MOLECULAR ASTROPHYSICS

Focusing on the organic inventory of regions of star and planet formation in the interstellar medium of galaxies, this comprehensive overview of the molecular Universe is an invaluable reference source for advanced undergraduates through to entry-level researchers. It includes an extensive discussion of microscopic physical and chemical processes in the Universe; these play a role in the excitation, spectral characteristics, formation, and evolution of molecules in the gas phase and on grain surfaces. In addition, the latest developments in this area of molecular astrophysics provide a firm foundation for an in-depth understanding of the molecular phases of the interstellar medium. The physical and chemical properties of gaseous molecules, mixed molecular ices, and large polycyclic aromatic hydrocarbon molecules and fullerenes and their role in the interstellar medium are highlighted. For those with an interest in the molecular Universe, this advanced textbook bridges the gap between molecular physics, astronomy, and physical chemistry.

A. G. G. M. TIELENS is a professor of Astronomy at both Leiden University and the University of Maryland. He has held appointments at the University of California in Berkeley, NASA Ames Research Center, the University of Groningen, and the Dutch Space Agency; was the project scientist of the HIFI instrument on board the Herschel Space Observatory; and was the NASA project scientist of the Stratospheric Observatory for Infrared Astronomy (SOFIA). He is a member of the Royal Netherlands Academy of Arts and Sciences and was awarded the Spinoza Prize in 2012.

MOLECULAR ASTROPHYSICS

A.G.G.M. TIELENS

Leiden University

CAMBRIDGE
UNIVERSITY PRESS

University Printing House, Cambridge CB2 8BS, United Kingdom

One Liberty Plaza, 20th Floor, New York, NY 10006, USA

477 Williamstown Road, Port Melbourne, VIC 3207, Australia

314–321, 3rd Floor, Plot 3, Splendor Forum, Jasola District Centre, New Delhi – 110025, India

79 Anson Road, #06–04/06, Singapore 079906

Cambridge University Press is part of the University of Cambridge.

It furthers the University's mission by disseminating knowledge in the pursuit of education, learning, and research at the highest international levels of excellence.

www.cambridge.org
Information on this title: www.cambridge.org/9781107169289
DOI: 10.1017/9781316718490

First published 2021

Printed in the United Kingdom by TJ Books Limited, Padstow Cornwall

A catalogue record for this publication is available from the British Library.

Library of Congress Cataloging-in-Publication Data
Names: Tielens, A. G. G. M., author.
Title: Molecular astrophysics / A.G.G.M. Tielens, Universiteit Leiden.
Description: Cambridge, UK ; New York, NY : Cambridge University Press, 2021. |
Includes bibliographical references and index.
Identifiers: LCCN 2020023795 (print) | LCCN 2020023796 (ebook) |
ISBN 9781107169289 (hardback) | ISBN 9781316718490 (epub)
Subjects: LCSH: Molecular astrophysics.
Classification: LCC QB462.6 .T54 2021 (print) |
LCC QB462.6 (ebook) | DDC 523.01/96–dc23
LC record available at https://lccn.loc.gov/2020023795
LC ebook record available at https://lccn.loc.gov/2020023796

ISBN 978-1-107-16928-9 Hardback

Additional resources for this publication at www.cambridge.org/tielens

Contents

Color plates can be found between pages 340 and 341

Preface

While interstellar molecules were first discovered some hundred years ago, over most of the intervening years, the molecular Universe lay dormant; the realm of a few brave pioneers that were undaunted by the prospect of the deep dive into molecular physics required to make sense of it all. Indeed, over much of this period, the Eddington quote in the Introduction sums up the attitude of the astronomical community, paraphrased to: *Molecules are not for astrophysicists*. However, over the last two decades, the opening up of the infrared and submillimeter spectral windows – driven by the rapid increase in detector technologies and ever-increasing telescope sizes – has provided us with a view of the richness of the molecular Universe. We are living in a molecular Universe where molecules are abundant and widespread and play an important role in the evolution of galaxies. We have also realized that regions of planet formation contain a rich organic inventory that may have provided the prebiotic roots of life. Finally, molecules provide an excellent tool to determine the physical conditions and probe the dynamics of many astronomically interesting objects and phenomena. Hence, molecular astrophysics has come into its own right as a key subdiscipline within astronomy. Conversely, future generations of astromomers will have to become familiar with all things molecular.

This book has grown out of lectures presented at a number of different summerschools on molecular astrophysics organized over the years. While preparing these lectures and discussing with students the molecular physics involved, I realized that a comprehensive introduction into molecular physics and its application to molecules in the interstellar medium of galaxies at the level of graduate students was sorely lacking. My earlier textbook on the physics and chemistry of the interstellar medium covers some of these aspects, and relevant chapters have been incorporated but updated with the latest developments and, in addition, their scope has been greatly expanded. The introduction to each chapter provides a short guide on what aspects would be particularly relevant for a course and which sections are, instead, provided for in-depth study by graduate students entering the field. Each chapter also provides a further reading and resource guide that will provide entry points for students into the rich literature of the field. Here, I have to appologize as, due to lack of space, it is impossible to do justice to all the relevant literature. The resource guide will typically provide references to some of the earliest studies and to some of the more recent developments. Together, this should enable students to trace back all relevant ideas and concepts. Finally, each chapter also contains a set of exercises that will allow students to test their comprehension of the material.

This book was largely written over a period of three years during extensive stays at the Astronomy Department of the University of California in Berkeley, the Astronomy Department of the University of Maryland in College Park, and the Astronomy Department of the University of Colorado in Boulder. I owe a deep debt of gratitude to these institutions for their hospitality and for providing a great atmosphere for science that was conducive to creative writing. Much of this book reflects a lifelong interest in the molecular Universe. I am deeply grateful to Harm Habing who, during the early stages of my career, taught me that graduate students are to be treated humanly. I wouldn't be where I am now if Harm hadn't stepped in at the right moment. I am also much in debt to Lou Allamandola and David Hollenbach who have been my guides through the molecular Universe over much of my career. Last, but not least, I greatfully acknowledge the many great students, postdocs, and collaborators who never tired of showing me the parts of the molecular Universe that inspired them.

Book Description

This work provides a comprehensive overview of our understanding of the molecular Universe, in particular the organic inventory of regions of star and planet formation in the interstellar medium of galaxies. It contains an extensive discussion of the microscopic physical and chemical processes that play a role in the excitation, spectral characteristics, formation, and evolution of molecules in the gas phase and on grain surfaces. Based on our current experimental, theoretical, and observational understanding of the molecular physics relevant for the interstellar medium of galaxies, this book includes the latest developments in this area of molecular astrophysics and provides a firm foundation for an in-depth understanding of the molecular phases of the interstellar medium. The physical and chemical properties of gaseous molecules, mixed molecular ices, and large polycyclic aromatic hydrocarbon molecules and fullerenes and their role in the interstellar medium are highlighted. This is an invaluable reference source for advanced undergraduate and graduate students and research scientists. Related resources for this book can be found at TBD.

Author

A. G. G. M. Tielens is a professor of Astrophysics at Leiden Observatory, the Netherlands, and at the Astronomy Department of the University of Maryland, College Park. Prior to this, he worked as an assistant researcher in the Astronomy Department of the University of California, Berkeley, a senior scientist at NASA Ames Research Center, California, professor of Astrophysics at the Kapteyn Institute in the Netherlands, and a senior scientist with the Dutch space agency, SRON. He was the project scientist of the HIFI instrument that flew on the Herschel Space Observatory, launched by the European Space Agency in 2009, and he was the NASA project scientist of the Stratospheric Observatory for Infrared Astronomy, SOFIA, from 2005 to 2007. He has published extensively on various aspects of the physics and chemistry of the interstellar medium of galaxies.

Cover Illustration

The tip of Orion's sword, where bright stars set their environment aglow in the light of large polycyclic aromatic hydrocarbon molecules. Figure courtesy of NASA/JPL-Caltech & T. Megeath (University of Toledo, Ohio).

1
Introduction

1.1 The Molecular Universe

Over the last 20 years, we have discovered that we live in a molecular Universe: a Universe with a rich and varied organic inventory; a Universe where molecules are abundant and widespread; a Universe where molecules play a central role in key processes that dominate the structure and evolution of galaxies; a Universe where molecules provide convenient thermometers and barometers to probe local physical conditions. Understanding the origin and evolution of interstellar and circumstellar molecules and their role in space is therefore key to understanding the Universe around us and our place in it and has become a fundamental goal of modern astrophysics.

After the discovery of the first diatomic molecules some 100 years ago, the preeminent astrophysicist of his time, Sir Arthur Eddington, lamented that "atoms are physics, but molecules are chemistry." Ever since, astrophysicists rued the moment that simple physical formulas had to give way to complex chemical solutions in a molecular Universe. To molecular astrophysicists, though, interstellar molecules provide a tool to probe macroscopic aspects of the Universe, whereas the harsh environment of space offers unique insight in microscopic processes controlling excitation and relaxation of isolated molecules. The chapters in this textbook introduce the reader to both aspects of molecular astrophysics.

The field is heavily driven by new observational tools that have become available over the last twenty years; in particular, space-based missions that have opened up the IR and submillimeter window at an ever accelerating pace. Furthermore, our progress in understanding the molecular Universe is greatly aided by close collaborations between astronomers, molecular physicists, astrochemists, spectroscopists, and physical chemists who work together in loosely organized networks. Together these groups provide the molecular data that is required to turn astronomers' pretty pictures into a quantitative understanding of the Universe around us. One focus is of course on understanding the unique and complex organic inventory of regions of star and planet formation that may well represent the prebiotic roots to life. A second focus is to appreciate the role of molecules in the evolution of the Universe. The third area to consider is the use of molecules to study the characteristics of the Universe, including physical conditions and the dynamics of, in particular, regions of star and planet formation but also stellar outflows and explosions, the interstellar medium of galaxies, and toroids around active galactic nuclei. In this chapter, we will introduce these three aspects in order to highlight the many uses of molecular astrophysics in astronomy. This will start with an overview of the rapid developments in astrobiology and the prebiotic origin of life, continue with examples of the use of molecules to study the Universe, and conclude with a bird's eye view of astrochemistry.

1.2 Astrobiology

1.2.1 Exoplanetary Science

Some 25 years ago, only nine planets were known and they were all part of the solar system. While some people have in the mean time mislaid one of those, at the moment of writing a variety of techniques have led to the discovery of some 3,800 additional planets (and counting) around distant stars. The field of exoplanetary studies was opened up by radial velocity studies in the mid 1990s that detected the presence of planets based on their small gravitational "tug" on the line position in the spectrum of the host star. The NASA Kepler mission revolutionized the field with the transit method that infers the presence of planets from small dips in brightness of the host star when planets pass in front of it. A number of other techniques, including direct imaging, have added a handful of planets. As a result, we now know that essentially every star has a planetary system. There are some hundred billion stars in the Milky Way and there are some hundred billion galaxies in the Universe accessible to us. That makes for a truly staggering number of ten thousand quintillion planetary systems. While the architectures of these planetary systems are not well known – as this is fraught with systematic issues – many of the detected planets are "hot-Jupiters": gas giants close to their host star that are quite inhospitable. We are, of course, particularly interested in planets like the Earth. The Kepler mission has taught us that one in every six stars like the Sun has a planet like the Earth at a distance where water would be liquid. That means that there are more than seven billion Earth-like planets in the habitable zone of their star and about one billion in the habitable zone around stars like the Sun. It is clear that the Universe provides ample opportunity for life to form.

1.2.2 Microbiology

While in the past we thought that life required very benign conditions, we have now discovered that life can thrive in a wide range of conditions. Indeed, extremophiles – organisms that thrive under extreme physical or chemical conditions – are very much part of the Earth's biosphere. Cyanobacteria give the Grand Prismatic Pool in Yellowstone National Park its much photographed color palette. These bacteria have no issue living in the scalding temperatures of this environment – or so we thought – yet they do. Likewise, the subglacial lakes in the Antarctic are home to a rich ecosystem. Bacterial colonies have built stromatolites reminiscent of billion-year-old fossils of early life on Earth.

These bacteria derive their energy not from the sun but from geothermally released chemicals. In the 1970s, tubeworm colonies were discovered living on and around hydrothermal vents on the deep ocean floor where mineral-rich, hot water is released through "chimneys." Life is made possible here by microbes that use chemosynthesis to convert chemical energy available in the disolved minerals to nutrients, and in the end this feeds the whole ecosystem. Life has also been discovered deeply buried in the driest spot on Earth, the Atacama desert in Chile. Here, bacteria have been discovered attached to hygroscopic salt pellets, a few meters deep in the ground. While it is not clear that these organisms developed here, for sure, they adapted well to these extreme conditions. Our definition of the habitable zone may well have been guided too much by the disclaimer "life as we know it."

1.2.3 Paleobiology

Some 4.567 billion years ago, dust particles coagulated in the early solar nebula and these aggregates were subsequently processed by energetic events to form the calcium-aluminum inclusions (CAI): one of the constituents of carbonaceous meteorites. These are the oldest objects known in the solar system: fossils of preplanet-forming events in the solar nebula. These CAIs were incorporated into planetesimals within a million years of their formation. The Earth (and other terrestrial planets) was put together through collisions of such planetesimals and larger embryonic bodies over a period of $\simeq$10 million years. During this assemblage process, the Earth was covered by a magma ocean. In this Hadean period, some 25–30 million years after the Earth was formed, the moon was created in a glancing collision of a Mars-sized impactor: the last one of these giant impacts. This resulted in very high temperatures, enduring magma oceans, and a silicate cloud cover. The runaway greenhouse effect lasted until the CO_2 in the atmosphere was subducted, perhaps some 100 million years later. Small (10's of microns) zircon crystals recovered from the Jack Hills in Australia are the oldest known crust and date back to 3.9–4.2 billion years ago. With the formation of a crust, a hydrosphere could develop and evidence for oceans dates back to this same period. Hence, life could start to take hold of the planet very early. The late heavy bombardment – when the solar system's architecture was rearranged due to a resonance between Jupiter and Saturn and a "flood" of planetesimals impacted the terrestrial planets – occurred 3.7–3.9 billion years ago, some 700 million years after the first rocks formed in the solar system. If life existed, the resulting impacts may well have "sterilized" the Earth.

Searches for the oldest fossils are highly contentious. Fossilized stromatolites discovered in Western Australia and dating back to $\simeq$3.5 billion years ago are widely accepted as genuine and attest to flourishing microbial life at that time. There are more controversial claims of "biological" activities preserved as structures in even older ($\simeq$3.7 billion years) rocks, which would have biology take off almost immediately after the end of the late heavy bombardment. But even $\simeq$200 million years is the equivalent of a blink of an eye in terms of starting life. It seems that as soon as conditions on the early Earth were conducive, life just took hold.

1.2.4 A New Paradigm

Thus, each of these fields – astronomy with the presence of countless planets like the Earth; microbiology with extremophiles everywhere; paleobiolology with fossiled life dating back to as far as we can trace – has undergone a revolution turning preconceived notions upside down. In essence, this is the logical conclusion of the Copernican and Darwinian revolutions and, together, they have changed our perception of the Universe and our place therein. The new paradigm is that the Universe teems with life. This has given rise to the new field of astrobiology where astronomy, planetary sciences, molecular physics, spectroscopy, chemical physics, geochemistry, geology, atmospheric sciences, biochemistry, and biology meet and mate.

1.3 The Prebiotic Origin of Life

1.3.1 The Earth's Volatile and Organic Inventory

From a molecular astrophysics point of view, one key aspect of astrobiology is the study of the inventory of volatiles such as water and organics available to newly formed terrestrial planets and our moons. This is a very active area of research and our insights are in a state of constant flux. Even the total water content of the Earth is controversial with estimates ranging from 0.5 to 2.5 times the water in the oceans. Presently, there is an active water cycle between the Earth's mantle and the surface where water is released by volcanic activity from the mantle and subducted in the form of hydrated sediments with a cycling timescale of $\sim 10^9$ yr. The carbon content of the Earth is equally uncertain with some researchers placing much of the Earth's inventory of carbon in the iron core. That carbon is of course not accessible and plays no further role in biochemistry. However, this does affect discussions on the carbon budget of the Earth and the processes that played a role in its origin. From the accessible carbon, some 90% may be stored in the mantle – as carbonates or graphite/diamond/carbides rather than dissolved in minerals – and cycles on a similar timescale as water.

Taking the solar system as our measure, the Earth could have derived its volatiles and organics from multiple sources. First, the last stages of the build up of the earth from smaller bodies are characterized by collisions of moon-sized bodies – embryos – and are highly stochastic in nature where "feeding" occurs over a wide range of initial radii (Figure 1.3). Hence, while, following the strong temperature gradient, embryos in the planetary disk may have started with a clear gradient in volatility and organic content, the "feeding-zone" of terrestrial planets in the habitable zone will include volatile- and organic-rich embryos. The heat released during this collisional formation process would have produced a steam atmosphere. Much of the water would dissolve in the magma ocean, only to be outgassed when the crust – mainly anhydrous minerals such as olivine and pyroxene – formed. For the Earth, this initial stage ended with the giant, moon-forming collision, which could have sheared off much of the initial water atmosphere but have left much water dissolved in the magma ocean.

Second, while this discussion has focused on planetesimal accretion as the main driver of planet formation, pebble accretion may provide a relevant alternative. Observations show that protoplanetary disks contain a large reservoir of millimeter- to centimeter-sized grains. Such pebbles may be the natural outcome of the dust coagulation process that is limited by bouncing or fragmentation. Observations show that pebble disks extend to some 100 AU and, as these pebbles will drift inward, they can provide a source of volatile- and organic-rich material for growing planets in the inner regions (Figure 1.4).

Third, the late heavy bombardment provides another scenario for the delivery of volatile and organic rich material to the inner terrestrial planets. The mean-motion resonance between Jupiter and Saturn rearranged the solar system, flipping Neptune and Uranus around and dislodging the cometary bodies in the outer solar system, some of which impacted the Earth–moon system, producing, among others, the prominent lunar basins and delivering volatiles to the Earth (Figure 1.5). This so-called late veneer delivered some 0.5% of the Earth's mass in the form of cometary bodies and much of this in the form of water ice and organics.

Hence, the volatile and organic inventory of the Earth has a diverse origin, where materials derived from a diverse set of locations were delivered by a variety of methods to the Earth (Figure 1.1). All may have contributed at some level. As will be discussed elsewhere in this book, the temporal evolution of a system may leave its imprint in the isotopic composition, reflecting, for example, a difference in zero-point energy that dominates the chemical exchange between different reservoirs in a kinetic (low temperature) setting. Alternatively, selective photodissociation may play an important role in the isotopic composition. These aspects have been used to assess the role of these different volatile/organic reservoirs for the Earth, and the isotopic composition of, e.g., H and N suggests that asteroidal delivery was a major contributor for the Earth (Figure 1.2) but this issue is not fully settled as we do not have a comprehensive inventory of the isotopic signature of relevant bodies.

1.3.2 Exogenous versus Endogenous Prebiotic Chemistry

Meteorites impacting the earth derive from collisional processes in the asteroid belt and hence their study may teach us much about the composition of their parent bodies. Of particular importance are CM and CI carbonaceous meteorites, which contain, by weight, some 5–20% of water and some 1–5 % of C mainly in the form of insoluble, poorly characterized, macromolecular material but also as well-defined chemical compounds. Hence, asteroidal bodies may have contributed to the volatile and organic inventory of the terrestrial planets. Sensitive analytical tools – originally developed for the analysis of moon rocks – have allowed detailed investigations of the composition of meteorites at the parts per million level. One spectacular result from this effort has been the identification and quantification of a very rich organic inventory, containing among others amino acids, carboxylic acids, purines and pyrimidines, and aliphatic and aromatic hydrocarbons (Table 1.1). While less well characterized, comets have a similar, diverse molecular composition. These rich and

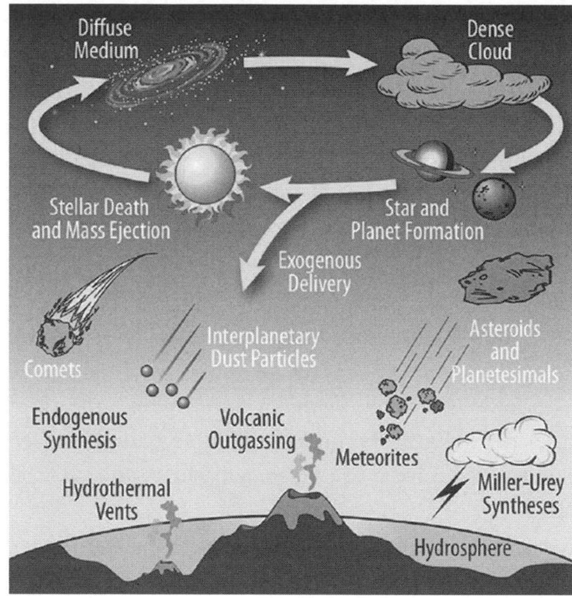

Figure 1.1 The cosmic history of prebiotic species will have left an imprint on the volatile and organic inventory of newly formed planets, including the Earth. As matter is injected by old stars and cycles between diffuse and dense star-forming clouds, a large number of processes contribute to a rich inventory of organic species in the interstellar medium. Some of this material may be incorporated into pebbles, cometesimals, and planetesimals where it can be further processed. These are the building blocks of planets and their organic inventory will be an amalgam of these sources and the processes involved. Further processing on the planetary surface, in its oceans, and in its atmospheres will drive further chemical complexity, leading possibly to the start of life. Figure kindly provided by J. Dworkin and adapted from [4]

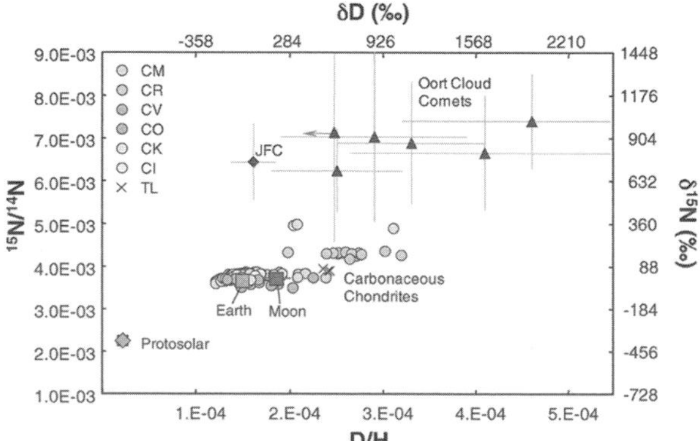

Figure 1.2 Isotopes are commonly used as tracers of the origin of molecular compounds. Here, the H and N isotopic composition of the Earth–Moon system is compared to that of different classes of meteorites and comets. JFC represents measurements for some Jupiter Family Comets. Note that N-isotopes derive from HCN/CN measurements which may not be representative of the nitrogen reservoir of comets. Figure taken from [24]

Table 1.1 *Organic compounds in meteorites*

Compound	Abundance[a] [ppm]	Compound	Abundance[a] [ppm]
CO_2	106	Monocarboxylic acids	332
CO	0.06	Dicarboxylic acids	25.7
CH_4	0.14	$\alpha-$Hydroxycarboxylic acids	14.6
NH_3	19	Amino acids	60
Alcohols	11	Diamino acids	0.04
Aldehydes	11	Sulphonic acids	67
Ketones	16	Phosphonic acids	1.5
Amines	8	Aliphatic hydrocarbons	12–35
$CO(NH_2)_2$	25	Aromatic hydrocarbons	15–28
Sugar-related compounds	$\sim$ 24	Basic N-heterocycles[b]	0.05–0.5
		Pyrimidines[c]	0.06
		Purines	1.2

Data taken from [31]. [a]Abundances in Murchison in parts per million. [b]Pyridines & quinolines. [c]Uracil and thymine

diverse inventories, and in particular the presence of a rich array of amino acids, have driven the point home that the organics delivered to the early Earth may well have prebiotic implications. This has brought the questions "has exogenous delivery given life a jump start on the early Earth and could it do so on exoplanets?" into focus in astrobiology. Some 5×10^{10} g/yr of organic carbon may have been delivered to the early Earth, including 3×10^6 g/yr of amino acids and 5×10^4 g/yr of nucleobases. Perhaps more relevant, taking Murchison as a model, a single meteorite could deliver 10 g of amino acids and 0.1 g of nucleobases and if trapped in a single warm pond could well be of importance. These amino acids are generally attributed to aqueous alteration on the meteoritic parent body and, specifically, the Strecker synthesis process where, for example, formaldehyde, ammonia, and hydrogen cyanide react to form aminoacetonitrile (NH_2CH_2CN). With water, this species reacts then on to glycine (NH_2CH_2COOH).

The alternative to exogenous delivery is endogenous synthesis. This is a logical consequence of Darwinian evolution and the idea can be traced back to a letter in 1871 from Charles Darwin to Joseph Hooker: "But if (and oh what a big if) we could conceive in some warm little pond with all sorts of ammonia and phosphoric salts, light, heat, electricity etcetera present, that a protein compound was chemically formed, ready to undergo still more complex changes." The best known example of endogenous synthesis is the Urey–Miller experiment where a gas mixture – simulating the early Earth's atmosphere – was subjected to electric discharges – representing lightning on the early Earth – and the products were collected. After many, many cycles of this process, the resulting mixture was analyzed and the presence of, among others, amino acids was demonstrated. The starting gas mixtures were very reducing, consisting of H_2O, NH_3, CH_4, and H_2.

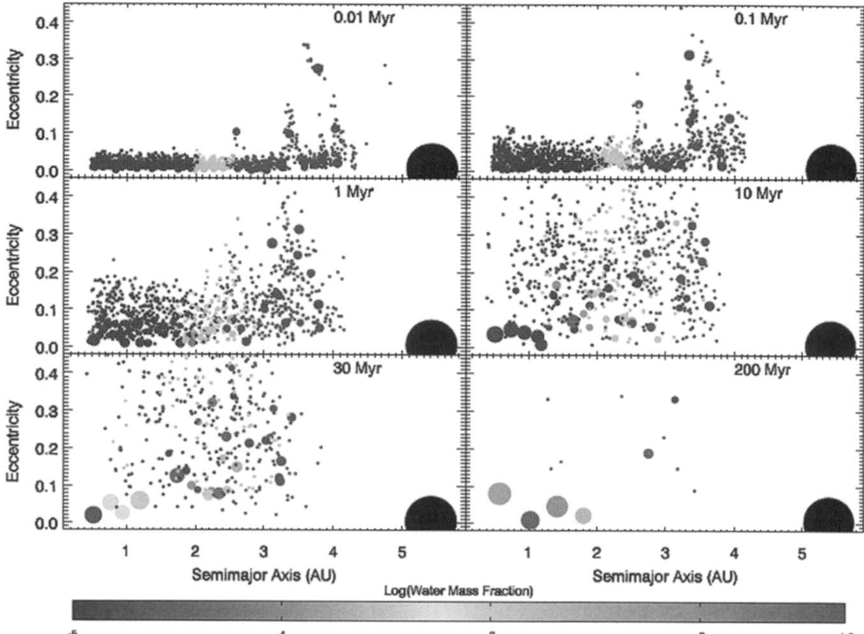

Figure 1.3 Snapshots in the evolution of disks of embryos from a simulation where Jupiter and Saturn are in the 3:2 mean motion resonance. The size of each body is proportional to its mass (but is not to scale on the x-axis). The interaction with Jupiter pumps up the eccentricity of the orbits of the embryos and they start to collide and grow, eventually leading to the formation of terrestrial planets. The temperature gradient in the protoplanetary disk has led to a compositional gradients in the embryos formed at different positions. The color of each body corresponds to its water content by mass (bottom scale, ranging from dry to 5% water). Jupiter is shown as the large black dot. Saturn is not shown. Figure taken from [21] (A black and white version of this figure will appear in some formats. For the color version, please refer to the plate section.)

In an aqueous setting, methane and ammonia are converted to formaldehyde and hydrogen cyanide. Strecker synthesis can then (again) take over to build amino acids. The Urey–Miller experiment has been repeated with many variations changing the composition of the initial mixture and/or the energy source. The conclusion is that, with the addition of energy, simple compounds can be converted into organic building blocks of proteins and other macromolecules in a reducing environment. In an oxidizing environment, on the other hand, this process is not very efficient.

1.3.3 Composition of the Early Earth's Atmosphere

The importance of the exogenous delivery versus endogenous synthesis debate hinges on the reducing versus oxidizing condition of the early Earth's atmosphere. This atmosphere resulted from outgassing and, taking current volcanic outgassing as a guide, would have

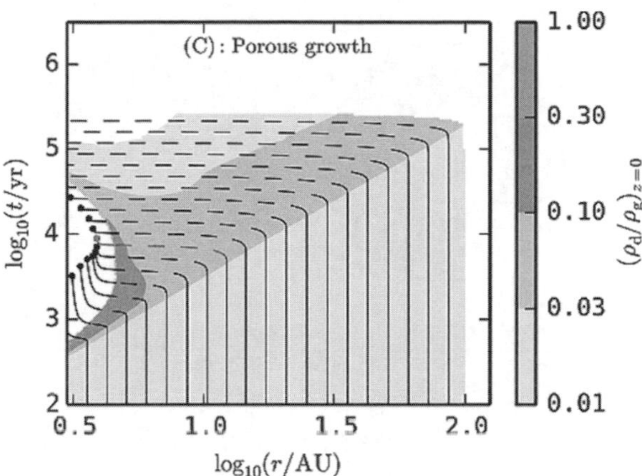

Figure 1.4 Growth from dust to pebbles to planetesimals in a protoplanetary disk. An initial phase of rapid coagulation and pebble formation (solid lines) is followed by a phase in which these pebbles decouple from the gas and drift inward (dashed lines). Pebbles formed in the inner disk will continue to grow and this growth is more rapid than the drift, culminating in the formation of a planetesimal (indicated by the black dot). The surface density in the outer disk is too low to allow planetesimal formation and the pebbles will drift inward where they can be "caught" by planets that have already formed. The gray scale indicates the dust-to-gas density in the mid plane. The details of these growth processes depend very much on the evolution of the porosity of the resulting aggregates. In this particular calculation, porosity is maintained to large sizes. Figure taken from [11]

been highly oxidized. However, as discussed in Section 1.3.1, outgassing is part of a cycle that balances with subduction, and on the early Earth this might have been different. If, on the early Earth, the O_2 fugacity of the magma was some two orders of magnitude lower, then ammonia and methane would be more important than carbon dioxide and molecular nitrogen. There is, however, little direct information on the mantle's fugacity. The composition of zircons seem to indicate that some 200 million years after the Earth's formation, the atmosphere was oxidizing in nature.

In a different approach, we can calculate the composition of an exoplanet's atmosphere resulting from outgassing. In thermodynamic equilibrium, the composition of the early Earth's atmosphere is a function of temperature, pressure, and (elemental) composition. The results reveal a strong dependence on the initial composition. Figure 1.6 shows exemplary calculation resulting in very reducing and very oxidizing atmospheres depending on whether H chondritic or CI chondritic compositions are adopted. H chondrites are thought to derive from S-type asteroids formed in the inner part of the asteroid belt ($\sim$2.2–3 AU). CI chondrites are thought to derive from asteroids in the outer ($\gtrsim$4 AU) asteroid belt. We do not really know the relative importance of these two potential sources for the early Earth let alone for exoplanetary systems. These results are only meant to illustrate the dependence on elemental composition. Real atmospheres are much more complex where, for example, penetrating UV radiation, cloud formation and settling, as well as

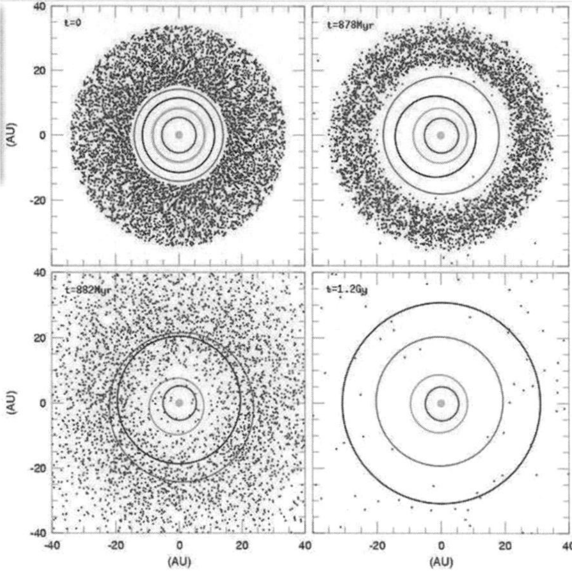

Figure 1.5 Four snapshots in the Nice simulation. The four giant planets were initially on nearly circular, co-planar orbits with semimajor axes of 5.45 (Jupiter), 8.18 (Saturn), 11.5 (Neptune), and 14.2 (Uranus) AU, indicated by the four circles. The outer cloud of dots indicates the cometary belt. The dynamically cold planetesimal disk was 35 $M_{\oplus}$, with an inner edge at 15.5 AU and an outer edge at 34 AU. The panels represent the state of the planetary system at four different epochs: (a) beginning of planetary migration (100 Myr); (b) just before the beginning of Late Heavy Bombardment (LHB; 879 Myr); (c) just after the LHB has started (882 Myr); (d) 200 Myr later, when only 3% of the initial mass of the disk is left and the planets have achieved their final orbits. The orbits of the planets evolve slowly through interaction of cometesimals with Jupiter, where the cometesimal is ejected and Jupiter's orbit shrinks. LHB is initiated by resonance between Jupiter and Saturn. As a result, Neptune leapfrogs over Uranus, creating havoc in the planetesimal disk. Most of the cometesimals are ejected but some are thrown into the inner solar system. Figure taken from [9]

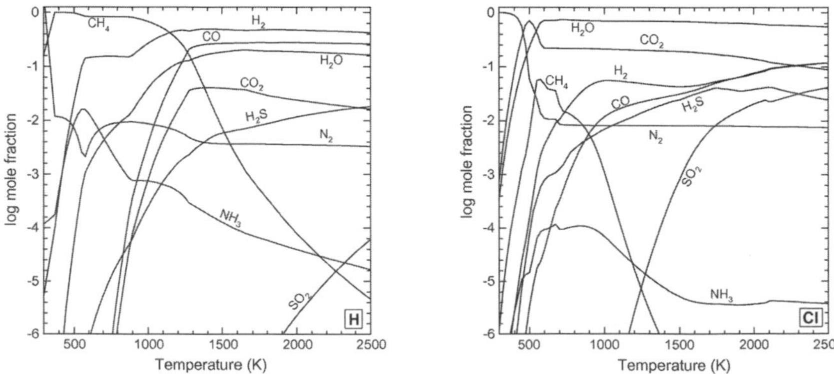

Figure 1.6 Atmospheric composition produced by outgassing calculated in thermodynamic equilibrium. The two panels are calculated for different elemental compositions corresponding to those of H chondrites (left) and CI chondrites (right). Calculations are performed at a pressure of 100 bars. Figure taken from [26]

circulatory mixing between hot and cold regions by strong winds can create large non-LTE effects. And so, while discussions can become quite passionate, this issue has not been settled.

1.4 Molecules and the Universe

Molecules are found in a wide range of objects, including diffuse and dense clouds in the interstellar medium, atmospheres of cool stars (later than K type) but also in sunspots, stellar environments such as outflows from Asymptotic Giant Branch (AGB) stars, red supergiants and luminous blue variables, post-AGB objects and planetary nebula, as well as supernovae and supernova remnants, atmospheres and surfaces of planets, moons and comets in the solar system, as well as exoplanets, photodissociation regions illuminated by bright stars, Hot Cores surounding massive protostars, Hot Corinos, protoplanetary disks and jets and outflows associated with low mass protostars, nuclei of galaxies including our own galactic center, as well as the nuclei of (Ultra)Luminous InfraRed Galaxies, and the molecular rings and toroids of active galactic nuclei. Also, molecules are abundant locally but have been detected as far back as red shifts of $\sim$7. These regions span a wide range in physical conditions (Table 1.2).

1.4.1 Molecules Tracing the Universe

The structure of the ISM can be well traced by molecules. Low level CO emission traces molecular gas and hence the spiral arm structure. An example of such a study is shown in Figure 1.7 where the CO emission traces the two curving arms in the grand-design

Table 1.2 *Physical conditions*

	n [cm^{-3}]	T [K]	Size scale [pc]
Diffuse clouds	$\simeq$100	$\simeq$80	3
Molecular clouds	$\simeq$10	$\simeq$1000	10
Photodissociation region	10^3–10^5	100–1000	0.01–0.1
Protoplanetary disk photosphere[a]	10^9	500	3×10^{-7}
Hot cores	10^6–10^{10}	150–1000	10^{-2}
Protostellar outflows	10^6–10^7	200–3000	$\sim 10^{-2}$
H_2O masers	$\simeq 10^9$	$\simeq$500	$\simeq 10^{-5}$
Circumnuclear disk	10^4	500	100
AGN toroid	up to $\simeq 10^9$	$\simeq$500	1

[a] Physical conditions vary strongly with location from very warm near the star to very cold far away as well as with height above the plane. The quoted values are indications for the photosphere at 1 AU.

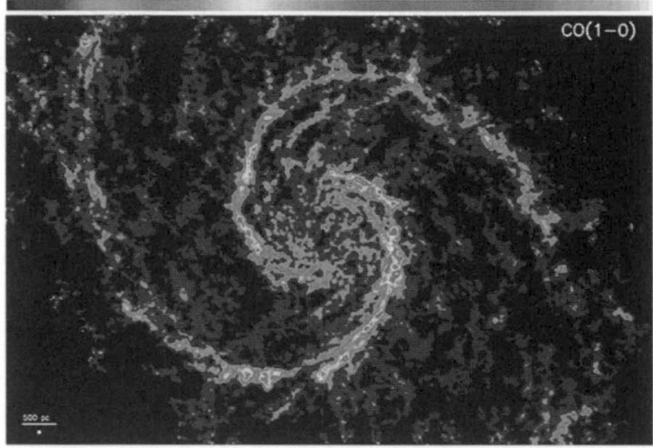

Figure 1.7 The nearby spiral galaxy, M51 – the whirlpool galaxy – in (integrated intensity of) the CO $J = 1 - 0$ line tracing molecular clouds in the spiral arms in the central 9 kpc. Observations obtained by the PdBI Arcsecond Whirlpool Survey (PAWS), using the Plateau de Bures interferometer at a resolution of $1.1 \times 0.9''$. A square-root scaling has been applied to enhance the faint emission. Figure taken from [27] (A black and white version of this figure will appear in some formats. For the color version, please refer to the plate section.)

spiral galaxy, M51. Because its energy levels are quite high, the most abundant molecule in the ISM, H_2, is hard to excite. Moreover, its rotational transitions are dipole forbidden. Emission from the CO molecule is then taken as a proxy for H_2, although it is recognized now that a substantial fraction ($\simeq$40%) of the molecular gas may be in so-called CO-dark molecular gas (where hydrogen is molecular but carbon is in the form of C^+ not CO). Molecular clouds are the sites of star formation and the CO luminosity has been used as a tool to determine the total gass mass with star-forming potential. Simple relationships have been derived linking the integrated CO intensity and the mass of molecular gas. This reservoir of gas with star-formation potential can then be linked to the global star-formation rate of a galaxy (Figure 1.8).

Polycyclic Aromatic Hydrocarbon (PAH) molecules are abundant and widespread in the ISM of galaxies. They also provide good tracers of the large-scale structure of the ISM in galaxies. Figure 1.9 shows the 8μm PAH emission of M51. Again, the spiral arms stand out well. In addition, the 8μm band also traces spurs trailing from the spiral arms. This, more diffuse, 8μm emission traces the general ISM, including HI and CO-dark molecular gas, which are particularly prominent in these spurs in the interarm region. PAHs are excited by absorption of FUV photons, emitted by massive stars. As massive stars live only briefly, bright 8μm emission is taken to trace current star formation and this relation has been calibrated against Hα, the total far-infrared emission, and the 24 μm dust emission (Figure 1.8).

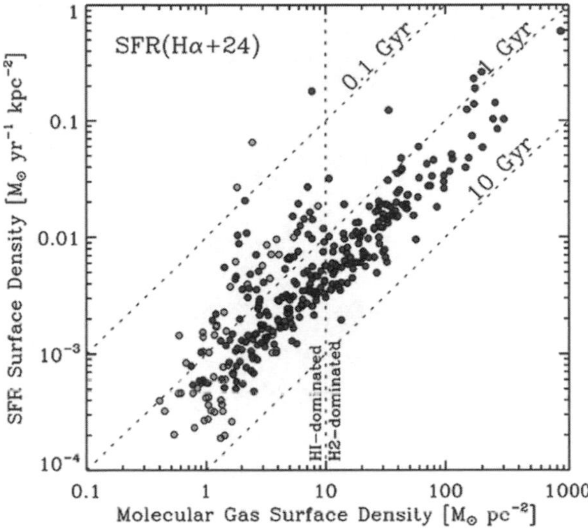

Figure 1.8 Modern forms of the Kennicutt–Schmidt law relate the star-formation rate – determined here from the observed Hα and 24μm intensities – to the molecular gas surface density – here determined from the observed CO intensity. The vertical dashed line indicates where the atomic (HI) and molecular (H_2) gas have equal surface density. The slanted dashed lines indicate different depletion timescales (i.e. timescale for complete conversion of the gas into stars). Figure taken from [29]

1.4.2 Molecules Measuring the Universe

Molecules are widely used to trace the characteristics including the physical conditions and dynamics of interstellar or circumstellar media and processes that regulate their evolution. Rotational transitions of molecules fall at millimeter wavelength and these lines can be studied using heterodyne techniques where routinely spectral resolutions of 10^6 – corresponding to sub-km/s – can be obtained. Molecules can thus be used to study the large scale dynamic response of gas in, e.g., spiral arms, feedback by jets and stellar winds on their environment, and gravitational collapse of molecular cloud cores. An example of the former is shown in Figure 1.10 taken from the CO survey of the Milky Way. A number of prominent features associated with the large-scale structure of the Milky Way are quite obvious. Comparison of the CO characteristics with tracers of other ISM components can provide insight into the relationship between different phases and the processes that regulate them.

An example of the use of molecular lines to trace the dynamics of star formation is shown in Figure 1.11, where the signature of collapse, the inverse P-Cygni profile, is quite obvious. Consider infall toward a protostar: For optically thick lines, the blue-side of the profile (at the backside of the protostar as viewed by us) is more absorbed than the

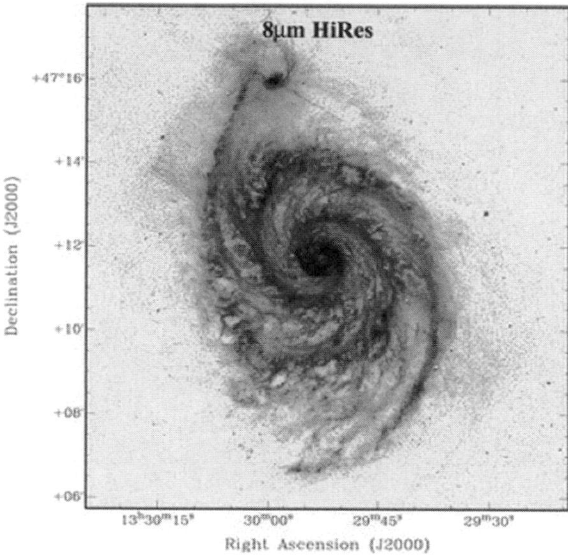

Figure 1.9 The nearby spiral galaxy, M51 – the whirlpool galaxy – in the PAH emission feature at 8 μm tracing the interaction of newly formed massive stars with their environment. Figure 1.7 shows the inner $4x4'$ of this galaxy in the CO $J = 1 - 0$ transition. Figure taken from [5]

red-side of the profile (in the foreground). Optically thin lines, on the other hand, will show a symmetric line profile. These observed profiles can be compared with detailed models for the so-called inside-out collapse of single cores to study the star formation process (c.f., Chapter 11).

Molecular line strengths are sensitive to the local physical conditions and this dependence on density and temperature links back to the excitation energy, E, of the energy level and the spontaneous emission coefficient, A, of that level. That will be discussed in more detail in Chapter 4 and used extensively in later chapters. Here, we will stress that with their multitude of energy levels – all differing in a systematic way in E and in Einstein A – molecules make for good barometers and thermometers. While there are pitfalls and drawbacks to this, molecules do provide powerful tools for the study of the physical conditions in the ISM. To first order, we can compare derived column densities in a so-called rotation diagram where the log of the column density – normalized by the statistical weight – is plotted versus the excitation energy of the level (Figure 1.12). For a Boltzmann distribution, $\log \left[N/g\right] \propto E/T$, the slope provides the (excitation) temperature. The total column density follows directly from the observed column densities where the unobserved level populations are estimated from the Boltzmann diagram or through the partition function at the estimated temperature. Non-LTE effects may make the derived temperature different from the kinetic temperature of the gas. The presence of a temperature gradient and of unequilibrated ortho-to-para ratios may complicate the interpretation and – tongue in cheek – provide fodder for models (see Chapter 4).

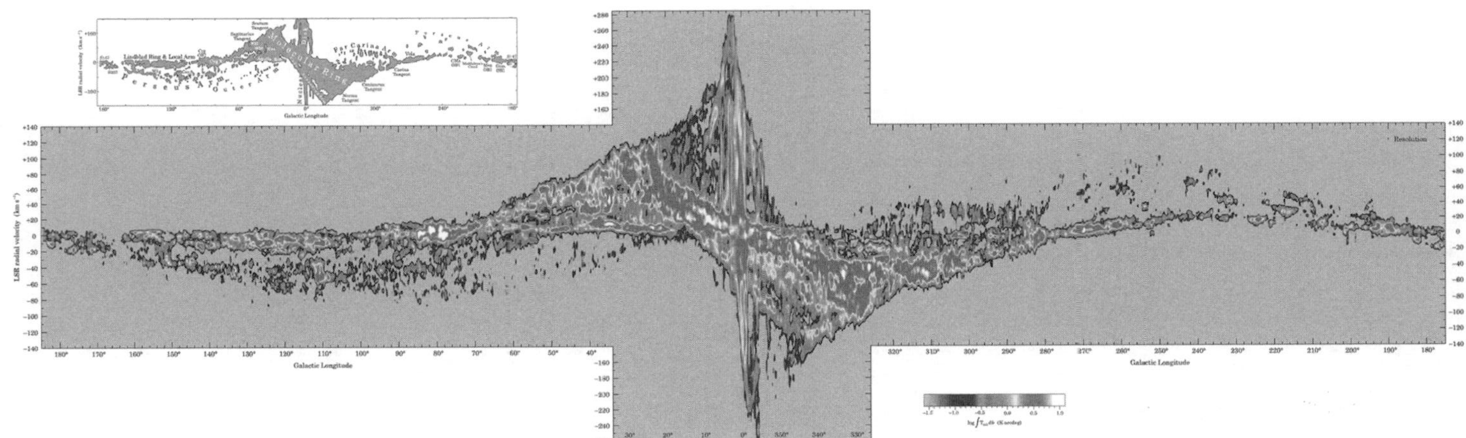

Figure 1.10 Longitude-velocity map of CO emission integrated over a $\sim 4^\circ$ wide strip centered on the Galactic plane. The map has been smoothed in velocity to a resolution of 2 km/s and in longitude to a resolution of $12'$. The inset identifies prominent structures in the molecular cloud distribution. Figure taken from [2] (A black and white version of this figure will appear in some formats. For the color version, please refer to the plate section.)

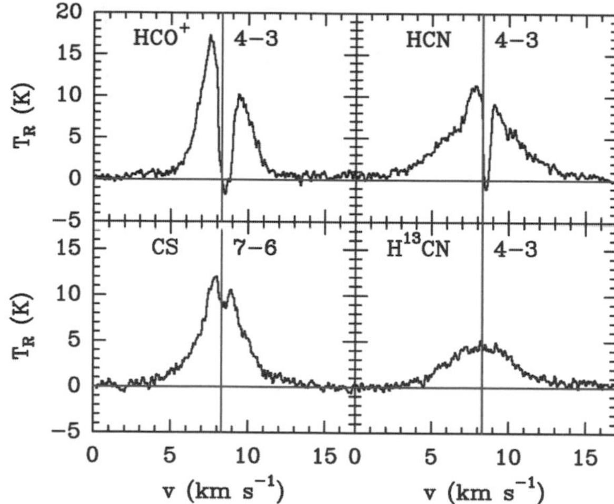

Figure 1.11 Continuum subtracted, molecular line profiles observed toward the central protostar in B335. The horizontal line indicates zero line intensity. The vertical line is the systemic velocity (8.3 km/s) of the source. The inverse P-Cygni profile is a clear signature of collapsing cloud. Figure taken from [6]

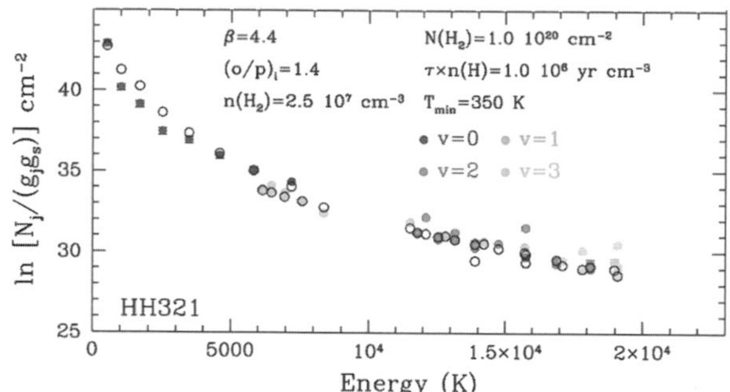

Figure 1.12 Rotation diagram of pure rotational and ro-vibrational transitions of H_2 in the HH 321 outflow. In this diagram, observed column densities for a given level – divided by the statistical weight of the level – are plotted as a function of the excitation energy. Filled and open circles refer to observational and model data, respectively. Emission lines coming from different vibrational states are indicated by different shades. The parameters of a best fit model through the data are indicated as well. The parameters describing this model include the temperature power law index, $dN/dT \propto T^{-\beta}$, the total column density, N, above the minimum temperature T_0, and the ortho-to-para ratio controlled by an initial value (o/p) and a rate coefficient describing its interconversion ($(\tau n)^{-1}$). Figure taken from [8]

1.5 Astrochemistry

The physical conditions in the ISM are very different (Table 1.2) from those in the Earth's atmosphere ($n \sim 10^{19}$ cm^{-3}, $T \simeq 300$ K) and ultra-high vacuum in a laboratory setting ($\sim 10^7$ cm^{-3}) approaches densities in only the densest cloud cores in the ISM.

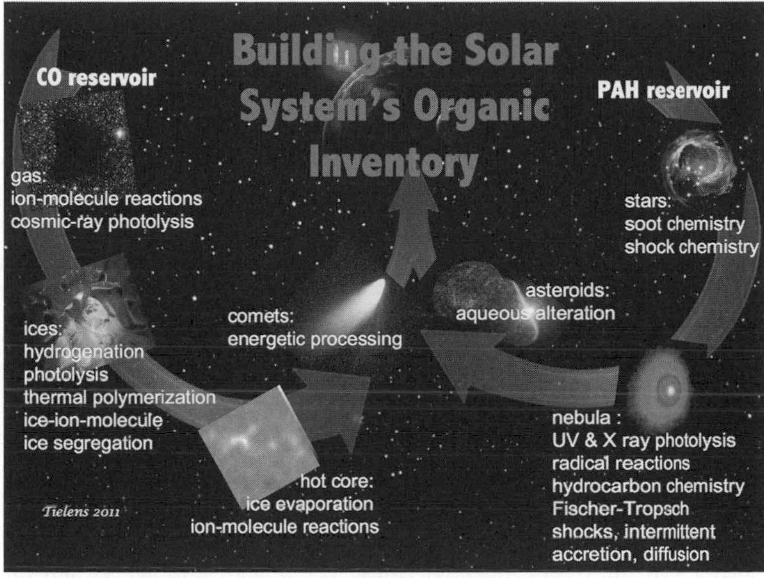

Figure 1.13 The organic inventory of the Solar System derives from a vast array of processes acting in a wide range of environments. Globally, two independent routes can be recognized. The first one builds up complex species from small radicals and starts with CO formation through ion–molecule reactions in dark clouds. Grain surface chemistry and photoprocessing of ices converts the main gaseous reservoir, CO, into complex species. The other route breaks down very complex species (e.g. PAHs) injected into the interstellar medium by stars into smaller and smaller species. Eventually, the species produced by either of these two chemical routes can become part of planetesimals and cometesimals in a protoplanetary disk environment that can deliver this organic inventory to planets in the habitable zone. Figure taken from [32] (A black and white version of this figure will appear in some formats. For the color version, please refer to the plate section.)

The interstellar medium is also far from thermodynamic equilibrium as the kinetic temperature of the gas is low but, yet, the medium is pervaded by highly energetic, cosmic ray particles and UV photons. These dissociate and ionize molecules and atoms. And while timescales are long, kinetics still controls the molecular composition at the low (10–100 K) temperatures of the ISM and, as a result, chemistry is quite different from that encountered on Earth.

Much of the carbon in regions of star and planet formation is locked up in very stable species that do not readily partake in chemical reactions. The story of interstellar organic chemistry is therefore by necessity a story that starts with breaking the carbon out of these species. The two main molecular reservoirs are CO and Polycyclic Aromatic Hydrocarbon (PAH) molecules. Figure 1.13 illustrates relevant chemical routes. In dark clouds, ion molecule chemistry converts gaseous carbon into predominantly carbon monoxide. Traces of hydrocarbon chains are also formed but on an overall "budget" scale, they are not important (Table 10.2). After accretion onto grains, hydrogenation reactions convert CO efficiently into formaldehyde and methanol as well as traces of other species in an icy grain mantle. Energetic processing of these ices by UV photons and cosmic rays can convert these molecules into more complex species, while heating by the protostar can initiate thermal polymerization reactions. The newly formed protostar will sublime these ices, and fast ions in shocks can sputter some molecules into the gas phase. Subsequent ion–neutral and neutral–neutral reactions in the warm dense gas of the Hot Core surrounding the protostar may convert, in particular, methanol into more complex species such as dimethyl ether and methyl formate. These can then accrete again into ices and other planetesimals. Thus, in this route, complex species are built up – in a bottom-up approach – by continued adding of small compounds, leading to a diverse organic mixture.

The other main molecular reservoir of carbon is Polycyclic Aromatic Hydrocarbon (PAH) molecules, which lock up $\sim$10% of the elemental carbon in the ISM. This carbon can be "tapped" in a top-down chemistry. This route starts with the formation of PAHs and other large, complex molecules in stellar ejecta through chemical processes akin to soot chemistry. The injected PAH family is further processed in the interstellar medium by energetic photons and particles, weeding out the less stable members. Further processing can occur in the photosphere of the protoplanetary disk by stellar UV and X-ray photons. Furthermore, in the inner, hot and dense (mid-plane) regions, H and OH radicals will breakdown these complex species into acetylene, methane, and carbon monoxide and carbon dioxide. Turbulence will diffuse these species and their daughter products – formed through neutral–neutral reactions – throughout the protoplanetary disk. Hence, this is a top-down chemistry where complex species are injected by stars and then broken down by (energetic) processing in the ISM to smaller and smaller species. Eventually, the species produced in these two chemical routes can become part of planetesimals and cometesimals where further processing may occur before this organic inventory is delivered to nascent planets.

1.6 Further Reading and Resources

Many of the molecular astrophysics and astrochemistry topics will be discussed in more detail in later chapters and relevant references that can serve as entry points to the literature will be provided there.

The properties of CAIs are reviewed in [13]. A study of the evolution of the Earth's atmosphere in the aftermath of the moon-forming impact is presented in [36]. The characteristics of the Jack Hills' zircon crystals, including their age, are discussed in [12]. The oldest known fossils are reviewed by [28]. Reference [35] provides a compendium of chapters on each of the subfields of astrobiology written by world-renowned experts for a broad scientific community.

The planetesimal growth scenario owes much to early pioneers, [25, 34]. Pebble accretion is described in [7, 20]. The Nice model is described in [9, 18]. The "Grand Tack" model describes the sculpting of the inner solar nebula through the migration of Jupiter and Saturn, resulting in the formation of the terrestrial planets. This model is described and reviewed in [21, 22].

The Darwin quotation is from [3]. The original Urey–Miller experiments were reported in [17] and the field has been recently reviewed by [16]. The composition of the soluble organic inventory of meteorites is summarized in [13], which should be consulted for original references. The composition of the atmosphere of the early Earth is discussed in [10]. The oxidation state of zircons has been studied by [33]. The calculations on the composition of the Earth's atmosphere are presented in [26].

There is a long history of searches for the infall signature of collapsing clouds. An early study is given in [14], while a more recent ALMA study is provided by [6].

Bibliography

[1] Bernstein, M., 2006, *Phil Trans Roy Soc B*, 361, 1689
[2] Dame, T. M., Hartmann, D., Thaddeus, P., 2001, *ApJ*, 547, 792
[3] Darwin Correspondence Project, Letter no. 7471
[4] Deamer, D., Dworkin, J. P., Sandford, S. A., Bernstein, M. P., Allamandola, L. J., 2002, *Astrobiology*, 2, 371
[5] Dumas, G., Schinnerer, E., Tabatabaei, F. S., et al., 2011, *AJ*, 141, 41
[6] Evans, N. J. II, Di Francesco, J., Lee, J.-E., et al., 2015, *ApJ*, 841, 22
[7] Johansen, A., Lambrechts, M., 2017, *Annu Rev Earth Planet Sci*, 45, 359
[8] Giannini, T., Nisini, B., Neufeld, D., et al., 2011, *ApJ*, 738, 80
[9] Gomes, R., Levison, H. F., Tsiganis, K., Morbidelli, A., 2005, *Nature*, 435, 466
[10] Kasting, J., 1993, *Science*, 259, 920
[11] Krijt, S., Ormel, C. W., Dominik, C., et al., 2016, *A & A*, 586, A20
[12] Maas, R., Kinny, P. D., Williams, I. S., et al., 1992, *Geochimica et Cosmochimica Acta*, 56, 1281
[13] MacPherson, G. J., 2003, in *Meteorites, Planets, and Comets*, ed. A. M. Davis, Vol. 1, *Treatise on Geochemistry*, eds. H. D. Holland, K. K. Turekian (Oxford: Elsevier-Pergamon), 201
[14] Mardones, D., Meyer, P. C., Tafalla, M., et al., 1997, *ApJ*, 489, 719
[15] Marty, B., 2012, *Earth Plan Sci Lett*, 313–314, 56
[16] McCollom, T. M., *Annu Rev Earth Plan Sci*, 2013, 41, 207
[17] Miller, S. L., 1953, *Science*, 117, 528
[18] Morbidelli, A., Levison, H. F., Tsiganis, K., Gomes, R., 2005, *Nature*, 435, 462
[19] Mumma, M. J., Charnley, S. B., 2011, *Annu Rev Astron Astrophys*, 49, 471
[20] Ormel, C. W., 2017, *Astrophys Space Sci Lib*, 445, 197
[21] Raymond, S. N., O'Brien, D. P., Morbidelli, A., Kaib, N. A., 2009, *Icarus*, 203, 644
[22] Raymond, S. N., Morbidelli, A., 2014, Complex Planetary Systems, *Proceedings of the International Astronomical Union*, 194
[23] Peltzer, E. T., Bada, J. L., 1978, *Nature*, 272, 443
[24] Saal, A. E., Hauri, E. H., Van Orman, J. A., Rutherford, M. J., 2013, *Science*, 340, 6138
[25] Safronov, V. S., Evolution of the protoplanetary cloud and formation of the Earth and the planets, NASA TT F-677
[26] Schaefer, L., Fegley, B., 2010, *Icarus*, 208, 438
[27] Schinnerer, E., Meidt, S., Pety, J., et al., 2013, *ApJ*, 779, 42
[28] Schopf, J. W., Kudryavtsev, A. B., 2012, *Godwana Research*, 22, 761

[29] Schruba, A., Leroy, A., Walter, F., et al., 2011, *ApJ*, 142, 37
[30] Sephton, M. A., 2002, *Nat Prod Rep*, 3, 261
[31] Sephton, M. A., Botta, O., 2008, *Space Sci Rev*, 135, 25
[32] Tielens, A. G. G. M., 2013, *Rev Mod Phys*, 85, 1021
[33] Trail, D., Watson, E. B., Tailby, N. D., 2011, *Nature*, 480, 79
[34] Wetherill, G. W., 1980, *Annu Rev Astron Astrophys*, 18, 77
[35] Woodruff, T. S. III, Baross, J., 2007, Planets and Life: The Emerging Science of Astrobiology (Cambridge University Press: Cambridge)
[36] Zahnle, K., et al, 2007, *Space Sci Rev*, 129, 35

2
Introduction to Chemistry

This chapter is meant as a refresher for astronomers in chemistry concepts that might be useful for their studies in molecular astrophysics. It discusses concepts such as energy levels, bonding, reactions, and classes of compounds. In addition, there is a brief overview of some common (laboratory) techniques in gas phase chemistry, surface science, and spectroscopy as well as different approaches to quantum chemistry. These sections are just meant as short overviews with the goal to alert you to the existence of these methods and their (dis)advantages as this may assist you in the rest of your studies. However, if you are seriously considering using these methods, you should consult a more specialized book. Excellent review articles and textbooks exist on each of these topics. The additional resources section at the end of this chapter provides possible entries into this literature. In a molecular astrophysics course, I would assign this chapter as reading and have a question/answer session where the students address their issues among themselves.

2.1 Chemistry Primer

Chemistry is the study of the characteristics of matter including its structure, composition, and interaction. Thus, chemistry investigates how atoms are put together to form chemical compounds and how the properties of these compounds depend on the arrangement of its constituent atoms. Each element is characterized by the number of protons in its nucleus (its atomic number, Z) and identified by a specific symbol. For the neutral element, the number of electrons equals the number of protons. Atoms use these electrons to participate in chemical reactions. The mass of an atom depends on the number of protons and neutrons in its nucleus and, hence, an element can have more than one isotope (e.g. atoms with the same number of protons but different numbers of neutrons). These isotopes will exhibit very similar chemical behavior but they will differ slightly in their spectroscopic properties.

2.1.1 Atomic Electronic Levels

The energy levels of an electron are quantised and are described by four quantum numbers. For a hydrogen atom (or any one electron hydrogenic system), the principal quantum number, n, describes its energy, $E_n = -2.18 \times 10^{-11}/n^2$ erg. For more electron systems,

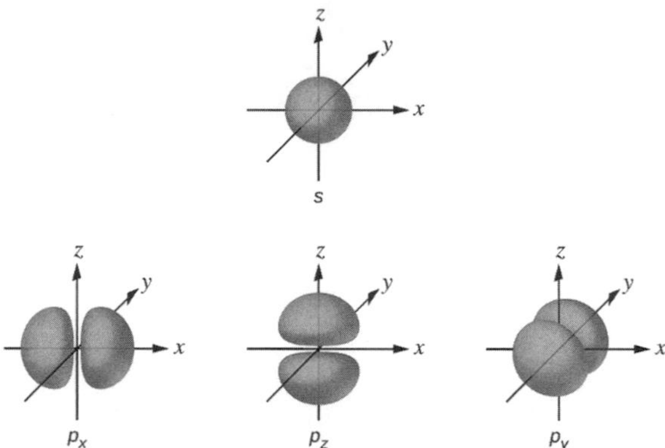

Figure 2.1 Representations of the electron density distributions in the s- and p-shells. The former is spherical while the latter has three equivalent dumbell shapes along each of the three axis. The 5 d- and 7 f-orbitals have more complicated geometrical shapes.

the principal quantum number broadly correlates with its energy where higher n corresponds to higher energies, larger orbit size, and less tightly bound electrons. A free electron is not bound (by definition) and can take on any energy.[1] The angular momentum of the electron orbital is described by the azimuthal quantum number, ℓ, and that sets the shape of the electron distribution. The value of ℓ is limited to $\leq n$ and traditionally they are indicated by letters; viz., $\ell = 0, 1, 2, 3, \ldots$ corresponds to $s, p, d, f \ldots$. The magnetic quantum number, m_ℓ, labels the directional dependence of the orbital electron distribution and can take any positive or negative integer between 0 and ℓ corresponding to $(2\ell + 1)$ values of m_ℓ. Finally, there is the spin quantum number, m_s, which can take values $\pm 1/2$. The orbit of an electron is then fully described by these four quantum numbers, n, ℓ, m_ℓ, and m_s with the restrictions on their possible values described above. As an example, the electron density distributions of the s- and p-orbitals are illustrated in Figure 2.1.

The electron configuration of an atom reflects the filling of these energy levels following the "aufbau" principle – lowest energy levels are filled first – and the Pauli exclusion principle – only two electrons (with opposite spin) in each level. To this we should add Hund's rule, which deals with the filling of levels within the same subshell and hence with the same energy: First, singly occupy sublevels before doubling any orbital so as to minimize repulsion due to electron overlap. Those electrons also assume the same spin so as to maximize the spin of the configuration. Physically, this maximizes the electron spin values and makes the electron spin wave function as symmetric as possible. As a result, the spatial wave function is antisymmetric and that reduces the Coulomb repulsion. Following these rules, carbon has six electrons and electron configuration of $1s^2 2s^2 2p^2$ while the eight electrons of oxygen result in $1s^2 2s^2 2p^4$. These electron configurations are illustrated in Figure 2.2.

[1] Limited by the size of the box it is in.

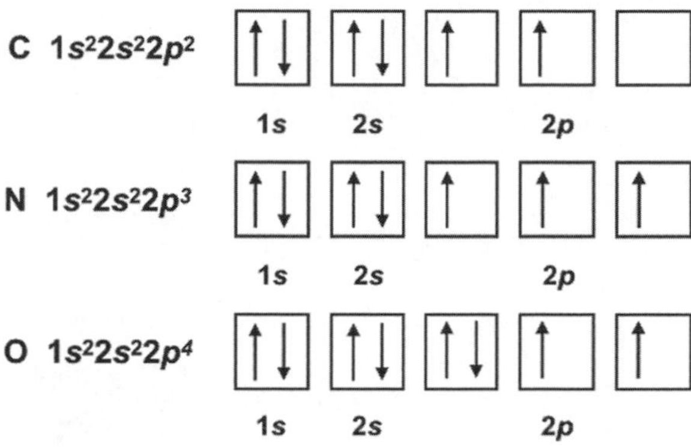

Figure 2.2 The electron configurations of C, N, and O. The arrows indicate the electrons and their (relative) spin orientation. For these atoms, the 1s and 2s (sub)shells are filled. The 2p subshells are filled following Hundt's rule. See text for details.

2.1.2 The Periodic Table

A chemistry primer starts off naturally with the periodic table of the elements (Figure 2.3). The periodic table of the elements was introduced by Mendeleev in 1869 and, in many ways, contains all of chemistry. In its modern form, the elements are arranged by atomic number and then stacked to reveal a clear pattern where elements in columns (called groups) have similar chemical properties (The horizontal rows are called periods). We recognize (Figure 2.3), for example, the alkali metals (group 1), the alkaline earth metals (group 2), the halogens (group 17), and the noble gases (group 18). This division can be linked to the electron configuration and we can recognize the s-block (groups 1 and 2), p-block (groups 13–18), d-block (transition metals, groups 3–12), and f-block (lanthanoids, or rare earth elements, and actanoids) elements. Broadly speaking, the elements can also be divided into metals, nonmetals, and metalloids. Metals are good conductors of heat and electricity, are ductile (can be drawn into a wire), are malleable (can be beaten into a thin sheet), and have a metallic lustre. All of these are consequences of metallic bonding and the presence of free electrons. Nonmetals have none of these properties and are generally gaseous at room temperature and pressure. Metalloids have intermediate properties and are mostly semi-conductors. B, Si, Ge, As, Sb, Te, and Po are metalloids. The noble gases and halogens as well as H, C, N, P, O, S, and Se are nonmetals. The other elements are metals.

The valence electron structure controls the properties of the atoms. The valence electrons are related to the atomic number and the periods are arranged in such a way that atoms with the same valence electron configuration are located in the same group. In moving from left to right across a period, the nuclear charge will increase. The radius of the atom will then decrease as the added electrons are not as effective in shielding the increased charge of the nucleus. Going down in a group, the added shell will be further away from the nucleus and hence the atomic size will increase. As a result, the ionization energy increases from left

IUPAC Periodic Table of the Elements

Key:

atomic number **Symbol** name conventional atomic weight standard atomic weight

1	2	3	4	5	6	7	8	9	10	11	12	13	14	15	16	17	18
1 **H** hydrogen 1.008 [1.0078, 1.0082]																	2 **He** helium 4.0026
3 **Li** lithium 6.94 [6.938, 6.997]	4 **Be** beryllium 9.0122											5 **B** boron 10.81 [10.806, 10.821]	6 **C** carbon 12.011 [12.009, 12.012]	7 **N** nitrogen 14.007 [14.006, 14.008]	8 **O** oxygen 15.999 [15.999, 16.000]	9 **F** fluorine 18.998	10 **Ne** neon 20.180
11 **Na** sodium 22.990	12 **Mg** magnesium 24.305 [24.304, 24.307]											13 **Al** aluminium 26.982	14 **Si** silicon 28.085 [28.084, 28.086]	15 **P** phosphorus 30.974	16 **S** sulfur 32.06 [32.059, 32.076]	17 **Cl** chlorine 35.45 [35.446, 35.457]	18 **Ar** argon 39.95 [39.792, 39.963]
19 **K** potassium 39.098	20 **Ca** calcium 40.078(4)	21 **Sc** scandium 44.956	22 **Ti** titanium 47.867	23 **V** vanadium 50.942	24 **Cr** chromium 51.996	25 **Mn** manganese 54.938	26 **Fe** iron 55.845(2)	27 **Co** cobalt 58.933	28 **Ni** nickel 58.693	29 **Cu** copper 63.546(3)	30 **Zn** zinc 65.38(2)	31 **Ga** gallium 69.723	32 **Ge** germanium 72.630(8)	33 **As** arsenic 74.922	34 **Se** selenium 78.971(8)	35 **Br** bromine 79.904 [79.901, 79.907]	36 **Kr** krypton 83.798(2)
37 **Rb** rubidium 85.468	38 **Sr** strontium 87.62	39 **Y** yttrium 88.906	40 **Zr** zirconium 91.224(2)	41 **Nb** niobium 92.906	42 **Mo** molybdenum 95.95	43 **Tc** technetium	44 **Ru** ruthenium 101.07(2)	45 **Rh** rhodium 102.91	46 **Pd** palladium 106.42	47 **Ag** silver 107.87	48 **Cd** cadmium 112.41	49 **In** indium 114.82	50 **Sn** tin 118.71	51 **Sb** antimony 121.76	52 **Te** tellurium 127.60(3)	53 **I** iodine 126.90	54 **Xe** xenon 131.29
55 **Cs** caesium 132.91	56 **Ba** barium 137.33	57-71 lanthanoids	72 **Hf** hafnium 178.49(2)	73 **Ta** tantalum 180.95	74 **W** tungsten 183.84	75 **Re** rhenium 186.21	76 **Os** osmium 190.23(3)	77 **Ir** iridium 192.22	78 **Pt** platinum 195.08	79 **Au** gold 196.97	80 **Hg** mercury 200.59	81 **Tl** thallium 204.38 [204.38, 204.39]	82 **Pb** lead 207.2	83 **Bi** bismuth 208.98	84 **Po** polonium	85 **At** astatine	86 **Rn** radon
87 **Fr** francium	88 **Ra** radium	89-103 actinoids	104 **Rf** rutherfordium	105 **Db** dubnium	106 **Sg** seaborgium	107 **Bh** bohrium	108 **Hs** hassium	109 **Mt** meitnerium	110 **Ds** darmstadtium	111 **Rg** roentgenium	112 **Cn** copernicium	113 **Nh** nihonium	114 **Fl** flerovium	115 **Mc** moscovium	116 **Lv** livermorium	117 **Ts** tennessine	118 **Og** oganesson

57 **La** lanthanum 138.91	58 **Ce** cerium 140.12	59 **Pr** praseodymium 140.91	60 **Nd** neodymium 144.24	61 **Pm** promethium	62 **Sm** samarium 150.36(2)	63 **Eu** europium 151.96	64 **Gd** gadolinium 157.25(3)	65 **Tb** terbium 158.93	66 **Dy** dysprosium 162.50	67 **Ho** holmium 164.93	68 **Er** erbium 167.26	69 **Tm** thulium 168.93	70 **Yb** ytterbium 173.05	71 **Lu** lutetium 174.97
89 **Ac** actinium	90 **Th** thorium 232.04	91 **Pa** protactinium 231.04	92 **U** uranium 238.03	93 **Np** neptunium	94 **Pu** plutonium	95 **Am** americium	96 **Cm** curium	97 **Bk** berkelium	98 **Cf** californium	99 **Es** einsteinium	100 **Fm** fermium	101 **Md** mendelevium	102 **No** nobelium	103 **Lr** lawrencium

For notes and updates to this table, see www.iupac.org. This version is dated 1 December 2018.

Figure 2.3 The periodic table of the elements. Credit: International Union of Pure and Applied Chemistry, IUPAC

to right in a period and decreases from top to bottom in a group. By adding an electron, an anion is formed and the electron affinity typically increases from left to right in the periodic table. The electrons in the outer shell are involved in bond formation and, thus, the chemical properties of an element are connected to the ground state configuration of the valence electrons. Atoms in the same group behave, therefore, chemically similar.

Finally, we should recall that masses of the atoms are given in atomic mass units, $1/12$ of the mass of the ^{12}C atom, 1.66054×10^{-24} g. Conversely, this implies that 12 g of ^{12}C atoms contains 6.022×10^{23} atoms (Avogadro's number). This is the definition of a mole. Any substance that contains this number of species will weigh its (total) atomic mass in grams.

2.1.3 Chemical Bonds

The electrons in the outer (partially filled) shell are involved in chemical bond formation. In covalent bond formation, one or more pairs of electrons are shared between atoms. In redox reactions, one or more electrons are transferred from one atom to another. Chemical reactions are then the reorganization of the electrons involved in bond formation. The reaction of a metal with a nonmetal generally results in an ionic compound. Common table salt (NaCl) is a good example. The reaction of a metal with a polyatomic ion or between a polyatomic cation and an anion will likewise result in an ionic bond; e.g. magnesium carbonate, $MgCO_3$, and ammonium cyanate, NH_4OCN. The reaction of nonmetals with each other or with hydrogen will generally lead to covalent bond formation.

Consider now covalent bond formation with the simplest example, the formation of molecular hydrogen. When two H atoms are brought together, each of the atoms will start to attract the electron of the other atom and this attraction will increase with decreasing distance. Maximum attraction occurs at about 0.74 Å and is then 4.5 eV. The electrons are located in between the nuclei, shielding their interaction. For closer distances, repulsion by the two nuclei will take over and the binding energy will decrease again. In chemistry, the binding energy of a molecule is typically reported in kJ/mole. For H_2, this is then $E_b = 4.5\,\text{eV/molecule} \times 1.6 \times 10^{-19}\,\text{J/eV} \times 6.022 \times 10^{23}\,\text{molecules/mole} = 435\,\text{kJ/mole}$; e.g., 1 eV equals 96 kJ/mole. Sometimes, binding energies are reported in kcal/mole and the conversion factor is 1 cal equals 4.2 J. It is also good to know the conversion to a temperature scale (1 eV equals 12,000 K) and to a wavenumber scale (1 eV equals 8100 cm^{-1}).

For H_2, each atom is equivalent and attracts the bonding electrons equally. The electron cloud is thus symmetric and H_2 is nonpolar. When one atom in a covalent bond has a higher electron negativity than the other, it will attract the electrons more, leading to a nonsymmetric electron cloud and a dipole moment. This is expressed in terms of the electronegativity: a dimensionless number that is a measure of the tendency of an atom to attract electrons in a covalent bond to itself. Its value ranges from about 0.7 (Cs) to 4 (F) and increases from left to right in a period, reflecting the increased nuclear charge. Thus, halogens have a high electronegativity while for metals it is low.

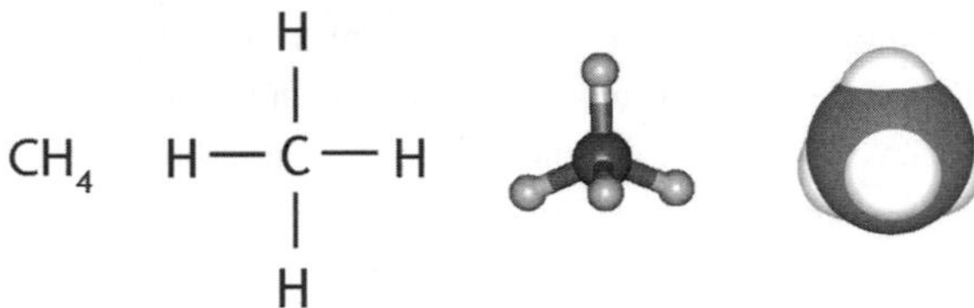

Figure 2.4 Different representations of a chemical species. (a) Chemical formula. (b) Structural formula. (c) 3-D ball and stick model. (d) Space-filling model. Figure kindly provided by Alessandra Candian

2.1.4 Molecular Representations

One way to represent a molecule is through its chemical formula that provides the number of each element in a compound (Figure 2.4). So, CH_4 (methane) contains four hydrogen atoms for each carbon (and no other elements). This by itself can be very confusing as isomers have the same chemical formula but distinct structures. In structural formulae (Figure 2.4), bonding between atoms is indicated, providing a convenient way to distinguish between isomers, but this has the drawback that it cannot be easily incorporated into text. In condensed structural formulae, atoms are arranged in bonded groups. For example, for the two isomers of C_2H_6O, dimethyl ether and ethanol, we can write CH_3OCH_3 and CH_3CH_2OH, respectively. The structural information of each group is then assumed to be understood by the reader. Molecules are sometimes also represented by line structures where bonds between atoms are indicated by lines (except for the C–H bond), with single bonds as one line, double bonds as two lines, and triple bonds as three lines, and all atoms are represented by their symbol except for C-atoms and H atoms connected to C atoms, both of which are not shown.[2] In a line structure, the saturated molecule cyclohexane (C_6H_{12}) would be a a hexagon. Benzene would also be a hexagon and the six electrons that are not involved in covalent bonds are often represented as a circle drawn within the hexagon. Sometimes, alternating double and single bonds are used to represent these so-called π-electrons. Molecules can also be represented by Lewis structures. Similar to the chemical formula, these are drawn by representing each atom by its symbol and each bonding pair of electrons with a line (so with up to three lines corresponding to a triple bond and six electrons). Nonbonding valence electrons are represented by dots next to the atom that they belong to.

Ball and stick and space-filled models provide a 3D visualization of the structural arrangement of the atoms (Figure 2.4). The shape of a molecule is an important factor for chemical reactivity and in the dipole moment of a molecule, and therefore the strength of its rotational transitions. Qualitatively, the geometry of molecules can be "predicted" by Valence Shell Electron Pair Repulsion (VSEPR) theory, in which the molecular shape is controlled by repulsion between pairs of electrons When considering the structure around a

[2] Anyone who thought that chemists cannot be the soul of a party should Google nanoputians.

central atom, VSEPR theory counts the number of sets of electron pairs where bonding and lone[3] electron pairs are counted equally. Here, it should be understood that single, double, or triple bonds are each counted as one set. The number of sets controls the geometry. Limiting ourselves to two, three, and four sets, these correspond to linear, trigonal planar, and tetrahedral geometries, respectively. This structure is then further modified according to whether bonding (B) or lone (L) pairs are involved. In this, it is considered that repulsion between L and L pairs is larger than the repulsion between L and B, which in turn is larger than between B and B.

Consider some examples: In CO_2, the central C atom is bonded by double bonds to each of the two O atoms. Hence, there are two sets and the structure is linear. For CH_4, we have four equal sets and the structure is tetrahedral with a bond angle of 109.5° ($\cos^{-1}\theta = -1/3$). For NH_3 there are three bonding pairs and a lone electron pair on the N atom. Again, there are four sets and a tetrahedral structure but because a lone pair is involved, the angle between the H-atoms is slightly smaller, 106.8°. H_2O also has four sets – two bonding and two lone pairs – and again a tetrahedral structure is implied. Once more, because of the increased lone pair repulsion, the bond angle between the H atoms is smaller; in this case, 104.5°. Finally, benzene has three sets and its structure is, therefore, trigonal planar with bond angles of 120°.

2.1.5 Valence Bond Theory

In valence bond theory, bonding electrons are localized in between the two atoms involved and bonds result from the overlap of the respective atomic orbitals. Consider the H_2 molecule again. When the two H-atoms approach each other, the $1s$ orbitals of the two atoms overlap in the region between them, constructive interference of the wave functions results in a new orbital with a high electron density between the two atoms, and a sigma (σ) bond is formed. In H_2, two s orbitals are mixed, but sigma bonds can also involve p orbitals as, for example, in F_2. In the F atom, the lone electron is located in a p orbital and these constructively interfere to produce enhanced electron density between the two atoms. In HF, s and p orbitals mix to a σ bond.

Consider now CH_4, which has four equivalent hydrogen with a bond angle of 109.5°. Now C has four electrons distributed as two in the $2s$ orbital and one electron in two of the p orbitals (Figure 2.2). Following the discussion above, we might naively expect that CH_2 rather than CH_4 were the preferred molecule or that, in CH_4, one $2s$ electron gets promoted to a $2p$ orbital, resulting in three σ bonds at 90° and one σ bond through the combined s orbitals. To understand the bonding in CH_4 we have to consider hybridization where orbitals on a single atom are combined – hybridized – into new orbitals, which are then used to bond to other atoms. Thus, for CH_4, the $2s$ and $2p$ atomic orbitals on the carbon atom hybridize into four sp^3 orbitals each containing one electron. The energy

[3] Nonbonding.

of the sp^3 orbitals is then $E_{sp^3} = (E_{2s} + 3E_{2p})/4$. These sp^3 orbitals will be arranged tetrahedrally to minimize their interaction. The energy required to promote the s electron to a hybidized orbital is more than compensated for by the energy gained by the additional molecular bonding allowed by the hybridization. Not all orbitals need to be hybridized. Hybridizing only two of the $2p$ orbitals with the $2s$ orbital results in three sp^2 orbitals and these are coplanar with a bond angle of 120°. The third $2p$ orbital will be at a 90° angle. So, for ethene ($H_2C{=}CH_2$), on each of the C atoms, we have two σ bonds to H atoms and a third σ bond to the other C atom. There is then one electron left in the $2p_z$ orbital. Overlap of the $2p_z$ orbitals of the two C atoms produces a π orbital with its maximum electron density above and below the plane of the three sigma bonds. So, the double bond consists then of one σ and one π bond. For acetylene, the $2s$ orbitals hybridize with one $2p$ level, giving rise to two sp orbitals and two remaining $2p$ orbitals. This leads to σ bonds between H and C and C and C and a linear arrangement. The singly occupied $2p_y$ and $2p_z$ levels on each of the C atoms sideways overlap and produce two π bonds – oriented perpendicular to each other and to the σ bond – resulting in a triple bond between the C atoms.

2.1.6 Molecular Orbital Theory

In valence bond theory, atomic orbitals are combined to produce hybrid orbitals, which then combine with orbitals on adjacent atoms to produce bond(s) localized to the two atoms. In molecular orbital theory, the resulting bonds may cover the entire molecule. Orbitals are really the solution of the Schrödinger wave equation and they have a phase. Consider again the H_2 molecule. Constructive (in phase) interference between the two $1s$ orbitals leads to enhanced electron density between the two nuclei and results in a σ orbital (Figure 2.5). Destructive (out of phase) interference results in reduced electron density between the two nuclei and a node and this bond is still symmetric, so, again it is a σ orbital. The constructive interference leads to bonding and this σ level is lower in energy than the atomic orbitals. The destructive interference will increase the energy relative to the atomic orbitals and

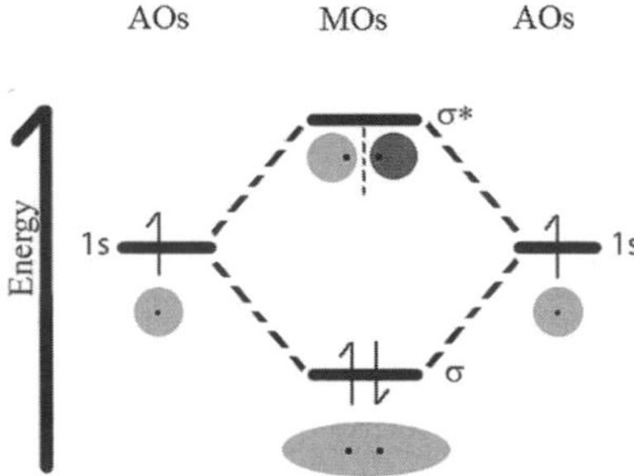

Figure 2.5 The atomic orbitals (AO) of two H atoms combine, according to phase, into bonding and nonbonding σ orbitals. The shape of the orbitals are indicated below the levels. Dark dots indicate the nuclei. The different shadings indicate different phases. Figure taken from Wikipedia, https://commons.wikimedia.org/wiki/File:H2modiagramCR.jpg

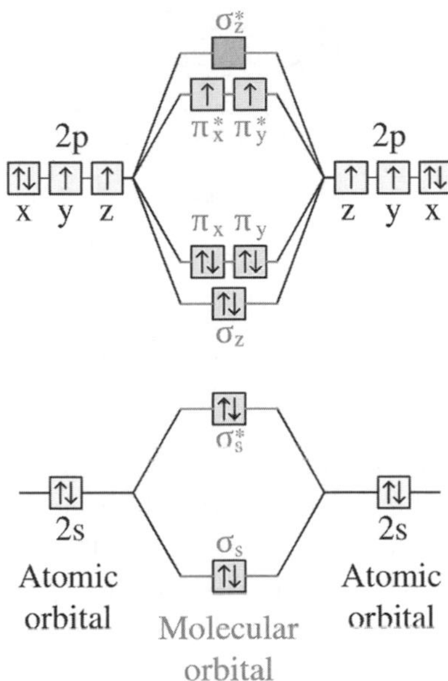

Figure 2.6 The atomic orbitals (AO) of two O atoms combine, according to phase, into bonding and nonbonding σ and π orbitals. Figure taken from Wikipedia, https://commons.wikimedia.org/wiki/File:Valence_orbitals_of_oxygen_atom_and_dioxygen_molecule_(diagram).svg

hence an antibonding sigma level ($\sigma^\star$). Following the aufbau principle, the two electrons are placed in the σ level.

Now, consider O_2 (Figure 2.6) where each O atom contributes four valence orbitals that will combine into eight molecular orbitals. The $2s$ orbitals combine (as for H_2) into bonding and antibonding σ_s orbitals. Taking z as the internuclear axis, the $2p_z$ atomic orbitals combine into a σ_p (constructive interference) and a $\sigma_p^\star$ (destructive interference). The $2p_x$ and $2p_y$ orbitals do not overlap but will combine into π and $\pi^\star$ molecular orbitals. These orbitals are then filled again following the aufbau principle, which leaves the two $\pi^\star$ molecular orbitals half filled and the $\sigma^\star$ empty (Figure 2.6).

2.2 Reactions

We can broadly distinguish between two reaction classes: electron transfer reactions and sharing of electrons, forming a covalent bond. Consider oxidation/reduction reactions: for example, oxidation of a metal atom by oxygen. This occurs through the transfer of two electrons to the oxygen atom. In this oxidation – loss of electrons by the metal atom – oxygen acquires a completely filled valence shell. The reverse process is called reduction – gain of electrons. Oxygen is in this the oxidizing agent but somewhat confusingly is actually reduced. Likewise, the metal atom is the reducing agent, but is oxidized. These redox reactions are electron transfer reactions. They do not need to lead to ions; for example, the oxidation of carbon leads to carbon dioxide formation.

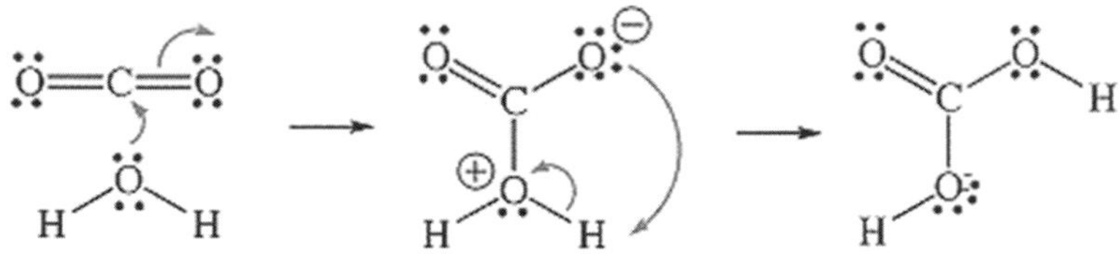

Figure 2.7 The formation of carbonic acid (H_2CO_3) from dissolved CO_2 involves acceptance of a lone electron pair from the O-atom in H_2O into the C orbital while the electrons in that orbital are transferred to an O atom. A proton is then transferred to internally neutralize this O^-.

Consider, now, an acid–base reaction. The common definition of an acid is a proton donor and of a base it is a proton acceptor. Start with water, which is both a proton donor – forming OH^- – and a proton acceptor – forming H_3O^+. In a solution, this is measured in terms of the pH given by $pH = -\log\left[H_3O^+\right]$ where $\left[H_3O^+\right]$ is the concentration of H_3O^+. The pH scale typically runs from 0 to 14. At 25°C, a solution with a pH less than 7 is acidic, while for a pH greater than 7, the solution is basic (or alkaline). Some examples of pH values are: hydrochloric acid (1 mol/ℓ) 0, formic acid (1 mol/ℓ) 1.9, lemon juice 2.0, carbonic acid (1 mol/ℓ) 3.2, coffee 4.5, milk 6.5, pure water 7.0, seawater 8.0, baking soda 9.0, antacids 10, bleach 13.0, sodium hydroxide (1 mol/ℓ) 14.0. Reactions between an acid and a base result in the formation of a salt (and water); viz., $HCl + NaOH \rightarrow NaCl + H_2O$. This is the Brønsted–Lowry definition of acid/base reactions.

In organic chemistry, a more general definition in terms of electron pair donors is useful. In the Lewis definition, an acid (base) accepts (donates) a pair of nonbonding valence electrons. These are called electrophile (nucleophile). Thus a Lewis acid–base reaction converts a lone pair on a base and an empty orbital on an acid into a covalent bond. Brønsted–Lowry acids/bases are retained in the Lewis definition as an acid is an electron pair acceptor (such as H^+) and a (Lewis) base is an electron pair donor (such as OH^-).

Consider now, as an example, the formation of carbonic acid (H_2CO_3) from dissolved CO_2 in water (Figure 2.7): The vacant orbital on the C atom in CO_2 accepts a nonbonding electron pair from the O-atom in H_2O to form a (two electron) covalent bond while transfering an electron to one of its O atoms. The resulting species is then internally "neutralized" by transfer of a proton from the water adduct to this O^-. Formulated this way, redox reactions and Brønsted–Lowry acids/base reactions are a class of Lewis acid–base reactions. Thus, species with lone electron pairs – NH_3 and H_2O – are Lewis bases. Benzene, with its electron-rich π-systems, is also a Lewis base. On the other hand, protons are Lewis acids and so are electron-poor π-systems such as $CH_2CHCH_2^+$.

2.3 Classes of Compounds

Functional groups determine the properties and reactivity of chemical compounds. Within a functional group, atoms have specific bonding arrangement. Some common functional

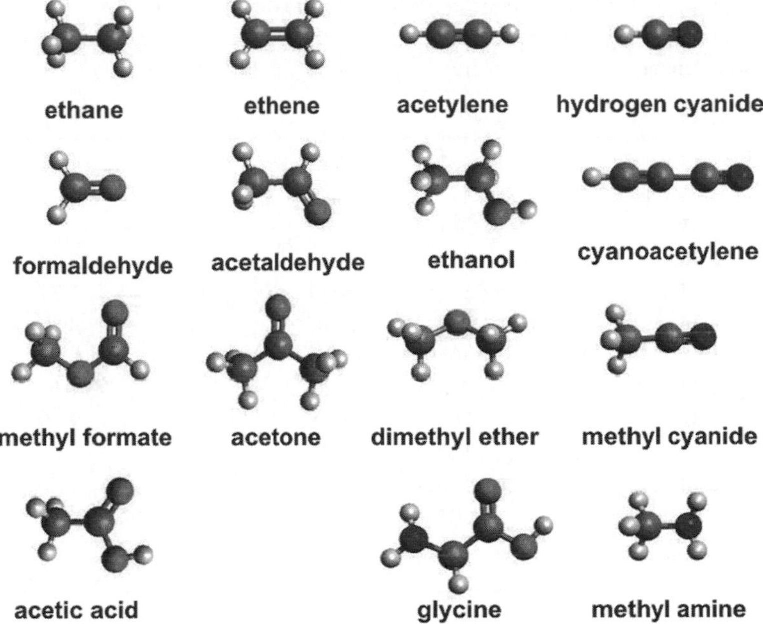

Figure 2.8 3-D ball and stick models of some molecules, illustrating the functional groups of classes of compounds. Dark gray, carbon atoms; light gray, hydrogen atoms; red, oxygen atoms; blue, nitrogen atoms. Note that methane is shown in Figure 2.4, PAHs are illustrated in Figure 8.1, and fullerenes are shown in Figure 8.25. (A black and white version of this figure will appear in some formats. For the color version, please refer to the plate section.)

groups are hydroxyl groups (R-OH) in alcohols, methyl groups (R-CH_3) in, e.g., alkanes, carbonyl groups (C=O) in aldehydes or ketones, carboxyl groups (COOH, consisting of a carbonyl and an hydroxyl group) in organic acids, nitrile groups ($C\equiv N$) in, e.g., cyano polyynes, and amino groups (NH_2) in, e.g., amino acids. Functional groups are the key structural elements that determine the electronic and vibrational spectra of a chemical compound as well as their reactivity, complexation, and solvation. A single compound can have more than one functional group and its associated character.

There is a well-developed formalism for naming chemical compounds that is based upon identifying the parent carbon backbone and the functional groups attached to it. The International Union of Pure and Applied Chemistry, IUPAC, has developed a systematic nomenclature for chemical compounds. Many small molecules also have common names. In the remainder of this section, the characteristics of classes of compounds are discussed. The structures of these classes of compounds are illustrated in Figure 2.8.

2.3.1 Alkanes, Alkenes, and Alkynes

These organic compounds only contain carbon and hydrogen connected by σ and π bonds, and as classes they differ in the presence of single (alkanes), double (alkenes), and triple

Table 2.1 *Average properties of chemical bonds*

Bond	Bond length	Bond energy		Bond	Bond length	Bond energy	
	[Å]	[eV]	[kJ/mole]		[Å]	[eV]	[kJ/mole]
H—H	0.74	4.48	432	C=C	1.33	6.36	614
H—C	1.09	4.28	413	C=N	1.38	6.37	615
H—N	1.01	4.05	391	C=O	1.20	7.72	745
H—O	0.96	4.84	467	O=O	1.21	5.13	495
H—S	1.34	3.60	347	N=O	1.21	6.29	607
C—C	1.54	3.60	347	C≡C	1.20	8.70	839
C—N	1.47	3.16	305	C≡N	1.16	9.24	891
C—O	1.43	3.71	358	C≡O	1.13	11.1	1072
C—S	1.82	2.68	259	N≡N	1.10	9.75	941
N—N	1.45	1.66	160				
O—O	1.48	1.51	146				
N—O	1.40	2.08	201				
Aromatic bonds							
C—C	1.40	5.37	518				
ϕ-H	1.08	4.9	472				
ϕ-C	1.52	4.43	427				
ϕ-N	1.42	4.45	429				
ϕ-O	1.52		464				

(alkynes) bonds. Some well-known alkanes are methane (CH_4; Figure 2.4) – the main constituent of natural gas – and propane (C_3H_8) – used as stove fuel. Ethene (ethylene, C_2H_4; Figure 2.8) is a familiar alkene that is widely used as a precursor to polymers. Note that the polymer polyethylene is formed by breaking the double bond. Alkenes are also known as olefins. Ethyne (acetylene, C_2H_2; Figure 2.8) is used as a fuel in welding torches because of its high temperature of combustion.

As these compounds have low polarity, they have low binding energies and the smallest members of these classes are gaseous at room temperature. For the same reason, they also do not react easily with either electrophiles or nucleophiles. The CH and CC bonds are quite strong and reactions require extreme conditions. Consider alkanes; the CC bond is the weakest (Table 2.1) and, in petrochemistry, cracking is used to convert long chain molecules in shorter ones that oxidize more easily in, e.g., car engines. Because of their low polarity, these compounds do not dissolve easily in water and are only weakly bonded in van der Waals clusters. The π bonds in alkenes inhibit rotations of the skeleton and when, e.g., methyl groups are attached to both sides of the double bond, they can be "locked" at the same or opposite sites. These trans and cis forms are, in some sense, separate chemical species.

2.3.2 Aromatic Compounds

Aromatic compounds, or arenes, are discussed in detail in Chapter 8 and we will only discuss some general properties here. We will focus here on aromatic compounds based on the benzene (C_6H_6) ring (Figure 8.2). The simplest polycyclic aromatic hydrocarbon, naphthalene ($C_{10}H_8$), consists of two fused rings and gives the mothball its odor.

Bonding in these hydrocarbons is through three σ bonds and one π bond. The former gives these compounds their planar structure while the latter provides an electron cloud above and below this plane. The carbon atoms form a regular hexagonal structure with bond angles of 120°. As for other hydrocarbons, the similarity in electron negativity of C and H leads to an apolar molecule that dissolves poorly in water and has no permanent dipole. Benzene is about 1.6 eV (150 kJ/mol) more stable than would be expected if it contained three isolated double bonds.[4] This increased stability is due to the delocalization of the π electron density over all the atoms of the ring. This hampers addition reactions as this energy would have to be supplied. Most reactions are therefore substitution reactions where an H atom is replaced by a functional group. Consider also azulene ($C_{10}H_8$), an isomer of naphthalene, which consists of a 7-ring bonded to a 5-ring where one of the π electrons of the 7-ring is transferred to the 5-ring to give additional stability and a dipole moment. In this case, the added stability is only about half of that in naphthalene.

2.3.3 Fullerenes

Fullerenes are discussed in detail in Section 8.7 and we will only discuss some general properties here. Fullerenes are an allotrope of carbon in the form of cages such as hollow spheres or ellipsoids. The truncated icosahedron with I_h-symmetry, C_{60}, is the smallest stable and, typically, also the most abundant fullerene when using laser vaporization synthesis methods. This fullerene has the shape of a soccer ball (Figure 8.25). The next stable homologue, C_{70}, has an elipsoidal shape reminiscent of an american football. Fullerenes are built up with fused hexagons and pentagons. The latter force curvature of the species as planes cannot be "tiled" with structures containing pentagons. Fullerenes where no two pentagons share an edge, e.g., each pentagon is completely surrounded by hexagons, are the most stable fullerenes. Because of its symmetry, each carbon atom in C_{60} is equivalent. Fullerenes are electron-deficient species and the bonds have an olefinic not an aromatic character. Hence, C_{60} reacts readily with nucleophiles, and addition reactions to the double bonds rather than substitution reactions rule, resulting in exohedral derivaties. Specifically, fullerenes are, thus, prone to H attack and the resonance energy is about 1.6 eV/C-atom. Endohedral species, with atoms embedded inside the hollow cage, can also be formed. Fullerenes can be dissolved in benzene.

[4] To be precise, the resonance energy is the energy released upon hydrogenation of the species minus the energy released when the same number of isolated double bonds are hydrogenated.

Fullerenes, in particular C_{60} and C_{70}, were serendipitously discovered by scientists searching for easy synthesis routes toward astronomically relevant carbonaceous molecules in the mid-1980s. "Milligram-production" of C_{60} was first achieved by scientists in search of analogue materials for interstellar dust some 5 years later. While it was immediately speculated that, in view of it high stability, the fullerene C_{60} could be expected to be very abundant in space, the first detection of C_{60} in space took until 2010 in the IR and 2015 in the visible.

2.3.4 Alcohols and Ethers

Alcohols contain an OH functional group bonded to a sp^3 C atom. The O atom is also sp^3 hybridized where two of the hybrid orbitals form bonds with the C and the H. The other two hybrid orbitals each contain a lone electron pair. Methanol (CH_3OH, wood alcohol) and ethanol (CH_3CH_2OH alcohol; Figure 2.8) are examples of simple alcohols. The phenol class of compounds consists of an hydroxyl group attached to a C-atom in an aromatic ring. Phenol is also the name of the simplest one, C_6H_5OH, which is also known as carbolic acid.

The OH group is highly polar and alcohols have a substantial dipole moment. Alcohols dissolve well in water, but as the carbon structure of the alcohol grows in size, the alcohol also developes alkane character. In phenols, the hydroxyl group has replaced an H atom on an aromatic structure. Phenols are more acidic than alcohols as the O^- anion can resonantly stabilize with the π electron system. Alcohols can react through cleavage of the O–H bond or the C–OH bond. However, alcohols are even weaker acids than water. In a sense, alcohols are like water with one of the H atoms replaced by a carbon chain, and not surprisingly the properties of alcohols are very similar to those of water.

In ethers, the second H atom on water is also replaced by a carbon structure and its properties are very similar as well. Dimethyl ether (CH_3OCH_3; Figure 2.8) is an example of a simple ether. In the past, ethyl ether ($CH_3CH_2OCH_2CH_3$) was widely used as a surgical anesthetic.

2.3.5 Aldehydes and Ketones

Aldehydes and ketones contain a carbonyl (C=O) group. If the carbon in the carbonyl group is connected to at least one H atom, the compound is an aldehyde. Otherwise it is a ketone. Some common aldehydes are formaldehyde (H_2CO), acetaldehyde (CH_3CHO), and benzaldehyde (C_6H_5CHO). Low levels of aldehydes occur naturally in many fruits and vegetables. The human liver breaks down ethanol into acetaldehyde. This toxic compound is then rapidly broken down to the salt anion acetate (CH_3COO^-), which is then decomposed into CO_2 and H_2O and readily eliminated. Acetone (CH_3COCH_3) – primary ingredient in nail-polish remover – is a common ketone.

In aldehydes and ketones, the C atom in the carbonyl group has three σ bonds, which gives this part of the compound a planar structure. The fourth electron makes a π bond to the O atom through sp^2 hybridization. The lone electron pairs on the O atom are therefore also in sp^2 orbitals. The carbonyl group is thus highly polar and hence these species have an appreciable dipole moment. The high electron density on the O atom of the carbonyl acts as a Lewis base and will accept protons. A nucleophile will attack on the carbon atom of the carbonyl to form, for example, an alcohol. As an aldehyde has at least one H atom attached, steric effects are less than for ketones and aldehydes are more reactive. Formaldehyde is prone to polymerization. In water, the lone electron pairs on the O atom of H_2O attack the C atom on the carbonyl group, and after transfer of a proton to the O atom on the carbonyl group, methylene hydrate is formed ($HOCH_2OH$). These methylene hydrate species react with each other to form (small) polymers, $[-CH_2O-]_n$, which in water are known as formalin, the embalming agent. When n gets large, the polymer paraformaldehyde is an insolubel white powder.

2.3.6 Carboxylic Acids

Carboxylic acids contain the carboxyl group; a carbonyl and hydroxyl group attached to the same C atom (COOH). The simplest carboxylic acids are formic acid (HCOOH) and acetic acid (CH_3COOH) and occur naturally in ants and in vinegar. Benzoic acid (C_6H_5COOH) is the simplest aromatic carboxylic acid and occurs naturally in fruits such as cranberries.

The carbon and the oxygen in the carbonyl group as well as the oxygen in the hydroxyl group are sp^2 hybridized, which gives the –COO– group a planar structure. Because of the sp^2 hybridization, one of the lone pairs on the oxygen in the hydroxyl group can conjugate with the π system of the carbonyl group. The OH group is a hydrogen bond donor while the C=O group can act as a hydrogen bond acceptor. Hence, they are soluble in water but as the chain length increases, the compound acquires more of the alkane character. Carboxylic acids are proton donors (e.g. Brønsted–Lowry acids) but they are only weak acids (e.g. they do not fully dissociate in water) compared to mineral acids such as HCl. Nevertheless, they are stronger than the corresponding alcohol, because the carboxylate ion ($-COO^-$) is more stabilized than the alkoxide ion ($-RO^-$) due to sharing of the π electron between the two oxygens.

2.3.7 Esters

An ester is obtained when the H atom on the hydroxyl group of a carboxylic acid is replaced by an alkyl or aryl group. Esters are typically obtained from a carboxylic acid and an alcohol through a dehydration reaction (e.g. $HCOOH + CH_3OH \rightarrow HCOOCH_3 + H_2O$). The simplest esters are methyl formate ($HCOOCH_3$) and methyl acetate (CH_3COOCH_3).

The carbonyl center gives the ester 120° C–C–O and O–C–O angles due to the sp^2 hybridization. The lone pair on the carbonyl oxygen can act as hydrogen bond acceptor and therefore esters are somewhat water soluble, but not as much as their "parents," the alcohols and carboxylic acids. For the same reason, their bonding is weaker and they more easily volatalize. Small esters give fruit and perfumes their fragrant odors. Esters are quite flexible and can easily rotate around the ester O atom. Nucleophiles can react with the carbonyl carbon. CH groups adjacent to the carbonyl are weakly acidic and the presence of a strong base can start condensation reactions.

2.3.8 Amines and Amides

Amines are derivatives of ammonia where one (or more) of the H atoms are replaced by a carbon-containing compound. When the N atom is attached to a carbonyl group, the compound is called an amide (e.g. R–CO–NR′R″). When there might be confusion on which C atom the amine group is attached, the prefix amino is used to name the compound, with a number indicating the location. Methyl amine (CH_3NH_2) is the simplest amine, where one of the H atoms of ammonia is replaced by a methyl group. Trimethyl amine gives fish its "fishy" odor. Aniline ($C_6H_5NH_2$, phenyl amine) is the prototypical aromatic amine and is used, for example, as the basis for dyes such as mauve.

The lone pair on the nitrogen atom gives the compound a tetrahedral structure around the amine group. Amines with at least one H atom still attached can hydrogen bond the H atom to the highly electronegative lone pair electrons of the N atom of another amine. The lone electron pair on the N can always accept a hydrogen bond and hence amines are soluble in water, but solubility decreases as the hydrocarbon chain length increases. Like ammonia, amines are bases as they can donate the lone electron pair of the nitrogen. The strength of the base depends on the substituent. In particular, when the electron pair on the N atom can partake in delocalization of an aromatic ring, the amine is a much weaker base. The lone electron pair on the N atom makes the amine a nucleophile that reacts easily with electrophiles.

Amino acids are in many ways considered to be the holy grail of the molecular universe given that they are the building blocks of proteins. In amino acids, the central C atom is bonded to an amino (NH_2) group, a carboxylic (COOH) group, and an H atom. The fourth sidechain provides the amino acid with its specific property. In glycine (NH_2CH_2COOH), this side chain is an H atom. In alanine ($NH_2CH_3CHCOOH$), it is a methyl group.

2.3.9 Nitriles

Nitriles – also called cyanides – have a C≡N group. The simplest cyanide, hydrogen cyanide (HCN) (prussic acid), is not considered an organic compound. The simplest organic nitrile, acetone nitrile (CH_3CN), is more generally known as methyl cyanide and is often used as a solvent, for example, to remove tars in the oil industry. Cyanopolyynes consist of C-chains with alternating triple and single CC bonds[5] ($(-C \equiv C-)_n$) with a terminal cyano group ($H(C_2)_nCN$). Benzonitrile (C_6H_5CN) is the simplest aromatic nitrile and is also used as a solvent.

[5] Note that polyacetylenes are polymers derived from acetylene and have alternating double and single bonds.

In the cyano group, both the C atom and the N atom are *sp* hybridized, providing a σ bond. The other electrons are in *p* orbitals and form two π bonds, resulting in a triple bond between the C and N atoms. The lone pair on the N atom is in an *sp* orbital. As nitriles do not have an H atom close to a strongly electronegative group, nitriles do not form hydrogen bonds among themselves. These species are linear and very rigid molecules due to the π bonding. However, the high electronegativity of nitrogen pulls the electrons from the triple bond to the nitrogen end of the bond. This gives these compounds a large dipole moment and hence strong intermolecular bonds. They are soluble in water as the lone electron pair on the N atom can act as an electron pair donor to an H atom of water. The *sp* character of the lone electron pair on the N atom makes nitriles less basic than amines. Analogous to the carbon in carbonyl groups, the C atom in the nitrile is electrophilic and therefore reacts with nucleophiles.

2.4 Laboratory Techniques

2.4.1 Spectroscopy

UV, Optical, and Infrared Spectroscopy

In absorption spectroscopy, a spectrometer shines UV/visible/infrared light through a sample, measuring the absorption spectrum by comparison to a reference spectrum of the light source without sample. The light source generally has a broad spectrum and the spectrum is measured by scanning the wavelength using, for example, a monochromator (which lets only only light of one wavelength through at a time). In the vacuum UV, the light of a tunable synchrotron source is used as the probe. Alternatively, a Fourier Transform Spectrometer is used (Figure 2.9), where light from the (broad) light source is split into two beams by a beam-splitter (which transmits and reflects half of the incident light). One beam is reflected off a standing mirror, while the other bounces off a moveable mirror, introducing a time delay. These mirrors reflect the two beams back to the beamsplitter where they are combined again and directed at the sample and then the detector. Depending on the time delay (e.g. difference in distance traveled by the beams) and the wavelength, constructive or destructive interference will occur. After transmission through the sample, the total intensity of the transmitted light is measured. This is repeated at different positions of the moveable mirrors, producing an interferogram (total intensity versus time delay). A spectrum is then constructed using a Fourier transformation from the time domain to the frequency domain.

In cavity ringdown spectroscopy, the species is brought into an optical cavity between two highly (99.9%) reflective mirrors (Figure 2.10). The probe laser light will bounce many times between these mirrors while slowly leaking out. This signal can be recorded as a function of time, providing a decay time constant. When the species in the cavity absorbs the laser light, the signal will drop faster. By scanning the probing laser over wavelength, an absorption spectrum can then be recorded. In essence, the highly reflective mirrors create a pathlength of several kilometers.

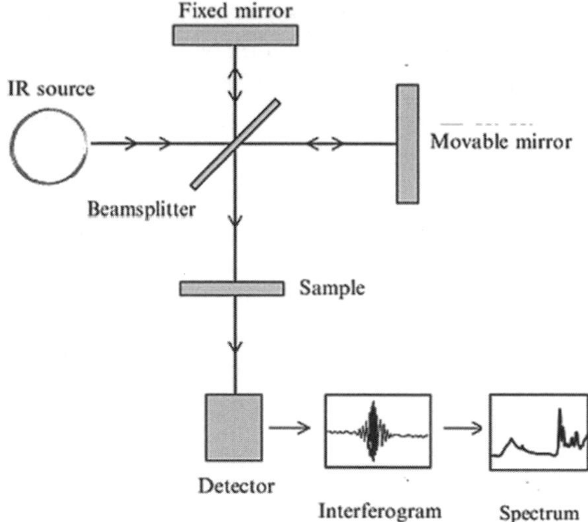

Figure 2.9 Fourier Transform Spectrometer. Light from the broadband source is split into two beams and recombined after bouncing off a fixed and a moveable mirror. After transmission through the sample, the detector measures the integrated light as a function of the position of the moving mirror. This results in an interferogram that is Fourier transformed into the spectrum

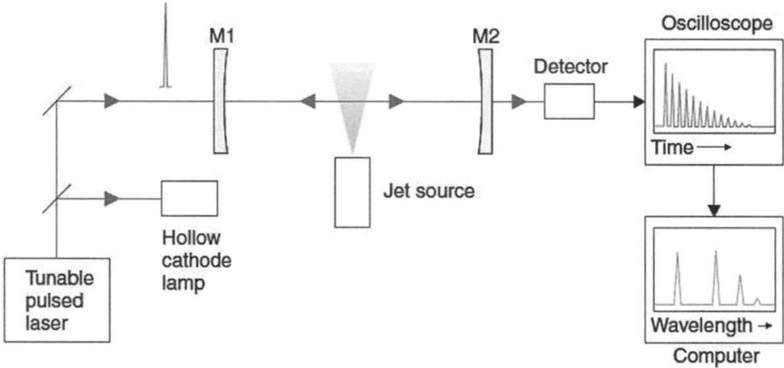

Figure 2.10 Cavity ringdown set up. A laser pulse is injected into an optical cavity between two high reflectance mirrors, M1 and M2. At each reflection, a small fraction of the light is transmitted through M2 and the decay of this signal is monitored by a photomultiplier detector. Comparing the decay time scales with and without the sample (the jet source) present measures the absorption. Scanning of the laser over a frequency range results then in an absorption spectrum. Figure taken from [6]

In Laser Induced Fluorescence, the species is excited by a laser and the fluorescent signal is recorded. A spectrum is then recorded by tuning the laser. This technique does require high fluorescence yields and is therefore not suitable for electronic levels that couple efficiently through interconversion of lower-lying electronic states or that have radiative long lifetimes such as vibrational states.

For simple molecules, these techniques are rather routine. However, for transient, highly reactive species – such as radicals and ions – or for species with low volatility, this technique does not suffice as it is difficult to to get enough species in the region probed to create a detectable absorption signal. Special techniques have been devised to measure the absorption spectra of these species. These generally rely on creating an ion signal that can be monitored using a mass spectrometer. Mass spectrometry provides a very sensitive detection technique, and the "trick" is then to convert an absorption signal into a mass signal.

One such technique is resonant two-color two-photon ionization. In this technique, neutral species are created and cooled down by supersonic jet expansion in a molecular beam. A first (tunable) laser excites the species to a resonant state and a subsequent pulse by a second laser ionizes it out of this selected state. Neither of the two lasers has enough energy to ionize the species but through this resonant state, ions are created that can be sensitively detected by a time-of-flight mass spectrometer. Scanning the first laser over the wavelengths produces an absorption spectrum.

Ion-dip techniques are a variant used to study, e.g., the IR spectra of neutral species. Once the electronic transitions are known, the system can be set up to lead to ionization from the ground state through two photons using lasers at fixed energies. A third laser is then scanned over the relevant (IR) wavelength range. Upon absorption of this laser light, the species will fall out of resonance, resulting in a disappearance of the ion signal. Recording this ion-dip signal as a function of the wavelength of the probe laser results in an absorption spectrum. As this technique uses state and mass selection, ground state properties are measured. Also, isotope effects are filtered out.

Messenger techniques can also be used. In this, the species is complexed with a weakly bonded noble gas atom. Following the procedure outlined above, the cluster cation is produced by the two-color scheme through an excited intermediate state. Careful tuning of the second laser is required to keep the internal energy of the ion cluster small. After production of the internally cold ion cluster, a tunable visible or IR laser is used to probe the absorption. In resonance with a low lying electronic or vibrational state, the increased internal energy will lead to loss of the noble gas atom and this can be recorded through mass spectrometry. For ions, the use of a cryogenically cooled 22-pole ion-trap can cool the species down to very low temperatures using a helium buffer gas.

The messenger technique has as its drawback that the complexed noble gas atom will slightly distort the energy levels involved. In the helium droplet technique, the species is trapped in a helium cluster and then absorption is monitored through evaporative cooling of the cluster as a function of the frequency of the light source. The effect of noble gas on the energy levels can then be assessed by comparing the spectra of droplets with different number of helium atoms.

Matrix Isolation Spectroscopy

Matrix isolation techniques (Figure 2.11) bring a species in the gas phase, mix it with an inert gas – typically a noble gas like argon or neon which is transparent to the probing

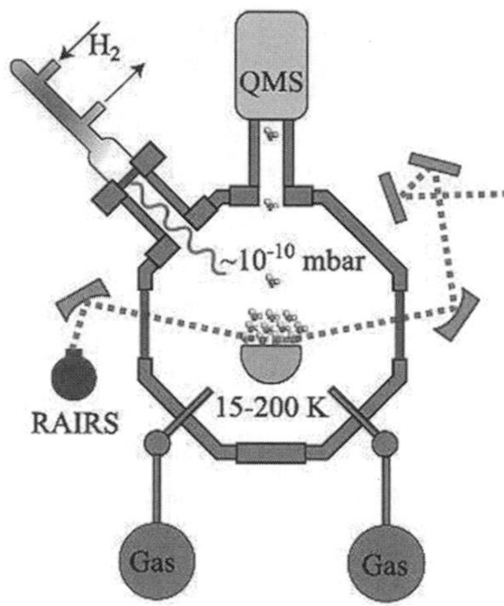

Figure 2.11 In matrix isolation spectroscopy, the species to be studied is codeposited with inert matrix material on a cold finger. The cold finger is rotatable. The ice can be irradiated by light from a (H_2 discharge) lamp. The chemistry can be followed in the IR. If a transparent substrate is used, the IR spectrum is measured in transmission. Otherwise a RAIRS technique can be used. A quadrupole mass spectrometer can be used to analyze sublimated gases after heat up of the sample through a resistive heater (or using spot heating by a laser). Surface chemistry can be studied by turning the sample to an inlet port hooked up to an atom or radical source. Figure taken from [24]

radiation – and freeze it on a cold finger kept at a temperature of, typically, $\simeq$10 K. An IR or UV/visible spectrometer is then used to record the spectrum. If the cold finger is transparent (e.g. saphire) to the probing radiation, a transmission signal can be measured. On a metalic surface (e.g. aluminum), a reflection spectrum is measured. At low concentration, species will have very weak absorptions and, in this case, the spectrum can be probed using Reflection Absorption InfraRed Spectroscopy (RAIRS; Section 2.4.3). Matrix isolation is particularly suited to studying transient species, e.g., radicals or ions. These species can be produced in the gas phase and then codeposited with an overabundance of inert gases. However, often, these species are produced in situ from a parent species using laser or Xenon lamp irradiation. The cold matrix environment allows for rapid quenching of excess energy and isolates the products. In this way, large concentrations can build up. To promote ion production, an electron donor (e.g. Na, K) or scavenger (e.g. CCl_4) can be added. Varying the concentration of these species can also demonstrate that the product is an ion.

The advantage of matrix isolation is that it is simple, can be used to measure a large wavelength range simultaneously, and uses relatively little material. One drawback, when parent species are involved, is to disentangle the absorption of the transient species from that of the parent or from other products. The main issue is, though, interactions of the species with the surrounding matrix, which can lead to substantial frequency shifts and broadening and affect band intensities. Electronic transitions in the optical can (red)shift by $\sim$100–1000 cm^{-1} ($\sim$25 to -250 Å; 0.5–5%), while the shift of vibrational bands in the IR are more modest, $\sim$2–6 cm^{-1}. In either case, this hampers detailed comparison of measured spectra with astronomical spectra.

Matrix isolation techniques are key to understanding the spectroscopic properties of, e.g., impurities trapped in ices. Precise peak positions and width of their vibrational (and electronic) transitions depend on the environment and this is key to analyzing astronomical infrared spectra of such species as CO and CO_2. Matrix isolation spectroscopy is often used in studies of electronic transitions to narrow the wavelength range that has to be measured, as molecular beam experiments typically only scan a small wavelength range at a time. Matrix isolation spectroscopy is more widely used in IR spectroscopy as the shifts are more modest – typically less than the observed width of, say, the Aromatic Infrared Bands. Moreover, trends in spectral properties with molecular structure, ionization, etc. can be trusted. Detailed identifications, though, do require gas phase confirmation.

Rotational Spectroscopy

Rotational spectroscopy is the premier tool to identify gas phase molecules present in space. Heterodyne techniques provide high resolution spectra ($\nu/\Delta\nu \sim 10^6$) and velocity resolve individual rotational transitions. Identification of species requires knowledge of the intrinsic rotational frequencies of astronomically relevant species. Rather than measuring all rotational transitions, this is done by measuring a set of transitions and fitting coefficients in polynomial expansions for the energy levels. Relevant transitions can then be calculated. Molecular rotational spectroscopy will be discussed in detail in Chapter 3. For a linear molecule, the energy levels can be described by the first three terms[6] in an expansion in $J\,(J+1)$, and a limited number of transitions have to be measured at high precision to be able to accurately extrapolate to higher frequencies. For an asymmetric top, on the other hand, there are three rotational constants, five centrifugal distortion terms, and seven higher order terms. Even for relatively simple molecules, some 20–30 constants have to be determined. When large amplitude motions are present – e.g., tunneling in NH_3 or internal rotors such as CH_3 in methanol – many more coefficients are required to accurately describe the rotational spectra. It is clear that measuring accurate rotational spectra is demanding.

2.4.2 Gas Phase Chemistry

Ion-flow Tube

A Selected Ion Flow Tube (SIFT) experiment is basically a tandem mass spectrometer (Figure 2.12). The primary reactant cation is produced, selected by a quadrupole mass filter, and then guided into a high volume flow of a buffer gas. The reagent is then introduced and the loss of reactant or formation of products is recorded downstream by a

6 The first two are the rotational constant and the centrifugal distortion parameter (cf. Chapter 3).

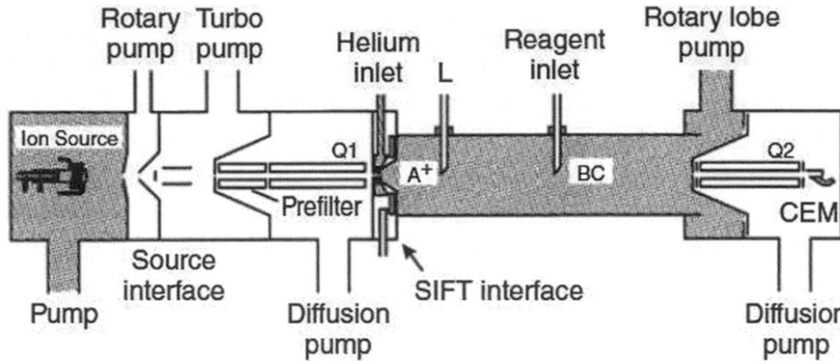

Figure 2.12 Schematic diagram of a selected-ion flow tube (SIFT) tandem mass spectrometer. Ions (A^+) are mass selected by a quadrupole mass filter and injected into the flow tube. Downstream, reactants (BC) are introduced and products (AB^+) of the reaction are measured with a mass spectrometer. Figure taken from [29]

mass spectrometer. Rate coefficients are then measured by varying the concentration of the reagent,

$$k = \Delta\,[\mathrm{AB}]\,/\,[\mathrm{A}]\,/\,(n\,(\mathrm{BC})\,\Delta t)\,, \tag{2.1}$$

where $\Delta\,[\mathrm{AB}]\,/\,[\mathrm{A}]$ is the relative change in the concentration of the reactant (A) and product (AB), n (BC) is the number density of the reagent, and Δt the reaction (flow) time. Heating or cooling (typically down to liquid nitrogen temperatures) of the flow tube addresses temperature dependence. SIFT has been extensively used to study ion–molecule reactions of atmospheric and astrophysical relevance.

Laval Nozzle Expansion

Flow tube set ups can also be used to study neutral–neutral reactions involving radicals. Radicals are produced by various techniques prior to injection into the buffer gas. After introduction of the reagent, the decay of the reactant or the formation of the product is followed downstream in a mass spectrometer. However, the system is much larger and it is difficult to cool the flow down in this case. Measurements at low temperatures require some adjustment in this technique. In the CRESU technique,[7] a Laval nozzle[8] replaces the heart of the flow reactor. The gas expands through this nozzle into a uniform, supersonic flow and this cools it down adiabatically. The design of the nozzle controls the temperature of the gas. This flow is then directed into the reagent background gas. The concentration of the reactants and products can be measured at various points through mass spectrometry or optical means. This technique requires large flow rates and hence heavy pumps. This can be mitigated by using pulsed Laval nozzle expansions.

[7] Cinétique de Réaction en Ecoulement Supersonique Uniforme.

[8] A Laval nozzle is the principle of a rocket. A carefully designed flow tube is pinched in the middle and this converts the thermal motion of highly pressurized gas on one side into a highly supersonic flow on the other side.

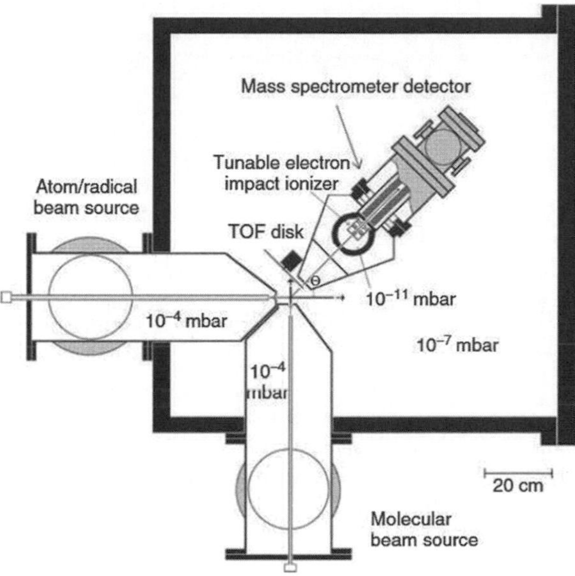

Figure 2.13 Schematic for a crossed molecular beam experiment. The products of the interaction of species in two supersonic beams are probed by a rotatable mass spectrometer. Figure taken from [29]

Crossed Molecular Beams

In crossed molecular beam experiments, two supersonic molecular beams, carrying the two collision partners, intersect at right angles (Figure 2.13). Concentrations and the interaction zone are limited to ensure single collision conditions. The supersonic expansion leaves the molecules internally very cold and with well collimated and unidirectional velocity vectors. The products of the scattering collision can then be measured at various scattering angles using a rotatable mass spectrometer and the differential scattering cross section can be evaluated.

This technique can be used to study reactive as well as inelastic collisions. In the latter, state-to-state excitation cross sections can be measured. The velocity of the products are set by energy and momentum conservation as the known energy and velocity of the collision partners is fixed by the velocity and crossing angle of the molecular beams. Products with different levels of internal excitation are then linked through velocities, following Newton's laws. By repeating the experiment numerous times under the same conditions, the full scattering process for each product (e.g. with different internal excitation) can be probed.

The crossed molecular beam set-up leads to high collision energies. In merged beams, collision energies can be manipulated to reach very small values. For charged particles (ions, electrons), electric fields can be used to merge the beams. As products are scattered out of the beam, it is easier to monitor a reaction by measuring the decay of the reagents (cf. Eq. (2.1)).

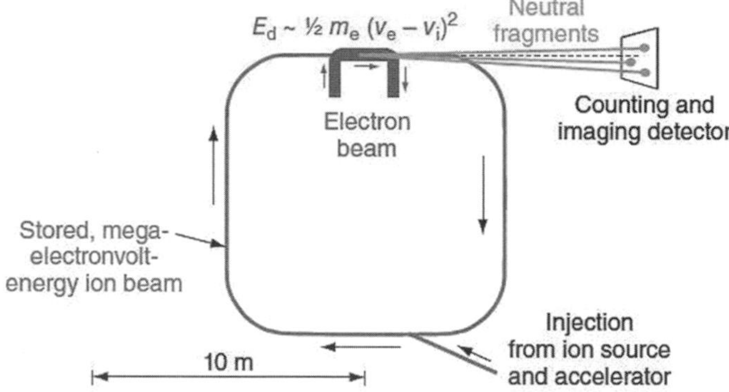

Figure 2.14 Schematic for a storage ring set up to measure dissociative electron recombination reactions. Ions are produced, accelerated, and stored in a storage ring. This ion beam interacts with electrons in a parallel beam, whose energies can be controlled, and the neutral fragments are detected downstream. Figure taken from [29]

Storage Rings

Thermal electron recombination rate coefficients are generally measured using a modification of the ion flow tube set up. In an FALP experiment[9] ions are produced by pulsed discharge, carried downstream by a buffer gas (e.g. helium), and a Langmuir probe is used to measure the electron density along the flow length. To reach low collision energies, storage rings can be used (Figure 2.14). One advantage over a merged beam is that the ions can be circulated many times and hence their internal excitation can be controlled much better. After introduction of the ion into the storage ring, the ion beam is merged – over a limited length – with an electron beam, and the interaction energy can be well controlled. Relative collision energies can now be brought down to as low as $\sim$1 meV and reaction rate coefficients down to 10 K can be determined. Neutral fragments decouple from the ring and can be detected using mass spectrometry to determine reaction rate coefficients, products, and branching ratios.

Ion Traps

Paul ion traps use static and oscillating electric fields to store ions for extended periods ($\sim 10^{-3}$–10^{2} s) in a small volume. Species are sublimated in an oven, ionized by an electron gun or UV laser, and then mass selected and transported into the ion trap through an iongate and a quadrupole mass filter. The ions are confined in the ion trap in the potential well created by applied electric fields. By applying a SWIFT pulse (Stored Waveform Inverse Fourier Transform) to the electrodes, unwanted species (masses) can be forced to absorb energy that ejects them from the trap. Helium background gas ($\simeq 10^{-6}$ bar) can cool the ions and helps keep the gyrating ion cloud confined. A high order multipole trap has

[9] Flowing Afterglow with Langmuir Probe.

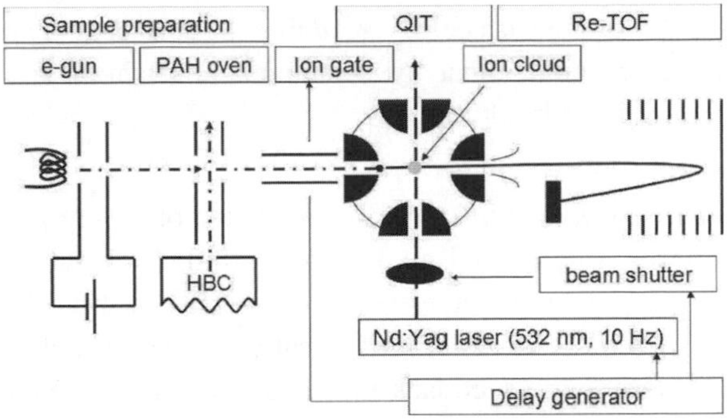

Figure 2.15 Schematic of an ion trap setup. From left to right, molecules are sublimated in an oven (in this case, the PAH, hexa-peri-hexabenzocoronene, HBC), ionized by an electron gun, and guided through an ion gate to a quadrupole ion trap (QIT), where the ions can be stored for $\sim$1 s. The ions can be irradiated by short (ns) Nd-Yag laser pulses. Afterward, the ion trap can be swept and the products analyzed using a time-of-flight (TOF) mass spectrometer. Figure taken from [35]

a large field free area where the ions can be effectively cooled down by collisions with helium buffer gas. In a 22-pole ion trap, temperatures as low as 10 K can be reached. The reagent can be cooled prior to introduction into the trap. The ions can also be irradiated by lasers to study their photochemical evolution. At set times, the ion trap is opened and reactants and products are extracted into a mass spectrometer. Reaction rate coefficients can be determined by varying the trapping time. As reagent densities of 10^{10} cm^{-3} can be reached in an ion trap and mass spectrometry is a very sensitive detection technique, relatively small reaction rate coefficients can still be measured (cf. Eq. (2.1)).

2.4.3 Surface Chemistry

Imaging Techniques

Scanning Electron Microscopy (SEM)

In scanning electron microscopy, a very narrow (4 − 40 Å) beam of energetic (0.1 − 50 keV) electrons from an electron gun is directed at a sample surface using lenses. This beam can be manipulated to scan the surface. The interaction of this beam with the surface can lead to reflection of the primary electrons, production of secondary electrons, and fluorescence, each of which can be recorded. Typically, secondary electrons are monitored as they originate $\simeq$30 Å in the surface. The number of electrons that are released depends on the orientation of the sample surface with respect to the beam and edges are, of course, highly limb-brightened. This gives the images a 3D feeling.

Scanning Tunneling Microscopy (STM)

In Scanning Tunneling Microscopy (STM), a very sharp conducting tip is brought to the surface. By applying a small bias between the tip and the surface, electrons can tunnel and the current will depend on the distance between the two, the voltage, and the local density of states (LDOS) of the surface. By scanning the surface, a map of the LDOS can be constructed. This can be done either by scanning at constant height or by keeping the current constant (through a feedback loop).

Atomic Force Microscopy (AFM)
In Atomic Force Microscopy, a very small tip is mounted on a spring-like cantilever. The free end of the cantilever is monitored by a detector. In this way, atomic scale variations on the surface are transformed into macroscopic variations. The tip is brought into contact with the surface and then the surface is methodically scanned. Height variations produce deflections of the cantilever and a feedback loop keeps the probe at a constant distance. In this way, the surface topology is mapped.

Both STM and AFMs can reach atomic-scale resolution without the use of energetic probing beams or optics, with their limitations. However, these techniques do not provide information on the nature of adsorbates. Theoretical modeling using, e.g., density functional theory can assist in this respect, though.

Analysis Techniques

Reflection Absorption Infrared Spectroscopy (RAIRS)
Transmission infrared spectroscopy is often used to identify the chemical character of species trapped in a sample (Section 2.4.1). This technique has been used extensively, for example, to monitor the chemical changes during processing of ices by UV photons or energetic ions. However, it is difficult to measure the absorption from the minute concentrations of adsorbates. In Reflection Absorption InfraRed Spectroscopy (RAIRS), the sample is mounted on a highly reflective metallic surface. The IR beam is directed under a highly glancing angle. The long pathlength makes this a more sensitive technique than transmission spectroscopy. However, sensitivity limitations remain an issue.

Temperature Programmed Desorption (TPD)
Species are adsorbed on a cryogenically cooled surface. These adsorbates can be trapped or, if mobile, can react. The sample is then heated in slow stages and at any time the flux of evaporating species is measured with a quadrupole mass spectrometer. The mass of the species of interest is then followed as a function of time as the temperature rises, typically, linearly. In this way, the mass of the species is measured as well as the temperature(s) at which it sublimates. The latter is a measure of the binding energies of the sites in which the species is trapped and, thus, probes the surface topology, including the presence of multiple binding sites, defects, kinks, and porosity. Typically, sites on the surface are characterized by a distribution of binding energies. This distribution can be probed by adsorbing inert species – e.g. helium atoms atoms or H_2 or N_2 molecules – and measuring their TPD spectra. By measuring the TPD signal after exposure of the surface to different fluences (e.g. different monolayer coverages), reactions that occur during the warm-up process can be studied as well. The temperature at which newly formed molecules are released will depend on the binding of the reacting species to the surface (setting the onset of migration) and the surface coverage (setting the distance over which reactants have to

migrate). Typically, with increasing coverage, newly formed species will be released at lower and lower temperatures (so-called second order behavior). This can be analyzed to derive migration and/or activation barriers. However, the effects of migration and a distribution of binding sites are often difficult to disentangle.

Laser Desorption
In laser desorption, a thin laser beam is directed at the sample that heats it locally to a high enough temperature to induce desorption. The ensuing plume of gas can then be analyzed by a quadrupole mass spectrometer. As only a small portion of the sample is thermally cycled, the chemical evolution during, say, a photolysis experiment can be monitored over time (e.g. fluence) with this technique. The internal excitation of the desorbing species can be probed using a resonant enhanced multiphoton ionization scheme.

Ice Chemistry

Thin layers of an ice are grown on a cryogenically cooled substrate in an ultrahigh vacuum ($p \sim 10^{-10}$ mbar) chamber. Typical temperatures of the cold finger are 10 K, comparable to dust temperatures inside dense clouds ($T_d \simeq 6-10$ K) but, with the proper cryostat, can go as low as 3 K. The ice composition is set by mixing relevant species in a gas bulb. The thickness of the ices is controlled by the flow rate and the exposure time and can be measured by monitoring laser interference fringes produced by the film during deposition.

Chemical reactions on the surface can be studied by directing a low energy atom beam through a separate beam line. Multiple beam lines can be used to study more complex chemistry. Atom or radical sources can be produced by thermal cracking or through a microwave discharge. Atom fluxes are in the range of $10^{10}-10^{14}$ atoms/cm^2/s, depending on the type and settings of the atom source. Reaction products are monitored with TPD or RAIRS. The chemical pathways can be elucidated by changing the atom exposure time of the ice surface. Or, when the ice and atoms are codeposited, by changing the relative flow rate of the atom beam. Calibration of the atom source is nontrivial as efficiency of atom production is not unity and recombination can occur on the walls of the beam line or – at high fluences – on the sample surface. Absolute reaction rates are very difficult to obtain. This is particularly true for reactions with H, as this atom is highly mobile on the surface and most of the deposited atoms are lost through H_2 formation. Relative rates – actually most relevant for astrochemistry – for competing reactions with different species are more easily determined, if the reaction rates are not too different. Extrapolation to different temperatures does need some care, though, and a proper description of the interaction (e.g. tunneling). Variations in the distribution of binding energies may also have an influence on the results.

Chemical reactions in the bulk of the ice can be promoted by UV photolysis or energetic ion bombardment. Products are analyzed – in situ – at set times by IR transmission or RAIRS techniques. At the end of the processing experiment, the ice can be heated to a set temperature and desorbed species can be analyzed using a quadrupole mass spectrometer. However, the products may also be formed during the warm up and desorption process.

The laser desorption method can be used to monitor the chemistry during the irradiation process, but still chemistry during the thermal processing may affect the results.

2.5 Quantum Theory

Quantum chemistry is an excellent tool to address key issues in chemistry and to give deep insight in the characteristics of species and their reactions, including the equilibrium geometry, the charge distribution, the electronic and vibrational spectral characteristics, the specific states involved in transitions, the characteristics of the transition states involved in a reaction, the structure of reaction intermediaries, and the reaction energetics and energy barriers.

In quantum chemistry, the Schrödinger equation is solved (generally with approximations),

$$\hat{H}\Psi_i\left(\vec{x}_1,\vec{x}_2,\ldots,\vec{x}_N,\vec{R}_1,\vec{R}_2,\ldots\vec{R}_M\right) = E_i\Psi_i\left(\vec{x}_1,\vec{x}_2,\ldots,\vec{x}_N,\vec{R}_1,\vec{R}_2,\ldots\vec{R}_M\right). \quad (2.2)$$

With $\hat{H}$ and Ψ, the Hamiltonian and the wave function describe the system consisting of N electrons and M atoms. The vectors $\vec{R}_j$ and $\vec{x}_l$ contain the coordinates of nucleus j and coordinates and spins of electron l, respectively. E_i is the energy of state i. The Hamiltonian is given by

$$\hat{H} = -\frac{1}{2}\sum_{j=1}^{N}\nabla_j^2 - \frac{1}{2}\sum_{l=1}^{M}\nabla_l^2 - \sum_{j=1}^{N}\sum_{l=1}^{M}\frac{Z_l}{r_{il}} + \sum_{j=1}^{N}\sum_{k>j}^{N}\frac{1}{r_{jk}} + \sum_{l=1}^{M}\sum_{f>l}^{M}\frac{Z_l Z_f}{R_{lf}}, \quad (2.3)$$

where the summations j and k run over the N electrons while l and f run over the M nuclei. In the Hamiltonian, the first two terms represent the kinetic energy of the electrons and nuclei, the third term is the electrostatic interaction between the electrons and nuclei, and the last two terms are the repulsive interaction between the electrons and between the nuclei. Note that to keep this expression simple, atomic units are used.[10] Realizing that the mass of the atoms is much larger than that of the electrons, we can assume that the electrons are moving in the background field of stationary nuclei, the Born–Oppenheimer approximation. The nuclear kinetic energy is then zero and their potential energy is constant and can be ignored. The Hamiltonian reduces then to three terms, the kinetic and potential energy of the electrons in the field of the atoms, and the electron interaction energy,

$$\hat{H}_{elec} = \hat{T} + \hat{V}_{Ne} + \hat{V}_{ee} \quad (2.4)$$

$$= -\frac{1}{2}\sum_{j=1}^{N}\nabla_j^2 - \sum_{j=1}^{N}\sum_{l=1}^{M}\frac{Z_l}{r_{il}} + \sum_{j=1}^{N}\sum_{k>j}^{N}\frac{1}{r_{jk}}. \quad (2.5)$$

[10] Atomic units: $e = \hbar = m_e = 1$. The length scale is then given by the Bohr radius, $a_0 = \hbar/m_e e^2 = 5.29 \times 10^{-9}$ cm = 0.529 Å and the energy scale is in Hartrees, $\hbar^2/m_e a_0^2 = 4.36 \times 10^{-11}$ erg, twice the ionization energy of atomic H.

For the electrons, the Schrödinger equation can be written in terms of the electronic wavefunction, $\hat{H}_{elec}\Psi_{elec} = E_{elec}\Psi_{elec}$, and energy, E_{elec}. The total energy is then $E_{tot} = E_{elec} + E_{nucl}$ where the latter is just the constant nuclear potential energy, $E_{nucl} = \sum_{l=1}^{M}\sum_{f>l}^{M}\frac{Z_l Z_f}{R_{lf}}$. Once Ψ_i is known, all the relevant properties can be computed. Specifically, if we have a trial electronic wavefunction, Ψ_{trial}, the expectation value for the energy is given by,

$$E_{trial} = \int \cdots \int \Psi^{\star}_{trial}\hat{H}\Psi_{trial}\, d\vec{x}_1 \ldots d\vec{x}_N \tag{2.6}$$

$$= \langle \Psi_{trial} | \hat{H} | \Psi_{trial} \rangle . \tag{2.7}$$

The $\Psi^{\star}$ indicates the complex conjugate of Ψ and, in the second line, the compact Dirac notation has been introduced. The energy, E_{trial}, is a functional as its value depends on the function adopted for the electronic wave function, Ψ_{trial}. This calculated "trial energy" will always exceed the actual energy of the ground state unless the ground state wave function is used. So, the strategy is to minimize E_{trial} by searching through "all" N-electron wavefunctions. Here, "all" is limited to a cleverly chosen set of wavefunctions that allow, for example, algebraic evaluation.

With the Born–Oppenheimer approximation, the problem has been subdivided in two separate steps. When the nuclei are specified (e.g. their position and charges), the external field the electrons are moving in is known and the Hamiltonian can be constructed. Then, the ground state electronic wavefunction can be determined for the specified nuclear structure. The nuclei can be considered to move in the potential generated by the electrons and their positions can now be updated. Next, the electronic wave function can be iterated upon with these new nuclear positions. After convergence, the properties of the system can be calculated.

Physical observables are always associated with the square of the wavefunction and, for example, the probability to find an electron in the volume, $d\vec{x}_1 \ldots d\vec{x}_N$, is $|\Psi_i(\vec{x}_1, \vec{x}_2, \ldots, \vec{x}_N)|^2 d\vec{x}_1 \ldots d\vec{x}_N$. Electrons are fermions and the Pauli exclusion principle requires that the wavefunction is antisymmetric with respect to the exchange of the (spatial and spin) coordinates of any two electrons,

$$\Psi_i\left(\vec{x}_1, \vec{x}_2, \ldots, \vec{x}_i, \vec{x}_j, \ldots, \vec{x}_N\right) = -\Psi_i\left(\vec{x}_1, \vec{x}_2, \ldots, \vec{x}_j, \vec{x}_i, \ldots, \vec{x}_N\right). \tag{2.8}$$

2.5.1 Hartree–Fock

We need to approximate the electron wave function, and for an N-electron system we can do this by expansion in N-electron basis functions. For a function of one variable, x_1, if we have a complete set of basis functions $\{\psi_i(x_1)\}$, then any function can be represented as a linear combination, $\Psi(x_1) = \sum_i a_i \psi_i(x_1)$. For a function of two variables, we can write for any fixed x_2, $\Psi(x_1, x_2) = \sum_i a_i(x_2)\psi_i(x_1)$. As $a_i(x_2)$ is a function of one variable, we can write, $\Psi(x_1, x_2) = \sum_{ij} b_{ij}\psi_i(x_1)\psi_j(x_2)$ and we can readily extend this to N variables. Following this, in the Hartree–Fock method, the electron wave function, Ψ_{HF}, is approximated as a single Slater determinant of the N one-electron spin-orbitals, ψ_j. So, each ψ_j depends on the coordinates of one electron,

$$\Psi_{HF} = \frac{1}{\sqrt{N!}} \begin{vmatrix} \psi_1(\vec{x}_1) & \psi_2(\vec{x}_1) & \dots & \psi_N(\vec{x}_1) \\ \psi_1(\vec{x}_2) & \psi_2(\vec{x}_2) & \dots & \psi_N(\vec{x}_2) \\ \vdots & \vdots & & \vdots \\ \psi_1(\vec{x}_N) & \psi_2(\vec{x}_N) & \dots & \psi_N(\vec{x}_N) \end{vmatrix} \tag{2.9}$$

where this form ensures that the electrons are indistinguishable and that the electron wave function is antisymmetric with respect to exchange of the coordinates of any two electrons. The one electron spin-orbitals consist of the product of a spatial orbital, $\phi_i(\vec{r})$ and one of two spin functions α and β representing spin up and spin down.[11] The spin orbital is proportional to the probability to find the electron in $[\vec{r}, \vec{r} + d\vec{r}]$, and typically these spin orbitals are chosen to be orthonormal ($(\langle \phi_i | \, | \phi_j \rangle = \delta_{ij}$ with δ_{ij} the Kronecker delta function. With these choices, we arrive at N-coupled equations for the spin-orbitals.

The one-electron wave functions are often replaced by a linear combination of atomic orbitals. For computational ease, they are actually often approximated by a combination of gaussian-type orbitals. These standardized sets of atomic orbitals are called basis sets. The coefficients in this combination are found by minimizing the energy. It should be realized that the orbitals are found by minimizing the energy but the energy depends on the orbitals. Hence, starting with guessed orbitals, the system is solved iteratively using numerical methods. Convergence is achieved when the calculated field is self-consistent with the assumed field. For that reason, this method is also called the self-consistent field method (SCF).

Without going into the algebraic details, in essence, the Hartree–Fock method boils down to replacing the complicated two-electron repulsion term in the Hamiltonian by the average electronic repulsion of each electron in the field of all other electrons, which reduces the problem to calculating the behavior of N non-interacting electrons in an effective potential. This simplifies the calculation but the electronic wavefunctions are only an approximation and will always yield an energy that is too high. That is, as the electrons are moving in an average field in the Hartree–Fock approximation, they are allowed to get too close together, resulting in too large an electrostatic repulsion term. Basically, the electron "creates" a hole around itself where, because of the Pauli exclusion principle, no other same-spin electrons are allowed.

Dealing with corrections for this correlation energy is at the heart of ab initio quantum chemistry. A number of approaches have been developed to account for electron correlation. These post Hartree–Fock methods include Configuration Interaction, in which the wave function is represented by a combination of Slater orbitals. The Hartree–Fock method describes the ground state of the system with one Slater determinant. In Configuration Interaction, electrons are kept apart by admixing in higher excitations, which are represented by

[11] These spin orbitals have the property that $\langle \alpha | \, | \alpha \rangle = \langle \beta | \, | \beta \rangle = 1$ and $\langle \alpha | \, | \beta \rangle = \langle \beta | \, | \alpha \rangle = 0$.

further Slater determinants. We can then truncate this after including only single electron excited states, double electron excited states, etc. If all excited states are included, the exact wave function is constructed.[12] The energy is then minimized with respect to the coefficients in the expansion. The Coupled Cluster post Hartree–Fock method also accounts for electron correlation effects by including excited electron states. However, instead of a linear expansion of the wavefunctions in terms of excited electronic states, the Coupled Cluster method uses an exponential expansion that is truncated after two terms (double excitation). This accounts in an approximate way for higher order terms as products of lower order terms.

Hartree–Fock by itself reaches accuracies of 0.02 Å in bond lengths, 10% in vibrational frequencies, and $1-1.75$ eV for bond dissociation energies. These accuracies can be improved upon by using correlation wave function methods but at great computational cost. Hartree–Fock itself has a formal scaling with N^4, but these improved methods scale with N^5 or higher. One advantage of Configuration Interaction and Coupled Cluster methods is that by increasing the number of terms included, the description of the system will be systematically improved. However, the computational requirements (e.g. storage as well as CPU time) increase rapidly with the system's size and only very small molecules can be studied this way.

2.5.2 Density Functional Theory

Because of its versatility, accuracy, and ease of use, density functional theory (DFT) has become the premier tool in quantum chemistry. Moreover, compared to experiments, DFT calculations are low cost and are often done in parallel to elucidate the actual workings of the experiments, including identifying reaction pathways and key intermediaries. Furthermore, any calculation can be very readily expanded to a wide range of related species to investigate trends. DFT calculations are good at characterizing molecular structures ($1-2\%$ in bond length), vibrational frequencies ($5-10\%$ but, after empirical corrections for anharmonic effects, ± 10 cm^{-1}), reaction enthalpies and free energies ($\simeq$0.2 eV, $\simeq$20 kJ/mole), and electronic transitions ($\simeq$0.2 eV, 500 Å). When anharmonicity is accounted for by including up to quartic force field terms in the Taylor expansion, then vibrational frequencies become as accurate as 0.5% without a correction factor.

In density functional theory, the N-electron wavefunction is not computed anymore. Rather, the focus is on the electron density distribution where it is realized that all the properties of the system are known if the electron density distribution is known. The electron density is given by the integral over the spin coordinates of all electrons and all but one of the spatial coordinates,

$$\rho(\vec{r}) = N \int \cdots \int |\Psi(\vec{x}_1, \vec{x}_2, \ldots, \vec{x}_N)|^2 \, ds \, d\vec{x}_2 \ldots d\vec{x}_N, \tag{2.10}$$

[12] Of course, within the underlying one-electron basis set used for expanding the orbitals.

where the integral represents the probability that one electron with arbitrary spin is located within $[r, r + dr]$ in the state represented by Ψ. The electron density distribution is actually an observable that can be determined through, e.g., X-ray diffraction or STM and AFM techniques. By reformulating the theory in terms of the electron density distribution, the N-body problem, which depends on $3N$ coordinates, has been reduced to a 1-body problem depending on three coordinates.

An energy functional can now be defined in terms of the electron density distribution that can be minimized iteratively to provide the correct energy density distribution function. Again, the theory considers N non-interacting electrons moving in an effective potential provided by the ions. But the energy functional includes an exchange correlation functional that depends on the electron density distribution function. As in the Hartree–Fock method, DFT solves N equations iteratively.[13] The orbitals are expanded over a basis set, but the basis set requirements are much more modest than in correlation methods. Essentially, the basis sets only need to represent the one-electron distribution as the inter-electron cusp (taking into account that no two electrons can occupy the same space) is included in the effective potential. That is the basic difference between DFT and Hartree–Fock. The exchange correlation part of the energy functional is approximated and often separate terms for the exchange and the correlation terms are used. For chemical applications, a variety of exchange-correlation functionals have been developed. B3LYP is a common functional employed, named after the authors (Becke for the exchange and Lee, Yang, and Parr for the correlation parts) where the exchange part combines the Becke exchange functional with the exact energy from the Hartree–Fock method. These functionals contain many adjustable parameters that are fitted to a number of well-studied molecules. It is therefore important to realize that validation against experiments is a key aspect for DFT as there is no systematic way to improve the functionals.

2.5.3 Molecular Dynamics

Numerical atomic simulations of the evolution of molecular systems can provide great insight in the behavior of excited molecules. In molecular dynamics studies, the characteristics of a molecular system are evaluated by generating realizations of the system through integration of the classical mechanics equations of motion on a potential energy surface. In this way, time averaged positions, velocities, and forces of the individual atoms in a molecule – potentially in a solvent, ice, or on a surface – are obtained and thus, e.g., the frequencies of vibrational modes, the movement of an excited species in or on the surface of an ice, or the fragmentation of a highly excited species can be followed. This allows calculation of absorption spectra and rate constants.

In molecular dynamics studies, the behavior of an excited molecule is studied by solving Newton's equations of motion. It requires a prescription for the interparticle interactions

[13] Iteratively, as the effective potential depends on the electron density distribution.

that provide the forces acting on the atoms in the molecule and these are used to update the positions and velocities of the atoms. That is,

$$\vec{F}(\vec{x}) = -\nabla V(\vec{x}) \tag{2.11}$$

$$\frac{d\vec{x}}{dt} = \vec{v} \tag{2.12}$$

$$\frac{d\vec{v}}{dt} = \frac{\vec{F}(\vec{x})}{m} \tag{2.13}$$

where $\vec{x}$, $\vec{v}$, and $\vec{F}$ represent the coordinates, velocity components, and forces of all N atoms. This provides a set of $6N$ ordinary differential equations that can be integrated numerically.

This is a microcanonical system with a fixed energy but to set up the system it may be connected to a thermal bath to keep the temperature constant. The thermal bath will act as a thermostat adding or subtracting thermal energy to keep the temperature constant. The temperature is given by,

$$T(t) = \frac{1}{3Nk} \sum_{i=1}^{N} m_i v_i^2. \tag{2.14}$$

At each temperature, the velocities are then scaled in such a way that,

$$\frac{dT(t)}{dt} = \frac{1}{\tau}\left(T_{bath} - T(t)\right), \tag{2.15}$$

where τ is a timeconstant over which equilibration will be achieved. A typical scaling factor for the velocities is,

$$f_v = 1 + \frac{\delta t}{2\tau}\left(\frac{T_{bath}}{T(t)} - 1\right). \tag{2.16}$$

After equilibrating to a set temperature, the thermal batch can be disconnected and the system evolved as a microcanonical ensemble.

In practise, then, starting off with a set of initial conditions for the positions of the atoms, velocities are chosen randomly from a Maxwellian distribution. The forces can then be calculated. The positions and velocities are evolved over a small time-step and the forces can be updated. Efficient and stable numerical algorithms exist to keep the accuracy up but the accumulation of errors is what limits the overall time a system can be evolved. Given enough time, the system will sample the Boltzmann distribution. Each time-step samples a new conformation of the system and these are averaged. The ergodic hypothesis allows the replacement of the ensemble-averaged properties with the time-averaged properties. As there are many initial realizations possible with this energy, these initial conditions have to be sampled through a Monte Carlo calculation.

The interparticle interactions can be specified by prescribing a potential energy surface – either calculated or semi-empirically – or by solving at each timestep the Schrödinger equations to update the interaction. The "quality" of these methods depend on the description of the interaction. The potential energy surface could be described by bonded terms – bond stretching, bending, and rotations – and nonbonding terms – van der Waals forces and electrostatic forces. Alternatively, the classical dynamics calculation could be combined with an electronic structure calculation – at each timestep – to provide updated interaction potentials. In contrast to molecular dynamics (MD) methods, such ab initio molecular dynamics methods (AIMD) do not require a prespecified potential energy surface and can follow "chemical" events. But, this comes at great computational costs. In MD, the nonbonding, electrostatic forces drop off slowly and dominate the computational cost as they scale with N^2. Truncation can then be helpful. Energy conservation is an issue, as small numerical round off errors will build up over time. Typical timesteps are a few femto seconds (10^{-15} s) and in MD the system can be evolved for nano- or even micro-seconds. For AIMD calculations, the electronic structure calculations are the time-consuming part and only nano-seconds are presently achievable.

2.6 Further Reading and Resources

References [5, 9] provide easily readible guides through the wondrous universe of the elements, and their basic properties and characteristics. Reference [28] presents a charming personal account of the fascination provided by the elements and their compounds, while the autobiographical writing of [26] weaves the elements through the darker experiences associated with fascist Europe in the Second World War. An entertaining discussion of the properties of molecules in every day life is presented in [1]. There are many excellent introductory textbooks on chemistry; c.f. [4]. Reference [25] is an in-depth look at the properties of chemical bonds.

There are web-based tools that provide help in naming chemical compounds; e.g. https://web.chemdoodle.com/demos/iupac-naming/, which allows you to draw a compound and provides its IUPAC name and www.chemspider.com, which translates chemical or condensed structural formulae into structures and names. The Avogadro program is an open source, multi-platform code that can visualize molecular structures [11]. Avogadro can be downloaded from http://avogadro.openmolecules.net/.

Fullerenes were first discovered in laser vaporization methods of graphite rods geared toward producing carbon clusters ([19]). Efficient production of C_{60} was first achieved by [18].

Laboratory techniques and their results are reviewed in [30], in particular [22, 27, 29]. Reference [31] provides an overview of the study of gas phase reactions of astrophysical relevance, while dissociative electron recombination reactions are reviewed by [10]. Surface chemistry techniques that have been widely used in astrochemistry have been described by [13].

There are many excellent textbooks on quantum chemistry, for example [32]. References [2, 8] review the popular post Hartree–Fock methods: configuration interaction and coupled clusters. The original references to DFT are [12, 17]. Reviews can be found in [14, 16].

Reference [15] provides an excellent in-depth introduction into density functional theory for chemists. Recent perspectives on DFT are provided by [3, 34]. Reference [23] provides an overview and critical assessment of density functionals commonly used in DFT.

Reference [7] is an excellent textbook on computation chemistry. References [20, 21] provide reviews of molecular dynamics studies. Ab-Initio Molecular dynamics studies are reviewed in [33]. There are a number of molecular dynamics software packages available through the web: The Amber package (http://ambermd.org) and the Gromacs package (www.gromacs.org) are widely used, particularly in the biochemistry community.

2.7 Exercises

2.1 Electron configuration

- Draw the electron configuration of atomic silicon.
- What is the spin?

2.2 Enjoy Tom Lehrer's the elements (www.youtube.com/watch?v=AcS3NOQnsQM) but only the compulsive neurotics will learn this by heart.

2.3 Ionization potentials

- Why is the ionization potential of nitrogen higher than for carbon?
- Why is the ionization potential of nitrogen higher than for oxygen?

2.4 What is the weight of a mole of glucose ($C_6H_{12}O_6$)?

2.5 Molecular geometry

- What is the shape of the hydronium cation, H_3O^+? Explain your answer.
- What are the bond angles in acetic acid (CH_3COOH)? Explain your answer.
- Why is water polar?

2.6 Chemical bonds

- What are the similarities and differences between ionic and covalent bonded species?
- Which of the following species are ionic and which are covalent compounds: NaCl, CCl_4, H_2O_2, Na_2CO_3? If the compound is ionic, write the symbols for the ions involved.
- Describe the bond of HF and provide a sketch of its electron cloud.

2.7 Valence theory

- Construct the bonding picture for HC≡N: (where : indicates the lone pair of electrons on the N atom).
- Sketch the orbitals.
- Explain your answer.

2.8 Molecular characteristics

- Describe the differences between alkanes, alkenes, and alkynes.
- Why is ethanol more soluble in water than butanol?
- Why is a carboxylic acid more polar than its respective alcohol?

2.9 Compare the advantages and disadvantages of different spectroscopic analysis techniques.

Bibliography

[1] Atkins, P. W., *Molecules* (New York: Scientific American Library)
[2] Bartlett, R. J., Musiał, M., 2007, *Rev Mod Phys*, 79, 291
[3] Becke, A. D., 2014, *J Chem Phys*, 140, 18A301
[4] Blackman, A., Bottle, S. E., Schnod, S., Mocerino, M., Wille, U., 2012, *Chemistry* (Australia: John Wiley and Sons)
[5] Callery, S., Smith, M., 2017, *The Periodic Table* (New York: Scholastic Inc.)
[6] Carpentier, Y., Rouillé, G., Steglich, M., et al., 2015 in [30], 29
[7] Cramers, C. J., 2004, *Essentials of Computational Chemistry: Theories and Models* (Chichester: John Wiley and Sons)
[8] Cremer, D., 2013, *Comput Mol Sci*, 3, 482
[9] Gray, T., Mann, N., 2009, *Elements: A Visual Exploration of Every Known Atom in the Universe* (New York: Black Dog & Lewenthal Publishers)
[10] Geppert, W., Larsson, M., 2008, *Mol Phys*, 106, 2199
[11] Hanwell, M. D., Curtis, D. E., Lonie, D. C., et al., 2012, *J Cheminformatics*, 4, 17
[12] Hohenberg P., Kohn, W., 1964, *Phys Rev*, 136, B864
[13] Hornekaer, L., 2015, in reference [30], 255
[14] Jones, R. O., Gunnarsson, O., 1989, *Rev Mod Phys*, 61, 689
[15] Koch, W., Holthauser, M. C., 2001, *A Chemist's Guide to Density Functional Theory* (New York: Wiley and Sons)
[16] Kohn, W., 1998, *Rev Mod Phys*, 71, 1253
[17] Kohn, W., Sham, L. J., 1965, *Phys Rev*, 140, A1133
[18] Kratschmer, W., Lamb, L. D., Fostiropoulos, K. , Huffman, D. R., 1990, *Nature*, 347, 354
[19] Kroto, H. W., Heath, J. R., O'Brien, S. C., Curl, R. F., Smalley, R. E., 1985, *Nature*, 318, 162
[20] Martinez, T., *Chem Rev*, 2018, 118, 3305
[21] Meuwly, M., 2019, *Comput Mol Sci*, 9, 1
[22] Müller, H. S. P., 2015, in reference [30], 68
[23] Mardirossian, N., Head-Gordon, M., 2017, *Mol Phys*, 115, 2315
[24] Öberg, K., 2009, *Complex Processes in Simple Ices*, PhD Thesis, Leiden University
[25] Pauling, L., 1960, *The Nature of the Chemical Bond*, 3rd edition (Ithaca, NY: Cornell University Press)
[26] Levi, P., 1984, *The Periodic Table* (New York: Schocken Books Inc.)
[27] Rice, C. A., Maier, J. P., 2015 in reference [30], 15

[28] Sacks, O., 2001, *Uncle Tungsten, Memories of a Chemical Boyhood* (New York: Knoff Inc.)
[29] Schlemmer, S., in reference [30], 109
[30] Schlemmer, S., Giesen, T., Mutschke, H., Jäger, C., 2015, *Laboratory Astrochemistry* (New York: Wiley and Sons)
[31] Smith, I. W. M., 2011, *Annu Rev Astron Astrophys*, 49, 29
[32] Szabo, A., Ostlund, N. S., 1982 *Modern Quantum Chemistry* (New York: MacMillan)
[33] Tuckerman, M. E., 2002, *J Phys: Condensed Matter*, 14, 1297
[34] Yu, H. S., Li, S. L., Truhlar, D. G., 2016, *J Chem Phys*, 145, 130901
[35] Zhen, J., et al., 2014, *Chem Phys Lett*, 592, 211

3
Molecular Spectroscopy

3.1 Introduction

Interstellar molecules – to be precise, diatomic radicals – were first discovered through their electronic transitions in the visual and near-UV regions of the spectrum in the 1930s. The diffuse interstellar bands (DIBs) – a set of some 400 interstellar absorption lines in the visible and far-red parts of the stellar spectra – date back even further. The DIBs also find their origin in molecular electronic transitions. In the solar system, the spectra of comets are dominated by emission bands due to simple molecular radicals and cations. Typically, molecular bond strengths are some 5–10 eV and electronic transitions occur in the far-UV. If there are low lying empty orbitals – as for radicals or ions – transitions shift toward longer wavelengths.

Vibrational transitions involve motions of the atoms in the molecule. They shift therefore by a factor $\sqrt{m_e/M}$ toward lower frequencies. As this is the mid-IR range of the spectrum, first observations of interstellar molecules in the infrared had to await the development of sensitive detectors and the opening up of this window in 1970s. Ro-vibrational transitions of molecules are routinely seen in absorption in the spectra of a wide variety of objects including cool giants (later than spectral type K), brown dwarfs, (exo)planet atmospheres, Hot cores associated with high mass protostars, and obscured galactic nuclei such as UltraLuminous Infrared Galaxies. In emission, the ro-vibrational transitions of H_2 pumped by UV photons are prominent in the spectra of photodissociation regions. These spectra also show strong emission bands due to vibrational fluorescence of UV-pumped polycyclic aromatic hydrocarbon molecules. The mid infrared spectra of all star-forming galaxies are dominated by these bands. Molecular vibrational emission bands are also present in the spectra of comets, protoplanetary disks, and shocks in molecular clouds. Vibrational transitions correspond to energies of typically a few hundred to a thousand degrees.

The discovery of (pure) rotational transitions due to interstellar molecules dates back to the late 1960s. As detectors improved, telescopes grew in size, and the sub-millimeter sky opened up, the list of molecules detected through their pure rotational transitions steadily grew over the years. When ALMA entered its operational phase, the pace of new molecular identifications increased manyfold. The energies associated with molecular rotations are typically a few degrees, depending on the moment of inertia, and occur in the (sub)-millimeter to far-infrared. Rotational transitions dominate the cooling of molecular clouds and in particular regions of star formation. The multitude of lines also provide powerful probes of the physical conditions in the emitting gases.

This Chapter introduces various aspects of molecular spectroscopy including energy levels, transition frequencies, selection rules, and transition strengths. There are separate sections on pure rotational transitions, on vibrational and ro-vibrational transitions, and on electronic spectra. Rotational spectroscopy has developed into a main staple of astronomy and a course would do well by discussing in depth the rotational spectra of linear molecules (Section 3.2.2). Symmetric top molecules (Section 3.2.3) present "merely" variations on a theme, while the spectra of asymmetric top molecules (Section 3.2.4) are too complex to be caught in simple rules (but are of great individual interest). These sections can be left for background reading. That also holds for the sections on hyperfine splitting, Λ-doubling, nuclear spin, and partition function (Section 3.2.5–3.2.8). The section on transitions strength (Section 3.2.9) is important for all students, though. In the vibrational spectroscopy section, Hooke's law and molecular identifications are important for all astronomy students while partition function and gas phase ro-vibrational spectra might be assigned to background reading. Electronic spectroscopy can also be left to the individual students interested in this aspect. The chapter ends with a discussion of the spectroscopy of three specific molecules, H_2, CO, and H_2CO (Section 3.5), and together they serve well as a framework to introduce students to relevant concepts introduced in this chapter. Finally, issues involving molecular excitation, emission intensities, and analysis of molecular observations are presented in Chapter 4.

3.2 Rotational Spectroscopy

3.2.1 Energy Levels

Rotational transitions arise from the rotation of the permanent dipole interacting with an oscillating electromagnetic field. Typically, this occurs in the microwave region of the spectrum. Rotational transitions are – to first order – set by the moments of inertia of the molecule. Working in the center of mass frame, we can define three orthogonal axes at the center of mass of the molecule, a, b, and c. The moments of inertia are then,

$$I_j = \sum_i m_i r_i^2 \quad j = a, b, c \tag{3.1}$$

where m_i is the mass of atom i at distance r_i from the center of mass and the summation is over all atoms. By convention $I_c \geq I_b \geq I_a$. As an example, for a diatomic molecule, the moment of inertia is $I = \mu R^2$ with μ the reduced mass and R the distance between the atoms. For a rigid rotor,[1] the rotational structure is set by the symmetry of the molecule.

[1] No distortion under rotation.

Table 3.1 *Classification of rotors*

Moment of inertia	Symmetry	Rotational constants	Example
$I_a = I_b = I_c$	Spherical top	$A = B = C$	CH_4[a], SF_6[a]
$I_a = 0, I_b = I_c$	Linear molecule	$A = \infty, B = C$	CO, CO_2[a]
$I_a < I_b = I_c$	Prolate symmetric top	$A > B = C$	C_2H_6
$I_a = I_b < I_c$	Oblate symmetric top	$A = B > C$	C_6H_6
$I_a < I_b < I_c$	Asymmetric top	$A > B > C$	H_2O

[a]These symetric molecules have no permanent dipole and therefore no pure rotational spectrum.

Table 3.1 summarizes the classification. For a rigidly rotating molecule, the rotational energy levels can be described classically by the rotations of a rigid body,

$$E_r = \frac{1}{2}\left(I_a\omega_a^2 + I_b\omega_b^2 + I_c\omega_c^2\right) = \frac{1}{2}\left(\frac{\mathcal{J}_a^2}{I_a} + \frac{\mathcal{J}_b^2}{I_b} + \frac{\mathcal{J}_c^2}{I_c}\right), \tag{3.2}$$

with ω_j the rotational frequency and $\mathcal{J}_j$ the angular momentum ($= I_j\omega_i$).

3.2.2 Linear Molecules

We will start with a simple system, a linear molecule. The quantum mechanical analog is then, $\mathcal{J} = \sqrt{J(J+1)}\hbar$ with J ($= 0, 1, 2, \dots$), the rotational quantum number, and we have,

$$E_r = \frac{J(J+1)\hbar^2}{2I}. \tag{3.3}$$

We can write this in terms of the rotational constant, B_e,

$$B_e = \frac{h}{8\pi^2 cI} \tag{3.4}$$

in wavenumbers, as,

$$\frac{E_r}{hc} = B_e J(J+1). \tag{3.5}$$

As a guide,

$$B_e \simeq \frac{17}{I\left(\text{amu Å}^2\right)}\ \text{cm}^{-1} \simeq \frac{500}{I\left(\text{amu Å}^2\right)}\ \text{GHz}. \tag{3.6}$$

Radiative rotational transitions are allowed for $\Delta J = \pm 1$, and the spectrum consists of a set of evenly spaced lines in frequency space (Figure 3.1); viz.,

$$\nu(J+1 \to J) = \frac{E(J+1) - E(J)}{hc} = 2B(J+1). \tag{3.7}$$

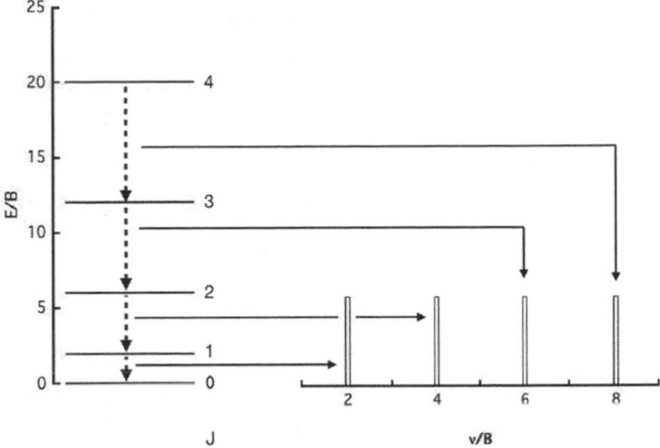

Figure 3.1 Left: Schematic rotational energy level diagram for a linear molecule. Energy – in units of the rotational constant, B – increases upward. The individual levels have been labeled with their J-value. Allowed transitions correspond to $\Delta J = \pm 1$. Right: To first order, the spectrum of a linear molecule consists of a set of equidistant lines. Frequency is in units of the rotational constant.

As B_e decreases with $1/I$, molecules consisting of heavier atoms will have transitions at lower frequencies. So, the lowest frequencies for CS and CO are at aproximately 49 and 115 GHz, respectively (cf. Table 3.3). Transitions of light hydrides occur in the far-infrared (e.g. HF at 1.2 THz), while the groundstate pure rotational transition of H_2 falls at 10.6 THz (28 μm).

A rotating molecule will pull apart and the moment of inertia will increase with increasing rotation. Allowing for this centrifugal stretching, the energy levels become

$$\frac{E_r}{hc} = B_e\, J\,(J+1) - D_e\, J^2\,(J+1)^2\,, \tag{3.8}$$

with D_e the centrifugal distortion constant. The line frequencies are then,

$$\nu\,(J+1 \rightarrow J) = 2B\ (J+1) - 4D_e\,(J+1)^3 \tag{3.9}$$

$$= 2B\ (J+1)\left(1 - \frac{2D_e}{B}\,(J+1)^2\right). \tag{3.10}$$

Centrifugal distortion will thus destroy the constant separation of rotational transitions in the spectrum of a rigid rotor. The way it is written above, centrifugal distortion represents a correction factor on B and this correction depends on the magnitude of the angular momentum. Obviously, for faster spinning molecules, the bond will lengthen more and the increased moment of inertia results in a decreased rotational constant. For a harmonic oscillator, $D_e = 4B_e^3/\nu_e^2$ with ν_e the vibrational frequency (in cm^{-1}). Centrifugal distortion is, thus, a "small" correction factor to the frequency of the rotation; i.e. $D_e/B_e = 4\,(B_e/\nu_e)^2 \ll 1$. Consider CO with $\nu_e = 2170\ \mathrm{cm}^{-1}$ and $B_e = 1.93\ \mathrm{cm}^{-1}$; the correction factor is 8.5×10^{-7}. In astronomy, for pure rotational transitions, this clearly has to be taken into account ($\Delta v \simeq 0.26\,(J+1)^2$ km/s). And even for ro-vibrational transitions (see Section 3.3.4), this correction factor amounts to 0.1 cm^{-1} on a level separation of 1 cm^{-1} at $J = 10$.

Anharmonic effects can be included as higher order correction terms. Vibrational motions occur at much higher frequency than rotations and hence a molecule will vibrate many, many times during one rotation. As rotations depend on the average of $1/R^2$, even for harmonic oscillators, the rotational constant will vary with vibrational excitation. For anharmonic oscillators, the rotational constant will decrease with increasing vibrational excitation as the average separation between the atoms increases. The dependence of the spectrum on the vibrational state is captured in small correction factors on the rotational constant as well as on the centrifugal distortion constant that depend on the vibrational quantum number of the modes involved,

$$B_v = B_e - \alpha_e \left(v + \frac{1}{2} \right) + \cdots \tag{3.11}$$

$$D_v = D_e + \beta_e \left(v + \frac{1}{2} \right) + \cdots . \tag{3.12}$$

These are small correction factors. If we include the cubic term in the potential expansion, we have

$$\alpha_e = \frac{24 B_e^3 R_e^3 g}{\omega_e^3} - \frac{6 B_e^2}{\omega_e}, \tag{3.13}$$

where g is the coefficient of the cubic term in the expansion (in units of cm^{-1}) and the fundamental vibrational frequency, ν_e, and the anharmonicity parameter, $x_e \nu_e$, are described in Section 3.3.1. Often this parameter is expressed using a Morse potential,

$$\alpha_e = -6 \left(\frac{B_e}{\nu_e} \right) \left(B_e - \sqrt{x_e \nu_e B_e} \right). \tag{3.14}$$

As the Morse potential is fully described by the bond energy, ν_e and $x_e \nu_e$, no new parameter is introduced and differences of this expression with measurements are then an indication of the deviation of the actual potential from the Morse potential. Expanding the potential again to cubic terms, we have for β_e,

$$\beta_e = D_e \left(\frac{8 \omega_e x_e}{\omega_e} - \frac{5 \alpha_e}{B_e} - \frac{\alpha_e^2 \omega_e}{24 B_e^3} \right). \tag{3.15}$$

However, β_e is a small correction on a small correction and is often neglected.

3.2.3 Symmetric Top Molecules

Let's now consider a symmetric top molecule (Table 3.1). Two moments of inertia are the same – say, $I_b = I_c$ (a prolate symmetric top) – so we can write,

$$E_r = \frac{\mathcal{J}_a^2}{2I_a} + \frac{\mathcal{J}_b^2 + J_c^2}{2I_b}, \tag{3.16}$$

with $\mathcal{J}^2 = \mathcal{J}_a^2 + \mathcal{J}_b^2 + \mathcal{J}_c^2$, we then have,

$$E_r = \frac{\mathcal{J}^2}{2I_b} + \mathcal{J}_a^2 \left(\frac{1}{2I_a} - \frac{1}{2I_b} \right). \tag{3.17}$$

We again make the classical to quantum mechanics transformation, $\mathcal{J}^2 \rightarrow J^2 (J + 1)\, \hbar^2$. As there are two main directions of rotation, there are two quantum numbers. Introducing K as the projection on the molecular axis – e.g. a for prolate molecules (and c for oblate species; Table 3.1) – we have $\mathcal{J}_a = K\hbar$. As J_a is quantized, the total angular momentum J can only take a few specific directions. With this, we arrive at,

$$\frac{E_r}{hc} = BJ\,(J+1) + (A-B)\,K^2, \tag{3.18}$$

where A and B are the rotational constants, $A = h/8\pi^2 c I_a$ and $B = h/8\pi^2 c I_b$. Again, the units for these rotational constants are wavenumbers. For an oblate symmetric top, we have to replace A by C $(= h/8\pi^2 c I_c)$. The rotational quantum number, J, is $J = 0, 1, 2, \ldots$. As K is the projection of J, it can attain the values, $K = 0, \pm 1, \pm 2, \ldots, \pm J$. Note that the energy is independent of the sign of K and each J-level has a $2J + 1$ fold degeneracy.

As for linear molecules, a correction term for centrifugal distortion can be included,

$$\frac{E_r}{hc} = BJ\,(J+1) + (A-B)\,K^2 - D_J J^2\,(J+1)^2 - D_{JK} J\,(J+1)\,K^2 - D_K K^4, \tag{3.19}$$

(cf. Eq. (3.8)).

The selection rules are $\Delta J = 1$ and, as there is no dipole moment along the symmetry axis (a), the angular momentum along the symmetry axis cannot change due to radiation, and we have $\Delta K = 0$. The rotational frequencies are,

$$\nu_{JK} = 2B\,(J+1) - 4D_J\,(J+1)^3 - 2D_{JK}\,(J+1)\,K^2. \tag{3.20}$$

If we write this as,

$$\nu_{JK} = \left(2B - 4D_J\,(J+1)^2 - 2D_{JK} K^2 \right)(J+1) \tag{3.21}$$

then we recognize again that the centrifugal distortion is a correction factor to B that depends on K. This represents a small change in the moment of inertia, I_B.

Figure 3.2 illustrates the energy level diagrams for prolate and oblate symmetric tops. We can consider two extremes: $K = J$ with rotation around the molecular axis and $E_{J,K=J}/hc \simeq AJ^2$ and $K = 0$ with rotation perpendicular to the molecular axis and

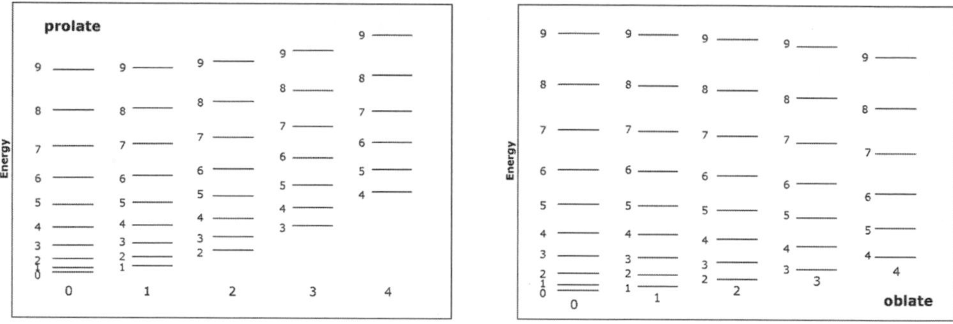

Figure 3.2 Schematic rotational energy level diagram for a prolate (left) and oblate (right) symmetric top molecule. Energy increases upward. The individual levels have been labeled with their J-value. The K-ladders have been shifted horizontally. Allowed transitions occur within each K-ladder. Because K is the projection of J on the molecular symmetry axis, we have $K \leq J$.

$E_{J,K=0}/hc = BJ(J+1)$. In general, for a given K, we recognize the energy level distribution for a linear rotor except that they start with $J = K$ rather than $J = 0$. Note, again, that the energy does not depend on the sign of K and hence, except for $K = 0$, each level is degenerate. Given the selection rule, $\Delta K = 0$, the spectrum is that of a linear rotor. Ignoring the centrifugal distortion, the different K-ladders would coincide but, in reality, transitions will split up. Also, an excited molecule will relax radiatively along its K-ladder and excitation will tend to bottle up in the meta-stable $J = K$ level.

3.2.4 Asymmetric Top Molecules

For asymmetric top molecules, all three moments of inertia are different and there is no simple general formula describing the energy levels. Specifically, $\mathcal{J}$ and its projection on a fixed space axis are constants of motion and J as well as M are good quantum numbers. However, the projection of $\mathcal{J}$ on any of the molecular axes is not conserved and hence none of the molecular axes carries out a simple rotation around $\mathcal{J}$. As a result K is no longer a good quantum number and there is no set of quantum numbers with a simple physical meaning that can be used to describe states. The energy of level, $J_{K_a K_b}$, is generally written as,

$$\frac{E_r}{hc} = \left(\frac{A+C}{2}\right) J(J+1) + \left(\frac{A-C}{2}\right) f(\kappa, K_a, K_b), \tag{3.22}$$

where the asymmetry parameter, κ is given by

$$\kappa = \frac{2B - A - C}{A - C}. \tag{3.23}$$

For a prolate top $\kappa = -1$ while for an oblate top $\kappa = 1$. Extensive tabulations for the function f exist. The spectrum of an asymmetric rotor is thus extremely complex. In

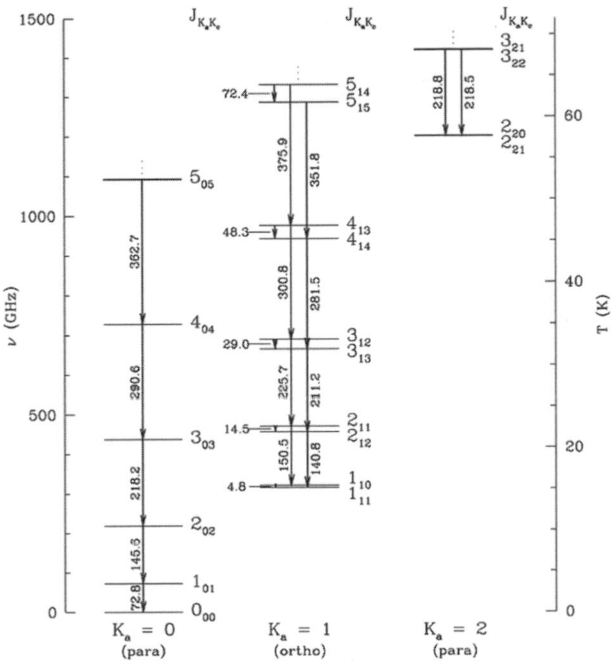

Figure 3.3 Energy level diagram for H_2CO, a near-prolate molecule. Note that formaldehyde has para (K_aK_b are ee or eo) and ortho (K_aK_b are oe or oo, where e=even and o=odd) levels. Figure reproduced with permission from [6]

addition to the irregular distribution of energy levels, the selection rules and transition probabilities between these levels are complicated as the dipole moment may lie in any arbitrary direction with respect to the principal axes of inertia.

When $I_b \simeq I_c \neq I_a$ (prolate), the levels can be represented by $J_{K_aK_b}$, where K_a and K_b are approximate quantum numbers. An analogous approach can be followed for an oblate-like case ($I_b \simeq I_a \neq I_c$). As an example, formaldehyde is a near-prolate molecule ($\kappa = -0.96$). The energy levels show the characteristic pattern of the prolate case (Figure 3.3). The slight deviation from the pure prolate symmetric top case lifts the degeneracy of the levels with $K > 0$ and each level is now split into two. Nevertheless, for the lowest levels, transitions between the different K-ladders are still not allowed. Formaldehyde has para and ortho states, characterized by antiparallel and parallel nuclear spins of the H atoms. These nuclear spin states combine with the rotational states (Section 3.2.7) and para states have K=even while ortho states have K=odd.

H_2O is another example of an asymmetric top molecule ($\kappa = -0.44$). The energy levels are labeled by $J_{K_aK_c}$, where the total angular momentum, J, is a good quantum number while the indices, K_a and K_c, refer to the corresponding prolate and oblate symmetric tops. Sometimes, the pseudo quantum number $\tau = K_a - K_b$ is introduced for labeling convenience. The index τ runs from $J, \ldots, 0, \cdots - J$ (c.f. K for a symmetric top) and

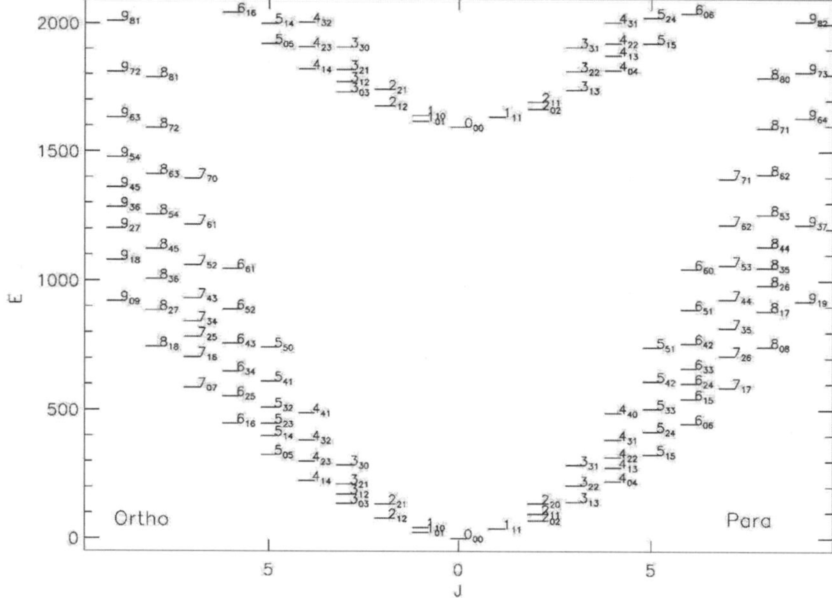

Figure 3.4 Energy level diagram for H_2O. Energies in cm^{-1}. Figure courtesy of F. Helmich

labels the levels in order of decreasing energy. The H_2O energy levels (Figure 3.4) can still be described by $J_{K_a K_c}$. From the symmetry, we have for the selection rules $\Delta J = 0, \pm 1$, $\Delta K_a = \pm 1, \pm 3 \ldots$ and $\Delta K_b = \pm 1, \pm 3 \ldots$. Hence, transitions between K-ladders are now allowed. Water also has para and ortho states, characterized by antiparallel and parallel nuclear spins. The nuclear para states have either K_a or K_c odd while the ortho states have K_a and K_c either both even or both odd.

3.2.5 Hyperfine Transitions

The rotational spectra of many astrophysically relevant molecules contain hyperfine splittings due to electric quadrupole and magnetic dipole interactions with atoms with nonzero nuclear spin, such as ^{13}C, ^{14}N, and ^{17}O. The nuclear spin, $\vec{I}$, will couple with the rotational angular momentum, $\vec{J}$ to $\vec{F} = \vec{I} + \vec{J}$. The number of hyperfine levels is the smaller of $2J+1$ and $2I+1$. This is most relevant for nitrogen-bearing molecules and levels with $J > 0$ are split in to three hyperfine transitions. The relative intensities of the hyperfine transitions of a molecular transition can be calculated and the reader should consult a specialized book on this topic.

3.2.6 Λ Doubling

Some species have nonzero electronic angular momentum and this can couple to the rotations of the molecule. This has an effect on the molecular spectrum. We will focus here on

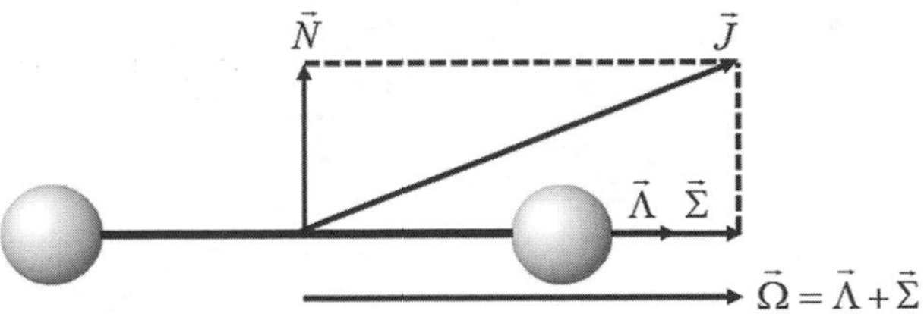

Figure 3.5 Λ-doubling of rotational levels in diatomic molecules. $\vec{N}$ is the nuclear rotational vector. $\vec{\Lambda}$ is the projection of the orbital angular momentum on the internuclear axis. $\vec{\Sigma}$ is the projection of $\vec{S}$ on the internuclear axis. $\vec{\Omega}$ is the vectorial sum of the latter two. When $\Lambda \neq 0$, the net current creates a magnetic field that can couple with the spinning electrons, splitting the energy levels.

a diatomic molecule with weak spin–orbit coupling and nonzero electron spin S. As discussed in Section 3.4.2, the term symbol for diatomic molecules is given by $^{2S+1}\Lambda_\Omega$ with $\vec{S}$ the total electron spin, $\vec{\Lambda}$ the projection of the orbital angular momentum on the internuclear axis, $\vec{\Sigma}$ the projection of $\vec{S}$ on the internuclear axis, and $\vec{\Omega} = \vec{\Sigma} + \vec{\Lambda}$ (Figure 3.5). The whole electronic shell can be considered to rotate around the molecular z axis. For a rotating molecule, we have the angular momentum of the rotation perpendicular to the z-axis, $\vec{N}$, and the total angular momentum of the electrons – composed of the orbital angular momentum, $\vec{L}$, and the spin angular momentum, $\vec{S}$ – precessing around the z-axis. The projections of the electronic components add to the total value (which is conserved), $\Omega\hbar = (\Lambda + M_s)\,\hbar$. As Ω is not zero, the total angular momentum of the molecule, J, is no longer perpendicular to the z-axis and the molecule will rotate around the space-fixed direction of $\vec{J}$.

The rotating molecule is then a symmetric top with the two moments of inertia, I_e, of the electrons rotating around the z-axis and I_m of atoms and electrons rotating around an axis perpendicular to the z-axis, with $I_e \ll I_m$. The rotational energy is then,

$$E_{rot} = \frac{J_x^2}{2I_x} + \frac{J_y^2}{2I_y} + \frac{J_z^2}{2I_z}. \tag{3.24}$$

We have $J_z^2 = \Omega^2\hbar^2$ and $J_x^2 + J_y^2 = J^2 - J_z^2 = \left(J\,(J+1) - \Omega^2\right)\hbar^2$ where we have made the classical to quantum replacement. This results in,

$$\frac{E_{rot}}{hc} = B_e\left(J\,(J+1) - \Omega^2\right) + A\Omega^2, \tag{3.25}$$

with the molecular rotational constant, $B_e = h/\left(8\pi^2 c I_m\right)$, and the electronic rotational constant, $A = h/\left(8\pi^2 c I_e\right)$. The latter term in the expression for the rotational energy does not depend on the rotation itself and is added to the electronic energy of the state Ω.

Generally, ground states of diatomic molecules are $^1\Sigma$ states with $\Omega = 0$ and we recover expression (3.5). But, as an example, consider OH with the electronic ground state $^2\Pi$ ($L = 1$ and $S = 1$) (Figure 3.6). Spin–orbit coupling provides then two sets of states

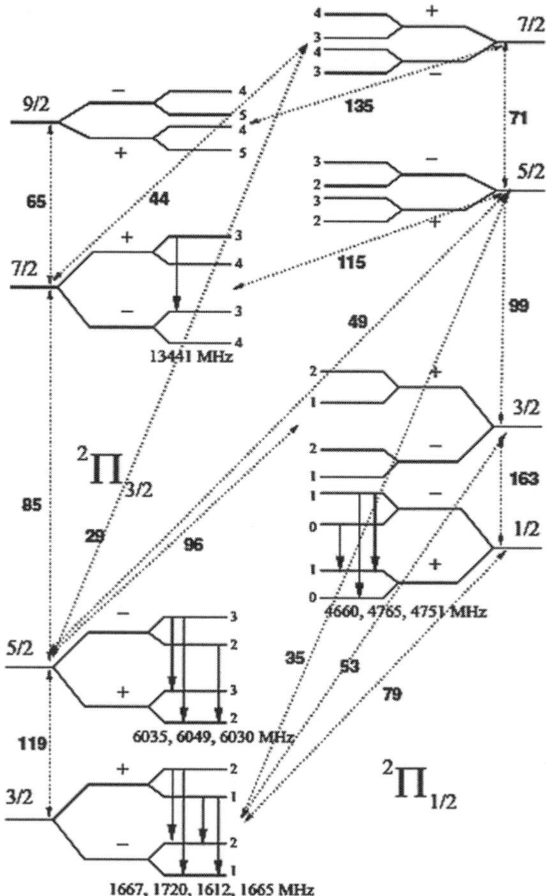

Figure 3.6 The rotational ladder of OH is split into two ($^2\Pi_{1/2}$ and $^2\Pi_{3/2}$) by spin–orbit coupling. Lambda doubling splits each level into half of opposite parity and hyperfine splitting results then in a total of four levels per J, labeled by the quantum number F. Splittings are not to scale. Known maser transitions (see Chapter 11) are indicated by solid arrows and labeled in MHz while relevant, pumping IR transitions are in μm. Figure taken from [10]

characterized by the electron orbital angular momentum, 1/2 and 3/2 ($\vec{J'} = \vec{L} + \vec{S}$), each containing a set of rotational ladders characterized by the rotational motion of the nuclei, J. Weak coupling between the electronic angular momentum and molecular angular momentum leads to lambda doubling of the rotational levels, depending on whether both are rotating in the same or in opposite sense. These Λ-doubles are then further split by coupling between the spin of the unpaired electron and the proton. These hyperfine levels have either parallel or antiparallel spins and are described by the total angular momentum quantum number, $F = J \pm I$. Each Λ-doublet can thus generate four lines; the transitions in which F does not change are called main lines. The other two are the satellite lines.

A magnetic field will split the $2F + 1$ states of each hyperfine level further, leading to a set of complex transitons with different polarization characteristics. Allowed transitions require a parity change and $\Delta F = 0, \pm 1$ but $F = 0 \rightarrow 0$ is forbidden.

In summary, for OH, the two possible orientations of the spin angular momentum relative to the internuclear axis give rise to multiplet splitting into $^2\Pi_{3/2}$ and $^2\Pi_{1/2}$ systems separated by approximately 140 cm^{-1}. The interaction between the electronic orbital motion and the nuclear rotation splits each rotational level into two spin doublet levels and this lambda doubling is of order $0.1-1$ cm^{-1}. Finally, each lambda-doublet component is doubled due to the contribution of the nuclear spin in its interaction with the electronic plus rotational motion with a separation of approximately 10^{-3} cm^{-1} (Figure 3.6).

3.2.7 Nuclear Spin: Statistical Weight and Ortho and Para Species

The rotational degeneracy factor, g_J, reflects the possible projections of the angular momentum vector on a spatial axis and equals, $g_J = 2J + 1$ (as m_j ranges from $-J, \ldots, 0, \ldots, J$). In addition, the statistical weight of rotational levels includes the statistical weight of the nuclei, g_n, given by

$$g_n = \frac{1}{\sigma} \prod_n (I_n + 1), \tag{3.26}$$

where I_n is the spin of nucleus n and the product runs over all nuclei. The factor σ – part of the rotational partition function – takes the symmetry of the molecule into account (c.f. Section 3.2.8). Nucleons are fermions and have an intrinsic spin, 1/2. The statistical weight of the nucleus can then be determined from the number of protons and neutrons. If the number of protons and neutrons are both even then the nucleus has no spin. If they are both odd then the nucleus has integer spin. If the number of protons plus the number of neutrons is odd, then the nuclear spin is 1/2. So, H, N, C, O have $I_n = 1/2, 1, 0, 0$, respectively. For relevant isotopes, we have D, ^{15}N, ^{13}C, ^{17}O, ^{18}O, equals $I_n = 1, 1/2, 1/2, 5/2, 0$. The statistical weight is then given by,

$$g(J) = g_n \ (2J + 1). \tag{3.27}$$

Many molecular species come in para and ortho species, which differ in their nuclear spins (para with antiparallel nuclear spins and ortho with parallel nuclear spins). The presence of separate para and ortho species is a direct consequence of the generalized Pauli principle: Wavefunctions have to be antisymmetric with respect to the exchange of two fermions (particles with half-integral spin) and symmetric with respect to exchange of two bosons (particles with integral spin). The total wavefunction is (approximately) the product of the electronic, vibrational, rotational, and nuclear wavefunctions, $\psi_{tot} = \psi_e \psi_v \psi_r \psi_n$. As an example, consider H_2 with a ground electronic state $^1\Sigma_g^+$, which is symmetric with respect to the exchange of the nuclei. The vibrational wave function depends only on the internuclear distance and is thus symmetric as well. This implies that product of the rotational and nuclear spin wavefunctions has to be antisymmetric.

Consider first the nuclear spin wavefunction. H atoms are fermions with $m_s = \pm 1/2$ (e.g. α and β). We have then four possible combinations for the two nuclei, 1 and 2: $\alpha(1)\,\alpha(2)$ and $\beta(1)\,\beta(2)$ and the two linear combinations, $(\alpha(1)\,\beta(2) + \alpha(2)\,\beta(1))\,/\,\sqrt{2}$ and $(\alpha(1)\,\beta(2) - \alpha(2)\,\beta(1))\,/\,\sqrt{2}$. The first three of these wavefunctions are symmetric with respect to exchange of the two nuclei. The latter is antisymmetric.

As mentioned above, we have to combine the symmetry of these states with those of the rotational states. The rotational wavefunction boils down to spherical harmonics, $Y_{J,M_J}(\theta,\phi)$, with J and M_J the rotational quantum numbers. Exchange of the two nuclei corresponds to a rotation by π and the symmetry properties of associated Legendre polynomials are given by $P(\pi - \theta) = (-1)^J\ P(\theta)$. Thus, even-$J$ levels are symmetric and have to combine with antisymmetric nuclear spin states, while odd-J levels are antisymmetric and have to combine with symmetric nuclear spin states to yield overall antisymmetric states.

This is also reflected in the statistical weight. Thus, for homonuclear diatomic molecules, with atomic spins, I_n, we have two sequences of statistical weights for rotational levels with J values differing in parity. The nuclear spin function is then given by, $g_n = g_n(i)$, according to Table 3.2, and,

$$g_n(1) = \frac{1}{2}[(2I_n + 1)^2 - (2I_n + 1)] \tag{3.28}$$

$$g_n(2) = \frac{1}{2}[(2I_n + 1)^2 + (2I_n + 1)] \tag{3.29}$$

Thus, for H_2 ($^1\Sigma_g^+$), we have $I_n = 1/2$ and for even (para) states, we have $g_n = 1$ while, for odd (ortho) states, we have $g_n = 3$. Half of the rotational states belong to the ortho form and half belong to the para form. The sum of the g_n's equals $(2I_n + 1)^2$.

Transitions between para and ortho states involve a change in nuclear spin and are thus forbidden. In terms of electro-magnetic allowed transitions, ortho and para forms are therefore "separate" species and, in space, exchange between them requires chemical reactions (e.g. $H_2 + H_3^+ \rightarrow H_2 + H_3^+$) or interaction with paramagnetic impurities (e.g. the lone electron pair on an oxygen atom) on a grain surface.

Table 3.2 *Nuclear spin statistical weight*[a]

	+	−	−	+
Electronic state Indices	g	u	g	u
Even J	$g_n(1)$	$g_n(1)$	$g_n(2)$	$g_n(2)$
Odd J	$g_n(2)$	$g_n(2)$	$g_n(1)$	$g_n(1)$

[a] For Fermi-Dirac statistics. For Bose-Einstein statistics, the entries for even and odd J in the same column are interchanged. The first two rows specify the symmetry of the electronic states (see Section 3.4). The last two rows indicate which function to use for even and odd J with the electronic state specified in the same column (Eq. (3.28)).

As another example, consider molecular oxygen, $^{16}O_2$, where each atom is a boson with $I_n = 0$. The electronic wavefunction of the ground state is asymmetric ($^3\Sigma_g^-$). Hence, the even J rotational states (and asymmetric nuclear spin states) have a weight equal to 0 and are missing in the spectra. The odd J states have a weight 1. In contrast, $C^{16}O_2$ has a symmetric ground electronic state ($^1\Sigma_g^+$). And now the symmetric rotational states couple with the symmetric nuclear spin states and have weight 1 while the asymmetric rotational states couple with the asymmetric spin states and have weight 0 and hence do not appear in the rotational spectra. Some other examples are, C_2H_2 with even (para) states and $g_n = 1$ and odd (ortho) states and $g_n = 3$. For water and formaldehyde, ortho and para states also have a ratio of the nuclear spin degeneracies of 3. For ammonia, we have two distinct species with ortho states with $K = 3n$ ($n = 0, 1, 2, \ldots$) and all spins parallel (3/2) and para states with $K \neq 3n$ and all spins not parallel (1/2). The ratio of the nuclear spin degeneracy factors is then 1.

3.2.8 Partition Function

Level populations are often expressed in terms of an excitation temperature (c.f. Sections 4.1.3, 4.2.1, and 4.2.2). In local thermodynamic equilibrium, we have for the level population, $n(J)$,

$$\frac{n(J)}{n} = \frac{g_J}{Z_r(T)} \exp\left[-E(J)/kT\right], \tag{3.30}$$

with T the temperature, $E(J)$ and g_J the energy and statistical weight of level J, and n the total density of the species. The rotational partition function, $Z_r(T)$, is given by,

$$Z_r(T) = \sum_J g_J \exp\left[-E(J)/kT\right]. \tag{3.31}$$

In this, the statistical weight factor includes the nuclear spin and hyperfine splitting factors. The rotational degeneracy factor reflects the projection on a spatial axis and equals, $2J+1$ (cf. Section 3.2.7). For symmetric top molecules, the K states (except for $K = 0$) are doubly degenerate and hence have a factor $g_K = 2$ in their rotational statistical weight.

For high temperatures, the summation can be replaced by an integral. These can be evaluated for rigid rotors. For a linear molecule, the partition function becomes,

$$Z_r(T) = \frac{1}{\sigma}\frac{kT}{hcB}. \tag{3.32}$$

Nonlinear molecules can rotate around three major axes and the partition function is,

$$Z_r(T) = \frac{\sqrt{\pi}}{\sigma}\left(\frac{1}{ABC}\right)^{1/2}\left(\frac{kT}{hc}\right)^{3/2}. \tag{3.33}$$

These approximations are only valid if kT/A, kT/B, $kT/c \gg 1$. In these expressions, σ is a molecule-dependent symmetry factor. To understand this factor, consider a homonuclear diatomic molecule. Rotation by 180° does not change the molecule. Hence, we are overcounting the number of states by a factor 2. In general, we have to correct the partition

Table 3.3 *Characteristics of molecular rotational transitions*

Species	Transition	ν_{ul} [GHz]	E_u [K]	A_{ul} [s^{-1}]	n_{cr} [cm^{-3}]
CO	1–0	115.3	5.5	7.2×10^{-8}	1.1×10^3
	2–1	230.8	16.6	6.9×10^{-7}	6.7×10^3
	3–2	346.0	33.2	2.5×10^{-6}	2.1×10^4
	4–3	461.5	55.4	6.1×10^{-6}	4.4×10^4
	5–4	576.9	83.0	1.2×10^{-5}	7.8×10^4
	6–5	691.2	116.3	2.1×10^{-5}	1.3×10^5
	7–6	806.5	155.0	3.4×10^{-5}	2.0×10^5
CS	1–0	49.0	2.4	1.8×10^{-6}	4.6×10^4
	2–1	98.0	7.1	1.7×10^{-5}	3.0×10^5
	3–2	147.0	14	6.6×10^{-5}	1.3×10^6
	5–4	244.9	35	3.1×10^{-4}	8.8×10^6
	7–6	342.9	66	1.0×10^{-3}	2.8×10^7
	10–9	489.8	129	2.6×10^{-3}	1.2×10^8
HCO^+	1–0	89.2	4.3	3.0×10^{-5}	1.7×10^5
	3–2	267.6	26	1.0×10^{-3}	4.2×10^6
	4–3	356.7	43	2.5×10^{-3}	9.7×10^6
HCN	1–0	88.6	4.3	2.4×10^{-5}	2.6×10^6
	3–2	265.9	26	8.4×10^{-4}	7.8×10^7
	4–3	354.5	43	2.1×10^{-3}	1.5×10^8
H_2CO	2_{12}–1_{11}	140.8	6.8	5.4×10^{-5}	1.1×10^6
	3_{13}–2_{12}	211.2	17	2.3×10^{-4}	5.6×10^6
	4_{14}–3_{13}	281.5	30	6.0×10^{-4}	9.7×10^6
	5_{15}–4_{14}	351.8	47	1.2×10^{-3}	2.6×10^7
NH_3	(1,1) inversion	23.7	1.1	1.7×10^{-7}	1.8×10^3
	(2,2) inversion	23.7	42	2.3×10^{-7}	2.1×10^3
H_2	2–0	1.06E4[a]	510	2.9×10^{-11}	10
	3–1	1.76E4[b]	1015	4.8×10^{-10}	300

[a] $\lambda = 28.2$ μm. [b] $\lambda = 17.0$ μm.

function by the (symmetry) factor equal to the distinct number of ways in which rotation brings a molecule into equivalent configurations. These symmetry factors can be evaluated through group theory but that is beyond the scope of this book. Simple symmetry considerations often suffice to show that, e.g., CH_4 has $\sigma = 12$.

We can define characteristic rotational temperature, $\theta_r = hcB/k$, and values for some astrophysically interesting species are summarized in Table 3.4.

3.2.9 Transition Strength

The excitation and deexcitation of molecular levels is described by the Einstein coefficients: The Einstein A_{ul} is the rate of spontaneous emission between an upper (u) and lower (l)

Table 3.4 *Characteristic rotational and vibrational temperatures*[1]

Species	Mode	θ_v [K]	θ_r [K]	Species	Mode	θ_v [K]	θ_r [K]
H_2	ν_1	6330	88	CO_2	ν_1	3360	0.561
CH_4	ν_1	4170	7.54		ν_2	954 (2)	
	ν_2	2180 (2)	7.54		ν_3	1890	
	ν_3	4320 (3)	7.54	CH_3OH	ν_1	5297	6.125
	ν_4	1870 (3)			ν_2	4315	1.185
NH_3	ν_1	4800	13.6		ν_3	4092	1.141
	ν_2	1360	13.6		ν_4	2125	
	ν_3	4880 (2)	8.92		ν_5	2093	
	ν_4	2330 (2)			ν_6	1928	
H_2O	ν_1	5360	40.1		ν_7	1546	
	ν_2	2290	20.9		ν_8	1487	
	ν_3	5160	13.4		ν_9	2834	
H_2CO	ν_1	4003	13.53		ν_{10}	2108	
	ν_2	2512	1.864		ν_{11}	1647	
	ν_3	2158	1.632		ν_{12}	389	
	ν_4	4091		CO	ν_1	3122	2.78
	ν_5	1797					
	ν_6	1679					

[1]Temperatures θ are expressed as $h\nu/k$ and hcB/k. Numbers in brackets indicate the degeneracy of the mode.

level. The Einstein B_{lu} coefficient is the absorption rate between these two levels while the B_{ul} coefficient is the rate of stimulated emission. For a two-level system controlled by radiation, the level populations are given by,

$$B_{lu} J_{ul} n_l = (A_{ul} + B_{ul} J_{ul})\, n_u. \tag{3.34}$$

In thermodynamic equilibrium, the level populations are given by the Boltzmann equation, $n_u/n_l = (g_u/g_l)\exp\left[-E_{ul}/kT\right]$ while the radiation field is given by Planck's law, $J_{ul} = B(\nu_{ul}) = 2h\nu_{ul}^3/c^2\left(\exp\left[h\nu_{ul}/kT\right] - 1\right)^{-1}$ where $E_{ul} = h\nu_{ul}$. Combining these equations gives rise to the well-known relationships between the Einstein coefficients,

$$g_u B_{ul} = g_l B_{lu} \tag{3.35}$$

$$A_{ul} = (2h\nu_{ul}^3/c^2)\, B_{ul}. \tag{3.36}$$

As the Einstein coefficients are properties of the species involved, they must be valid even when the system is not in thermodynamic equilibrium. In terms of the Einstein coefficients, the energy emitted is given by $j_{ul} = A_{ul} n_u h\nu_{ul}/4\pi$ and the energy absorbed from a pencil ray, corrected for stimulated emission, is given by, $\kappa_{lu} = (n_l B_{lu} - n_u B_{ul})\, h\nu_{ul}/4\pi$. In terms of the oscillator strength, f, the Einstein B coefficient is given by,

$$B_{lu} = \frac{4\pi}{h\nu_{ul}} \frac{\pi e^2}{m_e c} f_{lu}, \tag{3.37}$$

with m_e the mass of the electron.

Let us consider now the transition strength. A rotating dipole will emit an electromagnetic wave and the average power radiated is given by Larmor's formula,

$$\langle P \rangle = \frac{64\pi^4}{3c^3} \nu^4 \left|\mu_{ul}\right|^2, \tag{3.38}$$

with μ_{ul} the mean electric dipole moment associated with the transition. The Einstein A_{ul} is then,

$$A_{ul} = \frac{64\pi^4}{3hc^3} \nu^3 \left|\mu_{ul}\right|^2 \tag{3.39}$$

For a rotational transition, the dipole moment can be expressed as,

$$\left|\mu_{ul}\right|^2 = \mu_z^2 \frac{S_{ul}}{g_u}, \tag{3.40}$$

with μ_z the electric dipole moment along the molecular symmetry axis. The distribution of the line intensity over the different rotational bands is represented by the line strength of the transition, S_{ul}, the so-called Hönl–London factor. The calculation of the line strength can be quite complex. For linear molecules in the ground vibrational state, we have,

$$\left|\mu_{ul}\right|^2 = \mu_d^2 \frac{J+1}{2J+3}, \tag{3.41}$$

with μ_d the molecule's permanent dipole moment. For symmetric top molecules, we have,

$$\left|\mu_{ul}\right|^2 = \mu_d^2 \frac{J^2 - K^2}{2J+3}. \tag{3.42}$$

The $2J + 3$ factor in these expressions is the statistical weight of the upper level. Hence, when $J \gg 1$, we have $\left|\mu_{ul}\right|^2 \simeq \mu_d^2$. The transition probability increases rapidly with ν and hence with J; for a linear molecule, $A \propto (J+1)^4 / (2J+3)$. Which for large J simplifies to $\propto J^3$.

Recapping these results. The Einstein A scales with ν^3 times the transition dipole moment. The Einstein B scales with the transition dipole moment. The oscillator strength and absorption coefficient scale with the transition dipole moment times ν. In addition to these frequency dependencies, we also have to recognize the effect of mass of the particles on the intrinsic strength of transitions alluded to in Section 3.1. Finally, independent of this, when transitions are dipole forbidden, electric quadrupole or magnetic dipole transitions are a factor α^2 weaker with α the fine-structure constant (1/137).

Table 3.3 summarizes the characteristics of rotational transitions of some astrophysically relevant molecules. We note that homonuclear diatomic molecules (e.g. H_2, O_2, N_2), some symmetric, linear, heteronuclear molecules (e.g. CO_2, CS_2), some symmetric top molecules (e.g. C_6H_6), and spherical top molecules (e.g. CH_4) have no permanent dipole and, hence, no allowed rotational transitions. These species can still have magnetic dipole allowed transitions.

3.3 Vibrational Spectroscopy

3.3.1 Energy Levels

When a molecule vibrates, it may undergo a change in dipole moment and it will, then, couple to an electromagnetic field. Consider a diatomic molecule and small vibrations around an equilibrium position. The potential energy is then a quadratic function of the vibrational coordinates, and the essence of vibrational spectroscopy is then contained within Hooke's law for a harmonic oscillator,

$$\nu = \frac{1}{2\pi c}\sqrt{\frac{\kappa}{\mu}}, \tag{3.43}$$

with ν the fundamental frequency, κ the force constant, μ the reduced mass of the molecular units vibrating, and the factor c is included to transform the unit to wavenumbers (cm^{-1}) for cgs units, commonly used in spectroscopy. Molecular bond strengths are, of course, very similar to binding energies of electrons to an atom. However, the frequencies of the transitions of molecular vibrational levels are, thus, shifted to lower energies by $\sqrt{m_e/M}$, with m_e and M the mass of the electron and atom, respectively, and occur in the near- and mid-IR.

Real molecules are not harmonic oscillators and, as a result of anharmonicity of the bonding, hot bands (e.g. 2-1, 3-2) are generally shifted to lower frequencies. A Morse potential is a convenient description of the interaction of a diatomic molecule,

$$V(R) = D_e\left(\exp\left[-2a\left(R - R_e\right)\right] - 2\exp\left[-a\left(R - R_e\right)\right]\right), \tag{3.44}$$

where R and R_e are the internuclear distance and equilibrium distance, and D_e is the dissociation energy, and a equals,

$$a = \nu_e\sqrt{\mu/2D_e}, \tag{3.45}$$

with ν_e the fundamental frequency. We recognize the short range repulsive and long range attractive forces in equation (3.44). The Schrödinger equation can be solved exactly for the Morse potential, and the energy levels are,

$$E(v) = h\nu_e\left(v + 1/2\right) - h\nu_e x_e\left(v + 1/2\right)^2, \tag{3.46}$$

with $x_e\nu_e = \nu_e^2/4D_e$ the anharmonicity constant. The $1/2$ term accounts for the zero-point energy and implies that even when the molecule has no vibrational excitation, the atoms are still vibrating, in accordance with Heisenberg's uncertainty principle. The Morse potential has a maximum number of levels, $v_{max} \simeq 2D_e/\nu_e$. The frequencies of vibrational transitions are then,

$$\nu(v) = \nu_e - 2\nu_e x_e (v+1). \tag{3.47}$$

Now consider a polyatomic molecule and again study small vibrations around an equilibrium position that are represented by quadratic potentials. This results in a set of uncoupled harmonic oscillators. The total energy is then given by,

$$E = E_0 + \sum_k v_k h\nu_k, \tag{3.48}$$

with v_k and ν_k the vibrational quantum number and frequency of mode k. The zero-point energy, E_0, is given by,

$$E_0 = \frac{1}{2}\sum_k h\nu_k. \tag{3.49}$$

Anharmonicity – due to electronic repulsion at short distances, and dissociation at large distances – will play an important role. In polyatomic molecules, diagonal terms depend on the coordinates of a single oscillator while off-diagonal terms will link different oscillators and the energy (Eq. (3.48)) will contain cross terms. In principle, there are then $N_m(N_m+1)/2$ anharmonicity constants for a molecule with N_m normal modes.

For a molecule, translational degrees of freedom have been converted into vibrational and rotational degrees of freedom. In the harmonic approximation, the motions of the atoms in the molecule can be decomposed into the normal modes of vibration, which are linearly independent motions (i.e. mutually orthogonal). A molecule with N atoms will have $3N$ degrees of freedom, but three of those are associated with translational motion and three (two for linear molecules) with rotations, leaving $3N-6$ $(3N-5)$ vibrational modes. These modes will differ in their force constants and reduced mass and, hence, will occur at different frequencies. Some of the $3N-6$ modes may occur at the same frequency and hence these modes are degenerate. For N_{dg} degenerate modes, we have $n_v = 3N-5-N_{dg}$ for linear molecules and $n_v = 3N-6-\sum_{k=1}^{k=N_{dg}}(d_n-1)$ for nonlinear molecules with d_n the degeneracy of the nth mode. In principle, all atoms are moving for a given normal mode. However, it is sometimes advantageous to "classify" these modes as stretching (symmetric and asymmetric) and bending (scissoring, rocking, wagging, and twisting) vibrations of specific atoms (or groups of atoms) in the molecule. This can be particularly advantageous for CH stretching modes as these are relatively isolated. Thinking in terms of local modes can also be insightful when considering vibrations near the dissociation limit.

Modes will be infrared-active if the dipole moment changes during the vibration. Conversely, some vibrations can be infrared inactive; e.g. the symmetric stretching vibrations in C_2H_2. The symmetry of a mode is described using group theory and this is used to determine whether a mode is IR and/or raman active. Group theory is beyond the scope of this book. In general, the absorption strength will be stronger for more polar bonds.

Numbering of vibrational modes is governed by symmetry type set by group theory: Start with the completely symmetric modes and sort them by wavenumber. This yields the numbering of these modes. Do the same with subsequent symmetry types. In this, nondegenerate modes are numbered before degenerate modes. Degenerate modes receive the same number but are recognized by a superscript.[2] There is one (historical) exception: the bending modes of linear triatomic molecules is always labeled, ν_2. As an example, methane has nine fundamental modes: the symmetric stretch ($\nu_1 = 2917$ cm^{-1}), the doubly degenerate deformation mode ($\nu_2 = 1534$ cm^{-1}), the triply degenerate stretch ($\nu_3 = 3019$ cm^{-1}), and the triply degenerate deformation mode ($\nu_4 = 1306$ cm^{-1}). Only the ν_3 and the ν_4 modes are IR active. Vibrations in homonuclear molecules do not lead to changes in the dipole moment and hence are infrared inactive. Mixing of isotopes in such species will lead to infrared absorptions as the center of mass and the center of charge no longer coincide. Likewise, interactions with the environment in a solid will introduce weak infrared activity.

Anharmonicity introduces small shifts in the frequencies of the species. In addition, overtones ($\Delta v > 1$) become allowed. Besides these fundamentals, we can also expect combination bands (i.e. $\nu_i + \nu_j$ with $\Delta v_i = \Delta v_j = 1$ or $2\nu_i + \nu_j$ with $\Delta v_i = 2$, $\Delta v_j = 1$, etc.) and difference bands (i.e. $\nu_i - \nu_j$ with $\Delta v_i = \pm 1$ and $\Delta v_j = \mp 1$). In general, fundamental bands are much stronger than overtones or combination/difference bands. However, this can change due to resonance interaction. The near (accidental) resonances between a fundamental and an overtone/combination/difference band of the same symmetry will lead to mixing of the states. This will shift the frequencies of the modes away from each other and the weaker mode will borrow intensity from the stronger mode.

3.3.2 Partition Function

In general, the statistical weight of vibrational levels is $g_v = 1$ but, for degenerate modes, this has to be modified. The total vibrational statistical weight, including degeneracy, of the state, $\{v_1, v_2, v_3, \ldots\}$, is given by

$$g_{v_1, v_2, v_3, \ldots} = \sum_{k=1}^{k=n_v} \frac{(v_n + d_n - 1)\,!}{v_n!\,(d_n - 1)\,!}, \tag{3.50}$$

where each term represents the different way v_n quanta can be distributed over d_n (degenerate) modes. For a doubly degenerate vibration, this becomes $v_n + 1$, while for a triply degenerate vibration, we have $(v_n + 1)(v_n + 2)/2$.

For a single mode, the vibrational partition function is given by,

$$Z_{vib}(T) = \sum_v \exp\left[-E_{vib}(v)/kT\right], \tag{3.51}$$

[2] The bending mode of a linear polyatomic molecule is degenerate as the molecule can bend in the $x - z$ and/or $y - z$ plane. The two oscillators in the x and y direction are combined to yield circular motion and ℓ is the quantum number associated with this vibrational angular momentum. Level, v, in this mode has degeneracy $v + 1$ ($|\ell| = v, v - 2, \ldots 0$ or 1; $+$ and $-$ signs correspond to clockwise and anticlockwise rotation).

with $E_{vib}(v)$ the vibrational energy of level v of this mode. Ignoring anharmonic interaction, the partition function of a polyatomic molecule is given by the product of the vibrational partition function of each normal mode. For a single harmonic oscillator and measuring the energy relative to the zero-point energy, we have, $E_{vib} = vh\nu_e$, and

$$Z_{vib}(T) = \sum_{v} \left(\exp\left[-h\nu_e/kT\right]\right)^{v} = \left(1 - \exp\left[-h\nu_e/kT\right]\right)^{-1}. \tag{3.52}$$

Including degeneracy, this becomes,

$$Z_n(T) = \left(1 - \exp\left[-h\nu_e/kT\right]\right)^{-d_n}. \tag{3.53}$$

The total vibrational partition fruction is then,

$$Z_{vib}(T) = \prod_{k=1}^{n_v} \left(1 - \exp\left[-h\nu_e/kT\right]\right)^{-d_n}. \tag{3.54}$$

We can define characteristic vibrational temperatures as $\theta_v = h\nu_e/k$ and values for some astrophysically relevant species are summarized in Table 3.4. Typically, $\theta_v/T \ll 1$ and $Z_{vib} \simeq 1$ for a mode and hence, in essence, only the ground state is populated. However, large molecules have many, many modes and even if for each mode the vibrational partition function is close to 1, the product is still a large factor. Moreover, when excitation temperatures are high, partition functions can quickly become astronomically large. We will return to this in Chapter 8.3, where we discuss the excitation of large molecules.

3.3.3 Molecular Identification

Vibrational transitions are very characteristic for the motions of the atoms in the molecular group directly involved in the vibration but much less sensitive to the structure of the rest of the molecule. Characteristic band positions of various molecular groups are illustrated in Figure 3.7 and summarized in Table 3.5. Modes involving motions of hydrogen occur

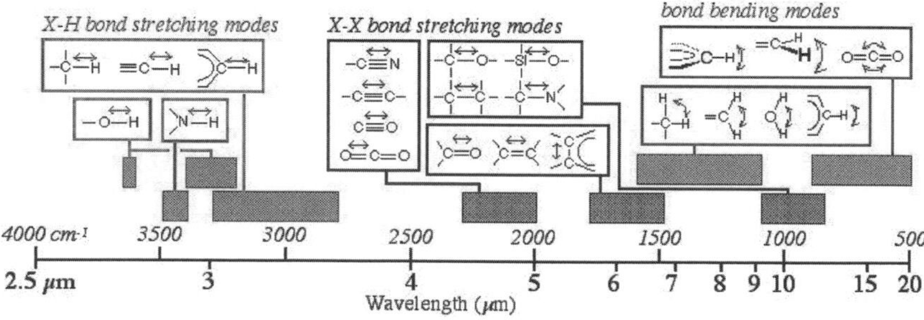

Figure 3.7 Summary of the vibrational frequencies of various molecular groups. The boxes indicate the range over which specific molecular groups absorb. The vibrations of these groups are schematically indicated in the linked boxes. Figure kindly provided by D. Hudgins

Table 3.5 *Characteristic vibrational band positions*

Group	Mode	Frequency range [cm^{-1}]	Note
OH stretch			
	Free OH	3,610–3,645[a]	Sharp
	Intramolecular[b]	3,450–3,600	Sharp
	Intermolecular[c]	3,200–3,550	Broad
	Chelated[d]	2,500–3,200	Very broad
NH stretch			
	Free NH	3,300–3,500	Sharp
	H-bonded NH	3,070–3,350	Broad
CH stretch			
	≡C–H	3,280–3,340	
	=C–H	3,000–3,100	
	CO–CH_3	2,900–3,000	Ketones
	C–CH_3	2,865–2,885	Symmetric
		2,950–2,975	Asymmetric
	O–CH_3	2,815–2,835	Symmetric
		2,955–2,995	Asymmetric
	N–CH_3	2,780–2,805	Aliphatic amines
	N–CH_3	2,810–2,820	Aromatic amines
	CH_2	2,840–2,870	Symmetric
		2,915–2,940	Asymmetric
	CH	2,880–2,900	
C≡C stretch			
	C≡C	2,100–2,140	Terminal group
	C–C≡C–C	2,190–2,260	
	C–C≡C–C–C≡C–	2,040–2,200	
C≡N stretch			
	Saturated aliphatic	2,240–2,260	
	Aryl	2,215–2,240	
C=O stretch			
	Non-conjugated	1,700–1,900	
	Conjugated[e]	1,590–1,750	
	Amides	1,630–1,680	
C=C stretch			
	–HC=C=CH_2	1,945–1,980	
	–HC=C=CH–	1,915–1,930	
CH bend			
	CH_3	1,370–1,390	Symmetric
		1,440–1,465	Asymmetric
	CH_2	1,440–1,480	
	CH	1,340	

Continued

Table 3.5 *Continued*

Group	Mode	Frequency range [cm^{-1}]	Note
CO–O–C stretch			
	Formates	∼1,190	
	Acetates	∼1,245	
C–O–C stretch			
	Saturated aliphatic	1,060–1,150	Asymmetric
	Alkyl,aryl ethers	1,230–1,277	
	Vinyl ethers	1,200–1,225	
Aromatic modes			
	C–H stretch	∼3,030	
	C–H deformation	∼1,160	In plane
	C–H deformation	900–740	Out of plane[f]
	C=C stretch	1,590–1,625, 1,280–1,315	

[a]The OH frequency of free water occurs at 3756 cm^{-1}. [b]H-bonded as dimer or polymer. [c]In a fully H-bonded network such as ice. [d]The OH is H-bonded to an adjacent C=O group. [e]Two double bonds separated by a single bond. [f]Pattern of bands whose position depends on number of adjacent C-atoms with H.

at considerably higher frequencies than modes involving similar motions of heavier atoms (again, Hooke's law, Eq. (3.43)). Thus, H-stretching vibrations occur in the 3 μm region, while stretching motions among (single bonded) C, N, and O atoms are located around 10 μm. Likewise, when the bond strength increases, the vibration shifts to higher frequencies and singly, doubly, and triply bonded CC vibrations shift from about 1000 to 2000 cm^{-1} (Figure 3.7).

Vibrational spectra can be used as fingerprints for the identification of the molecular groups of a species and, hence, provide a powerful tool to determine the class of molecules present (e.g. alkanes versus aldehydes). However, as a rule, vibrational spectra cannot easily distinguish the specific molecule within a class. The smallest molecule within a class often forms an exception to this rule. For example, the C–H stretching vibration of methane (3019 cm^{-1}) is shifted from that of other alkanes (2840–2975 cm^{-1}). For a pure substance, subtleties within the spectra can be used to identify the compound present. Thus, the spectrum of n-hexane (C_6H_{14}) will show absorptions at the positions of the stretching and bending vibrations of methyl (CH_3) and methylene (CH_2) groups in a relative strength commensurate with the intrinsic strength of the modes of these molecular groups and the relative number of CH_3 and CH_2 groups present in the molecule (2 versus 4 groups). The isomer isohexane (2-methylpentane with an additional methyl group replacing one of the H's on the second C-atom of the pentane molecule) will also show methyl and methylene absorptions but in a different relative strength (3 versus 3 groups). Furthermore, the methyl bands will be split due to interaction between the adjacent CH_3 groups. In the gas

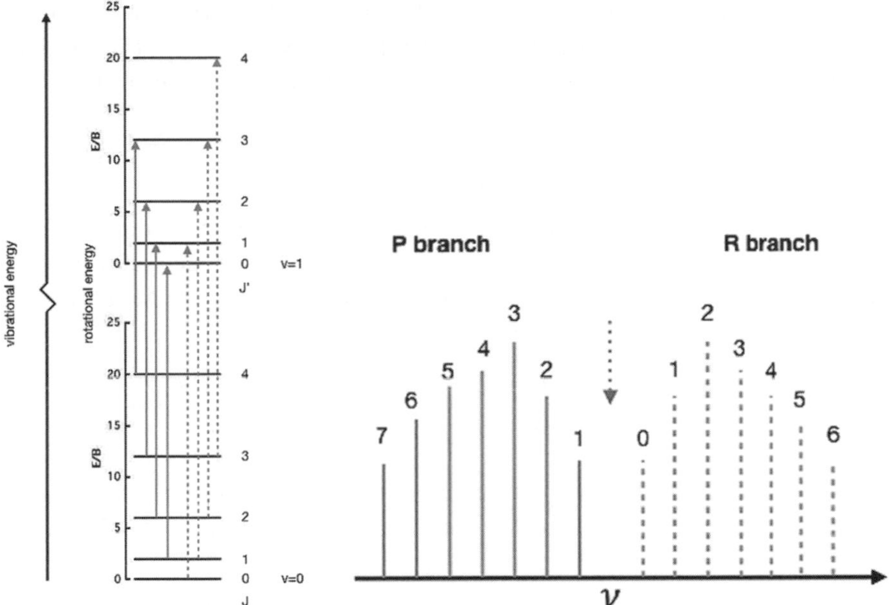

Figure 3.8 Ro-vibrational spectra of a diatomic rigid rotating harmonic oscillator. Left: Energy level diagram including the ground and first excited vibrational state and their associated rotational states. P-branch ($\Delta J = -1$) and R-branch ($\Delta J = +1$) transitions are indicated by full and dashed arrows. Q-branch ($\Delta J = 0$) transitions are missing. Right: The resulting spectrum consists of equidistant lines separated by $2B$ and centered around the fundamental frequency, ν. Q-branch transitions would pile up at ν. P and R branch transitions are labeled by the rotational quantum number from the level in the lower vibrational state. The relative strength of the transitions largely reflects the population distribution of the states involved. This will be discussed in Section 4.1.1.

phase, identification can be considerably aided by resolved P, Q, and R-branches (Section 3.3.4). In the solid state, when a mixture of species is present, identification of specific molecular species present within a class is often daunting.

3.3.4 Gas Phase Ro-vibrational Spectra

Molecules can vibrate and rotate simultaneously and that is reflected in the energy levels. For a diatomic molecule, we can write for the energy levels,

$$\frac{E}{hc} = \nu_e\left(v + \frac{1}{2}\right) - \nu_e x_e\left(v + \frac{1}{2}\right)^2 \tag{3.55}$$

$$+ B_e J\left(J+1\right) - D_e J^2\left(J+1\right)^2 - \alpha_e\left(v + \frac{1}{2}\right) J\left(J+1\right) \tag{3.56}$$

where the subscript e implies that the quantity is evaluated at the equilibrium position of the nuclei. The first two terms refer to vibrational energy levels of an anharmonic oscillator.

The next two are the rigid rotor plus the centrifugal distortion. The last term gives the vibration–rotation interaction. The first four terms have been described in Sections 3.2 and 3.3. For a Morse potential, we can write for the rotation–vibration interaction constant,

$$\alpha_e = 6\sqrt{\frac{x_e B_e^3}{\nu_e}} - 6\frac{B_e^2}{\nu_e}. \tag{3.57}$$

Ignoring for the moment the rotation–vibrational interaction, we recognize that the total energy is the sum of the rotational and vibrational energy. The selection rules are the same as for the individual motions, $\Delta v = \pm 1$, $\Delta J = \pm 1$. Now consider the transition from $v = 1 \leftarrow 0$. Simple algebra shows that the transition frequency is

$$\nu_{v=0\to 1, J\to J'} = \nu_e (1 - 2x_e) + B_e \left(J - J'\right)\left(J + J' + 1\right) \tag{3.58}$$

$$- D_e \left(J^2 (J+1)^2 - J'^2 \left(J'+1\right)^2\right), \tag{3.59}$$

where we have assumed the same rotational constants in the upper and lower state and indicated the rotational levels as J' and J'' for $v = 1$ and $v = 0$, respectively. For $\Delta J = 1$ – the so-called R-branch[3] – we now have $J' - J'' = 1$ and,

$$\nu_{v=0\to 1, J\to J+1} = \nu_e (1 - 2x_e) + 2B_e (J+1) - 4D_e (J+1)^3 \tag{3.60}$$

and for $\Delta J = -1$ – the so-called P-branch[3],

$$\nu_{v=0\to 1, J\to J-1} = \nu_e (1 - 2x_e) - 2B_e J - 4D_e J^3. \tag{3.61}$$

Note that in these expressions, J stands for the rotational quantum number in the lower vibrational state. There is no line at the band center, $\nu = \nu_e (1 - 2x_e)$. When the centrifugal distortion is small, we can ignore the D_e terms and we have two branches of equally spaced lines separated by $2B_e$ appearing on either side of the line center with the P-branch on the lower frequency side. Note that for, e.g., H_2 the selection rules are $\Delta J = \pm 2$ and these transitions are labeled as O(J) and S(J).

The ro-vibrational coupling expresses that as the bond length increases in a higher vibrational state, the moment of inertia also increases. As a consequence, the rotational constant will be smaller (i.e. $B_v \to B_e - \alpha_e (v + 1/2)$). We can write,

$$\nu_R \left(J''\right) = \nu \left(v''\right)_0 + 2B'_v + \left(3B'_v - B''_v\right) J'' + \left(B'_v - B''_v\right) J''^2 \tag{3.62}$$

$$\nu_P \left(J''\right) = \nu \left(v''\right)_0 - \left(B'_v + B''_v\right) J'' + \left(B'_v - B''_v\right) J''^2, \tag{3.63}$$

where the double prime refers to the ground state and the single prime to the upper state. Note that $B'_v - B''_v < 0$ and the spacing in the P-branch will be larger than in R branch. Also, the P-branch spreads out while the R-branch bunches up with increasing rotational quantum number. The bunching up of the R branch leads to the formation of a bandhead, which we can quantify,

[3] For the P-, Q-, and R-branches, transitions are labeled as P(J), QP(J), R(J), respectively, with J the rotational quantum number of the level in the lower vibrational state.

$$\frac{d\nu_R(J'')}{dJ} = \left(3B_v' - B_v''\right) + \left(B_v' - B_v''\right)J'' = 0. \tag{3.64}$$

This results in J'' (bandhead) $\simeq B_e/\alpha_e$. For CO, this is about 100.

Linear polyatomic molecules can be treated similar to diatomic molecules and their spectral patterns are the same. We can now recognize parallel bands where the vibration introduces a change in dipole moment along the molecular axis and perpendicular bands where the change in dipole moment is perpendicular to the molecular axis. For parallel bands, the selection rules are $\Delta v = 1$ and $\Delta J = \pm 1$ and transitions will show P and R branches. For perpendicular bands, the selection rules are $\Delta v = 1$ and $\Delta J = 0, \pm 1$. As the rotation–vibration constant is small, Q-branch[3] ($\Delta J = 0$) will occur at almost the same frequency but there is a small red shading as $\alpha_e < 0$. The amount of red shading is a function of the temperature.

For a symmetric top molecule, we have two rotational quantum numbers and two rotational constants. We can again recognize parallel and perpendicular bands now according to whether the change in dipole moment is parallel or perpendicular to the main axis of rotation of the molecule. The energy levels are (ignoring centrifugal distortion and rotational–vibrational coupling),

$$\frac{E}{hc} = \nu_e\left(v + \frac{1}{2}\right) - \nu_e x_e\left(v + \frac{1}{2}\right)^2 + B_e J(J+1) + (A - B)K^2. \tag{3.65}$$

The selection rules are $\Delta v = 1$, $\Delta J = \pm 1, 0$, and $\Delta K = 0$ (except for $K = 0$ where $\Delta J = \pm 1$) and there are P, Q, and R branches. For $K = 0$, the spectrum is that of a linear molecule without a Q-branch. In general, as I_A/I_B decreases, the intensity of the Q branch decreases. Consider first the parallel modes, which will have simple P, Q, and R branches except that – as for the pure rotational transitions – these split up into $J+1$ K-components at high spectral resolution. The perpendicular transitions are complicated by strong coriolis interaction. For a prolate and oblate top, the rotational energy levels are given as,

$$\frac{E}{hc} = BJ(J+1) + (A - B)K_A^2 \mp 2A\zeta K_A \tag{3.66a}$$

$$\frac{E}{hc} = BJ(J+1) + (C - B)K_C^2 \mp 2C\zeta K_C \tag{3.66b}$$

where ζ is the coriolis coupling constant.[4] The selection rules are $\Delta v = 1$, $\Delta J = \pm 1, 0$, and $\Delta K = \pm 1$. This large coriolis splitting shifts the K-structure and results in separate subbands ($K' \leftarrow K''$ transitions), where each subband has a P, Q, and R branch. The subbands' origins are approximately separated by $2(A(1-\zeta) - B)$. Depending on the magnitudes of A, B, and ζ these perpendicular subbands can be well separated or result in a massive congested spectra.

[4] The coriolis coupling constant, ζ, is the vibrational angular momentum but unlike for a linear molecule, ζ $(-1 \leq \zeta \leq 1)$ is not necessarily an integer.

Furthermore, the P and R branch will split because of a difference in $A - B$ between the upper and lower vibrational levels and the Q branch will split because of a difference between B in the upper and lower vibrational levels. For perpendicular bands, the selection rules are $\Delta v = 1$, $\Delta J = \pm 1, 0$, and $\Delta K = \pm 1$ and there are P, Q, and R branches. For the R branch, we have $\Delta J = +1$, $\Delta K = \pm 1$, and $\nu_R = \nu_o + 2B(J+1) + (A-B)(1 \pm 2K)$; for the P branch, we have $\Delta J = -1$, $\Delta K = \pm 1$, and $\nu_P = \nu_o - 2BJ + (A-B)(1 \pm 2K)$; for the Q branch, we have $\Delta J = 0$, $\Delta K = \pm 1$, and $\nu_Q = \nu_o + (A-B)(1 \pm 2K)$. Note that there are two sets of P, Q, and R branches for each lower state value of K.

For asymmetric top molecules, the spectra become very complex and have to be calculated numerically.

3.3.5 Accurate Line Lists

The potential of a molecule can be expanded into a Taylor series around the equilibrium position, which for a diatomic molecule reads as,

$$E(R) = \sum_n \frac{1}{n!} \left(\frac{\partial^n E}{\partial R^n} \right)_{R_e} (R - R_e)^n . \tag{3.67}$$

Such an approach can be the basis for the reproduction of the rotational–vibrational spectrum. The Dunham expansion is then often used for the term expansion,

$$E(v, J) = \sum_{k,l} Y_{k,l} (v + 1/2)^k (J(J+1))^l . \tag{3.68}$$

The coefficients, $Y_{k,l}$, in this expansion are then determined by comparison to experimental data. The first terms are readily assigned to ν_e & $-\nu_e x_e$ and B_e & D_e, and $-\alpha_e$ (cf. Eq. (3.46)) and can be linked to the coefficients in the expansion of the potential energy surface (Eq. (3.67)).

For many applications accurate line lists are a prerequisite for detailed modeling of terrestrial, (exo)planet, brown dwarf, or stellar atmospheres. Ab-initio potential energy surfaces – calculated, say, with coupled cluster methods – are used to predict line strengths and frequencies of transitions. These are then compared to high resolution spectroscopic measurements to establish their accuracies and, most importantly, assist in spectral assignments as well as make prediction for higher frequency regimes that are not as amenable to experimental study.

3.3.6 Solid State Vibrational Spectroscopy

Vibrational spectra of molecules frozen into an ice differ in several regards from gas phase spectra. In the solid state, rotations are generally suppressed and hence the rotational–vibrational bands collapse to one absorption band near the band origin. The precise band position will differ from that of the free (gas phase) species due to dispersive, electrostatic, induced, and repulsive interactions with neighboring molecules. For species that can form a

Table 3.6 *Integrated strength of ice bands*

Species	ν [cm^{-1}]	$\Delta\nu$ [cm^{-1}]	A [cm $molecule^{-1}$]
H_2O	3275	310	2.0 (−16)
H_2O	1670	160	1.0 (−17)
H_2O	750	240	2.8 (−17)
CO	2138	2.5	1.1 (−17)
CO_2	2340	18	7.6 (−17)[a]
CO_2	656	18	1.1 (−17)
CO_2	3708	12	1.4 (−18)
CO_2	3600	19	4.5 (−19)
CH_4	3010	7	1.0 (−17)
CH_4	1300	8	7.3 (−18)
NH_3	3375	45	1.3 (−17)
NH_3	1070	68	1.3 (−17)
CH_3OH	3250	235	1.3 (−16)
CH_3OH	2982	100	2.1 (−17)
CH_3OH	2828	30	5.4 (−18)
CH_3OH	1450	90	1.2 (−17)
CH_3OH	1026	29	1.9 (−17)
OCS	2042	45	1.5 (−16)
OCN^-	2160	25	1.3 (−16)[b]

[a]This band can be strongly affected by small particle scattering effects, which enhance its intrinsic strength (cf. Section 5.2.1). [b]Uncertain. Depends on the assumed efficiency of the production of this species in UV irradiation experiments.

hydrogen-bonded network such as – notably, H_2O but also CH_3OH and NH_3 – these shifts can be up to hundreds wavenumbers. The width of solid state absorption bands likewise depends on the matrix environment. For traces of weakly bonded species in an inert matrix, the width can be as narrow as 0.2 cm^{-1}. An amorphous ice will possess a distribution of binding sites. Variations in the interaction will then broaden the absorption features and width of 2–50 cm^{-1} are expected, denpeding on the mixture. For strong interactions – e.g. species in a H-bonding network – the width can be up to 300 cm^{-1}. For completeness, the far-IR spectral window is home to the phonon modes of solids and these can be used to study interstellar ices as well.

Despite these differences, the general rules for vibrational spectroscopy still hold and IR vibrational spectra provide convenient fingerprints for molecular identification (Figure 3.7, Table 3.5, Section 3.3.3). The absorption properties of some astronomically relevant ices are summarized in Table 3.6. These data refer to pure ices and there is an extensive literature on the effects of matrix variations.

3.4 Electronic Spectroscopy

Here, we will discuss the electronic spectra of simple molecules, with the emphasis on diatomic molecules. The resulting photodissociation cross sections are discussed in Chapter 6.3. The electronic spectra and photophysics of large molecules are discussed in Chapter 8.2.

3.4.1 Energy Levels and Notation

As discussed in Chapter 2, we can classify electrons according to the binding type involved (Figure 3.9): σ-electrons are localized between two atoms and are tightly bound. Transitions involving σ electrons occur therefore at short wavelength (1000–2000 Å). Delocalized π electrons are less strongly bound then σ-electrons and transitions for nonconjugated species are in the far-UV (1500–2500 Å). The energy levels of nonbonding electrons – say associated with lone pair electrons of O or N atoms or nonbonding π electrons – lie typically between those of bonding and antibonding orbitals. According to the Hund's rule, when there are degenerate orbitals, an electron will occupy an empty orbital before it starts to pair up. Typically, excitation of an electron does not affect the spin and, e.g., a singlet ground state (with $S = 0$, see Section 3.4.2) couples with a singlet excited state. For the excited electron, a spin flip can occur, converting, e.g., a singlet state in a triplet state (with $S = 1$). Following Hund's rule, these triplet states are at lower energy than the

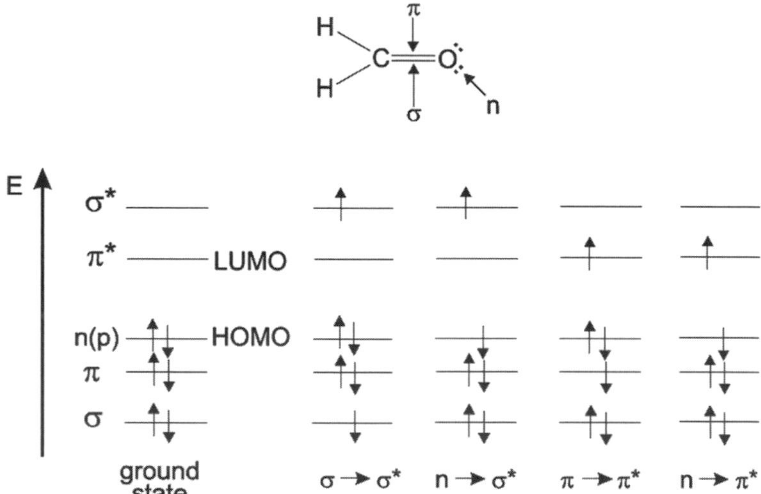

Figure 3.9 Schematic energy diagram for formaldehyde with the various types of electron molecular orbitals and relevant electronic transitions. The left-hand side illustrates the ground state while possible excitations of the different electrons are shown to the right. HOMO stands for highest occupied molecular orbital and LUMO for lowest unoccupied molecular orbital. Figure taken from [16]

corresponding singlet state. Triplet states can connect radiatively with singlet states through spin–orbit coupling,[5] but such transitions have small Einstein A's.

3.4.2 Diatomic Molecules

For atoms, the electronic states are labeled as $^{2S+1}L_J$ with L the orbital angular momentum of the electrons (the vector sum of the orbital angular momentum of the individual electrons), S the electron spin angular momentum (the sum of the spins of the individual electrons), and J the total angular momentum (the vector sum of the orbital and spin angular momenta). All of these are in units of $\hbar$. The different values for the total orbital angular momentum, $L = 0, 1, 2, \ldots$, are denoted by roman capitals, S, P, D, ... The electronic states of diatomic molecules are dealt with analoguously but now classified by the projection of their orbital angular momentum on the internuclear axis, $\Lambda = \sum \lambda_i$, their spin multiplicity, S, and the sum of projection of the orbital angular momentum and the spin on the internuclear axis, $|\Omega| = |\Lambda + \Sigma|$: e.g. $^{2S+1}\Lambda_\Omega$ with Σ the projection of S on the internuclear axis. The different values for the projected orbital angular momentum, $\Lambda = 0, 1, 2, \ldots$, are labeled as, Σ, Π, Δ, ... Electronic states of diatomic molecules are also labeled with letters. The ground state is labeled X and excited states with the same multiplicity are labeled A, B, C, ... while excited states with different multiplicity are labeled a, b, c, ...; both are in order of increasing energy. The Pauli exclusion principle requires an overall antisymmetry of the total wavefunction under exchange. The σ states of diatomics are labeled with superscript $+$ or $-$ to indicate symmetry with respect to reflection in a plane containing the internuclear axis. For homonuclear molecules, molecular orbitals are also labeled according to symmetry with respect to inversion through the center of symmetry and the electronic wavefunction is either even (labeled as subscript g for gerade) or odd (labeled as subscript u for ungerade) upon inversion.

3.4.3 Intensity and Franck–Condon Factors

Electronic transitions take place on timescales of femtoseconds ($\Delta t \sim \Delta r/v$ with $\Delta r \sim 10^{-7}$ cm and $v \simeq 10^8$ cm/s). The fastest atomic vibration is associated with CH stretching vibrations and occurs on a timescale of $\simeq 10^{-13}$ s. Hence, essentially, the nuclei are frozen during an electronic transition. This is the basis of the Born–Oppenheimer approximation. The wave function can then be expanded in its electronic (ψ), spin ($\mathcal{S}$), and vibrational (χ) components. The transition probability is then governed by the oscillator strengths,

$$f_e f_{so} f_v \sim \langle\psi|H_e|\psi\rangle \langle\mathcal{S}|H_{so}|\mathcal{S}\rangle \langle\chi|\chi\rangle, \tag{3.69}$$

where the first two terms describe the action of the electronic and spin–orbit coupling Hamiltonians on the wave functions and express the selection rules. The last term is

[5] In a mechanistic view, the magnetic moment associated with the orbital motion of the electron couples with the magnetic moment of the electron spin.

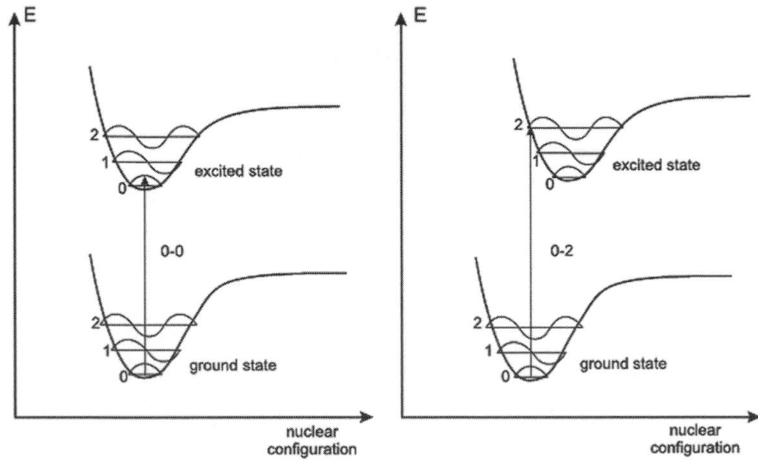

Figure 3.10 Schematic potential energy diagram with wavefunctions, χ, for the various vibrational levels in the ground and excited electronic state. Note that, for excited states, these have highest density near the turning points. Left: The bonding character does not change much between the two electronic states and overlap is largest for the 0–0 transition. Right: The bonding character changes much between the two electronic states and overlap is largest for the 0–2 transition. These two cases will give rise to very different vibrational progressions.

the vibrational overlap and, as there is no electronic action, there are no selection rules for vibrational transitions accompanying electronic transitions. However, the transition strength is governed by the overlap of the vibrational wave functions (Figure 3.10). If the bonding character between the two states is very similar, transitions in the vibrational ground state have the best overlap with the vibrational ground state in the excited electronic state. On the other hand, if the bonding character is very different, the potential energy curves are shifted and the ground vibrational state in the lower electronic state will couple best to an excited vibrational state in the upper electronic state through a vertical transition. Table 3.7 summarizes typical values for longest wavelength electronic transitions of molecular chromophores.

3.5 Specific Examples

Molecular Hydrogen

Following the discussion in Chapter 2, the two 1s orbitals of the H atoms combine into two molecular orbitals. Constructive interference between the two wave functions leads to the highest electron density between the two nuclei and an attractive bonding orbital, which is lower in energy than either atomic orbital. Destructive interference of the wave functions leads to a node (zero electron density) between the two nuclei. This antibonding orbital exceeds the energy of the atomic orbitals. The potential energy diagram of H_2 is shown in Figure 3.11. The configuration of the electronic ground state of H_2 molecule is $(1\sigma_g)^2$.

Table 3.7 *Long wavelength absorption bands of typical chromophores*[a]

Chromophore	λ [Å]	Absorption strength [cm^2/mol]	Transition
C–C	<1800	1000	$\sigma, \sigma^\star$
C–H	<1800	1000	$\sigma, \sigma^\star$
C=C	1800	10,000	$\pi, \pi^\star$
C=C–C=C	2200	20,000	$\pi, \pi^\star$
Benzene	2600	200	$\pi, \pi^\star$
Naphthalene	3100	200	$\pi, \pi^\star$
Anthracene	380	10,000	$\pi, \pi^\star$
C=O	2800	20	$n, \pi^\star$
N=N	3500	100	$n, \pi^\star$
N=O	6600	200	$n, \pi^\star$
C=C–C=O	3500	20	$n, \pi^\star$
C=C–C=O	2200	20,000	$\pi, \pi^\star$

[a]These correspond to the HOMO → LUMO transition.

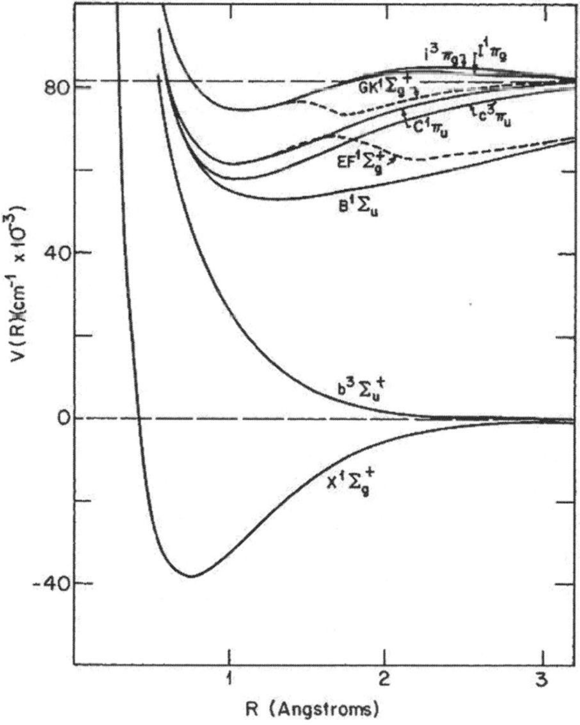

Figure 3.11 Energy level diagram for H_2. The horizontal dashed lines correspond to the energy of the atoms at "infinite" distance (1s and 1s or 1s and 2s). Figure taken from [25]

For both electrons, $\lambda = 0$ and hence $\Lambda = 0$, resulting in a Σ state. The electron spin is either $S = 0$ (singlet) or $S = 1$ (triplet). These are antisymmetric and symmetric under exchange. In the ground state, following the Pauli exclusion principle, the electrons will have opposite spins and $S = 0$. As the atomic orbitals are g, the ground state[6] is g. Thus, the ground state is $X\,^1\Sigma_g^+$ followed by the triplet state, $b\,^3\Sigma_u^+$. Higher electronic states couple two H-atoms with $n = 1$ and $n = 2$, for example. The electronic states of H_2^+ are at even higher energies. The ground electronic state of H_2 has 14 bound vibrational states and each of those has an "infinite" number of rotational states. As the nuclei are identical, the combined proton spin is either 0 or 1; e.g. antisymmetric and symmetric under exchange. Molecular hydrogen has para and ortho states (cf. Section 3.2.7). The odd rotational states are ortho ($I = 1$), the even ones are para ($I = 0$), and the statistical weights are 3 and 1 (Section 3.2.7). Selection rules are $\Delta\Lambda = 0, \pm 1$. $\Delta S = 0$, g states couple only to u states and vice versa, and parity is preserved. The change in vibrational quantum number is governed by the Franck–Condon overlap. Within the ground electronic state, there is no restriction on Δv. The pure rotational transitions have to preserve symmetry and occur either fully within the ortho or the para levels. In many ways, ortho and para H_2 behave therefore as separate chemical species that only exchange through chemical reactions (Section 3.2.7).

The $B\,^1\Sigma_g^+$ and $C\,^1\Pi_u^+$ are the next highest singlet states that can couple to the ground electronic state through allowed transitions (the Lyman and Werner band) with typical oscillator strengths of 10^{-2}. Figure 3.12 shows the rich spectrum of H_2 absorption in the Lyman and Werner bands observed toward a star in the Large Magellanic Cloud. Absorptions originate in the various rotational levels of the ground vibrational state in the ground electronic state and connect to ro-vibrational levels of the excited states according to the selection rules.

The vibrational frequency of H_2 in its ground electronic state is at $\nu_e = 4400$ cm^{-1}. As a very light hydride, anharmonicity is relatively important; viz., $\nu_e x_e = 121$ cm^{-1}. As H_2 is a homonuclear molecule, ro-vibrational and pure rotational transitions are dipole forbidden. The vibrational levels in the ground electronic state are still connected through weak quadrupole transitions with $\Delta J = 0 \pm 2$. These transitions are labeled by O(J''), Q(J''), and S(J''), with J'' the lower level rotational quantum number (i.e. 1 – 0 S(1) corresponds to the transition $v' = 1\,J' = 3$ to $v'' = 0\,J'' = 1$). The vibrational levels can be pumped by UV absorption in the Lyman–Werner bands followed by radiative decay as well as by collisions in warm gas. The former is important in diffuse clouds and in photodissociation regions associated with massive stars, planetary nebulae, or galactic nuclei (Section 11.6). The latter can be important in strong shock waves in molecular clouds (Section 11.7). Figure 3.13 shows the near-infrared spectrum observed toward the reflection nebulae, NGC 2023, illuminated by the B0.5 star, HD 37097, revealing strong ro-vibrational H_2 transitions. Because H_2 is such a light hydride, its pure rotational transitions occur at very high frequencies with ground state para ($J = 2$–0) and ortho (3–1) transitions at 28 and 17 μm, respectively.

[6] $g \times g = g$, $u \times u = u$, and $u \times g = g \times u = u$.

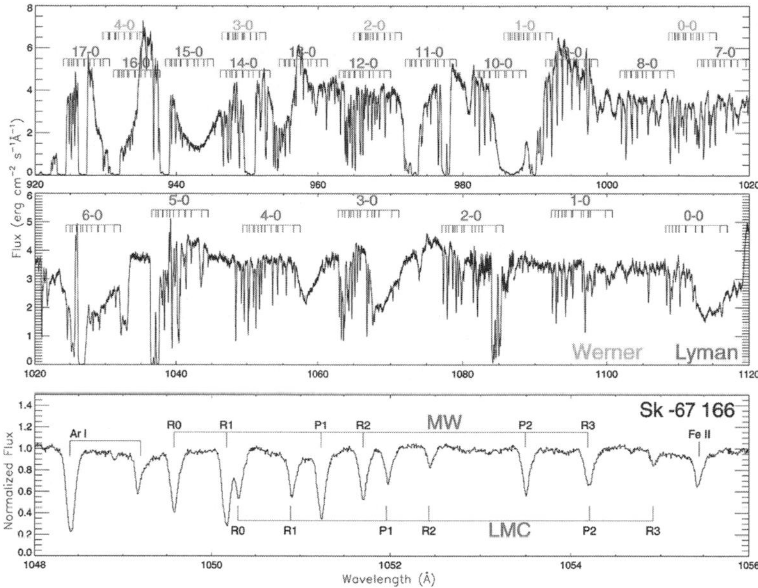

Figure 3.12 The 920–1120 Å absorption spectrum toward the star Sk -67°166 in the Large Magellanic Cloud measured by the Far Ultraviolet Spectroscopic Explorer, FUSE. Various Lyman (black) and Werner (gray) bands are indicated. This rich spectrum shows both LMC and Milky Way absorption components (bottom panel). Figure taken from [31]

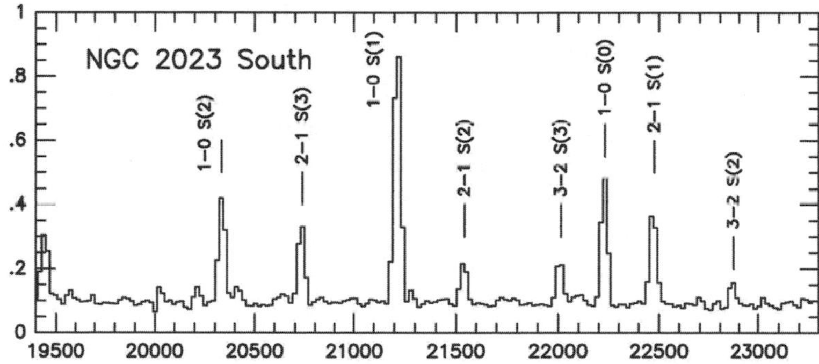

Figure 3.13 The near-IR emission spectrum of the reflection nebula, NGC 2023, revealing H_2 ro-vibrational emission lines associated with $v = 1 - 0$ and $v = 2 - 1$ transitions. Figure taken from [19]

Carbon Monoxide

For CO, we have to mix the atomic orbitals, which we will treat simplistically as one orbital from each center. The orbitals that we mix have to have the same symmetry and need to be close in energy. Thus, $2p_C$ mixes with $2p_O$ and $2s_C$ with $2s_O$ creating bonding and antibonding states (Figure 3.14). Also, s and p_z orbitals mix into σ orbitals (the z-axis is

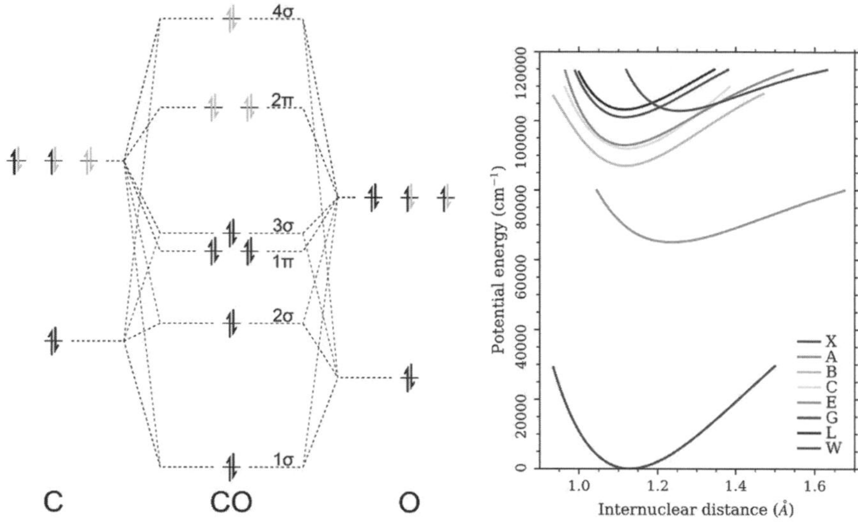

Figure 3.14 Left: Molecular orbitals of CO and their relationship to the atomic orbitals of C and O. As O is more electronegative than C, its orbitals are lower in energy. Right: Energy level diagram for CO. Figure courtesy of [13]

the bond axis) while the p_x and p_y orbitals mix into π orbitals. These states have to be filled and create the electron configuration, $(1\sigma)^2(2\sigma)^2(1\pi_x)^2(1\pi_y)^2(3\sigma)^2$. The wavefunctions of the electrons in the more stable orbits will have the highest electron density near the atom with the highest electron negativity (O). While the higher energy antibonding molecular orbitals will be closer to the less electronegative atoms (C). Thus, reactivity of CO will be concentrated on the C-atom. The total angular momentum and the total spin are both 0 and the ground state is $X\,^1\Sigma^+$. The energy level diagram of CO is shown in Figure 3.14.

Interstellar CO has been studied in the UV through the specific ro-vibrational transitions between the $X\,^1\Sigma^+$ and $A\,^1\Pi$, $B\,^1\Sigma^+$, $C\,^1\Sigma^+$, and $E\,^1\Pi$ states. Photodissociation of CO occurs, amongst others, through absorption into the E states – coupling to C(^3P) and O(^3P) – which predissociate because of an avoided crossing with the W state (Figure 3.14). UV transitions of CO occur in the far ultraviolet and have been measured using spectrographs on, e.g., HST and FUSE in diffuse clouds (Figure 3.15). These and comparable data has been analyzed through a curve of growth-type anaysis (c.f. Section 4.3.2).

The CO $v = 1 - 0$ fundamental vibration occurs at 2170 cm^{-1} (4.61 μm) and has only a modest anharmonicity (13.3 cm^{-1}). The ro-vibrational transitions of CO and its isotopes can be observed through the M-band window in the mid-IR (Figure 3.16) if measured at the right "time" to allow the interstellar transitions to shift out of the atmospheric bands by the Doppler effect due to the Earth's orbital motion. Solid CO is an important component of interstellar ices and its vibrational band is widely detected in absorption in sight lines traversing dense cold molecular clouds (Figure 3.16; see also Section 10.6). Overtone transitions, $\Delta v = 2$ of the CO fundamental are accessible in the K-band.

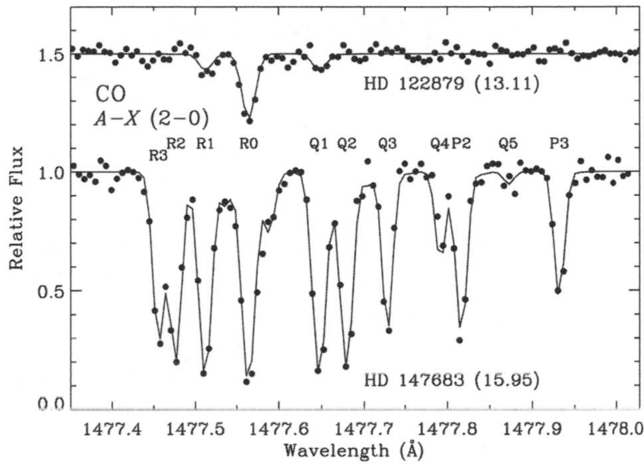

Figure 3.15 UV absorption spectrum measured toward the stars, HD 147683 & 122879, showing absorption lines due to the $A-X$ system. [24]

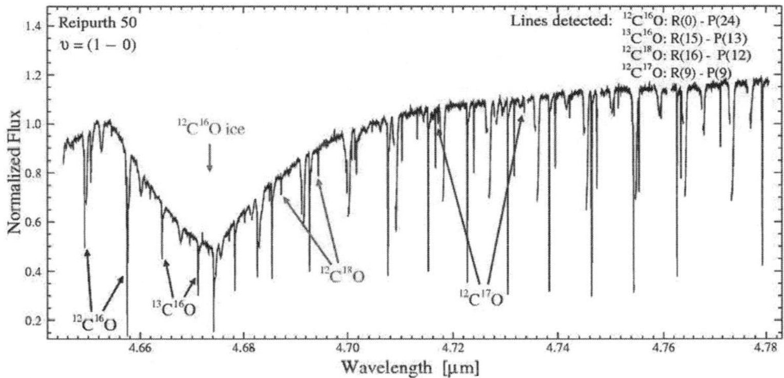

Figure 3.16 Ro-vibrational transitions of CO isotopes in the spectrum of the protostar Reipurth 50. Note also the broad solid CO band [26]

The pure rotational ladder of CO falls in the sub-millimeter and CO transitions are generally observed through the atmospheric windows at high spectral resolution (sub-km/s) and high sensitivity (at the quantum noise level) using heterodyne techniques on single dish telescopes (e.g. IRAM, APEX) or interferometers (e.g. plateau de Bure, ALMA). These instruments typically measure one CO line at a time (albeit that transitions from many other molecules may occur in the same spectral bandpass). Figure 3.18, actually, shows several transitions measured by the SPIRE Fourier Transform Spectrometer on board the Herschel Space Observatory. These observations lack the spectral resolution and have only limited sensitivity but they do obtain a large portion of the CO ladder. Observations of the rotational transitions of CO are widely used to study molecular clouds (Sections 10.1 and 10.2).

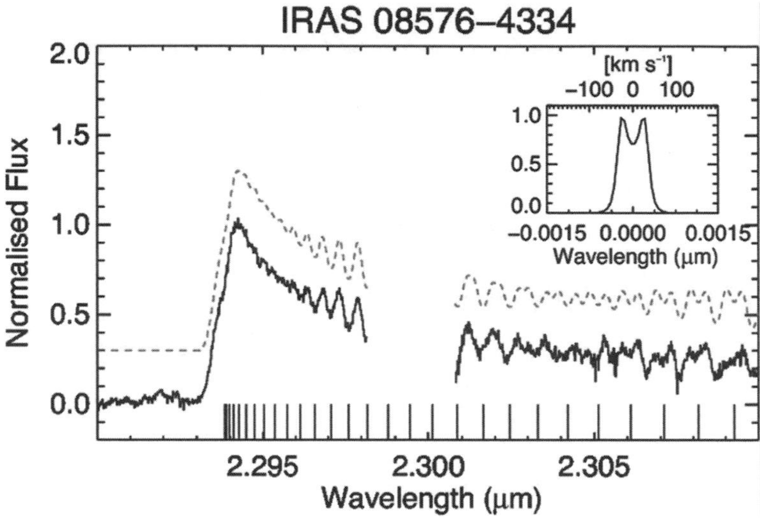

Figure 3.17 The near infrared spectrum of the massive protostar IRAS 08576-4334 reveals the $v = 2 - 0$ transition of CO. Positions of the individual ro-vibrational transitions – shifted to account for the Doppler effect – are indicated below. The dashed trace above the spectrum is a model fit of thee transitions convolved with the adopted line profile shown in the inset. Near the bandhead, the transitions pile up and that allows for an "easy" detection of this feature. Figure courtesy of [14]

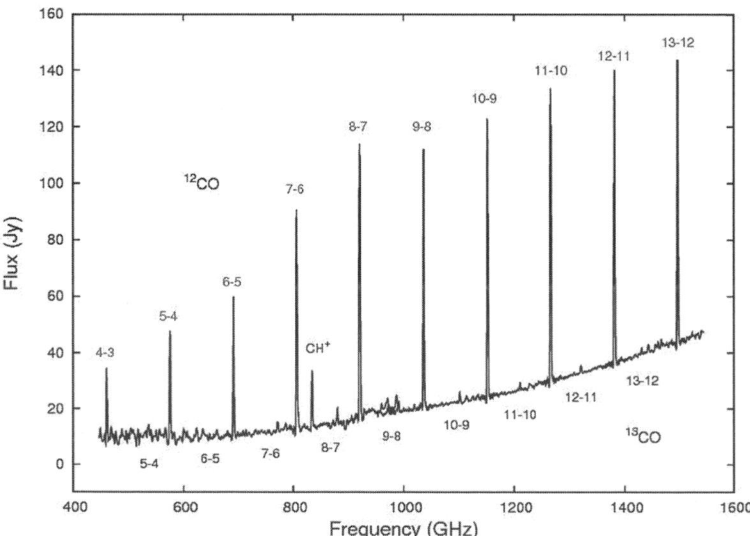

Figure 3.18 The pure rotational spectrum of CO of the planetary nebula, NGC 7027, measured by the SPIRE instrument on the Herschel Space Observatory. ^{12}CO lines are labeled at the top. ^{13}CO lines are labeled at the bottom. Figure courtesy of [32]

Polyatomic Molecules

The electron-level diagram for polyatomic molecules quickly becomes very complex. As an example, consider H_2CO (Figure 3.9). Following the discussion in Chapter 2 and ignoring the $1s$ electrons, the electron configuration of C is $(2s)^2(p_x)(p_y)$, where the p_x and p_y orbitals each have one electron with parallel spin and the $2p_z$ orbital is empty. One of the $2s$ electrons is promoted to the $2p_z$ orbital. The $2s$ orbitals hybridize with the two $2p$ orbitals to form three sp^2 orbitals (each with one electron). Overlap between two sp^2 hybridized orbitals of C with the $1s$ orbitals of the two H's gives rise to two σ bonds. The third sp^2 orbital of C overlaps with the oxygen sp^2 orbital, resulting in a σ_{CO} bond while overlap of the p_z orbitals on the C and O atoms gives rise to a π_{CO} bond. In addition, oxygen has two lone pairs of electrons, which are in nonbonding orbitals, n_O. Because of the higher electronegativity of O versus C, the energies of the molecular orbitals increase as, $1s_O < 1s_C < 2s_O < \sigma_{CH} < \sigma_{CO} < \pi_{CO} < n_O$ where the first three are essentially pure atomic orbitals, and the others are the σ, π, and nonbonding orbitals. The ground state configuration is then,

$$(1s_O)^2(1s_C)^2(2s_O)^2(\sigma_{CH})^2(\sigma'_{CH})^2(\sigma_{CO})^2(\pi_{CO})^2(n_O)^2(\pi^\star_{CO})^0 \tag{3.70}$$

In low-lying states, an n electron or a π electron can be promoted to a $\pi^\star$ orbital (Figure 3.9) and the electron spins can be parallel (triplet state) or antiparallel (singlet states). The ground state is $X\,^1A_1$ and excited states are $A\,^1A_2(n,\pi^\star)$ and $B\,^1A_1(\pi,\pi^\star)$. These states are labeled as S_0 ($^1A_1$) and S_1 ($^1A_2$) and T_1 ($^3A_2$).[7] Groundstate formaldehyde shows strong $\pi \rightarrow \pi^\star$ transitions around 1870 Å, while the $n \rightarrow \pi^\star$ transitions are much weaker (as there is little overlap between these essentially orthogonal orbitals) and occur around 2850 Å.

In the ISM, formaldehyde is observed through its pure rotational levels. The rotational level diagram of this near-prolate molecule is shown in Figure 3.3 and discussed in Section 3.2.4. The rotational transitions of formaldehyde provide a good tool for studies of physical conditions in dense molecular cloud cores (cf. Section 4.2.3) and are often used for this purpose, particularly for star-forming cores (Section 10.1). Gaseous formaldehyde has also been observed through its ν_1 and ν_5 symmetric and asymmetric CH_2 stretching vibrations around 3.6 μm in absorption in the spectrum of protostars and in emission in the spectrum of comets.

3.6 Further Reading and Resources

There are many textbooks that deal with atomic and molecular structure and transitions. For astrophysics, references [3, 22, 27, 30] provide comprehensive overviews. Conventions for line strength are detailed in [33] and line strengths of molecules are conveniently tabulated in appendix V of [30]. [1] provides a tutorial on the strength of (hyper)fine-structure lines. [18] treats in detail the conversion of observed line intensities into molecular column densities.

[7] For large molecules, nomenclature changes and the ground state, if singlet as for formaldehyde, is labeled S_0 while excited singlet states are labeled, $S_1 \ldots S_n$ and triplet states are labeled as $T_1, \ldots T_n$.

The PGOPHER program simulates rotational, vibrational, and electronic spectra and is accessible at http://pgopher.chm.bris.ac.uk/index.html. HITRAN (high-resolution transmission) database (https://hitran.org) provides a compilation of spectroscopic parameters that can be used to simulate infrared spectra. While HITRAN is developed for atmospheric research, the database is also very useful for astronomical purposes. Reference [23] provides more details on developing line lists. The ExoMol database, http://exomol.com/, provides accurate, evaluated line lists for relevant species. The field has been reviewed by [4].

Information on spectroscopy and excitation of atoms and molecules can be found at the website of the NIST (www.nist.gov/srd/atomic.htm) & https://physics.nist.gov/cgi-bin/micro/table5/start.pl. Rotational spectra of molecules can also be accessed through the website maintained by JPL (http://spec.jpl.nasa.gov/) and the Cologne Database for Molecular Spectroscopy (www.ph1.uni-koeln.de/vorhersagen/).

References [7, 12] provide early studies on the characteristics of IR spectra of astronomically relevant ice species. Studies on CO and CO_2 are particularly relevant [2, 9, 29]. Far-infrared properties of relevant ices are rarer. An early review is [21] and a recent study using THz timedomain spectroscopy is provided by [11]. Databases of laboratory spectra of interstellar ice analogs are available at: http://icedb.strw.leidenuniv.nl/, https://science.gsfc.nasa.gov/691/cosmicice/constants.html, and www.astrochem.org/databases.php.

3.7 Exercises

3.1 Transition frequencies

(a) Consider an electron in orbit around the nucleus. For a typical energy of 5 eV, what is the photon frequency required to move the electron out by 1 Å?

(b) In a similar fashion, atoms in a molecule can be considered connected by springs. Show that vibrational and electronic frequencies are then related roughly by $\sqrt{m_e/M}$.

(c) Show that, for rotations, the relationship is m_e/M.

3.2 Explain the difference between the rotational frequencies of CO and CS.

3.3 The $J = 2 \rightarrow 1$ rotational transition of $^{12}C^{16}O$ occurs at 230.538MHz. What is the bond length? What would you predict the $J = 2 \rightarrow 1$ transition of $^{13}C^{16}O$ to be?

3.4 Centrifugal distortion:

(a) A real molecule is not rigid and the contrifugal force will lead to an increased bond length when the molecule rotates faster. Consider a harmonic vibration where the restoring force is given by $-\kappa\,(R_e - R)$. The centrifugal force is given by $M\omega^2 R$. Balancing these two forces, use the quantized form of the angular momentum, to show that

$$R = R_e + \frac{J\,(J+1)\,\hbar^2}{M\kappa R^3}, \tag{3.71}$$

and thus that the nuclear separation increases with increasing rotation.

(b) Use Hooke's law to show that $R = R_e(1+\delta)$, where the small correction factor δ is given by,

$$\delta = \frac{J(J+1)\hbar^2}{M\kappa R_e^4}. \tag{3.72}$$

(c) The potential energy is then larger by $\kappa(R-R_e)^2/2$. Show that this results in,

$$E_{rot} = \frac{J(J+1)\hbar^2}{2MR_e^2} - \frac{J^2(J+1)^2\hbar^4}{2M^2\kappa R_e^6}. \tag{3.73}$$

(d) Show that this implies that for harmonic vibrations, D_e is given by $4B_e^3/\nu_e^2$.
(e) In this derivation, we have made the assumption that $\delta \ll 1$. Show that this is justified for CO.

3.5 Formaldehyde is a nearly prolate molecule with $A = 281970.37$, $B = 38835.42558$, and $C = 34005.73031$ all in MHz. Calculate the energy level diagram for the prolate molecule with $A' = A$ and $B' = C' = (B+C)/2$ and compare with the energy levels of H_2CO.

3.6 Rotational partition function.

(a) Linear rigid rotor:

(i) Derive the expression for the partition function of a linear rigid rotor (Eq. (3.32)) by making the transition from a summation to an integral, the substitution, $x = E(J)/kT$, and assuming that $T \gg hB/k$.
(ii) Adopting $T = 10$K, $B = 2\,\text{cm}^{-1}$ compare this approximate expression with the actual summed partition function. For what temperature is the approximation better than 10%?

(b) Derive the expression for the partition function of a rigid symmetric top following these steps:

(i) Write the partition function as a summation over the J and K states.
(ii) The energy is a function of J and K where J runs from 0 to ∞ and K from $-J, \ldots, 0 \ldots, J$ but this can also be seen as $K = 0, \ldots, \infty$ and $J = K, \ldots \infty$. Rewrite the equation for the partition function to reflect this.
(iii) Make the transition to integrals.
(iv) Do the integral over J first (see above).
(v) Realize that $K(K+1) \simeq K^2$. Do the integral over K.

(c) Consider the following intuitive derivation of the expression for the partition function of a rigid asymmetric top:

(i) Consider the partition function as a sum over all angular momentum states, L_a, L_b, and L_c, which we will assume, each, to run from $-\infty$ to ∞. Convert the summations to integrations, which yields for each axis, $\sqrt{2\pi I_i kT}$.

(ii) This has to be multiplied by $8\pi^2$. For a chosen axis, the angular integration yields a factor 2π. The integration over the orientation of this axis gives an additional factor of 4π.

(iii) Make the transition from classical to quantum mechanics.

3.7 Symmetry factor:

(a) What is the symmetry factor for H_2?
(b) What is the symmetry factor for H_2O?
(c) What is the symmetry factor for NH_3?
(d) What is the symmetry factor for CH_4?
(e) What is the symmetry factor for C_6H_6?

3.8 Consider a harmonic oscillator. Combine Newton's law with Hooke's restoring force and solve for a sinusoidal motion to derive the normal mode frequency.

3.9 The fundamental frequency of HCl is at 2886 cm^{-1}. What is its force constant? For CO, the fundamental frequency occurs at 2170 cm^{-1}. What two factors play a role in this difference with HCl?

3.10 The fundamental asymmetric stretching vibration of carbon monoxide occurs at 2143 cm^{-1}. The first overtone is at 4260 cm^{-1}.

(a) What are the harmonic vibrational frequency and the anharmonic constant?
(b) Predict the frequency of the second overtone.
(c) The second overtone occurs at 6350 cm^{-1}. What could be the origin of the discrepancy?
(d) Estimate the bond energy of this molecule.

3.11 The fundamental asymmetric CH stretching vibration of methane occurs at 3019 cm^{-1}. The first overtone is at 6006 cm^{-1}.

(a) What are the harmonic vibrational frequency and the anharmonic constant?
(b) Predict the frequency of the second overtone.
(c) Estimate the CH bond energy.

3.12 The vibrational partition function:

(a) Water has three vibrational modes at 3657, 1595, and 3758 cm^{-1}. Calculate the vibrational partition function at 300 K and at 1500 K.
(b) Carbon dioxide has vibrational modes at 1388, 667, and 2349 cm^{-1} where the 667 cm^{-1} mode is doubly degenerate. Calculate the vibrational partition function at 300 K and at 1500 K and compare to those of water.
(c) Consider now a 72-atom molecule (say, circumcoronene ($C_{54}H_{18}$)). Assume that each of the vibrational modes in this species has the (geometric) average partition function of a mode in the water molecule. Calculate the vibrational partition function of this molecule at 300 and 1500 K.

(d) Evaluate the partition function of this large molecule assuming the average CO_2 vibrational partition function.
(e) Compare these partition functions – marvel – and then draw reasoned conclusions on the importance of the temperature and the vibrational frequency spectrum.

Bibliography

[1] Axner, O., Gustafsson, J., Omenetto, N., Winefordner, J. D., 2004, *Spectro Chim Acta B*, 59, 1
[2] Baratta, G. A., Palumbo, M. E., 2017, *A & A*, 608, A81
[3] Bernath, P., 2005, *Spectra of Atoms and Molecules* (Oxford: Oxford University Press)
[4] Bernath, P., 2020, *J Quant Spectr Rad Trans*, 240, 1
[5] Cooksy, A. L., Blake, G. A., Saykally, R. J., 1986, *ApJ*, 305, L89
[6] Darling, J., Zeiger, B., 2012, *ApJL*, 749, L33
[7] d'Hendecourt, L. B., Allamandola, L. J., 1986, *A & A Suppl*, 64, 453
[8] DiSanti, M. A., Bonev, B. P., Magee-Sauer, K., et al., *ApJ*, 650, 470
[9] Ehrenfreund, P., Boogert, A. C. A., Gerakines, P. A., et al., 1996, *A & A*, 315, L341
[10] Gray, M., *Maser Sources in Astrophysics*, (Cambridge: Cambridge University Press)
[11] Giuliano, B. M., Gavdush, A. A., Müller, B., et al., 2019, *A & A*, 629, A112
[12] Hagen, W., Tielens, A. G. G. M., Greenberg, J. M., 1983, *A & A Suppl*, 51, 389
[13] Heays, A., private communication
[14] Ilee, J. D., et al., 2014, *MNRAS*, 429, 2960
[15] Klein, H., Lewen, F., Schieder, R., Stutzki, J., Winnewisser, G. 1998, *ApJ*, 494, L125
[16] Liao, Y.-Y., 2013, PhD Thesis, Ecole Normale Superieure de Cachan
[17] Mangum, J. G., Wootten, A., 1993, *ApJS*, 89, 123
[18] Mangum, J. G., Shirley, Y. L. 2015, *Publ Astron Soc Pacific*, 127, 266
[19] Martini, P., Sellgren, K., DePoy, D. L., 1999, *ApJ*, 526, 772
[20] Mina-Camilde, N., Manzanares, C., 1996, *J Chem Educ*, 73, 805
[21] Moore, M. H., Hudson, R. L., 1994, *A & A Suppl*, 103, 45
[22] Pradhan, A. K., Nahar, S. N., 2011, *Atomic Astrophysics and Spectroscopy* (Cambridge: Cambridge University Press)
[23] Schwenke, D. J., *J Phys Chem*, 1996, 100, 2867
[24] Sheffer, Y., Rogers. M., Federman, S. R., et al., 2008. *ApJ*, 687. 1075
[25] Shull, J. M., Beckwith, S., 1982, *Annu Rev Astron Astrophys*, 20, 163
[26] Smith, R., Pontoppidan, K. M., Young, E. D., et al., 2009, *ApJ*, 701, 163
[27] J. Tennyson, *Astronomal Spectroscopy* (World Scientific)
[28] Tennyson, J., Yurchenko, S. N., Al-Refaie, A. F., et al., 2016, *J Mol Spec*, 127, 73
[29] Tielens, A. G. G. M., Tokunaga, A. T., Geballe, T. R., Baas, F., 1991, *ApJ*, 381, 181
[30] Townes, C. H., Schalow, A. L., 1955, *Microwave Spectroscopy* (New York: McGraw-Hill)
[31] Tumlinson, J., Shull, J. M., Rachford, B. L., et al., 2002, *ApJ*, 566, 857
[32] Wesson, R., Cernicharro, J., Barlow, M. J., et al., 2010, *A & A*, 518. L144
[33] Whiting, E. E., Schadee, A., Tatum, J. B., Hougen, J. T., Nicholls, R. W., 1980, *J Mol Spectr*, 80, 249
[34] Yamamoto, S., Saito, S. 1991, *ApJ*, 370, L103

4

Molecular Emission and Absorption

Astronomy is largely an observational science and this chapter discusses the various aspects that enter into the analysis and interpretation of observations. Such a discussion has to start off by considering level populations and molecular excitation as that is at the bottom of much of the observational characteristics. Observations with ALMA as well as single dish telescopes bring us an in-depth view of many astronomical objects and phenomena and these observations concern rotational transitions of gas phase species. This chapter focusses, therefore, on excitation of rotational levels and subsequent emission in the microwave regime. The electronic excitation and vibrational fluorescence of large molecules is discussed at length in Chapter 8. The discussion starts with level populations and the concept of critical densities (Section 4.1.1). This is followed by a discussion of collisional (de-)excitation rates (Sections 4.1.4 and 4.1.5). For many species, rotational transitions are optically thick and this is often modeled and analyzed with the use of the escape probability formalism (Section 4.1.6). This chapter contains a lengthy discussion of this topic and the resulting emergent intensity of rotational transitions (Section 4.1.7). This discussion is followed by a section on the analysis of observations (Section 4.2) where the concept of rotation diagram analysis is discussed, including its many pitfalls. This is followed by a discussion on the use of molecules as barometers and thermometers of the gas (Section 4.2.3). The chapter closes with a short discussion on electronic excitation (Section 4.3.1) and the analysis of UV absorption lines of, e.g., H_2 using the curve of growth technique (Section 4.3.2). The chapter closes with resources that will guide a student into the wider literature.

As a guide for a course on molecular astrophysics, students would do well to focus on the sections on analysis of observations using rotation diagrams (Section 4.2) and the use of molecules to determine the physical conditions of the emitting gas (Section 4.2.3). In order to appreciate these discussions, students should also study carefully the two-level system both in the optically thin and the optically thick cases (Sections 4.1.5 and 4.1.7). The remainder of this chapter could be assigned for reading only.

4.1 Level Populations

4.1.1 Local Thermodynamic Equilibrium

In local thermodynamic equilibrium, the level populations are described by the Boltzmann distribution. The relative populations of two levels is then only a function of the temperature, T,

$$\frac{n_i}{n_j} = \frac{g_i}{g_j} \exp\left[-\left(E_i - E_j\right)/kT\right], \tag{4.1}$$

with E_i and g_i the energy and statistical weight of level i. Using the partition function, $Z(T)$ (Section 3.2.8), the level populations can be written as,

$$n_i = \frac{g_i n}{Z(T)} \exp\left[-E_j/kT\right], \tag{4.2}$$

with n the total density of the species under consideration. For a homogeneous cloud, similar expressions can be written for the column density.

In evaluating the partition function and level populations, care should be taken to include the statistics for nuclear spin (Section 3.2.7) when needed. For relative populations, this only plays a role for species with para and ortho states. For many molecules the nuclear spin is 0 and for rigidly rotating diatomic or linear polyatomic molecules, we can write, $g(J) = 2J + 1$ and $E(J) = hBJ(J+1)$ with J the rotational quantum number and B the rotational constant and we have ignored the small correction effect of centrifugal distortion on the energy levels (cf. Section 3.2.2).

The level populations are given by,

$$\frac{n(J)}{n} = \frac{g_J}{Z(T)} \exp\left[-E(J)/kT\right], \tag{4.3}$$

with g_J the statistical weight, $Z(T)$ the partition function, and $E(J)$ the energy of level J. The most populated level is then,

$$J_{max} \simeq \sqrt{\frac{kT}{2hcB}}. \tag{4.4}$$

This corresponds to an energy $kT/2$.

The strength of an emission line, $J+1 \rightarrow J$, is given by,

$$I_{J+1,J} = \frac{N(J+1)\, A_{J+1,J} h\nu_{J+1,J}}{4\pi}, \tag{4.5}$$

with $N(J+1)$, $A_{J+1,J}$, and $\nu_{J+1,J}$, the column density in the upper level, the Einstein A, and frequency of the transition, respectively. These are described in Section 3.2. If we adopt a homogenous cloud in LTE, the column density is given by eq. (4.2) times a length scale. The maximum emission occurs then for (see assignment 4.1),

$$J + 1 \simeq \sqrt{\frac{2.5kT}{hcB}}. \tag{4.6}$$

Maximum emission occurs thus for the level with an energy above ground of $\simeq 2.5kT$.

The strength of absorption lines scales with the absorption coefficient,

$$\kappa = \frac{h\nu_{J,J+1}\left(n_J B_{J,J+1} - n_{J+1} B_{J+1,J}\right)}{4\pi}. \tag{4.7}$$

The Einstein B depends only very weakly on the rotational quantum number through the line strength (cf. Section 3.2.9). Ignoring stimulated emission and assuming $J \gg 1$, maximum absorption will occur for the transition $J_{max} \rightarrow J_{max} + 1$ (see assignment 4.1),

$$J_{max} \simeq \sqrt{\frac{kT}{hcB}}. \tag{4.8}$$

Maximum absorption occurs thus for the level with an energy above ground of $\simeq kT$.

4.1.2 Critical Density

In general, the interstellar medium is not in thermodynamic equilibrium and we have to solve the statistical equilibrium equations to derive the level populations and thus the absorption and emission of the gas. Consider a two-level system separated by an energy difference, E_{ul}, connected by collisions with collisional rate coefficients, γ_{ij}, and radiative deexcitation with an Einstein coefficient, A_{ul}. The level populations can be found by solving the statistical equilibrium equation,

$$n_l n \gamma_{lu} = n_u n \gamma_{ul} + n_u A_{ul}. \tag{4.9}$$

Because of detailed balance, the collisional rate coefficients of the two levels are related, viz.,

$$\gamma_{lu} = \left(\frac{g_u}{g_l}\right) \gamma_{ul} \exp\left[-E_{ul}/kT\right], \tag{4.10}$$

where g_u and g_l are the statistical weights of the upper and lower level and E_{ul} is the energy difference between the two levels. Equation (4.9) can be rewritten as

$$\frac{n_u}{n_l} = \frac{\frac{g_u}{g_l}\exp\left[-E_{ul}/kT\right]}{1 + \frac{n_{cr}}{n}}, \tag{4.11}$$

with the critical density, n_{cr}, given by

$$n_{cr} = \frac{A_{ul}}{\gamma_{ul}}. \tag{4.12}$$

Thus, when the density is much larger than the critical density, collisions dominate the deexcitation process. We then recover LTE and the level populations are given by the Boltzmann expression at the kinetic temperature of the gas. When the density is low, every upward collisional excitation is followed by a downward radiative deexcitation and the population will "collapse" into the lower level.

This analysis can be generalized to multilevel systems where the critical density now compares radiative transitions to all lower levels with the collisional rates to all levels; viz.,

$$n_{cr} = \frac{\sum_{l<u} A_{ul}}{\sum_{l \neq u} \gamma_{ul}} \tag{4.13}$$

The principle is the same: LTE ensues when the density is larger than the critical density, and the levels fall out of equilibrium when the density is low.

For the intensity, we will need the emissivity, j_{ul}, for which we can write,

$$j_{ul} = \frac{n_u A_{ul} h\nu_{ul}}{4\pi} = \frac{\frac{g_u}{g_l} exp\left[-h\nu_{ul}/kT\right]}{1 + \frac{n_{cr}}{n} + \frac{g_u}{g_l} exp\left[-h\nu_{ul}/kT\right]} \frac{\mathcal{A}_j n \, A_{ul} h\nu_{ul}}{4\pi}, \tag{4.14}$$

where we have used the conservation equation for the species,

$$n_l + n_u = \mathcal{A} n, \tag{4.15}$$

with $\mathcal{A}$ the abundance of the species. In the high density limit, we recognize the denominator for the partition function of a two-level system and again recover LTE. In that limit, the emissivity scales with the density. In the low density limit, the emissivity simplifies to,

$$j_{ul} \simeq \frac{n^2 \mathcal{A}_j \, \gamma_{lu} \, h\nu_{ul}}{4\pi}. \tag{4.16}$$

Every upward collision now results in a photon and, ignoring radiative transfer effects, the emergent intensity scales with the density times the column density for a homogeneous cloud.

4.1.3 Excitation Temperature

As the ISM is often not in LTE, it is sometimes appropriate to define the excitation temperature as,

$$T_X \equiv \frac{E_{ul}}{k} \left[\ln\left(\frac{n_l g_u}{n_u g_l}\right)\right]^{-1}. \tag{4.17}$$

From equation (4.11), we find,

$$\frac{T}{T_X} - 1 = \frac{kT}{E_{ul}} \ln\left(1 + \frac{n_{cr}}{n}\right). \tag{4.18}$$

At high densities ($n \gg n_{cr}$), $T_X \simeq T$ as expected, while at low densities, ($n \ll n_{cr}$), T_X will be much less than T. Of course, formally, when $kT \ll E_{ul}$, T_X can also be close to T, even if $n \ll n_{cr}$. However, at such low densities and temperatures, excitation of the upper level will be exceedingly low and the transition will be largely irrelevant.

4.1.4 Collisional (De-)Excitation Rates

The collisional deexcitation rate coefficient can be written as,

$$\gamma_{ul} = \frac{4}{\sqrt{\pi}} \left(\frac{\mu}{2kT}\right)^{3/2} \int_0^\infty \sigma_{ul}\left(v\right)\, v^3\, exp\left[-\frac{\mu v^2}{2kT}\right] dv, \tag{4.19}$$

with μ the reduced mass of the system and $\sigma_{ul}(v)$ the collisional deexcitation cross section at the relative velocity, v, of the collision partners. The cross section will depend on the interaction potential of the collision partners. This will be discussed in more detail in the chapter on reactive collisions (Chapter 6). Here, we will briefly summarize these results. Collisional excitation and deexcitation rates are related by detailed balance (Eq. (4.10)).

Neutral–neutral interactions are regulated by short range van der Waals forces with cross sections typically $\sigma_{ul} = \pi\ (r_A + r_B)^2$ where the r_i's are the radii of the collision partners. Apart from resonances, this does not depend strongly on the interaction energy. Typically, $r_i \simeq 1$ Å. The deexcitation rate coefficient is then,

$$\gamma_{ul} = \sigma_{ul} \sqrt{\frac{8kT}{\pi\mu}} \simeq 7\times 10^{-11} \sqrt{\left(\frac{T}{100\ \mathrm{K}}\right)\left(\frac{\mathrm{amu}}{\mu}\right)} \left(\frac{\sigma}{5\ \text{Å}^2}\right) \quad \mathrm{cm^3\ s^{-1}}, \tag{4.20}$$

with μ the reduced mass.

The interaction of a charged particle with a neutral is assisted by the polarization induced by the incoming charge and this results in the Langevin rate,

$$k_L = 2\pi e \sqrt{\frac{\alpha}{\mu}}. \tag{4.21}$$

Typically, values for polarizabilities are 2 Å^3 and the Langevin rate for proton excitation is then $\simeq 3\times 10^{-9}\sqrt{\alpha/\text{Å}^3}\ \mathrm{cm^{-3}\ s^{-1}}$. Electron excitation rates are $\simeq 1.3\times 10^{-7}\sqrt{\alpha/\text{Å}^3}$ $\mathrm{cm^{-3}\ s^{-1}}$. Note that there is no dependence on temperature (cf. Section 6.4).

Electrons and ions interact through strong Coulomb forces. The deexcitation rate is is given by,

$$\gamma_{ul} = \sqrt{\frac{2\pi}{kT}}\, \frac{\hbar^2}{m_e^{3/2}}\, \frac{\Omega_{ul}}{g_u} \simeq 8.6\times 10^{-7} \sqrt{\frac{10^2\ \mathrm{K}}{T}}\, \frac{\Omega_{ul}}{g_u} \qquad \mathrm{cm^3\ s^{-1}}, \tag{4.22}$$

where Ω_{ul} is the collision strength, which contains all the (nasty) details of the interaction.

The different excitation rates are summarized in Table 4.1. Because of the Coulomb focussing, electron–ion interactions have the largest cross sections and rate coefficients at low temperatures. For neutral species, collisions with electrons can be very important when the electron fraction exceeds $\sim 5\times 10^{-4}$. For the diffuse ISM, this will occur when there is additional ionization above that due to photo-ionization of carbon – i.e., cosmic ray ionization of H. Table 3.2 summarizes critical densities for rotational transitions of molecular species of astronomical relevance. Perusing this data, it is clear that critical densities span a wide range of values covering relevant densities in diffuse and molecular clouds, photodissociation regions, regions of star formation, protoplanetary disks, and

Table 4.1 *Typical deexcitation rate coefficients*

Collision partners	$< \sigma_{ul} >$ [Å^2]	γ_{ul} [cm^3 s^{-1}]
Hydrogen–neutral	5	7×10^{-11} $(T/100\ \mathrm{K})^{1/2}$
Hydrogen–ion	250	4×10^{-9}
Proton–neutral	250	4×10^{-9}
Electron–neutral	200	1.5×10^{-7}
Electron–ion	1600	10^{-6} $(100\ \mathrm{K}/T)^{1/2}$

galactic nuclei. This means that molecular transitions will provide convenient probes of the physical conditions in the ISM (Section 4.2).

4.1.5 Molecular Excitation Studies

Most rotational transitions of interstellar molecules involve molecules that are subthermally excited; that is, collisional excitation is followed by radiative decay. The excitation of a species is then directly related to the collisional excitation rate coefficients. Analysis of molecular observations in terms of their physical conditions requires accurate (de)excitation collisional rate coefficients. Here, we will briefly discuss theoretical and experimental studies of this important topic.

As it is an impossible task to measure all state-to-state excitation collisional rate coefficients for astrophysically relevant species, analysis invariably relies on theoretical studies. These involve the evaluation of the collision cross section on the potential energy surface describing the interaction between the collider with the "target" species, generally in the ground electronic state. These are then converted into state-to-state collisional rate coefficients for use by astronomers in their analysis. Inside molecular clouds, H_2 is the main collision partner and para and ortho H_2 should be dealt with as separate collision partners. Typically, most of the ortho/para H_2 is in the rotational ground state. In PDRs, H_2 will be more highly excited, reflecting the higher kinetic temperature and the effects of UV pumping. Note also that ortho and para H_2 are in essence separate chemical species and the interconversion between these two states of H_2 – mediated by gas phase chemical reactions or through paramagnetic interactions on grain surfaces – is slow. Hence, the o/p ratio is often out of equilibrium with the local gas temperature. In diffuse clouds and PDRs, H_2 formation is a key process, driving the o/p ratio to 3 while, because of self-shielding, photodissociation favors a high p/o ratio. In dense, cold clouds, H_2 formation is of little importance and the o/p ratio tends to collapse to very low values. Besides H_2, H as well as electrons can be important collision partners. Sometimes, helium is evaluated as a collision partner as – like H_2 – it is a two electron system, but more easily calculated.

Theoretical molecular excitation studies start with calculating the potential energy surface. As rotational energies are small, accurate potential energy surfaces are required and

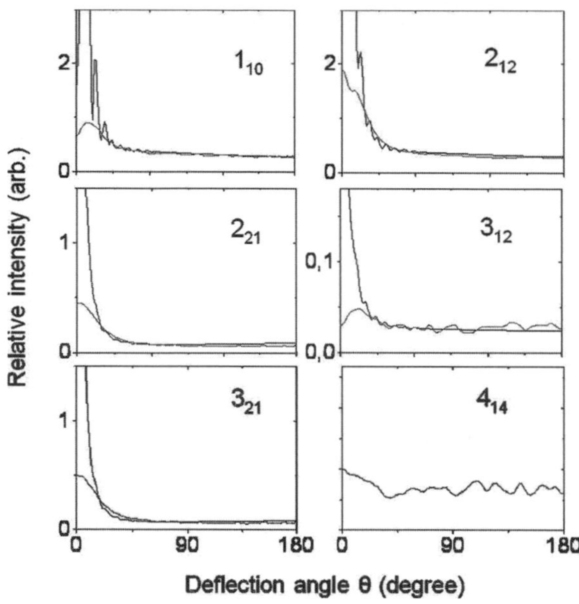

Figure 4.1 Comparison of laboratory measured (gray) and theoretical (black) differential scattering cross sections for para H_2 collisions with ortho H_2O at a collision energy of 71 meV. The experimental curves were scaled to match the theoretical curve at a deflection angle of 60°. Note that measurement errors become very large in the forward scattering (<5°) direction. Figure taken from [23]

high order theory calculations are a must. For small closed-shell systems, coupled cluster[1] calculations can reach accuracies of a few wavenumbers in the potential energy surface. For open shell systems, configuration interaction methods are required. As atomic H is a radical, those collisions will also require configuration interaction methods. The atomic orbitals used to build up the molecular orbitals have to be chosen carefully as the need for accuracy has to be balanced against computational costs. Once the potential energy surface has been calculated, it has to be fitted analytically to high precision to allow dynamical calculations. A good fit to the long range interaction needs care. With an accurate potential energy surface, the collisional excitation cross sections can be computed by solving the nuclear Schrödinger equation, which boils down to solving a set of coupled second-order differential equations. So, the calculation of state-to-state collisional rate coefficients is very involved and requires a substantial effort, and therefore deserves proper reference.

These theoretical results have to be validated against experimental measurements of selected transitions. Experimental studies can be done using the crossed molecular beam technique (Section 2.4.2). As an example, Figure 4.1 shows results for the excitation of para H_2O by para-H_2 with this technique. State-selected beams of H_2O and H_2 are crossed at

[1] Coupled-cluster and configuration interaction calculations are post-Hartree–Fock method (cf. Section 2.5.1).

90° angles. The production of excited state H_2O is monitored using the resonant enhanced multiphoton ionization technique. The H_2O^+ was extracted from the collision area, passed through a time of flight system, and velocity imaged. Crossed molecular beams do provide accurate relative differential scattering cross sections but not absolute values. In the comparison shown in Figure 4.1, the two have therefore been normalized. Double resonance techniques have been developed where one laser disturbs the distribution of the target species and, as collisions thermalize the distribution over time, a second laser is used to probe the depletion of the rotationally excited species through, e.g., laser-induced fluorescence. Combining this technique with the CRESU set-up (Section 2.4.2) allows low temperatures to be reached.

4.1.6 Optical Depth Effects

Consider now the effect of absorption on the level populations. Given a mean photon radiation field, J_{ul}, the statistical equilibrium equation for a two-level system can be written as,

$$n_l n \gamma_{lu} + n_l B_{lu} J_{ul} = n_u n \gamma_{ul} + n_u A_{ul} + n_u B_{ul} J_{ul}, \tag{4.23}$$

where the B's are the Einstein coefficients for absorption and stimulated emission. The level populations constitute a nonlinear problem: At any spatial point, the level populations depend on the radiation field while the radiation field depends on the level populations everywhere. Hence, the statistical equilibrium equations couple with the radiative transfer equations and these have to be solved simultaneously for the whole cloud. By introducing the concept of an escape probability – photons produced locally can only be absorbed locally – the local aspect can be recovered. Such a situation can, for example, result from the presence of a strong velocity gradient when photons are shifted outside of the line once they leave the "local" area. In calculating the optical depth, we will assume that the local-level populations hold globally. Both of these assumptions (locally determined optical depth and escape probability) are generally not correct. However, this decouples the statistical equilibrium equations from the radiative transfer equations and simplifies the problem considerably. In this approximation, we can write that the net absorptions, corrected for stimulated emission, are equal to those photons that do not escape,

$$\left(n_l B_{lu} - n_u B_{ul}\right) J_{ul} = n_u \left(1 - \beta\left(\tau_{ul}\right)\right) A_{ul}, \tag{4.24}$$

where $\beta(\tau)$ is the probability that a photon formed at optical depth τ escapes through the front surface. The level populations are now given by,

$$n_l n \gamma_{lu} = n_u n \gamma_{ul} + n_u \beta\left(\tau_{ul}\right) A_{ul}. \tag{4.25}$$

Comparing this to Equation (4.9), we note that when Equation (4.12) is replaced by

$$n_{cr} = \frac{\beta\left(\tau_{ul}\right) A_{ul}}{\gamma_{ul}} \tag{4.26}$$

Equation (4.11) is recovered. Thus the effect of photon trapping is to lower the density at which LTE is approached; i.e. after each absorption, there is a finite probability that the upper level is collisionally deexcited and the excitation energy is returned to the thermal energy of the gas.

The optical depth, corrected for stimulated emission, is given by,

$$\tau_{lu} = \frac{1}{4\pi}\int (n_l B_{lu} - n_u B_{ul})\,\phi(\nu)\,h\nu ds \tag{4.27}$$

$$= \frac{\pi e^2}{m_e c} f_{lu} N_l \left(1 - \frac{g_l}{g_u}\frac{N_u}{N_l}\right)\phi(\nu) \tag{4.28}$$

$$= \frac{c^2}{8\pi\nu^2} A_{ul} N_u \left(\frac{g_u}{g_l}\frac{N_l}{N_u} - 1\right)\phi(\nu), \tag{4.29}$$

with B_i, A_i, and f_i the Einstein B's, A, and oscillator strength of the transition and where we have assumed a homogeneous cloud with N_i the relevant column densities. The normalized profile function, ϕ, consists of a Doppler core and Lorentzian damping wings. The Doppler core is given by,

$$\phi_D(\nu - \nu_o) = \frac{1}{\sqrt{2\pi}\sigma_\nu}\exp\left[-\frac{(\nu-\nu_o)^2}{2\sigma_\nu^2}\right] \tag{4.30}$$

$$= \frac{1}{\sqrt{2\pi}\sigma_\upsilon \nu/c}\exp\left[-\frac{(\nu-\nu_o)^2}{2\sigma_\upsilon^2(\nu/c)^2}\right] \tag{4.31}$$

with σ_ν and σ_υ the dispersion in frequency space and velocity space, respectively. If turbulence also plays a role, the dispersion parameter is given by $\sigma_\upsilon^2 = \sigma_{th}^2 + \sigma_{tur}^2$ with σ_{th} and σ_{tur} the thermal and turbulent dispersion parameters, respectively. The velocity dispersion is often transformed into the Doppler parameter,

$$b \equiv \sqrt{2}\sigma_\upsilon = \sqrt{2kT/m} = 4.1\times 10^4 \sqrt{\left(\frac{T}{10\,\mathrm{K}}\right)\left(\frac{1\,\mathrm{amu}}{m}\right)}. \tag{4.32}$$

These are related to the full width half maximum of an optically thin line by,

$$\Delta\nu_{\mathrm{FWHM}} = 2\sqrt{\ln 2}\left(\frac{b}{c}\right)\nu \tag{4.33}$$

$$\Delta\upsilon_{\mathrm{FWHM}} = 2\sqrt{\ln 2}\,b. \tag{4.34}$$

The Lorentzian profile is given by,

$$\phi_L(\nu - \nu_o) = \frac{\gamma/4\pi^2}{(\nu-\nu_o)^2 + (\gamma/4\pi)^2}, \tag{4.35}$$

where γ is the damping parameter. Far from line center, the line profile will vary as $(\nu - \nu_o)^{-2}$. If natural line broadening dominates, $\gamma = \sum_l A_{ul}$. For radiation broadening or pressure broadening, γ will depend on the local conditions, but that is beyond the scope of

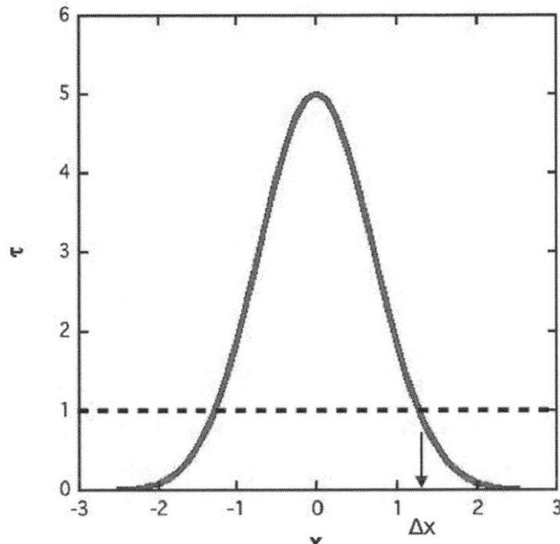

Figure 4.2 Gaussian line profile for a line with a peak optical depth of $\tau = 5$ as a function of the normalized frequency, $x = (\nu - \nu_0) / (\sqrt{2}\sigma_\nu)$. When the line center becomes optically thick, photons can still escape through the line wings. The escape probability is then approximately set by where the optical depth is unity. See text for details.

this book. Both natural broadening and Doppler broadening will apply and the convolution of the two gives rise to a Voigt profile,

$$\phi_V(\nu - \nu_o) = \int_{-\infty}^{\infty} \phi_D(\nu - \nu') \phi_L(\nu_o - \nu') d\nu' \tag{4.36}$$

$$\phi_V(u, a) = \frac{a}{\pi^{3/2}} \int_{-\infty}^{\infty} \frac{\exp[-y^2]}{a^2 + (u - y)^2} dy, \tag{4.37}$$

with $u = (\nu - \nu_o)/(\sqrt{2}\sigma_\nu$ the dimensionless frequency in units of the Doppler width and $a - \gamma/\Delta\nu_D$ the damping parameter. Apart from a slightly different definition, tables of the Voigt function are provided in [7]. As there is no simple analytical formula, we will only discuss the limits in the following.

With these definitions, we can evaluate the escape probabilities, which rules both emission and absorption.[2] For a slab with total optical depth τ, the fraction of the photons produced at depth τ' that escapes the slab is given by $\exp[-\tau']$ and we have to average this over the slab to get the escape probability,

$$\beta(\tau) = \langle \exp[-\tau'] \rangle = \frac{1}{\tau} \int_0^{\tau} \exp[-\tau'] d\tau' = \frac{1 - \exp[-\tau]}{\tau}. \tag{4.38}$$

For small optical depth, $\beta \to 1$ and for large optical depth, $\beta \to 1/\tau$. At this point, we have only considered radiative transfer in the line along a given direction characterized

[2] As photon pathways are fully reversible, absorption and emission are interchangeable. Hence, the concepts for the escape probability and curve of growth are related.

by an optical depth. As escape is a scattering process, we have to average this over all solid angles,

$$\beta(\tau,\vec{n}) = \frac{1-\exp\left[-\tau(\vec{n})\right]}{\tau(\vec{n})}, \tag{4.39}$$

where $\vec{n}$ is a directional vector. The escape probability averaged over all angles is,

$$\beta(\tau_o) = \frac{1}{4\pi}\int_{4\pi} \beta(\tau,\mathbf{n})\, d\Omega, \tag{4.40}$$

with τ_o a suitable reference optical depth. The escape probability depends now on the geometry and the distribution of absorbers. The escape probability has been evaluated under various assumptions and these have been fitted to simple formulae. For a homogeneous cloud, the escape probability is given by,

$$\beta(\tau_o) = \frac{1-\exp[-\tau]}{\tau} \quad \text{spherical geometry} \tag{4.41}$$

$$= \frac{1-\exp[-3\tau]}{3\tau} \quad \text{plane parallel geometry,} \tag{4.42}$$

$$= \frac{1-\exp[-2.34\tau]}{4.68\tau} \quad \text{semi-infinite slab,} \tag{4.43}$$

with τ_o the optical depth to the nearest surface.

Actually, photons will be emitted over the line profile, $\phi(\nu)$, and the escape probability is then,

$$\beta(\tau) = \int_0^{\tau} \frac{1-\exp[-\tau]}{\tau}\phi(\nu)\, d\nu. \tag{4.44}$$

For optically thick lines, escape occurs because the photon is reemitted after absorption at a frequency where the cloud is optically thin. For optically thin lines, the line will be dominated by the Doppler core and the line will look Gaussian. But for increasing optical depth (column density), the center gets saturated first and the profile will become flat topped. Additional emission will add intensity through the line wing (cf. Figure 4.2). The escape probability will then depend on the width, which is approximately given by,

$$\tau(\Delta\nu) \simeq N \exp\left[-\frac{\Delta\nu^2}{2\sigma_\nu^2}\right] \simeq 1, \tag{4.45}$$

or

$$\Delta\nu = \sqrt{2}\sigma_\nu\sqrt{\ln\left[\tau_p\right]} \propto \sqrt{\ln[N]}, \tag{4.46}$$

with τ_p the peak optical depth. Heuristically, we can then approximate the escape probability by assuming that all the photons emitted in the wing ($|\nu-\nu_0| > \Delta\nu$) escape while the photons in the core are absorbed. The escape probability is then given by,

$$\beta(\tau) = \frac{2}{\sqrt{\pi}}\int_{\Delta x}^{\infty} \exp\left[-x^2\right] dx = \left(\tau_p\sqrt{\pi\ln\left[\tau_p\right]}\right)^{-1}, \tag{4.47}$$

with $\Delta x = \Delta\nu/(\sqrt{2}\sigma_\nu)$. This is the logarithmic portion of the curve of growth. Eventually, the Lorentzian wings will take over. That is not important in interstellar emission lines but does play a role in strong absorption lines such as the H_2 Lyman–Werner bands. For Lorentzian-dominated lines, the intensity will scale again with the line width, which is now given by,

$$\tau(\Delta\nu) \propto N/\Delta\nu^2 \simeq 1, \tag{4.48}$$

or

$$\Delta\nu \propto \sqrt{N}, \tag{4.49}$$

the square root portion of the curve of growth. The transition from the logarithmic to the square root portion of the curve of growth will depend on the value of the damping parameter, a (Eq. (4.37).

The escape probability concept is often introduced in terms of the large velocity gradient method. Because of a systematic velocity gradient in a cloud, the Doppler effect will shift the emission line profile of a parcel of gas outside of the absorption line profile of a distant parcel of gas. If the velocity gradient is large then essentially all absorption will occur locally and we have reduced the statistical equilibrium equations to a local problem that is easily tractable. For the case of a large velocity gradient, the optical depth is given by,

$$\tau_{LVG} = \left(\frac{A_{ul}c^3}{8\pi\nu_{ul}^3}\right)\left(\frac{n_u}{|dv/ds|}\right)\left(\frac{g_u n_l}{g_l n_u} - 1\right), \tag{4.50}$$

with dv/ds the velocity gradient.

4.1.7 Emergent Intensity

Let us now consider a plane-parallel, homogenous, semi-infinite slab appropriate for a PDR. In this case, the escape probability is often approximated by,

$$\beta(\tau) = \frac{1 - \exp(-2.34\tau)}{4.68\tau} \qquad \tau < 7 \tag{4.51}$$

and

$$\beta(\tau) = \left[4\tau\left(\ln\left(\frac{\tau}{\sqrt{\pi}}\right)\right)^{1/2}\right]^{-1} \qquad \tau > 7, \tag{4.52}$$

where τ is the profile-averaged optical depth to the nearest surface,

$$\tau = \left(\frac{A_{ul}c^3}{8\pi\nu^3}\right)\left(\frac{n_u\Delta z}{b}\right)\left(\frac{g_u n_l}{g_l n_u} - 1\right). \tag{4.53}$$

These expressions are accurate to within 10%. In the limit of very small optical depth, the escape probability goes to $\frac{1}{2}$: in a semi-infinite slab, photons escape through half a hemisphere. At large optical depth, β scales with $1/\tau$.

The emergent intensity in the line is given by,

$$I = 2\int_0^z j_{ul}\left(\tilde{z}\right)\, d\tilde{z}, \tag{4.54}$$

where the factor 2 takes into account that photons are emitted only toward the front surface. Here, j_{ul} is given by,

$$j_{ul} = \frac{n_u A_{ul} h\nu_{ul} \beta\left(\tau_{ul}\right)}{4\pi}. \tag{4.55}$$

For a homogeneous cloud in thermodynamic equilibrium, we can integrate Eq. (4.54),

$$I = B\left(T\right)\left(\frac{b}{c}\right)\, \nu\, f\left(\tau\right), \tag{4.56}$$

where we have used the relations between the Einstein coefficients. The optical depth function, $f(\tau)$, is given by,

$$f\left(\tau\right) = 2\int_0^\tau \beta\left(\tilde{\tau}\right)\, d\tilde{\tau} = 0.428\left[E_1\left(2.34\tau\right) + \ln\left(2.34\tau\right) + 0.57721\right], \tag{4.57}$$

with E_1 the exponential integral. For $\tau \ll 1$, this simplifies to $f(\tau) = \tau$. The intensity scales then directly with the column density in the upper level, N_u, as

$$I = \frac{A_{ul}\, N_u\, h\nu_{ul}}{4\pi}, \tag{4.58}$$

e.g., the linear part of the curve of growth. For large optical depth, the logarithmic term in Eq. 4.57 dominates (i.e. the logarithmic portion of the curve of growth). Heuristically, for $\tau = 3$, $f(\tau)$ is $\simeq 1$ and,

$$I \simeq B\left(T\right)\frac{\nu b}{c}, \tag{4.59}$$

i.e., the integrated line intensity is the Planck function times the width of the line. For even higher optical depth, the damping wings of a line become important. This would give rise to the square root portion of the curve of growth, but this regime is of little importance in ISM studies and its effects are not included in the expression for the escape probability.

If we relax the thermodynamic equilibrium condition ($n \ggg n_{cr}\,\beta$), the intensity is still given by Eq. 4.58 in the optically thin limit. In this case, the population of the upper level is not given by the Boltzmann equation but rather by Eq. 4.11. We can then write for the intensity,

$$I = \frac{\gamma_{lu}\, n \mathcal{A}_j N\, h\nu_{ul}}{4\pi}, \tag{4.60}$$

where we have used the fact that the population of the upper level is small. In this limit, every upward collision leads to the emission of a photon. For an optically thick line, the level populations will vary through the cloud due to the effects of line trapping. The cooling law is now (cf. Eq. 4.14),

$$j_{ul} = \frac{\frac{g_u}{g_l}\exp\left[-h\nu_{ul}/kT\right]}{1+\frac{n_{cr}\beta(\tau)}{n}+\frac{g_u}{g_l}\exp\left[-h\nu_{ul}/kT\right]}\frac{\mathcal{A}_j n\, A_{ul}h\nu_{ul}}{4\pi}\beta(\tau), \tag{4.61}$$

which simplifies to,

$$j_{ul} = \gamma_{lu}\frac{\mathcal{A}_j n^2\, A_{ul}h\nu_{ul}}{4\pi}, \tag{4.62}$$

as long as $n \ll n_{cr}\beta(\tau)$. This integrates again to the optically thin limit (cf. Eq. 4.60). Thus, although the line is optically thick, because collisional deexcitation is unimportant, all upward transitions result in a photon that eventually escapes the cloud. This photon scattering is a random walk process. If we define ϵ as the probability that a photon is converted back into thermal energy of the gas,

$$\epsilon = \frac{n\gamma_{ul}}{n\gamma_{ul}+A_{ul}} \simeq \frac{n}{n_{cr}}, \tag{4.63}$$

then a line is "optically thin" in the subthermal regime when,

$$L < \frac{\lambda}{\sqrt{\epsilon}}, \tag{4.64}$$

with L the size of the cloud and λ the mean free path of a photon, which is equal to $1/\kappa$ with κ the line-averaged opacity. Thus, when the density is less than,

$$n < \frac{n_{cr}}{\tau^2}, \tag{4.65}$$

the cloud is effectively optically thin. For water lines, with their high critical densities, this can be a very important effect.

4.1.8 Multi-level Systems

In Section 4.1.2, we consider a two-level system connected by radiative and collisional transitions. A simple expression was derived for the level population (Eq. (4.11)) as a function of the ratio of the density to the critical density (Eq. (4.12)). Some further insight can be gained by considering multi-level systems. For rotational transitions in a linear molecule, we have the selection rule, $\Delta J = \pm 1$, and if we make the simplifying assumption that collisions only connect adjacent levels, we can show that the level populations are governed by the two-level solution,

$$\frac{n_u}{n_l} = \frac{g_u/g_l\exp\left[-\Delta E_{ul}/kT\right]}{1+n_{cr}/n}. \tag{4.66}$$

Of course, collisions do connect all levels. So, consider now a three-level system for which we can manipulate the statistical equilibrium equations (see assignment 4.2). The algebra is a little tedious but Figure 4.3 shows the result where we have used the molecular parameters

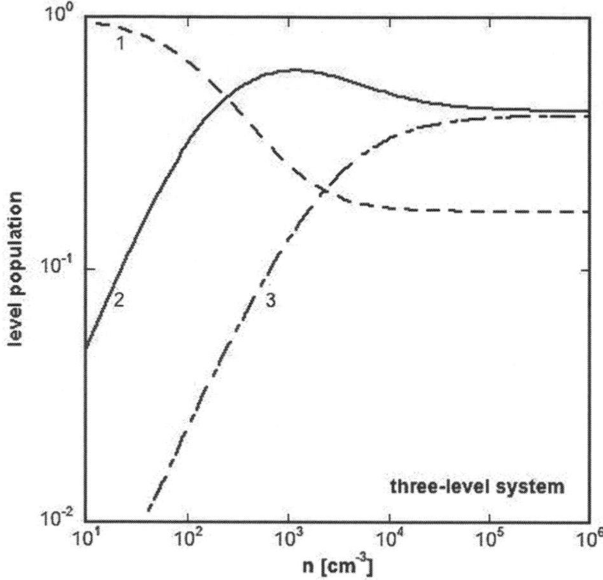

Figure 4.3 Illustrative example of the level populations for a three-level system. Molecular parameters have been taken for the three lowest rotational levels of CO. Curves labeled by J. The temperature is 30 K.

for the three lowest rotational levels of CO. As this figure illustrates, at low densities, the population drops down to the lowest level and the population of the excited levels is governed by the ratios,[3] $n/n_{cr,21}$ and $n/n_{cr,31}$. For the parameters adopted here, these are 1.1×10^3 and 8.9×10^3 cm^{-3}, respectively. At high densities, we recover LTE. For densities above $n_{cr,21}$, we find that the population of level 2 actually exceeds the LTE value. Essentially, the spontaneous decay rate increases rapidly with ν^3 and, hence, with increasing J. Thus, through the excitation of higher levels, the cascade process leads to a "pile-up" up at lower levels. This supra-thermal excitation is a common characteristic of molecular systems. Note that this does not correspond to an inverted population – the level population is higher than would be expected based upon the population of higher levels – and no laser action will ensue.

4.2 Analysis of Observations

4.2.1 Molecular Column Densities

We start this discussion with the radiative transfer equation for the specific intensity, I,

$$\frac{dI(\nu)}{ds} = -\kappa(\nu) I(\nu) + j(\nu), \tag{4.67}$$

[3] $n_{cr31} \equiv A_{32}/\gamma_{31}$.

where s measures the path of the radiation and where the absorption coefficient, κ, and specific intensity, j, are given by,

$$\kappa(\nu) = \frac{h\nu}{4\pi}(n_l B_{lu} - n_u B_{ul})\,\phi(\nu), \tag{4.68}$$

and

$$j(\nu) = \frac{h\nu}{4\pi} A_{ul} n_u \phi(\nu). \tag{4.69}$$

For a homogeneous cloud at excitation temperature, T_x, we have,

$$I(\nu) = I_0(\nu)\exp[-\tau(\nu)] + B(\nu, T_x)(1 - \exp[-\tau(\nu)]), \tag{4.70}$$

with I_0 the intensity of the background radiation field. As molecular observations are referenced against "blank" sky, we have,

$$I_{line}(\nu) = (B(\nu, T_x) - B(\nu, T_{BG}))(1 - \exp[-\tau(\nu)]), \tag{4.71}$$

where we have assumed that the background radiation field is a blackbody at T_{BG}. The line intensity is often expresed in terms of the radiation temperature,

$$T_R \equiv \frac{c^2}{2k\nu^2} I_{line}. \tag{4.72}$$

So, when we now introduce the Rayleigh–Jeans equivalent temperatures as,

$$J(T, \nu) \equiv \frac{h\nu/k}{\exp[h\nu/kT] - 1}, \tag{4.73}$$

we can write in the optically thin approximation, $\tau \ll 1$,

$$N = \left(\frac{3h}{8\pi^3 S\mu^2}\right)\left(\frac{Z(T)}{g}\right)\left(\frac{\exp[E_u/kT_x]}{\exp[h\nu/kT_x] - 1}\right)\left(\frac{\int T_R d\nu}{J(T_x) - J(T_{BG})}\right), \tag{4.74}$$

where we have to remember that the degeneracy is the product of the rotational (J and K) and nuclear spin (I) terms. If we assume $h\nu/kT_x \ll 1$, we can write this as,

$$N = \left(\frac{3k}{8\pi^3 \nu S\mu^2}\right)\left(\frac{Z(T)}{g}\right)\exp[E_u/kT_x]\,\frac{T_x}{T_x - T_{BG}}\int T_R d\nu. \tag{4.75}$$

This is typically a good approximation for low frequency transitions ($\nu < 25$ GHz) but is not warranted at sub-millimeter wavelengths unless the excitation temperature is large. If we assume $h\nu/kT_x \ll 1$ and ignore the background term, we have,

$$N = \left(\frac{3k}{8\pi^3 \nu S\mu^2}\right)\left(\frac{Z(T)}{g}\right)\exp[E_u/kT_x]\int T_R d\nu, \tag{4.76}$$

and this approximation is actually surprisingly good at sub-millimeter wavelengths as the neglect of the background radiation largely, but fortuitously, compensates for the approximation of the Rayleigh–Jeans limit.

4.2.2 Rotation Diagrams

We will focus our discussion first on linear molecules, whose behavior is very illustrative. Analysis of molecular observations often starts with the construction of the so-called rotation diagram: a plot of "measured" column densities versus excitation energy. The level populations are governed by (c.f. Eq. (4.2)),

$$N_J = \frac{gN}{Z(T_x)} \exp\left[-E_J/kT_x\right], \tag{4.77}$$

with T_x the excitation temperature and N the total column density. So, a plot of $\ln\left[N_J/g\right]$ versus E_J/k will give a straight line with a slope given by T_x and an intercept equal to $\ln\left[N/Z(T_x)\right]$. If the density is high compared to the critical densities of all the lines involved in the analysis, then T_x equals the kinetic temperature of the gas. That is, in general, not the case for rotational transitions. For example, compared to low-J transitions, high-J transitions in linear molecules occur at higher frequencies and, therefore, have higher Einstein-A values but similar deexcitation cross sections, resulting in higher critical densities. Thus, high-J transitions will fall out of LTE more quickly than low J levels. This is illustrated in Figure 4.4. This figure was constructed using RADEX to calculate the expected emission from a homogeneous cloud. Calculated intensities were then analyzed assuming optically thin emission,

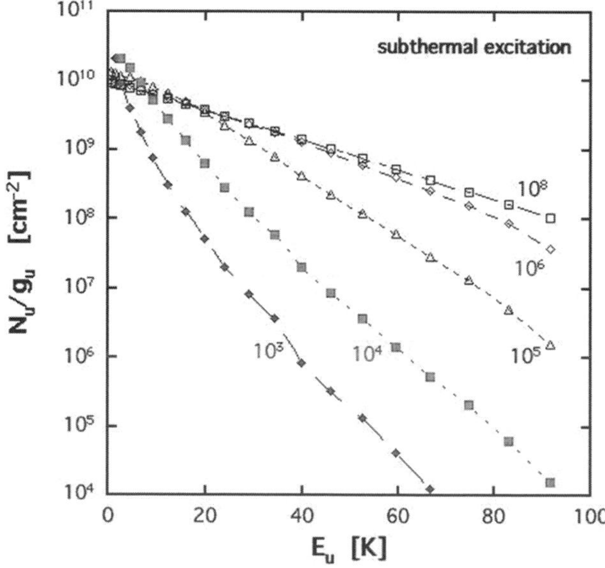

Figure 4.4 Rotation diagram for HC_3N at 20 K. A column density of only 10^{12} cm^{-2} has been adopted and all levels are optically thin. The labels associated with the different curves correspond to the adopted density (in cm^{-3}). At densities $n \gtrsim 10^7$ cm^{-3} all lines are in LTE, but at lower densities higher J states fall out of equilibrium. Figure constructed with RADEX [22] and adapted from [10]

$$N_J = \frac{4\pi I}{A_{J,J-1} h\nu_{J,J-1}}. \quad (4.78)$$

These column densities were used to construct the rotation diagram. At high densities, the level populations are given by the Boltzmann equation at the kinetic temperature. As the density drops below the critical density of the higher J states, population shifts toward the lower J states and the apparent temperature drops. At the lowest temperatures, clear positive curvature appears, reflecting suprathermal excitation (Section 4.1.8). At low densities, the population of level J is due to collisional excitation to all levels higher than J – all of which quickly relax radiatively – and produces an enhanced population of this level compared to higher ones. For CO, this behavior has been quantified and for a homogeneous cloud at temperature T, the excitation temperature of level J is related to the kinetic temperature by,

$$T_x(J) = \frac{E_J/k}{E_j/kT + 2.5\sqrt{E_J/kT} + 2}. \quad (4.79)$$

Optical depth effects also have to be kept in mind in rotation diagram analysis. For a moderately optically thick line emanating from a semi-infinite slab, the integrated line intensity is given by,

$$I = B(T)\, f(\tau)\, \frac{\nu b}{c} \simeq 0.428\, B(T)\, \ln(2.43\tau)\, \frac{\nu b}{c}. \quad (4.80)$$

This is essentially constant and the derived column density is independent of the actual column density. Formally, the ratio of the column density, N^*, derived from the observed intensity assuming optically thin emission (Eq. (4.77) to the true column density, N, is then,

$$\frac{N^*}{N} = \frac{\ln(2.34\tau)}{2.43\tau}. \quad (4.81)$$

In reality, the optical depth is a strong function of J for rotational transitions (c.f. Section 4.1) and this can give profound curvature in a rotation diagram. This is illustrated in Figure 4.5 where high column density clouds show strong curvature.

Beam filling factor is a third point to keep in mind in rotation diagram analysis. If the source does not fill the beam, the surface brightness and the derived column density are underestimated (by the beam filling factor). For a constant telescope aperture, low frequency transitions will be more affected than high frequency transitions and this too will introduce curvature in the rotation diagram. This is illustrated in Figure 4.6. For optically thick lines in LTE, T_R is constant, the surface brightness of a source scales with ν^2, and the beam scales with λ/D with D the telescope aperture. Hence, the inferred surface brightness from the observed flux scales with ν^4, resulting in inferred level populations independent of J. For optically thin transitions, substantial negative curvature is introduced. In principle, if the source size is known – say, the source is resolved at high frequencies – the effect of beam size can be corrected for by assuming that the source size does not change with J.

Positive curvature in a rotation diagram can also reflect the presence of a temperature gradient. If we adopt a powerlaw temperature distribution, $dN(T)/dT \propto T^{-b} db$, over a

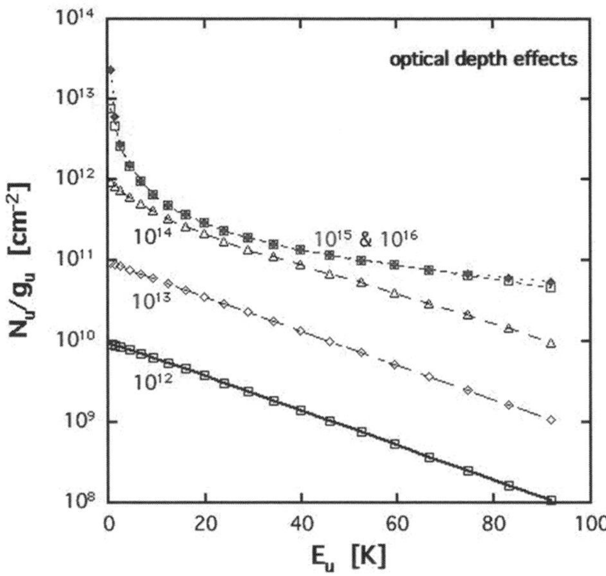

Figure 4.5 Rotation diagram for HC_3N at 20 K. A high density is assumed and all levels are in LTE. The different curves are labeled by the adopted column densities of HC_3N (in cm^{-2}) through the cloud. For low columns, all lines are optically thin and a straight line ensues (cf. Figure 4.4). At high column density, the lines are optically thick and the rotation diagram shows appreciable curvature. The derived column density is then independent of the actual column density. Figure constructed with RADEX [22] and adapted from [10]

range of temperatures, $[T_{min}, T_{max}]$, with $T_{min} \ll E_J/k$ and $T_{max} \gg E_J/k$, the excitation temperature is,

$$T_x\,(J) \simeq \frac{E_J}{b} \propto J^2, \tag{4.82}$$

and for the Herschel SPIRE or the PACS range of CO lines, this introduces a variation by a factor $\simeq$10.

We have focused here on linear molecules for which the rotational ladder is simple and rotational transitions obey simple relationships between transition strength and excitation energy. For asymmetric rotors, these relationships are more complicated and any particular level may be connected by several transitions to other levels, exhibiting a range of transition frequencies, Einstein A's, critical densities and optical depth and all the effects described above will come into play. Typically, this will lead to a large spread of derived column densities at the same excitation energy. This is illustrated in Figure 4.7.

Despite these drawbacks, rotation diagrams are a popular analysis tool but one that should be used with caution. All the effects discussed introduce curvature in the rotation diagram or large spreads. Conversely, a linear behavior in the rotation diagram does not necessarily imply optically thin emission as too small a range may have been probed. See, for example, the curves for high column densities in Figure 4.5, which are linear for

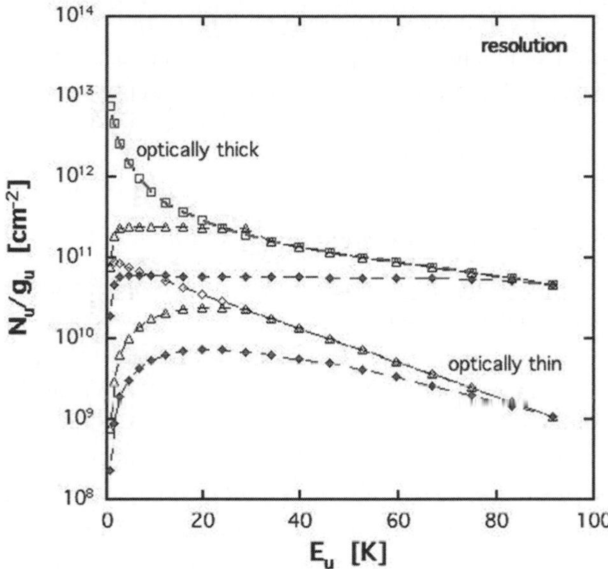

Figure 4.6 The effect of beam dilution on rotation diagram analysis is illustrated from the optically thick and optically thin HC_3N emission at 20 K. A high density is adopted and all lines are in LTE. The fully resolved cases correspond to the curves shown in Figures 4.4 and 4.5. For each case, three resolutions are shown: all lines are resolved and all lines originating from $J \geq 11$ and $J \geq 20$ are resolved. Figure constructed with RADEX [22] adapted from [10]

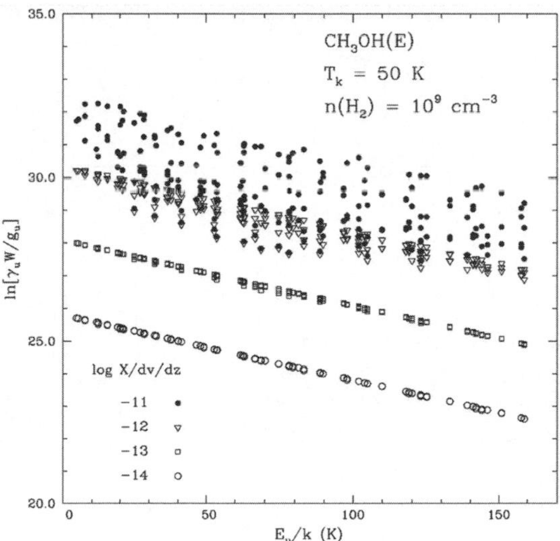

Figure 4.7 Rotation diagram for E-type CH_3OH at 50 K. A velocity gradient of 1 km/s/pc has been assumed. At low abundances, all transitions are optically thin and values derived from lines originating from the same level coincide. As the abundance increases, high frequency lines will progressively become more optically thick and their derived values drop below the optically thin values, and this leads to a large scatter in the rotation diagram. Figure adapted from [10]

$J \gtrsim 8$; when $h\nu/kT \gg 1$, the Planck function simplifies to the Wien law and the intensity will show linear behavior in a log-linear plot as long as $f(\tau)$ does not vary too much (e.g. for $J \gg 1$). Conversely, the presence of curvature may imply the presence of multiple temperature components but, equally well, it may just reflect the importance of optical depth or sub/superthermal excitation.

4.2.3 Physical Conditions in the Emitting Gas

The multitude of rotational lines due to a myriad of molecular species – each with their own characteristic critical density and excitation energy – provide a powerful tool to study the physical properties of molecular clouds. Figure 4.8 provides an overview of often used molecular diagnostics. Rotational lines of CO probe regions with densities of $\simeq 3 \times 10^3\ J^3$ cm^{-3} and temperatures of $\simeq 3\ J^2$. Thus, the 1–0 and 2–1 transitions are well suited for the study of general molecular clouds. The mid-J CO transitions, such as $J = 7-6$, are sensitive to luke-warm gas near YSOs or in PDRs, while 14–13 and higher transitions generally trace warm gas associated with shocks or XDRs. CO ro-vibrational emission in the fundamental (4.6 μm) or overtone region (2.3 μm) of the spectrum originate from

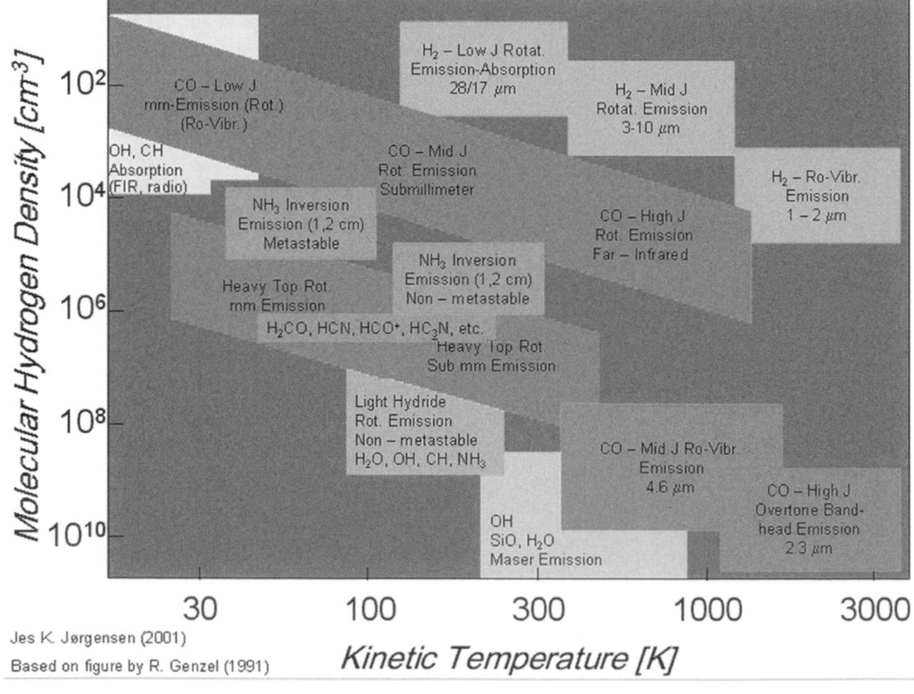

Figure 4.8 An overview of the molecular lines and the range of physical conditions in molecular clouds for which they are effective probes. Figure reproduced with permission from [9] (A black and white version of this figure will appear in some formats. For the color version, please refer to the plate section.)

dense, hot gas near a star or in a planet-forming disk. Because of the larger dipole moment, rotational lines of trace species such as CS, HCN, and HCO^+ have higher critical densities than CO and thus probe dense cores in molecular clouds. Molecular hydrogen has only electric quadrupole transitions and its low-J rotational transitions are readily excited in low density gas. However, the large energy spacing of H_2 does require high temperatures. Near-IR ro-vibrational transitions of H_2 are indicative of warm ($\sim$1000 K), dense ($\sim 10^4$ cm^{-3}) gas in shocks or are nonthermally excited through UV pumping in the Lyman–Werner bands. The metastable and non-metastable inversion lines of NH_3 are excited in medium density gas (10^4–10^6 cm^{-3}), while the far-IR rotational transitions of hydrides such as H_2O probe warm and dense (10^6–10^9 cm^{-3}) gas.

Molecular lines with their rotational ladders form ideal probes of the physical conditions in the emitting gas. The various rotational transitions originate from levels at different energies and have different critical densities. As a result, their relative intensities can be used to derive the density and the temperature of the gas. In order to extract this information from the observed lines, model calculations are required that take the molecular cloud structure, and the excitation and line radiative transfer properly into account. This is a formidable task and often simplifying assumptions are made to get a first impression of the implications of the observations, based upon diagnostic diagrams. A set of diagnostic diagrams is illustrated in Figure 4.9. In each of the panels, the ratio of two lines of a single species is shown in the form of contour plots. Line intensities have been calculated for a homogenous cloud of a given density and kinetic temperature. The results have then been converted into a contour plot in the $n - T$-plane. Perusing these diagrams and recalling the critical densities and energies of the levels involved, we realize that these line ratios are sensitive probes of the density and temperature of the gas in the range between the critical densities and excitation energies of the levels involved. The power of molecular rotational observations is evident. By properly selecting the species and lines involved, gas with any conditions can be traced. Of course, in general, molecular clouds will be highly inhomogeneous with spatial structure at all scales. Conversely, this implies that a line will trace the gas that best fits its critical density/energy level separation. Also, in using these diagrams, the optically thin approximation should be kept in mind. Often, main isotopes of molecules are optically thick and then less abundant isotopes have to be used to derive the physical conditions in the gas.

Consider the CO 3–2/6–5 ratio as an example. This ratio is a good measure of the density in the range of 3×10^3–10^5 cm^{-3} and a good measure of the temperature in the range of 20–70 K (for $n > 10^5$ cm^{-3}). Outside of this range, the levels of one or both of these lines is/are not effectively excited or the emission in one/both of these lines will be overwhelmed by more plentiful lower density gas present in the beam in a realistic interstellar environment.

4.2.4 Vibrational Absorption Bands

The discussion in the previous sections has focused on the excitation of rotational levels and analysis of (sub)-millimeter rotational transitions. The ro-vibrational transitions of simple

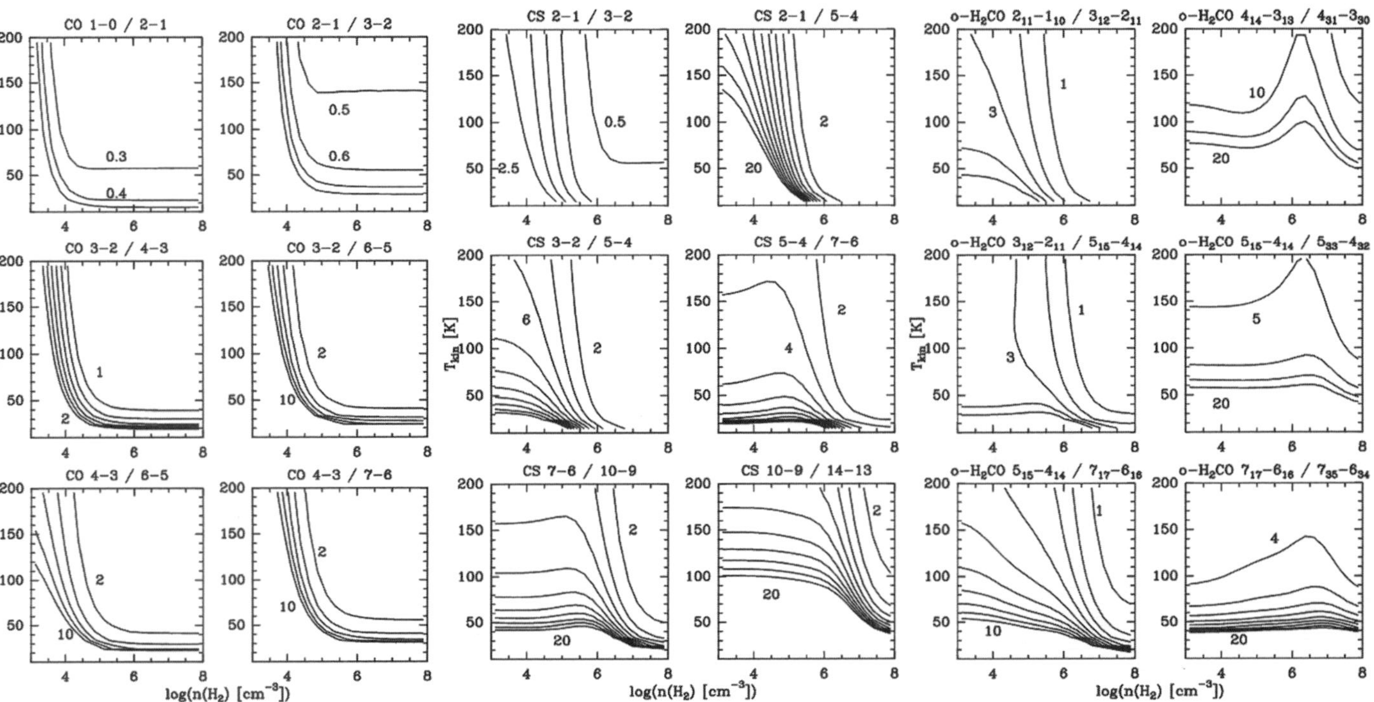

Figure 4.9 Contour plots of the expected ratio of integrated line intensities calculated in the optically thin limit using a large velocity gradient model. Contours are spaced linearly. The lines involved are indicated above each panel. Reproduced with permission from D. Jansen, 1996, PhD Thesis, Leiden

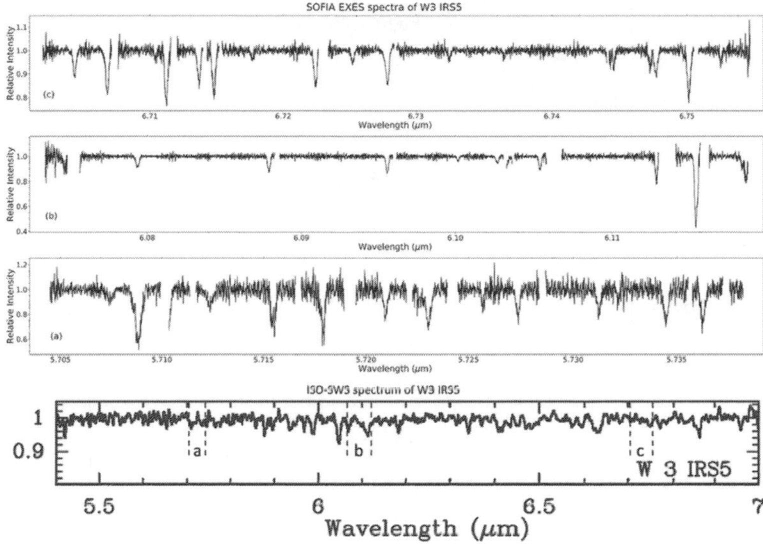

Figure 4.10 Ro-vibrational absorption bands of water in the mid-infrared spectrum of the massive protostar, W3-IRS 5. Bottom: The ISO/SWS spectrum obtained at a spectral resolution of 1500 (0.0042 μm, 200 km/s) reveals the presence of a large number of absorption bands. Taken from [4]. Panels a, b, and c: Spectra obtained by the SOFIA/EXES over small wavelength ranges at a resolution of 50,000 (0.00012 μm, 6 km/s) reveal that each of the bands in the low resolution spectrum contains a large number of ro-vibrational transitions of the ν_2 transition of H_2O, illustrating that high spectral resolution is a necessity for proper assessment of the characteristics of the absorbing gas. Figure reproduced with permission from [12]

molecules in the mid-IR also provide a wealth of information. In principle, a large number of transitions can be measured by the same instrument and the same telescope, which helps the analysis. In the K and L band, many of the largest telescopes have such high spectral resolution intruments available and a wide range of sources can be probed. The (T)EXES instrument on SOFIA (and on ground-based telescopes) can probe the 5–8 and 15–22 μm (and 8–13 μm) ranges, but only the brightest sources can be studied. As an example, Figure 4.10 shows ro-vibrational transitions of H_2O in the spectrum of the massive protostar, W3 IRS 5, illustrating the need for high spectral resolution. In these observations, the field of view is set by the background continuum source; at these wavelengths, typically the disk around the protostar. Sometimes, particularly in the main isotope of CO, multiple velocity components are revealed. Analysis of these absorption lines is relatively straightforward as the integrated optical depth can be measured and translated in column densities and these can be analyzed using a rotation diagram. For species with highly optically thick transitions, rather than curve of growth analysis (c.f. Section 4.3.2), overtones are often used to determine column densities. As overtones are only allowed because of anharmonic interaction, the transition strength is small, comparable to that of fundamental in the isotopic species. While this may imply that the main isotope and its

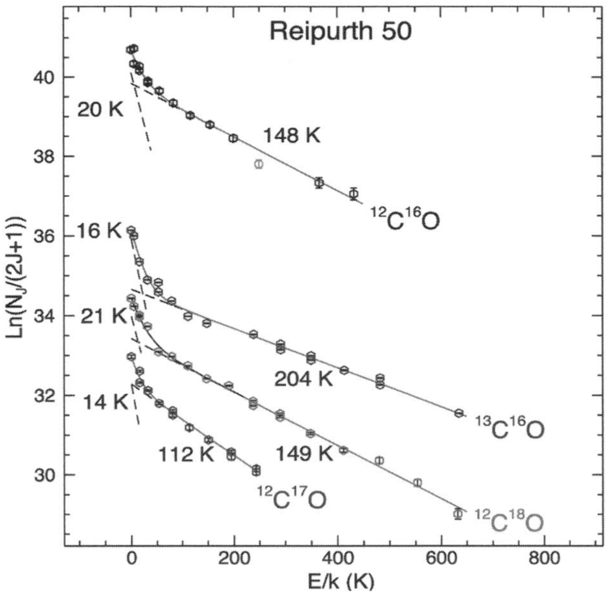

Figure 4.11 Boltzmann rotation diagram for different isotopes of CO derived from the mid-IR ro-vibrational transitions observed toward the low mass protostar, Reipurth 50. The corresponding spectrum is shown in Figure 10.29. Because of the high optical depth of the $\Delta v = 1$ transition, the $\Delta v = 2$ transitions in the K band window have been used for the main isotope. Figure reproduced with permission from [20]

isotopologues are colocated, it should be recognized tht the overtone occurs at much shorter wavelength and the background source may have a very different size and/or structure.

A rotation diagram analysis of the ro-vibrational transitions for the isotopologues of CO is shown in Figure 4.11. For CO, these rotation diagrams often reveal considerable curvature, which is generally interpreted as the presence of warm gas at a temperature of $\simeq$100–200 K in the Hot Core/Hot Corino near the young protostar and a much colder 10–20 K component associated with the surrounding molecular cloud. We note that this curvature is likely not caused by optical depth effects but a temperature gradient is a possibility.

4.3 Electronic Excitation

Electronic transitions can be an important excitation mechanism for some species. The key processes involved are discussed here for H_2, where pumping through the Lyman–Werner bands dominates the populations of the ro-vibrational states in the ground electronic state in diffuse clouds and moderately dense PDRs. That discussion is followed by a curve-of-growth analysis of UV absorption lines of H_2 in diffuse clouds. Further examples of this are described in Chapter 9 on diffuse clouds and Chapter 11 on PDRs associated with regions of massive star formation. Finally, UV pumping of large molecules, their subsequent electronic and vibrational cascade, and the resulting IR emission spectrum are discussed in Chapter 8, while the analysis of astronomical data is discussed in Chapter 12.

4.3.1 Ro-vibrational Excitation of H_2

Molecular hydrogen is readily electronically excited through allowed transitions from the ground vibrational and ground electronic state ($X^1\Sigma_g^+$) to various vibrational states of the excited electronic states, $B^1\Sigma_u^+$ and $C^1\Pi_u$ in the Lyman–Werner bands (cf. Figure 3.10). Most of the time ($\simeq$90%), H_2 will relax radiatively back to excited vibrational states in the ground electronic states. In the remainder, relaxation will be to the vibrational continuum of this state leading to dissociation (cf. Section 6.3.4). As H_2 is a homonuclear molecule, the excited vibrational states decay back to the ground state – typically, $\Delta v = 1$ – through (forbidden) quadrupole transitions. The Einstein A's for these transitions correspond to excited state lifetimes of $2{-}10 \times 10^5$ s. With typical collision rate coefficients (Section 4.1.4), this translates into critical densities of $\simeq 3 \times 10^4$ cm^{-3}. So, in the diffuse ISM, collisions do not play a role but, in dense PDRs, collisional redistribution may be important. For the lowest, pure rotational transitions, lifetimes are very long ($A_{ul} \simeq \left(\nu/10^{13}\,\text{Hz}\right)^{5.35}$ s^{-1}), resulting in low critical densities ($\simeq 2 \times 10^2$ & 3×10^3 cm^{-3} for H collisions in the para/ortho ground state transitions[4]). So, in diffuse gas, only the populations of the lowest rotational levels in the ground vibrational state are affected by collisions. The result of all of these considerations is that the population of all but the lowest rotational levels is determined by the cascade set up by the UV pumping process.

Let us consider this one step at a time and first evaluate the level populations solely based on the cascade. Consider that level $v_o J_o$ is populated at a rate $Q(v_o J_o)$. The level population is then given by $n(v_o J_o) = Q(v_o J_o)/\tau(v_o J_o)$, where τ is the lifetime of this level (the inverse of the sum of the Einstein A's of all transitions leading out of this level). The equilibrium population of level v, J, $n(v, J)$, produced by the cascade from level $v_o J_o$ follows then from,

$$n\left(v, J\right) = \tau\left(v, J\right) \sum_{v'=v}^{v_o} \sum_{J''=0}^{J_{max}} n\left(v'', J''\right) A\left(v'', J''; v, J\right), \tag{4.83}$$

which can be evaluated starting at level v_o down to level $v = 0$ for all J. The population of level, $n(vT)$, is defined by a set of cascade efficiency arrays, $a(v_o J_o; vJ) = n(vJ)A(vJ)$ for a given rate of entry into level $v_o J_o$. These cascade efficiency arrays can be calculated once and for all from the Einstein A's. The rate of entry into level vJ is then,

$$q\left(vJ\right) = Q\left(vJ\right) + \sum_{v'=v+1}^{14} \sum_{J'=0}^{J_{max}} Q\left(v'J'\right) a\left(v'J'; vJ\right). \tag{4.84}$$

4 For collisions with H_2, these critical densities are about two orders of magnitude higher.

In order to determine the Q's, the level populations of the ro-vibrational levels in the excited electronic states have to be determined,

$$n^{\star}\left(v', J'\right) A^{\star}\left(v', J'\right) = \sum_{v''} \sum_{J''}^{J_{max}} n\left(v'', J''\right) R\left(v'', J''; v', J'\right), \quad (4.85)$$

where the superscript $\star$ indicates that it pertains to the excited electronic state and $R\left(v'', J''; v', J'\right)$ is the rate of absorption from level $v''J''$ in the ground electronic state to level $v'J'$ in the excited electronic state, which is given by,

$$R\left(v'', J''; v', J'\right) = 4\pi \int_0^{\infty} \sigma(\nu) J(\nu) d\nu. \quad (4.86)$$

The cross section, σ is given by the oscillator strength and a Voigt line profile function. The mean intensity, J, of the radiation field depends on the incident radiation field from (nearby) stars and the radiative transfer through the cloud to the depth that is considered and thus on the dust properties as well as the abundances of H_2 in the relevant levels in the cloud. The radiation field couples different regions in the cloud and the level populations are therefore a nonlocal problem. We will come back to this in Section 6.3.4, where the photodissociation of H_2 is discussed. Here, we stress that once the radiation field is known, the local level populations can be evaluated through the cascade efficiency matrices.

Other processes than UV pumping can also influence the H_2 level populations and these have to be taken into account. This leads to a set of coupled statistical equilibrium equations,

$$n(v,J)\left(A(v,J) + k_{uv}(v,J) + \sum_i n(i) \sum_{v'} \sum_{J'} \gamma\left(i, v, J, v'J'\right) + \zeta_{CR}\right)$$
$$= \sum_{v'=0}^{14} \sum_{J'=0}^{15} n\left(v', J'\right)\left(A\left(v', J'\right) + k_{uv}\left(v'J', vJ\right) + \sum_i n(i)\gamma\left(i, v'J', v, J\right)\right)$$
$$+ n(H) n_o k_d (H_2, vJ) \quad (4.87)$$

where $A(v,J)$ refers to the summed Einstein A's of all transitions out of this level, $k_{uv}(v,J)$ is the total UV pump out of the level and includes thus also transitions that decay back to the vibrational continuum of the ground electronic state (and to dissociation), ζ_{CR} is the cosmic ray ionization rate, the collisional term includes vibrational and rotational terms of all relevant collision partners i, the first term on the right-hand side is the rate at which newly formed H_2 (on grain surfaces) is released into the gas phase in level vJ, and $k_{uv}(v'J', vJ)$ is the rate at which molecules are pumped from level $v'J'$ to the excited electronic state and then decay to level vJ in the ground electronic state. The k_{uv} rates can be related back to the R's (Eq. (4.86)), the abundance in the electronically excited state and the relevant Einstein A back to the ground electronic state, but that is left as an exercise for the reader.

4.3.2 Analysis of Observations

The H_2 population can be traced through UV absorption lines, and this has been done with instruments on board of Copernicus, the Hubble Space Telescope, and the Far Ultraviolet Space Explorer. The strength of absorption lines is generally measured in terms of the equivalent width in wavelength units,

$$W_\lambda = \int \frac{I_c(\nu) - I_o(\nu)}{I_c(\nu)} d\lambda = \frac{\lambda^2}{c} \int (1 - \exp[-\tau(\nu)])\, d\nu, \tag{4.88}$$

with I_c the (estimated) continuum intensity and I_o the observed intensity, and the optical depth is given by Eq. (4.27). We can follow the discussion on the escape probability to arrive at the curve of growth behavior. Ignoring stimulated emission, for small optical depth, all atoms are "seen" and the equivalent width is given by,

$$\frac{W_\lambda}{\lambda} = \frac{\pi e^2}{m_e c^2} f_{lu} N_l \lambda. \tag{4.89}$$

This is, approximately, the peak optical depth, τ_p, times the FWHM of the line. When the Doppler core of the line is saturated, absorption will increase slowly because of the increase in the width of the line. This leads to the logarithmic part of the curve of growth,

$$\frac{W_\lambda}{\lambda} = \frac{2b}{c} \sqrt{\ln[\tau_p]}, \tag{4.90}$$

where the peak optical depth is given by,

$$\tau_p = 7.5 \left(\frac{\text{km s}^{-1}}{b}\right) \left(\frac{N_l}{10^{12}\,\text{cm}^{-2}}\right) \left(\frac{5000\,\text{Å}}{\lambda}\right) f_{lu}. \tag{4.91}$$

For very high columns, the damping wings become important and the equivalent width is now given by,

$$\frac{W_\lambda}{\lambda} = \frac{2^{3/2}\lambda}{c} \sqrt{\left(\frac{\pi e^2}{m_e c} f_l\right) N_l \gamma}, \tag{4.92}$$

the square root portion of the curve of growth, as the equivalent width depends on the square root of the column density (as well as the square root of the damping parameter).

Observed line strength can now be fitted to curves of growth to determine column densities of the absorbers (Figure 4.12). The equivalent width is determined by assuming a suitable continuum. Often, a Doppler parameter is adopted from measurements of other (optically thin) species and this is used to construct a theoretical curve of growth, W_λ/λ versus $Nf\lambda$. The observations are then fitted against this curve to determine the column densities in the respective J levels. Three points should be made here: First, lines originating from the same level will only differ in their ordinate through their individual (and known) $f\lambda$ combination. That internal relationship will help locate them on the curve of growth; particularly if such a set straddles the linear-logarithmic or logarithmic-square root portions of the curve of growth. Second, for lines on the square root portion, the location

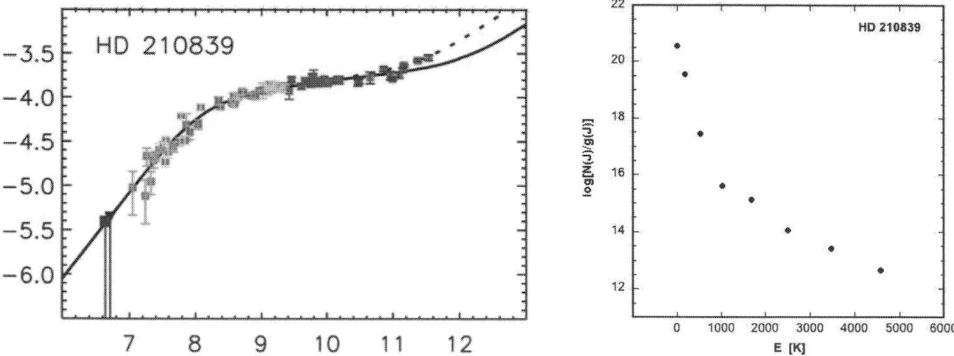

Figure 4.12 Left: Curve of growth analysis of absorption lines in the Lyman–Werner bands in the UV spectrum of HD 210839 taken from [13]. The data points are color coded (see the original paper) to indicate the J value. Full line is the curve of growth for a given b (and low γ). Variations in γ are captured through the solid and dashed lines on the right of the figure. Right: Rotation diagram of H_2 as determined from the curve of growth analysis. Two distinct populations can be recognized with excitation temperatures of $\simeq$70 and 550 K, respectively.

of the curve of growth will depend on the damping constant relevant for that transition and a set of curves will have to be taken into account; e.g. for relevant H_2 lines γ varies from 5×10^8 to 2×10^9 s^{-1}. Third, the logarithmic portion of the curve of growth depends on the Doppler parameter and this can be iterated upon, if necessary. Figure 4.12 shows the curve of growth fitting of the Lyman–Werner absorption bands observed toward the star HD 210839 with FUSE and the resulting rotation diagram (cf. Section 4.2.2).

4.4 Further Reading and Resources

Experimental and theoretical studies of molecular excitation have been reviewed by [19]. The monograph [8] discusses in depth molecular collisions in the interstellar medium.

Reference [15] provides a tutorial in converting observed intensities into column densities, including detailed examples for relevant molecules that illustrate K- and nuclear degeneracies, approximations for partition functions, line strength parameters, and hyperfine components. Reference [10] provides an in-depth analysis of the pitfalls in rotation diagram analysis, and much of the dicussion in Section 4.2.2 was derived from this paper. Suprathermal excitation was first discussed by [14] and its importance in rotation diagram analysis was stressed by [17]. Collisional data for excitation of astrophysically relevant molecules is available through the BaseCol website, http://basecol.obspm.fr maintained by the l'Observatoire de Paris. The RADEX program is available through the webpage http://var.sron.nl/radex/radex.php. This program provides a convenient interface to calculate expected line intensities for isothermal, homogenous, 1-D molecular clouds [22].

Reference [2] provides a pioneering study on the UV pumping of H_2 and the resulting population of vibrationally and rotationally excited states in relation to the analysis of

Copernicus UV absorption observations. Reference [3] extended this to higher densities and incident UV fields that are relevant to studies of the H_2 near-IR emission lines in reflection nebulae. Analysis of H_2 Lyman–Werner bands is described in [6, 13, 21]. Template molecular hydrogen curve of growth are provided in [16]. Oscillator strengths and transition probabilities of these lines are provided in [1]. These data can also be accessed through the SESAM website http://sesam.obspm.fr (SpEctroScopy of Atoms and Molecules) maintained by the Laboratoire dy Etudes du Rayonnement en de la Matiére at the l'Observatoire de Paris.

4.5 Exercises

4.1 Maximum intensity and absorption strength:

(a) Derive expression (4.4).
(b) Derive expression (4.6).
(c) Derive expression (4.8).
(d) Adopt $B = 2$ cm^{-1} and calculate the level population, absorption coefficient and (relative) emission intensity at $T = 10, 50, 100,$ and 300 K and compare this with the approximations.

4.2 Approximate the lowest rotational levels for a linear molecule by a three-level system connected by collisions and spontaneous emission.

(a) Rewrite the detailed balance equations as

$$a_1 n_2 - b_1 n_3 = c_1 \tag{4.93}$$

$$a_2 n_2 + b_2 n_3 = c_2, \tag{4.94}$$

and derive expressions a_ι, b_ι, and c_ι in terms of the relevant Einstein coefficients and collisional rate coefficients. Rewrite this in terms of relevant critical densities.

(b) Show that the level populations can be written as,

$$n_2 = \frac{c_1/b_1 + c_2/b_2}{a_1/b_1 + a_2/b_2} \tag{4.95}$$

$$n_3 = \frac{c_2/a_2 - c_1/a_1}{b_1/a_1 + b_2/a_2} \tag{4.96}$$

$$n_1 = 1 - n_2 - n_3. \tag{4.97}$$

(c) Take the low density and high density limits and compare these results with your expectations.
(d) Adopting the relevant molecular parameters and collisional rates ($\gamma_{21} = 6.55 \times 10^{-11}$, $\gamma_{31} = 8.42198 \times 10^{-11}$, $\gamma_{32} = 1.03 \times 10^{-10}$ cm^3 s^{-1}) for the three lowest levels of CO, calculate the level populations as a function of density.

4.3 Derive Equation (4.81).

(a) write down the expression for the peak optical depth of a line.
(b) Show that, for an optically thin line, equivalent width is approximately the peak optical depth times the full width at half maximum.
(c) Following the discussion in Section 4.1.6, derive Equation (4.90).

4.4 Ro-vibrational transitions:
Table 4.2 lists the measured lines of the overtone of the main isotope of CO toward the source, Reipurth 50. Here we will manipulate this data to derive the characteristics of the absorbing gas.

(a) Derive an expression for the energies of the rotational levels in the two vibrational states in terms of their rotational constants (see Section 3.3.4).
(b) Analyze the observed frequencies to derive these rotational constants.
(c) Calculate the rotational constants B_e and α_e and estimate the binding energy.
(d) The results differ slightly from established literature values. What could be the cause?
(e) Calculate the column densities from the observed strength of the lines assuming a FWHM of 3.3 km/s. Construct a rotation diagram and compare to Figure 4.11.
(f) Fit the observed column densities by a two-component model and derive the column density and excitation temperature of each component.

Table 4.2 *Measured optical depth of the overtone transitions of* $^{12}C^{16}O$

Transition	nu cm^{-1}	Tau	Error	A s^{-1}
(2,0) R(3)	4274.7407	1.351	0.075	3.90E-01
(2,0) R(2)	4271.1766	1.366	0.064	3.70E-01
(2,0) R(1)	4267.5421	1.076	0.061	3.40E-01
(2,0) R(0)	4263.8372	0.767	0.045	2.80E-01
(2,0) P(1)	4256.2171	0.783	0.047	8.30E-01
(2,0) P(2)	4252.3022	0.99	0.07	5.50E-01
(2,0) P(3)	4248.3176	1.014	0.05	4.90E-01
(2,0) P(4)	4244.2634	1.053	0.06	4.70E-01
(2,0) P(5)	4240.1398	0.964	0.064	4.50E-01
(2,0) P(6)	4235.947	0.838	0.045	4.40E-01
(2,0) P(7)	4231.685	0.772	0.046	4.30E-01
(2,0) P(8)	4227.354	0.623	0.048	4.20E-01
(2,0) P(9)	4222.9542	0.363	0.037	4.20E-01
(2,0) P(11)	4213.9486	0.275	0.035	4.10E-01
(2,0) P(12)	4209.3432	0.225	0.034	4.00E-01

Table 4.3 *Measured equivalent width toward HD210839*

λ [Å]	J	f_{lu}	W_λ [mÅ]	ΔW_λ [mÅ]	λ [Å]	J	f_{lu}	W_λ [mÅ]	ΔW_λ [mÅ]
927.017	2	2.3e-3	142.7	9.3	1031.191	3	1.1e-2	155.4	7.2
928.436	3	3.3e-3	109.5	21.5	1032.349	4	1.7e-2	139.2	6.4
933.578	3	2.e-2	145.8	13.1	1033.914	6	1.1e-2	32.7	4.4
933.788	4	1.1e-2	112.5	13.8	1035.181	4	1.1e-2	127.2	10.5
934.141	2	2.2e-3	138.2	11.4	1041.158	3	1.6e-2	165.8	3.6
934.789	3	7.1e-3	134.4	12.5	1041.729	6	1.7e-2	32.4	3.6
935.958	4	2.0e-2	118.3	11.8	1043.502	3	1.1e-2	160.7	10.0
941.596	2	3.3e-3	159.2	24.0	1044.542	4	1.5e-2	140.8	10.0
942.685	5	6.2e-3	85.8	9.5	1045.801	6	1.1e-2	25.3	4.1
942.962	3	5.7e-3	143.6	14.9	1047.550	4	1.1e-2	133.6	8.6
944.328	3	3.5e-3	150.3	9.3	1052.393	7	1.1e-2	20.3	3.8
944.718	5	2.0e-3	45.7	15.5	1052.496	5	1.1e-2	106.2	5.8
951.672	3	1.3e-2	137.3	7.7	1053.976	3	1.3e-2	161.6	4.7
956.578	2	9.1e-3	171.6	12.7	1056.472	3	9.6e-3	163.5	4.2
958.945	3	8.9e-3	140.5	13.9	1057.380	4	1.3e-2	135.5	5.2
960.263	5	2.3e-3	38.6	6.9	1058.316	6	1.1e-2	29.1	2.5
960.449	3	4.9e-3	128.9	10.1	1060.031	7	1.5e-2	26.5	2.0
962.151	4	8.7e-3	116.7	22.9	1060.581	4	9.8e-3	132.7	6.2
966.779	3	7.9e-3	168.5	29.3	1061.697	5	1.3e-2	122.5	7.9
968.292	2	7.2e-3	157.7	6.6	1065.596	5	1.0e-2	99.4	7.6
974.884	5	1.4e-2	107.8	11.4	1067.479	3	1.0e-2	171.9	10.0
975.344	2	6.6e-3	174.1	19.6	1070.141	3	7.6e-3	166.1	5.3
978.217	3	6.7e-3	137.7	7.5	1070.900	4	9.6e-3	140.1	8.9
979.803	4	1.3e-2	134.2	12.4	1071.497	6	1.0e-2	20.1	2.8
994.227	4	1.4e-2	130.0	12.5	1072.992	7	1.2e-2	23.0	2.9
994.517	7	1.1e-2	11.0	3.1	1074.312	4	7.8e-3	129.3	5.9
998.331	6	3.7e-2	77.0	6.9	1075.245	5	9.3e-3	111.0	8.2
999.268	4	1.6e-2	129.9	12.6	1080.492	6	8.9e-3	31.4	2.5
1003.982	2	1.7e-2	265.7	7.9	1081.267	2	4.7e-3	227.6	12.8
1005.390	2	9.9e-3	207.6	11.2	1081.713	3	6.4e-3	186.7	14.1
1010.938	2	0.24e-1	288.1	20.2	1084.562	3	5.0e-3	163.7	7.6
1016.458	2	1.0e-2	236.8	7.8	1085.146	4	6.0e-3	114.8	13.4
1017.001	5	1.1e-2	101.8	7.5	1085.382	6	7.9e-3	27.3	5.1
1017.831	5	2.4e-2	120.3	4.0	1086.630	7	8.6e-3	23.4	5.3
1018.214	7	8.8e-3	7.8	4.0	1088.796	4	5.2e-3	113.8	6.6
1019.013	6	2.0e-2	33.2	3.7	1089.515	5	5.7e-3	85.6	5.2
1019.500	3	1.1e-2	148.2	7.5	1099.787	3	2.5e-3	142.7	8.5
1023.434	4	1.1e-2	140.2	16.1	1100.164	4	2.9e-3	91.6	6.5
1028.801	7	2.3e-2	28.4	3.9	1100.982	7	5.2e-3	10.5	5.4
1028.985	3	1.7e-2	165.4	13.5	1104.083	4	2.6e-3	96.1	5.4
1030.071	6	1.8e-2	63.2	4.8	1109.313	5	2.6e-3	54.9	6.8

4.5 Curve of growth:

(a) Show that the peak optical depth is given by,

$$\tau_p = \sqrt{\pi} e^2/(m_e c) N_l f_{lu} (\lambda/b). \tag{4.98}$$

(b) Curves of growth have been evaluated numerically [7] and various approximations are available in the literature [5, 11, 18]. Here, the latter suffices to construct the curve of growth,

$$\frac{W_\lambda}{\lambda} \simeq \frac{\sqrt{\pi} b}{c} \frac{\tau_p}{1 + \tau_p / 2\sqrt{2}} \qquad \tau_p < 1.254 \tag{4.99}$$

$$\frac{W_\lambda}{\lambda} \simeq \frac{2b}{c} \sqrt{\ln\left[\tau_p / \ln 2\right] + \frac{\gamma \lambda}{4b\sqrt{\pi}} \left(\tau_p - 1.254\right)} \qquad \tau_p > 1.254\,. \tag{4.100}$$

Plot the curve of growth as a function of τ_p in the range $10^{-2} - 10^5$, assuming that $b = 10$ km/s, $\lambda = 1000$ Å, and $\gamma = 10^9$ s^{-1}, all appropriate for H_2.

(d) Compare these results to the simple expressions for the linear, logarithmic, and square root portion of the curve of growth.

(e) Equivalent widths for the Lyman–Werner bands of H_2 measured to the star HD 210839 are provided in Table 4.3. Fit these to the calculated curve of growth and derive column densities of hydrogen in the relevant rotational levels of the ground state.

(f) Examine the rotation diagram of H_2 and derive relevant excitation temperatures.

(g) Compare your results to Figure 4.12.

Bibliography

[1] Abgrall, H., Roueff, E., 1989, *A & A Suppl*, 79, 313

[2] Black J. H., Dalgarno, A., 1976, *ApJ*, 203, 132

[3] Black J. H., van Dishoek, E. F., 1986, *ApJ*, 322, 412

[4] Boonman, A. M. S., van Dishoeck, E. F., 2003, *A & A*, 403, 1003

[5] Draine, B. T., *Physics of the Interstellar and Intergalactic Medium* (Princeton: Princeton University Press)

[6] Federman, S. R., Cardelli, J. A., van Dishoeck, E. F., Lambert, D. L., Black, J. H., 1995, *ApJ*, 445, 325

[7] Finn, G. D., Mugglestone, D., 1965, *MNRAS*, 129. 221

[8] Flower, D., 2009, *Molecular Excitation in the Interstellar Medium* (Cambridge: Cambridge University Press)

[9] Genzel, R., 1991, in *The Physics of Star Formation and Early Stellar Evolution*, eds. C. J. Lada, N. D. Kylafis (Dordrecht: Kluwer), 155

[10] Goldsmith, P. F., Langer, W. D., 1999, *ApJ*, 517, 209

[11] Hansen C. F., McKenzie, R. L., 1971, *J Quant Spectros Radiat Transfer*, 11, 349

[12] Indriolo, N., 2020, private communication

[13] Jensen A. G., Snow, T. P., Sonneborn, G., Rachford, B. L., 2010, *ApJ*, 711, 1236

[14] Leung, C.-M., Liszt, H. S., 1976, *ApJ*, 208, 732
[15] Mangum, J. F., Shirley, Y. L., 2015, *PASP*, 127, 266
[16] McCandliss, S. R., 2003, *Molecular Hydrogen Optical Depth Templates for FUSE Data Analysis*, Publications of the Astronomical Society of the Pacific 115, 651
[17] Neufeld, D. A., 2012, *ApJ*, 749, 125
[18] Rodgers, C. D., Williams, A. P., 1974, *J Quant Spectros Radiat Transfer*, 14, 319
[19] Roueff, E., Lique, F., 2013, *Chem Rev*, 113, 8906
[20] Smith, R. L., Pontoppidan, K. M., Young, E. D., et al., 2015, *ApJ*, 813, 120
[21] Spitzer, L., Jr, Cochran, W. D., Hirsfeld, A., 1974, *ApJS*, 28, 373
[22] van der Tak, F. F. S., Black, J. H., Schöier, F. L., Jansen, D. J., van Dishoeck, E. F., A computer program for fast non-LTE analysis of interstellar line spectra, 2007, *A & A,* 468, 627
[23] Yang, C.-H., Sarma, G. , Parker, D. H., 2011, *J Chem Phys*, 134, 204308

5
Chemical Thermodynamics

5.1 Introduction

Interstellar and circumstellar media are far from chemical thermodynamic equilibrium as these low temperature, dilute regions are permeated by energetic radiation and particles. In addition, while dynamical timescales in space are long, chemical timescales can be truly "astronomical" and chemical equilibrium is difficult to attain. Nevertheless, it can be instructive to consider chemical thermodynamics as a state that the system strives to reach. This discussion starst with the introduction of the thermodynamic quantities: enthalpy, entropy, and Gibbs free energy. Starting with the Saha equation – familiar to astronomers from the ionization equilibrium in stellar atmospheres – the Gibbs free energy and the chemical potential are introduced and used to describe the system in thermodynamic equilibrium. We then consider specific examples of gas phase, solid–gas, and solid phase equilibria. This is followed by a discussion on methods in solving the resulting set of equations describing the chemical equilibria, and this chapter is concluded with a short discussion on the relevance of thermodynamics in space.

For a course on molecular astrophysics, this chapter is provided as background reading. The one concept that should be stressed is the microscopic reversibility of reactions, which results in a relationship between the forward and backward reaction rate coefficients (Eq. (5.33)).

5.2 Thermodynamics

Thermodynamics is concerned with the description of the system using macrosopic quantities; e.g., the internal energy, U, the total volume, V, the total number of species, N, the pressure, P, the density, ρ, and the temperature, T. As these average macroscopic quantities involve a large number of microscopic degrees of freedom, fluctuations will be small. Given enough time, a macroscopic system will evolve to an equilibrium state in which the macroscopic properties are determined by intrinsic properties of the system rather than by the external factors that have been brought to bear in the past.

In thermodynamical equilibrium, the state of a system is fully specified by the density, temperature, pressure of the system, and the number of moles of the components present.

This greatly simplifies model calculations. The thermodynamical properties of a system – free energy and entropy – can be described by the partition function, which describes how the total energy is shared among all the species. Collisions constantly redistribute the energy between the different species and between the different degrees of freedom of the species. As every state is assumed to be equally probable, the system will settle in the most probable population of states. Exercise 5.1 may help to refamiliarize yourself with this concept. Chemical thermodynamics is concerned with calculating the most probable chemical state.

If there is no heat exchange with the environment, the internal energy is a conserved quantity. Generally, we are also very familiar with the total number of species being conserved. However, in chemical thermodynamics that is not the case as reactions can transform species into others. In astrophysics, the ionization equilibrium in stellar structure studies is a familiar example of this issue. The Saha equation – used to describe this equilibrium for, for example, hydrogen – is just a special case of a more general formalism and we will start our discussion by considering the Saha equation. This equation is then generalized through the Gibbs free energy and chemical potential.

5.2.1 The Saha Equation

We will start this discussion by examining the simple reaction,

$$AB \overset{\Delta E}{\longleftrightarrow} A + B. \tag{5.1}$$

In thermodynamic equilibrium, the concentrations are given by the familiar Saha equation,

$$\frac{n(A)\,n(B)}{n(AB)} = \left(\frac{2\pi\,(m_A m_B/m_{AB})\,kT}{h^2}\right)^{3/2} \exp\left[-\Delta E/kT\right], \tag{5.2}$$

where we recognize the first term on the right-hand side as the quantum phase space of the coordinates of the particles, and the exponential represents the internal partition function. This equation is encountered in astrophysics when describing the degree of ionization in a stellar interior where AB, A, and B are the H-atom, proton, and electron, respectively, but it applies more generally for chemical equilibriums. Generally, it is then approached through the Gibbs energy of the system.

5.2.2 Enthalpy and Gibbs Free Energy

We will start the discussion on thermochemistry of a reaction describing the heat flow with the enthapy. The enthalpy (H) difference between products and reactants is given by,

$$\Delta H\,(\text{reaction}) = H\,(\text{products}) - H\,(\text{reactants}), \tag{5.3}$$

where a negative enthalpy corresponds to release of energy by a reaction (e.g. $\Delta E = -\Delta H$; for an exothermic reaction $\Delta H < 0$; for an endothermic reaction $\Delta H > 0$).

The enthalpy of a species is not just the bond energy but also involves the internal energy and the work done on the environment. Enthalpy is always measured relative to a standard state. In evaluating the enthalpy change of a reaction, the stochiometry of the reaction, the phases of the reactants and products, and the temperature of the reaction have to be kept in mind. As mentioned before, the enthalpy change does not depend on the reaction pathway and hence can be determined from the heat of formation and the thermodynamic properties of the species.

For stable molecules, the enthalpy (heat) of formation is generally a negative quantity, which implies that the formation of a compound from its elements is usually an exothermic process. For atoms, the enthalpy of formation can be derived from the bond strength if they are diatomic gasses in their standard state (e.g. H_2, O_2, N_2). For those elements that are solid in their standard state (e.g. C), the enthalpy of formation can be evaluated from vapor pressure data. For ions the enthalpy can be derived from their ionization potentials. The enthalpy for some astrophysically relevant species are listed in Table 5.1. These have to be transferred to the temperature of the gas under consideration using the specific heat of the species but that correction is generally small for astrophysical conditions.

The enthalpy measures the amount of energy involved in a reaction. A reaction will also change the entropy – disorder – of a system. Whether a reaction will occur is a balance between the energy released and the increase in entropy and that is measured by the Gibbs free energy.[1] Reactions will occur spontaneously when G is negative. Changes in the Gibbs free energy and entropy for many relevant species are also tabulated in Table 5.1. These two thermodynamic quantities, enthalpy and Gibbs free energy, as well as the entropy and the total internal energy are all state functions; e.g. they depend on the current state of the system not on how the system arrived in this state. Furthermore, H and G are relative quantities and a standard state has to be assigned: the pure solid, liquid, or gas at a temperature of 298.15 K and a pressure of 10^5 Pa.[2]

5.2.3 The Chemical Potential

In natural systems, pressure and temperature are generally kept constant and thermodynamic equilibrium is then best described using the Gibbs free energy, G, defined by,

$$dG = -SdT + VdP + \mu dN, \tag{5.4}$$

with S the entropy, and T, V, and P the temperature, volume, and pressure, respectively. The Gibbs free energy depends thus on the temperature, the pressure, and the number of particles. The chemical potential, μ, describes the change in internal energy, U, associated with a change in the number of particles, $\mu = (\partial U/\partial N)$. If we have a system consisting of

1 The term "free" reflects that the entropy has been removed from the internal energy and hence this amount of energy is available to do work.

2 Older literature may refer to the standard state at 1 Atm.

Table 5.1 *Thermodynamic quantities*[a]

Species	ΔH° [KJ/mole]	ΔS° [J/mole/K]	ΔG° [kJ/mole]
H_2	0	131	0
H	218	115	203.3
H^+	1536.2	109	1517
H^-	139	109	132
N_2	0	192	0
N	472.7	163	455.5
NH	376.6	181	3706
NH_2	190.4	195	199.8
NH_3	−45.9	193	−16.4
C^b	0	5.74	0
C	716.7	158	671.2
CH	594.1	183	560.7
CH_2	386.4	194	369.2
CH_3	145.7	194	148
CH_4	−74.9	186	−50.8
C_2H_2	226.7	201	209.2
C^+	1809.4	155	1758.8
CH^+	1626.7	172	1590.5
CN	435.1	203	405
HCN	135.1	202	124.7
O_2	0	205	0
O	249.2	161	231.7
OH	39.0	184	34.3
H_2O	−241.8	189	−228.6
OH^+	1317.1	183	1306.4
H_2O^+	975.4	189	989
H_3O^+	581.2	192	606.6
CO	−110.5	198	−137.2
CO_2	−393.5	214	−394.4
HCO	43.5	225	28.3
HCO^+	833	203	817.9
H_2CO	−115.9	209	−109.9
CH_3OH	−200.1	240	−166.2
$C_6H_6{}^c$	82.8	268	129.7
$C_{10}H_8{}^d$	150	245	252.4
$C_{24}H_{12}{}^e$	322.7	487	452.3

[a] Standard state is $T = 298.15$ K and $P = 10^5$ Pa. [b] Reference state for C is graphite. [c] Benzene. [d] Naphthalene. [e] Coronene.

multiple components, each characterized by a partial entropy, S_i, pressure, P_i, and chemical potential, μ_i, we have,

$$dG = -\left(\sum_i S_i\right) dT + Vd\left(\sum_i P_i\right) + \sum_i \mu_i dN_i. \tag{5.5}$$

Chemical equilibrium requires $dG = 0$ and at constant P and T, this results in,

$$dG = \sum_i \mu_i dN_i = 0. \tag{5.6}$$

Schematically, we can write reactions as,

$$\sum_i \nu_i \mathcal{A}_i = 0, \tag{5.7}$$

where $\mathcal{A}_i$ is the symbol for chemical species i and the ν_i's are the moles of species i involved in the reaction where ν_i is positive or negative according to whether we consider it a reactant or product, respectively. In terms of Reaction (5.1), we have, $-1\text{AB}+1\text{A}+1\text{B}=0$ and $\nu_{AB} = -1$, $\nu_A = \nu_B = 1$ and $\mathcal{A}_{AB} = \text{AB}$, $\mathcal{A}_A = \text{A}$, and $\mathcal{A}_B = \text{B}$. The changes in dN_i in Equation (5.6) are not independent and indeed Equation (5.7) implies that $dN_i \propto \nu_i$. This results in,

$$\sum_i \mu_i \nu_i = 0, \tag{5.8}$$

for the equilibrium condition. We recognize now that μ_i is essentially the molar Gibbs free energy of a compound (e.g. adding ν_i moles of species i adds $\nu_i \mu_i$ to the Gibbs free energy) or phrased differently the thermodynamic energy per unit amount (e.g. per mole).

So, we need to express the chemical potentials, μ_i in terms of phase space to close the description of the system. For fermions, we have the Fermi–Dirac distribution of states,

$$N(p) = \left(\exp\left[(\epsilon - \mu)/kT\right] + 1\right)^{-1}, \tag{5.9}$$

where ϵ is the energy consisting of the kinetic energy ($p^2/2m$, with p the momentum) and internal energy (rotation, excitation, ionization, and chemical bond energy) and we have ignored for the moment the subcript i. At low enough densities (or high enough temperatures), we can ignore the $+1$ in the denominator. The chemical potential follows then from the requirement that when we integrate the distribution function, we should arrive at the total number of species, n,

$$n = \frac{g}{h^3}\int N(p)\,4\pi p^2 dp = \frac{g}{Z_{int}}\left(\frac{2\pi mkT}{h^2}\right)^{3/2} \exp\left[\mu/kT\right], \tag{5.10}$$

where g refers to the number of equivalent states (i.e. the statistical weight), the factor in brackets is the quantum concentration (the translational partition function), and Z_{int} is the partition function for the internal degrees of freedom of the species, which we encountered already in Chapter 3. We have now,

$$\mu_i = kT \ln\left[n_i/g\left(2\pi m_i kT/h^2\right)^{3/2} Z_i\right], \tag{5.11}$$

where we have reintroduced the subscript i. So, the chemical potential is kT times the natural logarithm of the product of the spin, translational, and internal partition functions. The internal partition function is the product of the rotational, vibrational, and electronic partition functions. The former two are discussed in Chapter 3. For the electronic partition function, we generally only need to include the state of the lowest energy, $Z_{\text{electronic}} = \exp[-E_0/kT]$. Note that this energy is relative to a standard state and hence includes relevant molecular binding energies and ionization energies.

In general, we can write,

$$\mu_i = \mu_i^0 + kT \ln[P_i], \tag{5.12}$$

where μ_i^0 is the chemical potential at temperature T in a standard state (pure component at system temperature and standard pressure, generally taken to be 10^5 Pa), where it is implicitly taken that the P_i are in units of the standard state. Eq. (5.8) becomes,

$$\sum_i \mu_i \nu_i = kT \sum_i (\nu_i \ln[P_i]) + \sum_i \left(\nu_i \mu_i^0\right) = 0, \tag{5.13}$$

which we can rewrite as,

$$\prod_i P_i^{\nu_i} = K_p(T), \tag{5.14}$$

where the equilibrium constant, K_p,

$$\ln K_p(T) = -\sum \nu_i \mu_i^0 / kT, \tag{5.15}$$

is a function of temperature only and independent of the pressure and initial quantities of the reacting gases.[3] This is known as the law of mass action. The equilibrium constant is thus set by the partition functions of the substances involved in a reaction. In terms of Reaction (5.1), we have,

$$\frac{P_A P_B}{P_{AB}} = K_p = \frac{Z_A Z_B}{Z_{AB}} \exp\left[-\Delta E/kT\right], \tag{5.16}$$

with ΔE the binding energy of AB and Z_i the translational, rotational, and vibrational partition functions of species i. Thus, a large K_p means that most of AB is dissociated into A and B, while a small K_p means that the equilibrium lies on the AB side.

The change of free energy per mole, ΔG°, at standard conditions is given by,

$$\Delta G^\circ(T) = \sum \nu_i \mu_i^0. \tag{5.17}$$

[3] The subscript p reflects that the equlibrium constant refers to (partial) pressures.

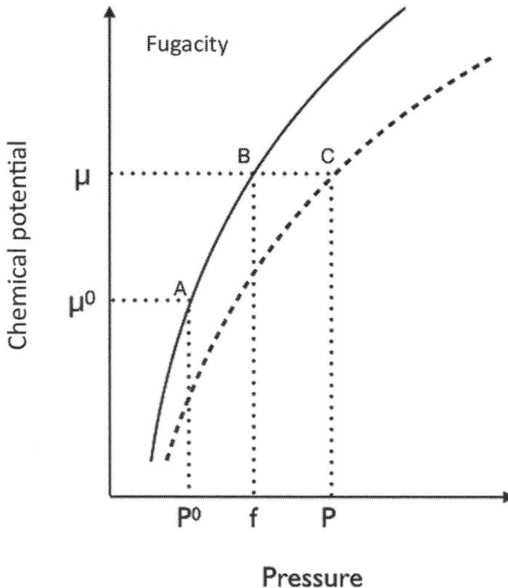

Figure 5.1 Schematic dependence of the chemical potential on pressure for constant temperature. Solid curve: an ideal gas. Dashed curve: a real gas. The standard state is labeled with A. Point B and C indicate the states of an ideal and a real gas with the same chemical potential. The fugacity of a real gas is defined as the pressure that an ideal gas would have with the same chemical potential as the real gas.

If we want to write the chemical equilibrium in terms of concentrations, $[i]$, we have to define a new equilibrium constant, K_C, given by,[4]

$$K_C = \exp\left[-\sum \nu_i \mu_i^0 / kT\right] (kT)^{-\sum_i \nu_i}, \tag{5.18}$$

to arrive at,

$$\prod_i [i]^{\nu_i} = K_C(T). \tag{5.19}$$

5.2.4 Fugacity

The chemical potential of an ideal gas is described by Eq. (5.12). Real gases show nonideal behavior due to attractive and repulsive interactions between the molecules. We can define the fugacity, f, as the pressure that a real gas would have if it behaved as an ideal gas (Figure 5.1) and, analog to Eq. (5.12), we can write,

$$\mu_i = \mu_i^0 + kT \ln\left[\frac{f}{P^0}\right], \tag{5.20}$$

4 The subscript C refers to concentration.

where we have now explicitly included the standard pressure. The fugacity is then,

$$f = P\gamma, \tag{5.21}$$

where the fugacity coefficient, γ, contains all nonideal aspects due to, e.g., attractive or repulsive interactions. The fugacity coefficient can be less than or larger than 1 but at very low presures it is 1 (because, at low temperatures, gases are ideal). Fugacity is, thus, the effective pressure of the material. Generally, in astrochemically relevant systems – stellar photospheres, inner protoplanetary disks, hot solar nebula, (exo)planetary atmospheres – the fugacity coefficient can be set equal to unity.

5.2.5 Solutions

We will now consider a solution where we define the solvent as the species with the largest concentration and the solute as traces in the solvent. While we tend to think of a liquid, a solution can also be a solid or a gas. Here, we consider a solid and a liquid, but only the nomenclature changes when we consider, say, a solid and a gas. In a solution, the chemical potential of a solute is given by,

$$\mu_i\,(\text{solution}) = \mu_i^0\,(\text{liquid}) + RT\,\ln[x_i], \tag{5.22}$$

where x_i is the mole fraction of the solute. As this is per mole, we have introduced the gas constant, R. As x_i is always less than 1, the chemical potential is reduced by dissolving in a solution. If the chemical potential of the solution is less than the chemical potential of the pure solute, the solute will dissolve in the solvent. Saturation (equilibrium) is reached when the chemical potential of the solution reaches that of the pure solid solute,

$$\mu_i\,(\text{solid}) = \mu_i^0\,(\text{liquid}) + RT\,\ln[x_i], \tag{5.23}$$

which we can rewrite as,

$$\ln[x_i] = -\frac{\Delta G^0_{\text{fusion}}}{RT}, \tag{5.24}$$

where we have introduced the molar Gibbs free energy for fusion, $\Delta G^0_{\text{fusion}}$. In terms of the enthalpy of fusion, we have,

$$\ln[x_i] = \frac{\Delta H^0_{\text{fusion}}}{R}\left(\frac{1}{T_m} - \frac{1}{T}\right), \tag{5.25}$$

with T_m the melting point of the solute.

This is again assuming ideal behavior where there is no interaction between solute and solvent or with other solutes. Nonideal behavior is captured by using the activity. The activity, a, of a material at a given temperature, pressure, and concentration is the ratio of the fugacity of that material relative to that of the standard state: the pure material at its saturated vapor pressure and the same temperature,

$$a = \frac{f}{f^\star} \simeq \frac{p}{p^\star}, \tag{5.26}$$

with $p^\star$ the saturated vapor pressure. By definition, the activity of a pure substance is equal to 1. For ideal (solid) solutions, we have,

$$\mu_i = \mu_i^\star + kT \ln x_i, \tag{5.27}$$

with $\mu_i^\star$ the chemical potential of the pure substance and x_i the mole fraction of i in the solution. As x_i is less than 1, the chemical potential of a solution is reduced compared to the pure substance. We can write equivalently,

$$\prod_i x_i^{\nu_i} = \exp\left[-\sum_i \nu_i \mu_i^\star / kT\right]. \tag{5.28}$$

If all the reaction partners are highly diluted in a solvent, we have,

$$\prod_i \rho_i^{\nu_i} = \rho_{\text{solvent}}^{\sum_i \nu^i} \exp\left[-\sum_i \nu_i \mu_i^\star / kT\right], \tag{5.29}$$

where ρ_i is the concentration of i in the solvent. In case of nonideal solutions, we have to multiply the mole fractions by activity coefficients γ_i, which contain the nonideal aspects.

5.2.6 Forward and Backward Reactions

As Equation (5.1) summarizes, for every forward reaction converting reactant(s) in product(s), there is a backward reaction converting the product(s) back into the reactant(s). In equilibrium, the rates of the two reactions involved balance. We have,

$$\text{forward rate} = J_f = k_f n\,(\text{AB}) \tag{5.30}$$

$$\text{backward rate} = J_b = k_b n\,(\text{A})\, n\,(\text{B}) \tag{5.31}$$

where k_f and k_b are the forward and backward rate coefficients. Balancing, rearranging, and introducing the equilibrium constant results in,

$$K_c = \frac{k_f}{k_b}. \tag{5.32}$$

We can write therefore,

$$\frac{k_f}{k_b} = \exp\left[-\Delta G^\circ / kT\right] = \exp\left[\Delta S^\circ / k\right] \exp\left[-\Delta H^\circ / kT\right], \tag{5.33}$$

where in the last step we have used $\Delta G = \Delta H - T\Delta S$. When the system is not in equilibrium, we have,

$$\Delta G = kT \ln\left[J_f / J_b\right]. \tag{5.34}$$

Equation (5.33) is an expression of microscopic reversibility and can be used to calculate the backward rate from measured or calculated forward rate (or vice versa).

5.3 Examples

At this point, it is useful to illustrate these concepts with some examples.

5.3.1 Gas Phase Equilibria

Consider, as an example, the dissociation of molecular hydrogen,

$$H_2 \leftrightarrow 2H. \tag{5.35}$$

If we ignore for the moment the rotational and vibrational degrees of freedom, we can write,

$$\frac{P^2\left(\mathrm{H}\right)}{P\left(\mathrm{H}_2\right)} = \frac{kT}{2^{3/2}}\left(\frac{2\pi m_H kT}{h^2}\right)^{3/2} \exp\left[-E_b/kT\right], \tag{5.36}$$

with E_b the binding energy of H_2. The atomic to molecular hydrogen ratio is then,

$$\frac{n\left(\mathrm{H}\right)}{n\left(\mathrm{H}_2\right)} = \frac{1}{2^{3/2}}\left(\frac{n_{q,H_2}}{n\left(\mathrm{H}_2\right)}\right)^{1/2} \exp\left[-E_b/2kT\right], \tag{5.37}$$

where we have introduced the quantum volume of H_2, n_{q,H_2} $(\equiv \left(2\pi m_{H_2} kT/h^2\right)^{3/2}$. Two points should be made here. First, the fraction of atomic hydrogen depends on $E_b/2$ rather than E_b as the binding energy per H-atom is half of the binding energy of H_2. Second, the factor $n\left(\mathrm{H}_2\right)/n_{q,H_2}$ is much smaller than 1 (in the classical limit considered here). As a result, the fraction of atomic hydrogen is much larger than expected from the Boltzmann factor alone. This reflects that two atoms have much higher entropy than a single molecule. This balance between energetics and entropy is at the heart of (chemical) thermodynamics.

Equation (5.36) has two unknowns and hence we have to provide one more relation. If we assume that initially all the hydrogen is molecular with density, n_o, we can write $n(\mathrm{H}_2) = (1-x)n_o$ and $n(\mathrm{H}) = 2xn_o$. We will now also include the rotational and vibrational partition functions and we have then from Eq. (5.14),

$$\ln\left[\frac{2x^2}{1-x}\right] = \frac{-E_b}{kT} - \frac{3}{2}\ln\left[\frac{2\pi\left(m_H/2\right)kT}{n_o^{2/3}h^2 Z_{rot}\left(T\right)Z_{vib}\left(T\right)}\right], \tag{5.38}$$

where E_b is 4.476 eV. In this expression, the species are assumed to be in the ground electronic state, leaving only the exponential factor containing the binding energy. We will approximate H_2 as a rigid rotor, harmonic oscillator to arrive at the rotational and vibrational partition functions (Chapter 3),

$$Z_{rot}\left(T\right) = \frac{T}{2\theta_r} \tag{5.39}$$

$$Z_{vib}\left(T\right) = \frac{\exp\left[-\theta_v/2T\right]}{1-\exp\left[-\theta_v/T\right]}, \tag{5.40}$$

with $\theta_r = hcB/k = 88$ K and $\theta_v = h\nu_v/k = 6215$ K (Table 3.4).

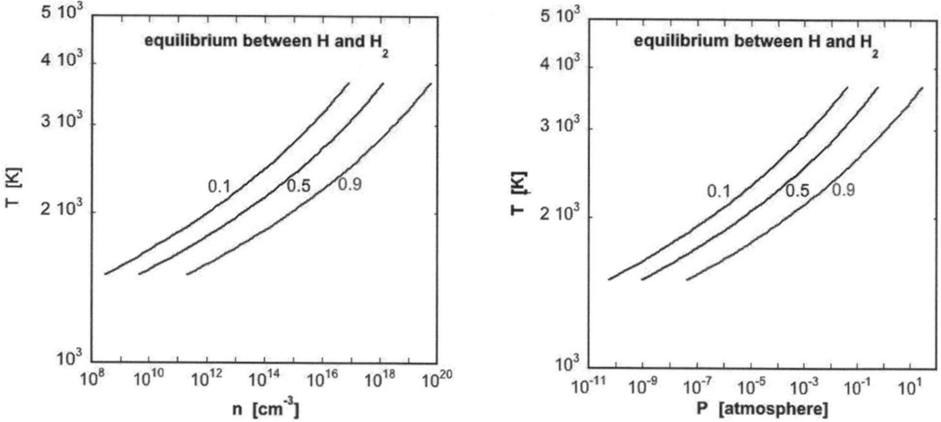

Figure 5.2 Lines of constant molecular hydrogen fraction in the temperature – density plane (left) and temperature – pressure plane (right). The curves are labeled by the molecular hydrogen fraction, x.

This relationship is shown in Figure 5.2. For a given temperature, hydrogen is largely atomic at low pressures/densities, while, at high pressures/densities, it is mainly molecular. This merely expresses the competition between the binding energy, which favors molecular hydrogen, and the entropy, which favors a larger number of particles (e.g. atomic H). The exponential dependence on temperature implies that the equilibrium is largely controlled by the temperature and shifts from atomic to molecular hydrogen over a small temperature range.

Besides dissociation, a pure hydrogen gas can also be ionized. The reaction describing this is,

$$\mathrm{H} \leftrightarrow p + e, \tag{5.41}$$

which results in,

$$\frac{P\,(\mathrm{H})}{P\,(\mathrm{p})\,P\,(\mathrm{e})} = K_p\,(H)\,. \tag{5.42}$$

The internal partition functions of the electron and proton are 2 (reflecting their spin states), while for the hydrogen atom, we will assume that only the lowest energy state is occupied (e.g. $Z\,(\mathrm{H}) = 4n^2 = 4$, where the factor 4 takes the electron and proton spin states into account). This equation can then be simplified to,

$$\frac{P\,(\mathrm{H})}{P\,(\mathrm{p})^2} = \left(\frac{h^2}{2\pi m_e kT}\right)^{3/2} \exp\left[E_I/kT\right]. \tag{5.43}$$

This is very similar to the equation describing the molecular hydrogen equilibrium and similar considerations apply. At low temperatures, the equilibrium will be to the left in Equation (5.41) and the gas will be neutral. At high temperatures, the gas will be fully ionized. Again, the transition between these two regimes is very sharp. As E_I is about three times larger than E_b, the transition occurs at about a three times higher temperature.

Of course, both sets should be solved simultaneously. However, as the transitions are so sharp and well separated in temperature, they could be dealt with separately. We have also ignored H^-, H_2^+ and H_3^+. In a pure hydrogen gas, H^- is never important as the binding energy of the additional electron is only 0.75 eV and so at temperatures at which electrons become available the equilibrium for H^- has shifted way to the left. However, if there is an additional source of electrons (i.e. metals with low ionization potentials), then H^- can become more important (e.g. in stellar photosheres such as the Sun). Likewise, the ionization potential of H_2 is 15.4 eV – much higher than the dissociation potential of H_2 – and by the time H_2 could be ionized, none is left. By the same token, the proton affinity of H_2 is only 4.38 eV and, when enough protons are around to shift the equilibrium to H_3^+, no H_2 is around and the temperature is too high for H_3^+ to be stable.

For later use, it is instructive to combine the dissociation and ionization balance. Ignoring the minor species, we can write an explicit expression accounting for H_2, H, p, and e in a pure hydrogen gas,

$$P_{gas} = P\,(\mathrm{H_2}) + P\,(\mathrm{H}) + P\,(\mathrm{p}) + P\,(\mathrm{e})\,. \tag{5.44}$$

Adopting charge neutrality we can write,

$$P_{gas} = P\,(\mathrm{H}) + 2\left(\frac{2\pi m_e kT}{h^2}\right)^{3/4}\sqrt{P\,(\mathrm{H})} + \frac{P\,(\mathrm{H})^2}{K\,(\mathrm{H_2})}, \tag{5.45}$$

which is a quartic equation in $\sqrt{P\,(\mathrm{H})}$, which, for a given P_{gas} and T, can be solved analytically or using numerical techniques.

5.3.2 Solid–Gas Equilibria

To illustrate the interaction of solids and gases in an astrophysical setting, we will consider a system dominated by hydrogen where oxygen is mainly locked up in H_2O. Consider now the equilibrium between solid corundum (Al_2O_3) and gaseous Al and O,

$$2\mathrm{Al} + 3\mathrm{O} \leftrightarrow \mathrm{Al_2O_3}. \tag{5.46}$$

For a solid, we use the activity in the mass action equation and we can write,

$$\frac{a_{\mathrm{Al_2O_3}}}{P_{Al}^2 P_O^3} = K_{eq}\,(\mathrm{Al_2O_3})\,. \tag{5.47}$$

As the activity of a pure substance is unity, we have,

$$-\ln\left[P_{Al}^2 P_O^3\right] = \ln\left[K_{eq}\,(\mathrm{Al_2O_3})\right]. \tag{5.48}$$

The partial pressure of atomic oxygen is set by the equilibrium with the main reservoir of oxygen, water,

$$2\mathrm{H} + \mathrm{O} \leftrightarrow \mathrm{H_2O}, \tag{5.49}$$

which is governed by,

$$\left[\frac{P_{H_2O}}{P_H^2 P_O}\right] = \left[K_{eq}\left(\mathrm{H_2O}\right)\right]. \tag{5.50}$$

We wll need the partical pressure of H for this, which is set by the equilibrium with the main reservoir of hydrogen, H_2,

$$2\mathrm{H} \leftrightarrow \mathrm{H_2}, \tag{5.51}$$

which we have already solved (Section 5.3.1). We can combine these to write,

$$P_O = \frac{P_{H_2O}}{P_{H_2}} \frac{K_{eq}\left(\mathrm{H_2}\right)}{K_{eq}\left(\mathrm{H_2O}\right)}. \tag{5.52}$$

With the approximations, the pressure of atomic oxygen is thus just set by the abundance of oxygen and the Gibbs free molar energies of the reactions involved.

The way this is now formulated, the conservation equations are built in by demanding that the partial pressure of H_2O is the abundance of oxygen – that is not locked up in CO (the main reservoir of carbon) – times the total pressure, and the H_2 pressure is the total pressure. It is instructive to solve this set of equations graphically. The Gibbs free energies per mole can be taken from tabulations. The left-hand side of Equation (5.47) is given by the Gibbs free energy of corundum. This gives the relationship between solid corundum and the vapor. We have to compare this to the partial pressures of atomic aluminum and of atomic oxygen set by the equations above. The intersection of the two curves sets the condensation temperature (Figure 5.3). When a cooling gas approaches this temperature, corundum is

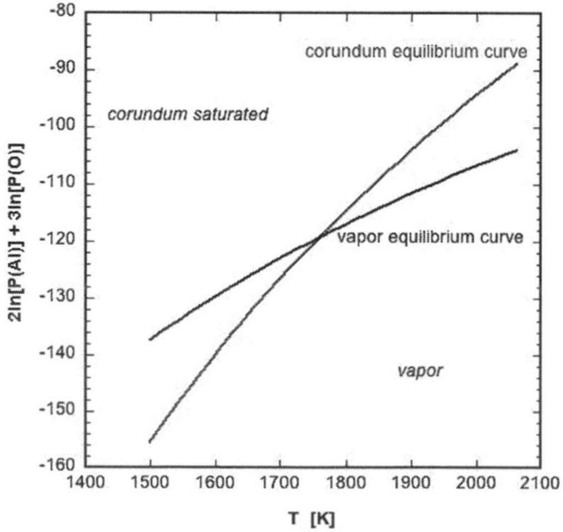

Figure 5.3 The equilibrium curve describing solid corundum (left-hand side of Eq. (5.47)) and the vapor equilibria of aluminum and oxygen (Eq. (5.52)) evaluated for solar system abundances and a total pressure of 10^{-3} Atm. The two curves intersect at the condensation temperature of corundum, 1758 K, for these conditions.

predicted to condense out in equilibrium. Above that temperature, aluminum will be in the gas phase, and the vapor curve describes the equilibrium. Below that point, the solid curve describes the equilibrium and thus also the equilibrium pressures of the aluminum and oxygen vapor. As a result, this also has direct influence on vapor condensation reactions that occur at lower temperatures. As an example, in the absence of corundum condensation, spinel ($MgAl_2O_4$) is predicted to condense out at 1685 K, but equilibrium corundum condensation lowers the partial pressure of aluminum and delays spinel condensation to almost 1500 K.

5.3.3 Solid Phase Equilibria

Here, we will consider the equilibrium between forsterite (Mg_2SiO_4), silica (SiO_2), and enstatite ($MgSiO_3$),

$$Mg_2SiO_4 + SiO_2 \leftrightarrow 2MgSiO_3, \tag{5.53}$$

where it is understood that silica is a melt. Starting with enstatite, at equilibrium we will have $1 - x$ enstatite, and $x/2$ forsterite and silica. We can then write,

$$\frac{4\,(1-x)^2}{x^2} = K_P, \tag{5.54}$$

with the equlibrium constant given by

$$K_P = \exp\left[-\mu/RT\right], \tag{5.55}$$

with R the gas constant. From the JANAF tables, we have $\mu \simeq 5$ kJ/mol at 1300 K and we find $x \simeq 0.7$. Thus, the fraction of forsterite and silica is 0.35 while the fraction of enstatite is 0.3.

5.4 Methods

In thermodynamic equilibrium, the system can be described by a set of equations describing mass balance. For hydrogen, ignoring ionization, this yields,

$$P_H = 2P\,(H_2) + P\,(H) + 2P\,(H_2O) + 3P\,(NH_3) + 4P\,(CH_4) + \cdots. \tag{5.56}$$

For each of the species involved, we can write a chemical reaction relating it back to its constituent atoms.[5] For example, for H_2O this is given by equation (5.52), which can be rearranged to an expression for $P\,(H_2O)$. These and similar equations for the other compounds involved can be substituted in Equation (5.56). Mass balance equations for all elements involved can be written and will contain similar terms. For elements with condensed phases, we have to take care to properly include condensation. Recall Equation (5.47); when $a_{Al_2O_3}$ equals unity, Al_2O_3 will condense out. At temperatures below the condensation temperature of a compound, the actual pressures of the constituent elements

[5] We make the choice to relate them back to atoms rather than the dominant component as Gibbs energies are generally given in terms of the atomic constituents. Recall that in thermodynamic equilibrium the chemical route leading to a species is irrelevant.

will be lower than the partial pressure indicated by the elemental abundance (cf. Section 5.3.2). If a fraction α_i of element i is condensed out, the partial pressure of this element has to be reduced by a factor $(1 - \alpha_i)$. Consider again Al_2O_3 as an example. Ignoring condensation, the partial pressure of Al, $\tilde{P}$ (Al), is given by,

$$\tilde{P}\,(\mathrm{Al}) = \mathcal{A}_{\mathrm{Al}}\beta\,(\mathrm{Al})\,P_{gas}, \tag{5.57}$$

where $\beta(\mathrm{Al})$ is the fraction of Al atoms in the gas phase ($\equiv P(\mathrm{Al})/(P(\mathrm{Al}) + P(\mathrm{AlO}) + \cdots)$). The actual pressure is $P(\mathrm{Al}) = (1 - \alpha_{\mathrm{Al}})\tilde{P}(\mathrm{Al})$. We will ignore for simplicity the effect of Al_2O_3 condensation on the oxgygen pressure (but it can easily be included). Considering that Al_2O_3 is in equilibrium with the gas phase, we have from Equation (5.47),

$$K_{eq}\,(\mathrm{Al_2O_3})\;P_{Al}^2 P_O^3 = K_{eq}\,(\mathrm{Al_2O_3})\;(1 - \alpha_{\mathrm{Al}})^2\tilde{P}^2(\mathrm{Al})P_O^3 = 1. \tag{5.58}$$

This provides an additional expression for α_{Al} that can be used to reduce the gas phase pressure of Al. Similar corrections can be made for all elements distributed between the gas phase and the solid phase. The gas phase and solid state compositions can then be solved by iterating on the mass balance and solid state fraction equations.

Solving for the chemical equilibrium requires the elemental abundances and relevant molar Gibbs free energies. For the solar nebula and protoplantary disks, solar abundances or interstellar medium abundances can be adopted. These abundances are less relevant for stellar atmospheres as other stars may have quite a different galactic chemical history. In addition, nucleosynthesis followed by 1st, 2nd, and 3rd dredge up as well as other mixing processes (e.g. rotation) during the red giants, and asymptotic giant branch phases may have profoundly altered the photospheric composition of these cool stars to the tune of even converting the star from O-rich (C<O) to C-rich (C>O) with a concomittant shift from oxides to carbon-bearing species. Nucleosynthesis also has a profound influence on the elemental composition of Wolf–Rayet stellar winds and nova and supernova ejecta. Chemical potentials or Molar Gibbs free energies can be obtained from the JANAF compilations or the NIST data base website.

5.5 The Role of Thermodynamics in Space

Chemical thermodynamics has been very successful in explaining the composition of meteorites in the solar system. Many of the minerals in meteorites follow the condensation sequence of a slowly cooling Solar nebula. Likewise, the measured composition of presolar dust isolated from meteorites with an isotopic composition betraying an origin in the ejecta from late-type stars – so-called stardust – follows the thermodynamic condensation sequence. Observationally, infrared spectra of dust around such objects also reveal the relevance of the condensation sequence.

Despite these successes, chemical thermodynamics should be used with care as chemistry is kinetics, and thermodynamics can at best only indicate the direction chemistry tends to flow. As already indicated in Section 5.3.2, the reactions describing chemical thermodynamics have no direct bearing on the actual chemical routes involved. Consider again the formation of Al_2O_3 analyzed in that section. The fraction of Al condensed in corundum is determined using the fiducial reaction (5.46). Actually, the main aluminum-bearing precursor molecule for corundum is AlO and growth of corundum involves reaction of this species with solid-Al_2O_3 where the additional oxygen required is made up through reactions with, likely, OH, where the latter is part of the (kinetic) H-H_2-OH-H_2O system. In thermodynamic equilibrium, this doesn't matter as all is regulated by the Gibbs free energies. But in reality, many of the relevant reactions may have activation barriers and therefore be frozen out at the temperatures and pressures involved. Abundances of key species are then much lower than expected. By the same token, the solid surface may first have to be activated before reaction can occur and that too may be inhibited by an activation barrier. Condensation of carbon dust presents a case in point as at relevant temperatures for stellar ejecta this involves the HACA mechanism: Hydrogen Abstraction from a molecular intermediary creating a radical site followed by Carbon Addition at that radical site. Typically, after two HACA steps, cyclization occurs, building up a graphitic hexagonal network. Given enough time, barriers involved in these reactions may not matter but then, as so often, time is of the essence.

Additionally, nucleation may inhibit attaining chemical thermodynamic equilibrium. The first steps in the condensation process involve nucleation of small clusters from simple gas phase species. The Gibbs free energies involved in cluster formation are in general different from those determined from atomization of solids. Simplistically, this is often caught by adding a macrosocopic quantity – surface free energy – to the Gibbs free energy: $\Delta G(r) = \Delta\mu\, 4\pi r^3/3 + \gamma\, 4\pi r^2$ with r the size of the cluster. The negative volume term drives the growth while the positive surface term describes the free energy penalty of the surface. It is easy to see then that there is a critical size, $d\Delta G/dr = 0$, which corresponds to a barrier to growth. Clusters smaller than this size will tend to dissolve while larger clusters will grow. In a slowly cooling gas, condensation will occur at a temperature well below the condensation temperature. In classical nucleation theory, a macroscopic approach is formulated to what is essentially a kinetic process. Really, the chemical routes involved in nucleation should be identified and these may well contain activation barriers that are not caught by a macroscopic surface free energy term.

Furthermore, we should examine the solid phase equilibria such as described in Section 5.3.3. For solid forsterite in contact with silica melt, reaction can proceed toward the left upon cooling. Equivalently, upon heating, an enstatite crystal will segregate into a forsterite crystal plus silica melt. However, under typical conditions in space, melts are not relevant, and reaction would have to occur between forsterite and gaseous SiO (e.g. out of equilibrium) or a forsterite grain would have to interact with a silica grain and collisions between grains might be very rare.

Finally, we will illustrate the relevance of chemical timescales based on the key reaction,

$$H_2 + H_2 \leftrightarrow H + H + H_2, \tag{5.59}$$

which has a rate of $3.7\times10^{-10}\,\exp\left[-48350/T\right]\,\mathrm{cm^3\,s^{-1}}$. Figure 5.4 provides the timescale as a function of temperatures for this reaction at two different pressures. These pressures

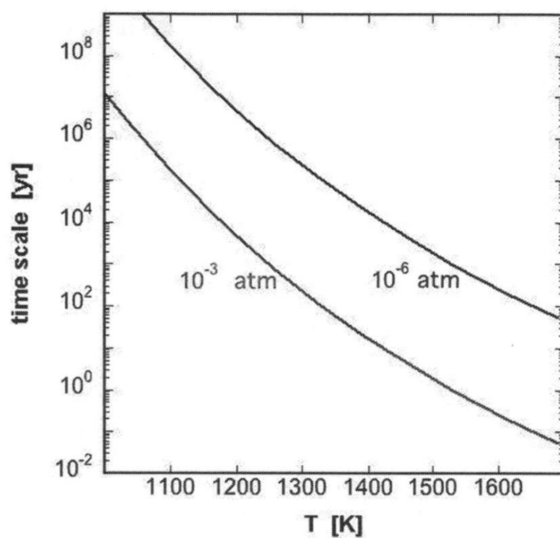

Figure 5.4 Chemical timescale for the collisional dissociation of H_2 for two different pressures.

are particularly relevant for the inner solar nebula or for inner protoplanetary disks and drives the H/H_2 ratio to equilibrium (Figure 5.2). So, with a dynamical timescale of $\sim 10^4$ yr, the atomic H fraction can be appreciable. The O/OH/H_2O fraction is set by the H/H_2 fraction (cf. Eq. (5.52)) and OH (and O) will burn carbonaceous solids to CO and CO_2. This will occur then on time scales set by the H/H_2 reaction.

5.6 Further Reading and Resources

Reference [5] provides a comprehensive text on chemical thermodynamics as applied to geosciences and planetary sciences.

An early review on chemical thermodynamic calculations relevant to the Solar nebula is provided by [8]. An early review on chemical thermodynamic calculations relevant to circumstellar dust formation is given in [12]. A more recent review – relevant to stardust – is provided by [11].

Over the years, various compilations of elemental abundances relevant to the solar nebula have appeared [1, 10]. Elemental abundances can have a profound influence on the chemical equilibrium [4, 7, 9].

Chemical potentials or Molar Gibbs free energies:
JANAF compilations (http://kinetics.nist.gov/janaf/janbanr.html),
NIST data base website (http://webbook.nist.gov/chemistry/form-ser.html).
A handy compilation in terms of simple polynomial fits is provided by [13].

The seminal paper [8] describes chemical thermodynamics and the composition of meteorites, while reference [11] highlights the condensation sequence of stardust in AGB ejecta. Observational evidence for the condensation sequence of dust in AGB ejecta is summarized in [14]. Chemical routes in nucleation reactions for carbon soot are described in [2, 6]. Oxide grain routes are described in [3].

5.7 Exercises

5.1 Partition function: consider the equilibrium AB $\leftrightarrow$ A+B+ΔE. Consider now three molecules AB that can dissociate into three atoms A and B. The partition function is given by,

$$Z_{tot} = \frac{Z_{\rm AB}^3}{3!} + \frac{Z_{\rm AB}^2}{2!}\frac{Z_{\rm A}}{1!}\frac{Z_{\rm B}}{1!} + \frac{Z_{\rm AB}}{1!}\frac{Z_{\rm A}^2}{2!}\frac{Z_{\rm B}^2}{2!} + \frac{Z_{\rm A}^3}{3!}\frac{Z_{\rm B}^3}{3!}, \tag{5.60}$$

with $Z_i = V\left(2\pi m_i kT/h^2\right)^{3/2} Z_i\,(int)$ where the first part is translational partition function and the second part describes the partition function of the internal degrees of freedom. Each of the terms in this expression describes one of the possible states of the system where the denominator takes the indistinguishable nature of the species into account.

(a) Generalize this expression for the total partition function when $N_{\rm AB}$, $N_{\rm A}$, and $N_{\rm B}$ species are present.

(b) When decreasing $N_{\rm AB}$ sequentially by one (and increasing each of $N_{\rm A}$ and $N_{\rm B}$ by one), the total partition function will first rise, reach a maximum, and then decrease again. At the largest likelihood, two successive probabilities will be approximately equal. Show that this condition is approximately given by,

$$\frac{Z_{\rm AB}}{N_{\rm AB}} = \frac{Z_{\rm A}}{N_{\rm A}+1}\frac{Z_{\rm B}}{N_{\rm B}+1}. \tag{5.61}$$

(c) Assuming $N \gg 1$, using the expression for Z_i, and recognizing that $N_i/V = n_i$, write this as,

$$\frac{n_{\rm A} n_{\rm B}}{n_{\rm AB}} = \left(\frac{2\pi m_{\rm A} m_{\rm B}/m_{\rm AB} kT}{h^2}\right)^{3/2} \frac{Z_{\rm A}\,(int)\, Z_{\rm B}\,(int)}{Z_{\rm AB}\,(int)}. \tag{5.62}$$

(d) Considering that A and B have no internal structure and that AB has only one (accesible) bound state, write the equivalent Saha equation and compare to Equation (5.2).

5.2 Hydrogen ionization equilibrium:

(a) Derive Equation (5.43).

(b) Assume initially all the hydrogen is atomic. Introducing $n\,(H) = (1-y)\,n_o$, rewrite this equation to an expression containing y and $1-y$. Make the equivalent of Figure 5.2 for hydrogen ionization and verify that the transitions temperature is about three times higher than for H_2 dissociation.

Table 5.2 *Gibbs free energies*[a,b]

Species	a	b	c	d	e
H_2	4.25E+05	−1.07E+05	2.70E+01	5.48E-04	−3.81498E-08
H_2O	8.66E+05	−2.28E+05	5.61E+01	7.63E-04	−4.95254E-08
Al_2O_3	0.00E+00	−7.33E+05	1.85E+02	−2.57E-03	0.00000E+00
AlO	3.55090E+05	−1.25068E+05	2.94347E+01	−3.97980E-04	2.26575E-08

[a] Relative to the atomic constituents and at 1 Atm. Units are cal/mole.
[b] $\Delta G = a/T + b + cT + dT^2 + eT^3$

(c) Cosmic ray ionization maintains an electron fraction of $\simeq 10^{-7}$ in dense ($n \simeq 10^4$ cm^{-3}) molecular clouds. Calculate the equivalent temperature of the gas if thermodynamic equilibrium were applicable.

(d) If the H ionization rate by electrons is given by,

$$10^{-9}\,(T/300)^{1/2}\,\exp\left[-13.6eV/kT\right], \tag{5.63}$$

would thermodynamic equilibrium be attained at this temperature?

(e) Derive Equation (5.45).

5.3 In Section 5.3.2, we have solved graphically for the condensation temperature of Al_2O_3. Derive an analytical expression for the condensation temperature using Equations (5.47) and (5.52) and calculate the temperature for a pressure of 10^{-3} atm. Relevant Gibbs free energies are provided in Table 5.2.

5.4 In Section 5.3.2, we have examined the formation of Al_2O_3 starting with the reaction (5.46). However, the main reservoir of aluminum is AlO. Demonstrate that this does not affect the derived condensation temperature. This is another manifestation that in thermodynamic equilibrium the actual pathway involved is not relevant (and that the Gibbs free energies refer to the atomic state).

Bibliography

[1] Anders, E., Grevesse, N., 1989, *Geochim Cosmochim Acta*, 53, 197
[2] Cherchneff, I., Barker, J. R., & Tielens, A. G. G. M., 1992, *ApJ*, 401, 269
[3] Cherchneff, I., Dwek, E., 2010, *ApJ*, 713, 1
[4] Cristallo, S., Piersanti, L., Straniero, O., et al., 2011, *ApJS*, 197, 17
[5] Fegley, B., jr., 2013, *Practical Chemical Thermodynamics for Geoscientists* (Amsterdam: Academic Press, Elsevier)
[6] Frenklach, M., Feigelson, E. D., 1989, *ApJ*, 341, 372
[7] Gallino, R., Bisterzo, S., 2011, *American Institute of Physics Conference Series*, 1386, 98
[8] Grossman, L, Larimer, J. W., 1974, *Rev Geophysics Spacephysics*, 12, 71

[9] Käppeler, F., Gallino, R., Bisterzo, S., Aoki, W., 2011, *Reviews of Modern Physics*, 83, 157
[10] Lodders, K., 2003, *ApJ*, 591, 1220
[11] Lodders, K., Fegley, B., Jr., 1999, *Asymptotic Giant Branch Stars*, 191, 279
[12] Salpeter, E. E., 1977, *Annu Rev Astron Astrophys*, 15, 267
[13] Sharp, C. M., Huebner, W. F., 1990, *ApJS*, 72, 417
[14] Tielens, A. G. G. M., Waters, L. B. F. M., Bernatowicz, T. J., 2005, in *Chondrites and the Protoplanetary Disk*, 341, 605

6
Gas Phase Chemical Processes

6.1 Introduction

Interstellar chemistry is dominated by kinetics rather than thermodynamics. This reflects the highly nonequilibrium nature of the ISM as low gas phase temperatures and densities combine with high fluxes of UV photons or energetic ions. In such a system, reactions and their rates are key to understanding the chemical composition of the medium. In this chapter, the different types of chemical reactions – relevant for the interstellar medium – are described and their reaction rates are discussed. For a general course, the discussion can center on the generic reactions and their rates, discussed in Section 6.2. These lead to a set of "simple" general rules that can be used to generate minimalistic sets of reactions that describe the "dance" of the atoms from one molecule to the next (Section 6.13.1). These general rules are key to understanding the chemistry of diffuse and dense molecular clouds and of regions of star formation (Chapters 9, 10, and 11). The rest of this chapter provides background for graduate students entering the field that will provide them with a deeper insight in the reactions and the factors that control their rates as well as in common methods employed to solve for the gas phase composition of molecular clouds. Further resources are given in Section 6.14.

6.2 Generic Reaction Processes and Their Rates

Gas phase reactions can be divided into different categories depending on their general effects. There are the bond formation processes, including radiative association, which link atoms into simple or more complex species. Reactions such as photodissociation, dissociative recombination, and collisional dissociation are bond destruction processes that fragment species into smaller species. Finally, there are the bond rearrangement reactions – ion–molecule exchange reactions, charge transfer reactions, and neutral–neutral reactions – that transfer parts of one coreactant to another one. In this section, we will describe simple approximations for the reaction rates. These reactions are discussed in greater detail in the remainder of this chapter. Generic examples of these reactions and typical reaction rates are

Table 6.1 *Generic gas phase reactions and their rates*

	Reaction	Rate[a]	Unit	note
Photodissociation	$AB + h\nu \rightarrow A + B$	10^{-9}	s^{-1}	1
Neutral–neutral	$A + B \rightarrow C + D$	3×10^{-11}	$cm^3\ s^{-1}$	2
Ion–molecule	$A^+ + B \rightarrow C^+ + D$	2×10^{-9}	$cm^3\ s^{-1}$	3
Charge–transfer	$A^+ + B \rightarrow A + B^+$	10^{-9}	$cm^3\ s^{-1}$	3
Radiative association	$A + B \rightarrow AB + h\nu$			4
Dissociative recombination	$A^+ + e^- \rightarrow C + D$	3×10^{-7}	$cm^3\ s^{-1}$	
Collisional association	$A + B + M \rightarrow AB + M$	10^{-32}	$cm^6\ s^{-1}$	3
Associative detachment	$A^- + B \rightarrow AB + e^-$	10^{-9}	$cm^3\ s^{-1}$	3
Cosmic ray ionization	$A + CR \rightarrow A^+ + e^- + CR$	4×10^{-16}	s^{-1}	

Notes: [a]Reaction rate coefficients evaluated at 300 K. [1]Rate in the unshielded radiation field. [2]Rate in the exothermic direction and assuming no activation barrier (ie., radical–radical reaction). [3]Rate in the exothermic direction. [4]Rate highly reaction specific.

summarized in Table 6.1. For a bimolecular reaction, $A + B \rightarrow C + D$, the corresponding rate of formation or destruction of a species is,

$$\frac{dn\,(A)}{dt} = \frac{dn\,(B)}{dt} = -k\,n\,(A)\,n\,(B) = -\frac{dn\,(C)}{dt} = -\frac{dn\,(D)}{dt}. \tag{6.1}$$

For a unimolecular reaction, $A \rightarrow C + D$, we have,

$$\frac{dn\,(A)}{dt} = -k\,n\,(A) = -\frac{dn\,(C)}{dt} = -\frac{dn\,(D)}{dt}. \tag{6.2}$$

Similarly, for a termolecular reaction, $A + B + M \rightarrow AB + M$, we have,

$$\frac{dn\,(A)}{dt} = \frac{dn\,(B)}{dt} = -k\,n\,(A)\,n\,(B)\,n\,(M) = -\frac{dn\,(AB)}{dt}. \tag{6.3}$$

We can make order of magnitude estimates for reaction rate coefficients using simple generic considerations.

For UV or cosmic ray processing, reaction rate coefficients are given by a cross section for the process to happen – as a function of energy – averaged over the energy dependent photon or particle flux. For bimolecular reactions, the collision cross section – as a function of the collision velocity – has to be multiplied by the velocity to turn it into a rate and then averaged over the velocity distribution of the reaction partners.

First, consider photochemical reactions. The reaction rate will be given by an average cross section, $\overline{\sigma}$ times a photon flux, $\mathcal{N}_{uv}$; viz.,

$$k_{uv} = \int \sigma\,(\nu)\,\mathcal{N}\,(\nu)\,d\nu \tag{6.4}$$

$$= \overline{\sigma}\,\mathcal{N}_{uv} = \left(\frac{\pi e^2}{m_e c}\right)\,f\,\left(\frac{\mathcal{N}_{uv}}{\Delta\nu}\right), \tag{6.5}$$

where the term in brackets is the classical expression for the integrated cross section of a harmonic oscillator (2.65×10^{-2} cm^2 Hz) and f is the oscillator strength.[1] Assuming absorption occurs over the range of 6–13.6 eV, the total photon flux in the interstellar radiation field is 10^8 photons cm^{-2} s^{-1}. With an oscillator strength of $f \sim 1$ typical for allowed transitions, the rate is then 10^{-9} s^{-1}. If dissociation occurs only over a small frequency range, then the photodissociation rate is concommitantly smaller as fewer photons are available for photodissociation. In a cloud, dust absorption (and scattering) will decrease the photon flux in a systematic way. For a homogeneous cloud with a given geometry, this can be readily modeled. For a clumpy cloud, this can be more involved. Sometimes, absorption by (other) gaseous species has to be taken into account. Finally, the interstellar radiation field is strongly attenuated deep in a cloud. Photodissociation is then dominated by photons produced by cosmic rays exciting gaseous species, which decay through photon emission. All of these aspects are discussed in more detail in Section 6.3.

Second, cosmic ray ionization can be dealt with in a very similar manner. This process is controlled by low energy (10–100 MeV) cosmic rays with a particle intensity of $\simeq$10 protons cm^{-2} s^{-1} sr^{-1}. The cross section for this process is $\simeq$0.03Å^2, resulting in a rate of $\simeq 4 \times 10^{-16}$ s^{-1}.

Third, for reactions between gaseous species, the reactants have to approach each other to roughly a Bohr radius. So, consider the reaction between a charged particle and a neutral. The interaction of an ion with a neutral is assisted by the polarization induced by the incoming charge. The field of the charged particle is $E \sim Ze/r^2$ and the induced field is $\alpha Ze/r^2$ with α the polarizability of the neutral species. So, the interaction energy will be $E \sim \alpha Z^2 e^2/r^4$. A deflection over an angle π will occur for an impact parameter, b_{crit}, which we can estimate by setting the interaction energy equal to the initial kinetic energy $E_{kin} = \mu v_\infty^2/2$;

$$b_{crit} \simeq \left(\frac{2\alpha Z^2 e^2}{\mu v^2} \right)^{1/4}. \tag{6.6}$$

The cross section scales, thus, with $\pi b_{crit}^2 \propto 1/v$. The reaction rate coefficient is given by the averaging this cross section times the velocity over the (Maxwellian) velocity distribution, $f(v)$,

$$k = \int \sigma(v)\, v\, f(v)\, dv, \tag{6.7}$$

and this is independent of velocity. The ion–molecule reaction rate is then approximately given by, $k_i \sim \pi e \sqrt{2\alpha/\mu}$. Besides electrostatic interaction, angular momentum will give rise to a centrifugal barrier that has to be overcome. A proper evaluation (Section 6.4) leads to the Langevin rate,

$$k_L = 2\pi e \sqrt{\frac{\alpha}{\mu}}. \tag{6.8}$$

[1] The effective number of classical oscillators associated with the absorption.

Typically, values for polarizabilities are 2 Å^3 and the Langevin rate is then $\simeq 10^{-9}$ cm^3 s^{-1}. Ion–molecule reactions are discussed in more detail in Section 6.4. Charge exchange reactions and electron detachment reactions are "special" cases of ion–molecule reactions.

For neutral–neutral reactions, we can adopt a hard sphere collision model where the cross section is given by $\pi\,(r_A + r_B)^2$ and the thermal velocity, $v_{th} = \sqrt{8kT/\pi\mu}$. With a cross section of 5 Å^2 and a thermal velocity of $\sim 10^4$ cm s^{-1}, we have a reaction rate coefficient of, $k_n \sim 3 \times 10^{-11}\,\sqrt{T/300\,\mathrm{K}}$ cm^3 s^{-1}. Most neutral–neutral reactions have substantial activation barriers as, during the reaction, bonds have to be broken and reformed. The reaction rate coefficient decreases then by the Boltzmann factor, detailing the fraction of collisions that are energetic enough to overcome this barrier. At 10 K, a barrier of $\simeq 1000$ K is prohibitive. As an aside, we note that for ion–molecule reactions and dissociative electron recombination reactions, the Coulomb energy can help to overcome reaction barriers. Neutral–neutral reactions involving radicals can occur at near the collision rate derived above. This is discussed in more detail in Section 6.5.

For electrons and ion interactions, we have to take Coulomb forces into account. Consider that energy exchange will only occur if the electron penetrates close enough to the electron cloud surrounding the ion; i.e. when $r \sim a_0$ with a_0 the Bohr radius. At that point, the potential energy is $e^2/a_0 \simeq 27$ eV. The Coulomb interaction will focus an electron to the charged species and this will lead to a cross section enhanced by a factor,

$$f_C = 1 + \frac{e^2/a_0}{kT}. \tag{6.9}$$

This is $\simeq 10^3$ at 300 K. At low temperatures the last term dominates by far and we can write for the reaction rate coefficient,

$$k_e \simeq \pi a_0 \frac{e^2}{kT}\,\overline{v} \sim 10^{-6} \left(\frac{300\,\mathrm{K}}{T}\right)\ \mathrm{cm^3\ s^{-1}}. \tag{6.10}$$

Measurements give typically a somewhat lower value, 3×10^{-7} cm^3 s^{-1}. The $T^{-1/2}$ dependence reflects that Coulomb interaction prefers low velocity interactions as the time "spent" near the interaction region is longer. Dissociative electron reacombination reactions are discussed in more detail in Section 6.8.

Fourth, for radiative association or radiative recombination reaction, the collisional rate coefficient is set by the type of interaction as discussed above. During the actual collision – when the reaction partners are in the interaction zone – stabilization has to occur through the emission of a photon. The collision time scale will be of the order of 10^{-14} s and, for an allowed electronic transition, the Einstein A will be $\sim 10^8$ s^{-1}, leading to a reaction probability of 10^{-6}. Hence, the resulting reaction rate coefficients are typically very small unless a long-lived collision complex can be formed as the collision energy is stored in internal degrees of freedom of the intermediary (e.g. vibrational excitation). We will discuss this issue further in Chapter 8, when we examine reactions of large molecules. Radiative association reactions for simple species are discussed in Section 6.6.

As ion–molecule rates are much faster than neutral–neutral rates and, when exothermic, generally do not posses activation barriers, even small levels of ionization can drive interstellar gas phase chemistry. In diffuse clouds or at the edges of molecular clouds, this ionization is provided either by penetrating UV photons from nearby masssive stars or by the interstellar radiation field. Inside dense cloud, ionization is provided by energetic cosmic ray particles. In the diffuse medium, the cosmic ray ionization rate of H is $\sim 4 \times 10^{-16}$ s^{-1}. Deeper in the cloud, the cosmic ray flux will be attentuated and the ionization rate will drop. Often, a generic rate of 3×10^{-17} s^{-1} is adopted for dense cores. As cosmic ray ionization is key to much of gas phase chemistry, molecular observations provide a handle on the cosmic ray flux in the ISM. This will be discussed in Section 9.5.

6.3 Photochemistry

6.3.1 Photodissociation Processes for Small Molecules

FUV photons permeating the diffuse ISM are a dominant destruction agent for small molecules. Typical bonding energies of molecules are in the range 5–10 eV, corresponding to wavelength of $\simeq$ 3000 Å and shorter. However, direct absorption into the dissociating continuum of the ground state is generally negligible. Figure 6.1 illustrates the various processes that can be involved in photodissociation of small molecules. The simplest dissociation occurs when the molecule is directly excited into a repulsive state. Spontaneous

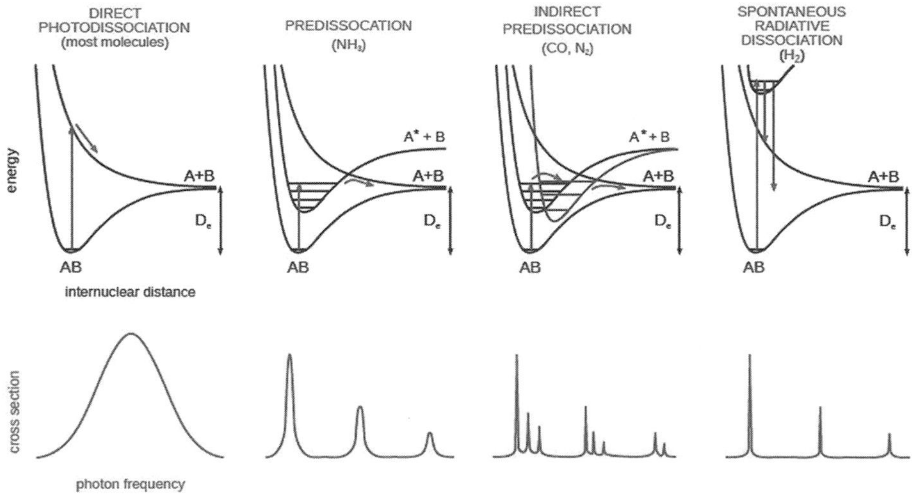

Figure 6.1 Schematic photodissociation processes (top) and cross sections (bottom). From left to right: Direct photodissociation into a repulsive state, predissociation through a bound state, indirect predissociation through two bound states, and spontaneous radiative dissociation. See text for details. Figure courtesy of A. Heays

radiative decay back to the ground state occurs on a timescale of $\simeq 10^{-8}$ s and that is much longer than dissociation, $\simeq 10^{-14} - 10^{-13}$ s (corresponding to a vibrational period). Hence, dissociation is very efficient. As, initially, the nuclei cannot respond to the electronic motion, the absorption cross section will peak at the vertical transition energy, which is not related to the dissociation energy of the molecule. The width of the absorption band is set by the steepness of the dissociative potential energy curve and the "width" of the vibrational potential well in the ground state. Direct predissociation occurs when a molecule is excited in a bound vibrational state of an electronic state and dissociates because this state mixes with the continuum of another (dissociative) electronic state (cf. Figure 6.1). Absorption occurs now in discrete lines, corresponding to state-to-state vibrational transitions. The width of these absorption lines depends on the coupling timescale to the dissociative state (e.g. Heisenberg's uncertainty principle). The strength reflects the oscillator strength of the initial electronic transition – and thus the Franck–Condon factors (cf. Section 3.4.3) – and the coupling between the states. Absorption can also occur into the dissociative (vibrational) continuum of the initial state. Indirect (or accidental) predissociation occurs when the excited electronic state crosses another state that is bound but predissociates by coupling to a dissociative state. Because multiple transitions are involved, this is a slower process and the absorption bands become very narrow. Spontaneous radiative dissociation corresponds to excitation into a bound electronic state, which decays radiatively into the vibrational continuum of the ground electronic state or to the continuum of a repulsive state. This process is important for H_2 and is discussed in more detail in Section 6.2.4.

As an example, Figure 6.2 shows the photodissociation cross section for H_2O and the electronic states involved. Direct absorption in the dissociative $\tilde{A}^1B_1$ state produces a broad

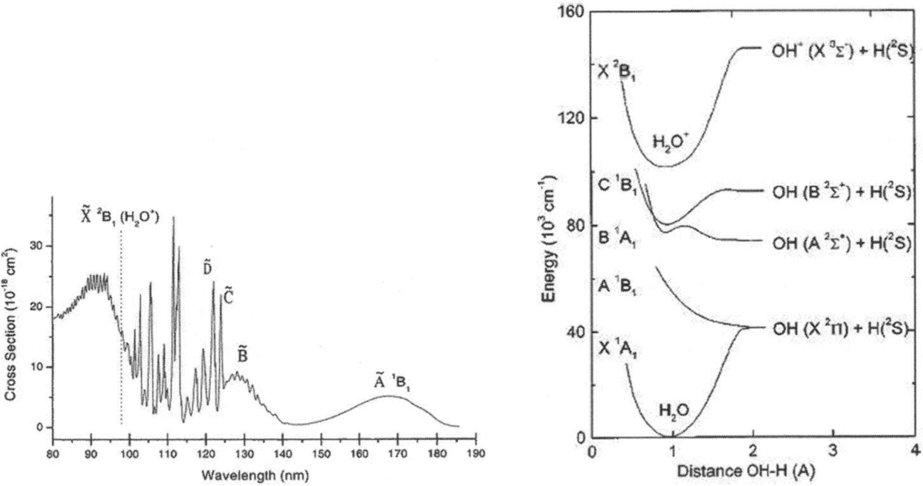

Figure 6.2 Photodissociation of H_2O. Left: Measured photodissociation cross section. Right: Energy levels involved in the photo-processes. See text for details. Figures courtesy of [25, 40]

absorption centered at $\simeq$170 nm. The second peak at $\simeq$130 nm is due to direct dissociation into the $\tilde{B}^1A_1$ state where the weak spectral structure reflects predissociation associated with excitation of the ν_2 bending and ν_1 symmetric modes. The sharp transition at 124 nm is due to predissociation through coupling of $\tilde{C}^1B_1$ with the $\tilde{B}^1A_1$ state. The narrow peak at 122 nm is due to predissociation of the $\tilde{D}^1A_1$ state (not shown) with the $\tilde{B}^1A_1$ state. The broad band around 90 nm is the ionization continuum associated with the $\tilde{X}^2B_1$ state. The spectral structure on this band is due to autoionization from "bound" levels in this state (of the neutral).

With increasing size, the density of states increases rapidly. The initial state can then couple rapidly to other electronic states, leaving the molecule eventually in a highly excited vibrational state of the ground electronic state. Dissociation can then still occur if enough energy accumulates into a specific mode to overcome the reaction enthalpy and any putative reaction barrier. This process is called unimolecular dissociation. We will discuss this in Section 8.2 when the photophysics of large molecules is reviewed.

For a limited number of simple, stable molecules of atmospheric and/or astronomical relevance, absorption cross sections have been measured in the laboratory and are available through electronic data bases (see resources). Photo-dissociation or photoionization yields and product distributions are less well characterized. Theoretical studies are particularly relevant for small radicals. Such studies require acurate electronic potential energy surfaces on which the nuclear motions can be evaluated. Density functional theory can be used to evaluate the ground state potential energy surface and time-dependent density functional theory to calculate the electronic absorption spectra of quite large molecules. Energy levels computed are accurate to about 0.2 eV. While that is not accurate enough for identification purposes of, e.g., diffuse interstellar bands, given the broad excitation spectrum of the interstellar radiation field, absorption rates are generally sufficiently accurate for astronomical purposes.

6.3.2 Photodissociation Rates

In the diffuse interstellar medium, the photodissociation rate is given by,

$$k_{pd} = \int_{\nu_i}^{\nu_H} 4\pi\, \mathcal{N}_{ISRF}(\nu)\, \alpha_{pd}(\nu)\, d\nu, \tag{6.11}$$

where $\mathcal{N}_{ISRF}(\nu)$ is the mean photon intensity of the interstellar radiation field, $\alpha_{pd}(\nu)$ the photodissociation cross section and the integration runs from the dissociation limit to the hydrogen photoionization limit. If there are allowed transitions to dissociative channels available, the cross section is typically 0.01–0.1 Å^2 for dipole allowed transitions in the wavelength region of interest.

Penetration of UV Photons

Inside a cloud, the radiation field will be attenuated and scattered by dust. Because dust absorption and scattering is not gray, the frequency distribution of the FUV radiation field will depend on the depth into the cloud. As a result, photodissociation rates will show a pronounced dependence on the depth in the cloud. For simple geometries, spherical or plane-parallel, this can be evaluated to arbitrary precision once the dust properties have been specified. Here, we will consider a plane parallel slab illuminated from the outside. Scattering is considered to be isotropic with albedo ω (e.g. nonconservative). Dropping for notational simplicity the dependence on the frequency, the radiative transfer equation for the intensity I is,

$$\mu \frac{dI(\tau,\mu)}{d\tau} = I(\tau,\mu) - S(\tau,\mu), \tag{6.12}$$

with τ the optical depth and μ the cosine of the angle. The source function, $S(\tau,\mu) = (\omega/2)\, J(\tau)$, can be written in terms of the mean intensity, $J(\tau) = \int_{-1}^{1} I(\tau,\mu)\, d\mu$. The solution of this equation is given by,

$$I(\tau,\mu) = g(\mu) \exp[-k\tau], \tag{6.13}$$

with

$$g(\mu) \propto (1 + k\mu)^{-1}. \tag{6.14}$$

This results in the following characteristic equation for the roots k,

$$\frac{k}{\log\left[(1+k)/(1-k)\right]} = \frac{\omega}{2}, \tag{6.15}$$

where only one of the roots is physical. This function is plotted in Figure 6.3. The intensity is then given by,

$$I(\tau,\mu) = \frac{C \exp[-k\tau]}{1 + k\mu}, \tag{6.16}$$

where the normalization constant, C, follows from the boundary condition. For an isotropic incident radiation field, we find for the mean intensity,

$$J(\tau) = J(0) \exp[-k\tau]. \tag{6.17}$$

As Figure 6.3 illustrates, the radiation field drops more slowly into the cloud than the extinction optical depth ($(1-\omega)\,\tau$) would indicate but – except for the pure absorption or conservative scattering case – faster than the absorption optical depth implies ($1 > k > 1 - \omega$).

The case for non-isotropic nonconservative scattering in plane parallel or spherical geometry can also be solved to arbitrary precision. In the asymptotic limit of high optical depth, an expression similar to Eq. (6.17) is recovered but the characteristic equation depends now also on the mean cosine of the scattering angle, g. Photo-rates are often presented as,

$$k_V(\lambda) = k(\omega,g)\, \frac{\tau(\lambda)}{\tau_V} = k(\omega,g)\, \frac{A(\lambda)}{A_V}, \tag{6.18}$$

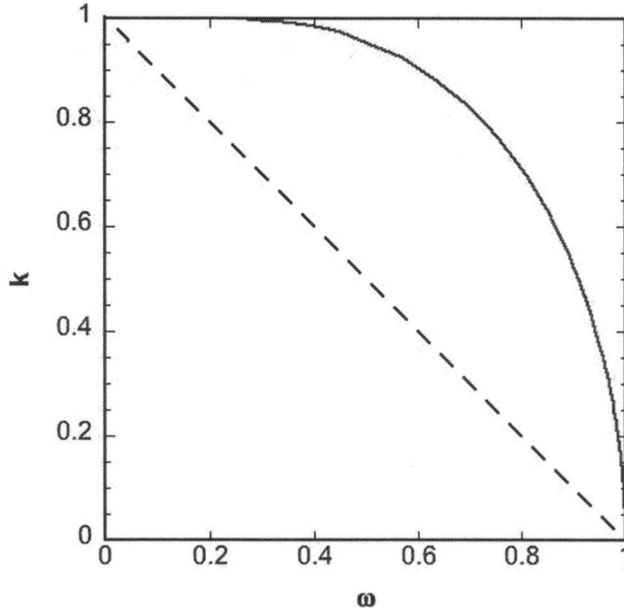

Figure 6.3 Roots, k of the characteristic equation (6.15) as a function of the scattering albedo, ω (full drawn curve). The dashed line gives $k = 1 - \omega$.

where A is the extinction in magnitudes. Results for the asymptotic limit case are presented in Figure 6.4 for often adopted scattering properties for dust in the diffuse interstellar medium. Curve B corresponds roughly to the commonly used model for interstellar dust (see also Section 6.3.2). As these results illustrate, while the extinction rapidly rises toward the far-UV, the mean intensity decays more gradually with wavelength.

Interstellar Radiation Field

The average interstellar radiation field in the diffuse ISM consists of three components, a far-red to near-IR component due to cool K-type giants, a visible component dominated by F0 stars, and a UV component originating in early type stars. Ignoring photospheric absorption bands and jumps, these components can be represented by three black bodies with $\left(W, T_{eff}\right)$ of (10^{-13},4000 K), (10^{-14},7500 K), and (6×10^{-17},25,000 K) with W the dilution factor and T_{eff} the effective temperature of the black body where $I(\nu) = \sum_i W_i B\left(\nu, T_{eff,i}\right)$. The latter component is of most relevance here and this so-called Habing field corresponds to 10^8 photons cm^{-2} s^{-1} with a typical energy of 10 eV or an integrated average intensity of 1.2×10^{-4} erg cm^{-2} s^1 sr^{-1}. More accurate – but fairly complex – representations are available. Often the following approximation is used,

$$\mathcal{N}(E) = 1.658 \times 10^6\, E - 2.152 \times 10^5\, E^2 + 6.919 \times 10^3 E^3 \quad \text{photons/cm}^2\text{/s/sr/eV}, \tag{6.19}$$

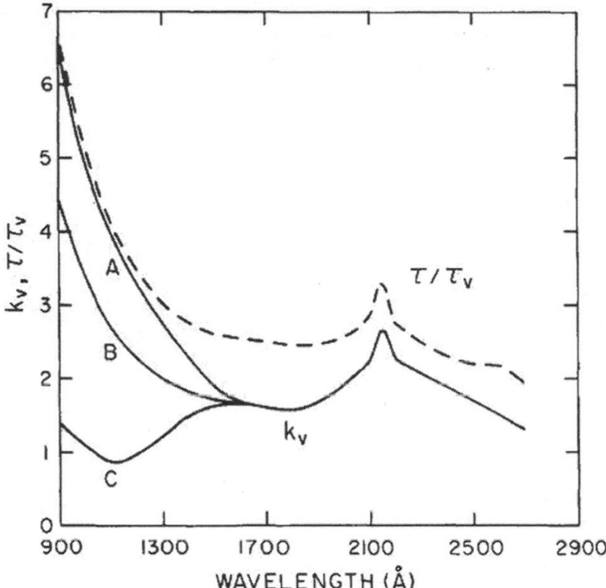

Figure 6.4 The interstellar extinction curve, τ/τ_V, and asymptotic decay rates, k, as a function of wavelength. The three curves labeled, A, B, and C, present three possible solutions for different assumptions concerning the albedo, ω, and mean cosine of the scattering angle, g. Figure taken from [26]

where E is the energy in eV. In radiation density and wavelength units (u_λ), this can be expressed as,

$$\lambda u_\lambda = 6.84 \times 10^{-14} \lambda_{1000}^{-5} \left(31.016 \lambda_{1000}^2 - 49.913\lambda_{1000} + 19.897\right) \quad \text{erg cm}^{-3} \quad (6.20)$$

where λ_{1000} is $\lambda/1000$ Å. In terms of energy density, this expression corresponds to 1.7 Habing fields. This spectrum is shown in Figure 6.5. Realistic stellar spectra are riddled with photospheric absorption lines but as, in general, absorption occurs in a continuum or a multitude of lines, these details are often less relevant.

Dust Properties and UV Penetration

Photo rates depend on the adopted extinction and scattering properties of the dust – which generally dominates the opacity. These dust properties are typically measured in the diffuse ISM and – in particular, the scattering properties – are actually quite uncertain (cf. Figure 6.4) and known to vary systematically from one region to the next. In particular, the properties of dust in dense clouds may have been modified considerably due to the effects of, for example, the growth of ice mantles and coagulation. As illustrated in Figure 6.6, for coagulated grains, extinction may be very gray throughout the visible and UV and photons may be able to penetrate much more deeply than expected for commonly adopted extinction properties.

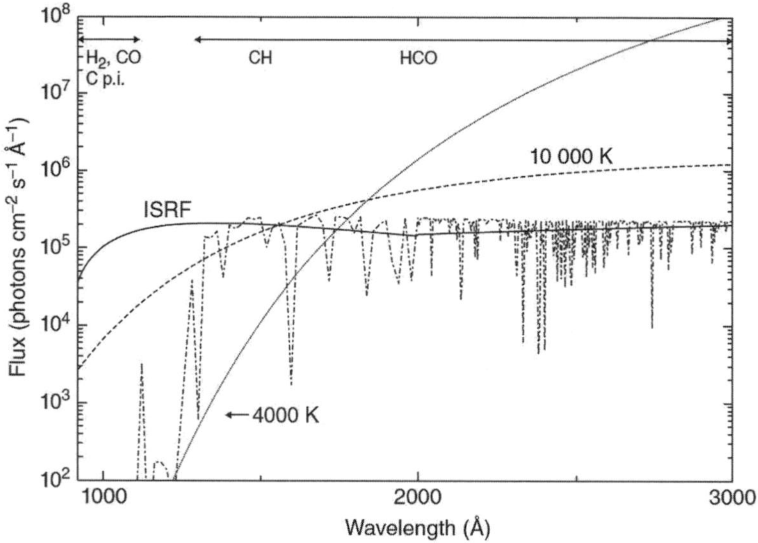

Figure 6.5 The average interstellar radiation field as a function of wavelength. The curve labeled ISRF is given by expression, Eq. (6.19), coupled with the cooler stellar component at longer wavelengths. The curves labeled 4000 and 10,000 K are black bodies with the same integrated intensities between 912 and 2070 Å as the ISRF. The dash-dotted curve is the stellar model for a B9.5 star. Absorption ranges for some relevant molecules are indicated near the top. Figure courtesy of [76]

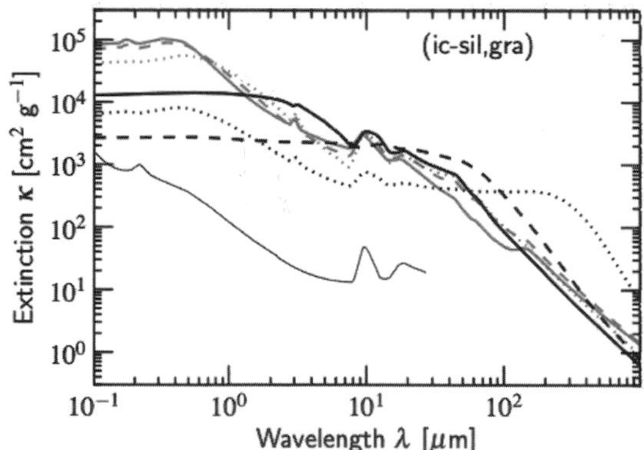

Figure 6.6 An illustration of the effects of coagulation on the interstellar extinction curve. Calculated extinction curves for porous aggregates grown in dense molecular clouds at different times. Coagulation was started using 0.1 μm ice-covered silicate and graphite monomers (in 2:1 ratio). For comparison, the interstellar extinction curve is also shown, arbitrarily shifted down. Figure courtesy [53]

In some cases, absorption of FUV photons by (abundant) atoms or molecules can also be important. This will lead to an extra reduction factor, $\beta(\tau)$, in the expression for the photodissociation rate, with β the probability that a UV photon will not be absorbed by these species along the way to the (optical) depth, τ, from the surface. In this regard, absorption by neutral carbon provides continuum opacity shortward of ≃1100 Å and can be of some importance for, e.g., the shielding of H_2. The optical depth is then given by,

$$\tau(C) = 1.6 \times 10^{-17} N(C), \tag{6.21}$$

with $N(C)$ the column density of neutral atomic carbon and β equals $\exp[-\tau(C)]$. Self-shielding by H_2 will be discussed in Section 6.3.4.

Clumpy Media

The photodissociation rates in Table 6.2 have been evaluated for homogenous plane parallel slabs but values for spherical geometry are also available. In reality, clouds are very inhomogenous and, except for dense cloud cores, any spot within a cloud may have some

Table 6.2 *Some important photo reactions and their rates*[1]

Ionization			Dissociation		
Reaction	k_0 [s^{-1}]	k_V	Reaction	k_0 [s^{-1}]	k_V
CH	7.6 (−10)	3.3	$CH \rightarrow C + H$	9.2 (−10)	1.5
CH_2	1.0 (−9)	2.3	$CH_2 \rightarrow CH + H$	5.8 (−10)	2.0
CH_4	6.8 (−12)	3.9	$CH_4 \rightarrow CH_3 + H$	1.2 (−9)	2.6
			$CH_3 \rightarrow CH_2 + H$	1.2 (−9)	2.6
			$CH^+ \rightarrow C^+ + H$	3.3 (−10)	2.9
			$CH_2^+ \rightarrow CH^+ + H$	1.4 (−10)	2.2
			$CH_3^+ \rightarrow CH_2^+ + H$	1.0 (−9)	1.7
			$CH_3^+ \rightarrow CH^+ + H_2$	1.0 (−9)	1.7
C_2	4.1 (−10)	3.8	$C_2 \rightarrow C + C$	2.4 (−10)	2.6
C_2H_2	3.3 (−10)	3.5	$C_2H_2 \rightarrow C_2H + H$	3.3 (−9)	2.3
			$C_2H \rightarrow C_2 + H$	5.2 (−10)	2.3
			$OH \rightarrow O + H$	3.9 (−10)	2.2
			$OH^+ \rightarrow O^+ + H$	1.1 (−11)	3.5
H_2O	3.1(−11)	3.9	$H_2O \rightarrow OH + H$	8.0 (−10)	2.2
O_2	7.6 (−11)	3.87	$O_2 \rightarrow O + O$	7.9 (−10)	2.1
			$CO \rightarrow C + O$	2.0(−10)	3.5[2]
			$HCO \rightarrow CO + H$	1.1 (−9)	1.1
H_2CO	4.8 (−10)	3.2	$H_2CO \rightarrow HCO + H$	1.0 (−9)	2.2
			$HCN \rightarrow CN + H$	1.6 (−9)	2.7
			$CN \rightarrow C + N$	2.9 (−10)	3.5
			$HC_3N \rightarrow C_2H + CN$	5.6 (−9)	2.2
			$NH_3 \rightarrow NH_2 + H$	1.2 (−9)	2.1

[1]Photoionization and dissociation rates are given by, $k_{pd} = k_0 G_0 \exp[-k_V A_v]$ with $G_0 = 1$ for the average interstellar radiation field in Habing units. [2]Self shielding can be important for CO photodissociation.

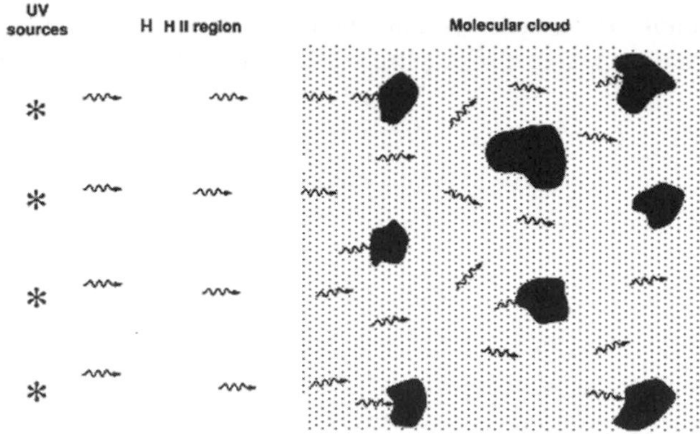

Figure 6.7 Schematic illustration of the penetration of UV photons from the interstellar radiation field into a medium consisting of dense clums embedded in a low density interclump medium. Figure courtesy [51]

sightlines with high optical depth while other sightlines are practically optically thin. FUV photons can then penetrate much deeper than naively expected from the average optical depth (Figure 6.7). The effects of this have been evaluated and the formalism developed in Section 6.3.2 for homogeneous clouds can be applied using effective dust properties.

Consider a cloud consisting of an interclump medium with randomly interspersed clumps. We will limit ourselves here to the case of highly optically thick clumps and a more diffuse interclump medium – $\kappa_0 \gg \kappa_1$ and $\kappa_0 \ell \gg 1$ with κ_i the dust opacity in phase i ($i = 0$ clump & $i = 1$ interclump) and ℓ the clump size. Define the effective opacity, κ_{eff} and albedo, ω_{eff},

$$\kappa_{eff} = \kappa_0 \frac{\kappa_1 \ell + p_0}{\kappa_0 \ell + 1} \tag{6.22}$$

$$\omega_{eff} = \omega \frac{1 + (1 - \omega)\, \kappa_0 \ell \,(\kappa_1 \ell / (\kappa_1 \ell + p_0))}{1 + (1 - \omega)\, \kappa_0 \ell} \tag{6.23}$$

When clumps dominate the penetration of radiation into the cloud, this can be further simplified to,

$$\kappa_{eff} = \kappa_1 + \frac{p_0}{\ell} \tag{6.24}$$

$$\omega_{eff} = \frac{\omega}{1 + (1 - \omega)\, \kappa_0 \ell}. \tag{6.25}$$

The mean intensity inside a clump, $\langle J_0 \rangle$, and in the interclump medium, $\langle J_1 \rangle$, are then given by the homogeneous solution at the effective opacity and albedo,

$$\langle J_i (z) \rangle = C_i \, J_{hom} (\omega_e, \kappa_e, z), \tag{6.26}$$

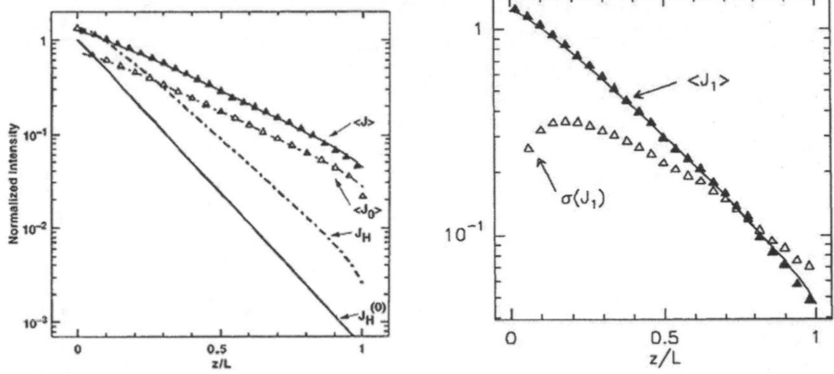

Figure 6.8 Calculated radiation field in a clumpy molecular cloud with an average optical depth of 7.5, a clump optical depth of $\kappa_0 \ell = 2.5$, and a clump filling factor of $p_0 = 0.1$ as a function of fractional depth z/L. The clumps contain 80% of the mass. Left: $< J >$ & $< J_0 >$ average intensity incident on and inside the clump. J_H solution for a homogeneous cloud with the same average optical depth. Right: $< J_1 >$ average intensity in the interclump medium. $\sigma\ (J_1)$ dispersion in the average intensity in the interclump medium. Figure courtesy [9]

where the homogeneous solution is give by Eq. (6.17) and the normalization constants are,

$$C_0 \simeq ((1-\omega)\,\kappa_0 \ell)^{-1} \tag{6.27}$$

$$C_1 \simeq \kappa_0 \ell / \left(\kappa_0 \ell - p_0\right). \tag{6.28}$$

In a clumpy medium, photons find a way deep inside by "flowing" around the clumps. The effects on the average intensity by penetration in a clumpy medium are illustrated in Figure 6.8. The average optical depth for the cloud is 7.5 in this calculation, quite typical for molecular clouds in the Milky Way. Clearly, the UV field incident on and inside a clump at fractional depth z/L is much higher than expected for a homogeneous cloud with the same average optical depth. Moreover, there are large fluctuations in the average radiation field inside the cloud, depending on whether there are "unobstructed" sightlines to the surface. Obviously, photoionization and dissociation rates can be much higher than naively expected. While this approach is relevant for generic studies, detailed Monte Carlo studies have been done for adopted, specific 3-D structures of molecular clouds.

6.3.3 Interstellar Rates

Tables 6.2 and 6.3 list some relevant photoionization and dissociation rates appropriate for semi-infinite, homogenous slabs, tabulated in the form described in Section 6.3.2 and Equation (6.18) transcribed to,

$$k_{pd} = k_0 \exp\left[-k_V A_v\right], \tag{6.29}$$

with k_0 the unshielded rate in the average interstellar radiation field (given by Eq. (6.20)) and A_v the visual extinction due to dust.

Table 6.3 *Some important atomic ionization reactions and their rates*[1]

Species	Ionization potential [eV]	k_0 [s^{-1}]	k_V
C	11.2	3.1 (−10)	3.3
S	10.4	6.0 (−10)	2.2
Mg	7.6	7.9 (−11)	2.1
Si	8.2	3.1 (−9)	2.3
Fe	7.9	2.8 (−10)	2.2
Na	5.1	1.5 (−11)	1.8
K	4.3	3.0 (−11)	1.7

[1]Rates are $k_{pd} = k_0 \exp[-k_V A_v]$.

Perusal of this data illustrates some clear trends. The typical rate of 10^{-9}–10^{-10} s^{-1} in the unshielded interstellar radiation field translates into a lifetime against photodissociation of 10^9–10^{10} s and, as reformation of molecules is generally much slower, molecular abundances are typically low. Only H_2 – where photodissociation occurs in very narrow bands and self-shielding is very important – as well as large species – that are particularly resistant against photodissociation because they can store excitation in vibrational modes and radiate them away (this will be discussed in Chapters 8 and 12) – can build up appreciable abundances in diffuse clouds. Species such as CO and CN, which are very strongly bound, have relatively low photodissociation rates – only the most energetic photons available ($h\nu \simeq 12$ eV) will be able to dissociate these species – and their photo rates drop rapidly into the cloud as high energy photons are most affected by dust extinction. Radicals such as HCO and CH have low binding energies and hence easily photodissociate even deep in a cloud – the binding energy of HCO is only 65 kJ/mol – and there is a weak dependence of the photodissociation rate on depth in the cloud. Photoionization typically takes more highly energetic photons than photo dissociation and, therefore, the rates are somewhat smaller and drop off faster. Finally, note that these rates are for a semi-infinite slab illuminated from one side. If radiation percolates through from the other side, the photo-rates depend also on the total optical depth of the cloud. Expressions for this situation have also been evaluated.

6.3.4 *Self Shielding of H_2*

Self-shielding is of particular importance for H_2 and, to a lesser extent, for CO and N_2, which are dissociated through line absorption, and these lines become readily optically thick. As an illustration, we will discuss here self-shielding of H_2 in some depth. Direct absorption from the singlet ${}^1\Sigma_g^+$ ground state into the repulsive triplet ${}^3\Sigma_u^+$ state of H_2 (Figure 3.11) is forbidden. Photodissociation of H_2 proceeds then through transitions between the ground state, $X^1\Sigma_g^+$, and the excited electronic states, $B^1\Sigma_u^+$ (Lyman) and

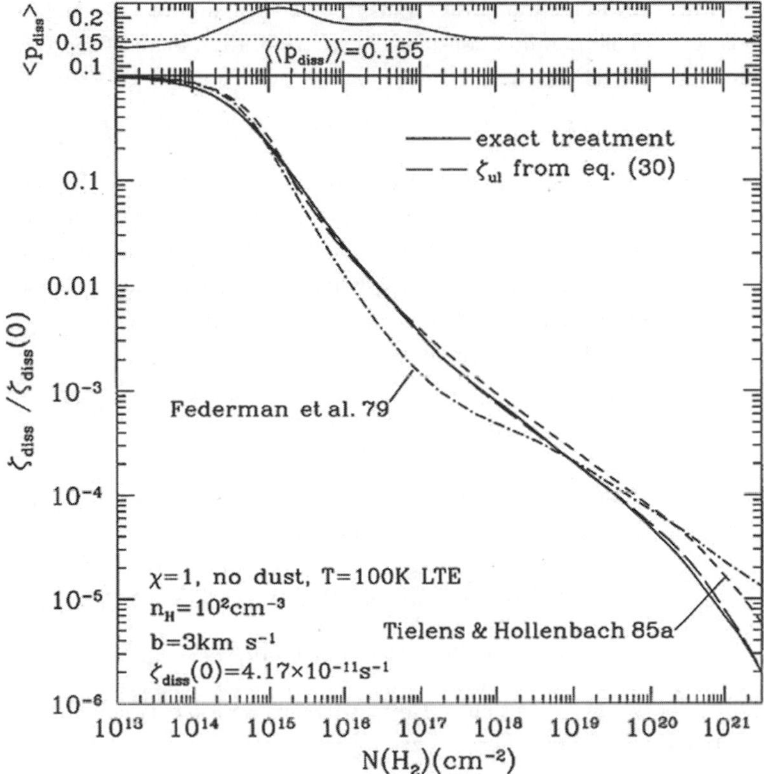

Figure 6.9 Calculated H_2 photodissociation rate – relative to the unshielded rate – as a function of H_2 column density in a cloud. The nomenclature in this figure uses ζ_{diss} rather than k_{uv}, where $\zeta_{diss}(0)$ is the unshielded rate. Simple approximations are taken from [23, 73]. The physical conditions used in these calculations are indicated in the lower left corner (note χ refers to a G_0 of 1.7 Habings, b is the Doppler broadening parameter, n_H is the density of H nuclei, and LTE has been assumed for the rotational population). The top panel gives the average photodissociation probability. Figure courtesy of [20]

$C^1\Pi_u$ (Werner), followed by fluorescence to the vibrational continuum of the ground electronic state – in about 10–15% of the time – which leads to dissociation. So, dissociation is initiated through line absorption in individual ro-vibrational transitions of these electronic states – the Lyman–Werner bands – and these occur in the 912–1100 Å range. The H_2 photodissociation rate follows then from a summation over all lines. When the H_2 column density in an individual level exceeds 10^{14} cm^{-2}, the FUV absorption lines become optically thick, and self-shielding becomes important. The photodissociation rate depends then on the H_2 abundance and level population distribution as a function of depth in the cloud and detailed model calculations have been performed. Results of such a calculation are presented in Figure 6.9.

It is instructive to evaluate the HI/H_2 transition using a simple self-shielding approximation. Consider a plane parallel, constant density slab, illuminated from one side. The photodissociation rate of H_2 is given by,

$$k_{UV}(H_2)\, n\,(H_2) = \beta_{ss}\,(N\,(H_2))\, \exp\left[-\tau_d\right]\, k_{UV}(0)\, n\,(H_2), \tag{6.30}$$

where $n(H_2)$ is the H_2 density, τ_d is the dust optical depth at 1000 Å, $N(H_2)$ is the H_2 column density into the cloud, $k_{UV}(0)$, is the unshielded photodissociation rate ($\simeq 4 \times 10^{-11}\, G_o$ s^{-1}), and $\beta_{ss}\,(N\,(H_2))$ is the self-shielding factor. This self-shielding factor is the reverse of the escape probability, which we encountered in Chapter 4 as the probability that a photon created at depth, τ, makes it to the surface and escapes. The self-shielding factor can be approximated by,

$$\beta_{ss} = \left(\frac{N\,(H_2)}{N_0}\right)^{-0.75}, \tag{6.31}$$

in the column density range $N_0 = 10^{14} \lesssim N\,(H_2) \lesssim 10^{21}$ cm^{-2}. This dependence of β_{ss} on the column is somewhat steeper than the dependence expected for heavily saturated lines on the square root portion of the curve of growth due to the effects of line overlap.

Assuming steady state and a H_2 formation per H-atom given by $k_d(H_2)n$, we can write,

$$k_{UV}(H_2)\, n\,(H_2) = k_d(H_2) n n_H. \tag{6.32}$$

Neglecting dust opacity ($N \lesssim 10^{21}$ cm^{-2}) and using the relation between total density, and the density of H and H_2, this yields,

$$n\,(H_2) = n \left(\frac{k_{uv}\, N_o^{3/4}}{k_d n\, N(H_2)^{3/4}} + 2\right)^{-1} \tag{6.33}$$

Realizing that $dN(H_2)/dz = n(H_2)$, we find,

$$\frac{4k_{uv}\,(0)\;\; N_o^{3/4}}{k_d\, n}\;\; (N\,(H_2))^{1/4} + 2N\,(H_2) = N, \tag{6.34}$$

where N is the column of hydrogen nuclei ($N = nz$). Ignoring the second term on the left-hand side, we find,

$$N\,(H_2) = \left(\frac{k_d\, n}{4k_{uv}\,(0)}\right)^4 \left(\frac{N}{N_0}\right)^4 N_0, \tag{6.35}$$

and for the abundance of H_2,

$$X\,(H_2) = \frac{dN\,(H_2)}{dN} = \frac{4N\,(H_2)}{N} = 4\left(\frac{k_d n}{4k_{uv}\,(0)}\right)^4 \left(\frac{N}{N_o}\right)^3. \tag{6.36}$$

Detailed evaluation of the H_2 abundance as a function of column density into clouds illuminated by different FUV fields is shown in Figure 6.10. Because,

$$\frac{k_d(H_2)\, n}{4k_{UV}\,(0)} = 1.9 \times 10^{-7}\, \frac{n}{G_o}, \tag{6.37}$$

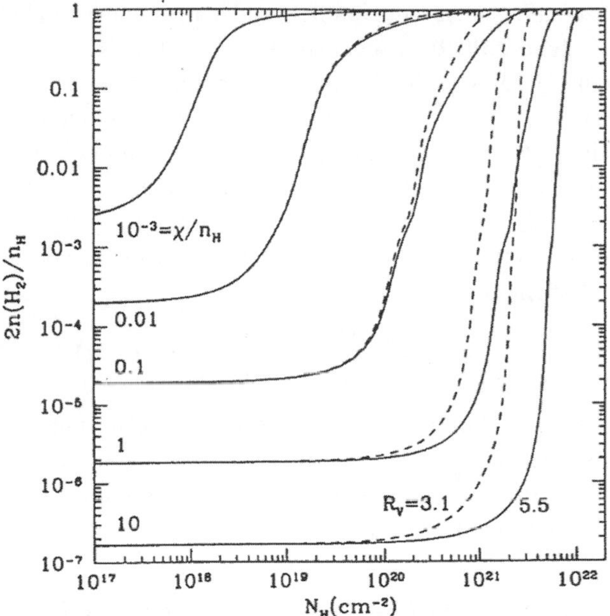

Figure 6.10 Calculated H_2 abundances as a function of total column density. The total gas density and temperature have been set equal to 200 cm^{-3} and 200 K, respectively. The different curves are labeled by the ratio of incident FUV field ($\chi = 1.7G_0$) and the density. The dashed ($\tau_d(1000Å) = 2 \times 10^{-21}N$) and solid ($\tau_d(1000Å) = 6 \times 10^{-22}N$) lines distinguish two different FUV dust extinction cross sections used. Figure courtesy of [20]

the structure of the HI/H_2 transition zone will be regulated by G_o/n: the equivalent of the ionization parameter in studies of ionized gas. While this derivation has limitations built in ($N_0 = 10^{14}$ cm$^{-2} \lesssim N_{H_2} \lesssim 10^{21}$ cm^{-2} and $N \lesssim 10^{21}$ cm^{-2}), it illustrates the sharp transition in the molecular hydrogen abundance with depth in the cloud (Figure 6.10) due to self-shielding (e.g. $n\ (N(H_2) \propto N^4$ and $X(H_2) \propto N^3$).

Equating the HI/H_2 interface or dissociation front with $X(H_2) = 1/4$, we obtain the hydrogen nucleus column density N_{DF} into the cloud where the gas is half molecular and half atomic,

$$N_{DF} \simeq 3.7 \times 10^{22} \left(\frac{G_0}{n}\right)^{4/3} \text{cm}^{-2}. \tag{6.38}$$

This corresponds to a column of 8×10^{19} cm^{-2} for a diffuse cloud with a density of 100 cm^{-3}. Dust opacity becomes important when $N_{DF} \gtrsim 5 \times 10^{20}$ cm^{-2}, or when,

$$G_0/n \gtrsim 4 \times 10^{-2}\ \text{cm}^3. \tag{6.39}$$

Self-shielding alone dominates H_2 dissociation and hence the location of the HI/H_2 transition when $G_0/n \lesssim 4 \times 10^{-2}$ cm^3 (Figure 6.10). This includes diffuse clouds and molecular clouds exposed to the interstellar radiation field and dense clumps in PDRs (cf. Chapter 9) with higher FUV fluxes. PDRs associated with bright FUV sources typically have $G_0/n \sim 1$ cm^3 and the location of the HI/H$_2$ transition is then dominated by dust absorption and typically occurs at $A_v \simeq 2$ ($N \simeq 4 \times 10^{21}$ cm^{-2}). At that point, dust has reduced the H$_2$ photodissociation rate sufficiently that an appreciable column of H$_2$ can build up, H$_2$ self-shielding takes over, and the HI/H$_2$ transition will still be very sharp.

The powerlaw fit to the self-shielding function (Eq. (6.31)) is very convenient and quite accurate (factor of 2) over the range $10^{14} < N\,(\mathrm{H_2}) < 5 \times 10^{20}$. A more accurate analytical function for the self-shielding factor is,

$$\beta_{ss} = \frac{0.965}{(1 + x/b)^2} + \frac{0.035}{(1 + x)^{1/2}} \exp\left[-8.5 \times 10^{-4}\,(1 + x)^{1/2}\right], \tag{6.40}$$

with $x = N\,(\mathrm{H_2})\,/5 \times 10^{14}\,\mathrm{cm^{-2}}$ and b the Doppler parameter in units of km/s. The self-shielding factor does depend to some extent on the excitation of H$_2$, but the main uncertainty resides in the poorly known dust extinction properties at the shortest wavelengths as illustrated by the dashed curves in Figure 6.10.

6.3.5 Cosmic Ray-Produced UV Radiation Field

Some FUV photons will be present even deep within dense cloud cores through processes following cosmic ray ionization of H$_2$. In particular, cosmic ray interaction with H$_2$ (or H) will produce (a first generation secondary) electrons with a typical energy of 19 eV. These electrons will lose their energy in collisions with the gas and, as the electron fraction is low and energy loss in an elastic collision with H$_2$ is small, mainly through ionization (producing further secondary electrons), dissociation, and through excitation of H$_2$ in the Lyman–Werner bands, viz,

$$H_2 + e \rightarrow H_2^+ + e + e \tag{6.41}$$

$$\rightarrow H_2^* + e \tag{6.42}$$

$$\rightarrow H + H + e, \tag{6.43}$$

where $\mathrm{H_2^*}$ refers to electronically excited H$_2$. Once the electron energy drops below some 10 eV, further ionization, dissociation, and electronic excitation are no longer important and at that point elastic collisions as well as vibrational and rotational excitation will take over. Here, we are concerned with the electronic excitations that are immediately ($A \sim 10^8$ s^{-1}) followed by radiative decay and produce an internal UV field. Because high energy cosmic rays have a long penetration depth into a molecular cloud, this leads to a "constant" source function and FUV photon intensity in the cloud. The excitation rate of the Lyman–Werner bands can be estimated by convolving the energy distribution of the secondary electrons produced by the primary cosmic ray with the collisional excitation cross sections. This yields a photon production rate of about $\epsilon_{CR}\zeta_{CR}$ with ζ_{CR} the primary cosmic ray ionization rate per H$_2$ molecule and ϵ_{CR} the number of photons per ionization ($\simeq$0.3). The mean intensity of the photon field, $\mathcal{N}_{CR}$, can then be found by equating this production rate to the photon destruction rate due to dust, $4\pi \, \overline{\kappa}_d \, \mathcal{N}_{CR}$, where $\overline{\kappa}_d$ is the mean dust absorption coefficient for these photons, for which we take the standard value in the diffuse ISM of $\overline{\kappa}_d \simeq (1-\omega)\, \sigma_d n_o \simeq 10^{-21} n_o$ with ω the albedo of the dust σ_d the dust extinction cross section per H-nucleus (these are averages) and n_o the density of hydrogen nuclei. This leads to,

$$4\pi \, \mathcal{N}_{CR} \simeq 1.5 \times 10^{3} \left(\frac{\zeta}{10^{-17}\, \mathrm{s}^{-1}} \right) \quad \text{photons cm}^{-2}\, \mathrm{s}^{-1}. \tag{6.44}$$

Note that this is independent of the density as production rate of photons scales with the density but the mean free path of the photons scales inversely with the density. We also ignored here direct radiative transitions of excited H_2 to low lying rotational levels in the ground vibrational state of the ground electronic state. Those photons will be resonantly scattered by H_2 and may eventually decay through transitions to the vibrational continuum of the ground electronic state (in about ten scatterings).

With a typical cosmic ray ionization rate of 3×10^{-17} s^{-1} per H_2 molecule, this mean intensity corresponds to about 10^{-4} of the average interstellar radiation field. Detailed calculations for the radiation field have been made. The spectrum shows numerous individual lines superimposed on a continuum due to transitions to the continuum of the ground electronic state and to transitions between the $a\,^3\Sigma_g^+$ and the $b\,^3\Sigma_u^+$ states (Figure 6.11). The details of the radiation field will depend somewhat on the rotational excitation in the ground electronic state but except for CO – whose dissociation will be controlled by line overlap – this detail will wash out in the derived photodissociation rates.

The photodissociation rate of species, i, due to these photons is,

$$k_{pd}\,(CR, m) = 4\pi \int \sigma_{pd,i}\,(\nu)\, \mathcal{N}_{CR}\,(\nu)\, d\nu \tag{6.45}$$

$$= \int \frac{\sigma_{pd,i}\,(\nu)}{(1-\omega)\, \sigma_d\,(\nu)} \epsilon\,(\nu)\, \zeta_{CR} d\nu, \tag{6.46}$$

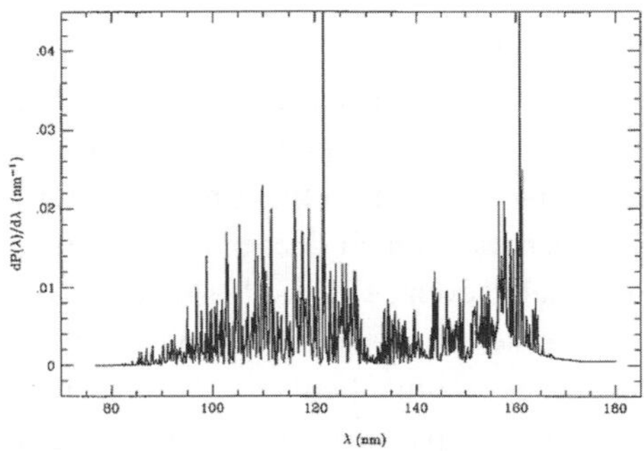

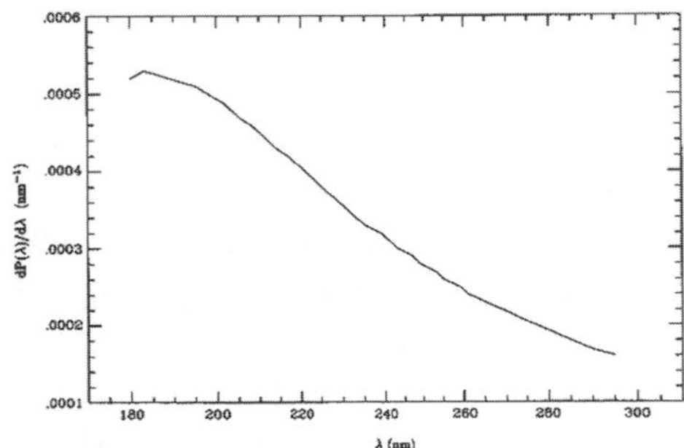

Figure 6.11 Calculated radiation field due to the excitation of H_2 by cosmic ray produced electrons. Left: The number of photons emitted in bins with a size of 0.1 nm between 75 and 175 nm. Right: The emission between 175 and 300 nm. A 3/1 $J = 1/J = 0$ rotational population has been assumed. Figure courtesy of [36]

Table 6.4 *Cosmic ray driven photo reaction rates*[1]

Species	p_i	Species	p_i
OH	1018	H_2CO	5320
H_2O	1942	CH_3OH	9344
O_2	1502	CH_3OCH_3	3428
CO	14[2]	C_2H_2	10310
CO_2	3416	HCN	6228
NH_3	3712	HC_3N	9512
CH_4	4678	CH_3CH_2OH	8614

[1]Photodissociation rate coefficients given as $k_{pd} = p_i\,\zeta_{CR}$.
[2]For CO, self shielding rather than dust is the limiting factor.

where $\sigma_{pd,i}(\nu)$ is the photodissociation cross section of species i and $\epsilon(\nu)$ is the production rate of photons with frequency ν per cosmic ray ionization. So, essentially, the photodissociation rate is set by the competition between the molecule under consideration and the dust for these cosmic-ray produced photons. We can then define the coefficients for dissociation of molecule m by cosmic ray-produced photons, p_m,

$$k_{pd}(CR,i) \equiv p_i\,\zeta_{CR}. \tag{6.47}$$

Table 6.4 provides some typical values for these coefficients. These are evaluated for dust properties appropriate for the diffuse ISM and they could be much larger if dust has coagulated inside dense cores (Figure 6.6).

6.3.6 Summary

Photodissociation is an important process in diffuse clouds with a rate that is comparable to ion–molecule reaction rates (see below). Dust shielding inside dense clouds will limit photodissociation, but the effects of this are difficult to quantify. In particular, the radiation field incident on clumps deep within a cloud can be much higher than naively expected. In dense cores, coagulation may also allow UV photons to penetrate much deeper than the (submm derived) dust column density would indicate. Finally, even deep inside shielded cloud cores, cosmic rays will produce a UV field. In this case, the resulting photodissociation rates depend inversely on the dust absorption properties and this field can be substantial. These effects are illustrated for H_2O in Figure 6.12. Thus, deep inside a homogeneous cloud ($A_V > 5$ mag), an H_2O molecule will be dissociated in about 5×10^5 yr. In a clumpy cloud, the extinction within the clump has to be of that order before the H_2O lifetime reaches a value of 5×10^5 yr. If dust coagulation is important, the lifetime of molecules against photo dissociation might be as short as 5×10^4 yr.

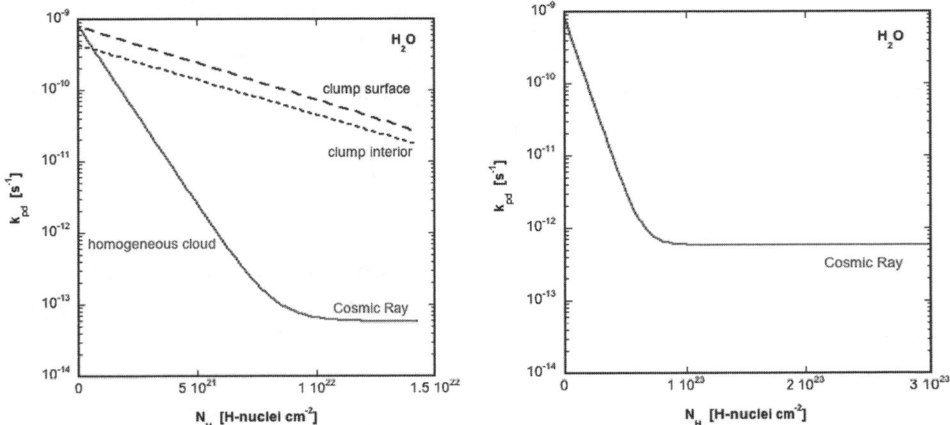

Figure 6.12 Calculated photodissociation rates for H_2O. Left: A comparison of the photo dissociation rate for a homogeneous cloud and a clumpy cloud (see Figure 6.8). Right: The photodissociation rate in cloud core where dust extinction cross sections have been reduced by a factor 10 due to coagulation (see Figure 6.6). Note the difference in column density scale between the two panels. The equivalent (average) visual extinctions through these two clouds are 7.5 and 16, respectively.

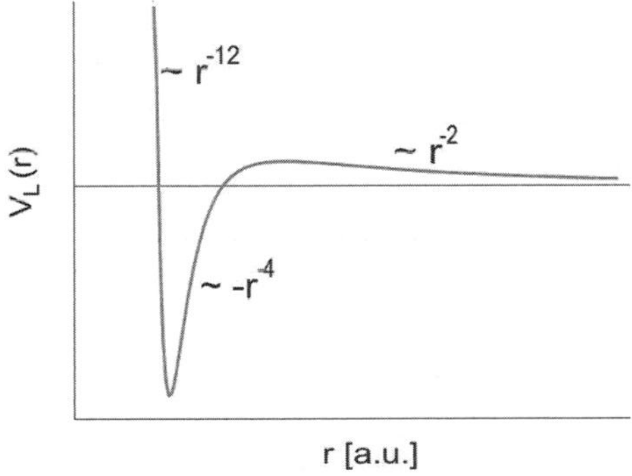

Figure 6.13 Schematic interaction potential for the collision of an ion with a neutral. The centrifugal barrier and the attractive potential due to the induced dipole moment are evident. Upon closest approach, the repulsive Lennard–Jones potential kicks in.

6.4 Ion–Molecule Reactions

Exothermic ion–molecule reactions as a rule occur rather rapidly because the strong polarization-induced interaction potential focuses the reactants together and the enhanced collision energy can be used to "overcome" any activation energy involved. The reaction rate is set by the rate of collisions where the reactants come close enough for a reaction to occur ($\simeq 2$ Å). The electric field, $\mathcal{E}$, of an ion with charge, q, induces a dipole moment, p, in the neutral,

$$p = \alpha \mathcal{E} = \frac{\alpha q^2}{r}, \tag{6.48}$$

with α the polarizablity of the neutral. In the ISM, we are typically concerned with single charges and $q = e$. The interaction potential is then,

$$V(r) = -\frac{1}{2}\frac{\alpha e^2}{r^4} + \frac{L}{2\mu r^2}, \tag{6.49}$$

where the second term represents the centrifugal barrier with L ($= \mu v_0 b$) the angular momentum, μ the reduced mass, v_0 the initial center of mass velocity, and b the impact parameter. The initial kinetic energy is thus, $E = \mu v_0^2/2$. The potential can also be written as

$$V(r) = -\frac{1}{2}\frac{\alpha e^2}{r^4} + E\left(\frac{b}{r}\right)^2. \tag{6.50}$$

This potential is sketched in Figure 6.14. To get some feeling for the values involved, the centrifugal barrier for an impact parameter of 10 Å with a velocity of 1 km/s is $\simeq$0.01 eV while the well depth associated with the induced dipole is $\simeq$1 eV. The barrier in this potential equals,

$$V_{max} = \frac{E^2 b^4}{2\alpha e^2}. \tag{6.51}$$

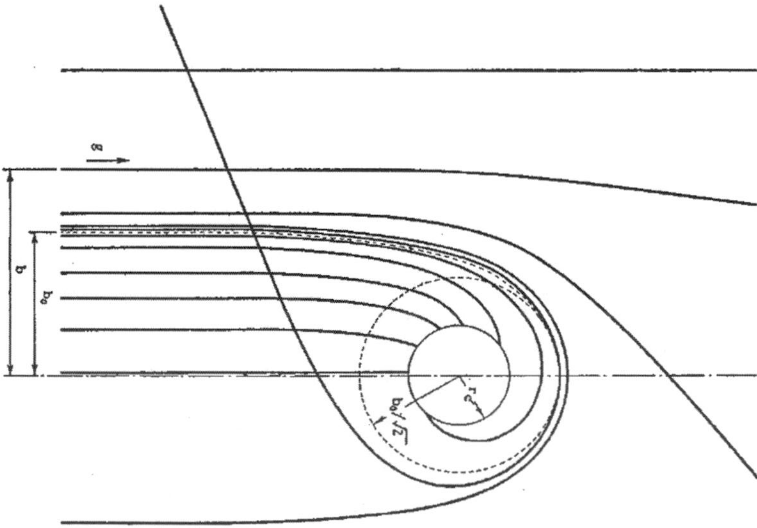

Figure 6.14 Trajectories as a function of impact parameter b. The dotted trajectory comes in at the critical impact parameter (here labeled as b_0). Only trajectories with an impact parameter less than this will get close enough for reaction to occur. Reproduced from [34], with the permission of AIP Publishing

Table 6.5 *Polarizabilities*

Species	α [10^{-24} cm^{-3}]	Species	α [10^{-24} cm^{-3}]
H	0.667	H_2	0.79 (0.93,0.72)
HD	0.79	D_2	0.78
C	1.76	CO	1.95 (2.60,1.63)
N	1.13	N_2	1.76 (2.38,1.45)
O	0.802	O_2	1.60 (2.35,1.21)
H_2O	1.49	NH_3	2.10
CH_3OH	3.32 (4.09,3.32,2.65)	CH_3CH_2OH	5.08 (5.76,4.98,4.5)
H_2CO	2.75 (2.76, 2-76, 1.83)	CH_3OCH_3	5.24 (6.38,4.94,4.39)
CH_4	2.6	C_2H_2	3.487 (5.12, 2.43)
HCN	2.59	CH_3CN	4.48 (5.74,3.85)
Pyrene	28.2	coronene	42.5
Circumcoronene	125 (168,168,37)	C_{60}	75-83

Thus, for a given collision energy E, when $b \leq b_{crit}$, the reactants will get close enough for reaction to (potentially) occur while when $b > b_{crit}$ the reactants will just scatter where b_{crit} is given by,

$$b_{crit} = \left(\frac{2\alpha e^2}{E}\right)^{1/4}. \tag{6.52}$$

In a hard sphere collision model, this results in a cross section,

$$\sigma\left(v\right) = \pi b_{crit}^2 = \pi \left(\frac{2\alpha e^2}{E}\right)^{1/2}. \tag{6.53}$$

The cross section scales inversely, with v canceling the larger collision rate at higher v and, hence, the so-called Langevin rate coefficient, k_L, will be independent of T; viz.,

$$k_L = \int_0^\infty v\sigma\left(v\right) f\left(v\right) dv = 2\pi e \sqrt{\frac{\alpha}{\mu}}. \tag{6.54}$$

Some relevant polarizabilities are given in Table 6.5 and, typically, $k_L \simeq 2 \times 10^{-9}$ cm^3 s^{-1} with no temperature dependence (cf. Table 6.6).

If the ion–molecule reactions involves a neutral species with a permanent dipole, then the ion–dipole interaction has to be taken into account and the reaction rate coefficient can be much larger than the Langevin rate coefficient. The potential energy is now,

$$V\left(r\right) = -\frac{1}{2}\frac{\alpha e^2}{r^4} - \frac{e\mu_D}{r^2}\cos\theta + \frac{L}{2\mu r^2}, \tag{6.55}$$

Table 6.6 *Ion–molecule reactions*

Reaction	α [cm^3 s^{-1}]	β
$H_2^+ + H_2 \rightarrow H_3^+ + H$	2.1 (−9)	
$H_3^+ + O \rightarrow OH^+ + H_2$	1.2 (−9)	
$H_3^+ + CO \rightarrow HCO^+ + H_2$	1.7 (−9)	
$H_3^+ + OH \rightarrow H_2O^+ + H_2$	1.5 (−9)	5.5
$H_3^+ + H_2O \rightarrow H_3O^+ + H_2$	1.7(−9)	5.4
$OH^+ + H_2 \rightarrow H_2O^+ + H$	1.1 (−9)	
$H_2O^+ + H_2 \rightarrow H_3O^+ + H$	6.1 (−10)	
$C^+ + OH \rightarrow CO^+ + H$	9.2 (−10)	5.5
$C^+ + H_2O \rightarrow HOC^+ + H$	2.1 (−9)	
$CO^+ + H_2 \rightarrow HCO^+ + H$	2.0 (−9)	
$He^+ + CO \rightarrow C^+ + O + He$	1.6 (−9)	
$He^+ + O_2 \rightarrow O^+ + O + He$	1.0 (−9)	
$He^+ + H_2O \rightarrow OH^+ + H + He$	5 (−10)	5.4
$He^+ + H_2O \rightarrow OH + H^+ + He$	5 (−10)	5.4
$He^+ + H_2O \rightarrow H_2O^+ + He$	5 (−10)	5.4
$He^+ + OH \rightarrow O^+ + H + He$	1.41 (−9)	5.5

[a]Reaction rate coefficients are either of the form, $k = \alpha$ or $k = \alpha \left(0.62 + 0.4767\beta\,(T/300)^{-1/2}\right)$.

with μ_D the dipole moment and θ the angle between the dipole and r. The evaluation of the interaction will now have to include an average over the angle. For a locked-in dipole, this can be solved and yields,

$$k_D = 2\pi e \left(\sqrt{\frac{\alpha}{\mu}} + \mu_D \sqrt{\frac{2}{\mu \pi k T}} \right). \tag{6.56}$$

At low tempertures, this can be as large as 10^{-7} cm^3 s^{-1} for a strong dipole. This is, of course, an extreme, as at large distances, the dipole will be freely rotating and the Langevin rate would be more applicable. In the literature, you will find,

$$k_D = 2\pi e \left(\sqrt{\frac{\alpha}{\mu}} + c\,\mu_D \sqrt{\frac{2}{\mu \pi k T}} \right), \tag{6.57}$$

where the parameter c (between 0 and 1) contains the details of the interaction. Defining $x = \mu_D / \sqrt{2\alpha k T}$, numerical studies give,

$$\frac{k_D}{k_L} = 0.4767x + 0.62000 \qquad x \geq 2 \tag{6.58}$$

$$= \frac{(x + 0.5090)^2}{10.526} + 0.9754 \qquad x < 2 \tag{6.59}$$

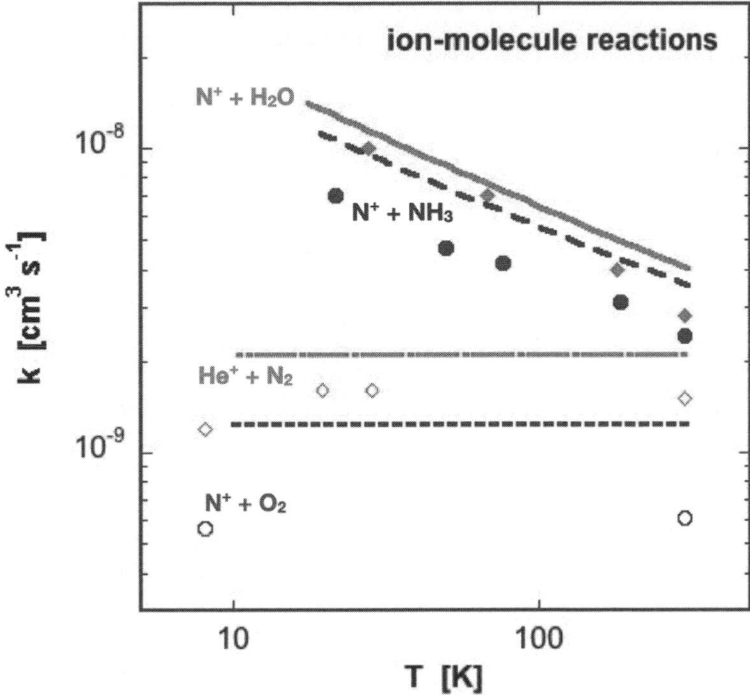

Figure 6.15 Comparison of measured ion–molecule rate coefficients with the theoretical estimates of Eqs. (6.54, 6.57, and 6.58). Experimental data taken from [45, 46, 61, 62]. Figure adapted from [70]

where the inverse temperature dependence reflects that at low temperatures rotation has less of an "averaging" effect.

Experimental techniques relevant for ion–molecule reactions have been discussed in Section 2.4.2. Some typical measurements are compared to these simple equations in Figure 6.15. The results show good agreement, including the temperature behavior of the reactions. One reason for this good agreement is that the centrifugal barrier is located at relatively large distances – at 10 K, $b_{crit} \simeq 20$ Å – much further out than where bond formation through electron exchange will become important ($\simeq 2$ Å) and the Langevin approach based on a centrifugal barrier seems reasonable.

Typically, $k_L \simeq 2 \times 10^{-9}$ cm^3 s^{-1} with no temperature dependence for neutral species with no permanent dipole moment and a weak inverse temperature dependence for neutral species with a permanent dipole moment (cf. Table 6.6). This is some two orders of magnitude faster than neutral–neutral reactions, even disregarding the activation barriers generally involved in the latter. It is clear then that a small amount of ionization can be very effective in driving interstellar chemistry.

Proton transfer reactions from one species to another are of particular relevance. These will occur if the proton affinity of the "receptor"-species is larger than that of

Table 6.7 *Proton affinities*[a]

Species	PA[a] [kJ mole^{-1}] [c]	Species	PA[b] [kJ mole^{-1}] [c]
N	405	H_2CO	712.9
O_2	421	HCN	712.9
H_2	422.3	CH	731
O	485	HCOOH	742
N_2	493.8	C_6H_6	750.4
CH_4	534.5	HC_3N	751.2
CO_2	540.5	C_2H	753
NH	589	CH_3OH	754.3
OH	593.2	NH_2	769
CO	594	HNC	772.3
C	623	CH_3CH_2OH	776.4
C_2H_2	641.4	CH_3CN	779.2
S	656	CH_3OCH_3	792
CH_3	673	Si	836
SH	690	CH_2	836
H_2O	691	NH_3	853.6
C_2	694	C_{60}	855–866
H_2S	705	$C_{24}H_{12}$	861.3

[a]In order of increasing proton affinity. [b]The value of the proton affinity is the negative of the enthalpy change in the proton tranfer reaction involving the given neutral species. [c]Conversion factor to eV/molecule and to K/molecule multiply by 1.04×10^{-2} and 1.2×10^{2}, respectively.

Table 6.8 *Some sample charge transfer reactions*[a]

Reaction	k [cm^3 s^{-1}]	Reaction	k [cm^3 s^{-1}]
$H^+ + O \rightarrow H + O^+$	4. (−10)	$He^+ + N_2 \rightarrow He + N_2^+$	2. (−9)
$H^+ + CO_2 \rightarrow H + CO_2^+$	1. (−10)	$He^+ + N_2 \rightarrow He + N + N^+$	8. (−10)

the "donor"-species. Table 6.7 provides a compilation of proton affinities for some astrophysically relevant species taken from the NIST data base. If the energetics are favorable, the rates for proton transfer reactions are well reproduced by the Langevin or average dipole orientation theory.

6.4.1 Charge Transfer Reactions

Charge transfer involving an atom can be important in setting the ionization balance of clouds. The charge exchange reaction between O and H^+ – which is nearly resonant – is particularly important in diffuse clouds because it jump starts oxygen chemistry. Charge exchange reactions are a special example of ion–molecule reactions. If the reaction only proceeds when the reactants approach each other closely, the centrifugal barrier has to be overcome and we arrive at the Langevin rate. In principle, electron transfer can occur at larger distances (up to 6 Å). Like other ion–molecule reactions, rates are of the order of 10^{-9} cm^3 s^{-1} in the exothermic direction, when there is an energy level in the product ion that is resonant (within ~0.1 eV) with the recombination energy of the incoming ion and the Franck–Condon factor connecting the upper and lower states is large.

The importance of the charge exchange reaction between O and H^+ has already been mentioned. Because the ionization potentials are exceedingly close, this reaction has a large rate coefficient. Charge transfer reactions involving other neutral atoms are rarely important. Charge transfer reactions between atoms and molecular species will have larger cross sections because of the larger number of electronic states available. The charge transfer process may leave the molecular cation in an excited electronic state from which it may dissociate in processes akin to photodissociation (of a cation) or dissociative recombination (of a doubly charged cation), possibly after radiative decay. As an example, He^+ is in accidental resonance with the $v = 3$, $K = 36$ and $v = 4$, $K = 4$ states of the $C^2\Sigma_u^+$ electronic state of N_2^+. These states can radiatively relax to form ground state N_2^+ or predissociate through the ${}^4\Pi_u$ state to N(^4S) and N^+(^3P).

6.5 Neutral–Neutral Reactions

Neutral–neutral reactions are reactions of the type,

$$A + B \rightleftharpoons C + D + \Delta E, \tag{6.60}$$

where it is assumed that the reaction is exothermic to the right. Thermochemistry has been discussed in Section 5.2. These reactions typically possess appreciable activation barriers because of the necessary bond breaking associated with the molecular rearrangement. The forward and backward reaction rates, k_f, and k_b, are related through the equilibrium constant, K (cf. Chapter 5, Eq. 5.33),

$$K = \frac{k_f}{k_b} \exp\left[\Delta E/kT\right]. \tag{6.61}$$

Ignoring for the moment the activation energy involved, the forward rate can be estimated from the classical "hard-collision" cross section; i.e. the collision where the impact parameter is small enough that the attractive interactive forces can overcome the angular momentum of the collision. For neutral–neutral collisions, the attractive interaction is due to van der Waals forces, and we will adopt here the London dispersion equation,

$$V_{vdw}(r) = -\frac{3}{2}\frac{\alpha_1\alpha_2}{r^6}\frac{I_1 I_2}{I_1 + I_2}, \tag{6.62}$$

with α_i the polarizability and I_i the ionization potential of the species involved. The potential also includes a centrifugal barrier (cf. Eq. (6.50)),

$$V(r) = -\frac{C_6}{r^6} + E\left(\frac{B}{r}\right)^2, \tag{6.63}$$

where the constant, C_6, is given by, $C_6 = 1.5\alpha_1\alpha_2/\left(1/I_1 + 1/I_2\right)$. For a given energy, this leads to a cross section,

$$\sigma(v) = \pi b_{crit}^2 = \frac{3\pi}{2^{2/3}}\left(\frac{C_6}{E}\right)^{1/3}, \tag{6.64}$$

and a rate,

$$k_{nn} = \int_0^\infty v\sigma(v) f(v)\, dv \simeq 8.6\frac{C_6^{1/3}}{\mu^{1/2}}(kT)^{1/6}. \tag{6.65}$$

Inserting some typical values ($\mu = 12$ amu, $I_i = 10$ eV, $\alpha_i = 10^{-24}$ cm^3), we find,

$$k_{nn} \simeq 10^{-9}\left(\frac{T}{300\,\text{K}}\right)^{1/6} \text{cm}^3\ \text{s}^{-1}. \tag{6.66}$$

The London C_6 coefficient in the Lennard–Jones potential tends to overestimate the interaction energy somewhat. Nevertheless, if there is no activation barrier, neutral–neutral can be very efficient. If there is an activation barrier involved, this reaction rate has to be multiplied by the Boltzman factor, $\exp\left[-E_a/kT\right]$. Even a modest barrier of 1000 K makes a reaction prohibitive at the 100 K typical for the diffuse ISM, let alone at the 10 K characteristic of molecular clouds. Such neutral–neutral reactions can be of importance, though, when the gas is warm; e.g. in stellar ejecta, in Hot Cores associated with protostars, in dense photodissociation regions associated with luminous stars, or in shocks.

Over the last decade it has been realized that neutral–neutral reactions between radicals or of radicals with unsaturated species can be fast even at low temperatures. At high temperatures, the rate coefficient shows the effect of an inner transition barrier associated with bond formation and a positive temperature dependence (Figure 6.16). At low temperatures, on the other hand, an outer, centrifugal barrier on the long range interaction potential surface dominates, leading to the formation of a van der Waals complex that can either dissociate or lead to reaction. At low temperatures, a negative temperature dependence may then result (Figure 6.16). As chemical reactions entail a "transfer" of an electron from the molecule to the radical, a small energy difference between the ionization potential of the molecule and the electron affinity of the radical is indicative of a potential fast reaction at low temperatures. Essentially, when this energy difference is small, the long range attractive forces may serve to "submerge" the activation barrier associated with avoided curve crossing. Data for some relevant reactions – summarized in Figure 6.17 – serve to illustrate this point. Perusing this data, when the difference in the ionization potential and electron affinity is less than about 9 eV, the rate coefficient increases with decreasing temperature [69].

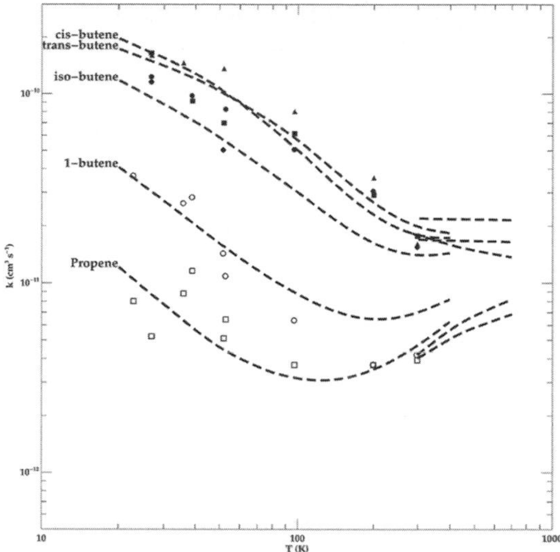

Figure 6.16 Measured rate coefficients for the reaction of O with alkenes. Solid lines indicate extrapolation using an Arrhenius law appropriate for high temperature. The dashed lines shows theoretical calculations, using transition state theory relevant at low temperature. Data taken from [3, 6, 10, 11, 12, 16, 18, 33, 57, 67, 68, 77]. Figure kindly provided by Jan Cami, adapted from [70]

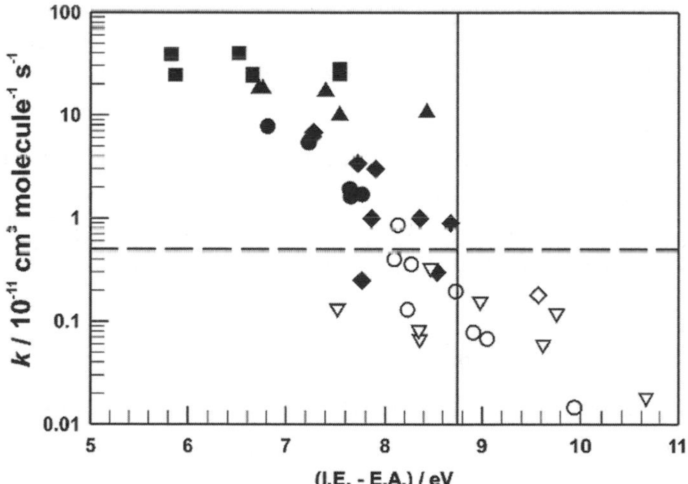

Figure 6.17 Rate coefficients for the reactions between alkenes and alkynes and H and O(^{3}P) atoms and OH, C_2H, and CN radicals. The rate coefficients are plotted against the difference between the ionisation energy (I.E.) of the unsaturated hydrocarbon and the electron affinity (E.A.) of the radical. Rate coefficients represented by open symbols increase with temperature around 300 K, whereas those represented by closed symbols decrease with temperature. The dashed horizontal line indicates a division between fast reactions at 300 K with a negative or zero temperature dependence and slow reactions a positive temperature dependence. The solid vertical line separates reactions with small and large differences in IE and EA. See text for detail. Figure courtesy of [69]

Table 6.9 *Neutral–neutral reactions*

Reaction	α	β	γ
$H_2 + O \rightarrow OH + H$	9.0 (−12)	1.0	4.5(3)
$H + OH \rightarrow O + H_2$	4.2 (−12)	1.0	3.5(3)
$H_2 + OH \rightarrow H_2O + H$	3.6 (−11)		2.1(3)
$H + H_2O \rightarrow OH + H_2$	1.5 (−10)		1.0(4)
$H + O_2 \rightarrow OH + O$	3.7 (−10)		8.5(3)
$OH + O \rightarrow O_2 + H$	4.0 (−10)		6.0(2)
$H_2 + C \rightarrow CH + H$	1.2 (−9)	0.5	1.4(4)
$H + CH \rightarrow C + H_2$	1.2 (−9)	0.5	2.2(3)
$C^+ + H_2 \rightarrow CH^+ + H$	9.4 (−12)	1.25	4.7(3)

[a]Reaction rates of the form $k = \alpha\,(T/300)^\beta\ \exp[-\gamma/kT]$.

Table 6.10 *Radical reactions*[a]

Reaction	α	β
$C + OH \rightarrow CO + H$	1.1 (−10)	0.5
$C + O_2 \rightarrow CO + O$	3.3 (−11)	0.5
$O + CH \rightarrow CO + H$	4.0 (−11)	0.5
$O + CH \rightarrow HCO^+ + e$	2.0 (−11)	0.44
$O + CH_2 \rightarrow CO + H + H$	2.0 (−11)	0.5

[a]Reaction rates are of the form, $k = \alpha\,(T/300)^\beta$.

Table 6.9 provides some examples of astrophysically relevant reactions. In general, the only neutral–neutral reactions that are of consequence for the cold conditions of dark clouds are those involving atoms or radicals, often with nonsinglet electronic ground states that do not have activation barriers. Some examples are listed in Table 6.10.

6.6 Radiative Association Reactions

In radiative association reactions, the collision product is stabilized through the emission of a photon; viz.,

$$A + B \xrightarrow{k_a} AB + h\nu. \tag{6.67}$$

The direct collision timescale is one vibrational period, $\tau_{col} \simeq 10^{-14}$ s for a bond involving H, which should be compared to the radiative lifetime of $\tau_{rad} \simeq 10^{-7}$ s for an allowed electronic transition and 10^{-3} for an allowed vibrational transition. The probability for the reaction is then 10^{-7} and 10^{-11}, respectively. With collision rate coefficients of 2×10^{-9} (ion–neutral, Section 6.4) and 2×10^{-10} (neutral–neutral reactions, Section 6.5) cm^3 s^{-1}, the reaction rate coefficient is expected to be very small. The reaction probability can be greatly enhanced if the collision partners have a resonance or form a long-lived activated complex, which stabilizes through the emission of a photon. Of course, most likely, the complex still redissociates. In laboratory settings, these rates are too slow to be measured but collisions with other species (rather than radiation) can lead to stabilization and ternary reactions are often used to estimate radiative association reaction rates. More generally, the rates of these reactions are estimated through theoretical calculations.

As a first application, consider the reaction,

$$C^+(^2P_{1/2,3/2}) + H(^2S_{1/2}) \longrightarrow CH^+(^1\Sigma^+) + h\nu. \tag{6.68}$$

Figure 6.18 shows the potential energy levels that connect to the dissociation limit of $C^+(^2P)$ and $H(^2S)$. At low temperatures, spin–orbit and rotational coupling will have a larger magnitude than the kinetic energy of the reactants and these have to be taken into account. Specifically, for the C^+–H system, the incoming species has to enter the $A^1\Pi$ state for an allowed electronic transition. Adiabatically, this state connects to the C^+ $(^2P_{3/2})$ fine-structure level, which is not much populated at the low densities of the ISM. However, collisions along the $a^3\Pi_1$ state can couple to the $A^1\Pi$ state at large distances where the energy difference is of the order of the fine-structure level separation in C^+ ($\simeq$60 cm^{-1} or $\simeq$8 meV). Also, quasibound, orbiting resonance states behind the $A^1\Pi$ centrifugal barrier (c.f. Section 6.4) will enhance the probability of tunneling and the formation of a long-lived collision complex. Detailed analysis shows that the potential energy curves are split into 12 states by spin–orbit and rotational coupling (Figure 6.18).

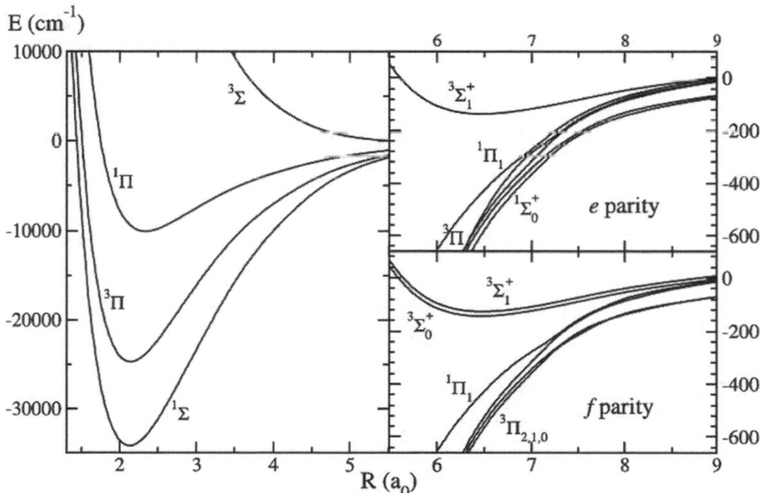

Figure 6.18 Left: Potential energy curves of the lowest electronic states of CH^+, connecting to the H and C^+ ground states. Relevant electronic states of $CH^+$are indicated. Right: Details of the fine-structure splitting of the adiabatic potential energy levels connecting near the dissociation limit. Note the difference in energy scale. Figure courtesy of [35]

The radiative association cross section is given by,

$$\sigma(E) = h \sum_{J'=J''-1}^{J'=J''+1} \frac{\pi}{k^2 p} (2J'+1) A(E, J'), \tag{6.69}$$

where k is the asymptotic wavenumber related to the kinetic energy and p is the degeneracy of the initial state. The Einstein A transition rate is given by

$$A(E, J') = \frac{64\pi^4}{3h} \sum_{v'', J''} \frac{1}{\lambda^3_{v''J''}} \frac{S_{J'', J'}}{2J'+1} M^2_{v''J'', J'}(E), \tag{6.70}$$

where λ is the wavelength of the emitted photon, $S_{J'', J'}$ is the Hönl–London factor, and the matrix element – connecting an initial continuum state (E, J') with the final bound state $(v', 'J'')$ – is given by,

$$M^2_{v''J'', J'}(E) = |\langle \psi(E, R)| \mu(R) |\psi_{v''J''}(R)\rangle|^2, \tag{6.71}$$

where the ψ's are the eigenfunctions of the $^1\Sigma$ state and $\mu = \langle ^1\Sigma | x | ^1\Pi_x \rangle$ is the electronic transition moment. Using potential energy surfaces, the cross section has been evaluated for this reaction and the energy-dependent radiative association rates are shown in Figure 6.19. The effect of resonances is quite evident but averaging over a Boltzmann energy distribution washes out these details. The overall effect is a higher rate.

If there are no excited electronic states available to the reactants or if internal conversion is rapid – due to the high density of states – relaxation occurs through IR emission in vibrational modes. In general, if there are no competing exothermic reaction channels possible for the intermediate reaction product, the reaction can be represented by,

$$A + B \underset{k_d}{\overset{k_a}{\rightleftarrows}} AB^* \xrightarrow{k_r} AB, \tag{6.72}$$

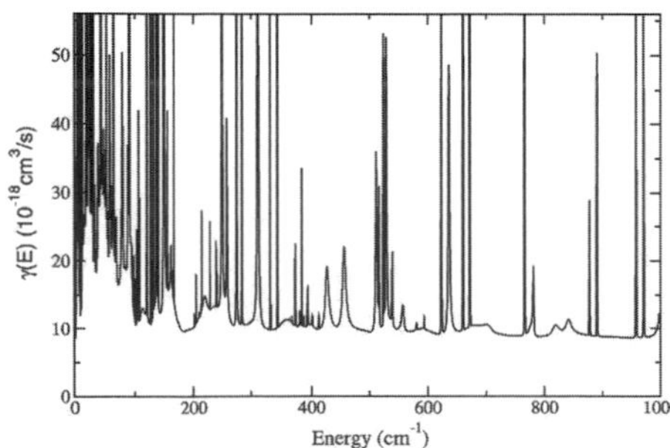

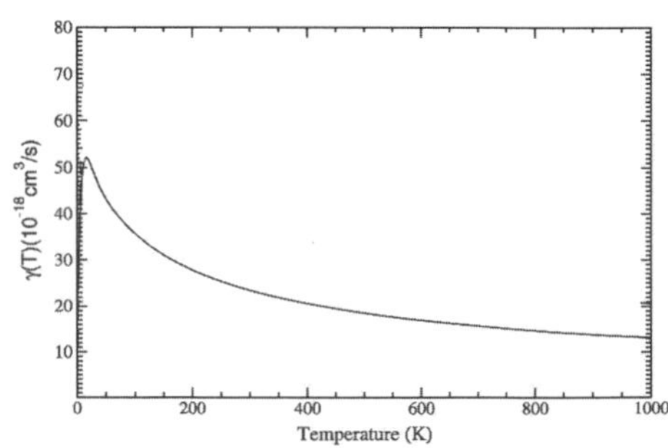

Figure 6.19 Calculated radiative association rates in the $C^+ + H \leftarrow CH^+$ reaction. Left: Energy-dependent radiative association rates. Right: Boltzmann averaged radiative association rate. Figure courtesy of [4]

with A and B the reactants and AB^* the activated complex. The overall radiative association rate, k_{ra}, for the reaction is then given by,

$$k_{ra} = k_a \left(\frac{k_r}{k_r + k_d} \right) \simeq k_r \left(\frac{k_a}{k_d} \right), \tag{6.73}$$

with k_a, k_d, and k_r the association, dissociation, and radiative stabilization rates of the activated complex. The collision complex will temporarily store the kinetic energy and the interaction energy in vibrational energy. Stabilization occurs then through emission in a vibrational transition with a radiative rate of typically 10^3 s^{-1}. The forward and the backward rates are related through detailed balance,

$$k_a\, Z_A\, Z_B = k_d\, Z_{AB^*}, \tag{6.74}$$

where the Z's are the partition functions per unit volume of the colliding species and the excited complex (see Sections 3.2.8 and 3.3.2). This corresponds to the ergodic assumption; i.e. the energy is divided over all available modes of the collision complex. At the temperatures we are concerned with, the partition functions of the colliding species are proportional to the rotational partition functions,

$$k_{ra} = k_r \left(\frac{h^2}{2\pi \mu k T} \right)^{3/2} \frac{Z_{AB^*}}{Z_A Z_B}, \tag{6.75}$$

where μ is the reduced mass. The first part represents the translational factors and here the Z_i's are the internal partition functions of species i. At low temperatures, vibrational degrees of freedom are not excited and, for A and B, the Z's are simply the rotational partition functions (see Section 3.2.8). For the activated complex, the vibrational degrees of freedom are key and the partition function is,

$$Z_{AB^*} = g_{en} \left(\mathrm{AB}^* \right) \int_0^\infty \rho\,(E) \exp\left[-E/kT \right] dE, \tag{6.76}$$

with g_{en} the electron–nuclear spin degeneracy factor and ρ the vibrational energy density of states. In this evaluation, we should only take states above the dissociation limit (D_0) into account and, ignoring possible barriers, we have,

$$\phi_{AB^*} = g_{en} \left(\mathrm{AB}^* \right) \int_0^\infty \rho\,(D_0 + E) \exp\left[-E/kT \right] dE \simeq g_{en} \left(\mathrm{AB}^* \right) \rho\,(D_0)\; kT, \tag{6.77}$$

where the last approximation reflects that $D_0 \gg kT$. The temperature dependence of the radiative association rate, $k_{ra} \propto T^{-n}$ reflects the temperature dependence of the (rotational) partition functions of the reactants and if the rotational spacing of the levels is small compared to the temperature, $n = (r_A + r_B + 1)/2$ with r_i the rotational degrees of freedom (cf. Section 3.2.8). Thus, for the reaction of two nonlinear molecules, this exponent is 3.5 and a very steep temperature dependence is predicted. Of course, this does assume that the vibrational degrees of freedom of the reactants are frozen out while the rotational degrees of freedom are not. At low densities, the latter are also frozen out (cf. Section 4.1).

The radiative association rate depends thus on the density of states of the activated complex and the dissociation energy of the complex. In Chapter 3, we already pointed out that the density of states is a steeply rising function of the number of internal degrees of freedom. We can make a simple-minded estimate of the density of states that brings out this dependence well. Consider a collision complex consisting of s harmonic oscillators of frequency ν and an internal energy equivalent to n quanta. Now, assuming that dissociation will occur when all these quanta are in any one mode, we can write for the probability that this occurs,

$$P = \frac{n\,!\,(s-1)\,!}{(n+s-1)\,!}. \tag{6.78}$$

Using the fequency of the oscillator, this can be converted into a lifetime for the activated complex,

$$k_{pd}^{-1} = \nu^{-1}\frac{(n+s-1)\,!}{n\,!\,(s-1)\,!}. \tag{6.79}$$

With $s = 9$, $j = 5$, and $\nu = 5 \times 10^{13}$ s^{-1}, we find, $k_{pd}^{-1} \simeq 2 \times 10^{-11}$ s, which is considerably longer than the direct collision timescale (10^{-14} s). Hence association can be greatly enhanced. For ion–molecule reactions, k_a is 2×10^{-9} cm^3 s^{-1} and the overall rate is $10^{-16} - 10^{-17}$ cm^3 s^{-1}. As activated complexes play an important role in unimolecular reactions, more sophisticated theories have been developed for the lifetime of such ergodic systems. These will be discussed in the chapter on interstellar PAHs and large molecules. Here, we merely emphasize that all these theories have in common that the density of states is a rapidly rising function of the internal energy of the species and of the number of atoms in the species (i.e. the number of modes over which this energy can be divided). Thus, a more tightly bound species or a larger species will lead to a longer lived complex and hence a higher radiative association rate.

Some insight on radiative association reactions can be obtained from ternary reaction studies, where the reaction is studied as a function of pressure – possibly with a buffer gas (e.g., He) – and stabilization of the activated complex can also occur through collisions. The reaction rate is then given by,

$$k_{eff} = k_a \frac{k_r + k_{col}n}{k_d + k_r + k_{col}n} \simeq k_{ra} + k_a \frac{k_{col}}{k_d} n, \tag{6.80}$$

with k_{col} the inelastic collisional rate coefficient and n the density of collision partners. At high densities, ternary reactions lead to the product while at low densities, radiative association dominates. So, high pressure data provides k_a/k_d (e.g. the lifetime of the activated complex) and, adopting a radiative stabilization rate ($\simeq 10^3$ s^{-1} for vibrational transitions), the radiative association rate coefficient can be estimated. The collisional rate coefficient depends on collision partner and temperature, and the efficiency of these collisions can range from 0.1 to 1. In general, using a geometric or Langevin collision rate coefficient leads then only to a lower limit on the rate coefficient. If the high pressure rate coefficient is measured as a function of density, a much better estimate results.

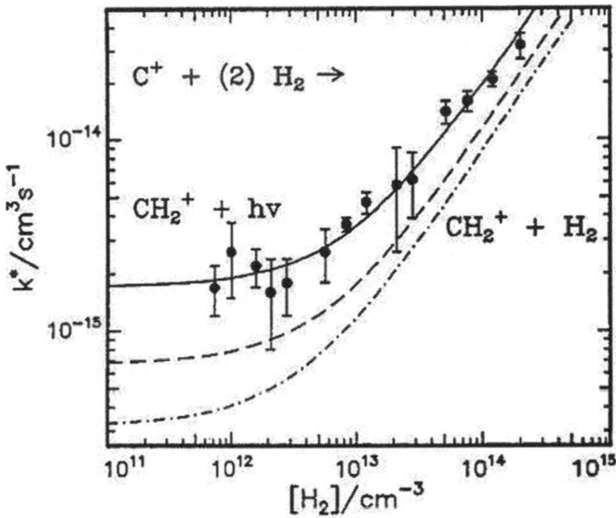

Figure 6.20 The effective rate coefficient for the reaction C^+ with para H_2 as a function of H_2 density at a temperature of 10 K. At high densities, the ternary reaction dominates. At low densities, radiative association takes over. The solid line is a fit combining the ternary rate (1.1×10^{-28} cm^6 s^{-1}) at high densities with the radiative association rate (6.7×10^{-16} cm^3 s^{-1}) at low densities. The dashed line shows the fit to the n-H_2 data. The dashed-dotted line is for pure H_2 $J = 1$. Figure courtesy of [29]

Figure 6.20 shows the measured rate constant for the reaction of $C^+ + H_2 \rightarrow CH_2^+ + h\nu$ at 10 K. The results are well fitted by Equation (6.80). With a Langevin collision rate and a radiative lifetime of 10^{-3} s, this corresponds to a collisional stabilization efficiency of 0.1 for H_2. The collisional stabilization efficiency for helium is about a factor 5 smaller. Conversely, for laboratory densities of 10^{12} cm^{-3}, ternary reactions will start to dominate over radiative association reactions.

Some important radiative association reactions are summarized in Table 6.11. Perusal of this table shows that, as the size of the system increases, the radiative association rate coefficient increases rapidly. The larger number of atoms implies a larger number of modes over which the energy can be distributed and hence a longer-lived activated complex. This then leads to a faster radiative association rate. This effect has been studied for the generic reaction system,

$$C_nH_m^+ + C_nH_m \rightarrow C_{2n}H_{2m} + h\nu, \tag{6.81}$$

and the results show that, for a typical bond energy of 3 eV, the reaction rate is 10^{-9} cm^3 s^{-1} even for N ($\equiv 2n + 2m$) as low as 10. For much smaller bond energies, the radiative association rate coefficient is much less. For example, for a characteristic physisorbed bond energy (0.1 eV), a rate coefficient of 10^{-9} cm^3 s^{-1} is only reached for N is $\simeq$100. Radiative association reactions and particularly their inverse, unimolecular photodissociation reactions, are also of great importance for the (photo)chemistry of large Polycyclic

Table 6.11 *Radiative association reactions*[a]

Reaction	k_{ra} [cm^3 s^{-1}]	Reaction	k_{ra} [cm^3 s^{-1}]
$H + C \rightarrow CH$	1.0 (−17)	$CH_3^+ + \text{n-}H_2 \rightarrow CH_5^+$	5.0 (−14)
$C^+ + H \rightarrow CH^+$	1.7 (−17)	$CH_3^+ + \text{p-}H_2 \rightarrow CH_5^+$	1.1 (−13)
$C^+ + \text{n-}H_2 \rightarrow CH_2^+$	6.70 (−16)	$CH_3^+ + \text{o-}H_2 \rightarrow CH_5^+$	3.1 (−14)
$C^+ + \text{p-}H_2 \rightarrow CH_2^+$	1.70 (−15)	$C_2H_2^+ + \text{p-}H_2 \rightarrow C_2H_4^+$	4.6 (−12)
		$C_2H_2^+ + \text{n-}H_2 \rightarrow C_2H_4^+$	1.3 (−12)
		$CH_3^+ + H_2O \rightarrow CH_3OH_2^+$	2.0 (−12)

[a] p-H_2, o-H_2, and n-H_2 indicate para ($J = 0$), ortho ($J = 1$), and "normal" molecular hydrogen (o:p=3:1).

Aromatic Hydrocarbon molecules in the interstellar medium and we will come back to this in Section 8.4.

6.7 Cosmic Ray Ionization

Ion–molecule reactions are ultimately driven by cosmic ray ionization. Cosmic ray interaction with H_2 leads to H_2^+ and subsequent reaction with H_2 leads to protonated H_2 (H_3^+). The proton can then be handed around[2] and "assist" in molecule formation. On the other hand, cosmic ray interaction with helium leads to He^+, which tends to be destructive. The relevant reactions are,

$$p + H_2 \rightarrow p + H_2^+ + e \qquad \text{ionization} \tag{6.82}$$

$$\rightarrow H + H_2^+ \qquad \text{electron capture} \tag{6.83}$$

$$\rightarrow p + H + H^+ + e \qquad \text{dissociative ionization} \tag{6.84}$$

$$\rightarrow p + 2H^+ + 2e \qquad \text{double ionization} \tag{6.85}$$

$$p + He \rightarrow p + He^+ + e \qquad \text{ionization} \tag{6.86}$$

$$p + He \rightarrow H + He^+ \qquad \text{electron capture.} \tag{6.87}$$

In addition, energetic electron interaction with H and He can also be important,

$$e_{CR} + H_2 \rightarrow e_{CR} + H_2^+ + e \qquad \text{ionization} \tag{6.88}$$

$$e_{CR} + He \rightarrow e_{CR} + He^+ + e \qquad \text{ionization.} \tag{6.89}$$

Cross section for these reactions are shown in Figure 6.21. At high energies, the data are well represented by the Bethe expression, $\sigma \propto \ln\left[\beta^2\right]/\beta^2$ with $\beta = v/c$. Perusing this figure, only the single ionization and electron capture reactions are relevant.

[2] With H, the chemistry is somewhat different but the point holds that ionization leads to molecule formation.

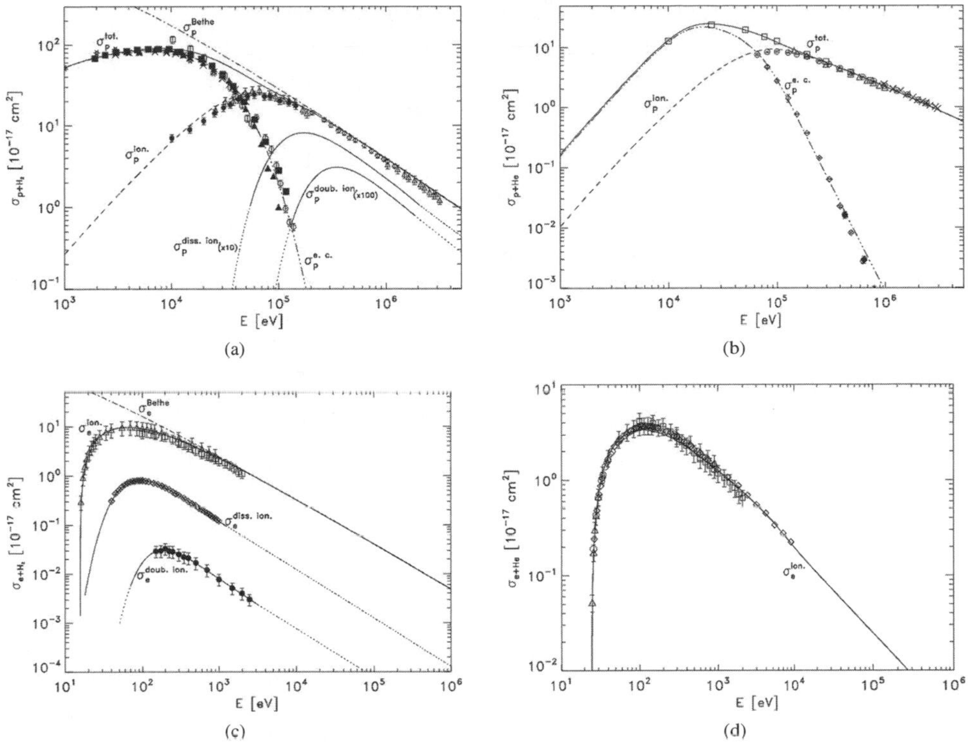

Figure 6.21 Cross section for the interaction of energetic protons (top) and electrons (bottom) with H_2 (left) and He (right). Separate curves are given for each of the processes in Eq. (6.82)–(6.88)): (a): Proton ionization ($\sigma_p^{ion.}$), electron capture ($\sigma_p^{e.c.}$), double ionization ($\sigma_p^{doub.ion.}$), and dissociative ionization ($\sigma_p^{diss.ion.}$). The drawn line indicates the total cross section ($\sigma_p^{tot.}$). (b): proton impacts on helium with the ionization (σ_p^{ion}) and electron capture cross section ($\sigma_p^{e.c.}$). (c) Electron impacts on H_2 showing ionization, dissociative ionization, and double ionization. (d) Helium ionization by energetic electrons. The curves labeled (σ_p^{bethe}) in panels (a) and (c) are twice the limiting, Bethe interaction cross section with atomic H. Figure courtesy of [55]

The cosmic ray ionization rate depends then on the cosmic ray flux and, given Figure 6.21, particularly the low energy flux, which is highly uncertain. The cosmic ray flux is shown in Figure 6.22. The high energy cosmic ray flux can be directly measured at the earth and is well represented by a power law. At low energies, modulation by the solar wind and its magnetic field prevents cosmic rays from penetrating the heliosphere. As cosmic ray ionization is at the bottom of gas phase ion–molecule reactions, molecular observations provide an indirect way to measure the cosmic ray flux in the ISM. This will be further discussed in Chapter 9.5. Here, we will parameterize the cosmic ray intensity, $I_{CR,i}$, of ion i as,

$$I_{CR,i}\,(E) = \frac{C E^{0.3}}{(E + E_0)^3} \;\; \text{cm}^{-2}\,\text{s}^{-1}\,\text{sr}^{-1}\,\text{GeV}^{-1} \tag{6.90}$$

Table 6.12 *Cosmic ray intensity parameters*

Ion	m_i [amu]	E_0^a [GeV]	C_i [$cm^{-2}\ s^{-1}\ sr^{-1}\ GeV^{1.7}$]
H	1	0.12	1.45
He	4	0.48	0.9
CNO	14	1.68	0.36
FeCoNi	58	6.95	0.24

[a] $E_0(i) = m_i E_0(H)$.

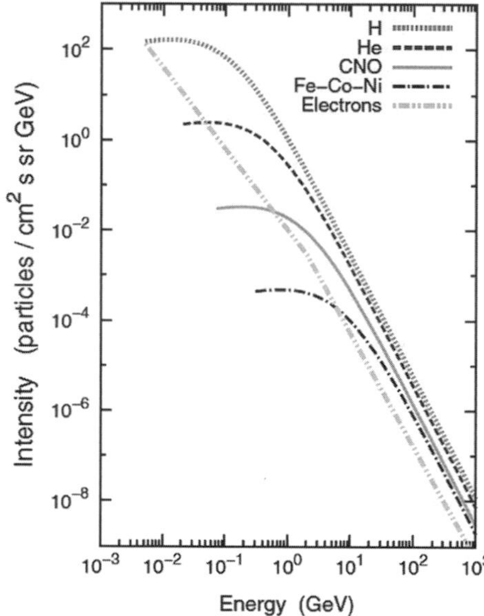

Figure 6.22 Cosmic ray intensity for electrons and different elements.

where the parameters in this expression are summarized in Table 6.12. For electrons we can adopt,

$$I_{CR,e}(E) = A\,E^{-\gamma} \tag{6.91}$$

with E in GeV, $A = 0.01$ and $\gamma = 1.8$ for $0.005 \leq E(GeV) \leq 2$ and $A = 0.016$ and $\gamma = 2.5$ for $2 \leq E(GeV) \leq 10^3$.

The ionization rate by cosmic ray protons is then,

$$\zeta'_{CR,p} = 4\pi \int_0^\infty I_{CR,p}(E)\,\sigma_{ion}(E)\,dE. \tag{6.92}$$

This is the primary cosmic ray ionization rate. The electrons produced by these ionizations have an energy of $\simeq$19 eV and, in turn, can produce further ionization (and excitation; cf. Section 6.3.5). The total ionization rate is then,

$$\zeta_{CR,p} = 4\pi\,(1+\eta_{sec}) \int_0^\infty I_{CR,p}(E)\,\sigma_{ion}(E)\,dE, \tag{6.93}$$

with η_{sec} ($\simeq$0.3) the number of ionizations produced by the first generation of electrons. We can now make an order of magnitude estimate of the cosmic ray ionization rate (see Section 6.2). Perusing Figures 6.21 and 6.22, we can surmise that ionization is dominated by 100 MeV protons, we then have,

$$\zeta_{CR} \simeq 4\pi I_{CR,e}\,(100\,\mathrm{MeV})\,\Delta E\,\sigma_{ion} \simeq 4\times10^{-16}\ \mathrm{s}^{-1} \tag{6.94}$$

A more proper evaluation of the cosmic ray ionization rate follows from convolving the ionization cross section with the cosmic ray intensity distribution given by Eq. (6.90) and this yields,

$$\zeta_{CR,p} = 6\times10^{-16}\left(\frac{100\,\mathrm{MeV}}{E_0}\right)^{2.56} \quad \text{ionizations s}^{-1}\ \text{H-nuclei}^{-1} \tag{6.95}$$

The cosmic ray ionization rate for helium and other species can be estimated in a similar way. In the high energy (Bethe) limit, the ionization cross section at energy E scales with $\sigma(E) \propto N_i/I_i^2 \ln(x_i)/x_i$ with N_i and I_i the number of electrons in subshell i and their binding energy, and $x_i = E/I_i$. Table 6.13 summarizes representative cross sections of some relevant species. At these high energies, cross sections behave similarly and these ratios can be used to determine the ionization rate.

Table 6.13 *Representative ionization cross sections at 100 MeV*[a]

Species	σ_i/σ_H
H_2	2
H	1
He	1.3
N_2	8.6
O_2	10.0
H_2O	5.6
CO	8.2
CH_4	6.9

[a] Ionization cross section, σ_i, of species i relative to that of H-nucleus (σ_H (100 MeV) = 2.9×10^{-19} cm^2). Primary ionization rates can be scaled from this.

This discussion centered on cosmic ray interaction with molecular gas. In atomic gas, the values for ionization are slightly different as, upon ionization of H, no energy is left as internal (vibrational) excitation and the electron is released with about 30 eV of energy rather than 19 eV. This also leads to a higher efficiency of further ionizations; e.g. the total cosmic ray ionization rate is about 1.5 times the primary rate.

Cosmic rays relevant for the ionization of gas in the interstellar medium are produced by supernova remants in a process called Fermi type I acceleration where atoms are scattered off magnetic fields back-and-forth across a shock front each time increasing the kinetic energy with $\Delta E/E \simeq \Delta v/c$. Some 10% of the supernova's energy is converted into cosmic ray energy density. The cosmic ray energy density will therefore scale with the star formation rate. Moreover, with an effective mean free path of $\simeq$0.15 pc, the cosmic ray flux will be higher near supernova remnants. With a Larmor radius $\ll$ 1 pc, cosmic rays are closely tied to the interstellar magnetic field. Cosmic rays will diffuse rapidly through the galaxy and will spend much of their "time" in the halo but eventually will leak out on a timescale of 10 Myr. With this timescale, the cosmic ray density will be enhanced by a factor of 10^3–10^4 relative to free streaming. As cosmic rays travel through the ISM and enter molecular clouds, they lose energy by interaction with the gas in ionization and dissociation processes. The range of cosmic rays – in terms of the equivalent H column that can be penetrated – is well described by,

$$N_H\,(E) = 2.4 \times 10^{20}\,E^{1.84}\ \mathrm{cm}^{-2}, \tag{6.96}$$

with E in MeV. This corresponds to a critical energy of 2 MeV for a diffuse cloud with a column density of 10^{21} cm^2. Molecular clouds have an average column density of 1.6×10^{22} cm^2 and a 10 MeV proton can penetrate the whole cloud. In addition, magnetic reflection of cosmic rays will act to decrease the cosmic ray density within clouds, but this is offset by the compression of the magnetic field. Scattering of cosmic rays off magnetic field inhomogeneities is not very relevant for the energies relevant to the ionization process.

6.8 Dissociative Electron Recombination Reactions

In dissociative electron recombination, an electron recombines with a positive molecule and as a result the molecule dissociates,

$$\mathrm{AB}^+ + \mathrm{e} \rightarrow \mathrm{A} + \mathrm{B}. \tag{6.97}$$

In the simplest form, the process starts with a dielectronic recombination step where the incoming electron recombines into a bound orbital and the excess energy is used to excite an electron in the ion core. The doubly excited species, AB** finds itself in a repulsive state and dissociates rapidly, before autoionization can occur (Figure 6.23). Alternatively, the first step may involve recombination into a Rydberg state of AB while the excess energy goes into vibrational excitation. This Rydberg state may then couple to the repulsive AB** state possibly through more than one step (Figure 6.23). Dissociative recombination is in

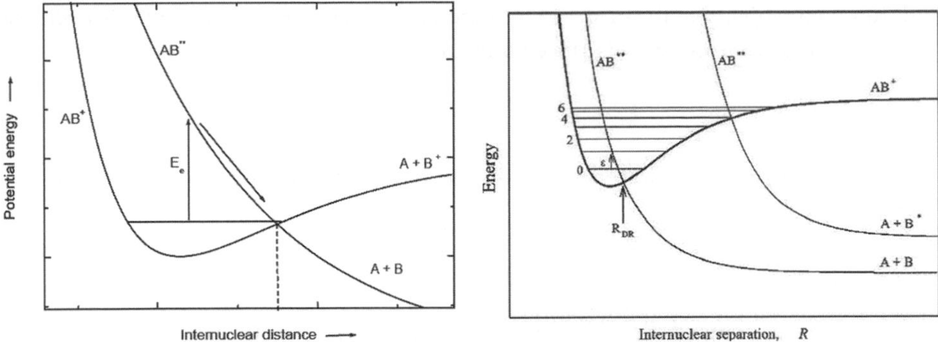

Figure 6.23 Schematic of dissociative electron recombination. Left: The direct process. Recombination of the ion AB^+ with an electron leads to the formation of a doubly excited species, AB^{**}, which slides down the potential, quickly passing the distance (indicated by a dashed line) where autoionization can occur. Right: The indirect process. Recombination of the ion with an electron leads to a vibrationally excited Rydberg state of the neutral that dissociates through a non-radiative transition to the dissociative electronic state, AB^{**}, either directly or through an intermediate state.

many ways akin to photodissociation with direct, predissociation, and indirect predissociation channels except that, instead of UV photons, excitation is due to the Coulomb energy (cf. Section 6.3.1). For atoms, only radiative processes are possible and recombination is a slow process ($\sim 10^{-8}$ s). For molecules, non-radiative processes are involved and these can be much faster ($\sim 10^{-12}$ s).

As discussed in Section 6.2, a simple estimate yields that the Coulomb interaction will increase the cross section by a factor e^2/a_0kT. Consider that a reaction will only occur if the electron and ion come close enough and a radiationless transition occurs with some efficiency, ϵ_{DR}. For the Coulomb interaction, there is no centrifugal barrier and hence no activation barrier if the reaction is allowed energetically. Consider a collision with velocity, v_∞, and impact parameter, b at large distances. Conservation of energy and angular momentum gives,

$$\frac{1}{2}\mu v_\infty^2 = \frac{1}{2}\mu v^2 + \frac{e^2}{r} \qquad \text{energy} \tag{6.98}$$

$$b v_\infty = r v \qquad \text{angular momentum.} \tag{6.99}$$

So, assuming that reaction occurs when a distance r_{min} is reached, we have for the collision cross section,

$$\sigma = \pi b^2 = \pi r_{min}^2 \left(1 + \frac{e^2/r_{min}}{\mu v^2/2}\right), \tag{6.100}$$

which we recognize as the hard sphere collision cross section multiplied by the Coulomb focusing factor. The latter dominates at low temperatures and we find then, typical for a

Coulomb interaction, that the cross section increases when the collision energy decreases, $\sigma \propto v^{-2}$, as the electron "spends" more time close to the nucleus,

$$k_e = \epsilon_{DR} \int_0^{\infty} \sigma(v)\, v f(v)\, dv, \tag{6.101}$$

with $f(v)$ the Maxwell–Boltzmann distribution. With our approximation, this becomes,

$$k_e = \pi r_{min} e^2 \epsilon_{DR} \left(\frac{8}{\pi \mu k T}\right)^{1/2} \tag{6.102}$$

$$\simeq 1.8 \times 10^{-6} \left(\frac{300\,\text{K}}{T}\right)^{1/2} \left(\frac{r_{min}}{1\,\text{Å}}\right) \epsilon_{DR}\ \text{cm}^3\ \text{s}^{-1}. \tag{6.103}$$

This is merely a collision rate and we have hidden all our ignorance about this process in the efficiency factor, ϵ_{DR}. This efficiency factor includes a Franck–Condon factor describing the overlap of the nuclear wavefunction of the initial ion state with the dissociating state (Section 3.4.3). In addition, a factor describing the competition with autoionization has to be included.

As many Rydberg states can be involved, dissociative recombination cross sections are difficult to evaluate theoretically. As discussed in Section 2.4.2, early experimental studies were based on ion-flow tubes. In more recent years, cryorings where ion and electrons can be merged at low relative velocities have been employed. Measured temperature dependencies range from the $T^{1/2}$ indicated by Eq. (6.102) to steeper dependencies when electronic or vibrational excitation can assist the recombination, in particular when the rate is not near the collisional limit or when other dissociation channels open up at higher energies. Theoretical estimates for product distribution are generally not reliable as the many Rydberg states available provide multiple curve crossings and a number of non-radiative steps open up. Experiments are thus needed and, for one, these show that, contrary to intuition, fragmentation to three fragments is the dominant channel. Rate coefficients and product distributions for some relevant reactions are given in Table 6.14.

6.9 Collisional Association and Dissociation Reactions

In laboratory settings, three body reactions generally dominate chemistry,

$$A + B + M \longleftrightarrow AB + M, \tag{6.104}$$

with rates $\simeq 10^{-32}$ cm^6 s^{-1}. In astrophysics, these reactions can become of importance for dense gas in stellar photospheres or the molecular layers directly surrounding them as well as in dense ($\simeq 10^{11}$ cm^{-3}) circumstellar disks. The reverse reaction, $H_2 + M \rightarrow H + H + M$ (M is H, He, or H_2), is particularly important in the warm inner regions of protoplanetary disks as well as in shocks where this reaction regulates the transition from C to J shock (Section 11.7.1). The rate for this reaction can be described by,

$$k = 1.85 \times 10^{-22} \sqrt{\frac{8kT}{\pi \mu}} \frac{a\,(kT)^{b-1}\,\Gamma(b+1) \exp\left[-E_0/kT\right]}{(1 + ckT)^{b+1}}\ \text{cm}^3\ \text{s}^{-1} \tag{6.105}$$

Table 6.14 *Electron recombination reactions*[a]

Reaction	α [cm^3 s^{-1}]	β
$H_3^+ + e \rightarrow H_2 + H$	3.8 (−8)	−0.45
$H_3^+ + e \rightarrow H + H + H$	3.8 (−8)	−0.45
$CO^+ + e \rightarrow C + O$	2.0 (−7)	−0.5
$HCO^+ + e \rightarrow CO + H$	1.1 (−7)	−1.0
$OH^+ + e \rightarrow O + H$	3.8 (−8)	−0.5
$H_2O^+ + e \rightarrow O + H + H$	2.0 (−7)	−0.5
$H_2O^+ + e \rightarrow OH + H$	6.3 (−8)	−0.5
$H_2O^+ + e \rightarrow O + H_2$	3.3 (−8)	−0.5
$H_3O^+ + e \rightarrow H_2O + H$	3.3 (−7)	−0.3
$H_3O^+ + e \rightarrow OH + H + H$	4.8 (−7)	−0.3
$H_3O^+ + e \rightarrow OH + H_2$	1.8 (−7)	−0.3
$CH^+ + e \rightarrow C + H$	1.5 (−7)	−0.4
$CH_2^+ + e \rightarrow CH + H$	1.4 (−7)	−0.55
$CH_2^+ + e \rightarrow C + H + H$	4.0 (−7)	−0.6
$CH_2^+ + e \rightarrow C + H_2$	1.0 (−7)	−0.55
$CH_3^+ + e \rightarrow CH_2 + H$	7.8 (−8)	−0.5
$CH_3^+ + e \rightarrow CH + H + H$	2.0 (−7)	−0.4
$CH_3^+ + e \rightarrow CH + H_2$	2.0 (−7)	−0.5

[a] Electron recombination rate coefficients are given as $k_e = \alpha\,(T/300)^{\beta}$.

Table 6.15 *Collisional dissociation of* H_2[a]

Collision partner	a	b	c	E_0^b
H	54.1263	2.5726	3.4500	0.180
H_2 (0,0)	40.1008	4.6881	2.1347	0.1731
He	4.8152	1.8208	−0.9459	0.4146
e^-	11.2474	1.0948	2.3182	0.3237

[a] Constants in the reaction rate coefficient for H_2 collisional dissociation (Eq. (6.105)). Taken from [47]. [b] In units of Hartrees (27.21 eV).

where μ is the reduced mass, Γ the gamma function, and the coefficients in this expression are given for relevant collision partners in Table 6.15.

6.10 Electron Attachment

Molecular anions are formed through radiative attachment reactions,

$$A + e^- \leftrightarrow \left(A^-\right)^* \rightarrow A^- \tag{6.106}$$

where $(A^-)^*$ is an excited molecular anion that can autoionize or radiatively stabilize. If an anion is formed, typically, the extra electron is captured in an antibonding valence orbital and the structure of the species is very different from the neutral. An electron affinity of 1–2 eV, depending on size, is required for stabilization to occur. The electron can also be bound to the neutral through long range interaction forming, e.g., dipole-bound states. The electron is now in a very diffuse orbital and the core "acts" as a proton. The extra electron has very little influence on the molecular structure. For a molecule to bind an electron this way, a dipole moment of $\gtrsim 2.5$ D is needed. A very high dipole moment ($\sim$9 D) is needed to have a second dipole bound state and, generally, stabilization has to occur through transfer to a valence bound anion.

The rate coefficient for electron attachment, k_{ea}, is given by,

$$k_{ea} = \left(\frac{k_r}{k_r + k_b} \right) k_f, \tag{6.107}$$

where k_f, k_b, and k_r are the forward (association), backward (autoionization), and radiative stabilization reaction rate coefficient, respectively, and the factor in brackets can be considered as an electron sticking efficiency. Evaluation of the latter requires knowledge on the vibrational transitions and their lifetimes as well as on the excitation of the species. The former can follow from quantum chemical calculations while the latter can be evaluated with the ergodic assumption for (large) species. The capture and autoionization rate coefficients are related through the principle of detailed balance (Section 5.2.6),

$$\frac{k_b}{k_f} = \frac{\rho^o}{\rho^-}, \tag{6.108}$$

with ρ^- and ρ^o the density of states of the anion and that of the neutral plus electron, respectively. Assuming that the neutral is in its vibrational ground state after loss of the electron, its density of states is that of the free electron,

$$\rho^o = \rho(e) = \frac{m^2 v}{\pi^2 \hbar^3}. \tag{6.109}$$

The density of states for the anion at the internal energy corresponding to the electron affinity of the neutral can be estimated from quantum chemically calculated vibrational frequencies. As the density of states is a rapidly rising function of number of modes available and of internal energy (Chapters 3 and 8), radiative attachment is promoted by large electron affinities and/or by a large size. For the open shell, carbon chain radicals, electron affinities are typically 3–4 eV. When the backward reaction is slow compared to radiative stabilization, the radiative attachment rate coefficient is given by the forward rate, which (for large species) is often assumed to be given by the Langevin rate (Section 6.4),

$$k_f = 2\pi \left(\frac{\alpha e^2}{m_e} \right)^{1/2}, \tag{6.110}$$

with α the polarizability of the neutral. However, that aproximation ignores the discrete character of the orbital momentum, which can enhance the electron attachment rate by a factor 2–4. For the larger hydrocarbon chains, reaction rate coefficients a few times 10^{-7} cm^3 s^{-1} at 10 K are indicated.

Radiative attachment rate coefficients for atoms are in general very small as there are few (electronic) states available to accept the collision energy, and radiative stabilization has to occur upon collision. With a collision timescale of 10^{-14} s and an Einstein A of $\sim 10^6 - 10^7$ s^{-1} for an IR to visible transition, the probability for reaction is $10^{-8} - 10^{-7}$. However, the reaction,

$$\mathrm{H} + \mathrm{e} \rightarrow \mathrm{H}^- = h\nu \tag{6.111}$$

is key to the abundance of H_2 in the early universe and its rate is therefore of importance. The rate of this reaction can be derived from the measured photodissociation cross section using the principle of detailed balance,

$$\sigma_f(v) = \frac{g_H}{g_{H^-}} \left(\frac{h\nu}{mvc} \right)^2 \frac{\sigma_b(\nu)}{\left(1 - \exp\left[-h\nu/kT\right]\right)}, \tag{6.112}$$

where the g_i are the statistical weights. The rate coefficient is then,

$$k_{H^-} = \int_0^\infty \sigma_f(v)\, v f(v)\, dv \tag{6.113}$$

$$= \frac{g_{H^-}}{g_H} \sqrt{\frac{2}{\pi c}} \left(\frac{h^2}{m_e ckT} \right)^{3/2} \exp\left[EA/kT\right] \int_{EA}^\infty \frac{\nu^2 \sigma(\nu)}{\exp\left[h\nu/kT\right] - 1} d\nu, \tag{6.114}$$

with $f(v)$ the Maxwell–Boltzmann distribution and EA the electron affinity of H. With the measured photodissociation cross section, the electron attachment rate can then be calculated (Figure 6.24).

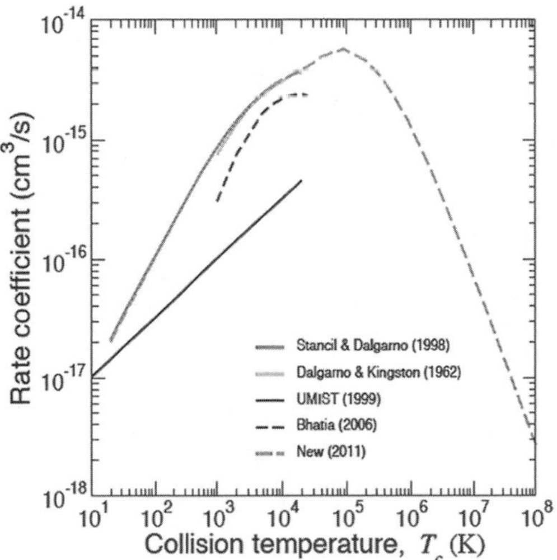

Figure 6.24 Calculated radiative attachment rate for the formation of H$^-$ using detailed balance (Eq. (6.113)). Figure adapted from [50]

Finally, electron attachment reactions with large molecules such as PAHs can be very rapid. This is discussed in Section 8.6.2.

6.11 Associative Detachment Reactions

With the realization that anions can be important in space, associative attachment reactions have come into play, where an anion and an atom react and the neutral product stabilizes through electron emission,

$$A + B^- \longrightarrow AB + e. \tag{6.115}$$

The simplest of these is the reaction,

$$H + H^- \longrightarrow H_2 + e, \tag{6.116}$$

which has long been recognized as an important route for H_2 formation in the early universe. The Langevin rate for this reaction is 2.8×10^{-9} cm^3 s^{-1}. The measured rate shows more complex behavior (Figure 6.25), reflecting deviations from the simple induced dipole potential energy surface.

Because carbon chain anions have been detected in C-rich stellar ejecta and in the ISM, their chemistry has been investigated in the laboratory. The reaction of C_n^- or HC_n^- with H indeed follows Equation (6.115) at about the Langevin rate. With increasing size of C_n^- (e.g. $n \geq 7$), radiative stabilization to the hydrogenated carbon chain becomes competitive. The reaction of C_n^- with N and O atoms leads to the formation of CN (or CN^-) and CO,

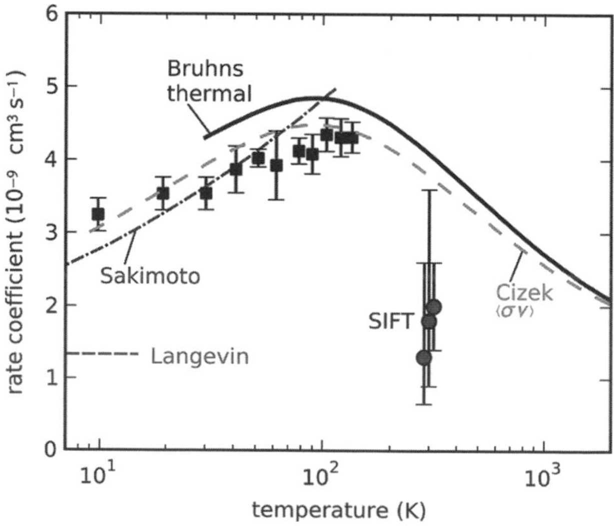

Figure 6.25 Low temperature 22-pole ion trap measurements of the associative detachment rate for the formation of H_2 from H^- are compared to quantum chemical calculations and flow tube measurements. Figure adapted from [31]

respectively. The reaction of HC_n^- with N leads to the formation of CN^- and with O, CO is the main product. The N-atom reactions occur at rates somewhat less than the Langevin rate while, with O, reaction proceeds at about the Langevin rate.

6.12 Non-LTE Effects

In this discussion, we have always tacitly assumed that the various degrees of freedom are in equilibrium and can be described by a single temperature. However, in general, the interstellar medium is far from equilibrium. With regard to the chemical processes, this can be of importance if a reaction with a high activation barrier is promoted by internal vibrational excitation of one of the species. For example, in photodissociation regions, molecular hydrogen can be highly vibrationally excited due to FUV pumping in the Lyman–Werner bands. The reactions $O + H_2 \rightarrow OH + H$ and $C^+ + H_2 \rightarrow CH^+ + H$ are notable examples that may then well go at the collision rate of these species with vibrationally excited molecular hydrogen. In contrast, in molecular clouds, the excitation of molecules may be much less than the kinetic temperature of the gas would predict if the density is below the critical density. In that case, reaction rate coefficients may be much less than determined in the laboratory. This should, for example, be kept in mind when evaluating the effects of neutral–neutral reactions in the warm post-shock gas.

Reactions may also be promoted by electronic excitation. Fine-structure excitation of O, C, and Si are of interest. The interaction of these species with neutral radicals is dominated by long range forces involving the atomic quadrupole. These interactions can suppress barriers in the potential energy surfaces that result from chemical interactions at intermediate separations. The reactivity is then determined by the quadrupole moment of the lowest spin–orbit states and, hence, is sensitive to the population of the fine-structure levels. The rate coefficients can then be enhanced by an order of magnitude compared to the usual extrapolations of laboratory studies (at 300 K) to low T using a $T^{1/2}$ dependence of neutral–neutral reactions.

The ortho-para ratio of molecular hydrogen is another example where non-LTE effects are often important. Various processes influence the abundance ratio of ortho to para H_2. These include chemical reactions in the gas phase and physical spin flipping reactions on grain surfaces. Because the gas and grain temperature are generally out of equilibrium, the ortho-to-para ratio is also not well described by the kinetic temperature of the gas. A notable reaction for which this may be of importance is the reaction of N^+ with H_2, which is the bottleneck in the formation of ammonia. This reaction has a 170K barrier for H_2 in its lowest vibrational and rotational state. For para H_2 this barrier is important but it is of no consequence for ortho H_2.

Finally, some reactions are promoted by translational energy. Now, generally, gas species have a Maxwellian velocity distribution that is rapidly established through collisions with the gas; e.g. H_2 molecules. The excess energy associated with exothermic reactions may leave the reaction products translational hot. If the reaction between this reaction product and H_2 is promoted by excess translational energy then this reaction may occur at a much faster rate than the kinetic temperature would indicate. As an example, the reaction $He^+ + N_2 \rightarrow N^+ + N + 0.29$ eV provides N^+ with 0.15 eV of kinetic energy and this helps the reaction $N^+ + H_2 \rightarrow NH^+ + H$, which has a small energy barrier (170 K). All of these effects have to be carefully evaluated for interstellar chemistry.

6.13 Gas Phase Chemistry Networks

6.13.1 General Chemical Rules

Based upon these chemical considerations, a few general rules governing gas phase chemical networks can be formulated. If an ion can react with H_2, those reactions will take precedence over all other reactions because of abundance considerations. If not, electron recombination generally dominates the loss channel for ions. If electron reactions are (also) inhibited, loss through reactions with neutrals will take over. For neutrals, reactions with ions will dominate (if not hindered by energy barriers). Because of abundance considerations, the important ions in diffuse clouds are C^+, H^+, and He^+, while in molecular clouds H_3^+, HCO^+, H_3O^+, and He^+ take over. If the neutral does not react with these species, neutral–neutral reactions with small radicals (atomic or diatomic) may become important, but again limited by energy barriers. The formation of neutrals is often dominated by electron recombination of an ion. Note that, in the general scheme of things, this will only be important if that ion does not react with H_2.

6.13.2 Gas Phase Models

Detailed gas phase chemical networks for various astrophysical environments will be discussed in Sections 9.3, 9.5, 10.5, 11.3.2, 11.4.2, 11.6.3, and 11.7.3. Here, we emphasize that involving ions in the reaction network carries a huge premium in terms of driving molecular complexity. For that reason, the small amount of ionization in molecular clouds due to penetrating cosmic ray ions takes on great significance for astrochemistry. Furthermore, molecular build up in the interstellar medium starts off with an atomic gas. Given the low rates of radiative association reactions, the first step combining two atoms into a diatomic species forms a bottleneck in the gas phase. Here, chemistry on grain surface comes to the rescue. The grains act as an extra body readily accepting excess energy and momentum. This is particularly true for molecular hydrogen formation. Once molecular hydrogen has formed on grain surfaces, gas phase chemistry can really take off. The physical and chemical principles involved in grain surface chemistry are discussed in Chapter 7.

6.13.3 Solving for the Chemical Composition

Once a set of chemical species and the reactions among them have been selected, differential equations for the abundance of each of the species can be written,

$$\frac{dn\,(i)}{dt} = -n\,(i)\left(\sum_j n\,(j)\,k_{ij} + \sum_j k_j\right) + \sum_{j,k} n\,(j)\,n\,(k)\,k_{jk} + \sum_j n\,(j)\,k_j, \tag{6.117}$$

The first term in brackets on the right-hand side includes all coreactants for species i as well as the unimolecular reactions (e.g. photodissociation or ionization, cosmic ray ionization) destroying species i while the second and third terms pertains to all bimolecular and unimolecular reactions resulting in the formation of species i. In steady state, this results in a set of quadratic equations,

$$\mathcal{F}(\vec{n}) = 0, \tag{6.118}$$

where the vector $\vec{n}$ contains the abundances of the species. Because in each reaction formation and destruction balance, this set of equations is dependent (e.g. each term appears as often with a plus sign as with a minus sign in this set of equations). However, for each element, we can replace one equation by a conservation equation. In addition, we can replace one equation by charge conservation. The resulting set of equations can be solved iteratively using standard Newton–Rhapson methods,

$$\vec{n}^{j+1} = \vec{n}^{j} - \mathcal{J}^{-1}\mathcal{F}, \tag{6.119}$$

where $\mathcal{J}$ is the Jacobian of this set of equations and $\mathbf{n}^j$ and $\mathbf{n}^{j+1}$ contain the current and an improved estimate of the abundances. Adopting initial guesses for the abundances, this set of equations is readily iterated upon for a given temperature, density, UV radiation field, cosmic ray ionization rate, and set of elemental abundances. Convergence depends, of course, on the initial guesses, but is generally rapid. Also, in chemistry, sometimes more than one (physical) solution is possible (see Section 6.13.4). Of course, many unphysical solutions are also available – e.g. where one or more abundance is negative – and care should be taken to avoid those in numerical schemes.

More generally, the time-dependent evolution of the chemical abundances is followed. In that case, the set of differential equations (Eq. 6.117) is integrated starting from some initial value – often taken as those appropriate for diffuse atomic clouds – and adopting values for the physical conditions. Of course, the evolution of the physical conditions must be followed simultaneously. It should be recognized that timescales for reactions in realistic chemical networks can be very different. In particular, cosmic ray ionization reactions and H_2 formation on grain surfaces are very slow compared to ion–neutral reactions. This results in a set of stiff differential equations, which have to be solved using special numerical methods such as the Gears method. Routines for this are available as standard in numerical libraries. Alternatively, and sometimes more practically, the chemical set of species can be divided into species whose abundances evolve slowly and those which are in rapid "equilibrium" with their environment. Cases in point are species whose abundances are controlled by proton transfer, reaction with H_2, or dissociative electron recombination reactions. In that case, the abundances of slowly evolving species can be explicitly followed by solving the relevant differential equations while the abundance of rapidly evolving species can be determined by solving a set of nonlinear equations. Numerical routines also exist for solving a set of differential equations and nonlinear equations simultaneously. In either case, the size of timesteps should be carefully chosen to keep the propagation of errors under control. Conservation equations provide helpful checks on the accuracy.

6.13.4 Chemical Bifucation

In nonequilibrium situations – and much of interstellar chemistry is in non-LTE – the complexity of chemical reaction systems with multiple feedback loops often leads to the coexistence of two or more, stable, chemical states. Such bistability is an example of hysteresis in which a system jumps back and forth between the two branches of stable states. This bistability occurs for a range of values of a control parameter, corresponding to two limit points. Under certain conditions, doubling bifurcations can evolve toward chaotic behavior. This is explored further in Exercise 6.5. In chemistry, the Belousov–Zhabotinsky reaction provides an illustrative example of this behavior (see Section 6.14).

In astronomy, the presence of multiple solutions can also occur. The coexistence of two stable phases in an ISM in thermal and pressure equilibrium presents a well-known example. Interstellar chemistry sometimes allows more than one physical solution. One such example is the chemical composition of dark clouds where the reactions that control the degree of ionization allow two distinct chemical equilibrium states to exist under certain conditions. In photodissociation regions, two stable solutions can occur controlled by the molecular heating processes. These are discussed further in Section 10.5.7 and 11.6.1. Time history provides the guide as to which solution will be realized starting from given initial conditions. As mentioned, when more than one solution is available, the system may start to oscillate between the possible states or, even, develop chaotic behavior. This aspect will not be covered in this book, but the reader should be beware that, in general, chemistry can be prone to chaotic behavior.

6.14 Further Reading and Resources

The monograph [52] provides further insight in the photodissociation behavior of small molecules. For large molecules, the monograph [74] can be consulted. A useful summary on photodissociation cross sections for small molecules is provided by [76]. Photoabsorption cross sections for families of PAHs are given in [44]. Depth-dependent photodissociation rates have been evaluated by [7, 60]. References [7, 26] are particularly relevant for the penetration of UV radiation into a cloud. A general monograph on radiative transfer relevant for the penetration of light into a medium is given in [14]. The UV interstellar radiation field has been studied by [37] while the wider wavelength range was taken from [48]. The detailed expression was taken from [19]. A widely used dust model is presented in [21] and values can be found on the website, www.astro.princeton.edu/~draine/dust/dust.html maintained by B. Draine. Dust extinction for aggregates has been described by [54]. The results shown in Figure 6.6 have been taken from the model study of the effects of coagulation in molecular clouds by [53]. The UV field inside dense clouds produced by cosmic rays was first discussed by [59]. A more recent and very detailed evaluation is presented in [36]. The penetration of radiation into a clumpy cloud has been investigated by [9] and detailed models for realistic clouds have been evaluated by [15, 71]. The photodissociation rate of H_2 has been evaluated by [1, 7, 20], the fluorescent emission spectrum has been calculated by [8, 20], and the transition from atomic to molecular hydrogen has been studied by [72]. The fit in Eq. (6.40) is taken from [20].

The website, http://satellite.mpic.de/spectral_atlas/index.html, provides an electronic data base for UV absorption spectra of many simple molecules. Photochemical rates relevant for atmospheric models are available through http://jpldataeval.jpl.nasa.gov. A useful compilation can also be found at http://home.strw.leidenuniv.nl/~ewine/photo/ maintained by E. van Dishoeck. Electronic absorption cross sections of PAHs and small carbon clusters are available on www.dsf.unica.it/~gmalloci/pahs/ maintained by G. Malloci.

The website http://iupac.pole-ether.fr maintained by the IUPAC Task Group on Atmospheric Chemical Kinetic Data Evaluation and the http://jpldataeval.jpl.nasa.gov maintained by the NASA Panel for Data Evaluation provide access to a chemical kinetics data base of neutral–neutral reactions relevant for atmospheric chemistry.

The theory of radiative association reactions and their rates has been discussed by [5]. The experimental measurements on radiative association rate coefficients have been taken from [29, 30]. The effect of size on the radiative association rate coefficient has been investigated by [38].

Ion–molecule reactions are discussed in the review by [70]. Various compilations of ion–molecule reactions have appeared [2]. The charge exchange reaction of He^+ with N_2 is discussed in [39].

Reactions of carbon chain anions and PAH anions have been investigated by [17, 22].

Cross sections for the interaction of high energy particles with gases are summarized in [63, 64]. The various processes relevant for the penetration of cosmic rays into molecular clouds have been discussed by [13, 55, 56].

Reviews on dissociative recombination are provided in [27] and [28, 43]. The KIDA (http://kida.obs.u-bordeaux1.fr) and UMIST (http://udfa.ajmarkwick.net) databases provide compilations of astrophysically relevant reaction rate coefficients. These databases are described in [49, 78].

The Belousov–Zhabotinsky reaction is a very illustrative example of a chemical bistability where the system oscillatees between two chemical states. A Youtube video can be found at www.youtube.com/watch?v=PpyKSRo8Iec.

6.15 Exercises

6.1 Reaction rate coefficients:

- Show that the centrifugal barrier in an ion–molecule reaction is given by Eq. (6.51) and that this leads to the Langevin rate (Eq. (6.54)).
- Calculate the centrifugal barrier for neutral–neutral reactions and derive the expression for the critical impact parameter (Eq. (6.64)) and the reaction rate coefficient (Eq. (6.65)).

- Derive the expression for the Coulomb focusing factor in Eq. (6.100) and the dissociative electron recombination rate coefficient (Eq. (6.102)).
- Compare and contrast the temperature dependence of these reaction rate coefficients.
- Compare and contrast the cross sections involved in these different types of reactions.

6.2 Cosmic ray ionization of molecular hydrogen leads to the formation of H_3^+. This species can transfer its "excess" proton to other species present. Compare CO and coronene; when the degree of ionization is low, where will this extra proton wind up? And where for high levels of ionization?

6.3 Consider the molecule AB formed through the following reactions,

$$A + B \longrightarrow AB + h\nu \qquad k_1 \tag{6.120}$$

and

$$A + BC \longrightarrow AB + C \qquad k_2 \, , \tag{6.121}$$

and destroyed through the reactions,

$$AB + D \longrightarrow A + BD \qquad k_3 \tag{6.122}$$

and,

$$AB + h\nu \longrightarrow A + B \qquad k_4 \, . \tag{6.123}$$

Derive expressions for the steady state abundance of AB in terms of the abundances of the other species and the rate coefficients of the reactions involved.

6.4 Compare and contrast the various chemical gas phase reactions. Describe the "general" rules controlling gas phase routes in the ISM and their "rational."

6.5 From deterministic behavior to chaos:

Consider a simple (biological) system where the population of bacteria in a culture is governed by $f(x) = 2x$ with x the current population and f the population, say, one hour later. So, starting with 1000 bacteria, after an hour there will be 2000, and then 4000, etc. Of course, growth cannot go on indefinitely and we could catch this resource limitation by, $f(x) = 2x(1 - x)$. This introduces a nonlinearity with interesting consequences. We are going to generalize this to the recursive relationship, $x_{n+1} = f(x, \alpha) = \alpha \, x_n (1 - x_n)$, where the growth rate, α, is the control parameter that relates the value at step $n + 1$ to that at n. We will restrict x to range $[0, 1]$; so, consider it as an (ill-defined) fractional population.

- Calculate 20 successive generations, starting at $x_0 = 0.5$, for different values of the control parameter, $\alpha = 0.1$, 0.5, 0.9, 1.0, 1.5, 2.0, and 2.5. Plot your results.
- Show that there are two equilibrium points, $x = 0$ and $x = (\alpha - 1)/\alpha$.
- Show that, for $\alpha < 1$, $x = 0$ is the stable equilibrium and for $1 < \alpha < 3$, $x = (\alpha - 1)/\alpha$ is stable. Check this with your numerical results above.

- Examine the results to understand why the value of $\alpha = 2$ corresponds to the replacement value.
- Calculate 100 successive generations, starting at $x_0 = 0.5$, for the following values of the control parameter, $\alpha = 3.0$ and 3.25. Plot your results. The system should now oscillate between two stable points. These two points are given by $x_{1,2} = \left(1 + \alpha \pm \sqrt{\alpha^2 - 2\alpha - 3}\right)/2\alpha$.
- Check that, when α increases to $1 + \sqrt{6}$, the system oscillates between four points.
- With increasing α the number of bifurcations rapidly doubles again and again, and for a value of 3.99, it has seemingly reached chaotic behavior. At this point, this deterministic system has a period of infinity.
- Recall that this is a full deterministic system. Chaos implies that as the initial value is slightly adjusted, the system quickly moves away from the previous solution. Check this.

Bibliography

[1] Abgrall, H., Le Bourlot, J., Pineau Des Forets, G., et al., 1992, *A & A*, 253, 525
[2] Anicich, V. G., Huntress, W. T., Jr., 1986, *ApJS*, 62, 553
[3] Atkinson, R., *Chem Rev*, 1986, 86, 60
[4] Barinovs, G., van Hemert, M. C., 2006, *ApJ*, 636, 923
[5] Bates, D. R., Herbst, E., 1988, in *Rate Coefficients in Astrochemistry*, eds. T. J. Millar and D. A. Williams, Dordrecht: Kluwer, p17.
[6] Biehl, H., Bittner, J., Bohn, B., Geers-Muller, R., Stuhl, F., 1995, *Int J Chem Kinet*, 27, 277
[7] Black, J., Dalgarno, A., 1977, *ApJS*, 34, 405
[8] Black, J. H., van Dishoeck, E. F., 1987, *ApJ*, 322, 412
[9] Boisse, P., 1990, *A & A*, 228, 483
[10] Boodaghians, R. B. , Hall, I. W., Toby F. S., Wayne, R. P., 1987, *J Chem Soc*, Faraday Trans. 2, 83, 2073
[11] Browarzik, R., Stuhl, F., 1984, *J. Phys. Chem.*, 88, 6004
[12] Carty, D., Le Page, V., Sims I. R. , Smith, I. W. M., 2001, *Chem Phys Lett*, 344, 310
[13] Chandran, B. D. G., 2009, *ApJ*, 529, 513
[14] Chandrasekhar, S., 1960, *Radiative Transfer* (New York: Dover)
[15] Cubick, M., Stutzki, J., Ossenkopf, V., Kramer, C., Röllig, M., 2008, *A & A*, 488, 623
[16] Cvetanovic, R. J., 1987, *J Phys Chem Ref Data*, 16, 261
[17] Demarais, N. J., Yang, Z., Martinez, O. Jr, et al., 2012, *ApJ*, 746, 32
[18] Donahue, N. M. , Clarke, J. S., Anderson, J. G., 1998, *J. Phys. Chem. A*, 102, 3923
[19] Draine, B. T., 1978, *ApJS*, 36, 595
[20] Draine, B. T., Bertoldi, F., 1996, *ApJ*, 468, 269
[21] Draine, B. T., Lee, H. M., 1984, *ApJ*, 285, 89.
[22] Eichelberger, B., Snow, T. P., Barckholtz, C., Bierbaum, V. M., 2007, *ApJ*, 667, 1283
[23] Federman, S., Glassgold, A., Kwan, J., 1979, *ApJ*, 227, 466
[24] Field, G. B., Somerville, W. B., Dressler, K., 1966, *Annu Rev Astron Astrophys*, 4, 207
[25] Fillion, J. H., et al., 2001, *J Phys Chem A*, 105, 11414

[26] Flannery, B. P., Roberge, W., Rybicki, G. B., 1980, *ApJ*, 236, 598.
[27] Florescu-Mitchell, A. I., Mitchell, J. B. A., 2006, *Phys Rep*, 430, 277
[28] Geppert, W., Larsson, M., 2008, *Mol Phys*, 106, 2199
[29] Gerlich, D., 1993, in *The Physics of Electronic and Atomic Collisions*, eds. T. Andersen et al., AIP Conference Proceedings, Vol. 295 (New York: AIP), 607.
[30] Gerlich, D., 1993, *J Chem Soc Faraday Trans*, 89, 2199.
[31] Gerlich, D., Jusko, P., Roucka, S., et al., 2012, *ApJ*, 749, 22
[32] Gillispie, G. D., Lim, E. C., *Chem Phys Lett*, 1979, 193
[33] Gingerich, K. A., Finkbeiner, H. C., Schmude, R. W., 1994, *J Am Chem Soc*, 116, 3884
[34] Gioumousis, G., Stevenson, D. P., 1958, *J Chem Phys*, 29, 294
[35] Graff, M. M., Moseley, J. T., Roueff, E., 1983, *ApJ*, 169, 976
[36] Gredel, R., Lepp, S., Dalgarno, A., Herbst, E., 1989, *ApJ*, 347, 289
[37] Habing, H. J., 1968, *Bull Astron Soc Neth*, 19, 421
[38] Herbst, E., Dunbar, R. C., 1991, *MNRAS*, 253, 341
[39] Inn, E. C. Y., 1967, *Planet Space Sci*, 15, 19
[40] Kanaev, A. V., Museur, L., Laarmann, T., Monticone, S., 2001, *J Chem Phys*, 115, 10248
[41] Kreckel, H., Bruhns, H., Čížek, M., et al., 2010, *Science*, 329, 69
[42] Kropp, J. L., Dawson, W. R., 1967, *J Chem Phys*, 71, 4499
[43] Larsson, M., Thomas, R., 2001, *PCCP*, 3, 4471
[44] Malloci, G., Cappellini, G., Mulas, G., Mattoni, A., 2011, *Chem Phys*, 384, 19
[45] Marquette, J. B., Rowe, B. R., Dupeyrat, G., Poissant, G., Rebrion, C., 1985a, *Chem Phys Lett*, 122, 431
[46] Marquette, J. B., Rowe, B. R., Dupeyrat, G., Roeff, E., 1985b, *A & A*, 147, 115
[47] Martin, P. G., Keogh, W. G., Mandy, M. E., 1998, *ApJ*, 499, 793
[48] Mathis, J. S., Mezger, P., Panagia, N., 1983, *A & A*, 128, 212
[49] McElroy, D., Walsh, C., Markwick, A. J., et al., 2013, *A & A*, 550, A36
[50] McLaughlin, B. H., Sadeghpour, H. R., Stancil, P. C., Dalgarno, A., Forrey, R. C., 2012, *J Physics: Conf Ser*, 388, 022034
[51] Meixner, M., Tielens, A. G. G. M., 1993, *ApJ*, 405, 216
[52] Okabe, H., 1978, *Photochemistry of Small Molecules* (New York: Wiley and Sons)
[53] Ormel, C. W., Min, M., Tielens, A. G. G. M., Dominik, C., Paszun, D., 2011, *A & A*, 532, A43
[54] Ossenkopf, V., Henning, T., 1994, *A & A*, 291, 943
[55] Padovani, M., Galli, D., Glassgold, A., 2009, *A & A*, 501, 619
[56] Padovani, M., Galli, D., 2011, *A & A*, 530, A109
[57] Perry, R. A., 1984, *J Chem Phys*, 80, 153
[58] Pino, T., Carpentier, Y., Féraud, G., et al., 2011, *EAS Publications Series*, 46, 355
[59] Prasad, S. S., Tarafdar, S. P., 1983, *ApJ*, 267, 603
[60] Roberge, W. G., Jones, D., Lepp, S., Dalgarno, A., 1991, *ApJS*, 77, 287.
[61] Rowe, B., Marquette J. B., Dupeyrat, G., Ferguson, E. E., 1985, *Chem Phys Lett*, 113, 403
[62] Rowe B. R., Canosa A., Le Page V., 1995, *Int J Mass Spectrom Ion Proc*, 149/150, 573
[63] Rudd, M. E., DuBois, R. D., Toburen, L. H., Ratcliffe, C. A., Goffe, T. V., 1983, *Phys Rev A*, 28, 3244
[64] Rudd, M. E., Kim, Y.-K., Madison, D. H., Gallagher, J. W., 1985, *Rev Mod Phys*, 57, 965

[65] Sakimoto, K., 1989, *Chem Phys Lett*, 164, 294
[66] Siebrand, W., 1966, *J Chem Phys*, 44, 4055
[67] Singleton, D. L., Cvetanovic, R. J., 1976, *J Am Chem Soc*, 98, 6812
[68] Singleton, D., Cvetanovic, R. J., 1976, *J Am Chem Soc*, 98, 6812
[69] Smith, I. W. M., Sage, A. M., Donahue, N. M., Herbst, E., Quanc, D., 2006, *Faraday Discuss*, 133, 137
[70] Smith, I. W. M., 2011, *Annu Rev Astron Astrophys*, 49, 29
[71] Spaans, M., 1996, *A & A*, 307, 271
[72] Sternberg, A., Le Petit, F., Roueff, E., Le Bourlot, J., 2014, *ApJ*, 790, 10
[73] Tielens, A. G. G. M., Hollenbach, D. J., 1985, *ApJ*, 291, 747
[74] Turro, N. J., 1991, *Modern Molecular Photochemistry*. (Sauselito: University Science Press)
[75] van Dishoeck, E. F., Black, J. S., 1986, *ApJ*,
[76] van Dishoeck, E. F., Visser, R., 2015, in *Laboratory Astrochemistry: From Molecules through Nanoparticles to Grains*, eds. S. Schlemmer, T. Giesen, H. Mutschke, C. Jaeger (Weinheim: Wiley-VCH) 229
[77] Van Orden, A., Saykally, R. J., *Chem Rev*, 1998, 98, 2313
[78] Wakelam, V., Herbst, E., Loison, J. C., et al., 2012, *ApJS*, 199, 21

7

Chemistry on Interstellar Grain Surfaces

7.1 Introduction

Gas phase species can accrete, migrate, and react with each other on cold grain surface in the interstellar medium, opening up the possibility of a rich and diverse chemistry. Indeed, chemistry on grain surfaces plays an important role in astrochemistry. Because the most abundant molecule in space, H_2, is very difficult to form in the gas phase, this molecule is generally thought to form on prevalent grain surfaces. Similarly, the presence of the peroxide species, HO_2 and H_2O_2 – which are difficult to make in the gas phase – in the molecular cloud, ρ Oph, suggests an active interchange between gas and solid phase. Also, observations of molecular clouds reveal a strong decrease in the abundance of many gas phase species toward their dense cores indicative of the importance of freeze out on cold interstellar grain surfaces. At the same time, infrared spectroscopy toward embedded protostars or background sources has revealed the widespread presence inside dense clouds of icy grain mantles, consisting of a variety of simple molecules – H_2O, CO, CO_2, CH_3OH, CH_4, NH_3 ... – (partially) mixed in low temperature molecular ices. As the composition of these ices is very different from that of the gas phase, this itself attests to a rich chemistry taking place on grain surfaces. The importance of freeze out is also inferred from observations of gaseous H_2O that show that this species is localized in a narrow region close to the surface of molecular clouds. In addition to this evidence that grains play an important role in the chemistry of molecular clouds, strong variations in the depletion of elements such as Fe, Si, and Mg between the cold neutral medium and the warm neutral medium reveal that there is also an active exchange of material between grains and gas phase in the diffuse ISM. Finally, it has been suggested that – if metallic grains are present – Fischer–Tropsch-like catalytic reactions can play a role in warm dense environments, such as the inner regions of protoplanetary disks or warm molecular zones close to the photosphere of stellar outflows. Here, we will discuss the chemical principles that play a role in grain surface chemistry, focusing on ice formation in molecular clouds in the interstellar medium.

The astronomical setting and the steps involved in interstellar grain surface chemistry are outlined in Section 7.2 and this can be the basis for the discussion in a molecular astrophysics course. The remainder of this chapter goes into much more detail and will be of interest to students entering the field or – with the references listed in Section 7.8 – could form a sound basis for an essay or presentation assignment in the class. The detailed chemical networks involved in interstellar ice formation are discussed and compared to the observations in Section 10.5. The role of grain surface chemistry and interstellar ices in the composition of hot cores and hot corinos is discussed in Section 11.2.

7.2 Characteristics of Surface Chemistry

In dense clouds, interstellar grains provide a surface on which accreted species can meet and react and to which they can donate excess reaction energy (Figure 7.1). Gas phase species will collide with grains, transfer some of their kinetic energy to the surface, and "stick" in physisorbed sites. From there, they can transfer to chemisorbed sites or desorb. Species with low barriers against migration will be mobile and can migrate searching for reaction partners. Besides this Langmuir–Hinshelwood process, molecules may also be formed through direct collision of a gaseous species with a trapped (and immobile) species on a grain surface (the Eley–Rideal process). Finally, an accreting species may come in and during the thermal accomodation process migrate on the surface and find a reaction partner (the Harris–Kasemo process). This migration process may also be initiated by a chemical reaction, producing a highly excited and therefore mobile reaction product. The chemical energy released upon reaction can heat the grain, can be transferred to center-of-mass translational motion and then desorption, and/or to excitation of internal degrees of freedom of the newly formed species. We will briefly discuss these "steps" (Figure 7.1) and the parameters that control them here. Further details can be found in subsequent sections.

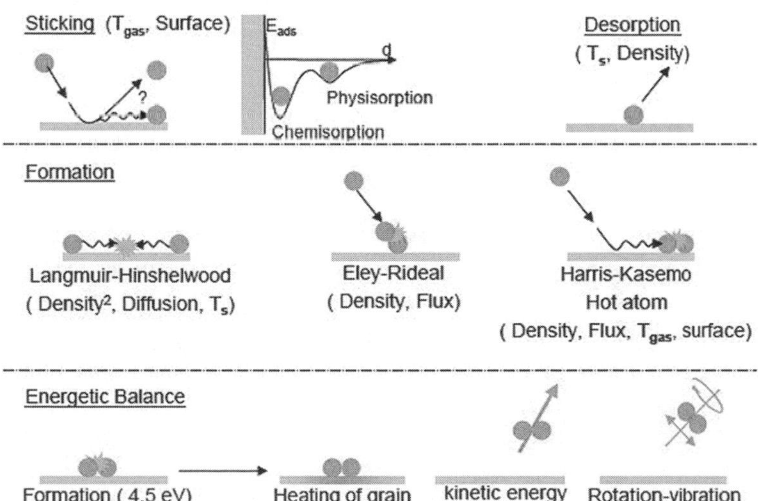

Figure 7.1 Relevant concepts in the formation of molecules on grain surfaces are schematically illustrated. Gas phase species accrete, diffuse, and react on an interstellar grain surface. Figure kindly provided by Lionel Amiaud, see [3]

7.2.1 The Astronomical Setting

Densities in diffuse clouds range from $\sim$50 to $\sim 10^3$ cm^{-3} while temperatures are in the range 20–100 K, corresponding to thermal pressures in the diffuse ISM of the order of 4000 K cm^{-3}. In molecular clouds, which are gravitationally dominated, thermal pressures are much higher and average densities range from 3×10^2 to 3×10^3 cm^{-3} while temperatures are typically $\simeq$10 K. Much higher values are attained in so-called hot corino's around low mass protostars ($n \sim 10^5 - 10^6$ cm^{-3} $T = 100-300$ K) and in so-called hot cores ($n \gtrsim 10^6$ cm^{-3}, 300–1000 K) around massive protostars. Hence, while much of the molecular cloud is at pressures of 10^4 K cm^{-3}, thermal pressures can be as high as 10^{10} K cm^{-3} in star-forming regions. For comparison, typical interstellar pressures ($\sim 10^{-15}$ mbar) are much less than pressures in even ultra high vacuum systems (UHV; 10^{-9} mbar). Molecular cloud temperatures are also very low and require cryogenic cooling or molecular beam expansion in the laboratory. Putting this in perspective, at an interstellar pressure of 10^4 cm^{-3} K, the incident flux is $\simeq 10^4$ particles/cm^2/s, while in an UHV setting, this is $\simeq 10^{10}$ particles/cm^2/s.

Consider that under these conditions, the mean free path of, e.g., an H-atom is $\simeq 10^{12}$ cm (a good fraction of the distance from the Earth to the Sun). Note that, in a laboratory setting under UHV conditions, the mean free path also well exceeds the experimental size scale. Gas particles interact thus mainly with the walls. In space, "walls" are provided by omnipresent dust grains. In the diffuse medium, these grains consist mainly of graphitic or silicate material but traces of other compounds are also inferred to be present, including silicon carbide, diamonds, aluminum oxide, spinel, iron or iron-sulfide, and probably many others. Typical sizes range from 30 to 3000 Å, corresponding to very small surface areas, 10^4–10^8 Å^2 or about $2 \times 10^3 - 2 \times 10^7$ sites per grain. The abundances of these solid compounds are quite low. The total mass density is about 1% of that of the gas and the total surface area is about 3×10^{-21} cm^2/H-atom.

7.2.2 Interstellar Surface Chemistry

Sticking on a grain surface will depend on the temperature of the gas and the grain as well as the grain material and surface structure. At 10 K, the sticking coefficient is close to unity even for the lightest species. The sticking coefficient will decrease slowly with temperature but even at 100 K, the sticking coefficient is still $\simeq$0.5. Residence time of a species on a grain surface is a much more limiting factor. Weakly bonded species such as H will sublimate at $\simeq$20 K from an "ideal" surface. They can remain trapped longer in deep wells or porous structures present on amorphous surfaces or on sufaces with many defects, kinks, or steps. But at 50–100 K, most simple species will quickly desorb from physisorbed sites and surface chemistry will then have to involve deeper chemisorbed sites.

Grain surfaces are the watering holes of ISM where species come to meet and mate; that is, in the ISM, grain surfaces function as regions that concentrate potential reaction partners in a limited space where collisions become frequent. Reactions on a grain surface occur then because some species on a surface are mobile and will "collide" with reactive species while scanning the surface. For amorphous H_2O ice, only H, D, and N will be mobile. On a CO, CO_2, N_2, and/or O_2 ice, atomic oxygen will also be mobile. As accretion rates are small – $\sim$1 particle/day – interstellar surface chemistry is limited by the rate at which species can be brought to the surface (the so-called diffusion limit). Conversely, an accreted species has a long time in which it can react with other species. This is particularly true for mobile species that can scan the surface for reaction partners. For the light species, H and D, the long timescales allow reactions with appreciable energy barriers to occur efficiently, largely assisted by tunneling. Also, at 10 K, an H_2O ice surface will be covered by about 0.5 monolayers of H_2 and reactions of H_2 with moderate activation barriers can occur readily as well.

These considerations lead to a set of key reactions that are very different from those controlling gas phase chemistry. In particular, chemistry will be dominated by mobile reaction partners that get to choose their reaction partners. Furthermore, as grains can absorb the reaction energy, addition reactions are much more important than (radiative) association reactions in the gas phase. This dominance of association reactions is further amplified by the "cage" effect: Reaction products are to some extent trapped in the well where they are created and may have multiple occasion to react. The rules of engagement are described in more detail in Section 7.5. However, because atomic H is very mobile, it is clear that hydrogenation will be a key characteristic of surface chemistry, converting O and O_2 to H_2O, N to NH_3, and CO to CH_3OH.

There are several processes that can return surface chemistry products to the gas phase. Of particular importance are thermal desorption in regions heated by nearby (proto)stars and photodesorption at the edges of clouds or in photodissociation regions. Sputtering is relevant in shocks while heating by cosmic rays can be of some importance deep inside dense cloud cores where few other "return" processes operate. Finally, the reation energy may lead to desorption of the reaction product or its neighbors but this is more controversial.

Finally, interstellar grain surface chemistry is in the diffusion limit, and typically the concentration of mobile radicals on the surface is either 0 or 1. Only in the latter case can a reaction occur. Hence, the chemical evolution of the system cannot be evaluated using "average" abundances of species and rate equations describing their interaction. Techniques have been developed to allow for this (Section 7.7), but combining gas phase with grain surface chemical networks is challenging.

7.3 Surface Processes

7.3.1 Accretion

The timescale for depletion of gas phase species on interstellar grains is given by,

$$k_{ac} = n_d \sigma_d \, v \, S\,(T, T_d) \simeq 10^{-17}\, S\,(T, T_d) \left(\frac{T}{10\ \mathrm{K}}\right)^{1/2} n \qquad \mathrm{s}^{-1}, \tag{7.1}$$

where n_d and σ_d are the number density and cross section of interstellar grains, $S(T, T_d)$ is the sticking coefficient of a species to a grain with T and T_d the gas and dust temperature, respectively. For the right-hand side, typical interstellar dust parameters ($n_d\sigma_d \simeq 10^{-21}n$ cm^{-1}, with n the density of H-nuclei) have been adopted and a mean mass appropriate for CO has been assumed.

The timescale at which gas phase species deplete out onto grains is $4 \times 10^9/n$ yr or as short as 4×10^5 yr in a dense core. This is much shorter than the dynamical evolution timescale of molecular clouds ($\simeq 3 \times 10^7$ yr for GMCs and $\sim 10^6$ yr for dense cores (the free-fall timescale)). This depletion timescale is quite comparable to the chemical evolution timecale of dense cores. We can also evaluate the timescale for species to arrive on a particular grain,

$$\tau_{ar} = (n_i \sigma_d v)^{-1} \simeq 3 \left[\frac{10^4\ \mathrm{cm}^{-3}}{n}\right] \left[\frac{1000\ \mathrm{\AA}}{a}\right]^2 \quad \text{day,} \tag{7.2}$$

with n_i and v_i the density and mean velocity of the accreting species, respectively, and where this equation has been evaluated for CO and a gas temperature of 10 K. So, in a dense core, this timescale is typically a few days.

7.3.2 Sticking

In the first step of grain surface chemistry, accreting species must "stick" to the grain. The sticking coefficient depends on the species accreting, the thermal velocity of the gas, the excitation of the phonon spectrum of the grain, as well as the interaction energy of the gas phase species and the surface. The approaching atom will be accelerated in the attractive potential well associated with the long range van der Waals interaction. As it collides with the surface, some of the kinetic energy will be transferred to the surface-collision partner. For hydrogen, typically, the energy transferred is small compared to the total (kinetic plus interaction) energy but, after the collision, some of the energy is "locked up" in horizontal motion along the surface. The atom will then "hop" along the surface and collide with another surface atom, losing more energy. This thermal accommodation process will continue until either the accreting atom has enough energy in the perpendicular direction to "escape" the potential well or its energy loss exceeds the initial kinetic energy and the species is trapped, although it may still rattle around on the surface before it has transferred all of its energy (kinetic plus binding) to the surface. Thus, under certain conditions, an accreting species may "visit" nearby sites during the thermal accommodation process.

When an incoming particle with kinetic energy, E_k, collides with a surface atom, the maximum energy transferred is,

$$\frac{\Delta E}{E_k + E_b} = \frac{4\mu}{M}, \tag{7.3}$$

for a central collision where μ and M are the reduced and summed mass of the projectile, m_{gas}, and the surface atom, m_{sur}, and E_b the binding energy. Typically, $E_k \ll E_b$ and

$\Delta E/E_b \simeq 4\mu/M$, which for an H atom on H_2O is $\simeq$0.25. The surface atom is, of course, connected to the solid and that affects the energy transfer. Consider the accommodation process as a collision between the species and one surface atom connected by a single spring to a wall (the rest of the grain) with a natural frequency, ω_D, the Debye frequency of the solid. The collision timescale is,

$$\tau_{col} \simeq \frac{b}{v} \simeq \sqrt{\frac{2b^2\mu}{E_k + E_b}}, \tag{7.4}$$

with b the size of the van der Waals well. When $\tau_{col}\omega_D \gg 1$, calculations show that the energy transferred can then be approximated as,

$$\frac{\Delta E}{E_k + E_b} \simeq \frac{m_{gas}}{m_{sur}} \frac{2\omega_D \tau_{col}}{\left((\tau_{col}\omega_D)^2 - 1\right)^2} \qquad \omega\tau_{col} \gtrsim 1.7 \tag{7.5}$$

$$\simeq \frac{m_{gas}}{m_{sur}} \qquad 1 \lesssim \omega_D \tau_{col} \lesssim 1.7. \tag{7.6}$$

The adsorbing particle will generally rebound at some angle to the surface and will transfer energy during each subsequent collision during the hopping process. The sticking probability is approximately given by,

$$P(\epsilon) \simeq \exp\left[-\epsilon^2/2\right], \tag{7.7}$$

with $\epsilon = E_k/E_c$ where E_c is approximately the geometric mean of the van der Waals interaction energy and the energy transferred, $E_c = \sqrt{E_b \Delta E}$, the average over a Maxwellian distribution of impact energies can be made to arrive at the sticking coefficient. Calculations show that the sticking coefficient, $S(T)$, is,

$$S(T) \simeq \frac{\gamma^2 + 0.8\gamma^3}{1 + 2.4\gamma + \gamma^2 + 0.8\gamma^3}, \tag{7.8}$$

where γ is defined as $\gamma = E_c/kT$. This has the limits,

$$S \simeq \gamma^2 \qquad \gamma \ll 1 \tag{7.9}$$

$$S \simeq 1 - 3/\gamma^2 \qquad \gamma \gg 1 \tag{7.10}$$

This assumes that, initially, surface atoms are stationary while in reality they will be vibrating with an average energy given by the dust temperature, T_d. Reflecting the phase of this vibration, the center of mass kinetics will vary and this has to be taken into account. Based upon noble gas thermal accomodation experiments and theoretical models, the sticking coefficient for H on grain surfaces has been evaluated to be given by,

$$S(T, T_d) = \left[1 + 4 \times 10^{-2}\,(T + T_d)^{1/2} + 2 \times 10^{-3}\,T + 8 \times 10^{-6}\,T^2\right]^{-1}. \tag{7.11}$$

The results show that for atomic H on an ice surface, the sticking coefficient is 0.8 at 10 K, dropping to 0.5 at 100 K. For accreting heavy species, matched better in mass to the surface atoms, the energy transferred is much higher and sticking is essentially assured.

In more recent years, detailed classical molecular dynamics studies have been made to evaluate the sticking coefficient. These arrive at a slightly different expression for the sticking fraction of impinging beam of species with energy E_k, viz.,

$$P\left(E_k\right) = \exp\left[-E_k/E_0\right], \tag{7.12}$$

with E_0 a fitting parameter derived from the numerical experiments, but which scales with the mass of the impactor. In that case, averaging over a Maxwell–Boltzmann distribution yield simply,

$$S\left(T\right) \simeq S_0 \frac{1 + 2.5\left(kT/E_0\right)}{\left(1 + \left(kT/E_0\right)\right)^{2.5}}. \tag{7.13}$$

For atomic hydrogen impacting crystalline or amorphous ice at 10 K, E_0 is evaluated to be 300 K. Experimentally, the sticking of molecular hydrogen has been studied by directing an effusive molecular beam on a cold surface and measuring the intensity of the reflected molecules using a quadrupole mass spectrometer. For H_2 sticking to ice and silicate surfaces, the results are in good agreement with Eq. (7.12) with the parameters summarized in Table 7.1. The dependence of E_0 on mass of the impacting species was confirmed in experiments using D_2. The sticking of atomic H on a surface is more difficult to measure experimentally as atoms adsorbed on the surface will diffuse and react to form H_2. Some constraints can be inferred from the recombination efficiency but this is difficult. Molecular dynamics studies on the other hand, can readily address this issue. The results of such a study are shown in Figure 7.2 and reveal good agreement between theory and experiments for H_2. However, the molecular dynamics results are not well fitted by Equation (7.13) with E_0 scaled by mass. Rather, an E_0 of $\simeq$100 K is indicated. A good fit is also obtained for H and H_2 using the expression given by Eq. (7.8) but then with $\gamma = E_0/T$ and multiplied by a prefactor S_0. The parameters for this fit are also given in Table 7.1. This latter fit is the preferred fit, except that dependence on surface temperature and composition are not fully explored.

Table 7.1 *Sticking coefficient for H_2*

Surface	Surface Temperature [K]	Species	$S_0{}^a$	$E_0{}^a$ [K]	$S_0{}^b$	$E_0{}^b$ [K]
Non-porous ASW[b]	10	H_2	0.76	87	0.62	226
Non-porous ASW[c]	10	H	1	52	1	244
Non-porous ASW[c]	70	H	1	52	0.5	244
Silicate[b]	10	H_2	0.95	56		

[a]Experimental study: Sticking coefficient given by eqn. (7.13). Taken from reference [17]. [b]Molecular dynamics study: Sticking coefficient given by Eq. (7.8) but with $\gamma = E_0/T$ and S multiplied by S_0. Taken from reference [105]

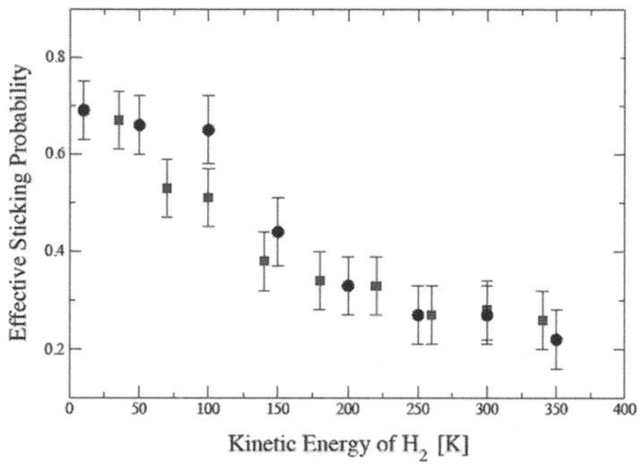

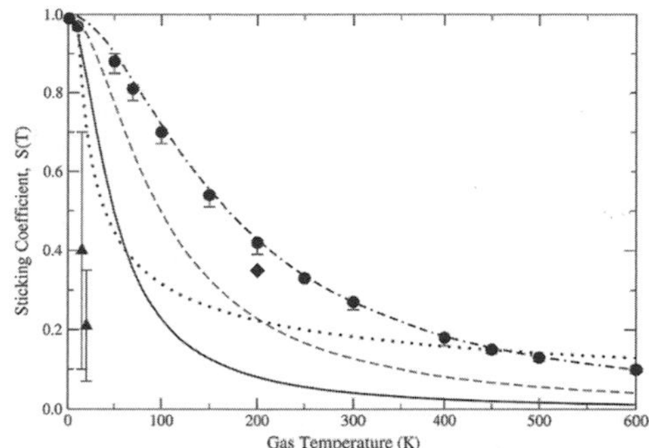

Figure 7.2 Left: A comparison of experimental (squares) and numerical (circles) sticking coefficients for H_2 on an ice surface [70, 105]. Right: The sticking coefficient of atomic H on ice surfaces calculated using molecular dynamics (circles) compared to Equation (7.8) with slightly different values for E_{vdw} (solid and dashed curves) and to Equation (7.8) with $\gamma = 244/T$ (dot-dashed curve). Dotted curve represents $T^{1/2}$ normalized at unity at 10 K. Figures courtesy of [104]

Energy transfer in the collision is controlled by the mass difference between impactor and the surface species (cf. Eq. (7.5)). In cold, dense clouds, grain surfaces may be partially covered by H_2, thereby facilitating energy transfer and increasing the sticking coefficient for atomic H. However, as the sticking coefficient under those conditions is close to unity anyway, this may not be too relevant.

Finally, the sticking coefficient of ions may be small. As a cation approaches a grain surface, it will be accelerated by the Coulomb potential of the grain (image) charge. Given the large difference between the ionization potential of gas phase species (e.g. 11.2 eV for C^+) and the work function of relevant grain materials (4.4 eV and 8 eV for graphite and silicates), electrons will likely tunnel out when the cation is near the surface and recombination will take place outside of the grain. The resulting neutral species will collide at relatively high energy with the grain and thermal accommodation will be inefficient.

7.3.3 Binding Energy

The physisorbed interaction between a species and a surface balances the long range van der Waals forces due to dispersion with the short range repulsive forces due to overlap of the electronic clouds of the species and the surface. The London dispersion forces have to be summed over all atoms in the surface and this transforms the leading contribution in the van der Waals interaction, $\propto d^{-6}$, to C_3/d^3. With some approximations, the constant, C_3, can be written as,

$$C_3 \simeq \frac{\pi n_s}{4} \alpha_s \alpha_i \frac{I_s I_i}{I_s + I_i}, \tag{7.14}$$

with n_s the density of the solid, α_s and α_i the polarizabilities of the solid and accreting species, and I_s and I_i the ionization potentials.

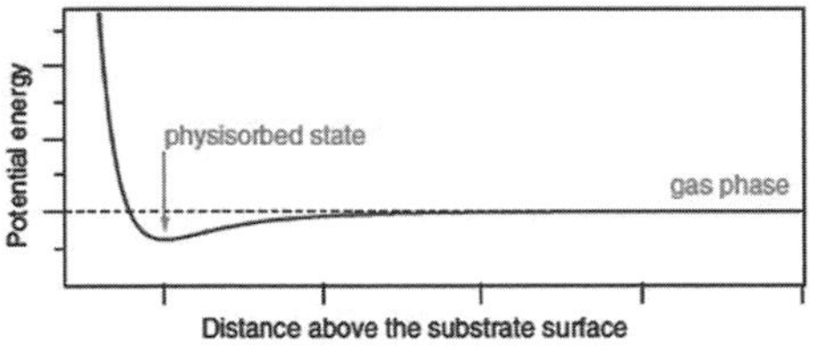

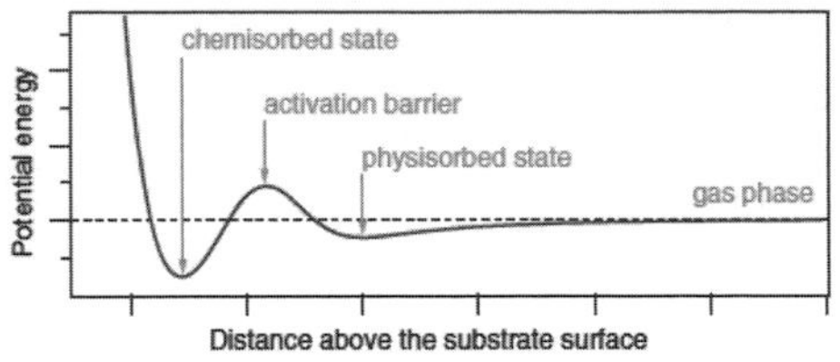

Figure 7.3 Left: Potential energy curve for physisorption on a surface. Typical physisorbed well energies are ~0.1 eV at a distance of ~2 Å above the surface. Right: Adsorption on a surface with a chemisorbed state. Chemisorbed wells are ~1 eV at a depth of ~0.2 Å. The relative location of the physisorbed and chemisorbed states determines the size of the activation barrier.

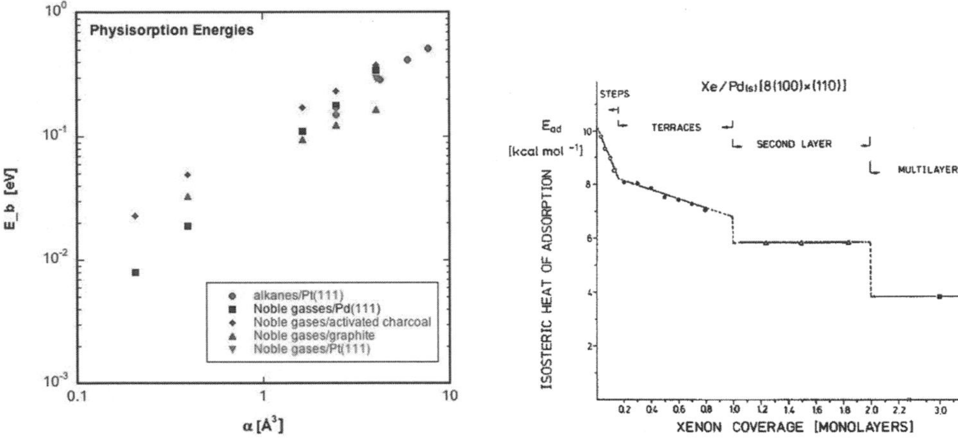

Figure 7.4 Left: Measured physisorption binding energies for noble gases and hydrocarbons on different surfaces as a function of polarizability of the adsorbed species. Data taken from [18, 29, 106, 112]. Right: Isosteric heat of adsorption (the negative of the adsorption enthalpy) for Xenon atoms on a Pd surface, revealing the presence of steps with enhanced binding. With increasing surface coverage, terraces are populated, followed by assemblage of a second layer, each characterized by its own binding energy. After the second layer, further layers are bonded by the bulk heat of adsorption for Xe films. A binding energy of 10 kcal/mol corresponds to about 430 meV or some 5000 K. Figure taken from [74]

To first approximation, the potential energy curve can be well represented by a simple, reduced potential, with a well depth D and a length scale $\ell = (C_3/D)^{1/3}$. As expected from Eq. (7.14), for a given surface, physisorbed binding energies scale well with the polarizability of the adsorbed species (Figure 7.4), albeit not always linearly. Experiments are in reasonable agreement with a single, heuristic, reduced potential given by the sum of a repulsive and a van der Waals attractive term,

$$\frac{V(z)}{D} = \left(\frac{3}{u-3}\right) \exp\left[-uz/a\right] - \frac{1}{(z+a)^3}, \tag{7.15}$$

where z is a normalized (to ℓ) coordinate above the surface, $a^3 = (1-3/u)$, and u is a fitting parameter. Graphite and noble gas solids have, $u \simeq 6.5$. For alkalide halides, $u \simeq 5.25$. Current theoretical studies focus on applying DFT to the adsorbate interaction. The difficulty resides into accounting for the dynamical correlation between fluctuating charge distributions that is at the basis of the van der Waals interaction. Recent attempts focus on calculating dispersion coefficients and have met with quite some success.

The physisorbed well is typically at a distance of some 2Å above the surface. Typical physisorbed energies range from 20 meV for Helium to up to 500 meV for highly polarizable species. Higher binding energies are obtained for species with large polarizabilities, e.g., PAHs. Likewise, species that can form hydrogen bonds will have a higher physisorbed energy. Near the surface, the continuum approach that underlies Eq. (7.14) will break down as the discrete nature of the atoms in the surface will lead to a potential energy landscape characterized by hills, valleys, and saddle points. Surface structure associated with kinks and steps on the surface can lead to further variations in the binding energy as, in essence, the adsorbant has additional nearest neighbors, increasing the binding. Figure 7.4 illustrates this for xenon adsorption on a stepped surface of palladium. Under these conditions, adsorbates first bind to the highest binding sites associated with the steps before filling up the terraces. The second layer still "feels" the metal surface but by the third layer, the van der Waals interaction is dominated by the xenon film. Initially, depending on the binding energy between adsorbates and the surface compared to the interaction between adsorbates, adsorbates may organize into islands rather than layers but eventually the influence of the surface will be lost and layers will form. Finally, with these kind of binding energies, species will only condense out at low temperatures ($\ll$ 100 K).

For some systems, the physisorbed site is a precursor state to a chemisorbed site where the adsorbate is bonded to the surface through overlap of one or more electronic orbitals (Figure 7.3). The binding energies involved are then much larger, $\simeq$1–5 eV). The chemisorbed potential wells are much deeper in the surface ($\simeq$0.3 1 Å) than the physisorbed sites. The two sites will be separated by an activation barrier, but, depending on the relative locations and depth of the physisorbed and chemisorbed sites, this barrier may be submerged. Figure 7.5 illustrates this for the interaction of H with graphene, which can be considered a model system for the interaction of H with graphitic surfaces, which is very relevant for H_2 formation in diffuse clouds. There is a physisorbed well with a depth of some 40 meV. Chemisorption requires a reconstruction of the carbon plane as the C-atom involved has to pucker out of the plane by some 0.5 Å. As a result, there is a substantial barrier of some 200 meV. The chemisorbed site has a depth of some 800 meV. As the H-atom approaches the surface, it is first trapped in the physisorbed site from which, it can thermally hop over or quantum mechanically tunnel through the activation barrier into the chemisorbed site.

The surface binding of a species can be much affected by surface structure as well as the presence of other adsorbates. Figure 7.6 show the result of Thermal Programmed Desorption study of D_2 to an amorphous solid water (ice) film. In these experiments,

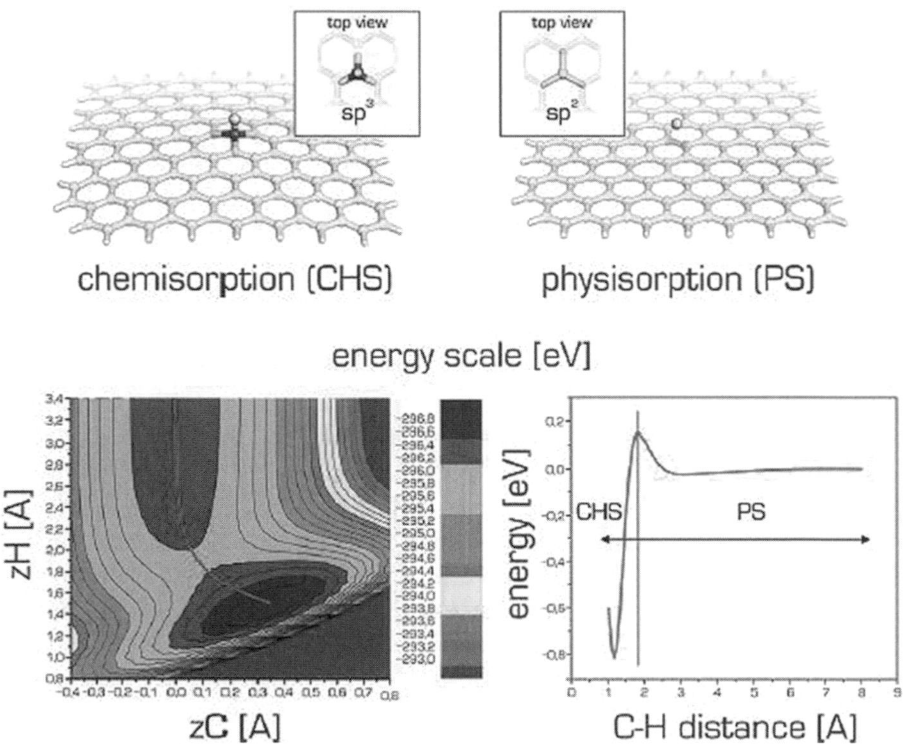

Figure 7.5 The calculated potential energy surface for the interaction of atomic H with graphene shows two minima associated with the physisorbed (PS) and the chemisorbed (CHS) site. A small activation barrier separates these two sites (bottom right). As the H-atom approaches the surface, it will first be trapped in the physisorbed site. For adsorption into CHS, the C-atom has to pucker out of the plane (bottom left; note the difference in scale for the x- and y-axis in this panel). The top two panels show a 3-D representation of the surface structure, illustrating the puckering of the plane. Reproduced from [87] with permission from the PCCP Owner Societies (A black and white version of this figure will appear in some formats. For the color version, please refer to the plate section.)

the ice was exposed to a beam of D_2, resulting in a low coverage (sub-monolayer) of physisorbed D_2. The ice was then slowly heated at a fixed rate (typically 10 K/min) and the amount of desorbing D_2 was measured as a function of temperature using a quadrupole mass spectrometer. At low surface coverages, D_2 evolves at much higher temperatures than at high surface coverage. For unreactive species, this type of TPD spectrum is characteristic for the presence of sites with a distribution of adsorption energies. D_2 is highly mobile at low temperatures and, at low coverages, will preferentially trap in the sites with the highest binding energies. As the exposure increases, more and more of the D_2 is "forced" into sites of lower binding energy and therefore released at lower temperatures. The inversion of this data into a distribution of binding energies is non-unique, but a possible distribution is shown in Figure 7.6. The range of adsorption energies is in good agreement with theoretical

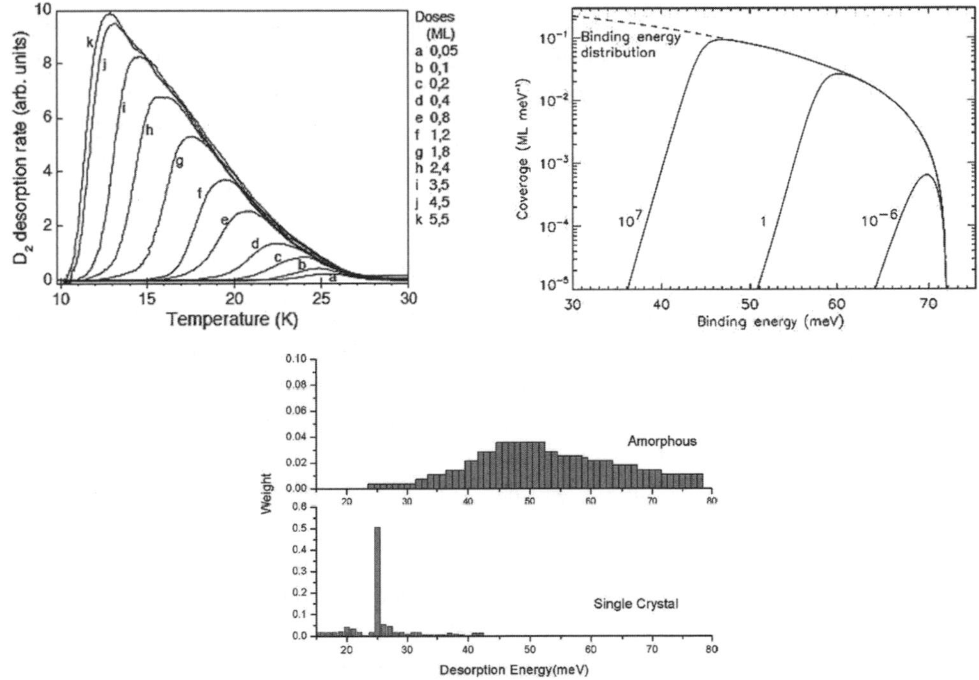

Figure 7.6 Top left: The D_2 desorption rate from an amorphous solid water surface as a function of temperature in Thermal Programmed Desorption experiments. The different curves correspond to different amounts of D_2 adsorbed on the surface (in units of monolayers (ML)). Top right: One possible adsorption energy distribution consistent with the TPD experiments. Curves are labeled by the gas phase H_2 density (in units of cm^{-3}) where the accreting and desorbing (at 10 K) flux are equal. Figure courtesy of [3, 63]. Bottom: Comparison of the binding site distribution for amorphous and single crystalline silicate. Figure taken from [52]

studies on well depth on amorphous solid water ice. Ignoring for the moment the small zero-point energy difference between H_2 and D_2, these results imply that, at 10 K, some 30–40% of the sites are occupied by H_2 in the ISM.

Other amorphous materials have similarly broad distributions of binding energies (Figure 7.6). As a result, H (and H_2) will bind to interstellar grain surfaces even at relatively elevated temperatures. In fact, even for single crystal surfaces there are a small number (5–10%) of enhanced binding sites. For a 1000 Å interstellar grain, this corresponds to $\simeq 10^5$ sites; a point that was well appreciated in the earliest H_2 formation studies.

In summary, binding energies to water ice can be reasonably well derived from the polarizability but there are exceptions. Notably, compared to their polarizability, atomic O is much more strongly bound while CH_4 is much less strongly bound. Species that can hydrogen bond – e.g. H_2O itself, as well as NH_3, CH_3OH, HCOOH – will also bond much more strongly to a water ice surface. Likely, the enhanced binding of atomic O on H_2O ice compared to its polarizability also reflects the formation of a hydrogen bond with the

O lone pair electrons. The binding of atomic O on a CO or O_2 ice is likely more modest, $\simeq$800 K. The origin of the low methane binding energy is not clear. True chemisorption may also occur for some species and quantum chemistry suggests that this is the case for atomic C. Finally, bonding will be affected by surface structure, including the presence of pores, in particular, for amorphous water ice. This will lead to a broad distribution of binding energies.

7.3.4 Surface Mobility

Surface migration from one site to the next is limited by a barrier against migration. In general, estimates for this barrier range from 0.1 to $0.5E_b$, depending on surface and adsorbate involved. For physisorption, a diffusion barrier of $0.3E_b$ is often adopted. Theoretical studies of H-atom mobility on amorphous solid water, indeed, predict values of $\simeq 0.3-0.4E_b$, but you should recall the existence of relatively deep physisorption wells on such surfaces (Figure 7.6). Assuming thermal hopping, the diffusion rate, k_{th}, is,

$$k_{th} = \nu_m \exp\left[-E_m/kT\right], \tag{7.16}$$

with ν_m an appropriate frequency factor. For a harmonic potential with depth E_b, we have,

$$\nu_m \simeq 5\times10^{12}\left(\frac{E_b}{10^3\,\mathrm{K}}\right)^{1/2}\left(\frac{1\,\mathrm{amu}}{m}\right)^{1/2}\quad \mathrm{s}^{-1}. \tag{7.17}$$

So, for atomic H bound by 500 K, the migration timescale is $\simeq 10^{-6}$ s^{-1}, increasing to 3×10^{-2} s^{-1} for a binding energy of 850 K (cf. Figure 7.6).

At low temperatures, quantum mechanical tunneling can become important. For a particle with energy E trapped in a well, tunneling to an adjacent site separated by a square barrier with height, E_m, and width, a, has a rate, k_{tun},

$$k_{tun} = \nu_m\left(1+\frac{E_m^2}{4E\left(E_m-E\right)}\sinh^2\left[\frac{2a}{\hbar}\sqrt{2m\left(E_m-E\right)}\right]\right)^{-1}, \tag{7.18}$$

with m the mass of the tunneling species. For particles with energy E above the barrier height, the rate can be found by replacing $E_m - E$ by $E - E_m$ and sinh by sin. The solution shows the well-known behavior of light reflection in a thin film. We can replace the argument of the sin and sinh functions by $(a/\lambda\!\!\!^{-})\sqrt{(E_m-E)/\pi kT}$, which relates the barrier width to the de Broglie wavelength, $\lambda\!\!\!^{-} = \hbar/\sqrt{2\pi mkT}$, and the energy to the mean energy of the system.

When the energy is much less than the barrier height, we recover the familiar WKB[1] approximation,

$$k_{tun} = \nu_m \exp\left[-\frac{2a}{\hbar}\left(2mE_m\right)^{1/2}\right]. \tag{7.19}$$

[1] Named after Wentzel, Kramers, and Brillouin.

Table 7.2 *Binding energies of astrophysically relevant species and surfaces*[a]

Species	α 10^{-24} cm^3	ASW[b] [K]	Graphite [K]	Silicate [K]
H	0.67	450–850[c]	480[h]	470[g]
D	0.63	545–980	530[h]	520[g]
C	1.8	720[d]		
N	1.1	720		
O	0.8	1300[e]	1400	
H_2	0.79	500–900[c]	530[h]	520[g]
HD	0.79	350–660[c]	550[h]	540[g]
D_2	0.78	350–660[c]	560[h]	550[g]
N_2	1.7	1435[f]		
O_2	1.6	1200	1300	
O_3	3.1	1800	2100	
CO	1.95	1575[f]		
CO_2	2.5	2500		
CH_4	2.45	1480[i]		
NH_3	2.1	5530		
H_2O	1.5	5700		
H_2CO	2.8	3260		
CH_3OH	3.2	5400		
HCOOH	3.3	5600		

Notes: [a]Taken from the compilation of [111] unless otherwise noted. [b]ASW: Amorphous solid water. [c]Broad distribution of binding sites with a few sites well above this range (see text for details) [3, 4]. [d]In analogy with atomic nitrogen. Quantum chemistry indicates the presence of a chemisorbed site with a binding energy of 14,100 K [92, 108]. [e]Binding energy to deepest wells. [f]Reference [36]. [g]Broad distribution of absorption sites measured for D_2 [52]. Other hydrogen isotopes scaled according to zero-point energy difference. [h]Measured for HD [61]. Other hydrogen isotopes scaled according to zero-point energy difference. [i]Measured on ice I_h [19].

We see that the rate depends exponentially on the barrier height and width and the particle mass. The quantum mechanical diffusion rate (Eq. (7.18)) has to be convolved with the distribution of thermal energies of the adsorbants to arrive at the actual rates. This introduces a temperature dependence but that dependence is much weaker than for the thermal hopping rates. The behavior of the migration rates is illustrated in Figure 7.7. At low temperatures, the tunneling rates are much larger than the thermal hopping rate but at high temperatures, the two rates converge. The dependence on barrier height and width and on particle mass are obvious. Other expressions for the potential barrier have been used and, while that can affect the detailed rate substantially, the key characteristic remains: at low temperatures, diffusion is dominated by tunneling while at high temperatures, thermal hopping takes over.

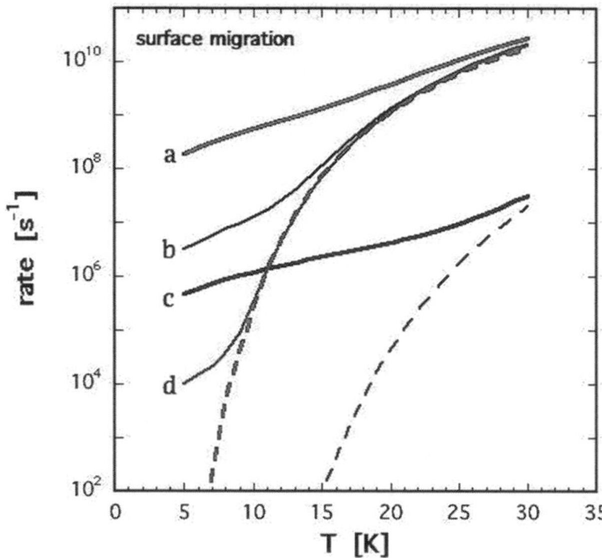

Figure 7.7 Migration rates on a grain surface. Solid curves correspond to quantum mechanical diffusion with $(E_m[K], a[\text{Å}], m[amu])$ given by $a = (165, 1, 1)$, $b = (165, 1, 2)$, $c = (370, 1, 1)$, $d = (165, 2, 1)$. The two dashed curves are the corresponding thermal hopping rates.

A number of experimental studies have been performed related to the mobility and recombination of hydrogen atoms to form H_2 on ice surfaces. While these studies have not always been interpreted in the same way, the consensus emerging from them is that surface mobility is much affected by the structure and surface morphology of the ice. Specifically, H_2O deposited at 10 K forms high density amorphous solid water and Thermal Programmed Desorption experiments (TPD; Section 2.4.3) reveal second order behavior for H_2 (and other inert species): the signature of a distribution of binding sites ($E_b \simeq$500–700 K, Figure 7.6). Such a distribution of binding sites hampers analysis of atomic H behavior and the mobilty of H has to be assessed from the molecular hydrogen behavior. For shallow binding sites on amorphous solid water ice, the site-to-site migration rate is estimated to be $\sim 10^9$ s^{-1} at 10 K. In deep sites, this rate is much slower, ~ 10 s^{-1}. Anticipating later discussion, it should be appreciated that the migration timescale is much faster than the accretion timescale on a grain and, in that limit, the exact timescale is less relevant for the ensuing chemistry.

Amorphous solid water is known to have a very porous open structure and this can influence the evaporation and diffusion process (Figure 7.8). In laboratory experiments, Knudsen-like diffusion through the gas phase in the pores can be important if,

$$D_g h(T) > D_s, \tag{7.20}$$

where D_g is the diffusion coefficient in the gas phase ($= \lambda v/3$, with λ the mean free path, v the velocity), h is a kinetic term relating the gas concentration to the total (gas

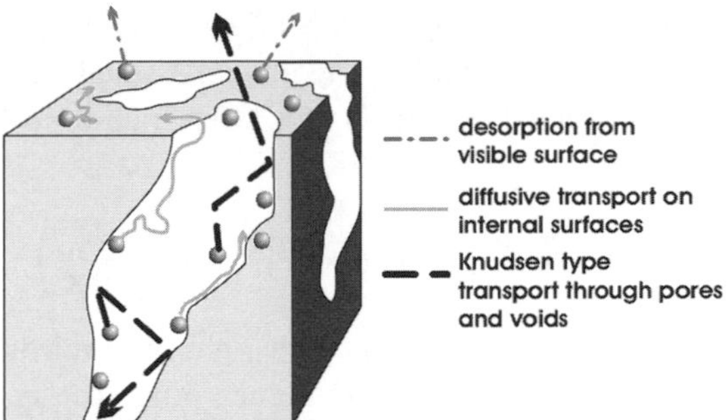

Figure 7.8 Schematic illustration of diffusion in a porous structure. The light-gray curly lines indicate transport on internal surfaces. Dashed black lines (Knudsen-like diffusion) indicate diffusion – following evaporation – in the voids and pores. Figure taken from [115]

plus surface) concentration ($\simeq \nu \exp[-E_b/kT]/(v/2\ell_{pore})$, and D_s the surface diffusion coefficient ($= D_0 \exp[-E_m/kT]$, with D_0 the pre-exponential factor ($= \nu a^2$ with ν an appropriate frequency factor and a the site dimension)). This leads to the condition,

$$\frac{E_b - E_m}{kT} < \ln\left[2\left(\frac{\ell_{pore}}{a}\right)^2\left(\frac{\lambda}{\ell_{pore}}\right)\right]. \tag{7.21}$$

The apparent surface area of amorphous water ice grown at low temperatures from a background gas ranges from 4×10^6 to 3×10^7 cm^2/g, which corresponds to pore radii of $\simeq$10 Å. As the binding energy is much larger than the migration energy, Knudsen diffusion is not very imporant and an H atom that enters one of the very many pores will have to migrate out through surface diffusion, and during that time evaporation is of no consequence. Pores are thus very conducive to molecular (hydrogen) synthesis as species will migrate without possibility of sublimation.

Diffusion of atomic O has been somewhat controversial. The earliest astrochemical models relied on luminescence studies of atomic O diffusing (and reacting) in low temperature noble gas matrices to conclude that atomic O is highly mobile inside these matrices and, by inference, also on interstellar ice surfaces. Recent studies imply O binding energies to amorphous ice surfaces of some 1250–1650 K, and migration barriers in excess of 600 K. This would imply that, on these surfaces, O could not effectively migrate until temperatures of $\sim$20 K (Figure 7.7). In contrast, studies on the reactions of O and O_2 mixtures indicate that atomic O is mobile even at temperatures as low as 10 K with migration rates of $\simeq 2$ s^{-1}. This difference may well reflect hydrogen bonding of atomic O with H_2O. For atomic N, lower binding energies and migration barriers have been derived (720 & 400 K).

7.3.5 The Interstellar Ice Case

In translating these results to the ISM, it has to be appreciated that – in cold molecular clouds – any deep potential wells will be blocked for H by adsorbed, abundant H_2.

Migration rates will then be controlled by the "open," lower binding energy sites; e.g. those with binding energies that allow efficient H_2 evaporation. A simple-minded estimate for this binding energy, E_0, can be obtained by equating (on a per-grain-basis) the H_2 accretion rate with the evaporation rate,

$$E_0 \simeq T \ln\left[4\mathcal{N}_{surf}\nu_m/n_0 v_{H_2}\right] \simeq 440\left(\frac{T}{10\,\mathrm{K}}\right), \tag{7.22}$$

with $\mathcal{N}_{surf}$ the number of sites per unit area and the right-hand side has been evaluated for a molecular hydrogen density of 10^4 cm^{-3}. Using the difference in polarizability and in zero-point-energy, the atomic H binding energy can then be estimated to be approximately $350\,(T/10\,\mathrm{K})$. Taking the migration barrier to be 1/3 of this, the migration rate (thermal hopping) would be $k_m \simeq 5\times10^7$ s^{-1}. So, on an interstellar grain in a dense, cold molecular cloud, almost all physisorbed sites will be occupied by molecular hydrogen. Every day, an atomic hydrogen is accreted which will "hover above the fray," rapidly migrating around and only now and then sinking down into a temporarily abandoned "deep" well but always quickly moving on – as it is being kicked out by an H_2 molecule – until it either reacts or evaporates. The H-atom residence time on a grain is dominated by the time spend in deeper wells, temporarily vacated by an H_2 molecule, and this will depend on the details of the surface morphology and H_2 coverage but all of that is directly tied to the accretion rate of H_2, and the evaporation timescale – k_{ev} (H) $\simeq 3\times10^{-4}$ s^{-1} – is not sensitive to the temperature. At high temperatures, the H_2 surface coverage is low, but then the atomic H thermal hopping timescale is very short anyway.

For heavier species, e.g., O and N atoms, the presence of adsorption sites with a range of binding energies also influences the analysis of migration and evaporation timescales. In this case, likely, H_2 would not block high binding energy sites as effectively. This aspect, too, has not been well sorted out.

The presence of molecular hydrogen also influences the hot atom diffusion during the accretion and thermal accommodation process. An H-atom accreting at 100 K from the gas phase onto a "clean" ice surface is calculated to transfer some 85 K of excess energy (thermal plus van der Waals energy) in each collision with an H_2O molecule and will then diffuse some 12 Å before being trapped in a surface site. For a surface with all sites filled with H_2, the energy transferred in a single collision of accreting H will be $\simeq$250 K and hot atom diffusion will not occur. This holds even more for an accreting heavier atom or a newly formed radical such as OH.

7.4 Surface Reactions

In this section, we will discuss the types of reactions that can play a role in interstellar grain surface chemistry. Reactions involving radicals are often barrierless and will occur upon collision. However, many reactions do possess appreciable activation barriers. As thermal hopping is often unimportant at the low grain temperatures, reactivity will be controlled by tunneling effects and this will be discussed in some depth. The emphasis will be on reactions involved in the conversion of CO into H_2CO and CH_3OH as well as the conversion of O and O_2 into H_2O. This section ends with a discussion of the chemistry involving H_2 and atomic O.

7.4.1 Radical–Radical Reactions

Reactions involving species with unpaired electrons can be expected to occur upon "collision" on the surface if one of the reactants is mobile on the surface. Thus, specifically, this includes reactions involving H atoms with themselves or with radicals such as OH, CH, etc. A number of relevant reactions are listed in Table 7.3. D-isotopic reactions can be included analogously to those of H-atoms. If migration timescales of C, N, O atoms are short as well, reactions involving these species may also have to be taken into account. The list in Table 7.3 is not complete as longer and longer carbon backbones can be made through sequential addition of, e.g., carbon atoms.

Table 7.3 *Radical–radical surface reactions*

Reactants				Products	Reactants				Products
H	+	H	⟶	H_2	O	+	O	⟶	O_2
H	+	O	⟶	OH	O	+	N	⟶	NO
H	+	OH	⟶	H_2O	O	+	C	⟶	CO
H	+	C	⟶	CH	O	+	CN	⟶	OCN
H	+	CH	⟶	CH_2	O	+	HCO	⟶	HCOO
H	+	CH	⟶	CH_3	C	+	N	⟶	CN
H	+	CH_3	⟶	CH_4	C	+	HCO	⟶	CCHO
H	+	N	⟶	NH	N	+	N	⟶	N_2
H	+	NH	⟶	NH_2	N	+	NH	⟶	N_2H
H	+	NH_2	⟶	NH_3	N	+	HCO	⟶	HNCO
H	+	O_2H	⟶	H_2O_2					
H	+	NO	⟶	HNO					
H	+	CN	⟶	HCN					
H	+	CN	⟶	HNC					
H	+	CNO	⟶	HNCO					
H	+	HCO	⟶	H2CO					
H	+	HCOO	⟶	HCOOH					
H	+	CH_3O	⟶	CH_3OH					
H	+	NCHO	⟶	NHCHO					
H	+	NHCHO	⟶	NH_2CHO					
H	+	CCHO	⟶	CHCHO					
H	+	CHCHO	⟶	CH_2CHO					
H	+	CH_2CHO	⟶	CH_3CHO					
H	+	N_2H	⟶	N_2H_2					

7.4.2 Radical–Molecule Reactions

Some radical-molecule reactions can occur as they have no or submerged barriers. Two key reactions are,

$$H + O_2 \rightarrow HO_2 \tag{7.23}$$

$$H + O_3 \rightarrow HO + O_2 \tag{7.24}$$

These will be further discussed when the chemical routes toward formation of water on grain surfaces are examined (Sections 7.4.4 and 10.6.2).

7.4.3 Reactions with Activation Barriers

The prolonged residence times of atoms on a grain surface allow reactions that are inhibited by activation barriers in the gas phase to proceed efficiently. Consider an H atom in a surface site where it can react with a species with a rate k_{reac}. This reaction will compete with migration to a neighboring site with a rate k_m. The probability for a reaction to occur is,

$$p_0 = \frac{k_{reac}}{k_m + k_{reac}} \simeq \frac{k_{reac}}{k_m}, \tag{7.25}$$

where we have assumed that the barrier against reaction is much higher than against migration. In general, the H-atom will migrate to the next site but a reaction may occur every time the H-atom enters a site with a possible coreactant. If the surface coverage of coreactants is θ_r, the maximum number of times a site associated with a coreactant is visited is given by,

$$n_r = \frac{k_m}{k_{ev}} \theta_r, \tag{7.26}$$

with k_{ev} the residence timescale of the H-atom on the grain. The probability that a reaction will occur on the kth visit to a site associated with the coreactant is,

$$p_k = (1 - p)^{k-1}\, p \tag{7.27}$$

The probability for a reaction before evaporation is,

$$p_r = \sum_{k=1}^{n_r} p_k = 1 - (1 - p)^{n_r} \simeq n_r p = \frac{k_{reac}}{k_{ev}} \theta_r. \tag{7.28}$$

As expected, the reaction probability scales directly with the total time spent in sites where the H-atom can react and is independent of the migration timescale. To get some feeling for the issue, we will evaluate this adopting $k_{ev} = 3 \times 10^{-4}$ s^{-1}. For a single coreactant on a grain with 10^6 sites, the 50% probability of reaction corresponds then to a reaction rate coefficient of, $k_{reac} \simeq 10^{-2}$ s^{-1}. With a frequency factor of $\nu_m = 5 \times 10^{12}$ s^{-1}, the probability to get over or through the barrier in a single try can be as small as 3×10^{-11} and the reaction will still occur readily. If the concentration of coreactants is as large as, $\theta_r = 10^{-1}$, this probability can be as small as 3×10^{-16}.

In general, an H-atom may choose from many possible coreactants. If we ignore evaporation, the probability for reaction with coreactant, i, is

$$\phi_i = \frac{k_{reac,i}\theta_i}{\sum_j k_{reac,j}\theta_j}. \tag{7.29}$$

We can also evaluate the competition between reactions with barriers and without barriers. Consider an H atom on a surface with one (other) radical present (e.g. reaction without a barrier) and species i with surface concentration, $\theta_i = 0.1$, where the reaction has a barrier. For a grain with 10^6 sites, the radical will "win" this competition if the reaction rate is less than $k_{reac} < 10^{-5}\,k_m$. For a migration rate of 10^9 s^{-1}, this corresponds to $k_{reac} < 10^4$ s^{-1} or a probability per "try" of $\sim 2 \times 10^{-9}$ to get over or through the barrier.

7.4.4 Surface Reactions Assisted by Tunneling

Hydrogenation (i.e. H + X → HX) is a key aspect of interstellar grain surface chemistry and, over the last 10 years, hydrogenation of astrophysically relevant species has been a focus of laboratory studies. If activation barriers are involved, these reactions will proceed slowly but as emphasized in Section 7.4.3, even if a reaction has a low probability per collision, it can still be important given the low accretion rates in the ISM. For such reactions involving H or D, tunneling will be very important. Tunneling corrections are notoriously difficult to determine. Estimates are often based upon the simple WKB expression for a square potential analog to Eq. (7.18), or a similar expression based on the asymmetric Eckart potential. We can get a crude estimate for when tunneling becomes important by comparing the tunneling probability with the Boltzmann probability to reach the top of the barrier (V). This allows us to define a critical temperature,

$$T_c = \frac{V\hbar}{2kS}, \tag{7.30}$$

with S the action integral,

$$p_{tun} = \int_a^b m\sqrt{V(s) - E}\,ds, \tag{7.31}$$

where a and b are the classical turning points ($V(a) = V(b) = E$, with E the kinetic energy of the particle) and s is the tunneling path. When $E \ll V$, this critical temperature simplifies to,

$$T = 75\sqrt{\frac{V}{1000\,\mathrm{K}}}\left(\frac{1\,\text{Å}}{a}\right)\mathrm{K}. \tag{7.32}$$

The importance of tunneling can be verified by isotope studies. Defining the kinetic isotope effect, IKE, as the ratio of the reaction rate coefficients of the light to the heavy isotope, we can write for the H-D system,

$$\ln\,[IKE] = \left(1 - \sqrt{\frac{m_H}{m_D}}\right)\frac{2S(D)}{\hbar}, \tag{7.33}$$

with m_i and $S(i)$ the relevant masses and action integral, respectively. Assuming a rectangular barrier and an energy much less than the barrier height, we arrive at,

$$\ln[IKE] \simeq 5.5 \left(\frac{a}{1\,\text{Å}}\right) \sqrt{\frac{V/k}{1000\,\text{K}}}, \tag{7.34}$$

which, with these parameters, results in a factor of 250 for IKE. Of course, this will also be affected by the difference in zero-point energy, which affects the barrier height but that is a much smaller effect than that caused by the mass difference. Analogous to the discussion around Figure 7.7, the dependence on E results in a modest temperature dependence for quantum tunneling.

This discussion assumes a rectangular barrier and the narrow width favors tunneling at low energies. That is OK as long as E is not too small compared to V (e.g. as for migration) but, for chemical reactions, the base of the barrier broadens and as a result the onset of tunneling is more gradual at very low temperatures. Calculated barriers are often approximated by the asymmetric Eckart potential where the energies, (V_f & V_b) at infinity are matched to the zero-point corrected energies as well as the second derivative of the potential along the reaction coordinate ($\partial^2 V/\partial q^2 = -m\omega^2$) at the barrier peak (the transition state). The equations describing the tunneling probability appear forbidding but they are essentially functions of $2\pi V_f/\hbar\omega$, $2\pi V_b/\hbar\omega$, and E/V_f. These expressions have to be integrated over the particle energy, introducing V_f/kT as a relevant parameter.

The mass to be used in these expressions deserves some further scrutiny. When considering the migration of a particle on a fixed lattice, the use of the mass of this particle is justified. However, when considering a reaction involving a polyatomic system, more care is required. Consider the colinear reaction,

$$\text{X} + \text{YZ} \rightarrow \text{XY} + \text{Z}. \tag{7.35}$$

Changes in the distance between X and Y and Y and X are related. We can represent the potential energy surface in 3-D space with axes, R_{XY}, R_{YZ}, and V. In such a representation, the path of the energy minimum reaction is curved in this space, following the potential energy minimum "valley," passing over a saddle point associated with the transition state, from reactants to products. For the tunneling reaction, we evaluate the action integral along this one-dimensional (but curved) path,

$$S = \int m_{eff} \sqrt{V(s) - E} ds, \tag{7.36}$$

with E the energy and V the potential evaluated over the dominant tunneling path, s, and we have introduced the effective mass, m_{eff}, to account for the change in reduced mass along this path as different particles become involved in the motion. For the colinear triatomic reaction X + YZ → XY + Z, the motion is fully characterized by the ratio, c, of the rate of change of the Y–X bond (dR_{XY}) to the rate of change of the Y–Z bond (dR_{YZ}),

$$c = \frac{dR_{XY}}{dR_{YZ}}. \tag{7.37}$$

And the effective mass is given by,

$$m_{eff} = \frac{m_X m_Z \, (1+c)^2 + m_Z m_Y c^2 + m_X m_Y}{(m_X + m_Y + m_Z)\left(1 + c^2\right)}. \tag{7.38}$$

When the particles going in and coming out have the same mass, $c = -1$ and the effective mass is given by,

$$m_{eff} = \frac{(m_Z + m_X)\, m_Y}{2\,(m_X + m_Y + m_Z)}. \tag{7.39}$$

The action integral is a compromise between pathways with small barriers and those with short tunneling lengths. In this, s is the (curvelinear) reaction coordinate, which couples the motion along the reaction coordinate to vibrational modes perpendicular to it that allows the system to "cut a corner" and tunnel through a reaction pathway that provides a shorter path than the reaction coordinate (increasing the transmission coefficient). In this formalism, this is accounted for through a (Jacobi) coordinate transformation expressed as an effective mass, which is related to the usual reduced mass, μ, by $m_{eff} = \mu(ds/dx)^2$ where dx is the reaction coordinate. This effective mass depends on the weight of each atom in the transition state from reactant to product as well as its relative contribution to the "transition motion." For the colinear tri-atomic reaction, this boils down to Eq. (7.38). This approximation is only valid near the transition state (but that is where most of the action is). In H (D) addition reactions, movement is essentially almost entirely associated with the incoming H (D) atom, and consequently, the effective mass is just 1 (2) amu. Abstraction reactions are also relevant and then the effective mass is 1 (2) amu if only H (D) is involved. For mixed isotopes, the effective mass is in between 1 and 2 amu.

Within the WKB approximation and assuming a square potential with height E_a and width a and that the kinetic energy is small, the tunneling rate is given by,

$$k = \nu \exp\left[-\frac{2a}{\hbar}\sqrt{2m_{eff}E_a}\right] \tag{7.40}$$

$$\simeq \nu \exp\left[-13.3\left(\frac{a}{\text{Å}}\right)\sqrt{\left(\frac{m_{eff}}{\text{amu}}\right)\left(\frac{E_a}{1000\,\text{K}}\right)}\right], \tag{7.41}$$

with ν the frequency factor in the well (Eq. (7.17)). Thus, the reaction probability depends on three factors: the size and width of the barrier and the effective mass of the (tunneling) particle. This could also be analyzed based on an asymmetric Eckart potential if quantum chemical calculations are available to determine good approximations to the actual barrier. There is a subtle point here as the Eckart potential is, among others, characterized by the curvature of the barrier at the top, which is often expressed in terms of a vibrational frequency. In this, it should be recognized that this frequency relates back to this effective mass.

To put this into context, let's adopt the WKB approximation and a width of 1 Å. A 50% probability for reaction of an H atom with a single coreactant on a surface with 10^6 sites requires a reaction rate coefficient of 10^2 s^{-1} (Section 7.4.3). With a pre-exponential factor of 2×10^{13} s^{-1}, the limiting barrier for this reaction is 3800 K. If the surface concentration is 0.1, then this limiting barrier for a 50% probability is 8000 K. It is clear that tunneling is a defining factor for grain surface chemistry in the diffusion regime (e.g. for low accretion rates). In that sense, interstellar ices are truly the watering holes of interstellar chemistry where species meet and mate.

In the next subsections, this formalism will be used to derive relevant parameters from experiments for a number of key reactions in interstellar grain surface chemistry. While the approximate nature of these expressions should be kept in mind, the results are closely tied to experiments. As interstellar grain surface chemistry is largely a competition between potential reactants for the highly mobile H (and D) atoms, this is a reasonable first approximation.

From Carbon Monoxide to Methanol

Laboratory techniques have been discussed in Section 2.4.3. Studies of hydrogenation reactions are performed under high fluence conditions and, as a result, H surface coverage is large and most H atoms are "lost" to reactions forming H_2. Deriving the factors controlling the tunneling rate is then difficult. Reaction rates can be derived using the kinetic isotope effect between H and the corresponding D reaction. As an illustration, consider H and D addition to CO. Reaction rate constants have been measured to be $R_{\rm H} = 6.8 \times 10^{-3}$ s^{-1} and $R_{\rm D} = 5.5 \times 10^{-4}$ s^{-1} but these contain the (unknown) surface concentrations of H and D atoms. These are set by the deposition rate, F_i, and the collision/reaction rate, $k_m(i)$, on the surface,

$$\frac{\theta(H)}{\theta(D)} = \left(\sqrt{\frac{F_H S_H}{F_D S_D}\frac{k_m(D)}{k_m(H)}}\right)^{1/2} \simeq 23, \qquad (7.42)$$

where we have assumed equal deposition rates and sticking coefficients and estimated the ratio of the migration rates assuming thermal hopping and a zero-point energy difference of 50 K, resulting in $k_m(H)/k_m(D) \simeq 500$. The barrier height for the reaction will differ because of the zero-point energy difference, and quantum chemical calculations for CO on an H_2O surface give 2740 and 2580 K for H and D, respectively. The effective masses are 1 and 2 amu. We can assume that the barrier width is the same and we have then,

$$a = \frac{\hbar}{2}\ln\left[\frac{\theta(H)\,k_{CO}(D)}{\theta(D)\,k_{CO}(H)}\right]\left(\sqrt{m_H E_a(CO,H)} - \sqrt{m_D E_a(CO,D)}\right)^{-1} \qquad (7.43)$$

$$\simeq 0.65\text{Å}, \qquad (7.44)$$

corresponding to a H plus CO rate of 6×10^6 s^{-1}. The reaction for D with CO is a factor 6×10^{-3} s^{-1} smaller. Note that the exposure timescales in these experiments imply all deep van der Waals sites are covered by H_2.

The next step in the hydrogenation of CO is reaction of H with H_2CO. Addition and abstraction reactions of H and D atoms with H_2CO and its fully deuterated analog have been studied in the laboratory as well,

$$H + H_2CO \rightarrow H_2 + HCO \quad (7.45)$$

$$\rightarrow CH_3O \quad (7.46)$$

Addition leads to the methoxy radical rather than the hydroxymethyl radical. Rates have been measured relative to the rate for the H + CO addition reaction. The resulting reaction parameters are summarized in Table 7.4. Note that abstraction is actually the preferred pathway. Methanol also reacts with H,

$$H + CH_3OH \rightarrow H_2 + CH_2OH, \quad (7.47)$$

and this reaction leads to the hydroxymethyl radical. This radical can react again with H_2 (see Section 7.4.5) or with H to reform methanol. These reactions have been studied in the laboratory using isotopically labeled species. Specifically, the relative rates for the reaction of D atoms with different isotopologues of methanol have been determined. Quantum chemical studies again provide reaction barriers and effective masses are 1.08 amu for these reactions. Analysis results then in a barrierwidth of 0.66 Å and this barrier width has been

Table 7.4 *Reactions with activation barriers*[a]

Reactants			Products		E_a [K]	a [Å]	k s^{-1}	Refs
H	CO	→	HCO		2740	0.65	6×10^6	[56, 114]
D	CO	→	DCO		2580	0.65	4×10^4	[56, 114]
H	H_2CO	→	CH_3O		2104	0.77	3.7×10^6	[45, 57]
H	D_2CO	→	CHD_2O		2063	0.77	4.2×10^6	[45, 57]
H	H_2CO	→	HCO	H_2	2959	0.51	7.6×10^7	[45, 57]
H	D_2CO	→	DCO	HD	3520	0.51	2.3×10^{6b}	[45, 57]
H	CH_3OH	→	CH_2OH	H_2	3224	0.66	1.4×10^6	[46, 47, 77]
H	CD_3OH	→	CH_2OH	H_2	4219	0.66	2×10^3	[46, 47, 77]
D	CH_3OH	→	CH_2OH	HD	3253	0.66	8.1×10^5	[46, 47, 77]
D	CH_2DOH	→	CHDOH	HD	3330	0.66	6.7×10^5	[46, 47, 77]
D	CHD_2OH	→	CD_2OH	HD	3430	0.66	4.2×10^5	[46, 47, 77]
H	H_2O_2	→	OH	H_2O	2508	0.45	7×10^{8c}	[81, 97]
D	H_2O_2	→	OH	H_2O	2355	0.45	3×10^{7c}	[81, 97]

Notes: [a]Energy barriers have been adopted from quantum chemistry studies and adjusted for zero-point energy differences. When available, results for species on a water ice surface have been adopted. The choice for the effective mass is discussed in the text. Barrier widths have been derived from the measured kinetic isotope effect or from a comparison to the CO + H reaction. [b]Note that this is not a colinear triatomic reaction. The effective mass for this reaction is 1.45 amu. [c]We have adopted effective masses for these reactions of 0.97 and 1.89 amu.

used to determine other relevant reaction rates in this system (Table 7.4). As abstraction of D from mixed isotopologues is not favorable, this is one way in which formaldehyde and methanol can be preferentially enriched in deuterium (Section 10.6.2) and is thought to be responsible for the very high deuterium fractionation of these species in Hot Corinos (Section 11.3.2).

Reactions Involved in Water Formation

There are several routes toward H_2O. Atomic O and H can directly react to form H_2O (cf. Section 7.4.1). Relevant here are the reactions of atomic H with O_2 – leading through H_2O_2 and OH to H_2O – and with O_3. The reaction of H with O_2 has a submerged barrier and occurs activationles in the gas phase. Laboratory studies reveal that this reaction occurs readily on ice surfaces as well. The product, the hydroperoxy radical (HO_2), reacts activationless with H,

$$H + HO_2 \rightarrow H_2O_2^{\star} \rightarrow H_2O_2 \tag{7.48}$$

$$\rightarrow OH + OH \tag{7.49}$$

$$\rightarrow H_2O + O \tag{7.50}$$

$$\rightarrow H_2 + O_2, \tag{7.51}$$

where $H_2O_2^{\star}$ is the highly excited complex, which can stabilize in various ways. There are no reaction barriers involved, although there are bottlenecks associated with the reorientation of the OH fragments along the minimum energy pathway. The excited intermediary has 3.7 eV of internal energy; $\simeq$2 eV above the 2OH dissociation channel. In the gas phase, OH is the favorite product but, on an ice surface, stabilization may occur through energy transfer to the phonon modes of the lattice or through dipole–dipole interaction with receptor modes of neighbouring species. Theoretical studies show that the reaction $H_2O_2 + H \rightarrow H_2O + OH$ occurs by simultaneous breaking of the OO bond and the formation of the OH bond. Energy barriers have been adopted from quantum chemical calculations and the barrier width is then estimated from isotope substitution experiments (Table 7.4).

The reaction of H with O_3 proceeds through the formation of a van der Waals complex,

$$H + O_3 \rightarrow OH + O_2. \tag{7.52}$$

In the gas phase, the long range potential steers and reorients the reactants and, along the minimum energy path where H attacks a terminal O in an out-of-plane approach, the van der Waals complex has a small submerged barrier, which is of no consequence. On an ice surface, the interaction with H_2O molecules could, in principle, affect the reorientation process and steric effects may hamper reaction. However, the reaction is known to occur rapidly on an ice surface and likely steric effects are minimal.

In many of these routes, OH is an intermediate product. As discussed in Section 7.4.5, OH will react rapidly with H_2 adsorbed on the grain surface to form H_2O, thereby releasing another H atom which hydrogenates other surface species. In a way, these oxygen-bearing species act thus as catalysts, $H + H_2 + XO \rightarrow H + H_2O + X$.

Other Relevant Reactions

Other reactions that have been suggested over the years are the hydrogenation of CS to H_2CS and CH_3SH. This is in complete analogy with the hydrogenation of CO, and gas phase barriers are known to be small enough to allow this reaction to proceed readily. The hydrogenation reactions of acetylene and ethene also have low activation barriers in the gas phase and are known to occur readily in low temperature matrices. Among the hydrogen abstraction reactions, H reacting with H_2S, N_2H_2, N_2H_4, and HNO are of interest. The radicals resulting from these reactions will react readily with (the next) H. This catalytic formation of H_2 may "siphon" off the H flux and can lead to H_2 formation, even at elevated ice temperatures. In this, the SH radical reacts readily with H_2 (Section 7.4.5) and hence the H-abstraction reaction is only relevant when the H_2 coverage is low.

7.4.5 Reactions Involving H_2

Reactions involving H_2 will have appreciable activation barriers. But because of the high surface coverage of H_2 on low temperature ($\lesssim$20 K) grain surfaces, such reactions may still occur. We can evaluate the probability for reaction of a radical with H_2 before another atom or radical arrives,

$$p_r = \frac{k_{reac}}{k_{ac}} \theta\,(H_2) \tag{7.53}$$

With an arrival rate of mobile coreactants of 10^{-5} s^{-1}, an H_2 coverage of 10^{-1}, and an effective mass of 2 amu, the limiting activation barrier is some 5000 K. Table 7.5 summarizes some relevant H_2 reaction on grain surfaces.

Table 7.5 *Reactions of H_2*

Coreactant		Products		$E_a{}^a$ [K]
NH_2	→	NH_3	H	3600
CH	→	CH_2	H	840
CH_2	→	CH_3	H	3600
CH_3	→	CH_4	H	4750
OH	→	H_2O	H	2959
CN	→	HCN	H	820
NCO	→	HNCO	H	2800
CH_3O	→	CH_3OH	H	2500

Notes: [a]Energy barriers for the corresponding gas phase reaction taken from the NIST database or from the combustion compilation by [110].

The reaction,

$$H_2 + OH \rightarrow H_2O + H, \tag{7.54}$$

is of particular relevance as OH is a common intermediary in all H_2O formation routes. This abstraction reaction proceeds through a transition state where the formation of the new OH bond proceeds simultaneously with the cleavage of the H_2 bond. This reaction is well studied in the gas phase – including isotope substitution – and has a substantial barrier, $\simeq$3000 K. On ice surfaces, laboratory studies also reveal this kinetic isotope effect for reactions involving H_2 and D_2 with OH (and no scrambling of the H's) with the H_2 reaction being about ten times more efficient than the D_2 reaction. Taken at face value, this would imply a rather small barrier ($a \simeq 0.4$ Å) and a very high rate ($\sim 10^{10}$ s^{-1}) for the H_2 reaction. However, these experiments were done under high H_2/OH deposition rates and actually the rate of H_2O_2 formation exceeded the H_2O formation rate. This would imply a more modest reaction rate 10^{-2} s^{-1}. Likely, some of the kinetic isotope effects measured in these ice reactions reflect differences in residence time of H_2 and D_2 on the surface and the actual barrier size and reaction rates are not well known. Nevertheless, these studies do reveal that the reaction proceeds rapidly at low temperatures in the presence of H_2 and as there is no "competition" the actual reaction rate is largely irrelevant.

Among other reactions, the reaction,

$$H_2 + CH_3O \rightarrow CH_3OH + H, \tag{7.55}$$

with a barrier of $\simeq$2500 K can be important. Note that, in contrast, the reaction involving the hydroxymethyl radical, CH_2OH, is endoenergetic by $\simeq$4400 K and will not play a role.

Finally, carbon chemistry on grain surface could be initiated by the H_2 insertion reaction with C and then quickly evolve toward CH_4, all through reactions with H_2. Similar insertion reactions with N are inhibited but NH_2 can react with H_2 on an interstellar grain surface. Other noted radicals are CN and NCO, but this list is far from complete.

7.4.6 Reactions of Atomic Oxygen

Atomic oxygen is more strongly bound to an H_2O ice surface than the van der Waals interaction would indicate (Section 7.3.3) and is therefore not mobile at 10 K. If O cannot react with an adjacent species on a timescale less than the accretion timescale of H ($\sim$1 day), then, likely, reaction with (the next accreted) H would lead to H_2O. Atomic O probably becomes mobile enough at temperatures $>$ 15 K to initiate a rich surface chemistry. On a CO, O_2, or N_2 surface, atomic O binds just by van der Waals forces and it can scan the surface for a reaction partner, opening up oxidation channels even at 10 K. Table 7.3 summarizes some relevant reactions involving atomic oxygen.

In addition to these radical reactions, we may have to consider oxidation reactions with activation barriers as timescales in the ISM can be long. Taking the accretion timescale ($\simeq 10^{-5}$ s^{-1}) as the limiting factor, atomic O can overcome an activation barrier of $\simeq 40\,T$ K. The reaction of atomic O with O_2 is known to occur at low temperatures but only if no radical–radical reaction channel is open (e.g. in the absence of H). Hence, there is a small activation barrier that has to be overcome. The reaction with HCO is another reaction that occurs at low temperatures, and in the gas phase, products are CO_2 + H as well as CO + OH. Two other reactions that may be relevant are O + CS → OCS and O + SO → SO_2 both of which occur in low temperature matrices. The reaction O + NO forming NO_2 may also be relevant.

The reaction of atomic O with CO has a puzzling history. Chemiluminescence studies reveal that $O(^3P)$ reacts with $CO(X^1\Sigma^+)$ in low temperature matrices to form $CO_2(^3B_2)$ in the triplet state from which it decays radiatively. However, gas phase studies as well as quantum chemistry calculations reveal a substantial barrier ($\simeq$2500 K) to this reaction pathway and calculated tunneling timescales are too long to make this reaction competitive in matrices or on ice surfaces. There is some experimental support for this reaction on an ice surface but the reaction is shown to be slow and if other reaction partners with low or without barriers are available (e.g. O_2 or a migrating radical (H)), then those are greatly favored. Nevertheless, given the long accretion timescales in the ISM, this reaction can still be important.

7.4.7 Reactions of "Hot" Radicals

The Harris–Kasemo reaction mechanism relies on hot atoms coming in, diffusing around on the surface and meeting a reaction partner before its energy is dissipated and the species is "trapped." Accreting species will be accelerated by the van der Waals potential and will "land" with appreciable kinetic energy. This may lead to some surface diffusion; e.g. on a bare surface, an H atom is calculated to move some 15 Å and visit several sites during the thermal accommodation process. However, at 10 K, interstellar grain surfaces will contain an apprciable amount of H_2 and energy transfer from an incoming species to this light molecule will be very efficient. Likely, this process may only become important at temperatures >20 K when thermal evaporation keeps the surface clean from H_2.

A modified version of the Harris–Kasemo reaction mechanism involves newly formed radicals. As a result of a reaction, a species may acquire a substantial amount of translational motion on the surface and diffuse some distance while colliding with potential reaction partners. As the newly formed species may also have appreciable internal excitation, reactions with appreciable barriers need to be considered as well. Experiments indicate that the reaction of OH with CO leading to CO_2 may be very relevant. The importance of this reaction will depend then on the competition between adsorbed CO and H_2 for the hot OH. At 10 K, H_2O formation may prevail while at higher temperatures CO_2 formation may take over. In this case as well, at low temperatures, diffusion of hot radicals over the surface may be limited by adsorbed H_2.

7.5 Chemical Networks: Rules of Engagement

The chemical network will depend on the chemical composition of the accreting gas, their accretion rates, and the migration timescale of accreted species on the grain surface.

Table 7.6 *Composition of the accreting gas in molecular clouds*[a]

Species	Gas phase abundance[b]
H_2	0.5
He	0.2
H[c]	$2/n$
D[d]	$2 \times 10^{-2}/n$
CO	8 (−5)
O[e]	2.4 (−4)
C	8 (−7)
N_2[f]	4 (−5)
H_2CO	2 (−8)
CH_3OH	2 (−9)
OH	3 (−7)
H_2O	5 (−9)
NH_3	2 (−8)
HCN	2 (−8)

Notes: [a]Taken from observations unless otherwise noted. a (-b) stands for $a \times 10^{-b}$. [b]Relative to H-nuclei. [c]Set by the balance between cosmic ray ionization of H_2 followed by dissociative recombination of the ions resulting, leading ultimately to H formation, and accretion of H on grain surfaces. n is the hydrogen nuclei density. [d]Atomic D abundance estimated from the observed DCO^+/HCO^+ ratio. [e]The main reservoir of oxygen in the gas phase is very uncertain. [f]Assuming all the gas phase nitrogen is in the form of molecular nitrogen.

Table 7.6 summarizes the composition of the gas phase inside dark cores of molecular clouds. Of course, H_2 (and He) is the dominant species in the accreting gas. The dominant accreting radicals are H, D, O, and C, while CO and N_2 are by far the key molecules accreting. The accretion timescale is much longer than in typically terrestrial laboratory settings. Interstellar grain surface chemistry is, therefore, in the "diffusion" limit where the reaction rate is essentially limited by the rate at which species are transported to the surface. At any time, at most two species that can migrate appreciably within an accretion or evaporation timescale (whichever is shorter) are present on any grain surface (compare Equations (7.2), (7.16), and (7.18)). Now, an accreted H (and D) will scan the whole surface very rapidly. On an H_2O surface, only N atoms can also scan an appreciable fraction of a grain surface, searching for a reaction partner before a next radical lands. On a CO, CO_2, N_2, and/or O_2 surface, O and C atoms will also be mobile. In principle, newly accreted species can "hop" around during the thermal accommodation process. However, the high surface concentration of H_2 will result in a rapid loss of kinetic energy and this is likely not important. Other species are trapped in the site where they originally land and are "forced" to wait for a reaction partner.

Thus, the reactions that need to be considered are reactions of migrating atoms (H, D, C, N, and O) with themselves and with other nonmobile species. The ability of atomic H to tunnel through even appreciable reaction barriers opens up hydrogenation as a key route toward chemical complexity in ices. In particular, reactions of H with O_2 and O_3 leading to H_2O and with CO and H_2CO to CH_3OH are key in this respect. On a surface where atomic O is mobile, oxidation of radicals can also be important, leading for example to formic acid (HCOOH) or methelyne glycol ($CH_2(OH)_2$) or methyl peroxide (CH_3OOH). Rapid gas phase reactions lock up most of the carbon in CO (rather than C), and formation of methane (CH_4) and more complex molecules (e.g. ethanol, CH_3CH_2OH) is inhibited. An appreciable fraction of the nitrogen might be in atomic form, leading to NH_3 formation as well as more complex species such as formamide ($HCONH_2$). Finally, at low temperatures, H_2 is an important surface species occupying some 50% of the available sites. A good fraction of this H_2 will actually be mobile, and newly formed or accreted radicals that can react with H_2 will preferentially do so.

7.6 Desorption

7.6.1 Thermal Desorption

The thermal desorption rate is set by the binding energy, E_b, of a species to a grain surface,

$$k_{des} = k_0 \exp\left[-E_b/kT\right], \tag{7.56}$$

where k_0 is the entropy factor, given by,

$$k_0 = \left(\frac{kT}{h}\right)\left(\frac{Z_{rot}}{Z_{vib}}\right), \tag{7.57}$$

where Z_{rot} is the rotational partition function of the species in the gas phase and Z_{vib} is the vibrational partition function of the adsorbed species. An adsorbed species has replaced (at least) one degree of translational freedom with motion of the surface bond. If this mode is fully excited, we have $Z_{vib} = kT/h\nu_z$ with ν_z given by,

$$\nu_z = \left(\frac{2N_s E_b}{\pi^2 m}\right)^{1/2} \simeq 5\times 10^{12}\left(\frac{E_b}{1000\,\mathrm{K}}\right)^{1/2}\left(\frac{\mathrm{amu}}{m}\right)^{1/2} \quad \mathrm{s}^{-1}, \tag{7.58}$$

where N_s is the surface density of sites. For a monatomic gas, $Z_{rot} = 1$, and we have $k_0 = \nu_z$. This makes intuitive sense as the evaporation rate is then the Boltzmann factor times the frequency at which the adsorbate exchanges energy with the surface. If the species is immobile on the surface, an additional factor has to be included, $2\pi mkT/N_s h^2$, which describes the ways the adsorbate can be distributed over the surface. The vibrational partition function is then 1. Rotational partition functions have been discussed in Section

3.2.8 and for a linear molecule, $Z_{rot} = kT/hcB$ with B the rotational constant. Finally, binding energies have been discussed in Section 7.3.3.

Temperature Programmed Desorption (Section 2.4.3) is an effective way to determine the thermal desorption behavior and binding energies of ices and species trapped in ices. The sublimation behavior of pure and mixed ices is illustrated in Figure 7.9. Pure ices show

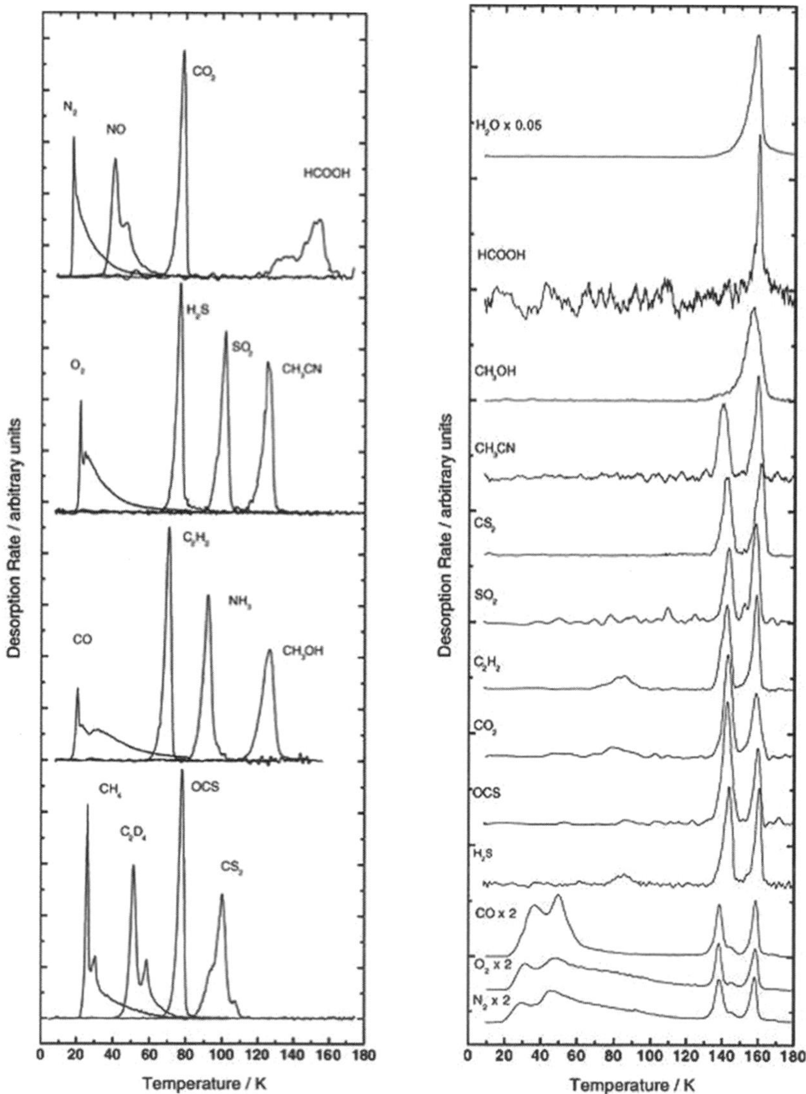

Figure 7.9 Left: Temperature Programmed Desorption trace of different pure ices illustrating the sublimation behavior and its dependence on binding energy. Right: Temperature Programmed Desorption trace of different species codeposited with an overabundance of H_2O ice. Figures taken from [22, 23]

a single peak corresponding to its binding energy. For mixed ices, a more complex behavior is seen. For apolar species (e.g. CO, N_2) present at substantial abundances, release first occurs at low temperatures where regions of high concentration (and low binding energy) sublimate. A second peak occurs around 120 K, when amorphous water ice crystallizes and the advancing crystalization front expresses contaminants out, which then sublimate. A small fraction of impurities can be trapped inside the ice even at high temperatures in clathrate-like structures. These are only released when H_2O itself sublimates. Species such as CH_3OH, HCOOH, and NH_3 partake in the H-bonding network of H_2O and are only released when H_2O sublimates. Layered ices show an even more complex behavior. Very volatile ices deposited on top of H_2O ice will sublimate largely when they are heated to their sublimation temperature. However, amorphous solid water deposited at low temperatures is very porous and some of the volatile outer ice may diffuse into these pores, and, as these pores collapse, be trapped (Figure 7.10). After this first release, these ices will thus behave like mixed ices.

At 10 K, all species (except H_2 and He) will stick on grain surfaces and will not sublimate over relevant timescales. For pure ices, CO and N_2 have a binding energy of $\simeq$800–900 K. Balancing accretion with sublimation, we find a sublimation temperature

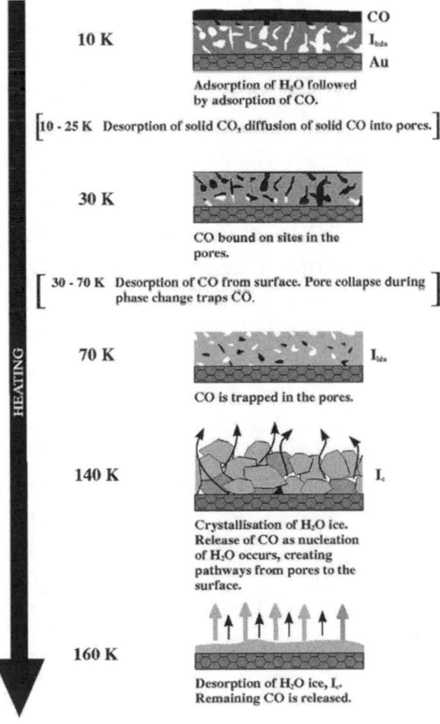

Figure 7.10 A cartoon describing the different processes relevant to the sublimation of mixed molecular ices. Figure taken from [22]

of $\simeq$15 K at a density of 10^4 cm^{-3}. Other pure ices will sublimate at concomitantly higher temperatures. The binding energy of many species is higher on an H_2O ice surface. For example, the sublimation temperature for CO on a H_2O ice is $\simeq$25 K. Measured dust temperatures in dark cloud cores are less than 10 K, slowly rising to about 14 K at an $A_v = 0.1$ magn. Hence, thermal desorption is never important in dark clouds. Once, a protostar is formed and heats up its environment, ices will desorb. H_2O ice will not desorb until $\simeq$90 K on interstellar timescales; e.g. only in Hot Cores/Hot Corinos (Section 11.3) directly surrounding a protostar. Traces of volatile species such as CO can remain trapped in H_2O ice to high temperatures. Interstellar timescales are much longer than in a laboratory setting. As a result, diffusion will tend to be more important in the ISM case. We note though that the propagation of the crystallization front will be driven by the latent heat of crystallization and hence will occur on a short timescale. The high temperature behavior of mixed interstellar ices may thus be very similar to that of laboratory samples.

7.6.2 Photodesorption

Absorption of a UV photon by an ice molecule will lead to electronic and vibrational excitation. This excitation will translate into translational motion through collisions with the neighbors. It may also lead to fragmentation where the fragmentation products acquire substantial kinetic motion. If the molecule is at the surface, it may well be ejected into the gas phase. Excited molecules created deeper in the ice mantle will have to "burrow" themselves to the surface before they can escape. During this process, they may transfer collisional energy to neighboring species and promote their ejection. Hence, this may lead to the ejection of species that do not, or poorly, absorb in the FUV range. The photodesorption yield is a convolution of the UV absorption cross section of the ice species with the incident UV radiation field and then mitigated/enhanced by the energy transfer to other other species. Results of photodesorption yields are summarized in Table 7.7.

Table 7.7 *Photodesorption yields*

Species	Yield[a]	E_{th}[b] [eV]	Reference
H_2O	3×10^{-3}	–	[5, 83, 113]
CO	$1 - 2 \times 10^{-2}$	8	[34, 84]
CO_2	10^{-3}	10.8	[37]
CH_4	$2 - 5 \times 10^{-3}$	9.5	[32]
NH_3	2×10^{-3}	–	[68]
CH_3OH	$< 3 \times 10^{-5}$[c]	–	[25]
N_2	0.025	12.4	[35]

Notes: [a] Yield per absorbed photon. [b] Absorption threshold. When no value is given, the experiments refer to broad wavelength irradiation. [c] No desorbed CH_3OH observed in experiments. Photodissociation products are released.

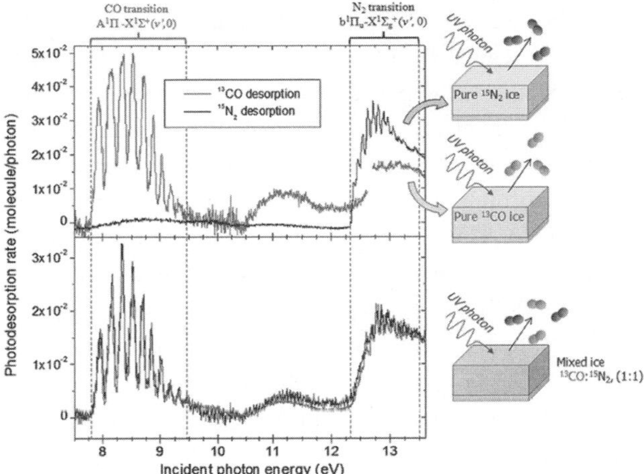

Figure 7.11 Photodesorption yields as a function of photon energy. Top: Yields measured for pure CO and pure N_2 ices. Bottom: Yields measured for a mixed CO/N_2 ice. Figure taken from [8]

Molecular dynamics studies of H_2O have provided insight into the relative importance of these processes. Photo absorption leads to fragmentation and, for absorption in the three top monolayers, the resulting H atom has a good chance of escaping. On the other hand, OH only escapes in $\simeq 2\%$ of the cases. Ejection of H_2O – either by knock out or by recombining H+OH radicals – occurs in less than 1% of the UV photon absorptions in the top two layers. Similar processes play a role for other species. The measurements shown in Figure 7.11 illustrate another important point. Pure CO and N_2 photodesorption spectra link the photoabsorption spectral structure to individual electronic transitions. The results for mixed CO:N_2 ices reveal that, even though only one species absorbs, both species desorb, providing direct evidence for the importance of energy transfer in the ice and knock out desorption. Similar results are obtained for C_6H_6:H_2O mixed ices excited through resonant absorption of benzene. In a way, this simplifies the issue of photodesorption as the yield of trace species can be related to absorption by the major ice component.

In summary, for species such as CO or N_2, which do not fragment for photon energies less than 13.6 eV, electronic excitation leads directly to excitation of the translational mode and ejection. Most species will, however, fragment and if that leads to an H-atom, the H will carry most of the excess energy. A light species with a low binding energy, such as CH_4, will have a high photodesorption yield through the knock-on process. However, energy transfer to neighboring species is of little importance if those species are very heavy (eg. large species) and/or have a high binding energy (e.g. CH_3OH). Photodesorption of these species will only occur if they are traces in a predominantly CO ice.

7.6.3 Chemical Desorption

Relevant chemical reactions on grain surfaces are typically highly exothermic, releasing some 4 eV of energy. Much of this may be in the form of internal vibrational and rotational excitation. The excited species will relax and as radiative relaxation involves long timescales ($\simeq 10^{-3}$ s), energy transfer to the bulk (e.g. heating of the grain) is likely the dominant process. As part of the process, some energy may wind up in center-of-mass translational motion of the newly formed species perpendicular to the surface; e.g. the molecule is ejected, possibly with substantial internal energy. In principle, energy transfer to the surface can occur through excitation of local phonon modes. Energy transfer is a quantum process and will only be efficient when donor and acceptor modes are closely matched in energy. Indeed, such energy transfer shows an energy gap law behavior with a rate that depends exponentially on the order of the multiphonon processes involved. The large energy gap between phonon modes of the ice ($\sim$200 cm^{-1}) and the internal molecular vibrations ($\simeq$3000 cm^{-1} if H is involved) implies that many phonons have to be excited simultaneously and this strongly inhibits this process. Likely, transfer to acceptor modes of nearby molecules will then be the dominant process, where small energy differences can be made up by simultaneous excitation of a phonon mode and/or of the van der Waals bond linking the species to the surface. This intermolecular coupling occurs through dipole–dipole interaction, and when the energies are well matched ($\sim$100 cm^{-1}) occurs on a timescale of nano seconds. Any excitation that may wind up in the van der Waals bond will couple readily to the phonon modes (which will have similar energies) and such excitation can quickly "leak" away. Whether or not molecule ejection will occur depends then on the rate at which the surface bond is excited by vibrational energy transfer to neighboring species as compared to the rate at which this mode is deexcited by transfer to the bulk phonons.

There are few experimental or theoretical studies addressing this process. Experimentally, the ejection of newly formed H_2 has been studied on graphitic surfaces and found to be very efficient. In this case, the energy gap between vibrationally excited H_2 and the CC modes of the surface is quite large and excited H_2 molecules are ejected. In the water formation routes on interstellar grain, the reactions of O + H and OH + H have high measured chemical desoprtion efficiencies ($\sim$0.3) but for the other astrophysically relevant routes, this efficiency is < 0.1. In contrast, in theoretical studies an efficiency of $\sim 10^{-3}$ is indicated for H_2O formation from OH + H. For H_2S, the inferred chemical desorption yield is measured to be $\gtrsim 10^{-3}$.

In the absence of clear experimental or theoretical guidance, astrochemical models often assume an arbitrary efficiency of release on formation. As the composition of interstellar ices and dark cloud cores are observed to be very different, this efficiency must be small and 10^{-2} is often adopted. But as the scant data indicates, chemical desorption is system-sensitive and no hard-and-fast rule has been formulated yet.

7.6.4 Sputtering

Interstellar shock waves are efficient in releasing adsorbed ices into the gas phase. In a J (jump) shock wave, the gas is quickly stopped but, because of their large inertia, grains will move relative to the gas at 3/4 of the shock speed until they have encountered their own mass in impinging gas atoms. As the stopping length is larger than the size of the shock front or the postshock cooling zone, charged grains will be betatron accelerated by the compression of the magnetic fluid. Hence, substantial velocity differences between gas and dust may persist for a long time. In a C (continous) shock, the ion fluid is smoothly compressed and accelerated and the neutral fluid is heated and accelerated by the drifting ions over a length scale that is long compared to the cooling length of the warm gas. So, in contrast to a J shock, the kinetic and thermodynamic characteristics of the gas do not change "instantaneously" to a high value from which the gas cools down, but rather the gas is heated continuously. As for J shocks, grains and gas will drift with respect to each other over a long length scale. Impinging species will have substantial energies and they can sputter ice species into the gas phase.

There is a vast literature on ion sputtering of materials. Energetic ion sputtering is a linear cascade in which an impinging atom undergoes binary collisions with substrate atoms until they are brought to a halt. Recoiling atoms may acquire enough momentum that they "escape." The sputtering yield will depend on the nuclear energy deposited per unit length relative to the binding energy, U_0, of the substrate; viz.,

$$Y(E) = 4.1 \times 10^{14} \frac{\alpha S_n(E)}{U_0}, \tag{7.59}$$

with $S_n(E)$ the nuclear stopping cross section of a projectile incident with energy, E, and α an energy independent function depending on the mass ratio between target and substrate atoms. Extensive models have been developed that fit the experimental results of a wide variety of materials. An example for H_2O ice is shown in Figure 7.12. Note that for high energy projectiles, electronic energy transfer dominates and this heats the target and drives a thermal evaporation wave. This results in a heat spike within a cylindrical region with a size, r_c, of typically some 50 Å. At 100 km/s, an incoming H_2 molecule will have an

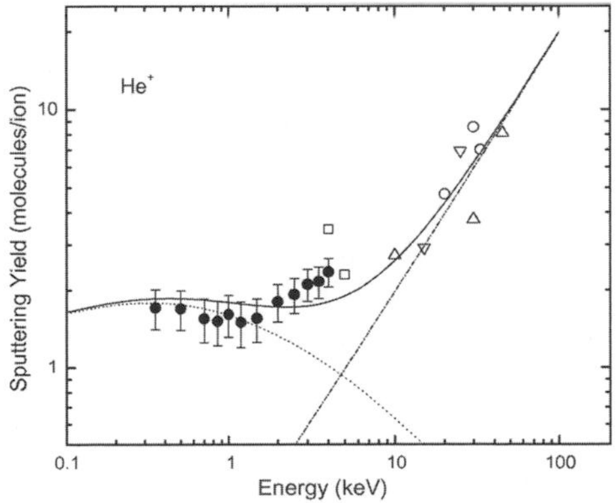

Figure 7.12 Measured sputtering yields of H_2O ice by He^+. The dotted line is proportional to the elastic stopping cross section. The dashed-dotted line is proportional to the electronic stopping cross section. The solid line is the sum. Figures taken from [33]

energy of 100 eV and relevant sputtering yields are substantial. He atoms will have a yield that makes them slightly more efficient in sputtering as the increased yield more than offsets the lower abundance.

As an order of magnitude estimate, we will ignore betatron acceleration and consider a grain slowed down by encountering its own mass. The number of sputtered species is,

$$N_{sput} = \frac{4\pi a^3 \overline{\rho}_s}{3m} \overline{Y}, \tag{7.60}$$

with $\overline{Y}$ the average sputtering yield, m the mass of the incoming particles, and $\overline{\rho}_s$ the average specific density of the grain material. The sputtered fraction of an ice mantle with a thickness Δa ($\Delta a \ll a$) is then given by,

$$f_{sput} = 2\,\frac{a}{\Delta a}\,\frac{\overline{\rho}_s}{\rho_{ice}}\,\overline{Y}. \tag{7.61}$$

Adopting $\overline{\rho}_s/\rho_{ice} = 2$, we can conclude that complete mantle removal requires $\overline{Y} \gtrsim 0.15\Delta a/a$ and, as $\Delta a/a$ is typically much less than unity, the corresponding threshold kinetic energy is modest and even a 10 km/s shock can strip a grain clean.

7.6.5 Cosmic Ray Driven Desorption

Cosmic ray desorption from interstellar ice grains is dominated by impacts of heavy atoms. A penetrating energetic ion deposits some 10–100 eV/Å (10–100 GeV/cm) in electronic excitation in its wake. This energy will couple to the lattice through nonradiative processes, driving a shock wave and heating the grain. Initially, particles will be ejected from around the entrance (and exit) channels but, once the energy spreads, a much larger surface area may reach the sublimation temperature and lead to quasi-thermal evaporation. Figure 7.13 illustrates the interaction of a heavy cosmic ray with a very small (130 Å) CO grain, which is completely vaporized by the deposited energy. For larger grains, the interaction can be captured through a so-called thermal-spike model where (part of) the grain cools down from an initially high temperature through evaporation of ice molecules.

Various models have been developed to estimate the sputtering of ice grains by Cosmic Ray ions. Figure 7.14 illustrates the results for a CO and H_2O ice, which differ in their binding energy. For comparison, a 100 Å thick ice mantle on a 1000 Å grain contains $\sim 4 \times 10^7$ species. With an Fe cosmic ray intensity of $\mathcal{N}_{CR} \simeq 10^{-4}$ ions cm^{-2} s^{-1} sr^{-1} at 100 MeV, the timescale between successive cosmic ray hits is,

$$\tau_{Fe-CR} = \left(4\pi\, a^2\, \mathcal{N}_{CR}\right)^{-1} \simeq 8 \times 10^4 \left(\frac{1000\,\text{Å}}{a}\right)^2 \text{yr}. \tag{7.62}$$

Hence, over the lifetime of a dense core, an icy grain will be hit multiple times by an energetic Fe cosmic ray and $\simeq 10^4$ species will be ejected from an H_2O ice mantle, and an order of magnitude more from a CO ice. Again, this is a small fraction of the total ice on a

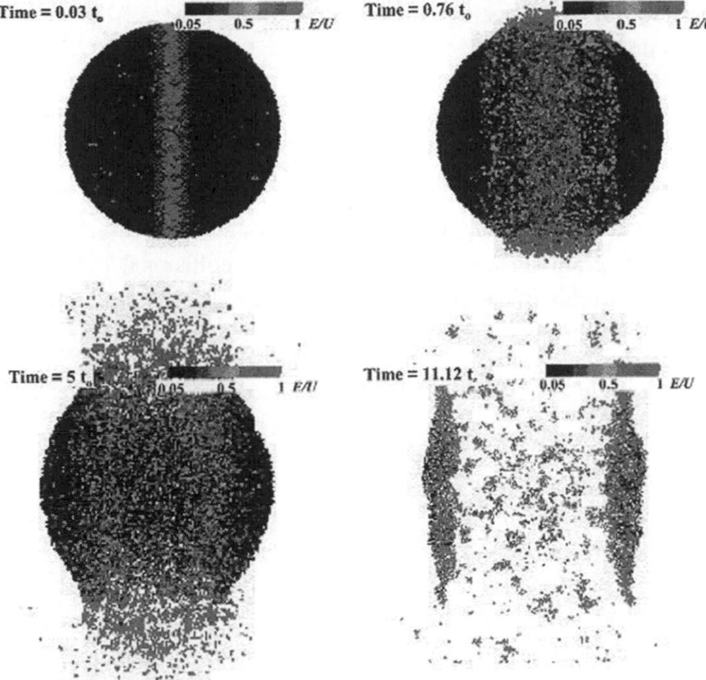

Figure 7.13 Molecular dynamics study of the interaction of a heavy cosmic ray with a small CO-ice grain. The total energy deposited is 8 keV. Each panel is a snapshot at a given timestep. The different colors correspond to energy in units of the binding energy, U_0. Figure taken from [12] (A black and white version of this figure will appear in some formats. For the color version, please refer to the plate section.)

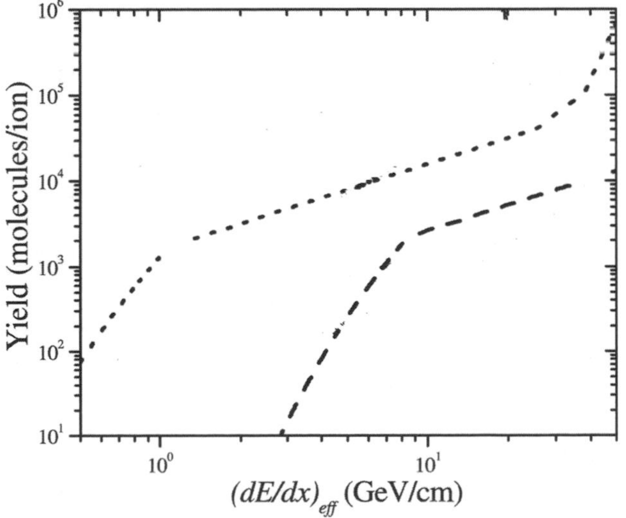

Figure 7.14 Sputtering yields as a function of deposited energy by Fe cosmic ray of CO (dotted line) and H_2O ice (dashed line). Figure taken from [12]

grain. This corresponds to an average injection rate of molecules from an H_2O ice mantle of $\simeq 5 \times 10^{-21}\, n\ \mathrm{cm}^{-3}\ \mathrm{s}^{-1}$.

7.7 The Stochastic Nature of Interstellar Surface Chemistry

Interstellar grain surface chemistry is in the diffusion limit where the chemical reactions are limited by the rate at which reactants can be brought to the surface. In contrast, in a laboratory setting, grain surface chemistry is in the collisional limit where collisions between reactants on the surface control the chemistry. The latter case can be evaluated using rate equations describing the rate of change of the (average) number density, $n\,(i)$, of the species present in terms of the reactive binary collisions between them,

$$\frac{dn\,(i)}{dt} = -n\,(i) \sum_j n\,(j)\, k_{ij} + \sum_{j,k} n\,(j)\, n\,(k)\, k_{jk} + k_{ac} n_g\,(i) - k_{sub} n\,(i)\,, \qquad (7.63)$$

where the first sum includes all coreactants for species i while the second sum pertains to all bimolecular reactions resulting in the formation of species i, and the last two terms refer to accretion of species i from the gas phase and sublimation into the gas phase. The reaction rate coefficient has to be treated with care. Considering a migrator "searching" the surface for a coreactant with which it can react upon "collision" (e.g. without activation barrier) – say, H plus H – the reaction is limited by migration on the surface and the reaction rate coefficient is given by the migration rate, k_m. The rate at which a species, say H_2, is formed is then,

$$R\,(H_2) = \frac{1}{2}\, k_m\, n^2\,(H)\,, \qquad (7.64)$$

where the factor 1/2 is to avoid counting each reaction twice. In case evaporation is not important, the average timescale to visit each and every site on the grain at least once scales with $N_{sites}\ln(N_{sites})$ where N_{sites} is the number of sites on the grain and this factor enters into the migration rate. In case evaporation is important, this correction factor to the rate at which species move from site to site has to be slightly adjusted. When relevant (hydrogenation) reactions possess an activation barrier, the probability for reaction, p_{ij}, in any collision is small and the reaction rate coefficient is then given by $k_{ij} = k_m p_{ij}$. This reaction probability will depend on how long the migrator is associated with the reaction site and as this is inversely proportional to the migration rate, the reaction rate is independent of the migration rate (see Section 7.4.3).

However, this formalism implicitly assumes an "infinite" surface where even small coverages are meaningful. In general, this is not the case in the interstellar medium. A typical reaction sequence may involve the accretion of a migrating atom on a coreactant-free surface; eg. at that point, the surface coverage of the accreted species is $N_{sites}^{-1} = (4\pi\, a^2\, N_s)^{-1}$ with N_s the number of sites per unit area while the surface coverage of coreactants is zero. The first coreactant that lands will react, setting all surface concentrations to zero until the next species lands. If a number of different coreactants could accrete, the reaction rate will then scale with the probability that the next accreting species is the specific one under consideration. For very mobile species (eg. H, D, and possibly C, N, and O), the surface migration/collision rate does not enter at all into this.

It is clear that using rate equations to calculate the reaction rate may lead to erroneous results, particularly when several species with very different surface mobilities accrete from the gas phase with very similar rates. Consider the accretion of species A and B at rates r_A and r_B (per site), which can react to form species A_2, AB, and B_2. In the diffusion limit where transport to the surface is the limiting factor, the formation rates of these species are,

$$R_{A^2} = r_A^2/2\,(r_A + r_B) \tag{7.65}$$

$$R_{B^2} = r_B^2/2\,(r_A + r_B) \tag{7.66}$$

$$R_{AB} = r_A\, r_B/\,(r_A + r_B)\,. \tag{7.67}$$

For the rate equation approach, we have in steady state,

$$r_A - 2\,k_A\,\theta_A^2 - k_{AB}\,\theta_A\,\theta_B \;= 0 \tag{7.68}$$

$$r_B - 2\,k_B\,\theta_B^2 - k_{AB}\,\theta_A\,\theta_B \;= 0, \tag{7.69}$$

where the k_i's are the migration rates (e.g. collision frequencies) on the grain surface. These two equations can readily be solved and we can calculate the formation rates,

$$R_{A^2} = k_A\theta_A^2 \tag{7.70}$$

$$R_{B^2} = k_B\theta_B^2 \tag{7.71}$$

$$R_{AB} = k_{AB}\theta_A\theta_B. \tag{7.72}$$

If A and B have the same mobility (eg. $k_{AB} = 2k_A = 2k_B$), it can be shown (see assignments) that these formation rates are equivalent to those calculated from the stochastic method. However, if the two species have very different mobilities, the rate equation method can lead to very erroneous results. This is illustrated in Figure 7.15. Essentially, in

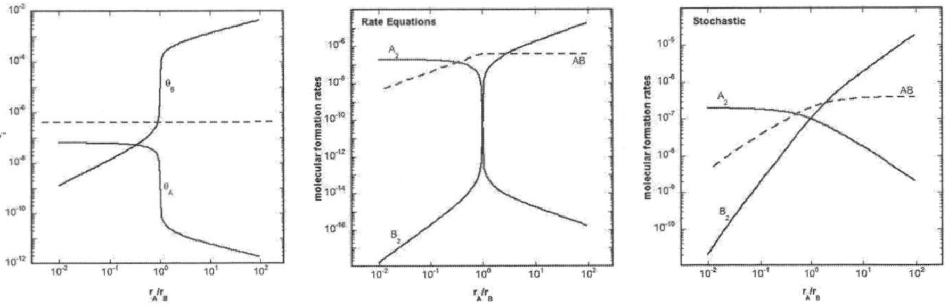

Figure 7.15 Grain surface reactions: Rate equations versus the stochastic aproach. Left: Surface concentrations for the rate equation aproach. The dashed line corresponds to one species on a 1000 Å surface. Middle: Molecular formation rates in the rate equation approach. Right: Molecular formation rates in the stochastic approach.

these calculations, the surface coverage of A is always $< 10^{-7}$ and, hence, meaningless for grains smaller than 3000 Å. When the migration rates are very different, the surface coverage develops a sharp transition at $r_A = r_B$ where the surface goes from A-dominated to B-dominated. The formation rates of the various species follow this flip-flop. This switch-over becomes more pronounced when the ratio of the accretion rate to the migration rates decreases.

In addition to these fluctuations driven by the accretion process, studies show that "cascade" effects can become important where – under conditions of low H fluxes – an appreciable abundance of a species is built up (e.g. O_3). Once an H lands and does react with O_3, the resulting OH can react with H_2, releasing a new H-atom that can react with another O_3 molecule. This process can then continue until the H-atom is lost to another reactant that does not react with H_2. The resulting avalanche leads to strong fluctuations in the abundance of even immobile species.

Various methods have been devised to deal with the discrete stochastic nature of the system under these conditions, each with their advantages and disadvantages. In the Monte Carlo approach, random number generators are used to explicitly evaluate the accretion history of the different species from the gas phase and their subsequent reaction sequences with coreactants present on the surface. Using a large number of drawings, the system will provide the surface concentrations of the species involved. The advantage of this method is that it takes the discrete nature of the problem fully into account. Moreover, Monte Carlo methods are very versatile and allow the inclusion of many different aspects that influence the reaction probability, such as the presence of multiple binding sites and porosity and/or morphology effects. A disadvantage is that it is very time consuming and it is difficult to integrate this approach with the rate equation method often used in evaluating chemical abundances in the gas phase.

In the Master Equation approach, the state of the surface is represented by a probability vector of the surface concentration of all species with elements $P(a,b,c,\ldots)$ where a species of type A, b species of type B, c species of type C ... are present. The sum of these probabilities is of course unity. The rate equations (Eq. (7.63)) are then replaced by time derivatives for these probabilities, which are constructed by properly accounting for the reactions involved. As an example, if only one type of species (H) can be present on a grain with N_{max} sites, there will be N_{max} equations describing the time evolution of the system. The rate of formation of H_2 is then given by,

$$R(H_2) = A\left(< N_H^2 > - < N_H >\right), \tag{7.73}$$

with A constant and the moment $< N_H^k >$ given by,

$$< N_H^k >= \sum_{N_H=0}^{N_{max}} N_H^k\, p(N_H), \tag{7.74}$$

and $p(N_H)$ is the probability to have N_H species on the surface. Comparing Equations (7.64) and (7.73), the inherent incompatability between these approaches when $< N_H >$ is small is clear. This set of equations can then be solved with appropriate boundary conditions using, for example, Runga Kutta techniques. Because of the computational effort involved, this method has only been applied to the simple system of H_2 formation. The stochastic evolution can also be sampled using a Monte Carlo method, and gas and grain surface chemistry can then be evaluated simultaneously if needed, but in practise this approach is too time consuming and has only been sparingly used in astrophysics. In an alternative approach, the master equation is rewritten in terms of the moments of the distribution. Again, time derivatives can be set up and solved using standard numerical techniques. Because the $(k+1)^{th}$ moment depends on the lower moments, the system has to be truncated in a clever way. Again, this method is very computer intensive and has not been used much.

The issues involved may become clearer when considering the average population of species i, $< n(i) >$, which is described by equations containing terms $< n(i)n(j) >$. This should be contrasted with Equation (7.63), where the terms contain $< n(i) >< n(j) >\equiv n(i)n(j)$. Hence, equations have to be added to evaluate the second order moments, $< n(i)n(j) >$. Now, the number of states grows exponentially with the number of species. Consider a system of only ten species which each can have only three states (zero, one, or two species on the surface), the number of equations is then $3^{10} \simeq 60,000$ and rigorous studies of this kind have to be limited to small systems. There are ways around this, though. First, only reactive species, which will have very low abundancies in interstellar grain surface chemistry, might be treated in a discrete way. This would include mobile species such as H, D, C, N, and O, as well as reactive radicals (e.g. HCO, O_2H, CH_3O). The concentration of abundant species can then be followed using rate equations. This boils down to replacing $< n(i)n(j) >$ by $< n(i) > n(j)$, where $< n(i) >$ is evaluated using the master equation approach and $n(j)$ follows from solving the system described by Eq. (7.63). This system could be reduced even further by realizing that at any one time there is only at most one mobile species on the surface. That is, in practice, the accretion rate of species is so low in the ISM (typically $\lesssim 1$ day^{-1}) that mobile species can always find a reaction partner (or evaporate). In any case, this approach leads to hybrid methods where rate equation and master equation approaches are intermixed. An automatic scheme can then be implemented whereby the method shifts from the master equation approach to the rate equation approach whenever given criteria are fullfilled on, for example, the average number of species present on the surface. This method is very versatile and can take the "discrete" effects associated with highly mobile species as well as "cascades" well into account.

Because of the difficulties in combining rate equations for gas phase reactions with the discrete approaches for surface chemistry, some authors have opted to modify the rate equations in a semi-empirical fashion to compensate for the inherent short comings of this method. Specifically, the expressions for the rate coefficients are altered to limit the probability that a reaction occurs in any given timestep to at most unity. In addition, other correction factors can be introduced to bring the results of the modified rate equation approach into agreement with those of stochastic theories for a few selected cases, which are considered representative. The resulting set of rate equations are then combined with the gas phase rate equations and integrated using standard ordinary differential equation integrators. The advantage of this method is that it is fast and designed to combine gas and grain surface chemistry. The disadvantage is that the system has to be validated semi-empirically a posteriori and this will always leave inherent uncertainties. A number of studies have compared results obtained with the modified rate equation approach to those from discrete methods for very limited systems. The results show substantial differences with the master equation approach when surface species are highly mobile and grain surface chemistry is in the diffusion limit. Clearly, modified rate equations have to be used with care. In a variation on this theme of modified rate equations, for rapidly migrating species, the reaction rate coefficient can be taken to be the accretion rate of a species times the probability (evaluated using Poisson statistics) that the other species is present on the grain surface whenever the average number of species present on the grain is less than unity. An adjustable switching function then allows a smooth transition from the rate equation to the modified rate equation system. This approach has been tested against large scale Monte Carlo studies with encouraging results.

7.8 Further Reading and Resources

Sticking and thermal accommodation are discussed in [14, 59]. Detailed molecular dynamics studies of the thermal accommodation process have been performed by [2, 104]. The simple reduced potential for physisorption is described in [106]. A more recent study on dispersion coefficients is reported in [1]. The presence of a wide range of binding sites on amorphous solid water ice has a long history in theoretical studies [2, 6, 91, 93]. Experimental evidence is more recent [49, 63]. Migration of atomic H on ice surfaces has been discussed by [6, 59, 64, 91, 93]. Diffusion of atomic oxygen has been reviewed by [101]. Studies on the binding of atomic oxygen to amorphous solid water are presented in [53, 72, 76] but that is at variance with the high mobility of atomic oxygen [71]. Atomic nitrogen binding energies and diffusion barriers have been measured by [72].

Tunneling has a long history in physics and chemistry. The monographs [7] and [78] provide good overviews of the field and the discussion given here leans heavily on the first of these references as well as [42, 66].

The hydrogenation of CO has been studied by several groups [41, 55, 57, 58, 60]. Reference [114] presents quantum chemical calculations on the energy barrier for the addition reaction of H and D to CO. Experimental studies on the kinetic isotope effect in this reaction have been reported by [56]. Experimental and quantum chemical studies of the addition and abstraction reactions of H and D with H_2CO have been reported by [45, 57]. Abstraction reactions for isotopically mixed methanol ice with H and D atoms have been studied by [46, 47, 77].

Laboratory studies on water formation on interstellar ice analogs have been reviewed by [111]. The reaction of H with O_2 in the gas phase has been studied by [102] and references therein. References [60, 75] have investigated this reaction on interstellar ice analogs. The peroxide reaction has been studied by [75, 81, 97]. Because of its importance in combustion chemistry, there is a long history on the reaction of H_2 with OH in the gas phase. A relative recent quantum chemistry study is [79]. The kinetic isotope effect in the reaction of H_2 with OH on ice surfaces has been studied by [80].The insertion reaction of C with H_2 has been well studied theoretically (viz., [51]) and was recently demonstrated to occur at low temperatures using the Helium droplet technique [62]. The reaction H_2 with CN has been recently studied by [11]. The chemiluminescence reaction of O plus CO in low temperature matrices has been reported in [39]. This reaction has been studied at low temperatures on a CO surface and shown to occur slowly [88, 89]. With excess O, O_2 wins the competition with CO [48]. The reaction of O and O_2 has also been studied by [90]. The reaction pathway of OH with CO to CO_2 has been critically reviewed by [111].

The desorption behavior of interstellar ice analogs has been reviewed by [24]. Relevant thermal desorption studies include [10, 22, 23, 36, 40]. Chemical desorption has been studied by [73, 82]. The importance of desorption of vibrationally excited H_2 formed through Langmuir–Hinshelwood mechanism on graphitic surfaces has been demonstrated experimentally by [65]. Sputtering of ices has been reviewed in an interstellar context by [100] and in a planetary ices by [13]. More recent studies include [33, 98]. The erosion of grain mantles in shocks has been investigated by [15, 38].

Early papers on interstellar grain surface chemistry [59, 99] identify many of the issues involved, including the importance of "enhanced" binding energy sites associated with kinks and steps on the surface, the importance of tunneling for migration on a surface and for reactions, and the reaction networks involved in ice chemistry. The field really opened up in the early 2000s as systematic laboratory studies were conducted for astrophysically relevant species. Some noteworthy papers are [69, 86] for H_2 formation and [55, 60] for interstellar ice chemistry. Comprehensive reviews on H_2 formation on interstellar grain surfaces are given in [107, 109] while reactions of importance for interstellar ices are reviewed in [50]. A review from an astronomer's perspective is provided in [101].

An early study using Monte Carlo techniques to study ice mantle composition is [99]. The versatility of Monte Carlo methods is quite obvious from [26, 27, 28]. The master equation approach has been used by [9, 96] for very limited systems. The master equation was solved using the moments of the distribution approach by [67]. Hybrid methods where rate equation and master equation approaches are intermixed have been developed by [30, 31, 95]. References [20, 21, 103] have approached merging gas phase and grain surface chemistry models from the opposite end. These studies solve both gas phase and grain surface chemistry using a Monte Carlo approach but this method can only handle limited systems. Modified rate equations have been developed by [16, 94] and validated by [44].

7.9 Exercises

7.1 Sticking coefficient:

- Calculate the sticking coefficient of an H atom on an H_2O ice surface at 10 and 100 K. Adopt a surface binding energy of 500 K.
- Calculate the sticking coefficient of an O atom on an H_2O ice surface at 10 and 100 K. Adopt a surface binding energy of 1300 K.
- Calculate the sticking coefficient of an H atom on an H_2O ice surface that is half covered with H_2 at 10 and 100 K. Adopt a surface binding energy of 500 K.

7.2 Consider a species physisorbed on a grain surface. Evaluate, as a function of binding energy (between 300 and 800 K), the sublimation timescale and the thermal hopping timescale at a temperature of 10 K and 30 K. Compare your results graphically with the rate of arrival of coreactants on a grain of 1000 Å for a gas phase density of coreactants of 1 cm^{-3}. CO is the main accreting species with a density of 10 cm^{-3}. If we assume that CO is chemically inert on a grain surface, evaluate (and compare) the rate at which newly accreted species are buried in the ice. Assume that the diffusion barrier is 30% of the binding energy

7.3 If the diffusion barrier is 30% of the binding energy, calculate the timescale for an O-atom to scan the whole surface of an H_2O ice and of a CO ice grain at 10 K and compare graphically with the accretion time and evaporation timescale as a function of grain size.

7.4 If the reaction barrier for H with CO is 1000 K and for H with H_2CO is 1200 K, calculate the relative probabilities for CO and H_2CO to react with H. Why could CH_3OH still be the dominant reservoir of carbon in interstellar ice?

7.5 The involvement of H_2 in interstellar grain surface chemistry:

- Calculate the limiting activation barrier for reactions of H_2 if the H_2 surface coverage is 0.5 and if there is only one H_2 molecule on a 1000 Å grain. You can assume that H_2 is highly mobile.
- Calculate the relative abundance of H_2 and HD on a grain surface at 10 K. Assume that all hydrogen and deuterium are in molecular hydrogen in the gas phase.
- What are the consequences for the deuterated fraction of ice molecules?

7.6 Thermal desorption:

- Calculate the sublimation rate for atomic H on an ice surface at 10 K.
- The sublimation temperature is defined as the temperature where accretion and thermal desorption balance. Calculate the sublimation temperature for a dark cloud core, assuming a CO abundance of 10^{-4} and a density of 10^4 cm^{-3}.
- Calculate the sublimation temperature for H_2O in a Hot Core ($n = 10^6$ cm^{-3}, $X\,(H_2O) = 10^{-4}$).

7.7 Assume a grain surface with one CO and one O_3 molecule. Evaluate the relative probability for reaction of a newly accreted H atom with either species. Do the same for a newly accreted D atom. Compare these probabilities. What does this imply for deuterium fractionation on grain surfaces?

7.8 Describe the various factors controlling surface reactions. Describe the general rules controlling grain surface routes in the ISM and their rational.

7.9 Describe the pros and cons of the various theoretical methods devised to describe grain surface chemistry.

Bibliography

[1] Amora-Barrios, N., Carchini, G., Błoński, P., López, N., 2014, *J Chem Theory Comp*, 10, 5002
[2] Al-Halabi, A., van Dishoeck, E. F., 2007, *MNRAS*, 382, 1648
[3] Amiaud, L., 2006, PhD thesis, Université de Cergy-Pontoise.
[4] Amiaud, L., Dulieu, F., Fillion, J.-H., Momeni, A., Lemaire, J. L., 2007, *J Chem Phys*, 127, 144709
[5] Andersson, S., van Dishoeck, E. F., 2008, *A & A*, 491, 907
[6] Ásgeirsson, V., Jónsson, H., Wikfeldt, K. T., 2017, *J Phys Chem C*, 121, 1648
[7] Bell, R. P., 1980, *The Tunnel Effect in Chemistry*, London: Chapman and Hall
[8] Bertin, M., Fayol, E., Romazin, C., et al., 2013, *ApJ*, 779, 120
[9] Biham, O., Furman, I., Pirronello, V., Vidali, G. 2001, *ApJ*, 553, 595
[10] Bisschop, S. E., Fraser, H., Öberg, K., et al., 2006, *A & A*, 449, 1297
[11] Borget, F., Müller, S., Grote, D., et al., 2017, *A & A*, 598, A22
[12] Bringa, E. M., Johnson, R. E., 2002, *Nucl Instr Meth Phys Rev B*, 193, 365
[13] Brown, W. L., Johnson, R. E., 1986, *Nucl Instr Meth Phys Rev B*, 13, 295
[14] Burke, J. R., Hollenbach, D. J., 1983, *ApJ*, 265, 223
[15] Burkhardt, A. M., Shingledecker, C. N., Le Gal, R., et al., 2019, *ApJ*, 881, 32
[16] Caselli, P., Hasegawa, T. I., Herbst, E., 1998, *ApJ*, 495, 309
[17] Chaabouni, H., Bergeron, H., Baouche, S., et al., 2012, *A & A*, 538, 125
[18] Chackett, K. F., Tuck, D. G., 1957, *Trans Faraday Soc*, 53, 1652
[19] Chaix, L., Dominé, F., 1997, *J Phys Chem B*, 101, 6105
[20] Charnley, S. B., 1998, *ApJ*, 509, L121
[21] Charnley, S. B., 2001, *ApJ*, 562, L99
[22] Collings, M. P., Dever, J. W., Fraser, H. J., McCoustra, M. R. S., Williams, D. A., 2003, *ApJ*, 583, 1058
[23] Collings, M. P., Anderson, M. A., Chen, R., et al., 2004, *MNRAS*, 354, 1133
[24] Collings, M. P., McCoustra, M. R. S., 2012, *EAS Publications Series*, 58, 315
[25] Cruz, -Diaz, G. A., Martín-Doménech, R., Munoz-Caro, G. M., Chen, Y.-J., 2016, *A & A*, 592, A68
[26] Cuppen, H., Herbst, E., 2005, *MNRAS*, 361, 565
[27] Cuppen, H. M., Herbst, E., 2007, *ApJ*, 668, 294
[28] Cuppen, H. M., van Dishoeck, E. F., Herbst, E., Tielens, A. G. G. M., 2009, *A & A*, 508, 275
[29] Da Silva, J., Stampfel, C., 2008, *Phys Rev B*, 77, 045401
[30] Du, F., Parise, B., 2011, *A & A*, 530, A131

[31] Du, F., Parise, B., Bergman, P., 2012a, *A & A*, 538, A91
[32] Dupuy, R., Bertin, M., Féraud, G., et al., 2017, *A & A*, 603, A61
[33] Famá, M., Shi, J., Baragiola, R. A., 2008, *Surf Sci*, 602, 156
[34] Fayolle, E., Bertin, M., Romanzin, C., et al., 2011, *ApJ*, 739, L36
[35] Fayolle, E., Bertin, M., Romanzin, C., et al., 2013, *A & A*, 556, A122
[36] Fayolle, E., et al., 2016, *ApJ*, 816, L28
[37] Fillion, J. H., Fayolle, E. C., Michaut, X., et al., 2014, *Faraday Discuss*, 168, 533
[38] Flower, D., Pineau des Fôrets, G., 1999, *MNRAS*, 268, 724
[39] Fournier, J., Deson, J., Vermeil, C., Pimentel, G. C., 1979, *J Chem Phys*, 70, 5726
[40] Fraser, H. J., Collings, M. P., McCoustra, M. R. S., Williams, D. A., 2001, *MNRAS*, 327, 1165
[41] Fuchs, G. W., Cuppen, H. M., Ioppolo, S., et al., 2009, *A & A*, 505, 629
[42] Garrett, B. C., Truhlar, D. G., 1979, *J Phys Chem*, 83, 2921
[43] Garrod, R. T., 2008, *A & A*, 491, 239
[44] Garrod, R. T., Vasyunin, A. I., Semenov, D. A., Wiebe, D. S., Henning, Th., 2009, *ApJ*, 700, L43
[45] Goumans, T. P. M., 2011, *MNRAS*, 413, 2615; errata: 2012, *MNRAS*, 423, 3775
[46] Goumans, T. P. M., Kästner, J., 2011, *J Phys Chem A*, 115, 10767
[47] Goumans, T. P. M., 2012, private communication
[48] Grim, R. J. A., D'Hendecourt, L. B., 1986, *A & A*, 167, 161
[49] Hama, T., et al., 2012, *ApJ*, 757, 185
[50] Hama, T., Watanabe, N., 2013, *Chem Rev*, 113, 8783
[51] Harding, L. B., Guadagnini, R., Schatz, G. C., 1993, *J Phys Chem*, 97, 5472
[52] He, J., et al., 2011, *Phys Chem Chem Phys*, 13, 15803
[53] He, J., et al., 2015, *ApJ*, 801, 120
[54] Herbst, E., Shematovich, V. I., 2003, *Astrophys Space Sci*, 285, 725
[55] Hidaka, H., Watanabe, N., Shiraki, T., et al., 2004, *ApJ*, 614, 1124
[56] Hidaka, H., Kouchi, A., Watanabe, N., 2007, *J Chem Phys*, 126, 204707
[57] Hidaka, H., Watanabe, M., Kouchi, A., Watanabe, N., 2009, *ApJ*, 702, 291
[58] Hiraoka, K., Miyagoshi, T., Takayama, T., et al., 1998, *ApJ*, 498, 710
[59] Hollenbach D. J., Salpeter, E. E., 1970, *J Chem Phys*, 53, 79
[60] Ioppolo, S., Cuppen, H. M., Romanzin, C., et al., 2008, *ApJ*, 686, 1474
[61] Katz, N., et al., 1999, *ApJ*, 522, 305
[62] Krasnokutski, S. A., Kuhn, M., Renzler, M., et al., 2016, *ApJ*, 818, L31
[63] Kristensen, L. E., Amiaud, L., Fillion, J-H., Dulieu, F., Lemaire, J-L., 2011, *A & A*, 527, A44
[64] Kuwahata, K., Hama, T., Kouchi, A., Watanabe, A., 2015, *Phys Rev Lett*, 115, 133201
[65] Lattimer, E., Islam, F., Price, S. D., 2008, *Chem Phys Lett*, 455, 174
[66] Leroy, R. J., Sprague, E. D., Williams, F., 1972, *J Phys Chem*, 76, 546
[67] Lipshtat, A., Biham, O., 2003, *A & A*, 400, 585
[68] Martín-Doménech, R., Cruz, -Diaz, G. A., Munoz-Caro, G. M., 2018, *MNRAS*, 473, 2575
[69] Matar, E., Congiu, E., Dulieu, F., et al., 2008, *A & A*, 492, L17
[70] Matar, E., Bergeron, H., Dulieu, F., et al., 2010, *J Chem Phys*, 133, 104507
[71] Minissale, M., et al., 2013, *Phys Rev Lett*, 111, 053201
[72] Minissale, M., Congiu, E., Dulieu, F., 2016, *A & A*, 585, A146
[73] Minissale, M., Dulieu, F., Cazaux, S., Hocuk, S., 2016, *A & A*, 585, A25
[74] Miranda, R., Daiser, S., Wandelt, K., Ertl, G., 1983, *Surface Sci*, 131, 61

[75] Miyauchi, N., Hidaka, H., Chigai, T., et al., 2008, *Chem Phys Letters*, 456, 27
[76] Murray, B. J., 2003, PhD thesis, (University of East Anglia).
[77] Nagaoka, A., Watanabe, N., Kouchi, A., 2007, *J Phys Chem A*, 111, 3016
[78] Nakamura, H., Mil'nikov, G., 2013, *Quantum Mechanical Tunneling in Chemical Physics* (Boca Raton, FL: Taylor & Francis)
[79] Nguyen, T. L., Stanton, J. F., Barker, J. R., 2011, *J Phys Chem A*, 115, 5118
[80] Oba, Y., Watanabe, N., Hama, T., et al., 2012, *ApJ*, 749, 67
[81] Oba, Y., Osaka, Y., Watanabe, N., et al., 2014, *Faraday Discuss*, 168, 185
[82] Oba, Y., Tomaru, T., Lamberts, T., et al., 2018, *Nature Astronomy*, 2, 228
[83] Öberg, K. I., Linnartz, H., Visser, R., van Dishoeck, E. F., 2009, *ApJ*, 693, 1209
[84] Paardekooper, D. M., Fedoseev, G., Riedo, A., Linnartz, H., 2016, *A & A*, 596, A72
[85] Perets, H. B., et al., 2005, *ApJ*, 627, 850
[86] Pirronello, V., Biham, O., Liu, C., Shen, L., Vidali, G., 1997, *ApJ*, 483, L131
[87] Pykal, M., Jurečka, P., Karlický, F., Otyepka, 2016, *Phys Chem Chem Phys*, 18, 6351
[88] Raut, U., Baragiola, R. A., 2011, *ApJ*, 737, L14
[89] Roser, J. E., Vidali, G., Manicò, G., Pirronello, V., 2001, *ApJ*, 555, L61
[90] Rosu-Finsen, A., McCoustra, M. R. S., 2018, *Phys Chem Chem Phys*, 20, 5368
[91] Senevirathne, B., Andersson, S., Dulieu F., Nyman, G., 2017, *Mol Astrophys*, 6, 59
[92] Shimonishi, T., Nakatani, N., Furuya, K., Hama, T., 2018, *ApJ*, 855, 27
[93] Smoluchowski, R., 1983, *J Phys Chem*, 87, 4229
[94] Stantcheva, T., Caselli, P., Herbst, E., 2001, *A & A*, 375, 673
[95] Stantcheva, T., Shematovich, V. I., Herbst, E., 2002, *A & A*, 391, 1069
[96] Stantcheva, T., Herbst, E., 2004, *A & A*, 423, 241
[97] Taquet, V., Peters, P. S., Kahane, C., et al., 2013, *A & A*, 550, A127
[98] Teolis, B. D., Plainaki, C., Cassidy, T. A., Rau, U., 2017, *J Geo Res: Planets*, 122, 1996
[99] Tielens, A. G. G. M., Hagen, W., 1982, *A & A*, 114, 245
[100] Tielens, A. G. G. M., McKee, C. F., Seab, C. G., Hollenbach, D. J., 1994, *ApJ*, 431, 321
[101] Tielens, A. G. G. M., 2013, *Rev Mod Phys*, 85, 1021
[102] Troe, J., Ushakov, V. G., 2008, *J Chem Phys*, 128, 204307
[103] Vasyunin, A. I., Semenov, D. A., Wiebe, D. S., Henning, Th., 2009, *ApJ*, 691, 1459
[104] Veeraghattam, V. K., Manrodt, K., Lewis, S. P., Stancil, P., 2014, *ApJ*, 790,1
[105] Veeraghattam, V. K., 2014, PhD, University of Georgia
[106] Vidali, G., Cole, M. W., Klein, J. R., 1983, *Phys Rev B*, 28, 3064
[107] Vidali, G., 2013, *Chem Rev*, 113, 8762
[108] Wakelam, V., Loison, J. C., Mereau, R., Ruaud, M., 2017, *Mol Astrophys*, 6, 22
[109] Wakelam, V., Bron, E., Cazaux, S., et al., 2017, *Molecular Astrophys*, 9, 1
[110] Wang, H., Frenklach, M., 1997, *Combust Flame*, 110, 173
[111] Hama, T., Watanabe, N., 2013, *Chem Rev*, 113, 8783
[112] Wellendorf, J., Silbaugh, T. L., Garcia-Pintos, D., et al., 2015, *Surf Sci*, 640,36
[113] Westley, M. S., Baragiola, R., Johnson, R. E., Baratta, G. A., 1995, *Nature*, 373, 405
[114] Woon, D. E., 2002, *ApJ*, 569, 541
[115] Zacharia, R., 2004, PhD thesis, Freie University Berlin

8
Physics and Chemistry of Large Molecules

8.1 Introduction

It is now well established that large molecules containing 10's of atoms are prevalent in the ISM of galaxies, locking up some 10% of the elemental carbon. IR emission features at 3.3, 6.2, 7.7, 8.6, 11.2, and 12.7 μm dominate the emission spectra of most objects with associated interstellar or circumstellar material. These emission features have generally been attributed to IR fluorescence from UV pumped Polycyclic Aromatic Hydrocarbon (PAH) molecules containing between 50 and 100 C-atoms. A few special objects show an IR emission spectrum dominated by the bands of C_{60} at 7.1, 8.6, 17.4, and 18.9 μm. All of these objects are planetary nebulae but only very few planetary nebulae have IR spectra dominated by C_{60}. The IR band at 18.9 μm is distinctly displaced from that of the PAHs and this band can be – weakly – discerned in (some) PAH dominated spectra, demonstrating that this fullerene is present in the ISM as well but generally "overpowered" by emission from the much more abundant PAHs.

The Diffuse Interstellar Bands (DIBs) are a set of absorption features in the visible wavelength range. They were recognized as due to interstellar material in the earliest spectra and some 400 DIBs have been catalogued and new ones are reported every year. The strongest and best-known bands occur at 4430, 5778, 5780, 5796, 6177, 6196, 6284, and 6614 Å. While most of the bands fall in the visible, they actually stretch from the near-UV through the visible to the far-red and the near-IR. It is now generally accepted that the DIBs are due to electronic absorptions by molecules (rather than due to absorption by impurities in dielectric grains). Of course, the recent, unambiguous assignment of two far-red absorption bands – at 9632.7 and 9577.5 Å – to electronic transitions in C_{60}^{+}, makes this link uncontroversial. But even before this identification, the molecular origin (rather than a grain-origin) of the DIBs was well accepted. Among the observational evidence is the absence of large variations in the peak position and profile of these bands between sight-lines with very different dust properties (e.g. grain size). Also, the detailed profiles of some DIBs show substructure that resembles the rotational substructure of molecular transitions, which varies (weakly) in peak position and relative strength, indicative of varying excitation conditions. Finally, the enigmatic object, the Red Rectangle, shows *emission* bands at wavelengths near to, but not exactly at, the position of some (absorption)

DIBs, suggesting that these transitions may be involved in a fluorescence process when the conditions are right.

This chapter describes the principles of the chemistry relevant for large molecules, which is distinctly different from that of small molecules discussed in Chapter 6. The interaction with UV photons is of particular relevance as, for one, it is at the basis of the IR emission of these species; that is, a molecule absorbs a UV photon and becomes highly excited with an excitation temperature of $\sim$1000 K for a very short time ($\simeq$1 s) while it cools back down to a typical interstellar temperature of $\sim$10 K until it absorbs another UV photon perhaps a day, a month, or a year later depending on how far away it is from bright stars. This process can repeat itself some hundred million times before the molecules succumbs to the harsh conditions of the ISM and is destroyed. While the species is highly excited, other processes can also take place, including ionization as well as fragmentation (with, e.g., H or C_2H_2 loss). The molecule can also interact with gas phase species. Reactions with atomic H and electrons are of particular relevance.

The concepts relevant to the chemistry of large molecules are described with PAHs in mind. However, many of these principles apply with some modification to the chemistry of fullerenes and that is discussed as well. A course on molecular astrophysics could leave Sections 8.2.3, 8.4.2, 8.4.3, 8.5.2, 8.5.3, 8.6.1, and 8.7 to self study.

8.2 The Physics and Chemistry of Interstellar PAHs

The structure of aromatic compounds has been summarized in Section 2.3.2. Polycyclic Aromatic Hydrocarbon molecules (PAHs) are a family of planar molecules, consisting of carbon atoms arranged in a characteristic honeycombed lattice structure of fused six-membered rings decorated by H-atoms at the edges (cf. Figure 8.1). PAHs are a subset of the group of Polycyclic Aromatic Compounds that also comprises heterocyclic

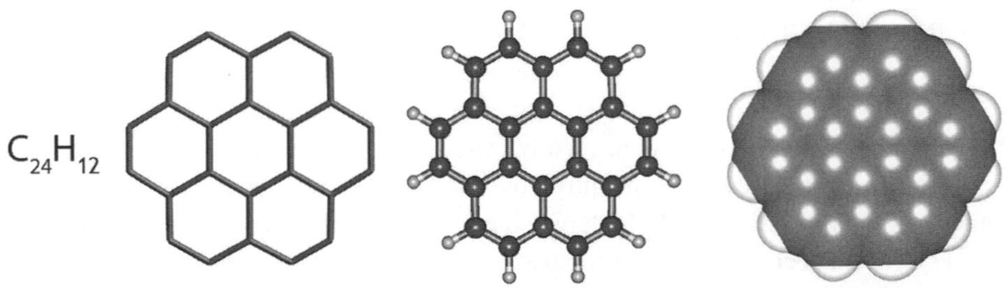

Figure 8.1 Four common representations of the structure of the PAH molecule, coronene ($C_{24}H_{12}$). Left: Structural formula. Center left: Only the hexagonal structure of the C atoms is represented where each covalent bond is indicated by a line – a so-called Kekulé structure. As the π electrons are delocalized, 20 equivalent Kekulé forms exist. Center right: Stick and ball, where black and white balls indicate C and H atoms, respectively. Right: 3-D model where the size of the atoms and their distances are representational. Black and white are C and H atoms. Image courtesy of Dr Alessandra Candian.

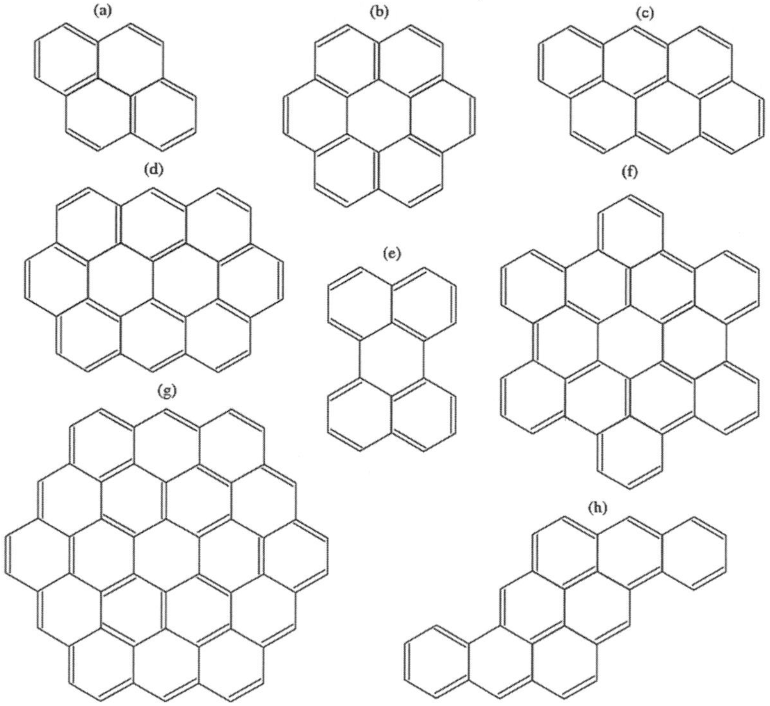

Figure 8.2 Molecular structure of some representative Polycyclic Aromatic Hydrocarbon molecules. (a) pyrene ($C_{16}H_{10}$), (b) coronene ($C_{24}H_{12}$), (c) anthanthrene ($C_{22}H_{12}$), (d) ovalene ($C_{32}H_{14}$), (e) perylene ($C_{20}H_{12}$), (f) hexabenzocoronene ($C_{42}H_{18}$), (g) circumcoronene ($C_{54}H_{18}$), (h) pyranthrene ($C_{30}H_{16}$).

aromatic compounds (heterocycles). With the explosive expansion of graphene research, it is apparent that there are an infinite number of PAHs conceivable. There are strict naming rules introduced by IUPAC (International Union of Pure and Applied Chemistry) but those are not always adhered to. The IUPAC rules are very complex and have many exceptions and, moreover, official names can get very cumbersome. This does not encourage strict adherence, and nicknames are then readily introduced, particularly in a subfield such as astrochemistry dominated by a healthy dose of anarchy. That does not make for easy exchange of data and ideas with the larger chemistry discipline, though. Compounds have a unique CAS number, but their use too has not been adopted by astrochemists. Naming of PAHs starts by recognizing the largest parent basic unit such as naphthalene, phenanthrene, anthracene, pyrene, chrysene, or coronene (Figure 8.2) and then rings or functional groups are added to the name.

The ground state of atomic carbon consists of $2s^2\,2p^2$. The s and p orbitals will hybridize to three sp^2 orbitals and one (remaining) p orbital and each of these orbitals will contain one of the four C valence-electrons. The sp^2 orbitals will combine to form σ orbitals with neighboring (C or H) atoms. Each sigma bond will contain two (localized)

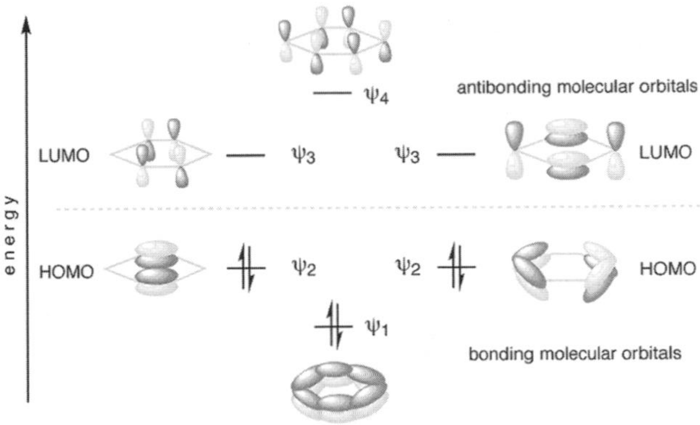

Figure 8.3 Molecular orbital representation of the π energy levels of benzene. The dashed line indicates the energy of an isolated p orbital and separates the bonding from the anti-bonding π levels. The shades of the electron clouds indicate the phase. When the phases match, the orbitals overlap to generate a common region of the same phase and the orbitals with the largest amount of overlap are lowest in energy.

electrons. The remaining atomic p orbital will be located perpendicular to the plane defined by the sigma bonds. Together these orbitals will form a delocalized, conjugated π electron system containing the six remaining electrons in benzene. So, the six atomic p orbitals form six molecular orbitals, symmetrically distributed in energy around the isolated p-orbital level (c.f. Figure 8.3). Three of these are bonding and the other three are anti-bonding. The lowest, most stable π orbital consists of all p-orbitals overlapping around the ring. The next two π orbitals each have one node. These two (ψ_2 and ψ_3) levels are degenerate. The three bonding molecular orbitals can accommodate the six electrons. This is in a way similar to the filled electron shell of noble gasses and accounts for the stability of benzene. The energy level system of larger PAHs is built up in a similar way (cf. Figure 8.4). Note that aromaticity depends on the number of electrons in the conjugated system not on the size of the ring or its charge. So, the 1,3-cyclopentadiene anion ($C_5H_6^-$) and the tropylium cation ($C_7H_7^+$) have a cyclic conjugated system of π electrons and are aromatic.

Generally, the more extensive the delocalized electron cloud is, the more stable the PAH molecule. Essentially, for these species to react, the σ bonding has to be broken and the extended aromatic π system has to be disrupted. Indeed, benzene (C_6H_6) is highly unsaturated but it is very stable against, i.e., addition reactions. This is in contrast to alkenes that readily undergo addition reactions (e.g. $C_2H_4 + Br_2 \rightarrow C_2H_4Br_2$). This high stability resides in the resonance energy associated with the π-bonding system. Estimates of the resonance stabilization yield about 150 kJ/mole or $\simeq$1.6 eV per benzene. As a general rule, planar, monocyclic rings with a continuous system of p orbitals and $(4n + 2)$ π electrons (with $n = 0, 1, 2, \ldots$; the so-called Hückel's rule) are aromatic and the π electrons provide additional stability (c.f. Figure 8.4). Strictly speaking, the term aromaticity is connected to monocyclic rings. Moreover, a species such as pyrene ($C_{16}H_{10}$) does not fullfil

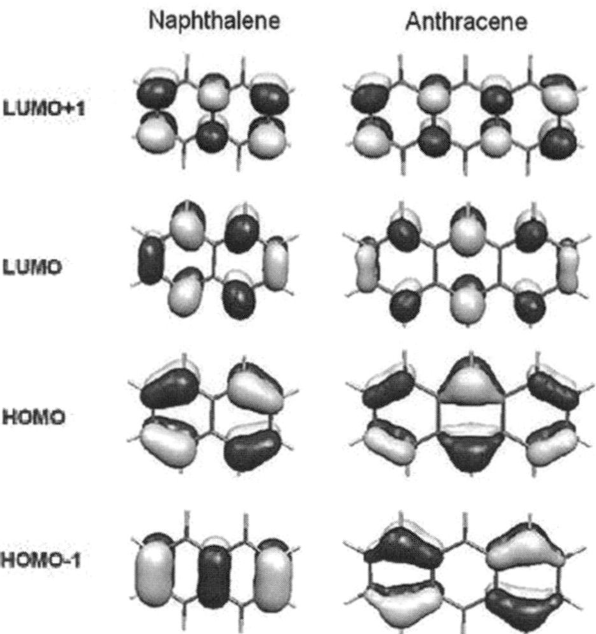

Figure 8.4 Molecular orbital representation of the π energy levels of the two ring PAH, naphthalene (left; $C_{10}H_8$) and the three ring PAH, anthracene (right; $C_{14}H_{10}$).

Hückel's rule. Yet, pyrene is conjugated, cyclic, planar, and very stable and this reflects its extensive conjugated π electon system. Various extensions of the aromaticity criteria for polycyclic systems have been devised based on theoretical and experimental criteria. Here, we will sidestep this issue and "define" polycyclic aromatic hydrocarbons as polycyclic, π-electron delocalized systems in which the C–C bond length is characteristically $\simeq$1.4 Å. The ease with which DFT allows molecular structures to be evaluated nowadays makes this a workable definition for all but the pure chemists.

The high stability of PAHs can be assessed from the average binding energy of carbon atoms in C_nH_m,

$$E_C = (E_{atom} - mE_{C-H})/n, \tag{8.1}$$

where the atomization energy E_{atom} is corrected for the average bond C–H energy, E_{C-H}. Figure 8.5 shows this average binding energy calculated for a set of PAHs in the NASA Ames PAH data base. There is a rapid increase in average binding energy with number of carbon atoms, which eventually levels off to the asymptotic value of the average binding of C-atoms in graphite, 7.3 eV. For a given size, the "circular" or "oval" PAHs in the coronene and ovalene families[1] are the most stable structures. Adding protruding rings

[1] Molecules consisting of a coronene or ovalene core surrounded by succesive concentric rings of hexagons.

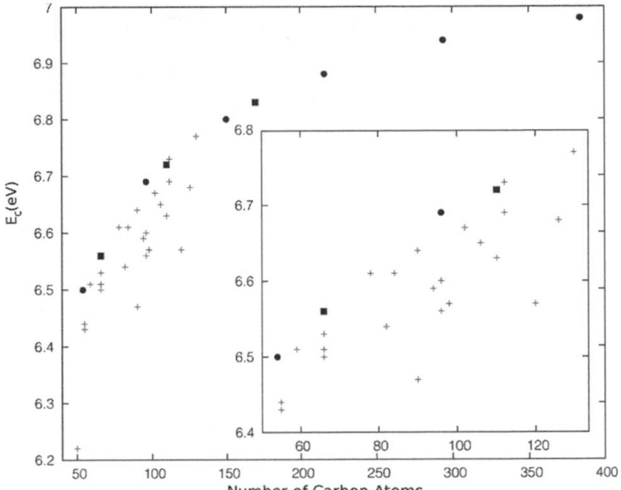

Figure 8.5 Binding energies per carbon, E_C (as defined in Equation (8.1)), as a function of the number of carbons. The coronene family is plotted with filled circles, the ovalene family with filled squares, and the remaining PAHs are plotted with plus symbols. Figure courtesy of [86]

to such superPAHs[2] makes them less stable. The average binding energy for the C-atom in such pendant rings has been estimated to be some 0.6 eV per C-atom less than for C-atoms in the "core." For comparison, the binding energy for aromatic H is substantially less, 4.5 eV, and other functional groups are typically bound by even less than this.

The six overlapping π electrons in benzene form one orbital that "loops" around the entire molecule, and this provides stabilization to this molecule.[3] Aromaticity as defined here pertains then to molecules that have at least one of such "loops." Qualitatively, the relative stability of PAH species can be assessed by Clar's sextet rule. This empirical finding indicates that the most stable PAH structure is the one that has the largest number of disjoint aromatic sextets; e.g., benzene-like rings connected to adjacent rings through single CC bonds. A molecule such as tetracene has only one sextet, while chrysene has two and triphenylene has three. As will be discussed in Section 12.3.5, interstellar PAHs show large and systematic variations in their edge structure ranging from zig-zag edges to armchair edges (cf. Figure 12.15). Following Clar's sextet rule, a PAH with zig-zag edges has fewer sextets than a PAH with armchair edges and is expected to have lower stability and greater reactivity.

[2] The name superaromatics has been introduced to classify these compact, stable PAHs [92] and the nickname "superPAHs" seems quite appropriate. The term "GrandPAH" has also been introduced but then from a slightly different perspective; e.g. to describe a very limited set of PAHs that dominate the interstellar PAH family under the most extreme conditions of space (cf. Section 12.3.9). As GrandPAHs are kinetically rather than thermodynamically defined, the two terms are not necessarily synonymous.

[3] This is called a sextet and contains three double bonds in one ring; often indicated by drawing a circle in the molecular structure.

8.2.1 The Astronomical Setting

The general structural formula of centrally condensed superPAHs is $C_{6r^2}H_{6r}$ with $3r^2 - 3r + 1$ hexagonal cycles arranged in $r - 1$ rings around the central cycle. The first three are coronene ($C_{24}H_{12}$), circumcoronene ($C_{54}H_{18}$), and circumcircumcoronene ($C_{96}H_{24}$). With a C–C bond length of $\simeq$1.4 Å, the surface area of one aromatic cycle is $\simeq$5 Å^2. The total surface area of a compact PAH is then approximately,

$$\sigma_{PAH} \simeq 3 \times 10^{-15}\, r^2 \simeq 5 \times 10^{-16}\, N_c \text{ cm}^2, \tag{8.2}$$

with N_c the number of C atoms of the PAH. The radius of this approximate circular PAH is then,

$$a \simeq 0.9 \times 10^{-8}\, N_c^{1/2} \text{ cm}. \tag{8.3}$$

So, a "typical" 50 C-atom interstellar PAH will have some 20 fused cycles, some 20 H-atoms at its periphery, a radius of $\sim$6 Å, and a surface area of 200 Å^2.

The various timescales involved in the excitation and relaxation processes are very different. UV photon absorption in the ISM occurs on a timescale of,

$$\tau_{uv} = k_{uv}^{-1} = (4\pi\, \sigma_{uv}\,(PAH)\, \mathcal{N}_{uv})^{-1} \simeq \frac{1.4 \times 10^9}{N_c\, G_o} \text{ s}, \tag{8.4}$$

with $\sigma_{uv}\,(PAH)$ the UV absorption cross section of the PAH (approximately equal to 7×10^{-18} cm^2 per carbon atom (N_c)) and $\mathcal{N}_{uv}$ the mean photon intensity of the radiation field – here expressed in terms of the Habing field, G_o (10^8 photons cm^{-2} s^{-1}). So, a typical interstellar PAH with $N_c = 50$ absorbs a UV photon once a year in the diffuse ISM and every 10 minutes in a PDR such as the Orion Bar. In contrast, as we will discuss in Section 8.2.2, conversion of the electronic excitation energy into the vibrational manifold (nonradiative decay) is very rapid; $\sim 10^{-12}$ s for internal conversion and $\sim 10^{-9}$ s for intersystem crossing. Internal vibrational redistribution occurs also on a very rapid timescale, $\sim 10^{-9}$ s. Radiative vibrational relaxation through mid-IR transitions, on the other hand, is much slower, $\sim$1 s and through the lowest lying vibrational states in the far-IR is even much slower than this (e.g. $A \sim \nu^3$). For species with large energy gaps between the ground state and the first excited electronic state, electronic fluorescence and phosphorescence can be an important deexcitation process as well (cf. Section 8.2.2). For very high internal excitation, fragmentation processes can compete with radiative relaxation. For coronene ($C_{24}H_{12}$), this occurs at an internal energy of some 12 eV and this threshold scales approximately with the size (cf. Section 8.5). Other processes that are of importance in the ISM include ionization, which will scale with the UV absorption rate once the photon energy well exceeds the ionization potential ($h\nu \gg I_p$) and for a neutral PAH would of the order of $10^9/G_0$ s. The recombination timescale in neutral atomic regions (with an electron fraction of $\simeq 10^{-4}$) is given by $\simeq 2 \times 10^8/n_H$ s with n_H the H density. Collisional deexcitation is similarly a slow process, $\sim 10^9/n_o$ s, with n_o the density of collision partners.

The large difference between the FUV absorption rate and the IR emission rates implies that the vibrational excitation temperature of the species will fluctuate widely. Indeed, a small PAH may reach temperatures in excess of 1000 K immediately after FUV photon absorption but, after a second, it will have cooled down to $\sim$10 K and remain that cold until it absorbs another FUV photon after a day or so (depending on the local FUV flux). These excitation fluctuations are characteristic for all molecular-sized species, including fullerenes, diamandoids, and clusters of molecules; and this will be described more fully in Section 8.4.

8.2.2 Photophysics of Large Molecules

For small molecules, the density of background states is very small compared to the decay width of the excited level. Coupling to other states then becomes very directed, leading to either directly or through predissociation to dissociation channels (cf. Section 6.3.1). For large molecules, the density of states is very large and other energy decay processes open up. Figure 8.6 – a so-called Jablonski diagram – shows a schematic energy level diagram for large molecules, specifically geared toward neutral PAHs. They consist of a series of singlet (paired-up electron spins) and triplet (parallel electron spins) states. The ground state is a singlet state, S_0, and the first triplet state, T_1, lies below the first excited singlet state, S_1. For odd-electron systems (cationic or other open shell PAH species), the electronic system consists of doublet states and quartet states with D_0 the ground state. For the hydrocarbons of concern here, the lowest quartet state (Q_1) lies above the first excited doublet state (D_1). For large molecules, photon absorption into an excited electronic state is followed by a variety of radiationless and radiative processes (Figure 8.6). For neutral aromatic molecules, these commonly include internal conversion (IC) from a highly excited singlet state to the vibrational manifold of lower-lying (singlet) states ($S_n \rightarrow S_{n-1}$), intersystem crossing (ISC) from the (first) excited singlet state to the vibrational manifold of a lower triplet state and its inverse process ($S_1 \leftrightarrow T_1$). These compete with radiative processes; visible fluorescence mainly from S_1 to S_0, phosphorescence from T_1 to S_0, and infrared vibrational relaxation within the singlet (and triplet) states. For odd-electron systems, the D_1 state only couples to the D_0 state and ISC followed by phosphorescence is not important.

Competition between these different processes is controlled by their relative rates. The first step, electron excitation, occurs on a timescale of $\simeq$1 femto second (1 femto s $= 10^{-15}$ seconds). For aromatic hydrocarbon species, internal conversion from highly excited electronic states to lower lying states typically happens on a timescale of tens to hundreds of femto seconds but internal conversion from the first excited singlet state to the vibrational manifold of the ground state ($S_1 \rightarrow S_0$) is much slower. The rate constant for radiationless transitions in large molecules, $k_{i\rightarrow f}$, is controlled by the energy gap as expressed in Fermi's golden rule as originally formulated by Dirac,

$$k_{nr}\left(i \rightarrow f\right) = \frac{2\pi}{\hbar}\left|V_{if}\right|^2 \rho_f\left(E\right), \tag{8.5}$$

where ρ_f is the density of excited vibrational (final) states that match the energy of the initial state.[4] The matrix element of the perturbation that couples the two states is given

[4] Strictly speaking, the density of states capable of mixing the two electronic states.

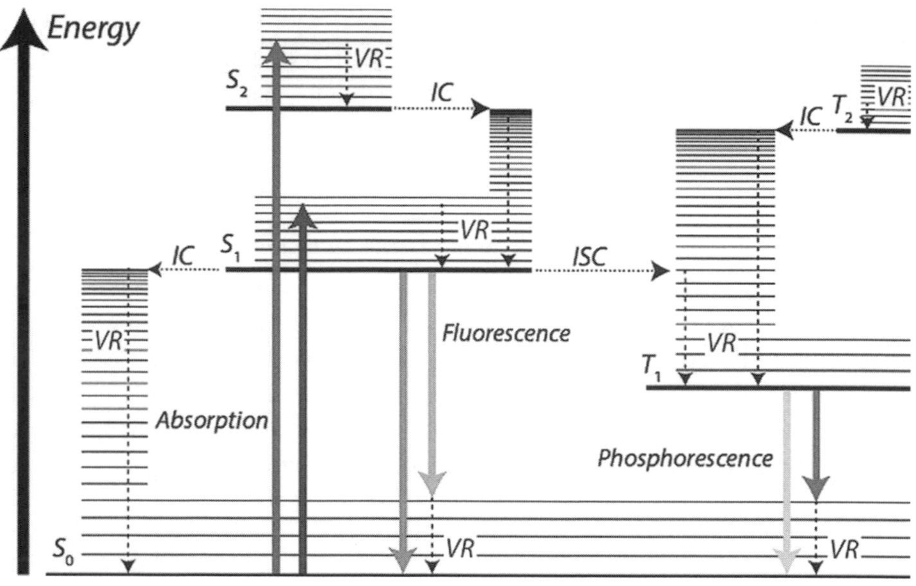

Figure 8.6 The flow of excitation in a schematic energy diagram for large neutral PAH molecules (Jablonski diagram). Photon absorption (dark upward arrows) will take the molecule up to a higher lying electronic state with the same symmetry. This process can be followed by internal conversion (IC) to the vibrational manifold of a low lying electronic state or through intersystem crossing (ISC) to a electronic state with a different symmetry (e.g. singlet to triplet or doublet to quartet) (both indicated by dotted arrows). The system can relax down through emission in electronic bands (fluorescence for allowed transitions (light downward arrows) or phospherescence for forbidden transitions (light downward arrows)), IR emission through the vibrational modes, or through unimolecular dissociation. This diagram is appropriate for laboratory experiments where collisions typically bring the species down to the ground vibrational state (dashed arrows) of the excited electronic state and fluorescence/phosphorescence occurs from that level. Both IC and ISC are reversible processes and all states can be revisited in accordance to their phase space importance. At high photon energies, ionization and fragmentation are effective competitors. Note that a molecule can also arrive in an excited electronic state through electron recombination or a chemical reaction. See text for details.

by $V_{if} = \langle \Psi_i | \hat{h} | \Psi_f \rangle$ with $\hat{h}$ the perturbation operator that couples nuclear with electronic motion. For singlet–singlet transitions, the matrix element is controlled by the vibrational component and essentially the rate scales with the square of the overlap of the vibrational wavefunctions in the two states, the Franck–Condon factor, $\langle \chi_i | | \chi_f \rangle^2$ (cf. Figure 3.10; Section 3.4.3). For intersystem crossing, this operator also includes the coupling of electron spin with orbital angular momentum. Calculating these rates is beyond the scope of this book. Empirically, when a series of related species is considered, the rate constant decreases exponentially with the energy gap as illustrated for PAHs and related compounds in Figure 8.7. This is a convolution of two factors: The density of states increases exponentially with the energy, facilitating the transition. However, the Franck–Condon overlap

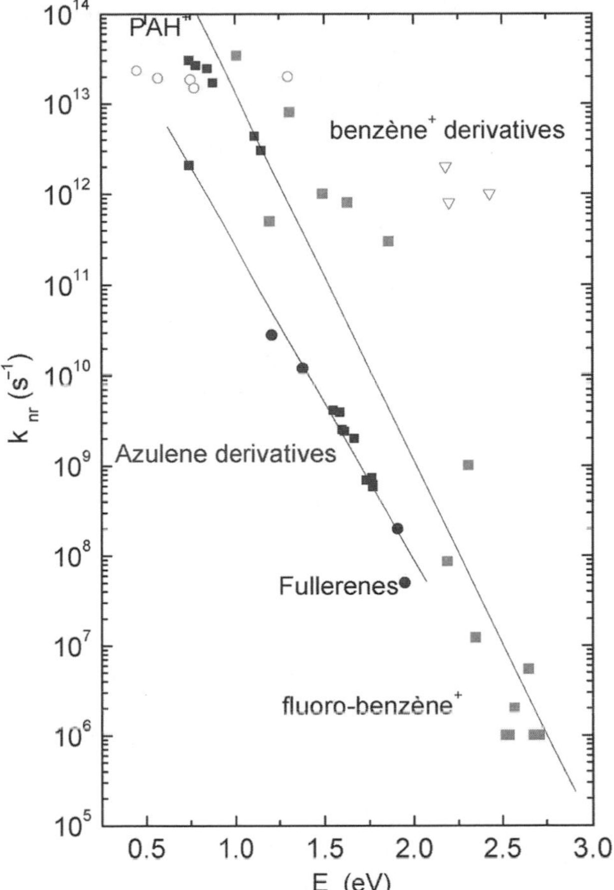

Figure 8.7 Nonradiative electronic relaxation rate of neutral and ionized PAHs and related compounds dominated by interconversion [80]. The lines – guiding the eye – signify the energy gap law. Figure kindly provided by Thomas Pino, see reference [80]

integral decreases even faster with increasing energy. Best overlap is obtained for the highest frequency vibrations as fewer vibrational quanta are needed to match the energy of the excited electronic state. The empirical relations can be represented by,

$$k_{IC} = 7.6 \times 10^{14} \exp\left[-E\left(\text{cm}^{-1}\right)/1025\right] \qquad \text{neutrals} \tag{8.6}$$

$$k_{IC} = 1.7 \times 10^{15} \exp\left[-E\left(\text{cm}^{-1}\right)/1500\right] \qquad \text{cation} \tag{8.7}$$

$$k_{ISC} = 2.6 \times 10^{4} \exp\left[-E\left(\text{cm}^{-1}\right)/1850\right] \qquad \text{neutrals,} \tag{8.8}$$

where the last rate constant refers to intersystem crossing. The spin flip that has to occur is caused by spin–orbit coupling where, in a classical picture, a change in the angular momentum describing the electron movement around the nuclei compensates the electron spin flip. As in the case of internal conversion, intersystem crossing is favored when a good overlap exists between vibrational wave functions of the excited singlet state, S_n, with vibrational wave functions of the triplet state, T_n. Empirically, the spin-forbidden triplet–singlet coupling factor is $\simeq 10^{-7}$ for relevant energies.

These conversion processes are reversible. As Eq. (8.5) indicates, the forward and backward rates are related by, $k_b/k_f = \rho_S(E_S)/\rho_T(E_T)$, where the densities of states are at the energy in that state. The system may then transfer back and forth between the S_1 and T_1 states – establishing a steady state abundance equilibrium between the two states – until eventually it decays to the ground state through (delayed) electronic fluorescence, vibrational fluorescence, or phosphorescence. Delayed fluorescence is only significant for large energy gaps between S_1 and S_0 (e.g. low IC rates to the ground state), small energy gaps between the S_1 and T_1 states (rapid ISC and reverse-ISC), and long triplet lifetimes.[5] Delayed fluorescence was first studied through excitation of the triplet state followed by collisional activated excitation in the triplet state to a state commensurate with the first excited singlet state.[6] In general, delayed fluorescence is much weaker than direct fluorescence but it has been observed for aromatic species. Fluorescence from higher electronic states can also occur if the associated band gap is appreciable. Besides direct fluorescence, delayed fluorescence can also be important in this case as the lifetime of the S_1 state is often considerable. In a similar fashion to reversible ISC between the triplet and the singlet system, reversible IC leads to revisits of higher electronic states and the ratio of the quantum yields between adjacent states relates to the relative density of states. For PAHs, fluorescence from $S_2 \rightarrow S_1$ has been observed where the smaller band gap than for $S_1 - S_0$ is partly compensated for by the larger oscillator strength. In the high energy limit, when all energy gaps are, in some sense, small, all states will be visited many times before decay happens. In that case, Fermi's golden rule implies that the system will spend most of its time in the lowest state(s) but all processes are open. That includes (auto)ionizing states (cf. Section 8.6.1).

Consider, as an example, coronene with S_2, S_1, and T_1 energies of $\simeq 30{,}000$, $\simeq 23{,}000$, and $\simeq 19{,}000$ cm^{-1}, respectively. The $S_1 \rightarrow S_0$ IC rate is 1.6×10^5 s^{-1} while the ISC rate to T_1 is 2.1×10^6 s^{-1} and the triplet–singlet coupling factor is more than compensated for by the much smaller energy gap. The radiative transition $S_1 \rightarrow S_0$ has a rate of $\simeq 8.4 \times 10^5$ s^{-1} and the fluorescence yield is 0.27. The phosphorescence rate from level T_1 is much slower ($\simeq 0.017$ s^{-1}). The nonradiative transition rate from T_1 to S_0 is also slow, $\simeq 0.086$ s^{-1} and the phosphorescence yield is 0.12. The reverse rate from T_1 to S_0 depends on the internal excitation. Consider coronene excited to the S_2 level. After IC to the S_1 state, and ISC to the T_1 state, delayed fluoresence has a yield of $\simeq 0.08$, only slightly less

[5] This delayed fluorescence is sometimes called Poincaré fluorescence in the astronomical literature as the earlier chemical physics literature on this topic was overlooked.

[6] In a liquid or solid, the vibrational energy is thermally coupled to the surrounding medium through collisions, driving the excitation down to the ground state of the (excited) electronic state and transfer from T_1 back to S_1 becomes sensitive to the temperature of the bath. In the molecular physics literature, this is therefore also called Thermally Activated Delayed Fluorescence and there is an extensive literature related to the development of efficient diodes.

than the phosphoresence yield (0.11) and implies a reverse-ISC rate of $\simeq 6 \times 10^5$ s^{-1}. The IC and ISC rates refer to the species in the ground vibrational state of the excited electronic state as they are measured in the solid phase. In the gas phase, collisional deactivation is not important and – promoted by the better Franck–Condon overlap for vibrationally excited species – the IC rate will increase much more rapidly with vibrational excitation than the ISC rate. As PAHs increase in size, the energy gap between the S_1 and S_0 states decreases and nonradiative coupling from S_1 to S_0 becomes competitive with fluorescence and intersystem crossing. This occurs around an energy gap of about 2 eV or a compact PAH of some 25 rings (e.g. a PAH slightly larger than circumcoronene). For ions, the radical nature implies that the energy gap between the ground and first excited doublet state is much smaller than for the corresponding neutral and the excitation will more quickly couple down to the ground state (cf. Figure 8.7). Also, the lowest quartet state lies above the first excited doublet state and hence temporary storage of the electronic energy in the quartet system through ISC and release through reverse-ISC has little effect. Delayed fluorescence can still occur when the excited molecule "revisits" the D_1 state.

Summary

Summarizing now the energy flow in large aromatic molecules (Figure 8.6): After electronic excitation (1 fs) and internal conversion in high S states, the excitation will wind up in highly vibrationally excited states. The vibrational states initially populated this way will couple to other vibrational modes on a timescale of tens of picoseconds. Full equilibration of the available energy over all vibrational states takes much longer (nanoseconds). For relatively small molecules, IC to the first excited electronic state may be followed by (visible) fluoresence to the ground electronic state on a nanosecond timescale. When the band gap is small or the excitation is very high, electronic fluorescence in the visible/near-IR is important. For large neutral molecules or radical species, the band gap is typically small and the density of states is so high that relaxation through the vibrational modes tends to dominate and visible fluorescence is inhibited. On a much longer timescale (seconds), the molecule will relax radiatively through sequential vibrational transition(s). Finally, intersystem crossing (ISC) may transfer the excitation to an electronic state with a different multiplicity (singlet to triplet or doublet to quartet). The "excess" energy in this state will lead to vibrational excitation that will be emitted in the vibrational modes. The molecule will return to the electronic ground state through the emission of a visible photon – in a process called phosphorescence – and, as this is a spin forbidden transition, this occurs on a timescale of $10^{-4} - 10^2$ s. At high internal energies, dissociation in a unimolecular reaction can compete if enough energy is localized into one specific bond. This dissociation process is generally evaluated with the ergodic approximation: The time average of a specific property (e.g. fragmentation) is equal to the average over the full (phase) space. The excited system has lost all memory of where it originated. Conversely, the system can be described by only its total energy. Of course, angular momentum is a conserved quantity as well but that is often ignored in (fragmentation) studies. In order to understand radiative relaxation and fragmentation processes, we need to understand first the excitation of a molecule with a given energy, E, and that is discussed in Section 8.3. Finally, at photon energies well above the ionization potential, ionization can compete effectively with radiative processes either through direct photo ionization channels (Section 8.6) or through thermoionic emission (Section 8.6.1) when during the ergodic evolution of the system, the (auto)ionizing state has a finite probability to be visited.

8.2.3 PAH Electronic Absorption Cross Section

The absorption spectrum of benzene is linked to its electronic structure in Figure 8.8. The HOMO and the LUMO[7] are each a pair of degenerate molecular orbitals. The four possible transitions from $\pi_{2,3}$ to $\pi^{\star}_{4,5}$ lead to transitions from the ground state ($^1A_{1g}$) to the singlet $^1B_{2u}$ and $^1B_{1u}$ and the degenerate $^1E_{1u}$ states. This results in three bands – the α, p and β bands – with substantial substructure. The first two of these correspond to symmetry forbidden transitions as they link states with the same electron configuration. The electronic spectra of PAHs show a gross similarity to that of benzene. Due to configuration interaction, the HOMO and LUMO are not degenerate anymore and this results in levels above and

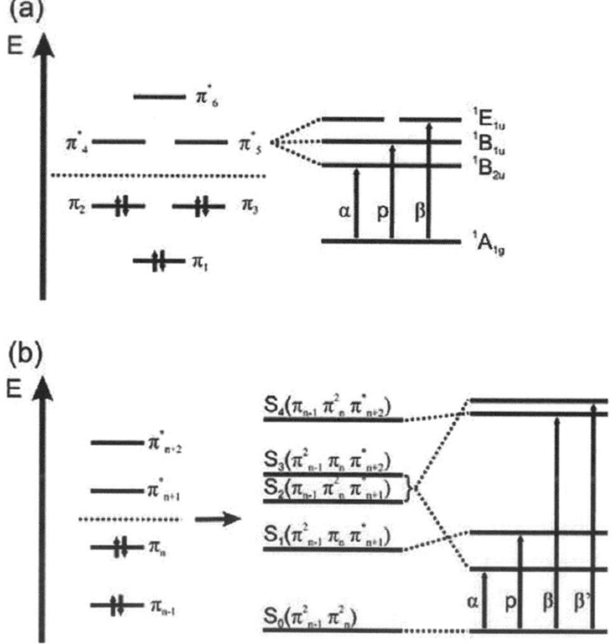

Figure 8.8 Lowest electronic transitions: (a) benzene linked to the electronic levels system. (b) PAHs. Configuration interaction (indicated by the arrow) breaks the degeneracy of the HOMO–LUMO levels. Figure taken from [87]

[7] Highest Occupied Molecular Orbital and Lowest Unoccupied Molecular Orbital.

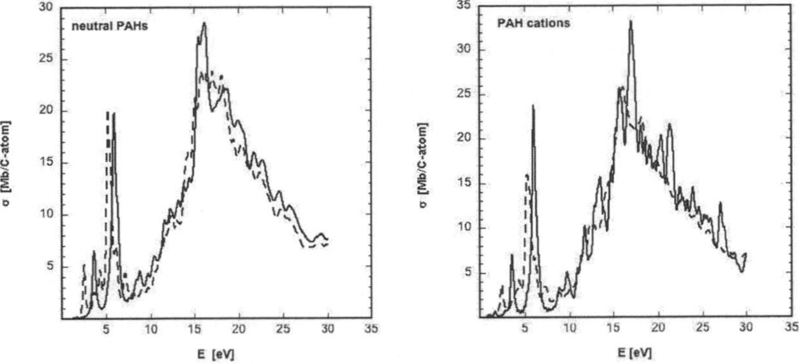

Figure 8.9 Calculated UV absorption cross sections of neutrals (left) and cations (right). Solid and dashed lines represent coronene and circumcoronene, respectively. The units are megabarn (10^{-18} cm^2 per C-atom). Figure adapted from [68]

below the original level. There are four transitions possible (Figure 8.8) and the relative location of the α and p transitions depends on the extent of the configuration interaction and hence the detailed molecular structure.

Electronic absorption spectra have been measured for a few small PAHs. Systematic studies were performed using time-dependent DFT for a number of PAHs ranging from naphthalene ($C_{10}H_8$) to circumcoronene ($C_{54}H_{18}$) and hence spanning a range in sizes and molecular structure (Figure 8.9). The accuracy of these calculations is about 0.3 eV and they should be considered indicative, helped by the forgiving scale of the figure. Only bound–bound transitions are included in these calculations but it should be noted that over this range of energies, bound–free absorption is important as well (cf. Table 8.3). PAH absorption spectra are characterized by two broad absorption systems around 6 and 16 eV, respectively. The former is due to $\pi^\star \leftarrow \pi$ transitions. Higher energy absorption corresponds to $\sigma^\star \leftarrow \pi, \pi^\star \leftarrow \sigma, \sigma^\star \leftarrow \sigma$, and Rydberg transitions. In these calculations, the discrete transitions above the ionization potential correspond to superstates that couple well to the ionizing continuum. Indeed, photo ionization studies show that the ionization yield above $\simeq$15 eV is unity.

Electronic spectra are very molecule specific and ideal for identification purposes. The Diffuse Interstellar Bands carry therefore the promise of much insight in the large molecule component of the ISM and it is of interest to consider the electronic structure and spectra of these species in some detail. At low energies ($\lesssim$10 eV), individual PAH spectra do show identifiable structural fingerprints and the bands essentially relate to well resolved one-electron π-transitions. In the astronomical literature, benzene – or any catacondensed PAH – is sometimes represented by a perimeter model. Consider a ring of radius R on which the π electrons are localized. This is a variant of the particle in a box or the energies of a rigid rotor in a plane and can be solved to give for the energy levels,

$$E_m = \frac{m^2\hbar^2}{2I}, \tag{8.9}$$

where $I = m_e R^2$ is the moment of inertia and the angular momentum can take on values, $L_z = m\hbar$, with $m = 0, \pm 1, \pm 2, \ldots$. So, the energy levels are doubly degenerate. For benzene, the highest occupied and lowest unoccupied molecular orbital (HOMO and LUMO) correspond to $m = 1$ and $m = 2$. The transition frequency is,

$$\nu = 30{,}000\ (2m+1)\left(\frac{\text{Å}}{R}\right)\ \text{cm}^{-1}, \tag{8.10}$$

which for $R = 1.39$ Å – the CC bond length – is $\simeq 50{,}000$ cm^{-1} or about 2000 Å. This falls in between the two most intense bands of benzene (at 54,500 and 48,000 cm^{-1}). Analysis of naphthalene with 10 π electrons would lead to very similar results as now $m = 0$, 1, and 2 are occupied. In general, this simple model does not reproduce the electronic structure of aromatic systems well as the energy of the levels is calculated to rapidly increase with m. In reality, the electron levels occur in pairs symmetrically arranged around the zero-level. Figures 8.3 and 8.4 illustrate this for benzene and naphthalene. Similar to an infinite sheet of graphite, the energy levels are spread over an energy range of $\pm 3\beta$ around this zero-level where β is the resonance integral between two neighbors, 2.9 eV. As the size of the species increases, more π states are present but the range over which they are distributed is very similar. This behavior is illustrated in Figure 8.10. As for graphene and graphite, the highest density of (degenerate) energy levels occurs at $\pm\beta$ for all sizes, but the energy gap – the "HOMO–LUMO" distance – decreases rather smoothly with PAH size approximately as $E_g \propto M^{1/2}$ (Figure 8.11). Hence, as the molecule increases in size, the first absorption will shift toward the red. For compact PAHs, experiments give for the first

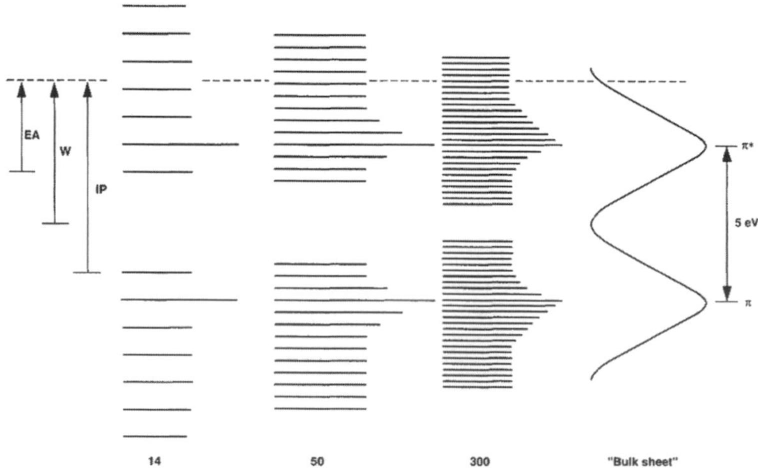

Figure 8.10 Schematic of the influence of size on the energy level structure of PAH species. The arrows on the left-hand side indicate the electron affinity, ionization potential and work function for the smallest PAH. Figure taken from [11]

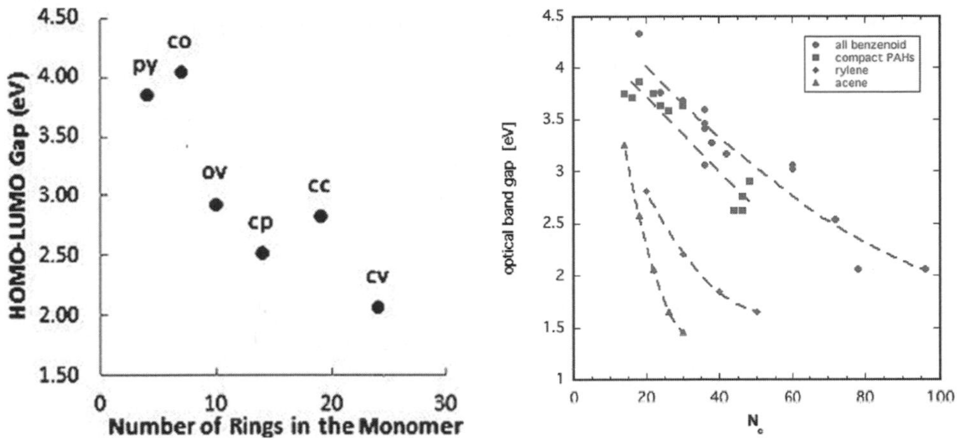

Figure 8.11 Left: The HOMO–LUMO band gap for a series of small PAHs calculated using DFT. The PAHs shown are pyrene (py), coronene (co), ovalene (ov), circumpyrene (cp), circumcoronene (cc), and circumovalene (cv). Figure adapted from [4]. Right: Optical bandgap measured from the p-band transition in several series of PAHs, illustrating the dependence on size (number of π electrons) and molecular structure. Dashed curves represent least square fits to the different series. Adapted from [87]

absorption, $\nu \simeq 30,000/m^{0.5} + 10,000\ \mathrm{cm}^{-1}$ with m the number of hexagon rings, in good qualitative agreement with this simple model. For an "infinite" graphene sheet, the energy gap disappears and for graphite the $\pi, \pi^\star$ states actually slightly overlap due to interaction between the sheets.

At high energies, collective (plasmon) excitation of electrons become important. Above $\simeq$10 eV, ionization spectra are contaminated by shake-up bands where more than one electron is involved in π band transitions. For σ transitions, the one electron picture holds up to energies of $\simeq$13 eV. The challenge to assign these bands requires high level theory and has only been partly met.

8.3 Statistical Physics

8.3.1 Microcanonical and Canonical Ensembles

It is appropriate to start this discussion on the excitation of large molecules in space with a primer on statistical mechanics. In contrast to most terrestrial and laboratory settings, large molecules in space can be considered as isolated systems, as exchange of energy with the surroundings through photons or collisions is very infrequent. Large molecules in space are thus best described as microcanonical ensembles. A system completely isolated from its surroundings has a constant energy, E, and constant number of particles, N (and a constant volume, V). The probability to observe the system with energy E_s is then, $P_s \propto \delta\left(E_s - E\right)$. If the degeneracy of the quantum state associated with energy, E, is

W (e.g. there are W states with energy E) and each state has the same a priori probability, then we have $P_s = 1/W$. The entropy, S, is then given by,

$$S = k \ln [W]. \tag{8.11}$$

The microcanonical ensemble can be contrasted to the canonical ensemble, which is a system in contact with a heat bath. Because of the thermal contact, energy is not constant but all realizations of the system are described by the same average energy, kT, where T is the temperature of the heat bath. In terms of an isolated molecule in space, we can consider a specific mode as a canonical ensemble with the remainder of the molecule as the heat bath. The probability, P_s, to find the system (e.g. specific mode) with energy E_s is then given by the Boltzmann distribution,

$$P_s = \frac{1}{Z} \exp\left[-E_s/kT\right], \tag{8.12}$$

with Z the (canonical) partition function,

$$Z = \sum \exp\left[-E_s/kT\right]. \tag{8.13}$$

The entropy is,

$$S = -k \sum P_s \ln [P_s]. \tag{8.14}$$

The canonical ensemble approaches the microcanonical ensemble when root mean square fluctuations in energy, ΔE_{rms}, are small. Consider the fluctuations in energy, E, relative to the mean, $\overline{E}$,

$$\Delta E = E - \overline{E}. \tag{8.15}$$

We can square these fluctuations and average them to arrive at the root mean square fluctuations in energy,

$$\Delta \overline{E^2} = \overline{E^2} - \overline{E}^2, \tag{8.16}$$

where we have used the fact that the average of a sum is the sum of the average. The rms fluctuations in energy are then,

$$\Delta E_{rms} = \sqrt{\Delta \overline{E^2}}. \tag{8.17}$$

In the canonical system, we have,

$$\overline{E} = \sum P_s E_s \tag{8.18}$$

$$\overline{E^2} = \sum P_s E_s^2. \tag{8.19}$$

Keeping in mind that $\partial \exp\left[-E/kT\right]/\partial T = \left(E/kT^2\right) \exp\left[-E/kT\right]$ and using that the heat capacity is given by, $C_V = \partial \overline{E}/\partial T$, it is easy to show that,

$$C_V = \frac{\overline{E^2} - \overline{E}^2}{kT^2}, \tag{8.20}$$

and

$$\Delta E_{rms}^2 = kT^2 C_V. \quad (8.21)$$

For the relative energy fluctuations, we have,

$$\frac{\Delta E_{rms}}{\overline{E}} = \sqrt{kT^2}\,\frac{\sqrt{C_V}}{\overline{E}}. \quad (8.22)$$

As both C_V and $\overline{E}$ scale directly with the size, N, of the system, we have

$$\frac{\Delta E}{\overline{E}} \propto \frac{1}{\sqrt{N}}. \quad (8.23)$$

Hence, as $N \rightarrow \infty$, the energy fluctuations will go to 0 and that is the microcanonical limit.

8.3.2 Density of States

In order to proceed further, we need to specify the number of states at energy E or, equivalently, the number of states per unit energy in the interval $[E, E + \Delta E]$ (the density of states). A microcanonical system with s degrees of freedom can be described by the s coordinates, q_i, and the associated momenta, p_i. Its evolution is governed by its Hamiltonian, $H(\mathbf{p}, \mathbf{q})$. Any state of the system can be described by a point in the $2s$-dimensional phase space spanned by these coordinates and momenta. The number of states, $\Phi(E)$, with energy less than E is,[8]

$$\Phi(E) = \frac{1}{h^s} \int \cdots \int dp_1 \cdots dp_s\, dq_1 \cdots dq_s. \quad (8.24)$$

The density of states, $\rho(E)$ is, $\rho(E) = d\Phi(E)/dE$. The partition function is given by the Laplace transformation of the density of states,

$$Z(T) = \int \rho(E) \exp\left[-E/kT\right] dE. \quad (8.25)$$

As an example, consider a classical harmonic oscillator with frequency $\omega = \sqrt{k_r/m}$, with k_r the restoring force and m the reduced mass. States with energy E are found on an ellipse with axes, $\pm\sqrt{2E/m\omega^2}$ and $\pm\sqrt{2mE}$ for q and p, respectively. With $\omega = 2\pi\nu$, the total number of states with energy less than E is, $\Phi(E) = E/h\nu$ and the density of states, $\rho(E) = 1/h\nu$. This results in,

$$Z(E) = \frac{kT}{h\nu}, \quad (8.26)$$

the average energy divided by the energy spacing. The partition function of a collection of s distinguishable, harmonic oscillators is $Z = \prod_{i=1}^{s} Z_i$, which yields,

$$Z_{cho}(T) = (kT)^s \prod_i^s \frac{1}{h\nu_i}. \quad (8.27)$$

[8] Note, $d\Phi = W dE$.

For a collection of, s, indistinguishable classical oscillators with frequencies, ν_i, the density of states is given by,

$$\rho(E) = \frac{E^{s-1}}{(s-1)!\prod_i^s h\nu_i} \tag{8.28}$$

where the indistinguishable character gives rise to the factorial term. Defining the geometric average frequency as $\overline{\nu}^s = \prod \nu_i$ and using Stirling's approximation, $\ln[n!] = n\ln[n]$, we can write,

$$\rho(E) = \frac{1}{\overline{\nu}}\left(\frac{E}{(s-1)\,h\overline{\nu}}\right)^{s-1}, \tag{8.29}$$

where the density of states is now given in units of per wavenumber.

While this analytical expression for the classical density of states is very handy, it is only accurate at very high energies and at low energies the quantum nature has to be taken into account. For the quantum harmonic oscillator, energy levels are discreet,

$$E_n = (n + 1/2)\,h\nu, \tag{8.30}$$

and the degeneracy is 1 ($g_n = 1$). As discussed in Section 3.3.2, the partition function is particularly easy to calculate,

$$Z_{qho} = \sum_{n=0}^{\infty} g_n \exp\left[-E_n/kT\right] \tag{8.31}$$

$$= \frac{\exp\left[-h\nu/2kT\right]}{1-\exp\left[-h\nu/kT\right]} \tag{8.32}$$

$$= \left[2\sinh\left(h\nu/2kT\right)\right]^{-1}. \tag{8.33}$$

For high energies ($kT \gg h\nu$ or equivalently when $h \to 0$), we recover the classical result. The probability to find the system with n excitations is,

$$p_n = \frac{\exp\left[-E_n/kT\right]}{Z_{qho}} = a^n\,(1-a), \tag{8.34}$$

with $a = \exp\left[-h\nu/kT\right]$. The average occupation number is,

$$\langle p_n\rangle = \sum_{n=0}^{\infty} np_n = \frac{a}{1-a} = \left(\exp\left[h\nu/kT\right]-1\right)^{-1}, \tag{8.35}$$

which at low energies is just the Boltzmann factor. The mean energy in the oscillator is then,

$$E_1 = h\nu\left(\left(\exp\left[h\nu/kT\right]-1\right)^{-1} + 1/2\right), \tag{8.36}$$

which at low energies is just the zero-point energy and at high energies is $kT + 1/2\,h\nu$. The heat capacity is given,

$$C_V = \frac{dE_1}{dT} = k\left(\frac{h\nu}{kT}\right)^2 \frac{\exp\left[h\nu/kT\right]}{\left(\exp\left[h\nu/kT\right] - 1\right)^2}, \tag{8.37}$$

which goes to the classical limit, k for high energies. For s distinguishable quantum harmonic oscillator, we have,

$$\ln\left[Z_{qho,s}\right] = s\ln\left[Z_{qho}\right] = s\,(h\nu/2kT) - s\ln\left[1 - \exp\left[-h\nu/kT\right]\right], \tag{8.38}$$

and the mean energy, $U_s = s\,U_1$.

In general, finding a simple expression for the density of states is difficult, essentially because of the restriction imposed on the energy of the states that should be counted (e.g. only those with energies in the range $(E, E + dE)$). Nevertheless, Equation (8.29) does demonstrate three essential features of the density of states: 1) As expected, the natural variable is the total energy in the system divided by the characteristic energy of the system, $E/h\overline{\nu}$. 2) For a given number of modes, the density of states increases rapidly with internal energy. 3) For a given internal energy, the density of states increases rapidly with the number of modes. We have encountered this already when we considered the vibrational partition function in Chapter 3 (see Exercise 3.12).

The difficulty in evaluating the density of states for real molecules lies in the quantum nature of vibrations and the coupling between them (and with rotational states) through anharmonic interactions. To illustrate the effect of anharmonic interactions, let us compare the density of states for a Morse potential and a harmonic potential. The energy states for a Morse potential are given by (cf. Section 3.3.1),

$$E = \left(n + \frac{1}{2}\right)h\nu - \left(n + \frac{1}{2}\right)^2 \frac{(h\nu)^2}{4D_e}, \tag{8.39}$$

with D_e the dissociation energy. This leads to,

$$\left(n + \frac{1}{2}\right) = \frac{2D_e}{h\nu}\left(1 - \sqrt{1 - \frac{E}{D_e}}\right). \tag{8.40}$$

The density of states is,

$$\rho\,(E) = \frac{dn}{dE} = \left[h\nu\sqrt{1 - \frac{E}{D_e}}\right]^{-1} \tag{8.41}$$

$$= \frac{1}{h\nu}\left(1 + \frac{1}{2}\frac{E}{D_e} + \frac{3}{8}\left(\frac{E}{D_e}\right)^2 \dots\right), \tag{8.42}$$

where the first term is the harmonic result and subsequent terms can be considered correction factors. It is readily seen that when E approaches D_e, the density of states diverges.

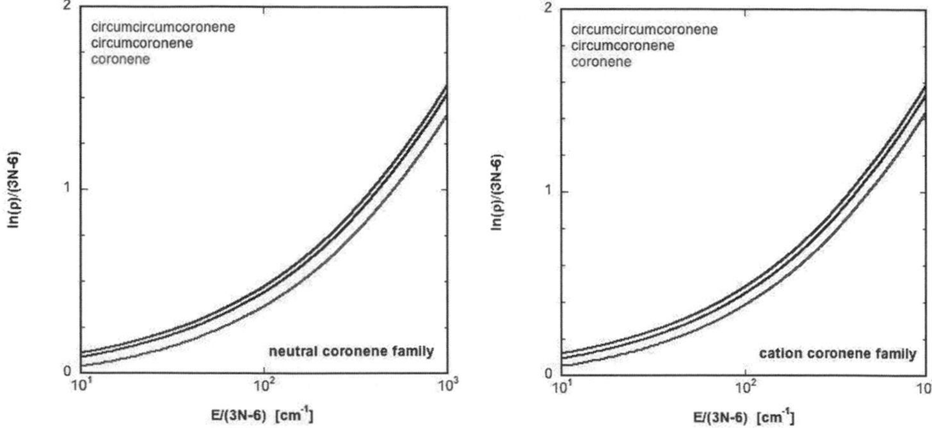

Figure 8.12 Density of states calculated in the harmonic approximation for the first three molecules in the coronene family. The energy (in cm^{-1}) and natural logarithm of the density of states (in cm^{-1}) have been scaled with the number of degrees of freedom. Note the very similar behavior. The small shifts reflect small differences in the geometric mean of the vibrational frequency (cf. Eq. (8.29)).

8.3.3 The Density of States of PAHs

While in the past, the density of states has been calculated using a variety of approximations, with the advent of fast computers, it has become feasible to calculate the (harmonic) frequencies of large molecules such as PAHs using DFT. In this approximation, the density of states can be evaluated using direct counting algorithms and, despite its limitations, these have been widely used. The caveat of the harmonic approximation should, however, be kept in mind and these approximations should be used with caution. The density of states for the first three members of the coronene family are shown in Figure 8.12. Except for a slight offset, the behavior of the density of states is very similar for different PAHs. The microcanonical temperature is given by,

$$\frac{1}{kT_m} = \frac{1}{k}\frac{dS}{dE} = \frac{d\ln\left[\rho\left(E\right)\right]}{dE}. \tag{8.43}$$

The behavior of this derivative can be glanced from Figure 8.12. For the same level of internal excitation ($= E/(3N-6)$), the microcanonical temperatures of different PAHs are very similar, reflecting their similarity in binding and hence vibrational modes. For constant internal energy, the larger PAH will have, as expected, a much lower temperature.

One particularly useful representation for the density of states is,

$$\ln\left[\rho\left(E\right)\right] = 2.84\times10^{-2}\ (3N_c-6)\left(\frac{E}{3N_c-6}\right)^{0.6} - 6.15, \tag{8.44}$$

with E and ρ in units of cm^{-1}. This approximation is valid over the range 0.025–200 cm^{-1}per mode (0.00001–0.075 eV per atom) to typically 10%. Note that this is an approximation for $\ln[\rho(E)]$ and the error in $\rho(E)$ is large. This can readily transformed into a relationship between the internal energy and the microcanonical temperature. A slightly better representation is provided by the following simple powerlaws,

$$T = 3750\left(\frac{E\,(\mathrm{eV})}{3N-6}\right)^{0.45} \qquad 100 < T < 1000 \tag{8.45a}$$

$$T = 11000\left(\frac{E\,(\mathrm{eV})}{3N-6}\right)^{0.80} \qquad 1000 < T < 10000 \tag{8.45b}$$

which fit the temperature to typically better than 5% over this range. Again, for applications that depend exponentially on the temperature (Boltzmann factor, Arrhenius law, emission on the Wien side of the black body curve), the error is large and it is better to work with an accurate evaluation of the density of states.

Because of the (small) skewness of the canonical distribution, there is a small difference between the energy of the system, E, and the average energy, $< E(T_m) >$. The probability, $P\left(E', T\right)$, that, in a canonical system with mean energy kT, an individual mode has an internal energy, E', is,

$$P\left(E', T\right) dE' = \frac{\rho\left(E'\right)}{Z\left(T\right)} \exp\left[-E'/kT\right] dE'. \tag{8.46}$$

A Taylor expansion of this equation provides the correction factor due to the skewness of the distribution,

$$E \simeq < E\left(T_m\right) > -kT_m. \tag{8.47}$$

We can consider the last term as a finite heat bath correction. When the system becomes large, this correction term becomes very small. We can introduce the microcanonical heat capacity, C_m,

$$C_m = \frac{dE}{dT_m} = 2.2\,\frac{E}{T_m}, \tag{8.48}$$

where the right-hand expression is derived using Equation (8.45a) for the density of states of PAHs. Thus, this energy difference can also be seen as a relationship between the microcanonical and canonical heat capacities, C_m and C_c; viz.,

$$C_m \simeq C_c - k. \tag{8.49}$$

Again, this correction factor is small for large systems.

In the ergodic approximation – when energy exchange between the modes is very rapid – the probability of excitation of mode i with v quanta is given by,

$$P_i\left(v\right) = \frac{N_{v,i}\left(E\right)}{N\left(E\right)} = \frac{\rho_r\left(E - vh\nu_i\right)}{\rho\left(E\right)} \tag{8.50}$$

with $N_{v,i}(E)$ the number of species with v quanta in mode i with frequency ν_i, $N(E)$ the total number of species with energy E. The total density of vibrational states, $\rho(E)$, is the number of ways the energy E can be divided over all available states, while the reduced density of states, $\rho_r(E - vh\nu_i)$, represents the number of ways the energy, $E - vh\nu_i$, can be divided over all modes except the mode, i, under consideration.

8.4 The Excitation of Interstellar PAHs

We have discussed the relationship between the internal energy and the excitation of a species (c.f. Figure 8.12) as well as derived approximate expressions for this excitation at a given internal energy either through the density of states (Eq. (8.44)) or through the excitation temperature (Eq. (8.45a)). We have to put this into an interstellar context by considering excitation and deexcitation processes on timescales that are relevant. Let us first consider the limit where excitation is a very slow process. We will describe then the excitation as instantaneous followed by decay to the ground state before the next photon is absorbed. Assuming that these processes are Poisson distributed, the energy distribution function, $G(E)$, is given by,

$$G(E)\,dE = \frac{\overline{r}}{dE/dt}\exp\left[-\overline{r}\,\tau_{min}(E)\right]dE. \tag{8.51}$$

This is basically the probability that, after absorption of a single UV photon, the species can cool to an internal energy, E (linked to the cooling timescale $\tau_{min}(E)$) before the next photon is absorbed. The average photon absorption rate, $\overline{r}$, is given by,

$$\overline{r} = 4\pi\,\sigma_{\rm PAH}(E_{uv})\,\mathcal{N}(E_{uv}), \tag{8.52}$$

with $\mathcal{N}(E_{uv})$ the mean photon intensity of the radiation field at energy, E_{uv}. The time it takes to decay from the initial energy after photon absorption, E_1 to E, $\tau_{min}(E)$, is given by the integral expression,

$$\tau_{min}(E) = \int_E^{E_1}\frac{1}{dE/dt}\,dE. \tag{8.53}$$

The energy immediately after photon absorption is given by,

$$E_1 = E_0 + E_{uv}, \tag{8.54}$$

with E_0 the initial energy in the system. In the evaluation of Equation (8.53), we have to take all processes into account,

$$\frac{dE}{dt} = -4\pi\sum_i \kappa_i\,B(\nu_i, T(E)) - \sum_j k_{frag,j}(E)\,\Delta E_{frag,j} - \ldots, \tag{8.55}$$

where the first term on the right-hand side is the IR emission with κ_i the intrinsic strength of vibrational mode i and the summation is over all vibrational modes. The other term describes the energy loss associated with fragmentation where $k_{frag,j}$ and $\Delta E_{frag,j}$ are the rate for fragmention and the energy involved,[9] and the summation is over all fragmentation processes, j. In evaluating the energy decay rate through IR emission, we have implicitly assumed that the modes are harmonic, which allows us to simplify this equation greatly (see Section 8.4.2).

The IR emission of PAHs can be readily described in the canonical approximation characterized by a mean energy, kT, and the temperature distribution function and energy distribution function of a PAH can be readily related through,

$$G(T)\,dT = G(E)\,dE. \tag{8.56}$$

The emission temperature is related to the internal energy through the heat capacity,

$$E(T) = \int_0^T C_V(T)\,dT \tag{8.57}$$

For harmonic oscillators, we can write for the heat capacity (cf. Section 8.3.1),

$$C_V = k \sum_{i=1}^{s} \left(\frac{h\nu_i}{kT}\right)^2 \frac{\exp\left[h\nu_i/kT\right]}{\left(\exp\left[h\nu_i/kT\right] - 1\right)^2}, \tag{8.58}$$

where the summation is over all s degrees of freedom of the species. Sometimes, the heat capacity of PAHs is approximated by that of graphite. While this neglects the CH modes, this is a good approximation if the internal energy well exceeds the energy of the lowest vibrational modes ($\sim$50 cm^{-1}).

These two equations ((8.57) and (8.58)) provide the canonical relation between the temperature and the internal energy of species. Figure 8.13 shows this relation for the three first members of the coronene family. As expected from Figure 8.12, the relations are very similar, reflecting the similarities in their binding and hence vibrational spectrum. Small deviations at low temperatures/energies reflect that larger species have lower frequency modes that dominate the energy content at low energies. Similar results are obtained for their ionized brethrens. We have now two expressions for the relation of the internal energy and the temperature: The microcanonical relation through the density of states given by Eq. (8.43) and the canonical relation through the heat capacity, Eqs. (8.57) and (8.58). These two are compared in Figure 8.13 for neutral circumcircumcoronene. There are small differences at low temperatures but for astrophysical purposes these are not very relevant.

Because of the similarity in the vibrational properties of PAHs, to first order, the time dependent evolution of the temperature is independent of the size of the PAH (e.g. $dE/dt = C_V dT/dt$ and E and C_V are both extensive quantities and scale linearly with size) and a reasonable approximation for small ($\simeq$ 50 C-atoms) neutral PAHs is given by

$$\frac{d\ln[T]}{dt} \simeq -\,1.1\times10^{-5}\,T^{1.53} \qquad \mathrm{K\ s^{-1}} \qquad T > 250\ \mathrm{K}. \tag{8.59}$$

For 2-dimensional materials, the Debye law for the heat capacity scales approximately with T^2 and to zeroth order that is also a good approximation for PAHs. The energy decay rate

[9] Including the bond energy and kinetic energy carried away by the fragment.

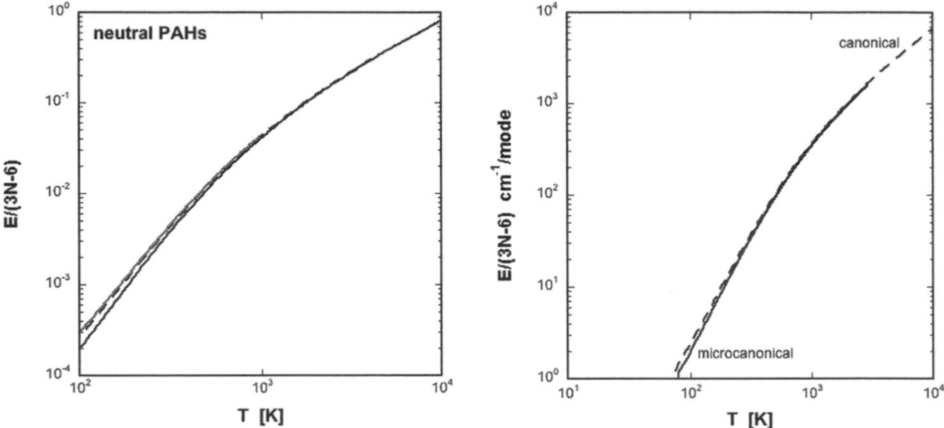

Figure 8.13 Left: Canonical relation between the internal energy (in eV) and the temperature of a species for neutral coronene, circumcoronene, and circumcircumcoronene. Right: Comparison between the canonical and microcanonical temperature-internal energy relations for circumcircumcoronene. See text for details. Note the different units between the two panels (1 eV equals 8100 cm^{-1}).

scales then with a high power of the temperature ($\sim T^{4.5}$). At a temperature of 1000 K, this corresponds to a cooling timescale of $\simeq 2$ s. Again, to first order, this is independent of the PAH size as at a given temperature the emission rate as well as the energy content scale similarly with the number of modes. Ionized PAHs are somewhat (factor 3–4) more effective coolers due to the larger intrinsic strength of the C–C modes.

In reality, an interstellar PAH will be exposed to a spectrum of exciting UV photons. This set of equations should then be averaged over the UV spectrum; e.g. each UV absorption frequency yields a distinct E_1 (Eq. (8.54)). The energy distribution function, ($G(E)$ (or, equivalently, the temperature distribution function, $G(T)$) has to be properly averaged over this range in E_1 (or T_1) and the rate of UV photon absorption (Eq. (8.52)) has to be averaged over the excitation spectrum as well. The energy distribution function averaged over the energy spectrum of the photon field, $\mathcal{N}(E)$, is,

$$\langle G(E)\rangle = \frac{4\pi}{\overline{r}} \int_{h\nu_1}^{h\nu_2} G\left(E, E'\right) \sigma_{\mathrm{PAH}}\left(E'\right) \mathcal{N}\left(E'\right) dE', \tag{8.60}$$

where the ν_i's correspond to the limiting energies of the radiation field.

For small species, the cooling timescale is much faster than the UV absorption timescale and the initial temperature, T_0 – corresponding to E_0 in Eq. (8.54) – can be set equal to the microwave background radiation temperature. Calculation of the temperature distribution function is then rather straightforward. Higher internal energies than 13.6 eV can be reached through multiphoton processes, which become important when $\tau_{uv} < 0.1\tau_{cool}$ where $\tau_{cool} = T/(dT/dt)$. Combining Equations (8.4), (8.45a), and (8.59), this yields $G_0 > 10^6\ (50/N_c)^{1.61}$ for a typical absorbed FUV photon energy of 10 eV. In this case, T_0

and T_1 have to be determined in an iterative fashion. One way to do this is by iterating on $G(T)$; viz.,

$$G_{n+1}(T_0, T) = \int G_n(T_0, T')\, G(T', T)\, dT' \tag{8.61}$$

where the term in the integral is the probability that, starting at temperature, T_0, the species has a temperature T' after n photon events multiplied by the probability that the absorption of an additional photon takes it to temperature T. If the species can absorb over a range of photon energy, $\langle G_n(T_0, T)\rangle$ has to be used. The resulting $G_{n+1}(T_0, T)$ has to be averaged over the energy spectrum of the UV radiation field. Starting with a delta function at the microwave background temperature, this system can then be iterated until convergence is attained. Multiphoton processes are particularly important when considering fragmentation for PAHs larger than $\simeq$24 C-atoms.

8.4.1 Temperature Distribution Function

The temperature distribution function of interstellar PAHs depends strongly on the size of the species and the strength of the radiation field (Fig. 8.14). As the cooling timescale is so much shorter than the UV absorption timescale, each species is very cold for almost all of the time. Once a species absorbs a UV photon, its temperature rises to a maximum of $\simeq$1000 K. This maximum temperature increases with decreasing size, reflecting the reduced heat capacity for smaller species. As larger PAHs have lower frequency modes, they cool down to lower temperatures than small PAHs. In the calculations reported in Figure 8.14, the microwave background radiation temperature was not included and hence

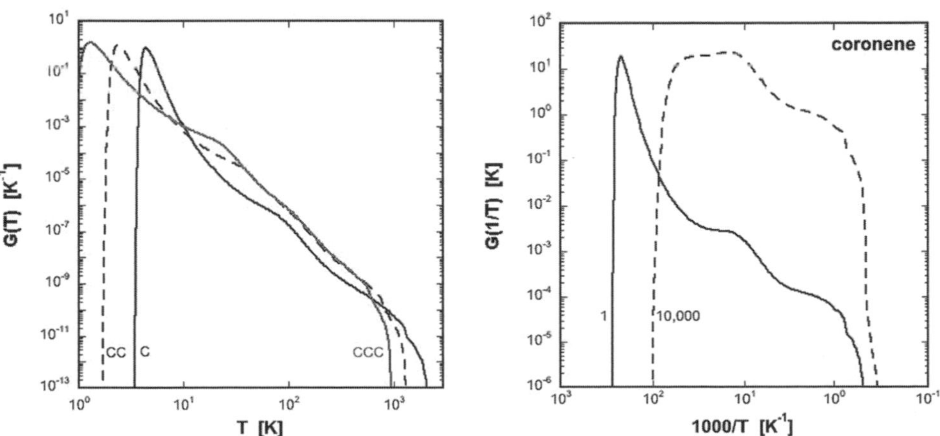

Figure 8.14 The temperature distribution of interstellar PAHs. Left: Calculated temperature distribution, $G(T)$, for coronene (dotted), circumcoronene (dashed), and circumcircumcoronene (solid) averaged over the interstellar radation field with $G_0 = 1$. Right: The temperature distribution, $G'(1/T)$, for circumcoronene for two different radiation fields.

there is no lower limit on the temperature. In reality, the temperature drop will stop at $\sim$5 K. Figure 8.14 also displays the temperature distribution as $G'(\beta)$ with $\beta = 1/T$, where it should be understood that $G'(\beta)\,d\beta = G(T)\,dT$. A typical Arrhenius law process or emission on the Wien side of the Planck function is governed by $\int \exp[-a\beta]\,G'(\beta)\,d\beta$. This representation also brings out the weak shoulder at high temperatures due to two-photon processes. This shoulder becomes much more prominent when the strength of the UV radiation field increases. Processes with particularly high thresholds will benefit enormously from two- or multi-photon processes. As the strength of the radiation field increases, the PAH does not cool down to very low temperatures and multiphoton processes become concommitantly more important.

It is instructive to compare the behavior of large molecules with that of dust grains (Figure 8.15). As the species increases in size, the energy content in radiative equilibrium increases as well and the temperature distribution function changes from a function peaking at low temperature with a long tail to a Gaussian distribution around a mean internal energy. Consider the energy distribution as a canonical process with mean energy, $\bar{E}$, which is related to the temperature through the heat capacity. The photon absorption and emission processes are stochastic. Focus first on the absorption process. Define $< h\nu_{uv} >$ as the average photon energy absorbed. Then, N_{abs} photons have to be absorbed in order to reach this steady state value with $N_{abs} = \bar{E}/ < h\nu_{uv} >$. For a stochastic process, the dispersion

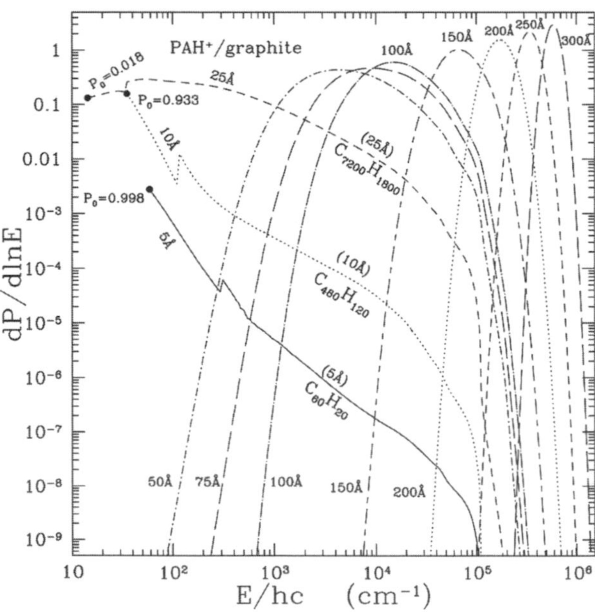

Figure 8.15 The internal energy distribution function for different sized species. The kinks in the distribution functions for the smallest sizes are artifacts due to approximations in the adopted properties. Figure taken from [64]

in the number of photons absorbed is then $N_{abs}^{1/2}$ or in terms of the dispersion in energy, $\sigma_{abs} = (\bar{E} < h\nu_{uv} >)^{1/2}$. When N_{abs} is large, the distribution function is then given by the gaussian,

$$P(E) = \frac{\exp\left(-(E-\bar{E})^2 / (2\bar{E} < h\nu_{uv} >)\right)}{\sqrt{2\pi\bar{E} < h\nu_{uv} >}} \tag{8.62}$$

So, as the average internal energy becomes larger and larger compared to the typical absorbed photon energy, the energy distribution function will become more and more sharply peaked around this mean value. The IR emission process will, in the same way, give rise to a Gaussian distribution with a dispersion set by the number of photons emitted; e.g. $\sigma_{em} = (\bar{E} < h\nu_{ir} >)^{1/2}$ with $< h\nu_{ir} >$ the average energy of the IR photons that are emitted. The combined absorption and emission distribution function is then also a Gaussian with a dispersion given by, $\sigma^2 = \sigma_{uv}^2 + \sigma_{ir}^2$ and because the average IR emission energy is so much less than the UV absorption energy, $\sigma \simeq \sigma_{uv}$. As this figure well illustrates, for PAHs, single photon events dominate and the distribution function is given by Eq. (8.51). As the species increases in size, the distribution function transits to a Gaussian distribution and when $\bar{E} \gg h\nu_{uv}$ becomes very sharply peaked at $\bar{E}$.

8.4.2 IR Emission Spectrum

The emission intensity due to the vibrational transition $v \rightarrow v-1$ in mode i is given by,

$$I(E,i,v) = \frac{N_{v,i}(E)\ A_{v,i}\ h\nu_i}{4\pi}, \tag{8.63}$$

with $A_{v,i}$ the Einstein A emission coefficient associated with this transition and $N_{v,i}(E)$ all species with internal energy E with v quanta in mode i. Summing over all v in the harmonic approximation, we can write for the intensity in mode i,

$$I(E,i) = \kappa_i\ B(\nu_i, T), \tag{8.64}$$

with κ_i the absorption coefficient,

$$\kappa_i = B_i\ N(E), \tag{8.65}$$

where B_i is the Einstein coefficient for absorption in mode i. For the expected intensity, we then have to integrate over the energy cascade,

$$I(i) = \int_{E_o}^{E_1} I(E,i)\ G(E)\ dE, \tag{8.66}$$

or equivalently,

$$I(i) = \int_{T_o}^{T_1} I(E(T),i)\ G(T)\ dT, \tag{8.67}$$

Finally, the interstellar PAH family may be very diverse and occur in different ionization stages, each with its own intrinsic properties, including distinct vibrational modes. The IR intensity of such a family can then be found by a proper summation; viz.,

$$I(\nu) = \sum_k \sum_j \sum_i n(k)\ f(j)\ I_{k,j}(i)\ \phi_{k,j}(i,\nu), \tag{8.68}$$

where the summation is over the modes, i, of ionization stage, j, of molecule, k. Here, $n(k)$ is the density of the specific PAH molecule, k, and $f(j)$ the fractional abundance of ionization stage, j. The integrated intensity, $I_{k,j}(i)$, of mode i of molecule k in ionization stage j is given by Equation (8.67), and $\phi_{k,j}(i,\nu)$ is the intrinsic line profile of this transition.

8.4.3 Emission Profile

The classical oscillator model for a dipole transition leads to the Lorentz emission profile,

$$\phi(\nu) = \frac{1}{\pi} \frac{\gamma/4\pi}{(\nu-\nu_0)^2 + (\gamma/4\pi)^2}, \tag{8.69}$$

with ν_0 the central frequency and γ the classical damping constant,

$$\gamma = 8\pi^2 e^2/3m_e c\lambda^2. \tag{8.70}$$

The width of the Lorentzian damping profile is equal to $\Delta\nu = \gamma/2\pi$. The quantum mechanical oscillator also gives rise to the Lorentzian profile but the damping constant is now set by the finite lifetime of the levels involved. This is a consequence of Heisenberg's uncertainty principle, $\Delta E \Delta t > \hbar$. With $\Delta t = A^{-1}$ (A is the Einstein coefficient for spontaneous transitions), this translates into an uncertainty in the frequency of $\Delta\nu = A/2\pi$. For a typical ro-vibrational Einstein A of 10^2 s^{-1}, we then expect a linewidth of $\Delta\nu = 6 \times 10^{-10}$ cm^{-1}, which is much less than the turbulent Doppler width of the line, $\sim$0.01–0.03 cm^{-1}.

However, emission by a highly vibrationally excited molecule is not well represented by an isolated harmonic oscillator, and anharmonicity effects dominate the emission profile of highly excited species. In a way, during the emission process, the constant "tugging" of the other excited modes on the atoms in the molecule leads to a changing potential associated with the emitting mode. This is akin to the collisional broadening process that governs the broadening of stellar absorption lines. The interaction of two vibrational modes, k and l, is caught in anharmonic coefficients, $\chi_{k,l}$, that describe shift in the fundamental vibration frequency due to the interaction with other (excited) modes. For a given excitation of a molecule, the total energy is,

$$\frac{E(\vec{n})}{hc} = \sum_k \nu_k (n_k + 1/2) + \sum_{k \le l} \chi_{k,l} (n_k + 1/2)(n_l + 1/2), \tag{8.71}$$

where $\vec{n}$ is the vector of occupation numbers, n_i, describing the population of the fundamental vibrational modes. The energy of the transition corresponding to $n_k + 1 \rightarrow n_k$ of mode k with this set of occupation numbers is,

$$\frac{E^{(k)}(\vec{n})}{hc} = \nu_k + 2\chi_{k,k} n_k + \sum_{i \neq k} \chi_{i,k}(n_i + 1/2)\,. \tag{8.72}$$

Here, the second term on the right accounts for overtone emission within the mode and the last term is the anharmonic interaction with the other modes. This expression is for a given distribution of the internal energy, E, over all available modes. As the energy sloshes around on a very short timescale compared to the emission timescale, this expression has to be averaged over all possible realizations of the system (i.e. all possible $\vec{n}$ with an energy equal to the total energy in the molecule).

Anharmonicity typically leads to red shifts of lines (cf. Section 3.3.1) and, as the internal excitation of the molecule increases, the occupation numbers will increase, this redshift will become more pronounced, and the transition will broaden. Figure 8.16 illustrates this behavior for the CH out-of-plane bending mode of tetracene, where the anharmonic interaction between the modes have been calculated using DFT. The redshift and broadening of the line profile is quite apparent. As this is a statistical process, the calculated line profiles are quite symmetric at any internal energy. As the molecule cools down during the emission process, the frequency shift decreases. The observed emission profile has to be determined by following this energy cascade from the initial energy all the way down,

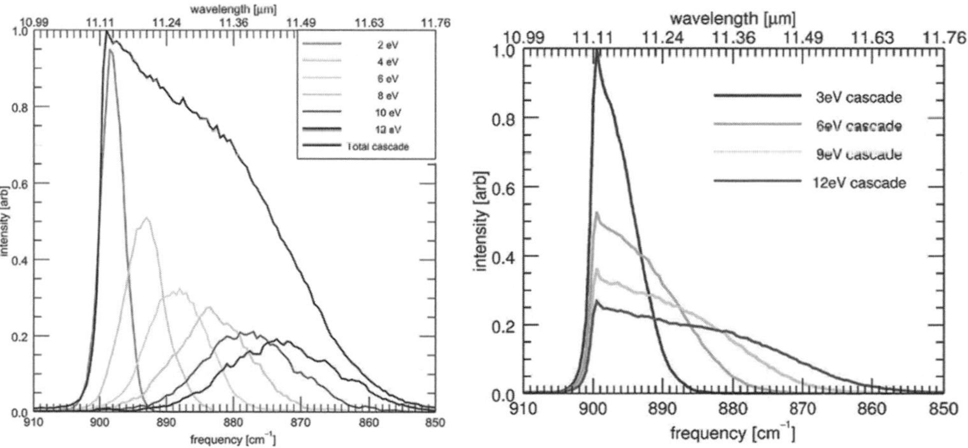

Figure 8.16 Left: Emission profiles of the CH out of plane bending mode of tetracene calculated for different internal energies. The anharmonic constants have been calculated using DFT. Due to anharmonicity, the line shifts to the red and broadens as the internal energy of the species increases. The black line is the spectrum obtained following the full cascade from an initial internal energy of 12 eV. Right: Calculated emission profiles following the full energy cascade starting at increasing initial internal energies. Figure taken from [67]

using the convolution of Eq. (8.66). This will lead to a characteristic red-shaded profile (Figure 8.16) and the "redness" of the profile is a measure of the internal excitation of the species.

8.5 Photochemistry

UV photon absorption leaves a molecule highly vibrationally excited and it can cool through IR emission as described in Section 8.4.2 but for high enough excitation, the internal energy can also be used to fragment the species. In the simplest approximation – so-called RRK theory after its originators, Rice, Rampsperger, and Kassel – the species is treated as a system of s identical oscillators of frequency ν, and a molecule with internal energy E will dissociate if one of these oscillators has an energy in excess of a critical energy, E_o. Defining,

$$n \equiv E/h\nu, \tag{8.73}$$

and

$$m \equiv E_o/h\nu, \tag{8.74}$$

the probability for dissociation can be found from combinatorial statistics,

$$P = \frac{(n-m+s-1)!\,n!}{(n-m)!\,(n+s-1)!}. \tag{8.75}$$

Generally, n and m are not integers. Moreover, a molecule will have a range of oscillator frequencies. In the quantum RRK version of this theory, this expression is rewritten to,

$$P = \frac{\Gamma(n-m+s-1)\,\Gamma(n)}{\Gamma(n-m)\,\Gamma(n+s-1)}, \tag{8.76}$$

where Γ is the gamma function and ν is now the geometric mean of the vibrational frequencies. If the oscillator frequency is very small (e.g. n and m are both very large) and the number of quanta is much larger than the number of oscillators (e.g. $n-m \gg s$), this probability reduces to,

$$P \simeq \left(\frac{n-m}{n}\right)^{s-1}. \tag{8.77}$$

An additional factor ν transforms this probability then to the rate constant,

$$k(E) = \nu\left(\frac{E-E_b}{E}\right)^{s-1}. \tag{8.78}$$

We encountered this expression already in Section 6.6 on radiative association reactions. The use of this simple expression has its limitations. Nevertheless, it predicts correctly that the rate rapidly decreases with increasing number of oscillators and increases with the excess energy in the system. Of course, typically, for interstellar PAHs, m and $n-m \simeq 30 < s$ and hence use of this expression is not really justified.

This early RRK theory has led to the development of RRKM theory where the reaction is described as dissociation through a transition state, or more properly through a saddlepoint that separates the reactant from the products. In RRKM theory, it is assumed that all species that encroach upon the phase space area of the transition state will dissociate and that this is a one-way street (products cannot go back to the parent molecule). Also, it is assumed that dissociation takes place through one separable coordinate perpendicular to all other coordinates. The RRKM rate is then given by,

$$k_{RRKM}(E) = \frac{W^{\star}(E - E_o)}{h\rho(E)} \tag{8.79}$$

where $W^{\star}(E - E_o)$ is the sum of states at the transition state between 0 and $E - E_o$ and $\rho(E)$ is the density of states of the parent species. For a set of classical oscillators (cf. Eq. (8.29) in Chapter 8.3.2), this results in the classical RRKM rate constant,

$$k_{RRKM}(E) = \left(\frac{E - E_o}{E}\right)^{s-1} \frac{\prod_{i=1}^{s} \nu_i}{\prod_{i=1}^{s-1} \nu_i^{\#}} \tag{8.80}$$

where the ν_i's and the $\nu_i^{\#}$'s are the vibrational frequencies of the species and of the transition state, respectively. This ratio of frequencies is itself a frequency and hence we recover an expression akin to Equation (8.78).

We can also consider the unimolecular dissociation reaction as the reverse of the association reaction of the products forming the parent upon collision. In detailed balance, the rates of these forward and backward reactions are related (Section 5.2.6). The unimolecular rate is then given by,

$$k(E,\epsilon) = A(\epsilon)\,\frac{\rho_d(E - E_o - \epsilon)}{\rho_p(E)}, \tag{8.81}$$

where E is the initial excitation energy of the species, E_o is the binding energy of the fragment, ϵ is the energy carried away by the fragment (-R), and ρ_p and ρ_d are the densities of states of the parent (PAH-R) and the daughter (PAH-), respectively. The factor A is proportional to the cross section for the reverse reaction. If we assume that ϵ is small compared to E, then we can expand the density of states in ϵ and only retain the first term. After integration over ϵ, the total rate becomes,

$$k(E) = k_o(T_d)\,\frac{\rho_d(E - E_o)}{\rho_p(E)}, \tag{8.82}$$

where $k_o(T_d)$ depends on the interaction potential (in the reverse reaction). As long as there is no activation barrier for the association reaction, the dependence on temperature will be small. This microcanonical daughter temperature T_d is defined analogously to Equation (8.43) as,

$$\frac{1}{kT_d} = \frac{1}{k}\frac{dS}{dE} = \frac{d\ln[\rho_d(E)]}{dE}. \tag{8.83}$$

This expression can be evaluated using the density of states as given by Equation (8.44) but the equation becomes a bit cumbersome because of the difference in degrees of freedom for the parent and daughter. When the parent and daughter differ by only a few degrees of freedom, we may replace the densities of states by a single function (and include the slowly varying correction factor in the expression for k_o). The unimolecular dissociation rate can then be written in Arrhenius form,

$$k\,(E) = k_o\,(T_e)\,\exp\left[-E_o/kT_e\right], \tag{8.84}$$

where T_e is an effective temperature defined as,

$$\frac{1}{kT_e} = \frac{d\ln\left[\rho_d\,(E^\star)\right]}{dE}, \tag{8.85}$$

with $E^\star$ some energy between E and $E - E_o$. It can be shown that, to first order, this effective temperature is given by,

$$T_e \simeq T_m - \frac{E_o}{2C_m}, \tag{8.86}$$

where the second term is the finite heat bath correction. With the expression for the microcanonical heat capacity for PAHs (Eq. (8.48)), this becomes,

$$T_e = T_m\left(1 - 0.23\,\frac{E_o}{E}\right). \tag{8.87}$$

For a given internal energy, the correction factor increases when the (binding) energy involved becomes a larger fraction of the energy in the system. The dependence on size is somewhat hidden: A larger species will require a larger internal energy to attain the same microcanonical temperature (cf. Eq. (8.45a)) and hence, the correction factor will be smaller. For PAHs absorbing a given photon energy, $h\nu$, this becomes,

$$T_e \simeq 3750\left(\frac{h\nu\,(\mathrm{eV})}{N_c}\right)^{0.45}\left(1 - 0.23\,\frac{E_o\,(\mathrm{eV})}{h\nu\,(\mathrm{eV})}\right) \tag{8.88}$$

and, of course, for a given binding energy and a given photon energy, the correction factor does not change. However, the rate will rapidly decrease with the size of the species. In the limit $E \gg E_o$ and ignoring the difference in number of oscillators between daughter and parent, Expression (8.84) can be shown to be formally equivalent to Equation (8.81) with $T_e = T_m$.

Finally, the microcanonical RRKM rate can also be obtained from the canonical rate equation using an inverse Laplace transformation. Recall that the partition function is the Laplace transformation of the density of states (Eq. (8.25)). Conversely, the density of states is the inverse Laplace transformation of the partition function. Likewise, the sum of states is the inverse Laplace transformation of $kT\,Z\,(T)$. Without derivation, we can then use the convolution properties of inverse Laplace transformations to arrive at,

$$k_c = A_\infty\,\frac{\rho\,(E - E_\infty)}{\rho\,(E)} \tag{8.89}$$

with A_∞ and E_∞ the high pressure pre-exponential and activation energy, respectively. The latter is related to the 0 K activation energy E_o through,

$$E_\infty = E_o + kT + \left(E^\# - E\right), \tag{8.90}$$

where the term in brackets refers to the difference in thermal energy between the transition state and the molecule. There is a difference with the RRKM rate in that the density of states refers to the molecule and not the transition state. The attractive part of this is that it allows the calculation of the rate even when the transition state properties are not known.

The pre-exponential factor is the equivalent of k_o and its value depends on the interaction potential. If we set it equal to,

$$k_o = \frac{kT_e}{h} \exp\left[1 + \frac{\Delta S}{R}\right] \tag{8.91}$$

with ΔS the change in entropy associated with the transition state which depends on T_e, we recover the canonical Arrhenius expression for a gas in thermal equilibrium at temperature T_e. The canonical expression for the entropy change is given by,

$$\Delta S = k \ln\left[\frac{\prod \Phi_i^\#}{\prod \Phi_i}\right] + \left(\frac{E^\# - E}{T}\right) \tag{8.92}$$

where the Φ's are the molecular vibrational and rotational partition functions of the transition state and the parent molecule. These can be evaluated from the vibrational frequencies and moments of inertia (Sections 3.2.8 and 3.3.2).

8.5.1 The Astrochemical Rate Coefficient

Figure 8.17 shows, as an example, the calculated dissociation rate for H/H_2 loss from two PAHs (see Section 8.5.2 for details) as a function of the internal energy. The dissociation rate increases rapidly with internal energy above the threshold as it becomes increasingly likely that the system enters the transition state region where dissociation will occur. In evaluating the photochemistry of *interstellar* PAHs, the rate for dissociation has to be compared to other decay channels in order to arrive at the probability for dissociation upon photon absorption. Emission of IR photons is often the dominant competing channel, and the photodissociation probability after absorption of a FUV photon of energy, E, can be written as,

$$p_d(E) = \frac{k(E)}{k(E) + k_{IR}(E)}, \tag{8.93}$$

where strictly speaking, the unimolecular dissociation rate should have been evaluated over the cascade. However, because of the steep dependence of this rate on internal energy, it can be evaluated at the internal energy before cooling sets in. We can now define the total dissociation rate, k_t, as,

$$k_t \equiv \int k(E)\, G(E)\, dE, \tag{8.94}$$

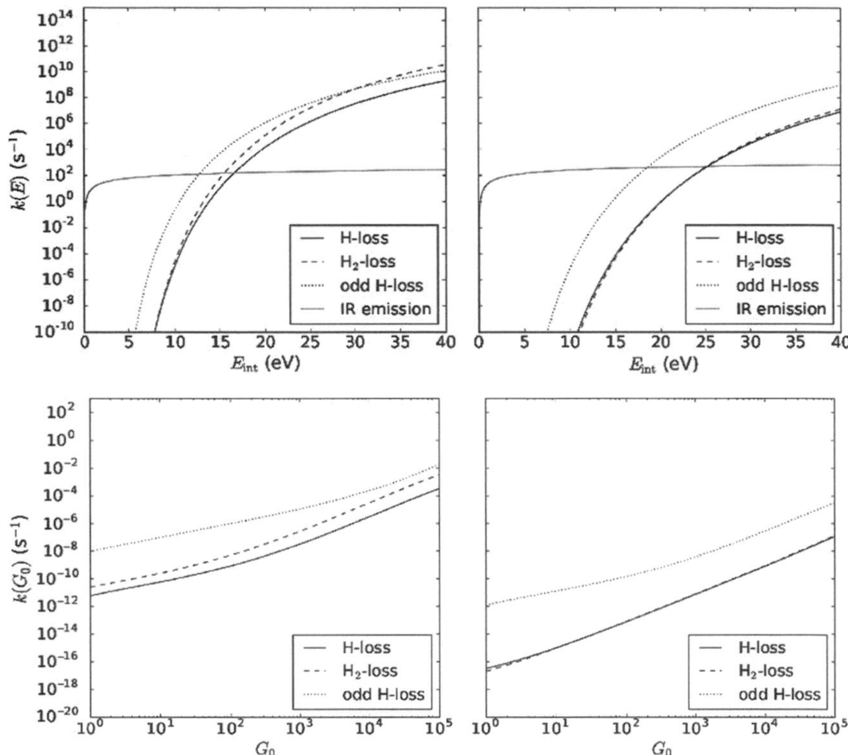

Figure 8.17 Top: Comparison of IR emission and photodissociation rates for H and H_2 loss from ovalene (top left) and circumcoronene (top right). The rate for H-loss after a first H-has already been removed is also shown. Bottom: Calculated photodissociation rates as a function of the strength of the incident radiation field for ovalene (bottom left) and circumcoronene (bottom right). Figure adapted from reference [26]

where it should be remembered that in the evaluation of $G(E)$ all energy loss processes are taken into account (cf. Eq. (8.55)).

Figure 8.17 shows, as an example, the calculated dissociation rate for H/H_2 loss from two PAHs (see Section 8.5.2 for details) as a function of the strength of the radiation field. A typical PAH can absorb some 10^8 photons over its lifetime in the ISM. Hence, even probabilities as small as 10^{-8} per photon event can still be relevant if there is no balancing process reforming the parent species.

8.5.2 Experimental Studies

Fragmentation studies have almost exclusively been performed on PAH cations and their derivatives as ions are readily stored in ion-traps on seconds time scales and/or products are easily characterized using mass spectrometry (Section 2.4.2). Relevant results are summarized in Table 8.1 where kinetic parameters have been determined from time-resolved

Table 8.1 *Kinetic parameters in the fragmentation of PAH and fullerene cations*

Reaction	E_o [eV]	ΔS [J K^{-1} mole^{-1}]	E_{app} [eV]
$C_{16}H_{10}^+ \rightarrow C_{16}H_9^+ + H$	4.6	10.7	9.06
$C_{16}H_{10}^+ \rightarrow C_{16}H_8^+ + H_2$	3.5	−12.7	11.18
$C_{10}H_8^+ \rightarrow C_8H_6^+ + C_2H_2$	4.4	3.5	10.35
$C_{60}^+ \rightarrow C_{58}^+ + C_2$	8.3	156[a]	39.4
$C_{60} \rightarrow C_{58} + C_2$	10	131[a]	

Reference: [65, 84] [a]Note that these have been calculated from the Arrhenius factor (6×10^{21} and 8×10^{20}).

measurements. More generally, results are presented in terms of the appearance energy, E_{app}; e.g. the internal energy at which the reaction occurs on the timescale relevant to the detection method. For a time-of-flight mass spectrometer, this timescale implies a rate of $\simeq 10^4$ s^{-1}. In the ISM, the relevant timescale is the IR cooling timescale of ~ 10 s^{-1} (cf. Section 8.5.1), which is many orders of magnitude longer. For probabilities as small as 10^{-8}, the relevant rate is 10^{-7} s^{-1}. Hence, by necessity, laboratory determined rates have to be extrapolated over many orders of magnitude.

The kinetic parameters for H and H_2 loss have been determined for the pyrene cation ($C_{16}H_{10}$) through time-resolved mass spectrometry measurements (Table 8.1). This corresponds to an appearance energy of $\simeq$9–11 eV for the pyrene cation. The values for H-loss are in good agreement with DFT calculations. Based on DFT studies followed by kinetic Monte Carlo simulations, both H and H_2 losses start with H-roaming, which has a relatively low barrier ($\simeq$3.2 eV), and the formation of a CH_2 group. Most of the time, the H will move back to the emptied site (with a barrier $\simeq$0.8 eV) but H or H_2 can be eliminated from this (short-lived) aliphatic intermediary ($\simeq$2.5 eV). The exact barriers involved for these two processes – and therefore their relative importance – seem to depend on the particular molecular structure involved. If a single H is lost, the PAH is left with an odd number of H's and loss of the second H has a much reduced barrier ($\simeq$3.8 eV compared to a total of $\simeq$5.7 eV). Monte Carlo simulations show that this two-step process – a roaming H forms an sp^3 group followed by H/H_2 elimination from this aliphatic side – is preferred over the direct loss of an sp^2 H.[10]

Adopting the values in Table 8.1 as "generic," the unimolecular dissociation rate for H and H_2 loss from ovalene and circumcoronene cations are shown in Figure 8.17. As expected, these rates rise sharply with the internal energy of the cation and depend strongly on the size of the PAH cation. This merely reflects the dependence of the density of states

[10] Compare this to walking up a steep slope which is much assisted by steps.

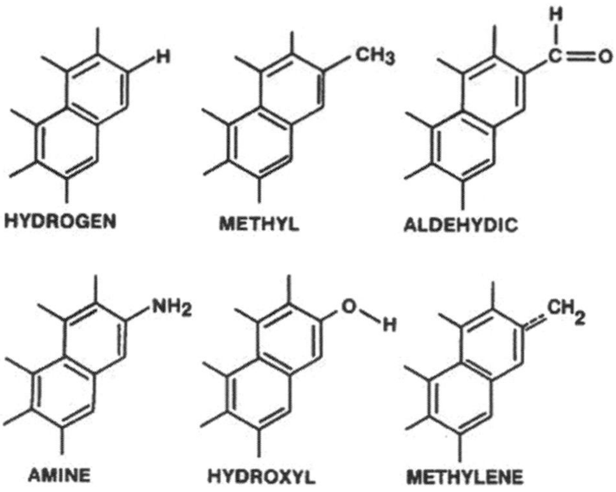

Figure 8.18 Structural formulae of some sidegroups on PAHs. The sidegroups replace an H atom on the periphery of the carbon backbone outlined by the hexagonal network. Figure reproduced with permission from [37]

on these values (cf. Fig. 8.12). These results also illustrate that IR relaxation dominates at low energies, while fragmentation takes over at high internal energies. The internal energy corresponding to this transition depends on PAH size. Thus, while for ovalene ($C_{32}H_{14}$), fragmentation takes over at about 13 eV, for circumcoronene ($C_{54}H_{18}$), it takes 25 eV. As a corrolary, molecules smaller than ovalene (e.g. coronene, $C_{24}H_{12}$), fragment readily by a single photon in the neutral ISM, but larger species require absorption of multiple photons. As a result, the photodissociation rates for these species in the ISM scales with G_0^{α} with $\alpha > 1$ (Figure 8.17).

Estimates for the relevant kinetic parameters in the fragmentation of neutral PAHs and PAH derivatives (Figure 8.18) are summarized in Table 8.2. Calculated RRKM dissociation rates for these groups are compared in Figure 8.19. These are based upon binding energies rather than direct measurements and should, thus, be considered as guides. The results summarized in Figure 8.19 and Table 8.2 indicate that sidegroup fragmentation is dominant over aromatic H loss and this is well supported by experiments. Indeed, because of their very different binding energies, the rates are very different for different sidegroups. Following the weakest link approach, these sidegroups should be sorted according to their binding energies. We conclude that, in a "competition," the methyl ($-CH_3$) group of an ethyl ($-CH_2CH_3$) sidegroup is the first to go, followed by H-loss from a methyl, hydroxyl, or amine group, the complete methyl or amine group, and finally, the aromatic hydrogen and the hydroxyl group. Loss of an acetylene group is predicted to be the slowest channel. Often, side group fragmentation involves isomerization pathways. For example, fragmentation of highly excited, methoxylated PAH cations (PAH–OCH_3^+) can proceed through several pathways, which typically lead to" loss" of a C atom from the ring together with

Table 8.2 *Unimolecular dissociation of neutral PAHs*

Sidegroup[a]	Bond[b]	E_b [eV]	p_d^c
Hydrogen	PAH – H	4.47	10^{-10}
Methyl	$PAHCH_2$ – H	3.69	10^{-2}
	PAH – CH_3	4.0	10^{-4}
Ethyl	$PAHCH_2$ – CH_3	3.1	0.9
Hydroxyl	PAHO – H	3.69	10^{-2}
	PAH – OH	4.5	10^{-10}
Amine	PAHNH – H	3.47	10^{-1}
	PAH – NH_2	4.0	10^{-4}
Acetylene	PAH′ – C_2H_2 [d]	8.0	negligible

[a]Structural formulae of these sidegroups are illustrated in Figure 8.18. [b]Bond denoted by –. [c]Estimated photodissociation probability for a 50 carbon atom PAH, assuming $E = 10$ eV and $k_{IR} = 1\ s^{-1}$ (cf. Eq. (8.93). [d]Loss of acetylene group from PAH skeleton.

the attached methoxy functional group in conjunction with the conversion of a hexagon to a pentagon. Photolysis of PAH-quinones (PAH=O) cations similarly results in pentagon formation (and loss of a CO group). For methylated PAH cations, photolysis results in the loss of an H-atom from the methyl group and, at low energies, rearrangement to a tropylium[11] structure. At high energies, formation of the benzyl-like structure (PAH–CH_2) is competitive.

The rates in Figure 8.19 and Table 8.2 are evaluated at a fixed energy. Proper evaluation of the astrochemical rates requires averaging these rates over the energy distribution function (cf. Eq. (8.94)). The total dissociation rate is then dominated by the highest energies that are typically achieved. While for small PAHs these are achieved upon absorption of a single photon, for a 50 C-atom PAH, two or more photon processes are important. The relative importance of the different channels will strongly depend on PAH size, despite theory predicting little dependence of the kinetic parameters on size. The similarity in kinetic parameters implies that fragmentation occurs at similar internal excitation temperatures but that corresponds to much higher internal energies for larger PAHs (cf. Eq. (8.45a)). Hence, small energy differences, δE, between the various channels are then amplified to large internal energy differences, $\Delta E \propto (3N_c - 6)\,\delta E$ or in terms of the appearance energy, $\Delta E_{app}/E_{app} \sim (N_{c,1}/N_{c,2} - 1)$. This is borne out by experiments where C-loss channels are competitive with H-loss channels (for derivatives of) small PAHs but large PAHs exclusively fragment through complete H-stripping before C-loss channels open up.

[11] The tropylium ion is the heptagonal cyclic species ($C_7H_7^+$) with six delocalized π electrons in three molecular orbitals distributed over the seven carbon atoms. This aromatic structure is very stable.

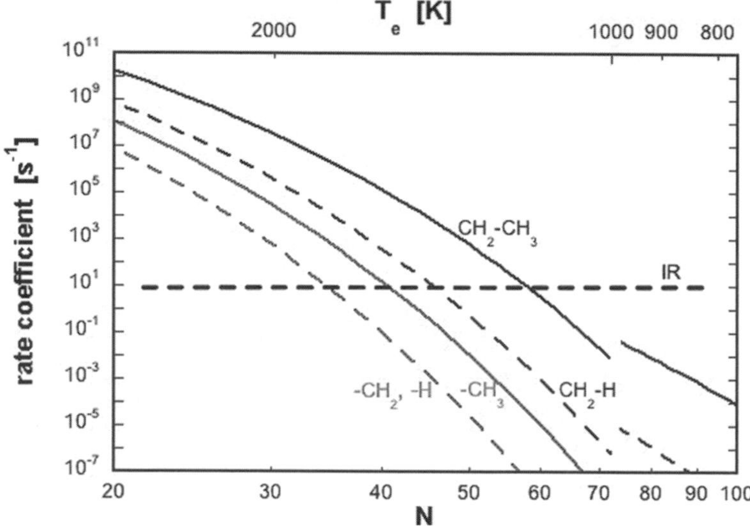

Figure 8.19 Calculated RRKM unimolecular dissociation rates (Eq. (8.84) with Eqs. (8.87) and (8.45a)) for different side groups as a function of PAH size for an internal energy of 10 eV. The critical energies have been set equal to the binding energies of the group Eq. (8.93). The pre-exponential factor has been set equal to 3×10^{16} s^{-1}. The break in the curves reflects the change over between the two $E-T$ relations. The horizontal dashed line indicates the IR relaxation rate. Figure adapted from [46]

In the above analysis, we have tacitly assumed that IC is very rapid and leaves the species highly vibrationally excited in the lowest electronic state. As discussed in Section 8.2.2, that is not a good assumption for small molecules, which have relatively large energy gaps between the lowest electronic states, and energy can become trapped in excited electronic states from which radiative electronic (rather than vibrational) transitions are important. Take coronene as an example: The energy flow goes rapidly from highly excited electronic states down to S_1. Because of the large energy gap with S_0 (23,000 cm^{-1}) only $\simeq$5% of the times, the system will transfer to the S_0 state nonradiatively (which would leave the molecule with all the photon energy in vibrational excitation). In $\simeq$27% of the time, relaxation will be radiatively to the S_0 state and some 23,000 cm^{-1} of energy is emitted in a near-UV photon. In the remainder of the cases (2/3), coronene will transfer to the T_1 state, storing 19,000 cm^{-1} in electronic excitation. Radiative and nonradiative relaxation rates from the T_1 state are much slower than IR emission, and vibrational deexcitation to the ground vibrational state of the T_1 will be the most important next step in the ladder. Eventually, electronic relaxation will take the system to the S_0 state in 11% of the cases while nonradiative relaxation occurs in the remaining $\simeq$55% of the cases. Thus, fragmentation in the electronic excited states will occur at "reduced" internal vibrational excitation. Furthermore, fragmentation in the S_1 state will compete as a deexcitation process with radiative and nonradiative electronic transitions, which have a much higher rate than vibrational fluorescence. As the species increases in size, the energy gap, S_1-S_0, decreases, nonradiative relaxation to the ground electronic state will dominate (cf. Figure 8.7), and these considerations are of lesser importance. Similar considerations apply to other closed shell species.

For PAH ions (and other open shell species), we have to consider doublet and quartet states. In this case, the lowest quartet state, Q_1, is above the first excited doublet state, D_1, and hence ISC will be preferentially from the former to the latter. Hence, storage of the system in the quartet ladder is not very important. Also, states will be lower in energy and so relaxation to the ground state will be much faster. As a result, small PAHs tend to dissociate much faster when ionized than when neutral.

8.5.3 Breakdown of the Statistical Limit

The ergodic approximation plays a key role throughout this discussion on the excitation and the resulting IR emission and fragmentation. When the energy in the molecule is low, the states will decouple and the energy will be localized in specific modes. At that point, the energy cascade will have to be described on a per mode basis. This transition from a delocalized to a localized phase in the energy cascade occurs when the energy width of the states is less than the energy spacing of the states to which they couple. For isolated oscillators, the width of the state is set by the radiative decay rate. However, PAHs consist of a large number of coupled modes and the width of the state is set by the anharmonic coupling between levels. At the onset of coupling, the IVR rate is then,

$$k_{IVR} \simeq \frac{2\pi c V^2 \rho(E)}{\hbar}, \tag{8.95}$$

with V the coupling matrix element and ρ the density of states. The timescale for IVR is measured to be $\sim$10 ps. In evaluating the breakdown of the statistical limit, it should be kept in mind that states do not couple directly to the entire bath of modes but rather couple anharmonically to doorway states that couple anharmonically to other states and eventually after several "tiers" couple to the vibrational quasi continuum. The IVR timescale relevant for this discussion is then determined by the coupling matrix element and the local density of states that can couple to the relevant state through anharmonic interaction. As anharmonic couplings are linked to the properties of the mode and modes behave very similar for all species in the PAH family, we expect that the IVR limit does not depend much on the size of the aromatic system. For anthracene, the onset of IVR is measured to occur around an internal energy of $\sim$1800 cm^{-1} or a density of states of $\sim$120 cm^{-1}. The internal energy to reach this density of states will scale roughly with,

$$E_{IVR} \simeq 1800 \left(\frac{24}{N}\right)^{2/3}. \tag{8.96}$$

This limit on IVR has two consequences. First, when the modes decouple the excitation is trapped in specific modes. Hence, in the energy cascade, the last $\simeq$1000 cm^{-1} – some

1% of the total energy – will be emitted by these modes and these are by necessity the lowest energy modes. The profile of this emission will not be broadened by IVR but rather be dominated by P(Q)R rotational structure. The rotational population of the PAHs is dominated by the IR fluorescence and collisional/radiative deexcitation processes. As will be discussed in Sections 9.6.4 and 12.6, this will lead to $J \simeq 150$ in the diffuse ISM and $J \simeq 300$ in PDRs. The spectroscopic implication of this will be discussed in Section 12.3.7.

Second, limited IVR has implications for unimolecular fragmentation processes. RRKM is predicated on a comparison of the number of states in the transition region leading to dissociation to the total number of states accessible to the system ($W^\star$ and ρ in Eq. (8.79)) and localization diminishes W. Because of localization, the flow into the transition region will also be reduced. Both of these will lead to smaller rate constants. The number of states accessible to the system also decreases and this will tend to increase the rate coefficient. Fragmentation of interstellar large molecules – by necessity – involves relatively low levels of excitation: typically, < 1000 cm^{-1} per mode, which is small compared to the C-H stretching frequency. Hence, these aspects are particularly relevant.

As H-loss is the first step in the photochemical evolution of interstellar PAHs, consider the excitation in the C–H stretching modes. Anharmonicity couples these modes strongly to each other and excitation in these modes will slosh around from one C–H stretching mode to another and only very slowly leak away to other modes in the system. Following Fermi's golden rule, the C–H stretching modes will couple best to the initial electronic excitation and this localization may well enhance the reaction rate coefficient. Eventually, once the flow of the energy establishes itself in the system – say on a timescale of tens of nano seconds – ergodicity will set in. Hence, fragmentation rates may be characterized by an initial high rate set by the direct coupling to the reaction coordinate followed by a much slower rate set by the energy flow in the molecule. This non-ergodic behavior may be particularly relevant when extrapolating laboratory measurements to the small rates (long timescales) relevant to space.

8.6 Photo Ionization

Ionization is a key process taking place upon photo irradiation if the absorbed photon energy is above the ionization threshold (I_p). The photo ionization rate is given by,

$$k_{ion} = 4\pi \int_{i_p}^{\infty} \sigma\left(\nu\right)\, Y_{ion}\left(\nu\right)\, \mathcal{N}\left(\nu\right)\, d\nu, \tag{8.97}$$

with σ the absorption cross section, Y_{ion} the ionization yield, and $\mathcal{N}$ the average photon intensity all at frequency ν. In HI regions and PDRs, this can be written as,

$$k_{ion}\left(Z_d\right) \simeq 2.8 \times 10^{-12}\, \left(13.6\, - IP\left(Z_d\right)\right)^2\, N_c\, G_o\, \mathrm{s}^{-1}, \tag{8.98}$$

Table 8.3 *Ionization potentials*[a]

Z^b		−2	−1	0	1	2	3
Species	Chemical formula						
coronene	$C_{24}H_{12}$		0.47	7.3	11.4		
Ovalene	$C_{32}H_{14}$		1.2	6.7	9.5	12.3	
Circumpyrene	$C_{42}H_{16}$		1.5	6.0	9.2	12.4	
Hexabenzacoronene	$C_{42}H_{18}$		0.9	6.9	9.2	11.5	
Circumcoronene	$C_{54}H_{18}$		1.4	6.1	9.0	12.9	
Circumovalene	$C_{66}H_{20}$		1.9	5.7	8.3	10.9	
Circumcircumcoronene	$C_{96}H_{24}$	0.6	3.1	5.7	8.2	10.8	13.4
	C_{60}		2.65	7.6	11.5		
	C_{70}		2.72	7.5	11.4		

[a]Electron affinities and ionization potentials in units of eV. Data taken from experiments, DFT calculations or estimated from conducting disk equivalence. [b]Consider the reaction: $PAH^z + h\nu \rightarrow PAH^{Z+1} + e$. No value implies a negative electron affinity or an ionization potential exceeding the H-ionization potential, 13.6 eV.

where the ionization potential, IP is in units of eV and G_0 is the strength of the UV photon field in Habing units. The value 13.6 eV reflects the ionization potential of H, which limits the photon radiation field.

Ionization potentials for a few representative PAHs are collected in Table 8.3. These have been taken from experiments, quantum theory, or estimated by comparing to the ionization potential of a conducting disk,

$$IP_z = W + \left(Z + \frac{1}{2}\right)\frac{e^2}{C} \simeq 4.4 + \left(Z + \frac{1}{2}\right)\frac{25.1}{N_C^{1/2}}\ \text{eV}, \tag{8.99}$$

with W the work function (4.2 eV for a graphene sheet) and C the capacitance of the species where $C = 2a/\pi$ has been used (appropriate for a circular disk) and the size has been approximated by $a = 0.9N_C^{1/2}$ Å. This refers to the process,

$$\text{PAH}^Z + h\nu \rightarrow \text{PAH}^{Z+1} + e. \tag{8.100}$$

When Z is negative, we are really considering electron attachment and the energy involved is the electron affinity. When the electron affinity is negative (as for example for benzene and naphthalene), the electron is unbound. Typically, the ionization yield increases slowly with energy above the ionization potential ($Y = (h\nu - IP_Z)/9$ for $h\nu < I_p + 9$, with $h\nu$ and IP_Z in eV, and 1 otherwise).

Photo ionization can compete well with fragmentation when the photon energy is well above threshold. This competition is illustrated for the coronene cation in Figure 8.20. The measured appearance energy of the coronene dication (10.95 eV) is close to the threshold

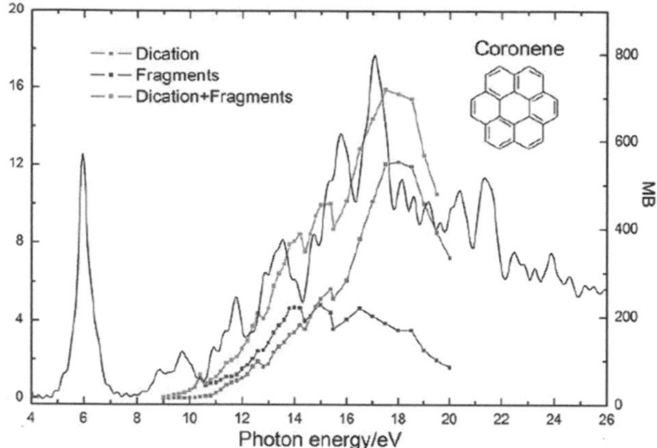

Figure 8.20 The relative intensity (left axis) of the photo-products (dications and singly ionized fragments) of the coronene cation irradiated with VUV photons as a function of photon energy. The sum of both processes is also shown. The solid line is the computed photo absorption cross section in Mb (10^{-16} cm^2, right axis). Figure reproduced with permission from [99]

ionization potential of 10.58 eV. For a large PAH, where the internal energy required for fragmentation is very high, a single photon will lead to ionization rather than fragmentation unless the ionization potential is above 13.6 eV.

8.6.1 Thermionic Emission

Ionization can also occur in a delayed process when a highly excited species releases its excess energy through ejection of an electron. When a well-defined bound state above the ionization potential couples directly to the ionization continuum, this molecular auto ionization of a Rydberg state is described by the quantum states involved and this is generally included in the ionization studies described in Section 8.6. However, in a large molecular system such as PAHs and fullerenes, delayed ionization can occur where the species is highly excited and ergodically "visits" an ionizing electronic state. For a cation this is thermionic emission of an electron; for an anion this is autodetachment. These are the reversible process of electron recombination and attachment reactions and the rates of these processes are related through detailed balance (cf. Section 5.2.6). Delayed emission of an electron can be described in a way completely analogous to the unimolecular description of the fragmentation process discussed in Section 8.5. Using reversibility, this can be written as,

$$k_{tie}\left(E,\epsilon\right) = \frac{2m}{\pi^2\hbar^3}\sigma\left(\epsilon\right)\epsilon\exp\left[-\epsilon/kT_f\right]\exp\left[-I_p/kT_i\right], \tag{8.101}$$

with E the internal energy of the initial state, ϵ the kinetic energy of the ejected electron, σ the electron capture cross section, T_f the (final) temperature of the product, T_i the (initial) temperature of the parent, and I_p the ionization potential. For the Coulomb capture cross section we can use,

$$\sigma(\epsilon) = \frac{\pi e^2 a}{\epsilon}. \tag{8.102}$$

The thermionic emission rate follows from integration over all electron energies,

$$\begin{aligned} k_{tie}(E) &= \frac{2me^2a}{\pi^2\hbar^3} kT_f \exp\left[-I_p/kT_i\right] \\ &\simeq 2.3\times10^{15}\left(\frac{N_c}{50}\right)\left(\frac{T_f}{1000\,\mathrm{K}}\right)\exp\left[-12\left(\frac{I_p}{\mathrm{eV}}\right)\left(\frac{1000\,\mathrm{K}}{T_i}\right)\right]\mathrm{s}^{-1} \end{aligned} \tag{8.103}$$

It should be understood that the T's in these expressions are the appropriate microcanonical temperatures.

Thermionic emission is in competition with other energy decay channels and then it is quite obvious that this process is only relevant when the ionization potential is comparable to the binding energies involved in fragmentation processes. In practise, thermionic emission is important for C_{60} with an ionization potential of 7.6 eV and an E_0 for C_2 loss of 10 eV (Figure 8.21). For neutral PAHs with ionization potentials well above relevant binding energies, thermionic emission plays little role. Hence, while the carbon backbone of PAHs is protected against fragmentation by peripheral H's, for C_{60} thermionic emission plays that role. The thermionic process is also much more important for PAH anions as electron affinities tend to be much lower than ionization potentials.

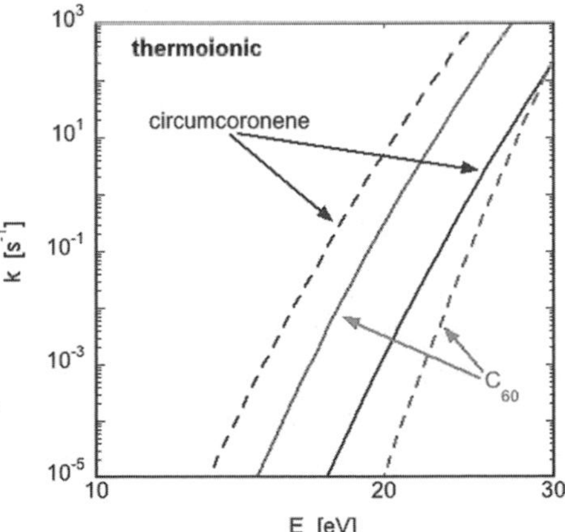

Figure 8.21 The competition between thermionic emission (solid) and fragmentation (dashed) for the PAH, circumcoronene ($C_{54}H_{18}$), and C_{60}. The excitation has been calculated using the approximations of Eq. (8.45a).

8.7 Gas Phase Chemical Processes

8.7.1 Electron Recombination

Recombination of a PAH cation with an electron leads to a neutral intermediary with an internal energy of some 6–7 eV (the ionization potential of the neutral PAH) that can relax through radiation and, in principle, through fragmentation,

$$\mathrm{PAH}^+ + e \rightarrow \mathrm{PAH}^\star \rightarrow \mathrm{PAH}_{-H} + H \tag{8.104}$$

$$\rightarrow \mathrm{PAH} + nh\nu, \tag{8.105}$$

where PAH* is the excited intermediary, $nh\nu$ indicates the emission of multiple IR photons, and in this example it is assumed that fragmentation takes place through the lowest energy loss channel (H-loss). If we compare the ionization potential of a neutral PAH with the appearance potential for H-loss (cf. Section 8.5.2),[12] then for small PAHs, fragmentation can compete with IR relaxation. However, already for pyrene ($C_{16}H_{10}$) and coronene ($C_{24}H_{12}$), the appearance potential is higher than the ionization potential and relaxation is preferred through IR emission (or other channels).

For the recombination rate, we can use the results from Section 6.8 (e.g. Eq. (6.103) with $r_{min} = 0.9N_C^{1/2}$ Å (the geometrical size appropriate for a compact PAH). As this essentially assumes a spherical species, a correction factor of 0.8 is often included to account for the planar shape of PAHs. This leads to,

$$k_e \simeq 1.3 \times 10^{-6}\, N_C^{1/2} \left(\frac{300\,\mathrm{K}}{T}\right)^{1/2} \quad \mathrm{cm^3\ s^{-1}}. \tag{8.106}$$

Electron recombination rates have been measured at room temperature for the small PAH cations, naphthalene, azulene, acenaphthene, anthracene, phenanthrene, fluoranthene, and pyrene, using a modified flowing afterglow Langmuir probe technique (see Section 2.4.2). Measured electron recombination rates are compared to the results of Eq. (8.106) in Figure 8.22. This comparison suggests that the measured values approach the theoretical estimate with increasing size but experiments on larger PAHs would be very valuable.

8.7.2 Electron Attachment

Electron attachment reactions for atom and small molecules have been discussed in Section 6.10. When the electron affinity is $\simeq$1 eV, IC and IVR followed by radiative stabilization is fast. For PAHs, the electron affinity increases with PAH size and for $N_C > 30$ the electron affinity typically exceeds 1 eV and auto ionization is of little importance. The (Langevin) rate coefficient can then be estimated from the PAH polarizability, which for PAHs scales well with the surface area, and can be approximated by $\alpha \simeq 1.6 \times 10^{-24} N_C$ cm^3.

$$k_{ea}\,(\mathrm{PAH}) = k_L\,(\mathrm{PAH}) \simeq 1.3 \times 10^{-7}\, N_C\ \mathrm{cm^3\ s^{-1}} \tag{8.107}$$

[12] These appearance potentials have been measured for cations but should be very similar for neutral PAHs.

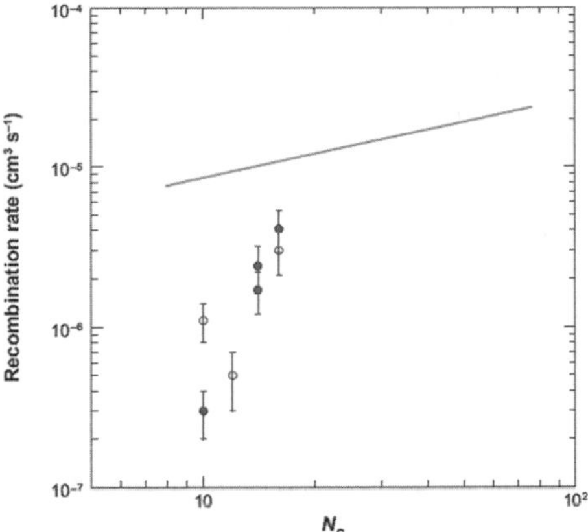

Figure 8.22 Experimentally measured electron recombination rates for some simple PAH cations (filled circles: PAHs with only hexagons; open circles: PAHs containing one pentagon) are compared to the theoretical recombination rate for a thin conducting disk. Figure adapted from [16]

Electron attachment rates have only been measured for a few small PAHs with measured rates of $10^{-9}-10^{-8}$ cm^{-3}, in good agreement with quantum chemical calculations. This is much less than Eq. (8.107) would indicate but then the electron affinities of the PAHs investigated were only $\simeq$0.5 eV and a very small sticking efficiency is expected. No electron attachment experiments have been performed on PAHs with astrophysically relevant sizes but they are calculated to have high electron affinities.

A high electron-attachment rate is contingent on the existence of accessible low energy negative ion states in which capture can occur (s-wave capture). Some species can efficiently attach electrons at elevated temperatures ($T > 2000$ K) through the p-wave capture process, but a small energy barrier ($\hbar^2/m_e R^2$) inhibits the formation of anion s states and results in a low electron-attachment rate at low temperatures. Not all PAHs show low lying anion resonances in their electron-scattering cross section; hence these few experimental results cannot be confidently extrapolated to all (large) PAHs. Indeed, some PAHs in the astronomical family may efficiently form anions, whereas others do not.

8.7.3 Ion–Molecule Reactions

The interaction of neutral PAHs with cations is expected to occur at the Langevin rate (cf. Section 6.3),

$$k_L = 2\pi e \sqrt{\frac{\alpha}{\mu}}, \tag{8.108}$$

e.g. regulated by the polarizability α of the PAH and the reduced mass, μ, which is typically dominated by the ion. Polarizabilities of a few representative PAHs are given in Table 6.6. Polarizabilities can be calculated using semi-empirical molecular orbital quantum theory to 6% accuracy, while time-dependent density funtional theory yields results at the 10% level. For small PAHs, we can approximate the polarizability by $\alpha \simeq 1.5 \times 10^{-24} N_C$ cm^{-3} (e.g. scaling with N_C or volume) and we have,

$$k_L = 6 \times 10^{-9} \sqrt{\frac{N_C}{50}} \sqrt{\frac{12\,\text{amu}}{m_x}}\ \text{cm}^3\ \text{s}^{-1}. \tag{8.109}$$

For very large PAHs, the polarizability will approach that of an infinitely thin disk (e.g. scaling with $N_C^{3/2}$) but that limit is generally not obtained for astrophysically relevant sizes. For noncompact PAHs, the polarizability may actually be localized on a limited number of edge groups but the effect of this has not been explored. Of particular interest is proton transfer, and proton affinities of PAHs – typically $\simeq$850 kJ/mole (Table 6.7) – are larger than for many astrophysically relevant species. Such proton transfer reactions can be expected to be fast.

Charge-exchange reactions involving neutral PAHs,

$$\text{PAH} + \text{M}^+ \rightarrow \text{PAH}^+ + \text{M}. \tag{8.110}$$

have only been measured for anthracene with C^+, Ar^+, and He^+. These reactions proceed at the Langevin rate ($2 - 6 \times 10^{-9}$ cm^3 s^{-1}, depending on the species involved). For most species, the exothermicity of the reaction is a few electron volts, and charge exchange is the only channel open at low temperatures. The reaction with C^+ can also lead to carbon insertion, and a branching ratio of 0.6:0.4 has been estimated for charge transfer versus insertion for anthracene. Insertion probably proceeds through addition in the π-system followed by the formation of a cyclopropa-PAH peripheral group. He^+ provides another possible exception because the excess energy leaves the cation highly excited and may even lead to double ionization. The doubly ionized PAH may then fragment.

For reactions involving atomic ions with low ionization potentials, adduct formation is the only channel open and measured binding energies for potasium and sodium to benzene are in the range 1 – 1.2 eV. In analogy to the discussion involving the interaction of electrons with neutral PAHs (cf. Section 8.7.2), at these energies, "sticking" is very likely. The ionization potential of Fe is less than that of benzene and, with a binding energy of 2.4 eV, adduct formation is also expected to be rapid. The ionization potential of larger PAHs will be less than that of Fe but nevertheless adduct formation is expected to be prevalent, forming a complex with a Fe atom located on one of the outer rings where the charge would be (delocalized) in the aromatic ring system. The (neutral) metal atom on the PAH surface is tightly bound and available for further reactions with other collision partners. Measured reaction rates for small PAHs are close to the Langevin rate and the same is expected for large PAHs.

For neutral species interacting with PAH cations, again, the Langevin rate can be used to derive the collision rate. The reaction of atomic H with PAH cations is particularly relevant, leading to hydrogenated PAH cations. This reaction has been measured to have a rate of 1.4×10^{-10} cm^3 s^{-1} for small PAHs, about a factor of 10 less than the Langevin rate. For H interacting with interior C atoms, a barrier of $\simeq$0.2 eV has been calculated as the C atom has to pucker out of the plane. For small PAH cations, reactions with atomic O leads to both association as well as loss of CO. For pyrene, the latter channel has a branching ratio of only 0.05 while, for coronene, CO formation is not observed at all. For small PAH cations, reactions with atomic N lead to association and to HCN formation. However, in this case, the reaction rate of N with PAH cations decreases rapidly with size: The association rate with the pyrene cation is small ($\simeq 10^{-12}$ cm^3 s^{-1}) and N is not observed to react at all with the coronene cation. H_2 does not react with PAH cations. The singly dehydrogenated PAH cation, $C_{10}H_7^+$, reacts (slowly) with H_2 (5×10^{-11} cm^3 s^{-1}). In contrast, the singly dehydrogenated pyrene cation does not react with H_2 (but does react with atomic H). This may be related to its triplet ground state and could be a more common characteristic of larger, dehydrogenated PAH cations.

8.7.4 Reactions of Anions

Reactions of dehydrogenated aromatic anions, PAH_H^-, can be quite relevant to interstellar chemistry. These species react with atomic H through associative detachment reactions,

$$\mathrm{PAH}_H^- + H \rightarrow \mathrm{PAH} + e, \tag{8.111}$$

at close to the Langevin rate, 5×10^{-10} cm^3 s^{-1}. The large number of states available to the free electron make this a very efficient process compared to regular H-attachment to these anions. Treating large PAHs as conducting disks, their gas phase acidity is $\simeq PA + 25.1/N_C^{1/2} \simeq 12.5$ eV and dehydrogenation increases this by only 0.1 eV. This is much less than the gas phase acidity of H_2 and hence the reaction of dehydrogenated PAH anions with H_2 is endothermic and will not proceed in the ISM. Reactions of dehydrogenated PAH anions with a variety of species – including acetylene, methanol, and methylcyanide – lead to proton transfer at, essentially, the collision rate. With other species, association (CO_2) or atom transfer reactions (O_2) or both (N_2O) are important. We note that the electron affinity of Fe bonded to coronene is about 1eV, to be contrasted to the coronene and iron electron affinities of 0.5 and 0.15 eV, respectively. For the complex, the charge is largely localized on the iron atom as the high polarizability of coronene provides the remainder of the interaction energy. This seems to be a general property of metal–PAH anion interactions. As the electron affinity is larger than the Fe-coronene binding energy, no detachment reaction is expected between Fe and coronene anions or PAH anions in general.

8.7.5 Mutual Neutralization Reactions

For mutual neutralization reactions, we can use – by lack of better – the same expression as for molecular ion–electron recombination (Eq. 6.102) except that we use the cation mass. This leads to a rate of,

$$k \simeq 3 \times 10^{-8}\, N_C^{1/2} \left(\frac{300\,\mathrm{K}}{T}\right)^{1/2} \left(\frac{\mathrm{amu}}{\mu}\right)^{1/2}. \tag{8.112}$$

8.7.6 Neutral–Neutral Reactions

Neutral, dehydrogenated PAHs are free radicals and are therefore intrinsically reactive. The reaction of atomic H with dehydrogenated PAHs is of key importance in the rehydrogenation of PAHs. No activation barrier is expected, leading to a rate of (cf. Section 6.4),

$$k_{H,PAH} \simeq 3 \times 10^{-9} \left(\frac{N_C}{50}\right)^{1/3} \left(\frac{T}{100\,\mathrm{K}}\right)^{1/6} \mathrm{cm^3\ s^{-1}}. \tag{8.113}$$

This is, in magnitude, very comparable to assuming a geometric collision cross section and a thermal velocity (albeit that the dependence on N_C and T are somewhat different). Reactions of PAH radicals with unsaturated or radical species can be rapid as well. The presence of a centrifugal barrier on the attractive van der Waals potential leads to the formation of a complex that can stabilize or dissociate again. If the difference between the electron affinity of the PAH and the ionization potential of the molecule is relatively small ($\lesssim$9 eV), the barrier associated with bond rearrangement may be submerged and reactions can occur even at low temperatures (cf. Section 6.4). Figure 8.23 shows, as an illustrative example, the reaction of the phenyl radical C_6H_5 with vinyl-acetylene (C_4H_4) forming naphthalene ($C_{10}H_8$). As this example illustrates, the long range van der Waals interaction indeed "submerges" the barrier in the entrance channel below the energy of the reactants. The long-lived complex is then stabilized by shifting groups around and by H loss. These types of reactions are thus promoted by high polarizabilities of the reactant molecule. Reaction rates have not been measured but could be as large as $\simeq 10^{-10}$ cm^3 s^{-1}.

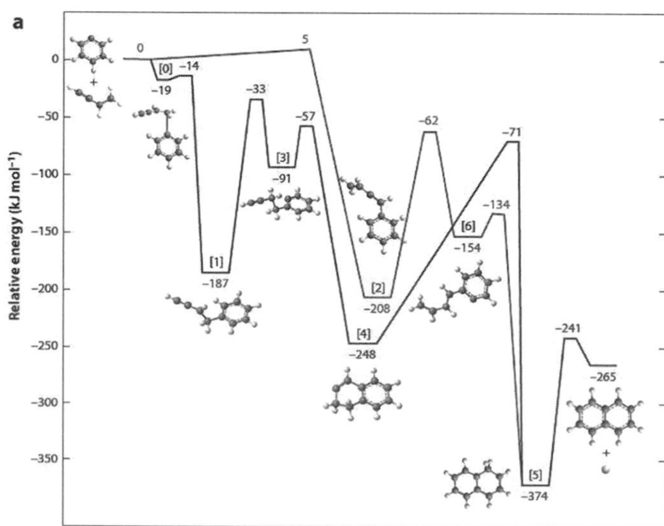

Figure 8.23 Relevant sections of the potential energy surface of the reaction of phenyl radicals (C_6H_5) with vinylacetylene (C_4H_4) leading to naphthalene. The proposed pathway leading to naphthalene in cold molecular clouds only has submerged barriers. Figure taken from [53]

Reactions of neutral atoms with neutral PAHs may lead to interesting complexes. Neutral Fe atoms are bound to coronene by about 0.6 eV. The difference with the cationic interaction (binding energy $\simeq$2.5 eV) resides in the charge polarization and the induced electrostatic interaction. Because of the mass difference, the reaction rate coefficient will be somewhat slower than indicated by Eq. (8.113).

8.7.7 PAH Clustering

PAHs have a propensity to cluster. For small PAHs, attractive electrostatic and dispersion interactions favor T-shaped dimers. For the dimers of larger PAHs, the dispersive interaction becomes dominant and the preferred structure is parallel. For larger clusters, thicker stacks will develop. When the thickness of the stack approaches the diameter of the monomer, a cluster will grow further as a packing of separate stacks or more complex geometric arrangements. Data on the binding energy of PAHs are summarized in Figure 8.24. They scale linearly with the polarizability, α, of the PAH and, hence, with the number of carbon atoms. The heat of sublimation of solid PAH samples is somewhat larger than the, somewhat uncertain, exfoliation energy of graphite ($\simeq$50 meV per C-atom). As fewer atoms are involved, dimer binding energies will be somewhat less than the measured heat of sublimation for solid PAH samples or for PAHs binding to graphene. This difference will become less for larger PAHs. Commonly used functionals in DFT calculations have difficulty in dealing with long range dispersion interactions, and special functionals have been developed specifically to deal with this interaction. These were not

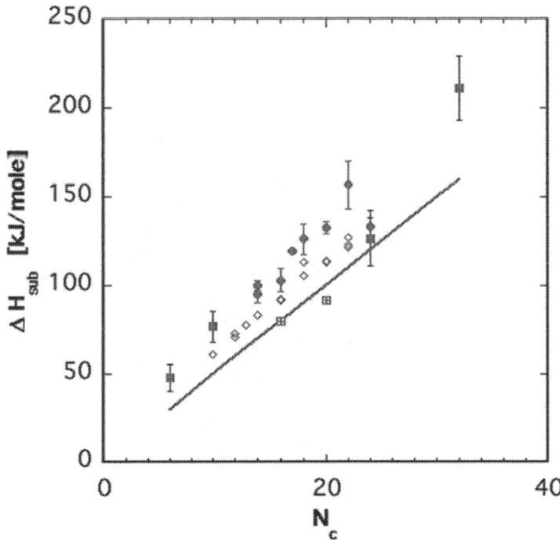

Figure 8.24 Heat of sublimation for PAHs. Experimental and quantum chemical data are indicated by filled and open symbols, respectively. Circles: ΔH^0_{subl} data from sublimation of solid PAH samples [77]. Squares: Data taken from sublimation of PAHs on graphene surface [98]. Diamonds: DFT calculations of dimerization energy for neutral PAHs [63]. Plussed squares are DFT calculations of cationic PAH dimers [85]. The solid line indicates the exfoliation energy of graphite (5 kJ/mol).

used in the calculations reported in Figure 8.24 but the agreement with the measurements is, nevertheless, gratifyingly good. Charging is expected to enhance the binding in PAH dimers and this is supported by measurements on the benzene and the naphthalene dimer. But those dimers are (apparently) not parallel stacked, and quantum chemical calculations for larger, parallel stacked dimers do not show much of an effect. From this, it is reasonable to assume that two coronene ($C_{24}H_{12}$) molecules will have a binding energy of $\simeq$1 eV, while two circumcoronene molecules are bonded by $\simeq$2.8 eV. This approaches covalent binding values.

The collision rate coefficient for neutral PAHs is $\simeq 4 \times 10^{-11}\,(T/10\,\mathrm{K})^{1/2}\,(N_c/50\ \mathrm{C\text{-}atom})^{1/2}$ cm^3 s^{-1}, while for ions we can use the Langevin expression, $\simeq 6 \times 10^{-10}$ cm^3 s^{-1} or about an order of magnitude faster. As the collision energy, 10 K or about 10 meV, is small compared to the interaction energy, a long-lived complex will form during which the PAHs can reorient themselves to their energetically most preferred arrangement. With a binding energy in excess of $\simeq$1 eV, this complex can live a millisecond; e.g. long enough to radiate an IR vibrational photon and become bound.

8.8 Fullerenes

Fullerenes consist of a closed structure of fused hexagons and pentagons (cf. Section 2.3.3), forming hollow spheres or ellipsoids, of which the "soccer ball" C_{60} is the best known (Figure 8.25). Many stable fullerenes exist – C_{24}, C_{28}, C_{32}, C_{36}, C_{50}, C_{60}, C_{70}, ... – and all require 12 pentagons for closure (C_n with $n = 2\,(10 + m)$ corresponding to 12 pentagons and m hexagons). We will focus here on C_{60} as its presence in space has been unambiguously established. The C atoms in C_{60} are placed on the vertices of a truncated icosahedron structure, analogous to a soccerball. C_{60} is built up of 20 hexagons and

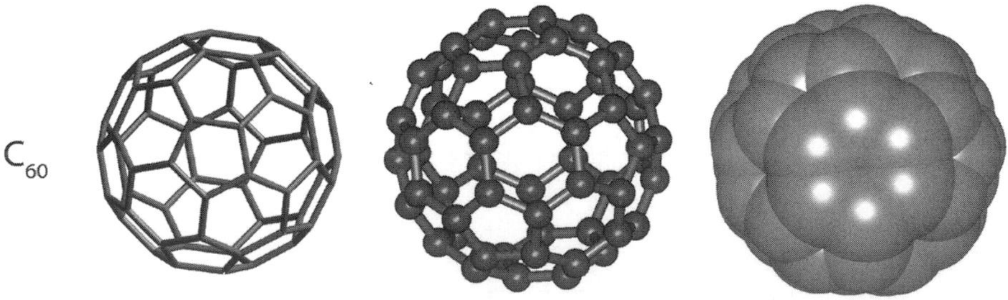

Figure 8.25 Four common representations of the structure of the fullerene molecule, C_{60}. Note that all C-atoms are equivalent. Each of them is part of two hexagons and one pentagon. Left: Structural formula. Center left: The structure of the C atoms is represented where each covalent bond is indicated by a line. This right image has been oriented in such a way to bring out that each pentagon is the center of a ring of hexagons. Center right: Stick and ball model where the black balls indicate C atoms. This image is centered around a hexagon. Right: 3-D model where the size of the atoms and their distances are representational. Image courtesy of Dr Alessandra Candian

12 pentagons, where each pentagon is surrounded by 5 hexagons and each hexagon touches on 3 pentagons. C_{60} is the first fullerene in which the pentagons are isolated, which is at the basis of its enhanced stability. The bonds in common between two hexagons have a double bond character and are shorter (1.38 Å) than the single bond involving a pentagon (1.45 Å). The diameter of C_{60} is 7.1 Å but when the size of the electron cloud is included it is 10.3 Å. There is an internal cavity with a diameter of 4 Å. The binding energy is 7.4 eV/C-atom. All the double bonds are associated with the six membered rings and C_{60} is not superaromatic but rather has a polyenic structure. While for PAHs, the chemistry is dominated by substitution reactions that preserve their planar structure, for fullerenes substitution reactions are not relevant. Rather, addition reactivity centers on attacking the C-bonds in common between hexagons. The van der Waals binding energy of C_{60} to graphene is 0.85 eV, as expected much less than for a PAH of comparable size (e.g. circumcoronene, $C_{54}H_{18}$, $E_{vdw} = 2.5$ eV).

Figure 8.26 illustrates the energy level diagram of C_{60}. The "spherical" structure of C_{60} allows the energy levels to be expanded in spherical harmonics of the angular momentum, l. C_{60} has a HOMO–LUMO gap of $\simeq$1.5 eV. The HOMO level is 5-fold degenerate while the LUMO level is threefold degenerate. As these levels are both ungerade, transitions between them are forbidden. The photophysics of C_{60} parallels that of PAHs with rapid interconversion to S_1 (at an energy of 1.99 eV) and is followed by intersystem crossing to the triplet state which lies about 0.4 eV below the S_1 state. The fluorescence yield is small as most of the excited C_{60} decays through phosphorescence with a quantum yield of

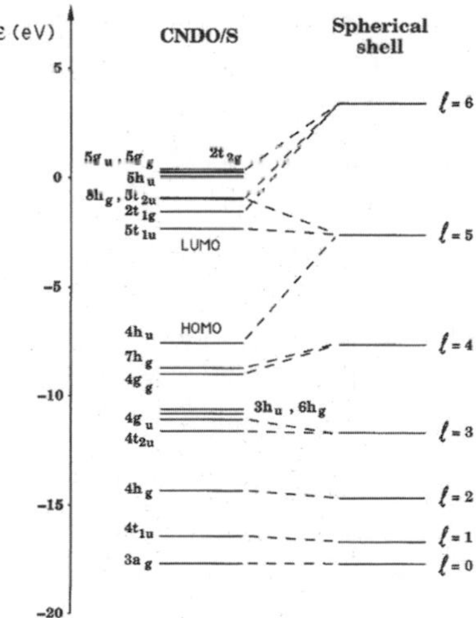

Figure 8.26 Energy level diagram for C_{60}. Except for the 2 σ levels, $3h_u$ and $6h_g$, all levels are bonding and antibonding π levels. The HOMO (h_u) and LUMO (t_{1u}) levels are separated by $\simeq$1.5 eV. Left: Quantum chemical calculation using CNDO/2. Right: Spherical shell model. Figure adapted from [22]

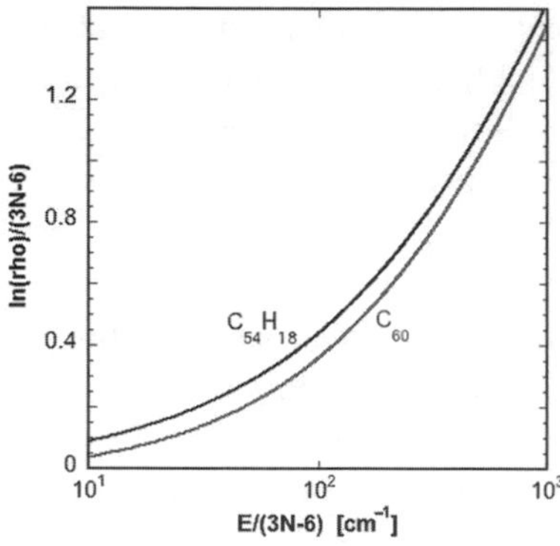

Figure 8.27 The C_{60} normalized density of states, ρ (on a natural log scale) as a function of the normalized energy, E. Both E and ρ are in units of cm^{-1}. For comparison, the density of states of circumcoronene is also shown.

0.96. As for PAHs, the fragmentation chemistry of fullerenes is controlled by the density of states and the kinetic parameters involved. As Figure 8.27 illustrates, the density of states of C_{60} is very similar to that of PAHs and for the same internal excitation (same energy per mode), the microcanonical temperature – given by the slope, $dE/d\ln[\rho]$ (cf. Eq. (8.43)) – is very similar. The kinetic parameters are, however, very different (Table 8.1) and C_{60} is much less prone to photo fragmentation. The activation energy required for C_2 loss from C_{60} is 7.1 eV (Table 8.1) and, using Eq. (8.89) this results in an internal energy required for C_{60} fragmentation of $\simeq$26 eV, corresponding to absorption of about 3–4 photons. The other difference with PAH fragmentation is that PAHs are "protected" by their peripheral H's as, for large PAHs, first all H have to be stripped off before the C-skeleton starts to fragment. In regions of high atomic H density, rehydrogenation can, in principle, prolong the lifetime of the PAH skeleton against fragmentation. Because of its high binding energy, C_{60} is inherently very stable. Neutral C_{60} is further "protected" by thermionic emission (Section 8.6.1) and by rapid fluorescence from low lying electronic states (Section 8.2.2).

The ionization potentials of C_{60} and C_{70} are summarized in Table 8.3. The radius of a spherical fullerene molecule can be approximated by $a \simeq 0.46 N_C^{1/2}$ Å. Within the framework of an electrostatic model (cf. Eq. (8.99)), the ionization potential is then,

$$IP_z = W + \left(Z + \frac{1}{2}\right)\frac{e^2}{C} \simeq 5.7 + \left(Z + \frac{1}{2}\right)\frac{31}{N_C^{1/2}} \text{ eV}, \tag{8.114}$$

with W the work function and we have used the capacitance of a sphere ($C = a$). For the same number of C-atoms, the fullerene ionization potentials are slightly higher as their size is slightly smaller than for PAHs. The opposite holds for the electron affinity of fullerenes; for the same number of C-atoms, the electron affinity will be slightly smaller. C_{60} and C_{70} have high electron affinities and recent studies reveal an electron attachment rate of $\simeq 10^{-6}$ cm^3 s^{-1}. The measured photo ionization yield of C_{60} increases approximately linearly between the ionization potential and 13.6 eV. The photo ionization rate is estimated to be $\sim 10^{-8} G_0$ in the (neutral) ISM, which is very comparable to that of PAHs of the same size. The electron recombination rate has not been measured but is likely also quite similar to that of PAHs of similar size. Hence, the ionization behavior of C_{60} will resemble that of PAHs of a similar size.

The polarizability (76×10^{-24} cm^3) and geometric size of C_{60} are very comparable to that of PAHs of the same number of C-atoms and hence collision rates involving them will be similar. The interaction with atomic H is, of course, of quite some interest. Chemical attack will focus on the polyenic bonds asociated with the hexagons, which have the higher electron density. Indeed, atomic H will preferentially bind to C-atoms in common between two hexagons. The first one is bound by $\simeq$2.1 eV while the second one is slightly more strongly bound ($\simeq$2.9 eV) if adsorbed on an adjacent C-atom and slightly less strongly bound ($\simeq$1.9 eV) if adsorbed on a next nearest neighbor. Further H-atoms can be added with an average binding energy of $\simeq$2.9 eV up to 36 H-atoms, after which the average C–H bond energy slowly decreases to 2.6 eV for $C_{60}H_{60}$. Note, however, that, for the latter, the Gibbs free energy makes this reaction endothermic above 220 K. While the reaction has not been studied in detail, a small activation barrier might be involved as the molecular structure has to rearrange somewhat. In analogy with PAHs, a collision rate of $\simeq 3 \times 10^{-9}$ cm^3 s^{-1} can be expected. Figure 8.28 compares the H-loss rates as a function of internal energy wth the IR cooling rate, illustrating that photodissociation will be rapid in the ISM, except in the most shielded environments.

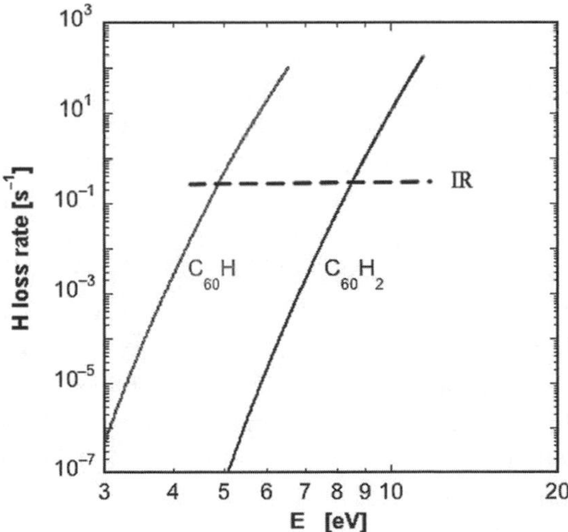

Figure 8.28 Comparison of the IR emission rate and the photodissociation rate for $C_{60}H$ and $C_{60}H_2$. Binding energies of 2.1 and 2.9 eV and a preexponential factor of 2×10^{14} s^{-1} have been adopted in this calculation.

Reaction with H^+ leads to charge transfer with a rate – estimated based upon polarizability – of 2×10^{-8} cm^3 s^{-1}. No adduct formation is observed, probably because of the large amount of excess energy available. Reflecting the high ionization potential of He, the reaction with He^+ leads to C_{60}^{2+} as well as C_{60}^+. Fullerenes have a high proton affinity (cf. Table 6.8) and proton transfer reactions will be rapid.

Ionized fullerenes will react with neutrals through charge exchange if the energetics allow this (e.g. with Na, K, Al, Ca) and through adduct formation (cf. the discussion in Section 8.7.3) otherwise. With hydrogen, adduct formation is observed to occur with a rate in excess of 1×10^{-10} cm^3 s^{-1} and the calculated binding energy (3.3 eV) is somewhat higher than for the neutral. As for the neutral, multiple H's can be attached to C_{60}^+. Reaction of H_2 with C_{60}^+ is inhibited by a calculated energy barrier of 1.7 eV and does not occur under interstellar conditions. Physisorbed binding energies of molecular hydrogen to C_{60} and its cation are calculated to be about 50 meV.

8.9 Further Reading and Resources

Early reviews on interstellar PAHs are [6] and [73]. A more recent review is [94]. The proceedings, [49, 59], provide entries into the physical and chemical literature.

PAH nomenclature in the IUPAC system is described in [32]. The photophysics of large molecules is well described in the textbooks [17, 18, 95]. Early measurements on interconversion and on intersystem crossing rate coefficients are reported in [89], while more recent results are summarized in [80]. References [19, 33, 72, 90] report theoretical studies on the energy gap law and the Franck–Condon factors involved. Delayed fluoresencence related to reverse ISC has been measured for PAHs and for fullerenes by [13, 36, 55] and introduced into the astronomical literature by [61]. A recent review is presented in [35]. Fluorescence from higher states has been reviewed by [45]. An early review on the photochemistry of large molecules relevant to an astronomical setting is [58]. The photophysics of PAHs and the relation between molecular structure and optical spectra of PAHs has more recently been reviewed by [87]. The importance of electron correlation and multi-electron excitation for PAHs has been demonstrated in the femto-second XUV excitation studies by [71] and the theoretical studies on shake-up ionization spectra by [28]. The high stability of superPAHs was originally introduced by [92] based upon group additive rules and further worked out by, among others, [86] using DFT quantum theory.

The reader is referred to textbooks on statistical mechanics for details on (micro) canonical ensembles, for example [56]. The discussion on this subject here gained much from [10], a series of papers on the microcanonical excitation by Klots with the main one, [54], and by [8]. Various aspects involved in the astrochemical problem of loss of hydrogen and various sidegroups are discussed in [5, 46, 62, 74, 93]. Relevant photochemical studies have been published by [51, 52, 99, 100]. The breakdown of ergodicity has been studied experimentally by Zewail and coworkers [34, 43] and reviewed from a theoretical perspective by [39]. The implications for the emission spectrum has been investigated by [75].

The formalism for the calculation of the temperature distribution function of vibrationally excited PAHs is based upon a pioneering study in this field, [81]. A particularly insightful description is given in [1]. This was put in the context of interstellar PAHs by [12]. Different approaches were taken by [31, 48, 60]. The anharmonic line profile of vibrational emission bands was first discussed by [14] and further investigated experimentally by [47] and modeled by [27, 78]. Recent experimental studies on anharmonicity have been reported in [70] and analyzed using DFT by [66]. Line profiles have been systematically investigated by [67].

Various aspects of the ionization of large molecules were first discussed by [57], including a discussion of the then available experimental data. More recent measurements of electron attachment and electron recombination cross sections were made by [24] and [3, 16]. Photo ionization yields of PAHs have been measured by [96] and [50]. Electron attachment rates for PAHs have been measured by [2, 25, 30]. The ionization balance of interstellar PAHs has been extensively discussed by [11, 62, 88]. Quantum chemical calculations of the polarizability of PAHs have been performed by [7]. Charge exchange reactions involving neutral PAHs and involving dehydrogenated PAH anions have been studied by [24, 29], respectively. van der Waals binding between PAHs and graphene has been studied by [63, 77, 82, 85, 98]. The properties of coronene clusters and their photochemical evolution have been experimentally investigated by [23]. Clustering of PAHs in an astrophysical setting has been studied by [83].

There is a vast literature on the chemistry of fullerenes and its derivatives. An early review is [42] and one useful textbook is [44]. Ion–molecule reactions relevant to interstellar PAHs and fullerenes have been reviewed by [15, 21, 79], while relevant neutral–neutral reactions are reviewed in [53]. Radiative cooling of fullerenes has been discussed by [40]. The importance of thermionic emission in excited C_{60} has been discussed by [41, 97]. An early paper on the photophysics of C_{60} is [76].

The database, http://astrochemistry.oa-cagliari.inaf.it/database/, maintained by G. Malloci [69] provides electronic spectra of a set of small to moderately sized PAHs calculated using DFT. The database www.astrochemistry.org/pahdb/ provides vibrational spectra of a set of PAHs either calculated by DFT or measured using matrix isolation spectroscopy [20]. The NIST database provides much relevant data on individual molecules (http://webbook.nist.gov/chemistry/form-ser.html).

8.10 Exercises

8.1 PAHs and resonance structures:

(a) Draw the three resonance structure of naphthalene.
(b) Draw example resonance structures of coronene having one, two, and three delocalized electron rings.
(c) Evaluate the resonance energy of benzene. Resonance energy is defined as the added stability associated with delocalized double bonds as compared to the

compound with localized double bonds. Compare the enthalpy for hydrogenation of cyclohexene (a cyclic 6 C-atom compound with one localized double bond; -120 kJ/mole) and benzene (-208 kJ/mole) to estimate that the benzene resonance energy is 150 kJ/mole.

(d) We can also calculate the resonance energy of benzene from the following data: C–H, C–C, C=C bond strengths of 416.3, 331.4, 591.1 kJ/mole. Heat of vaporization of graphite is 718.4 kJ/mole, dissociation energy of H_2 is 435.9 kJ/mole, and heat of formation of benzene is 82.9 kJ/mole. Compare the two answers.

8.2 Energy fluctuations:

(a) Derive Equation (8.20).

(b) Consider an ideal monatomic gas at high temperatures where $E = 3/2NkT$. Derive an expression for the relative fluctuations in energy.

(c) Calculate the magnitude of energy fluctuations of the air in your study. Use $C_V \simeq 10^7$ erg/g for air.

(d) In the limit of high temperature, show that you expect 5.5% energy fluctuations for a 75 C-atom PAH.

8.3 Canonical and microcanonical systems. Derive the microcanonical expression for the entropy (Eq. (8.11)); Realize that the microcanonical ensemble is a special case of a canonical ensemble and use Eq. (8.14).

8.4 Average occupation number. Derive Eq. (8.35) (hint: Derive first d/da of the geometric series $\sum a^n$).

8.5 Use $\langle E \rangle = d \ln (Z) / d\beta$ to show that the mean energy of s oscillators is s times the mean energy of one oscillator.

8.6 IR emission intensity. For harmonic oscillators, all transitions in mode i occur at the same frequency and the total intensity at this frequency can be obtained by summing over v. Also, for harmonic oscillators, $A_{v,i}$ is equal to $v\, A_{1,i}$. Show now that in the canonical approximation (e.g. use the Boltzmann expression for the excitation), the intensity in mode i i given by Eq. (8.64) with Eq. (8.65).

8.7 Describe the difference between a canonical and microcanonical ensemble. Each of these can be used to describe the vibrational excitation of PAHs. Discuss the pros and cons of each approach.

8.8 Draw the Jablonski diagram for coronene and use it to describe the photophysics of this species.

8.9 Describe the radiative heating and cooling of interstellar PAHs and make a comparison with the heating and cooling of large interstellar dust grains. In this, focus on understanding Figure 8.15.

8.10 Discuss the photochemistry of interstellar PAHs. Pay attention to the different channels that could be open and the dependence on excitation/size.

8.11 Discuss the different processes linking the ionization states of PAHs.

8.12 Compare and contrast the physics and chemistry of PAhs and fullerenes.

Bibliography

[1] Aannestad, P. A., 1989, in *Evolution of Interstellar Dust and Related Topics*, eds., A. Bonetti, J. M. Greenberg, S. Aiello (Amsterdam: Elsevier), 121

[2] Abouaf, R., Diáz-Tendero, S., 2009, *Phys Chem Chem Phys*, 11, 5686

[3] Abouelaziz, H., Gomet, J. C., Pasquerault, D., Rowe, B. R., 1993, *J Chem Phys*, 99, 237

[4] Adkins, E. M., Giaccai, J. A., Houston Miller, A., 2017, *Proc Comb Inst*, 36, 957

[5] Allain, T., Leach, S., Sedlmayr, E., 1996, *A & A*, 305, 616

[6] Allamandola, L. J., Tielens, A. G. G. M., Barker, J. R., 1989, *ApJS*, 71, 733

[7] Alparone, A., Librando, V., Minniti, Z., 2008, *Chem Phys Lett*, 460, 151

[8] Andersen, J. U., Bonderup, E., Hansen, K., 2001, *J Chem Phys*, 114, 6518

[9] Andrews, H., Candian, A., Tielens, A. G. G. M., 2016, *A & A*, 595, 216

[10] Baer, T., Hase, W. L., 1996, *Unimolecular Reaction Dynamics: Theory and Experiments* (Oxford: Oxford Press)

[11] Bakes, E. L. O., Tielens, A. G. G. M., 1994, *ApJ*, 427, 822

[12] Bakes, E. L. O., Tielens, A. G. G. M., Bauschlicher, C. W., 2001, *ApJ*, 556, 501

[13] Baleiz ao, C., Berberan-Santos, M. N., 2008, *Ann NY Acc Sci*, 1130, 224

[14] Barker, J. R., Allamandola, L. J., Tielens, A. G. G. M., 1987, *ApJ*, 315, L61

[15] Bierbaum, V. M. 2014, *The Diffuse Interstellar Bands*, IAU symp, 297, 258

[16] Binnier, L., Alsayed-Ali, M., Foutel-Richard, A., et al., 2006, *Discus Faraday Soc*, 133, 289

[17] Birks, J. B., 1970, *Photophysics of Aromatic Molecules* (London: Wiley & Sons)

[18] Birks, J. B. (ed.), 1973, *Organic Molecular Photophysics, Vol I & II* (London: Wiley & Sons)

[19] Bixon, M., Jortner, J., 1969, *J Chem Phys*, 50, 4061

[20] Boersma, C., Bauschlicher, C. W. Jr., Ricca, A., et al., 2014, *ApJS*, 211, 8

[21] Bohme, D. K., 1992, *Chem Rev*, 92, 1487

[22] Braga, M., Larsson, S., Rosén, A., Volosov, A., 1991, *A & A*, 245, 232

[23] Bréchignac, P., Schmidt, M., Masson, A., et al., 2005, *A & A*, 442, 239

[24] Canosa, A., Laubé, S., Rebrion, C., Pasquerault, D., Gomet, J. C., Rowe, B. R., 1995, *Chem Phys Lett*, 245, 407.

[25] Carelli, F. Grassi, T., Gianturco, F. A., 2013, *A & A* 549, A103

[26] Castellanos, P. Candian, A., Andrews, H., Tielens, A. G. G. M., 2018, *A & A*, 616, A167

[27] Cook, D., Saykally, R., 1998, *ApJ*, 493, 793

[28] Deleuze, M. S., 2002, *J Chem Phys*, 116, 7012.

[29] Demerarais, N. J., Yang, Z., Martinez, O., Wehres, N., Snow, T., Bierbaum, V. E., 2012, *ApJ*, 746, 32

[30] Denifl, S., Ptasińska, S., Sonnweber, B., et al., 2005, *J Chem Phys*, 123, 104308

[31] Draine, B. T., Li, A., 2001, *ApJ*, 551, 807

[32] Ehrenhauser F., 2015, *Polycyclic Aromatic Compounds*, 35, 161.

[33] Englman, R., Jortner, J., 1970, *Mol Phys*, 18, 145

[34] Felker, P. M., Zewail, A. H., 1983, *Chem Phys Lett*, 102, 113

[35] Ferrari,P., Janssens, E., Lievens, P., Hansen, K., 2019, *Int Rev Phys Chem*, 38, 405

[36] Fetzer, J. C., Zander, M., 1990, *Z. Naturforsch*, 45a, 727

[37] Geballe, T. R., Tielens, A. G. G. M., Allamandola, L. J., et al., 1989, *ApJ*, 341, 278

[38] Gillispie, G. D., Lim, E. C., *Chem Phys Lett*, 63, 193

[39] Gruebele, M., Wolynes, P. G., 2004, *Acc Chem Res*, 37, 261

[40] Hansen, K., Campbell, E. E. B., 1996, *J Chem Phys*, 104, 5012
[41] Hansen, K., Hoffman, K., Campbell, E. E. B., 2003, *J Chem Phys*, 119, 2513
[42] Hebard, A. F., 1993, *Annu Rev Mater Sci*, 23, 159
[43] Heikal, A., Bañares, L., Semmes, D. H., Zewail, A. H., 1991, *Chem Phys*, 157, 231
[44] Hirsch, A., Brettreich, M., 2005, *Fullerenes: Chemistry and Reactions* (Weinheim: Wiley VCH Verlag)
[45] Itoh, T., 2012, *Chem Rev*, 112, 4541
[46] Joblin, C., Tielens, A. G. G. M., Allamandola, L. J., Geballe, T. R., 1996, *ApJ*, 458, 610
[47] Joblin, C., Boissel, P., Leger, A., et al., 1985, *A & A*, 299, 835
[48] Joblin, C., Toublanc, D., Boissel, P., Tielens, A. G. G. M., 2002, *Mol Phys*, 100, 3595
[49] Joblin, C., Tielens, A. G. G. M., 2011, *EAS Publications Series*, 46
[50] Jochims, H. W., Rühl, E., Baumgärtel, H., Tobita, S., Leach, S., 1997, *Int J Mass Spectrom Ion Proc*, 167/168, 35
[51] Jochims, H. W., Rühl, E., Baumgärtel, H., Tobita, S., Leach, S., 1994, *ApJ*, 420, 307
[52] Jochims, H. W., Baumgärtel, H., Leach, S., 1999, *ApJ*, 512, 500
[53] Kaiser, R. I., Parker, D. S. N., Mebel, A. M., 2015, *Annu Rev Astron Astrophy*, 66, 43
[54] Klots, C. E., 1989, *J Chem Phys*, 90, 4470
[55] Kropp, J. L., Dawson, W. R., 1967, *J Chem Phys*, 71, 4499.
[56] Landau, L. D., Lifshitz, E. M., 1988, *Statistical Mechanics* (Oxford: Pergamon Press)
[57] Leach, S., in *Polycyclic Aromatic Hydrocarbons and Astrophysics*, eds. A. Léger, L. d'Hendecourt, N. Boccara (Dordrecht: Reidel), 89
[58] Leach, S., 1996, in *The Diffuse Interstellar Bands*, eds. A. G. G. M. Tielens, T. P. Snow (Dordrecht: Kluwer), 281
[59] Léger, L., d'Hendecourt, N., Boccara, N., eds., 1997, *Polycyclic Aromatic Hydrocarbons and Astrophysics* (Dordrecht: Reidel)
[60] Léger, A., d'Hendecourt, L., Défourneau, D., 1989, *A & A*, 216, 148
[61] Léger, A., D'Hendecourt, L., Boissel, P., 1988, *Phys Rev Lett*, 60, 921
[62] Le Page, V., Snow, T. P., Bierbaum, V. M., 2003, *ApJ*, 584, 316
[63] Li, B., Ou, P., Wei, Y., et al., 2018, *Materials*, 11, 726
[64] Li, A., Draine, B. T., 2001, *ApJ*, 554, 778
[65] Lifshitz, C., 2000, *Int J Mass Spectrom*, 200, 423.
[66] Mackie, C. J., Candian, A., Huang, X., et al., 2015, *J Chem Phys*, 143, 224314
[67] Mackie, C. J., Chen, T., Candian, A., et al., 2018, *J Chem Phys*, 149, 134302
[68] Malloci, G., Mulas, G., Joblin, C., 2004, *A & A*, 426, 105.
[69] Malloci, G., Joblin, C., Mulas, G., 2007, *Chemical Physics*, 332, 353
[70] Maltseva, E., Petrignani, A., Candian, A., 2015, *ApJ*, 814, 23
[71] Marciniak, A., Despré, V., Barillot, T., et al., 2015, *Nature Comm*, 7, 7909
[72] Medvedev, E. S., 1982, *Chem Phys*, 73, 243
[73] Puget, J. L., Léger, A., 1989, *Ann Rev Astron Astrophys*, 27, 161
[74] Montillaud, J., Joblin, C., Toublanc, D., 2013, *A & A*, 552, A15
[75] Mulas, G., Malloci, G., Joblin, C., Toublanc, D., 2006, *A & A*, 460, 93
[76] O'Brien, S. C., Heath, J. R., Curl, R. F., Smalley, R. E., 1988, *J Chem Phys*, 88, 220
[77] Oja, V., Suuberg, E. M., 1998, *J Chem Eng Data*, 43, 486
[78] Pech, C., Joblin, C. Boissel, P., 2002, *A & A*, 388, 639
[79] Petrie, S., Bohme, D. K., 2000, *ApJ*, 540, 869

[80] Pino, T., Carpentier, Y., Féraud, G., et al., 2011, *EAS Publications Series*, 46, 355
[81] Purcell, E. M., 1979, *ApJ*, 206, 685
[82] Rapacioli, M., Calvo, F., Spiegelman, F., et al., 2005, *J Phys Chem A*, 109, 2487
[83] Rapacioli, M., Joblin, C., Boissel, P., 2006, *A & A*, 460, 519
[84] Reinköster, A., et al., 2004, *J Phys B: At Mol Opt Phys*, 37, 2135
[85] Rhee, Y. M., Lee, T. J., Gudipaty, M. S., et al., 2007, *PNAS*, 104, 5274
[86] Ricca, A., Bauschlicher, C. W., Jr., Boersma, C., Tielens, A. G. G. M., & Allamandola, L. J., 2012, *ApJ*, 754, 75.
[87] Rieger, R., Müller, K., 2010, *J Phys Org Chem*, 23, 315
[88] Salama, F., Bakes, E. L. O., Allamandola, L. J., Tielens, A. G. G. M., 1996, *ApJ*, 458, 621
[89] Siebrand, W., 1966, *J Chem Phys*, 44, 4055
[90] Siebrand, W., 1967, *J Chem Phys*, 46, 440
[91] Smith D., Spanel P., Mark, T. D., 1993, *Chem Phys Lett*, 213, 202
[92] Stein, S. E., 1978, *J Phys Chem*, 82, 566
[93] Tielens, A. G. G. M., et al., 1987, in *Polycyclic Aromatic Hydrocarbons and Astrophysics*, eds., A. Léger, L. d'Hendecourt, N. Boccara (Dordrecht: Reidel), 99
[94] Tielens, A. G. G. M., 2008, *Annu Rev Astron Astrophys*, 46, 352
[95] Turro, N. J., 1991, *Modern Molecular Photochemistry* (Sauselito: University Science Press)
[96] Verstraete, L., Léger. A., d'Hendecourt, L., Defourneau, D., Dutuit, O., 1990, *A & A*, 237, 436
[97] von Helden, G., Holleman, I., Meijer, G., Sartakov, B., 1999, *Opt Express*, 4, 46
[98] Zacharia, R., Ulbricht, H., Hertel, T., 2004, *Phys Rev B*, 69, 155406
[99] Zhen, J., Castillo, S. R., Joblin, C., et al., 2016, *ApJ*, 822, 113.
[100] Zhen, J., Castellanos, P., Paardekoper, D., et al., 2014, *ApJ*, 797, L30

9
Diffuse Clouds

Diffuse atomic hydrogen clouds are a major repository of gas in the ISM and play an important role in the ecology of galaxies. They are a natural "'weather" phenomena in the ISM. A medium that is in pressure and thermal equilibrium allows two coexisting phases: diffuse clouds ($T \simeq 80$ K, $n \simeq 50$ cm^{-3}) embedded in a more tenuous intercloud medium ($T \simeq 8000$ K, $n \simeq 0.5$ cm^{-3}). The interchange from intercloud to cloud material can be brought about by pressure disturbances connected to turbulence or by supernova shock waves sweeping up the ISM. While the column densities of diffuse clouds are insufficient to start collapse under their own gravity, they are the main reservoir from which star-forming molecular clouds form. Hence, their conversion into molecular clouds, in some sense, regulates star formation.

The low dust extinction of diffuse clouds allows optical absorption line studies and, in the early parts of the twentieth century, diffuse atomic clouds were discovered through their atomic (e.g. CaII, NaI, KI) absorption lines in the optical spectra of stars. The first molecules (e.g. CH, CH^+, CN) followed in the 1930s. The launch of satellites such as Copernicus and the International Ultraviolet Explorer in the 1970s, the Hubble Space Telescope in the 1990s, and the Far Ultraviolet Spectroscopic Explorer in the early 2000s extended the study of such absorption lines into the ultraviolet with, in particular, the Lyman α line of HI and the Lyman–Werner bands of H_2, molecules such as CO, OH, HCl, as well as a host of important atomic and ionic lines. These observations are limited to pinhole studies toward nearby stars and/or high lattitude, extragalactic background sources.

With the 21 cm HI line, the radio sky opened up in the 1950s for studies of diffuse clouds throughout the Milky Way and other galaxies. This revealed a varied interstellar medium of cloud and intercloud gas and a wide variety of structures at all scales. While the cloud–intercloud phases reflect the radiative interaction of massive stars with their surroundings, much of this structure is driven by mechanical energy input in the form of stellar winds and supernova explosions, which leads to a turbulent medium. Indeed, the effects of turbulent energy dissipation on the energy balance can linger for thousands of years. Understanding these structures, their characteristics, their origin and evolution, and their relationship with massive stars is at the base of studies of the interstellar medium of early type galaxies and their evolution.

Molecules play an important role in diffuse clouds. In particular, photo-electrons from large molecules dominate the heating of the gas. Cooling, however, is dominated by the C^+ fine-structure line emission at 158 μm. Large molecular anions also provide a route toward neutralization of atomic cations that is faster than radiative recombination. Molecules also provide convenient probes of diffuse clouds. While the 21 cm line of HI has long been used to trace atomic gas in the ISM, IR fluorescence of large molecules pumped by UV photons acts as "a dye" delineating diffuse clouds and their interaction with stellar photons. And this can be done at a much finer spatial scale than radio observations allow. Also, the 21 cm line observations do not provide insight into the physical conditions of the gas. Indeed, it is even difficult to separate the emission from diffuse clouds from that of the warm intercloud medium. In contrast, rotational transitions of small molecules are good thermometers and barometers and much of our insight into the physical conditions in diffuse clouds results from molecular observations, particularly in the visible and UV.

Diffuse molecular clouds have played an important role in the development of astrochemistry. The first molecules – the radicals CH, CH^+, and CN – were discovered through their optical absorption lines in the spectra of bright stars and these sightlines were by necessity diffuse clouds, characterized by low dust optical depths ($A_v < 1$ magn). Conversely, this implies that the UV radiation field is strong throughout the cloud, keeping molecular abundances low. This strong radiation field simplifies the reaction networks involved considerably and makes diffuse clouds a testbed for the study of the first chemical steps that convert atomic gas into molecular gas. The recent "explosion" of simple hydride detections in diffuse clouds with the opening up of the sub-millimeter sky through the HIFI instrument on Herschel and GREAT on SOFIA – supported by ground-based observations from APEX – have really opened up diffuse clouds as interstellar laboratories for astrochemistry. In this respect, reactions with H_2 are a cornerstone of hydride gas phase chemistry and diffuse clouds offer the opportunity to observe this molecule directly through absorption in the Lyman–Werner bands. Such observations have revealed a steep increase in the H_2 abundance with column density of diffuse clouds and this can be used to probe the formation mechanism of this key molecule.

In this chapter, we will discuss the molecular astrophysics of diffuse clouds. We will start with a short discussion on the characteristics of diffuse clouds as derived from observations (Section 9.1). This is followed by an in-depth discussion of the chemistry in these regions. First, the formation of H_2 is discussed in Section 9.2 with an emphasis on catalysis on grain surfaces and a comparison between simple models and the observations. The reaction networks of other elements are described in depth in Section 9.3. The abundances of some species – notably CH^+ – have been difficult to explain by chemical networks relevant for diffuse clouds as their formation requires elevated temperatures. In Section 9.4, we discuss the chemistry of turbulent dissipation zones in the light of these observations. Much of the chemistry of diffuse clouds is linked to cosmic ray ionization of, e.g., H and H_2, and molecular observations have long been used to infer cosmic ray ionization rates of interstellar gas (Section 9.5). The Diffuse Interstellar Bands are a set of relatively broad, optical absorption lines due to large molecules. Their characteristics and the molecular physics involved are described in Section 9.6. Photo-electric heating by large molecules, such as PAHs, plays an important role in the thermal behavior of diffuse interstellar gas. This process will be discussed in Section 12.8.

A course on molecular astrophysics would benefit much from this chapter. Analysis of observations in terms of column densities and physical conditions (Section 9.1) provides relatively simple examples of the theory encountered in Chapters 3 and 4. The section on H_2 formation goes perhaps a little deeper than a course would need but Figure 9.3 is easy to understand, linking the UV dissociation rate and the grain surface formation rate of H_2. Relevant aspects of these can be pulled in from Sections 6.3.4 (e.g. Eq. 6.35 and Figure 6.10) and 7.3 (e.g. Eq. 7.9). The detailed chemistry description in Section 9.3 illustrates well the importance of kinetics over thermodynamics in interstellar chemistry. As the number of key reactions for each species is really very limited, students can easily grasp the essentials and develop simple analytical models that can provide much insight. The chemistry of photodissociation regions (Section 11.6) is very much related to that of diffuse clouds. This discussion also provides a good training ground for the more complex chemistry relevant for dark clouds, shocks, and protoplanetary disks. This section also provides insight into how molecular observations can be used to track the impact of turbulence and determine the cosmic ray ionization rate. Students who want to be challenged by a century-old problem should digest Section 9.6 on the Diffuse Interstellar Bands.

9.1 The Characteristics of Diffuse Molecular Clouds

9.1.1 Cloud Properties

Clouds in the interstellar medium are broadly divided into diffuse atomic clouds and molecular clouds, which differ in their physical conditions. In this chapter, we will discuss largely atomic clouds, which are characterized by relatively low densities ($\sim$50 cm^{-3}), moderate temperatures ($\sim$80 K), column densities up to 5×10^{21} H-nuclei cm^{-2}, and low molecular abundances ($n_{H_2}/n_H < 1$. These are typical conditions expected for clouds in thermal and pressure equilibrium with their environment. In lower or higher pressure environments (e.g. outer or inner galaxy), these conditions will be somewhat different. Turbulence wil also cause pressure variations. In regions of high UV fields, pressures also tend to be higher with concomitant effects on the temperature and density. Photodissociation regions associated with HII regions or reflection nebulae are characterized by high incident UV fields. The physical and chemical processes in PDRs are very similar to those in diffuse clouds and we will discuss this in Section 11.6.

The class of diffuse clouds is sometimes further divided into diffuse atomic clouds, diffuse molecular clouds, and translucent clouds depending on their column density or H_2/H and CO abundance ratios (see Figure 9.1). Typical sizes of diffuse clouds are $\sim$0.5 pc, masses are $\sim$3 $M_\odot$, and column densities are $\sim 10^{20}$ H-nuclei cm^{-2}, but in reality they span a range in these parameters. These column densities correspond to a visual extinction of 0.05 magn in the visual and 0.1 in the FUV. Hence, they are optically thin to the interstellar

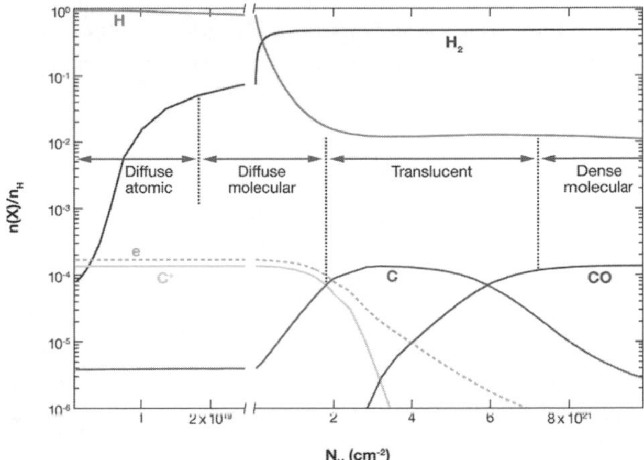

Figure 9.1 Classification of diffuse clouds according to their column density. The curves are the results of a calculation illustrating the transition from H to H_2 and from C^+ to C and CO that occur over this column density range. Figure courtesy of [105]

radiation field and the penetration of FUV photons controls their chemistry. In contrast, UV photons play no role in dense molecular clouds and we will discuss them in Chapter 10.

9.1.2 Column Densities and Abundances

Molecular absorption lines in the optical and UV can be readily analyzed to determine the column densities of the absorbing level involved. This has been discussed in Section 4.3.2 for far-UV H_2 lines, but a similar procedure is applicable to all absorption lines. If the line is optically thin, the observed equivalent width can be converted directly into a column density using the oscillator strength of the transition. The latter has to be either measured in the laboratory or calculated using quantum chemical methods. Often the lines are saturated in the core and a curve of growth method has to be used to analyze the equivalent width (see Section 4.3.2 and Exercise 4 in Chapter 4). Once the column densities of the different levels have been derived, a Boltzmann rotational diagram analysis can be used to derive the total column density of the species (Section 4.2.2). Abundance determinations require accurate column densities of atomic and molecular hydrogen along the same sight-line. Molecular hydrogen is done through a curve of growth analysis of the many Lyman–Werner absorption lines. For atomic hydrogen, the column density is derived from a fit to the wings of the observed profile of the Lyα line, assuming that it can be accurately represented by a pure atomic damping profile.

The molecular composition of diffuse clouds has also been studied at (sub)-millimeter wavelength through absorption against strong – often extragalactic – background sources. Analysis of the observed (typically ground state) line is then done very similarly using the known molecular parameters (e.g. oscillator strength in terms of the dipole moment) and adopting an excitation temperature (generally set equal to the 2.73 K microwave

Table 9.1 *The composition of diffuse clouds*

Species	Abundance[a]	Ref	Species	Abundance[a]	Ref
H	1.2	[98]	N	2.5 (−4)	[77]
H_2	1	[98]	N_2	1.4 (−7)[b]	[55]
HD	2.2 (−6)	[13]	NH	2.1 (−9)	[12]
H_3^+	5 (−8)[c]	[73]	NH_2	4.0 (−9)[d]	[86]
O	9.5 (−4)	[8]	NH_3	4.0 (−9)[d]	[86]
OH	1.2 (−7)	[93]	NH^+	< 4.0 (−10)[d]	[86]
H_2O	< 1.2 (−8)[i]	[104]	CN	6.9 (−9)	[11]
OH^+	1.6 (−8)[d,k]	[49]	HCN	9.5 (−9)[e]	[69]
H_2O^+	4.0 (−9)[d,k]	[49]	HNC	1.6 (−10)[e]	[69]
H_3O^+	2.4 (−9)[d,k]	[64]	HC_5N	< 4.2 (−11)[e]	[69]
C^+	4.3 (−4)	[9]	S	2.0 (−7)	[78]
C	7.6 (−6)	[16]	S^+	2.9 (−5)	[78]
C_2	3.8 (−8)	[107]	SH	1.1 (−8)[d]	[79]
C_3	3.8 (−9)	[72]	H_2S	4.4 (−9)[d]	[79]
CH	6.0 (−8)	[101]	SH^+	7.0 (−9)[d]	[36]
CH_2	7.9 (−8)	[71]	CS	1.8 (−9)[d]	[79]
CH^+	7.6 (−8)	[11]	SO	8.0 (−10)[d]	[79]
CH_3^+	< 2.0 (−9)[j]	[48]	Cl	2.6 (−7)	[78]
C_2H	6.5 (−9)[e]	[69]	Cl^+	6.2 (−8)	[78]
c-C_3H	3.4 (−11)[e]	[67]	HCl	6.5 (−10)	[26]
C_3H_2	1.1 (−9)[e]	[69]	HCl^+	9.0 (−9)	[17]
CO	6.1 (−6)	[57]	H_2Cl^+	3.0 (−9)	[80]
HCO	1.8 (−10)[e]	[67]	HF	1.4 (−8)[h]	[108]
HCO^+	4.2 (−10)[e]	[69]	CF^+	8.4 (−10)[e]	[67]
HOC^+	7.8 (−12)[e]	[68]	ArH^+	3.0 (−10)[k,l]	[99]
H_2CO	3.1 (−9)[e]	[69]	C_{60}^+	3.0 (−9)[f]	[5]
CH_3OH	< 1.1 (−10)[e]	[69]			

[a]Abundance relative to H_2 measured toward ζ Oph through UV absorption lines (cf. [103], $N(H_2) = 4.2 \times 10^{20}$), toward B2200+420 in millimeter emission lines [69], or from HIFI/Herschel studies [34], unless otherwise indicated. Analysis of the C_2 rotational population results in $n/G_0 \simeq 90$ cm^{-3} for ζ Oph. The physical conditions for the other sightlines are not well constrained. [b]Measured toward the star HD 124314 [55]. [c]Measured toward the star, ζ Per. [d]Measured through sub-millimeter line absorption against G10.6-0.4 and normalized through CH to H_2 (3.5 (−8); [102]). [e]Measured through millimeter line absorption against B2200+420 and normalized against CN (6.9 (−9)). [f]Measured toward the well-known "DIB-star" HD 183143. [h]Average abundance measured through sub-millimeter line absorption. [i]H_2O abundance (relative to H_2) in translucent clouds is measured to be 5.0 (−8) through sub-millimeter absorption lines [28]. [j]Upper limit measured toward Cyg OB 2 no 12. [k]Abundance relative to H-nuclei. The H_2 fraction corresponding to the gas with these species has to be very small (see text for details). [l]Measured through sub-millimeter line absorption against Sgr B2(N) and (M).

background temperature). As there is no direct measure of HI and H_2 possible this way, abundances are derived by scaling to other species. Based upon an observed rougly linear correlation of CH column densities with the dust extinction, CH is often used as a proxy. There is, however, considerable scatter in this relation. A much tighter relation is present in CH versus H_2 but even then variations are present. Furthermore, the conditions along the sightlines toward these background (sub)millimeter sources may well be very different than those along the optical/UV sightlines.

Diffuse clouds show a wide variety of simple molecules but most of the elements are predominantly in atomic form. Typical abundances are listed in Table 9.1. Unless otherwise noted, these have been measured toward the star ζ Oph, which – for diffuse clouds – is a relatively dense (300 cm^{-3}) and extincted ($A_V = 0.75$ magn) sight-line. Optical and UV absorption line studies typically require column densities of 10^{12} cm^{-2} and hence derived abundances are limited to 10^{-9} and up. Pure rotational transitions are somewhat more sensitive and derived abundances can be slightly lower. When comparing optical/UV with sub-mm derived abundances, the differences in normalization should be kept in mind. Moreover, the sub-millimeter sightlines may probe a quite different part of the parameter space of diffuse clouds. There are clear compositional differences. For example, in the sub-millimeter studies, OH is less abundant while the H_2O/OH ratio is higher than in the UV studies.

9.1.3 Physical Conditions

Molecular transitions also provide convenient thermometers. Rotational diagrams derived from optical or UV absorption lines provide the excitation temperature of the molecule. The pure rotational transitions of homonuclear molecules (e.g. H_2, C_2) are forbidden and have low critical densities. These provide, therefore, good determinations of the kinetic temperature of the gas. Figure 3.11 shows a portion of the FUV spectrum of the star Sk $-67°$ riddled with H_2 Lyman–Werner transitions. Figure 4.10 shows, as an example, the rotation diagram for H_2 as measured toward λ Cep (HD 210839). Two distinct excitation regimes can be recognized. For low J, the critical densities of the levels involved are low enough that this represents the kinetic gas temperature. For the higher H_2 levels, other processes come into play. First, UV pumping in the Lyman–Werner bands will be followed by electronic fluorescence, leaving the molecule in an excited vibrational state of the ground electronic state. Subsequently, the molecule will cascade down through ro-vibrational transitions to the ground-vibrational state (Section 4.3.1). The excitation temperature of the high rotational levels in the ground vibrational state is then a measure of the strength of the UV pump and thus the incident radiation field. Second, dissipation of turbulence leads to highly localized regions of warm, dense gas. The population of states with $J \gtrsim 3$ will be enhanced in these intermittent turbulent dissipation zones (Section 9.4). In this case, these higher states measure the dissipation of turbulent energy rather than the UV field.

Given their low critical density, the lowest rotational levels of C_2 also provide good estimates of the kinetic temperature. The populations of the higher levels are affected by

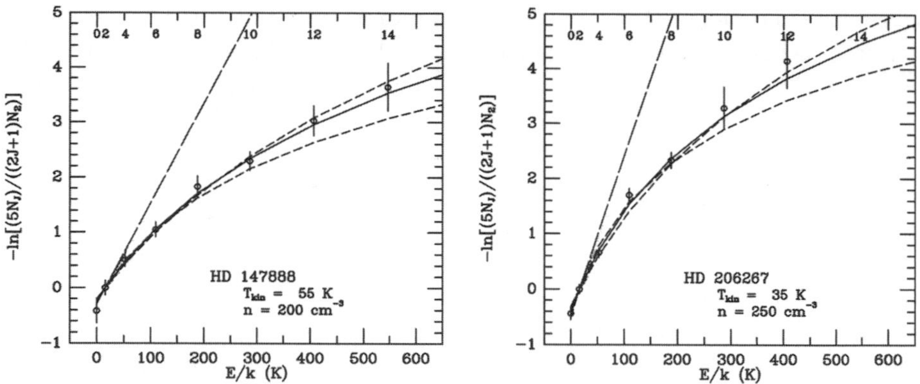

Figure 9.2 The Boltzmann rotational diagram for C_2 (relative to the $J = 2$ population) toward the stars, HD 147888 and HD 206267 as measured through the F–X (0–0) and (1–0) bands at 1341 8 and 1314 Å and the D–X (0–0) band at 2313 Å. The rotational level, J is indicated near the top. The long-dashed line is the Boltzmann distribution at the kinetic temperature estimated from the lowest levels. The solid line is the best non-LTE fit to the data. Parameters for these model are indicated in the panels. The short-dashed lines indicate models with ($T_k - 10$ K, $n - 50$ cm^{-3}) and ($T_k + 10$ K, $n + 50$ cm^{-3}) and illustrate that the physical conditions are well determined. The near-IR and UV radiation fields that pump these levels have been set equal to the average interstellar radiation field. Strictly speaking, the model fits depend on n/G_0. Figure courtesy of [106]

electronic pumping through the near-infrared Phillips systems and the UV Mulliken system. Deexcitation is through quadrupole and intersystem transitions to the rotational levels in the ground state and through collisions. The populations of the higher levels are thus a measure of n/G_0. This can be modeled in the same way as described for H_2 in Section 4.3.1. The results can be plotted in a Boltzmann diagram (cf. Figure 9.2). The curvature in this diagram reflects the importance of suprathermal excitation (Section 4.1.8 and 4.2.2). As this example illustrates, observations of C_2 provide a powerful way to probe the physical conditions (e.g. T_k and n/G_0) in the absorbing gas.

Abundances of many molecules – notably H_2 and CH – are sensitive to the physical conditions. In diffuse clouds, molecules are, typically, formed through bimolecular reactions and destroyed through photodissociation. Abundances are then sensitive to the ratio of the density to the strength of the UV field, n/G_0 (Sections 9.2 and 9.3).

In summary, molecular observations can be used to determine the physical conditions of the absorbing gas. As the abundances and rotational level populations are controlled by an interplay of UV pumping and collisional processes, only the combined parameter, n/G_0, can be reliably determined. The degeneracy between different models can only be broken if an independent measure of the radiation field or density were available. In the past, the high J H_2 lines have been used to determine G_0. However, the column densities of these levels may also be influenced by regions of turbulent dissipation (Section 9.4). The fine-structure level populations of C provide a measure of the density, independent of the strength of the radiation field. The [CII] fine-structure at 158 μm (1.9 THz) is the dominant cooling line of diffuse clouds and acts as a thermostat, regulating the temperature to the level separation of these fine structure levels (92 K) in such a way that the cooling balances the heating. These levels are typically very subthermally excited[1] and hence the population of the upper level is sensitive to density and temperature (cf. Section 4.1). However, absorption lines from the upper level are very weak and have only been measured for a few sightlines.

For typical diffuse clouds, the ratio n/G_0 is $\simeq$50 cm^{-3}. However, optical and UV observations of diffuse clouds focus on gas along the line of sight toward late type O stars or early B stars. The n/G_0 ratio can be much higher near a bright star as the thermal expansion of the overpressurized ionized gas in the HII region surrounding the star will sweep up and compress the neutral medium in a shell. Eventually, this expansion will stop when the HII region reaches pressure equilibrium with the ISM; e.g. at a density of $\simeq$0.25 cm^{-3}. For an O9 ionizing star, the radius of the swept-up shell is then $\simeq$100 pc and the density in the shell is $\simeq$50 cm^{-3}. During this evolution, the parameter n/G_0 is $\simeq 250\left(100\,\mathrm{cm}^{-3}/n\right)^{1/3} \simeq 280\,(R/100\,\mathrm{pc})^{1/2}$. Of course, this expansion takes time and n/G_0 will typically be less. More importantly, the actual structure of the ISM near massive stars is generally much more complicated. Early O stars have strong stellar winds and their mechanical energy input will dominate the expansion over the thermal pressure of the ionized gas. The winds of later type O stars are more feeble. However, short-lived, higher mass stars in an association may already have exploded as supernova creating a very tenuous cavity filled with hot gas. Because of their random motion, later type O stars may then find themselves in this cavity and away from their birth sites. This complex interplay of energetic and radiative energy input in their surrounding medium drives the evolution of the ISM. Molecular observations of diffuse clouds provide one way of studying the effects of feedback of massive stars on the ISM.

Finally, as an aside, the excitation of molecules such as CN and CO is strongly affected by the cosmic microwave background radiation (CMB) and indeed early studies of the temperature of the CMB were performed by analyzing CN data. Such measurements can also provide a view of the background radiation temperature over cosmic history.

9.2 The Formation of H_2

The results of UV studies of the H_2 abundance in diffuse clouds are summarized in Figure 9.3. The abundance of molecular hydrogen is set by a balance among accretion, diffusion, and reaction on grain surfaces (see Chapter 7) and dissociation by UV photons. The UV photo-dissociation process has been well characterized (cf. Section 6.3.4) and, hence, these observations can be used as a probe of H_2 formation. We will first review these observations and derive the H_2 formation rate in diffuse interstellar clouds empirically and then examine the chemical process involved in the latter.

[1] For the well-known diffuse cloud along the line of sight toward ζ Oph, the population of the upper level is only about 0.01 of that of the ground state and yet the kinetic temperature of the gas is 75 K.

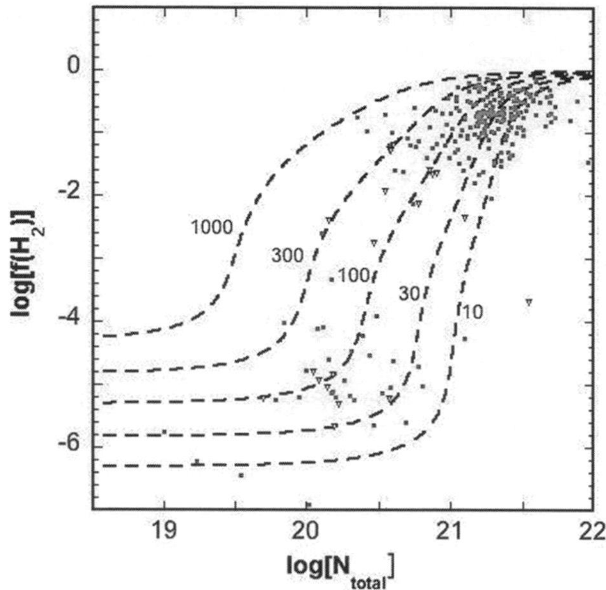

Figure 9.3 Observed H_2 fraction, f_{H_2}, versus the total column density, $N_{H_{tot}} = N_{HI} + N_{H_2}$. Upper limits are indicated with inverted triangles. The curves are equilibrium models for constant density (as indicated) and an assumed $G_0 = 1$. See text for details. Figure adapted from [120]

9.2.1 UV Observations and the Rate of H_2 Formation

Consider a constant density diffuse cloud with the interstellar UV radiation field incident from one side. Balancing the formation of H_2 on grain surfaces with the UV photodissociation, we have,

$$k_d\, nn\,(\mathrm{H}) = k_{uv}\,(\mathrm{H}_2)\, n\,(\mathrm{H}_2) = k_{UV}\,(0)\,\beta_{ss}\,(N\,(\mathrm{H}_2))\exp\left[-\tau_d\right] n\,(\mathrm{H}_2) \tag{9.1}$$

where k_d is the H_2 formation rate on dust surfaces, n is the density of H-nuclei ($n = n\,(H) + 2n\,(H_2)$), $k_{UV}\,(0)$ the unshielded photodissociation rate, and $\beta_{ss}\,(N\,(H_2))$ the self-shielding factor. The photodissociation of H_2 and the role of self-shielding has been discussed in Section 6.3.4; we can rewrite Eq. (9.1),

$$k_d\, nn\,(\mathrm{H})\exp\left[\sigma_d N\,(\mathrm{H})\right] = k_{UV}\,(0)\,\beta_{ss}\,(N\,(\mathrm{H}_2))\exp\left[-\sigma_d N\,(\mathrm{H}_2)\right] n\,(\mathrm{H}_2)\,, \tag{9.2}$$

with σ_d the UV cross section per H-nuclei. This separates the H and H_2 dependencies and, realizing that $n\,(H)/n\,(H_2) = dN\,(H)/dN\,(H_2)$ (cf. Section 6.3.4), we evaluate both sides separately. For the left-hand side,

$$k_d\, nn\,(\mathrm{H}) = k_d\, n\int_0^{N(H)}\exp\left[\sigma_d N'\,(H)\right] dN'\,(H) = \frac{k_d n}{\sigma_d}\left(\exp\left[\sigma_d N\,(H)\right] - 1\right). \tag{9.3}$$

This provides the column density of HI in terms of the column density of H_2,

$$N(HI) = \frac{1}{\sigma_d} \ln\left[\alpha G\left(N\left(H_2\right)\right) + 1\right], \tag{9.4}$$

where α is given by,

$$\alpha = \frac{k_{UV}(0)}{2k_d n}, \tag{9.5}$$

and G is,

$$G\left(N\left(H_2\right)\right) = \sigma_d \int_0^{N(H_2)} \beta_{ss}\left(N\left(H_2\right)\right) \exp\left[-2\sigma_d N'\left(H_2\right)\right] dN'\left(H_2\right). \tag{9.6}$$

The self-shielding factor is accurately given by Eq. 6.40. For given dust properties, the G function can be evaluated and the relationship between the HI and H_2 column can be calculated.

Adopting $k_d = 3 \times 10^{-17}$ cm^3 s^{-1}, $k_{UV}(0) = 2.7 \times 10^{-11}$ s^{-1}[2], we have $\alpha = 1.33 \times 10^3/n$ in the average interstellar radiation field. The results for this simple model are compared to the observations in Figure 9.3 for different densities. The structure in the curves reflects the detailed behavior of the self-shielding function (Figure 6.10). With this choice for the H_2 formation rate on dust, the observed fraction of H_2 is well explained for densities in the range 10–1000 cm^{-3}. These are typical densities for diffuse clouds. Note that points to the right of a curve may actually refer to lines of sight with several, lower column density, diffuse clouds with those characteristics. Also, time-dependent effects may play a role as the timescale for H_2 formation is appreciable, $\tau(H_2) \simeq 10\ (100/n)$ Myr, which would tend to increase the HI fraction for a given density. This may well be relevant for some of the stringent upper limits. Positive density fluctuations associated with turbulent dissipation would tend to increase the H_2 fraction, which would survive in an unshielded environment for a few hundred years but in a well-shielded environment for up to a few million years (Section 9.4).

The H-nuclei column density toward the dissociation front where $n(H_2) = 1/4$ is of particular interest (Section 6.3.4). Analysis shows that this column density is well represented by,

$$N(H) = \frac{3}{4\sigma_d} \ln\left[\left(\alpha\overline{G}\right)^{4/3} + 1\right], \tag{9.7}$$

where we have adopted the simple expression for the dependence of the self-shielding factor scaling with the 3/4 power of the H_2 column density (Eq. 6.31 in Section 6.3.4). In this expression,

[2] As this is evaluated at the surface of a diffuse cloud, an unshielded rate equal to half that in the diffuse ISM is appropriate. The value used here is evaluated for the Draine average interstellar radiation field.

$$\alpha\overline{G} = G_0 \left(\frac{\sigma_d}{1.9 \times 10^{-21}\ \text{cm}^2\ \text{H-nuclei}^{-1}}\right)\left(\frac{10^2\ \text{cm}^{-3}}{n}\right)$$
$$\times \left(\frac{3\times 10^{-17}\ \text{cm}^3\ \text{s}^{-1}}{k_d}\right)\left(1 + 8.9\left(\frac{\sigma_d}{1.9\times 10^{-21}\ \text{cm}^2\ \text{H-nuclei}^{-1}}\right)\right)^{-0.37}, \quad (9.8)$$

where we have gone back to the Habing normalization of the radiation field.[3]

9.2.2 H_2 Formation on Grain Surfaces

Direct radiative association of ground-state H atoms forming H_2 (Section 6.6) has a very low rate as the molecule can only stabilize through emission of a vibrational quantum which is a dipole forbidden transition (Section 3.5). Reaction of a ground-state H atom with an H-atom in $n = 2$ leads to an H_2 molecule in an excited electronic state (cf. Figure 3.11 and 6.9) and radiative relaxation can occur through allowed electronic transitions but the rate is still very small (k_{ra} (H_2) $\simeq 5 \times 10^{-15}$ cm^3 s^{-1}) given the low emission probability (3×10^{-5} with $A_{ul} \simeq 10^9$ s^{-1}). As the abundance of atomic H in $n = 2$ is very small, this route is essentially closed. The associative detachment (Section 6.11) of H^- with H also leads to H_2 formation, where the former is formed through electron attachment (Section 6.10). The associative detachment rate is quite rapid ($\simeq 4 \times 10^{-9}$ cm^3 s^{-1}) but electron attachment is slow ($\simeq 10^{-18} T$ cm^3 s^{-1}) and H^- is rapidly photodissociated ($2.4 \times 10^{-7} G_0$ s^{-1}). Hence, the overall rate is slow in the diffuse ISM (see Exercise 9.5), but this reaction is important in the early universe.

As gas phase routes toward H_2 are effectively closed in the diffuse ISM, it is generally accepted that H_2 formation occurs predominantly on the surfaces of interstellar grains. The H_2 formation rate per unit volume, $R_d(H_2)$, can then be expressed as (cf. Eq. 7.1),

$$R_d\,(H_2) = \frac{1}{2} S\,(T, T_d)\,\eta\, n_d \sigma_d\, n\,(H)\, v_H, \quad (9.9)$$

where $S(T, T_d)$ is the sticking probability of an H atom with temperature T colliding with a grain of temperature T_d, η is the probability that an adsorbed H atom will migrate over the grain surface, find another H atom, and form H_2 before evaporating from the grain surface, $n_d\sigma_d$ is the total grain surface area per unit volume, n_H is the H atom density, and $v_H = 1.5 \times 10^4\, T^{1/2}$ cm s^{-1} is the thermal speed of the H atoms. Typically, $n_d\sigma_d \simeq 10^{-21}\, n$ cm^{-2} and Equation (9.9) becomes,

$$R_d\,(H_2) \simeq 3 \times 10^{-17} \left(\frac{T}{100\,\text{K}}\right)^{1/2} S(T, T_d)\eta(T_d)\, n\, n_H\ \text{cm}^{-3}\,\text{s}^{-1}. \quad (9.10)$$

For sufficiently low temperatures, S and η are ~ 1 (see below) and the rate coefficient of H_2 formation on interstellar grain surfaces is $k_d = R_d/nn_H = 3 \times 10^{-17}$ cm^3 s^{-1}: the

[3] In the Draine definition, $G_0 = 1.7$ for the average interstellar radiation field.

value used in constructing the curves in Figure 9.3. Hence, with standard values for the parameters, H_2 formation on grain surfaces can explain quantitatively the observed H_2 abundances.

Because S and η tend to decrease with increasing temperature as atoms bounce from the grain surfaces or evaporate before reacting, this rate will decrease with T and T_d and the details of this dependence will depend on the interaction of atomic H with astrophysically relevant surfaces. This is particularly relevant for photodissociation regions where the strong irradiation by the stellar radiation field will lead to substantially higher temperatures ($T_g \simeq 100-1000$ K, $T_d \simeq 30-100$ K; Section 11.6) as well as in dissociative shocks (T_g can be several hundreds of degrees in the cooling gas where molecules reform, $T_d \simeq 10$ K; Section 11.7).

The sticking coefficient of H on interstellar grain surfaces has been discussed in Section 7.3.2. Experimental studies have been performed under thermal conditions ($T_g \simeq 100$ K, $T_d \simeq 15$ K) that are close to those expected in diffuse clouds and we can confidently adopt a value of 0.5.

The reaction probability, η, has been the subject of a number of experimental and theoretical investigations. The earliest studies analyzed migration and reaction of atomic H physisorbed on ice surfaces and realized that, in space, H_2 formation is inhibited on perfect surfaces at temperatures in excess of some 13 K because an H atom will evaporate before the next one accretes. However, it was also realized that H_2 formation is efficient up to a critical temperature between 25 and 50 K with the presence of a small number of sites with enhanced adsorption energies associated for example with kinks or surface defects. Experimental studies on (poly)crystalline silicate surfaces, indeed, indicate a very limited range of temperatures ($\simeq 8-15$ K) over which H_2 formation can proceed efficiently. However, on amorphous silicate surfaces, H_2 formation may proceed over a much wider range of temperatures given the much higher binding energy (700 versus 300 K). For interstellar dust temperatures relevant for diffuse clouds, H_2 formation should be efficient. This is not the case for the elevated dust temperatures relevant for photodissociation regions and we will reexamine these issue with H_2 formation in Section 11.6.3.

9.3 Chemistry

Chemistry in the ISM is controlled by kinetics not thermodynamics and this is very clear for diffuse clouds. The chemistry in diffuse clouds is an interplay of grain surface chemistry, forming H_2, ion–molecule reactions that drive molecule formation, and UV photochemistry that breaks these molecules down again to atoms. As there are four temperatures involved – the kinetic gas temperature, the dust temperature, the (dilute) radiation field temperature, and the temperature that characterizes the level of ionization – that are all different, it is clear that thermodynamic equilibrium is not attainable. The chemistry involved in the H/H_2 ratio has been discussed in Section 9.2. Here, we will focus on chemistry involving other elements.

As the main reservoirs and the kinetics involved are very different, the chemical networks of O, C, and N are very different as well. In atomic gas, there are two initiating steps: (1) the formation of H_2 on grain surfaces and (2) cosmic ray ionization of H. The former has been discussed in Section 9.2 and the latter will be discussed below. When, due to self-shielding, the H_2 abundance builds up, cosmic ray ionization of H_2 can become important to jump start the chemistry. In any case, molecule formation is strongly counteracted by FUV photo-destruction in diffuse clouds and – except for H_2 – atoms and atomic ions dominate the composition of the gas. The role of grains is two-fold: First, grains provide UV opacity, which shields molecules in the deeper portions of the cloud – but in contrast to dark clouds, never fully. Second, grain surface reactions play a key role in the formation of H_2. For other species, the role of grain surface chemistry is not as well established. N-chemistry may demand a dominant role by surface chemistry as well.

9.3.1 The Chemistry of O

Atomic oxygen has an ionization potential slightly higher than H and hence will be largely neutral in diffuse clouds. Ion-molecule routes toward O-hydrides start then with cosmic ray ionization of H (Figure 9.4). As radiative recombination is slow, the main destruction pathway for H^+ is charge exchange with neutral oxygen atoms, which is nearly resonant

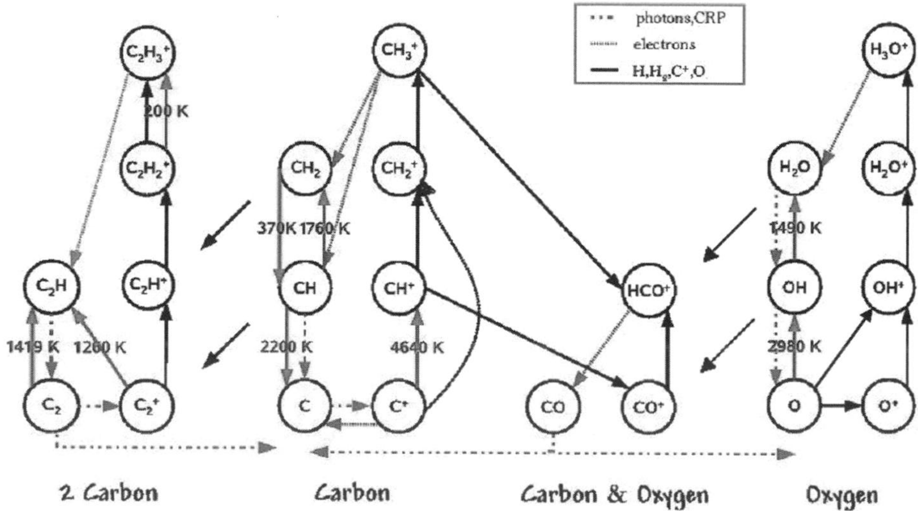

Figure 9.4 Key reactions in the build-up of molecules involving oxygen and carbon. Gas phase chemistry is initiated by ionization, followed by reaction with H_2 (if rapid) and electrons. These reactions convert C and O into small hydride radicals and then CO, which is very stable. The code indicates reactions with H_2, C^+, H, and H^+ (solid), reactions involving UV photons or cosmic rays (dashed), and electron recombination (slanted dotted). Reactions with barriers are labeled with the barrier height in degrees K. See text for details. Figure adapted from [35]

with a rate of $\simeq 4\times 10^{-10}\,\exp\left[-227/T\right]\,\text{cm}^3\,\text{s}^{-1}$ for ground state O ($^3P_2$) atoms (but much faster for $J = 1$ or 0 O atoms). Once O^+ has formed, reaction with H_2 then leads quickly to OH^+, H_2O^+, and then H_3O^+ (Figure 9.4). The latter dissociatively recombines to OH and H_2O. Reactions with C^+ then lead to CO^+ and HCO^+ and, eventually, CO. Mutual neutralization reactions of H^+ with PAH anions are important as the fraction of PAH anions is $\sim$0.3 (Section 9.5.3). There is an alternative channel to start O-hydride formation: Cosmic ray ionization of H_2 followed by reaction with H_2 produces H_3^+, which then rapidly proton transfers with atomic O to form OH^+. However, this is only a minor channel compared to the H CR-ionization route in diffuse clouds as dissociative recombination is the major destruction channel for H_3^+.

As this discussion illustrates, oxygen chemistry is closely tied to the cosmic ray ionization rate (as well as other processes) (but with limitations) and we will discuss this in Section 9.5. Here, we will discuss the abundance of O-hydride cations. The abundances of OH^+ and H_2O^+ have been measured by GREAT/SOFIA and HIFI/Herschel in the submillimeter and in the UV from the ground. Typical abundances are given in Table 9.1. The abundance of OH^+ and H_2O^+ are related through H-abstraction from H_2 and dissociative electron recombination,

$$\frac{n\left(\mathrm{OH^+}\right)}{n\left(\mathrm{H_2O^+}\right)} = 0.64 + 0.1\left(\frac{100\,\mathrm{K}}{T}\right)^{1/2}\left(\frac{X\left(\mathrm{e^-}\right)}{1.4\times 10^{-4}}\right)\left(\frac{1}{X\left(\mathrm{H_2}\right)}\right), \tag{9.11}$$

where standard reaction rate coefficients have been used. The second term on the right-hand side describes the competition between electron recombination and further reaction with H_2. The observed OH^+/H_2O^+ ratio of $\simeq$5 (Table 9.1) toward background sources implies that the electron recombination channel dominates. Hence, these sightlines are characterized by a low f_{H_2} fraction of $\simeq$0.02. A similar expression can be derived for the H_2O^+/H_3O^+ fraction, and the observations in Table 9.1 imply a somewhat higher f_{H_2}. In any case, these sightlines probed by HIFI are quite diffuse (cf. Figure 9.3).

9.3.2 The Chemistry of C

There are distinct differences in the initial chemical steps involved in the formation of oxygen- and carbon-bearing species in diffuse clouds. First, most of the oxygen is neutral and the chemistry starts by charge transfer between O and H^+ (see above). In contrast, photoionization keeps carbon largely ionized. Second, the build up of oxygen-bearing species is driven by the reaction of O^+ with H_2, which proceeds rapidly. The analogous reaction of carbon-bearing species – H_2 reaction starting with C^+ – is inhibited by a high activation barrier (Figure 9.4). Instead, small hydrocarbon radicals have to be formed through the slow radiative association reaction of C^+ with H_2 forming CH_2^+ (only 1 in $\sim 10^6$ collisions are reactive). This ion can react quickly with H_2 to CH_3^+ but further reaction with H_2 is again inhibited by an activation barrier. Radiative association to CH_5^+ is slow (1 in $\sim 10^4$ collisions) but can still be relevant. However, CH_5^+ dissociatively recombines to CH_3 rather than CH_4. Thus, in contrast to H_2O, the direct gas phase route to CH_4 is effectively closed. The small hydrocarbon ions that do result from this route can react with atomic O, eventually resulting in CO, or they may dissociatively recombine to small, neutral, hydrocarbon radicals. Reaction of these with C^+ will then lead to acetylenic species and longer chains can be built up from there through self-reaction of acetylenic radicals/ions and through C^+ or C insertion reactions (Figure 9.4).

The column density of CH is often used as a proxy for the H_2 density. Referring back to Figure 9.4, every reaction of C^+ with H_2 will eventually lead to CH and the CH to H_2 ratio is approximately given by,

$$\frac{X\,(\mathrm{CH})}{X\,(\mathrm{H_2})} \simeq \frac{k_{ra}}{k_{uv}\,(\mathrm{CH})}\, n\left(\mathrm{C^+}\right) \simeq 5\times 10^{-10}\,\frac{n_o}{G_o}, \tag{9.12}$$

with k_{ra} the radiative association rate (6.7×10^{-16} cm^3 s^{-1}), k_{uv} the photodestruction rate of CH ($\simeq 9.1\times 10^{-10}\exp\left[-1.5A_v\right]$ s^{-1}), and an abundance of 1.4×10^{-4} has been adopted for C^+ and an A_v of 1 magn. For a typical diffuse cloud in the Solar neighborhood, $n_o/G_0 \simeq 100$ cm^{-3} and the predicted CH abundance is in good agreement with the observed one (Table 9.1). For diffuse clouds in thermal and pressure equilibrium, n_o/G_0 is approximately constant. The CH column density will then indeed track the H_2 column density reasonably well, although variations of a factor of 3 are observed, presumably implying higher densities due to pressure variations or higher UV fields due to the proximity of an illuminating star. We note that the route toward CH^+ is effectively closed at low temperatures due to the high activation barriers. The chemistry involved in CH^+ will be discussed in Section 9.4 on turbulent dissipation regions.

CO is not a good tracer of H_2 in diffuse and translucent clouds as CO needs an appreciably larger column density of gas to be shielded against dissociating UV photons than H_2. Observational studies reveal a rapid rise in the CO abundance with column density of H_2 (Figure 9.5) and only for gas phase column densities approaching 3×10^{21} does the CO abundance approach the gas phase carbon-abundance in the diffuse ISM. There is some indication of two distinct regimes in the relationship between these two species corresponding to H_2 columns smaller and larger than 2.5×10^{20} cm^{-2}. Models link these two regimes to the two different formation routes of CO, the CH^+ with O route at low columns and the C^+ with OH route at high columns, both balanced by CO photodissociation (Figure 9.4). Models for the chemistry of diffuse and translucent clouds are in good agreement with the observations (Figure 9.5). We note that CO self-shielding plays some role in the dependence of the CO abundance on depth in the cloud, and this really kicks in between a column density of 1 to 2×10^{21} cm^{-2}. At high column densities, photodissociation is no longer relevant and CO destruction is due to reaction with cosmic ray produced He^+.

9.3.3 The Chemistry of N

As the ionization potential of N is higher than for H, nitrogen cannot be photo-ionized or charge exchange with H^+ and N will be neutral in diffuse clouds. Reactions of N with H_2 are inhibited by strong activation barriers and this route toward nitrogen-bearing molecules

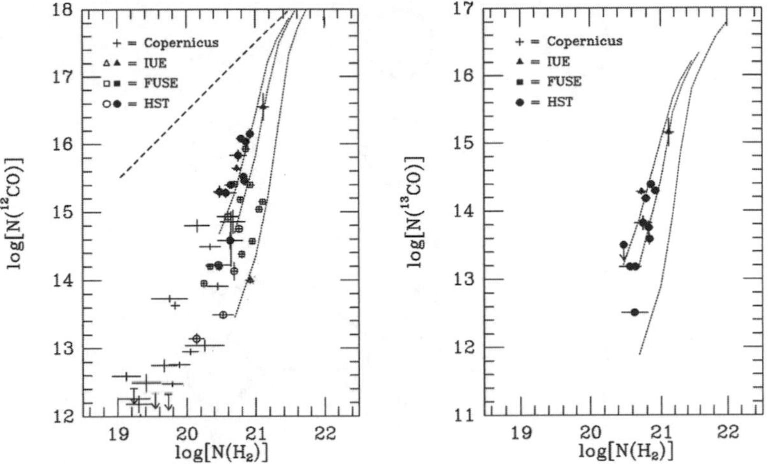

Figure 9.5 The observed relation of ^{12}CO (left) and ^{13}CO (right) with H_2 column densities in diffuse and translucent clouds. Data obtained using Copernicus, IUE, HST, and FUSE. Filled circles refer to objects with data on both CO isotopes. The dotted curves show the results of model calculations [116] for different I_{uv} ($= 1.7G_0$) and density combinations (from left to right: $(I_{uv}, n) = (0.5, 500)$, $(1, 500-1000)$, $(10, 2000)$). The dashed line is the expected relationship if all the hydrogen is in H_2 and all the gas phase carbon in CO. Figure adapted from [107]

is effectively closed (Figure 9.6). In diffuse clouds, nitrogen chemistry is initiated, then, by cosmic ray ionization of N. The reaction of N^+ with H_2 has a small, somewhat uncertain endothermicity of 10–20 meV ($\simeq$200 K). The rate of this reaction depends strongly on the initial fine-structure state of the ion[4] and the rotational state of the molecule.[5] In the ISM, N^+ tends to be in the ground state and, at low temperature, the reaction rate coefficient depends critically on the ortho-to-para ratio of H_2. At 100 K, however, both o- and p-H_2 react rapidly ($\simeq$3–4 $\times$ 10^{-10} cm^3 s^{-1}) but at 50 K the rate coefficient for p-H_2 has already decreased by almost a factor of 10. A variety of processes are involved in setting the H_2 o/p ratio, which we will touch on in the chapter on molecular clouds (Section 10.5). Here, we note that observed column density ratios of H_2 $J = 0$ to $J = 1$ range from -0.2 to $+0.3$ in dex and hence this reaction can be expected to be quite fast in diffuse clouds. Subsequent reactions with H_2 are fast until NH_3^+ is produced. The activation barrier of 1000 K for the reaction of NH_3^+ with H_2 leads to a very slow reaction rate coefficient,[6] severely hampering the formation of NH_4^+. Dissociative recombination of the N-hydride cations leads to the formation of, e.g., NH and NH_2. In contrast to O, the alternative route toward N-hydrides, involving H_3^+, is effectively closed as the proton affinity of atomic N is less than that of H_2 (Table 6.7). Finally, photodissociation of nitrogen-hydrides is very rapid with rate coefficients of k_{uv} $\sim$5–10 $\times$ 10^{-10} cm^3 s^{-1}.

[4] The two excited fine structure levels are 70 and 190 K above ground.
[5] The first ortho state is 170 K above the ground para state.
[6] Accounting for tunneling, $k \simeq 2 \times 10^{-13}$ cm^3 s^{-1} at 100 K.

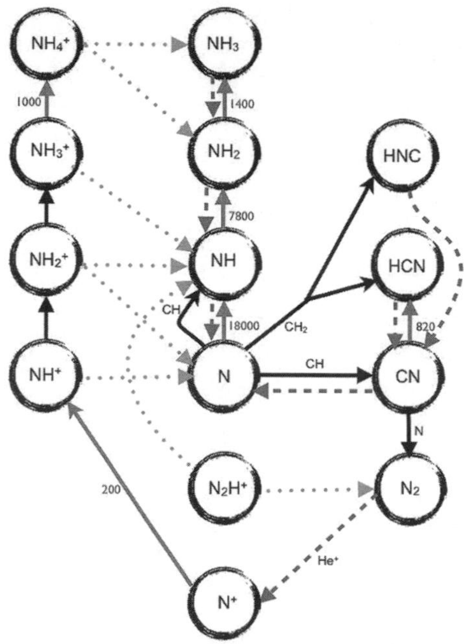

Figure 9.6 First chemical steps in gas phase routes for nitrogen-bearing species. The solid lines indicate reactions with H_2, or if indicated, CH and CH_2. Dashed lines are reactions involving UV photons or cosmic ray produced He^+. Dotted lines are electron recombination reactions. Reactions with barriers are labeled by their barrier height. The initiating reaction, cosmic ray ionization of N, is not marked. See text for details. Figure adapted from [35]

Despite the difficulties in initiating N-chemistry in diffuse clouds, nitrogen-hydrides are observed to be quite abundant (Table 9.1). Exact abundances are somewhat difficult to determine: Measurements refer to absorption line studies against strong sub-millimeter background sources and there is no direct measurement of the H or H_2 column density. Using the CH column as a proxy, typical abundances are 2×10^{-9}. As these measurements refer to the inner galaxy, the elemental N abundance – and thus the atomic N abundance – may be slightly higher than in the local solar neighborhood, perhaps 0.1–0.2 dex higher (e.g. $X(N) \simeq 10^{-4}$), or an abundance ratio, $X(\mathrm{NH}_n)/X(\mathrm{N}) = 2 \times 10^{-5}$. This can be compared to the simple chemistry outlined above. In essence, gas phase N-chemistry is initiated by cosmic ray ionization of atomic N and the hydrides formed at the end of the reaction chain are eventually photo-dissociated. Balancing these processes leads to the following relation,

$$\frac{X(\mathrm{NH}_n)}{X(\mathrm{N})} = \frac{\zeta_{CR}}{k_{uv}} \simeq 1.8 \times 10^{-6} \left(\frac{\zeta_{CR}}{2 \times 10^{-16}\,\mathrm{s}^{-1}} \right), \tag{9.13}$$

where we have adopted a high cosmic ray ionization rate per H-atom appropriate for the diffuse ISM (cf. Section 9.5). This is an order of magnitude lower than the observed NH_n

to N abundance ratio estimated above. Moreover, there is no simple way to form NH_3 and this species should be much less abundant than NH and NH_2; this is also contrary to the observations. This discrepancy with the observations has been taken to imply that grain surface reactions are an important source of ammonia. The importance of surface chemistry for NH_3 formation in dense clouds is well accepted, as ammonia is observed to be abundant in interstellar ices and is very abundant in the gas phase of hot cores whose chemistry is dominated by evaporated ices (Sections 10.5 and 11.3). In diffuse clouds, we can assume that accretion of atomic N on grain surfaces followed by hydrogenation will be counterbalanced by photodesorption into the gas phase, where photodissociation will complete the cycle. If every accreted N atom is converted into NH_3, released into the gas phase, and then broken down along the chain $NH_3 \rightarrow NH_2 \rightarrow NH \rightarrow N$, we have,

$$\frac{X\,(\mathrm{NH}_n)}{X\,(\mathrm{N})} = \frac{k_d}{k_{uv}} \simeq 5.4 \times 10^{-6} \left(\frac{n/G_0}{10^2\ \mathrm{cm}^{-3}}\right) \tag{9.14}$$

This is a factor of 3 higher than pure gas phase models could produce but still falls short of the observations. Attenuation of UV radiation in the cloud may increase this estimate somewhat (as long as all accreted N still photodesorbs). We mention again that grain surface chemistry would alleviate the NH_3 issue. Finally, we note that whether surface chemistry could explain the observed abundance ratios of the N-hydrides depends on the photodesorption process, which is not well understood for ammonia on bare silicate/carbon surfaces.

There are several pathways to CN, which involve reactions of hydrides with either the main nitrogen or carbon reservoir. The reaction of CH with atomic N dominates with a small (20%) contribution from C_2 with N. The alternative route of NH with C^+ is quite important too ($\simeq$50%). Balancing the first route with CN photodissociation yields for the abundance ratio,

$$\frac{X\,(\mathrm{CN})}{X\,(\mathrm{CH})} = \frac{k_n X\,(\mathrm{N})\, n_o}{k_{uv}\,(\mathrm{CN})} \simeq 0.15 \left(\frac{n_o/G_0}{100\,\mathrm{cm}^{-3}}\right), \tag{9.15}$$

where we have adopted for the neutral–neutral reaction, $k_n = 1.7 \times 10^{-10}\ \mathrm{cm}^3\ \mathrm{s}^{-1}$ and for the UV rate, $k_{uv}\,(\mathrm{CN}) = 5.2 \times 10^{-10} \exp\left[-3.5A_v\right]\ \mathrm{s}^{-1}$ and a typical H-nuclei column of $10^{21}\ \mathrm{cm}^{-2}$. The relationship is well fit by a ratio of 0.15 (Figure 9.7). For the sightline to ζ Oph, we have $n_0/G_0 \simeq 90$ and the total $A_V = 0.8$ (Table 9.1); we find that the calculated CN-to-CH abundance ratio is $\simeq$0.1, which is much less than observed ($\simeq$1). Probably, there are other routes toward CN but we do expect that they all scale the same way with n_o/G_0.

9.3.4 The Chemistry of S

With an ionization potential of 10.4 eV, sulfur is predominantly in the form of S^+ in diffuse clouds. Because sulfur–hydrogen bonds are so weak, the chemistry of sulfur is unique (Figure 9.8): Reactions of S, S^+ and all their hydrides with H_2 are endothermic and will not proceed in diffuse clouds. Hence, sulfur chemistry has to be initiated by the slow

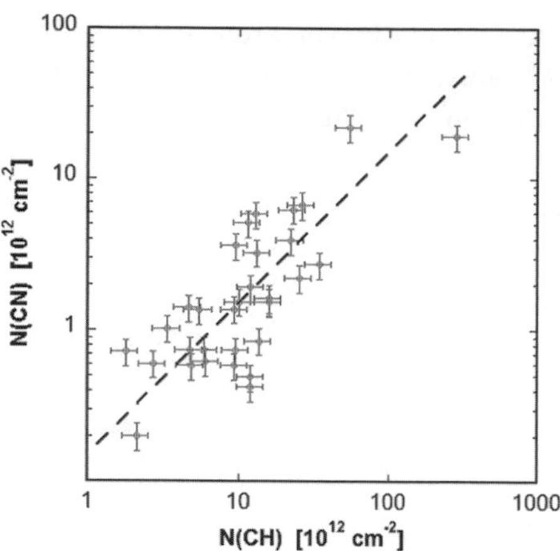

Figure 9.7 The relation between the CN and CH abundance. Typical error bars of 20% have been assumed. The curve is an eyeball fit to the data ($X\,(\mathrm{CN})\,/\,X\,(\mathrm{CH}) = 0.15$). Figure adapted from the compilation of [102]

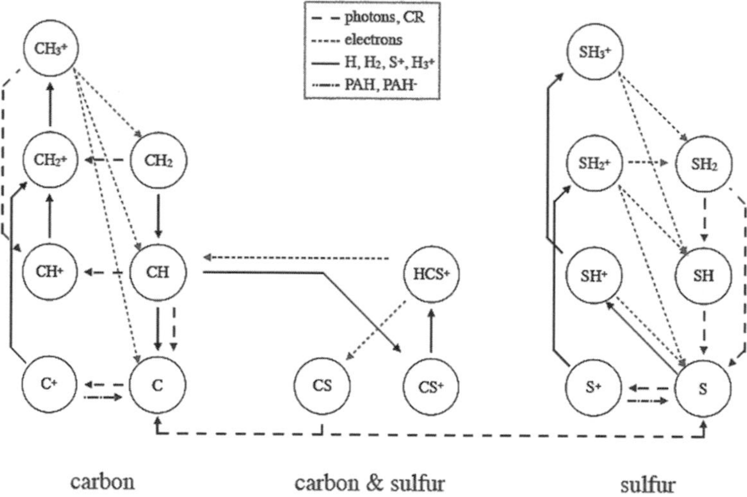

Figure 9.8 First chemical steps in gas phase routes for sulfur-bearing species. The reaction coding is indicated in the inset. See text for details. Reproduced by permission from [37]

radiative association reaction of S^+ with H_2 ($k_1 = 1.2 \times 10^{-17}$ cm^3 s^{-1}). The other route involving SH^+ to form eventually H_3S^+ has two bottlenecks as it has to start with proton transfer from H_3^+ to the trace amount of neutral atomic sulfur present followed by the slow radiative association process of SH^+ with H_2 (5.8×10^{-16} cm^3 s^{-1}). Dissociative electron recombination forms the neutral hydrides, H_2S & SH, and of course S. We then find that the abundance ratio of SH to S^+ is given by,

$$\frac{X(\mathrm{SH})}{X(\mathrm{S^+})} = \frac{k_1 n(\mathrm{H_2})}{k_{uv}(\mathrm{SH})} \tag{9.16}$$

$$\simeq 6 \times 10^{-7} \left(\frac{f_{\mathrm{H_2}}}{0.5}\right)\left(\frac{n/G_0}{100\,\mathrm{cm^{-3}}}\right), \tag{9.17}$$

where k_{uv} (SH) is the photodissociation rate of SH (9.8×10^{-10}) and $f_{\mathrm{H_2}}$ is the molecular hydrogen fraction. For ζ Oph, we calculate X (SH) $X\left(\mathrm{S^+}\right) \simeq 5 \times 10^{-7}$, which is some three orders of magnitude less than observed (Table 9.1). So, ion–molecule chemistry in diffuse clouds predicts low abundances of sulfur-bearing species. As sulfur is predominantly ionized, it will not accrete on grains and grain surface chemistry followed by photodesorption is not an effective source of sulfur hydrides. Given the high activation barriers, the presence of small amounts of warm gas will have a profound influence on the abundances of sulfur-bearing compounds. We will come back to this in the discussion on the chemistry of turbulent regions (Section 9.4).

9.3.5 Halogen Chemistry

The high ionization potential of fluorine (17.4 eV) implies that this atom is predominantly neutral in diffuse clouds. The reaction of F with H_2 is exothermic by 432 kJ/mol (4.5 eV) but has a small energy barrier of 800 – 1000 K depending on orientation. The NIST recommended rate coefficient is $1.1 \times 10^{-10} \exp\left[-450/T\right]$ cm^3 s^{-1} ($190 < T < 380$ K) but this is affected by excitation of the F fine-structure level and rotational excitation of H_2.[7] Carefully controlled studies yield reaction rates of 2×10^{-13} and 2×10^{-12} cm^3 s^{-1} at 10 and 100 K, respectively. Photodissociation is an important loss channel for hydrogen fluoride with an unshielded rate of 1.2×10^{-10} s^{-1}. At moderately high densities ($n > 10^2$ cm^{-3}), reactions of HF with C^+ forming CF^+ take over as the main destruction agent. Taking the 100 K value for the reaction of F with H_2 and balancing this with UV photodissociation results in,

$$\frac{\mathrm{HF}}{\mathrm{H_2}} \simeq 3 \times 10^{-8} \left(\frac{n/G_0}{100\,\mathrm{cm^{-3}}}\right) \tag{9.18}$$

where it is assumed that most of the F is atomic (a reasonable approximation) and an F elemental abundance of 1.8×10^{-8} has been adopted. As for so many species – notably CH – the abundance of HF scales with n_o/G_0 but theory and experiments indicate a stronger dependence on physical conditions (e.g. T). The good correlation between the column densities of HF and CH is then somewhat surprising (Figure 9.9). Analysis of the observations suggests that HF is a major ($\simeq$40%) reservoir of the elemental F in diffuse clouds.

Neutral chlorine does not react readily with H_2 at the low temperatures of the ISM (the activation barrier is about 2200 K). The chlorine ionization potential (12.97 eV) is very

[7] The F $^2P_{3/2} - {}^2P_{1/2}$ fine-structure splitting is 581 K. Rotational excitation of H_2 may range from 170 to 510 K.

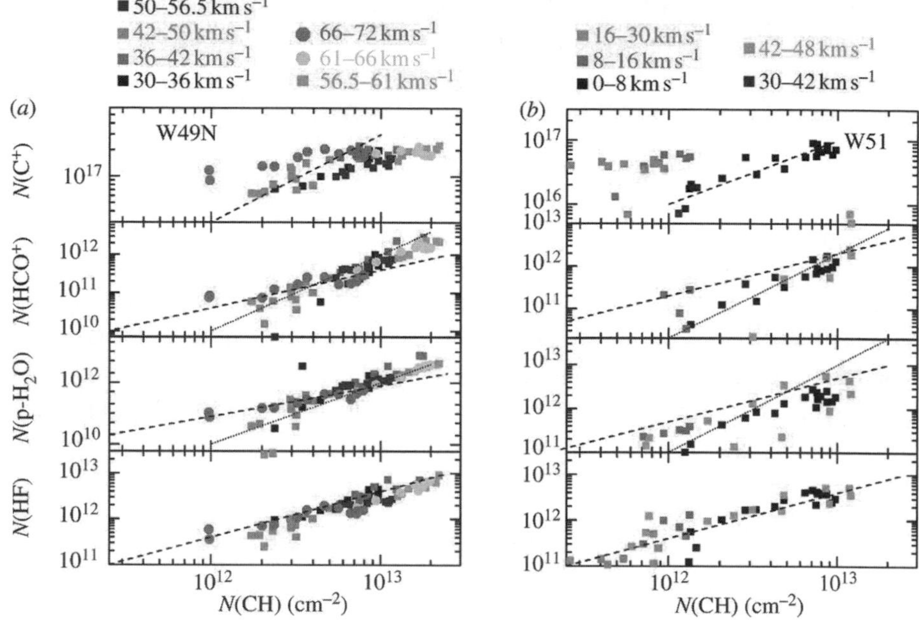

Figure 9.9 Comparison of the HF, p-H_2O, HCO^+, and C^+ column densities with that of CH along the W49N (a) and W51 (b) line of sight. Shading indicates different velocity intervals. Each point corresponds to a velocity channel of 0.6 km s^{-1}. Dashed lines show linear trends, while dotted lines indicate quadratic trends. Reproduced by permission from [33]

close to the Lyman limit but there can be an appreciable abundance of Cl^+, particularly at the cloud edge, and this ion reacts at the Langevin rate with H_2 forming HCl^+, which reacts with H_2 to H_2Cl^+, also at the Langevin rate. The latter dissociatively recombines with a minor channel to HCl (0.1). Photodissociation is the dominant destruction agent (1.7×10^{-9} s^{-1}) for HCl. Despite this relatively simple chemistry, there are a number of poorly understood aspects in the observed abundance ratio of chlorine-bearing species. These include the observed low H_2Cl^+/HCl^+ ratio ($\simeq 0.3$), which implies that dissociative recombination dominates over reactions with H_2 for HCl^+. This would imply a largely atomic gas with a very low H_2 fraction (10^{-2}). Likewise, the abundance ratio of HCl to its parent H_2Cl^+ is puzzling.

9.4 The Chemistry of Turbulent Regions

CH^+ was one of the first molecules discovered in the ISM. Yet, understanding its chemical origin has long remained enigmatic. For diffuse clouds, direct pathways to CH^+ are closed by the high energy barrier involved (Figure 9.4; Section 9.3.2). As the abundance of CH is well matched by models, the basic carbon chemistry seems understood. Hence the high abundance of CH^+ is generally taken to imply the presence of warm gas.[8] First models relied on heating by shock waves to produce the required high temperatures, but such models have largely been discounted as there are no obvious velocity shifts with respect to other molecular lines and the CH^+ column density shows a rough correlation with total H column.

It is now thought that CH^+ traces Turbulent Dissipation Regions, as viscous dissipation of turbulence is not uniform in time and space but rather occurs localized and in bursts. As regions of large velocity shear (rather than (compressible) shocks) dissipate the suprathermal energy stored in supersonic turbulence, chemistry in the resulting warm gas can drive the formation of CH^+ – as well as other species. In addition, a small drift velocity will develop between ions and neutrals and this can also help overcome energy barriers in reaction between atomic ions and H_2. This can be "caught" by using an effective temperature, T_{eff}, in the relevant reaction rate coefficients given by,

$$\frac{3}{2}kT_{eff}^2 = \frac{3}{2}kT_r^2 + \frac{1}{2}\mu\left(u_i - u_n\right)^2, \tag{9.19}$$

where T_r is the weighted kinetic temperature,

$$\frac{T_r}{\mu} = \frac{T_n}{m_n} + \frac{T_i}{m_i}, \tag{9.20}$$

and the subscripts i and n refer to ions and neutrals, respectively. To guide our thinking, for the typical model shown in Figure 9.10, the turbulent heating rate is $\Gamma_{tur} \simeq 3 \times 10^{-23}$ erg cm ^{3} s ^{1}, which exceeds the photo-electric heating rate of interstellar gas by about a factor 30. As the 158 μm C^+ fine-structure line cannot keep up with this additional heating, cooling is dominated by H_2 rotational emission. The resulting temperature, $\simeq$750 K, and the ion–neutral drift both help to drive the reaction,

$$C^+ + H_2 \rightarrow CH^+ + H \qquad k = 10^{-10} \exp\left[-4640/T\right] \text{ cm}^3 \text{ s}^{-1}. \tag{9.21}$$

The methylidine cation is lost through reactions with H_2 at the Langevin rate, resulting in a CH^+/C^+ fraction of $\sim 10^{-3}$ (Figure 9.10). Other chemical signatures of warm gas along the line of sight are SH^+ (Section 9.3.4) and OH (Section 9.3.1),

$$S^+ + H_2 \rightarrow SH^+ + H \qquad k = 10^{-10} \exp\left[-9860/T\right] \text{ cm}^3 \text{ s}^{-1} \tag{9.22}$$

$$O + H_2 \rightarrow OH + H \qquad k = 3.1 \times 10^{-13} \left(\frac{300\text{ K}}{T}\right)^{2.7} \exp\left[-3150/T\right] \text{ cm}^3 \text{ s}^{-1}. \tag{9.23}$$

The chemical (and thermal) signature of turbulent dissipation persists for some 3×10^3 yr or equivalently, the size of the chemically enhanced regions are $\simeq$100 AU for typical parameters.

[8] Vibrationally excited H_2 may play a role in PDRs but their abundance is too low in diffuse clouds to be important.

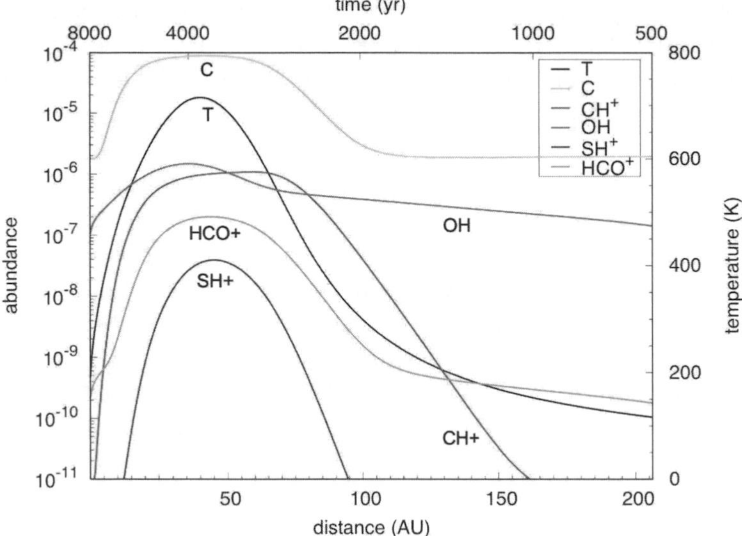

Figure 9.10 The structure of turbulent dissipation regions. Gas enters the region at turbulent velocities of 3.5 km/s. The dissipation of the turbulent energy heats the gas to ~700 K. The increased temperature as well as the ion–neutral drift drive the endothermic reaction of, e.g., C^+ and S^+ with H_2 forming CH^+ and SH^+ and these chemical signatures of turbulent dissipation last for a few thousand years. Reproduced by permission from [37]

The total column density of a chemical tracer of turbulence is then set by the number of active dissipating vortices along the line-of-sight, $\mathcal{N}_{vor}$, which is set by comparing the typical dissipation rate of a vortex with the average turbulent energy dissipation rate, ϵ_{tur}, in the ISM,

$$\mathcal{N}_{vor} \simeq \frac{\epsilon_{tur}}{\Gamma_{tur}} \frac{L}{\ell_{vor}}, \tag{9.24}$$

where L is the size of the cloud(s) along the line of sight, and ℓ_{vor} the scale-size of a vortex. Observations of turbulence in the ISM yield $\epsilon_{tur} \simeq 2 \times 10^{-25}$ erg cm^{-3} s^{-1} but with a large scatter, which results then in $\mathcal{N}_{vor} \simeq 10^2$. So, the number of active vortices along any line-of-sight is very large but the volume filling factor is small.

Models for turbulent dissipation regions (TDR) are in good agreement with the observations of CH^+ with $n \sim 50$ cm^{-3}, sizes of $\sim 10^2$ AU and strain rates of $\sim 10^{-11}$ s^{-1} (Figure 9.11). Likewise, the models predict the SH^+ to CH^+ ratio well. Turbulent dissipation regions will also influence the abundances of other species (Figure 9.10), notably OH and H_2, as reactions of the main O-reservoir with H_2 have substantial energy barriers. The enhanced molecular abundances in TDRs can percolate through to other species such as HCO^+ and CO, whose chemistry involves small O— or C—hydrides. This is illustrated in Figure 9.12 for models of different densities but with constant total column density.

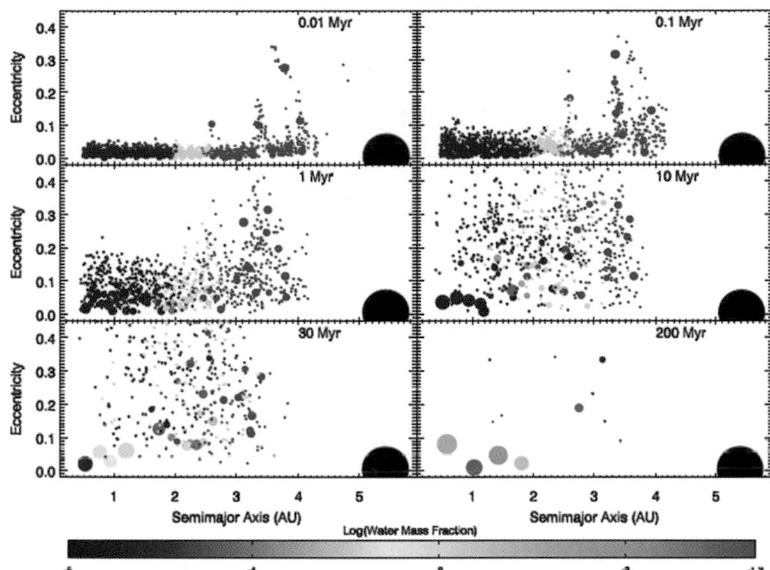

Figure 1.3 Snapshots in the evolution of disks of embryos from a simulation where Jupiter and Saturn are in the 3:2 mean motion resonance. The size of each body is proportional to its mass (but is not to scale on the x-axis). The interaction with Jupiter pumps up the eccentricity of the orbits of the embryos and they start to collide and grow, eventually leading to the formation of terrestrial planets. The temperature gradient in the protoplanetary disk has led to a compositional gradients in the embryos formed at different positions. The color of each body corresponds to its water content by mass (bottom scale, ranging from dry to 5% water). Jupiter is shown as the large black dot. Saturn is not shown. Figure taken from [21].

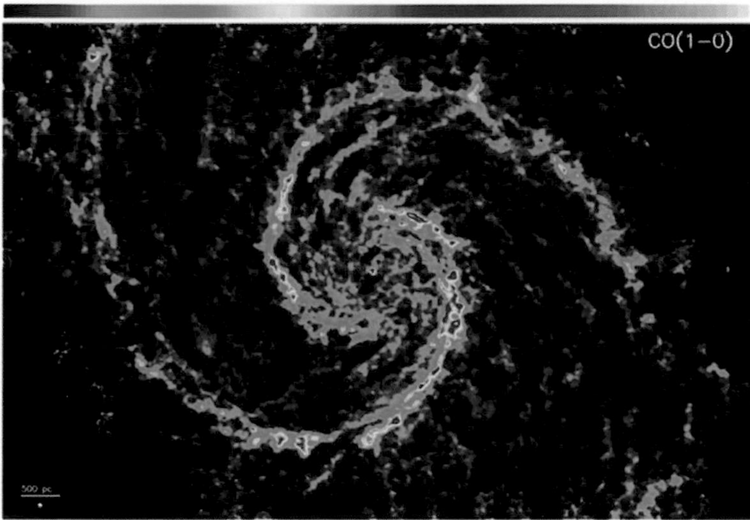

Figure 1.7 The nearby spiral galaxy, M51 – the whirlpool galaxy – in (integrated intensity of) the CO $J = 1 - 0$ line tracing molecular clouds in the spiral arms in the central 9 kpc. Observations obtained by the PdBI Arcsecond Whirlpool Survey (PAWS), using the Plateau de Bures interferometer at a resolution of $1.1 \times 0.9''$. A square-root scaling has been applied to enhance the faint emission. Figure taken from [27].

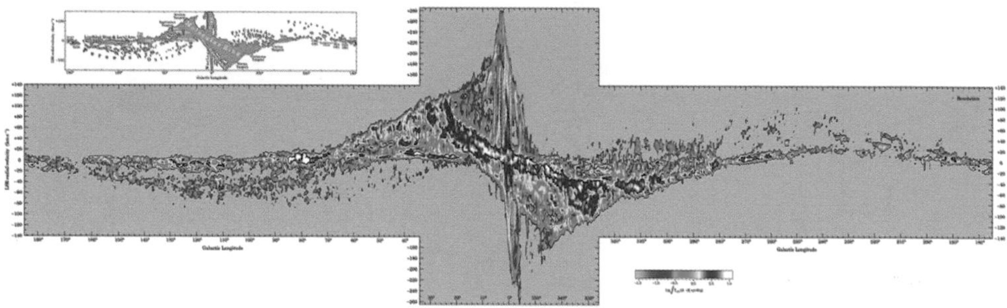

Figure 1.10 Longitude-velocity map of CO emission integrated over a $\sim 4^\circ$ wide strip centered on the Galactic plane. The map has been smoothed in velocity to a resolution of 2 km/s and in longitude to a resolution of $12'$. The inset identifies prominent structures in the molecular cloud distribution. Figure taken from [2].

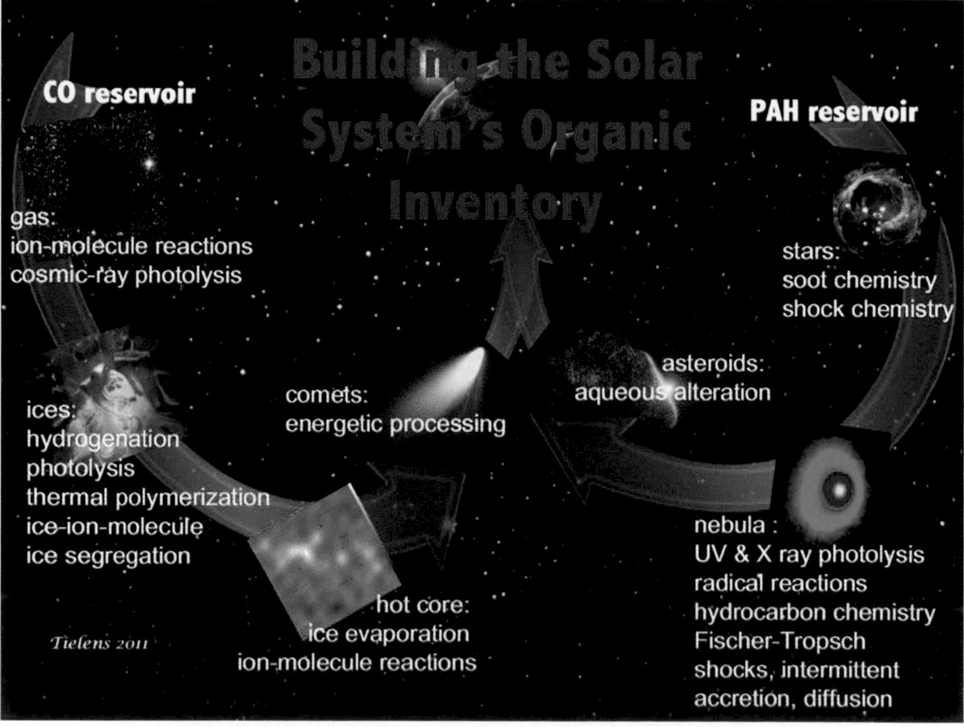

Figure 1.13 The organic inventory of the Solar System derives from a vast array of processes acting in a wide range of environments. Globally, two independent routes can be recognized. The first one builds up complex species from small radicals and starts with CO formation through ion–molecule reactions in dark clouds. Grain surface chemistry and photoprocessing of ices converts the main gaseous reservoir, CO, into complex species. The other route breaks down very complex species (e.g. PAHs) injected into the interstellar medium by stars into smaller and smaller species. Eventually, the species produced by either of these two chemical routes can become part of planetesimals and cometesimals in a protoplanetary disk environment that can deliver this organic inventory to planets in the habitable zone. Figure taken from [32].

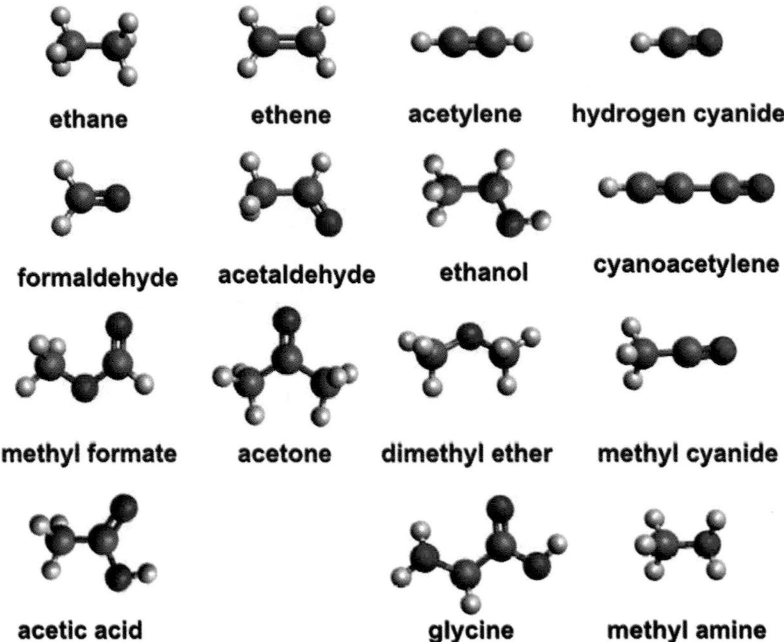

Figure 2.8 3-D ball and stick models of some molecules, illustrating the functional groups of classes of compounds. Dark gray, carbon atoms; light gray, hydrogen atoms; red, oxygen atoms; blue, nitrogen atoms. Note that methane is shown in Figure 2.4, PAHs are illustrated in Figure 8.1, and fullerenes are shown in Figure 8.25.

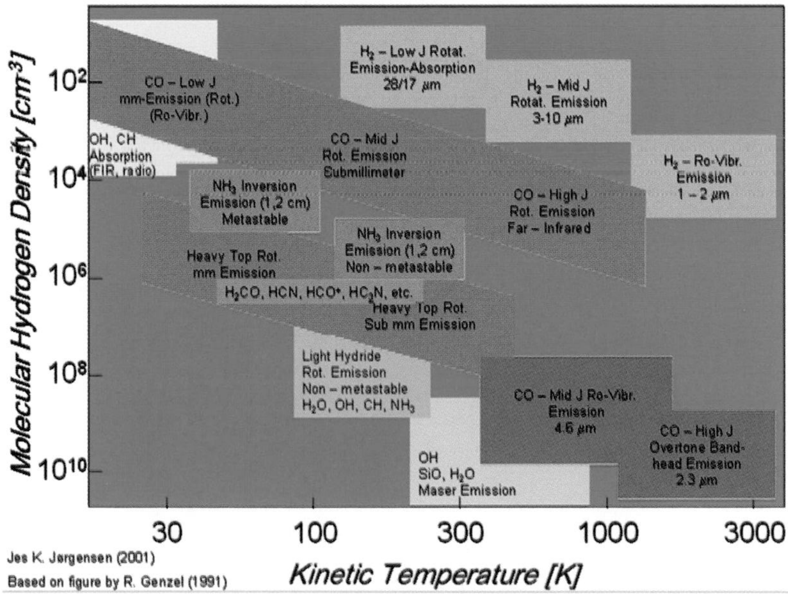

Figure 4.8 An overview of the molecular lines and the range of physical conditions in molecular clouds for which they are effective probes. Figure reproduced with permission from [9].

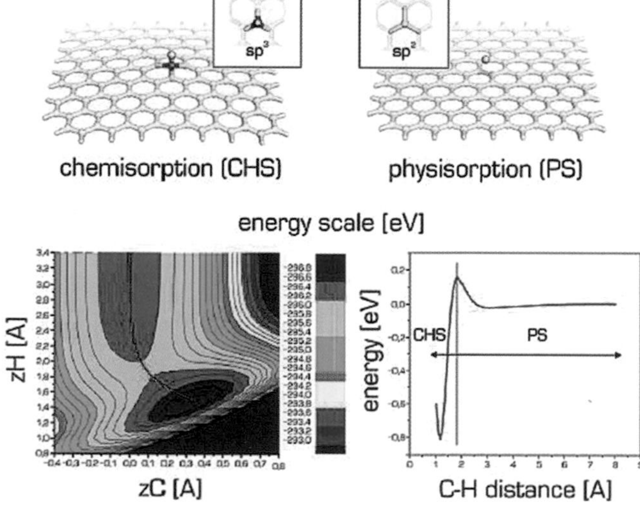

Figure 7.5 The calculated potential energy surface for the interaction of atomic H with graphene shows two minima associated with the physisorbed (PS) and the chemisorbed (CHS) site. A small activation barrier separates these two sites (bottom right). As the H-atom approaches the surface, it will first be trapped in the physisorbed site. For adsorption into CHS, the C-atom has to pucker out of the plane (bottom left; note the difference in scale for the x- and y-axis in this panel). The top two panels show a 3-D representation of the surface structure, illustrating the puckering of the plane. Reproduced from [87] with permission from the PCCP Owner Societies.

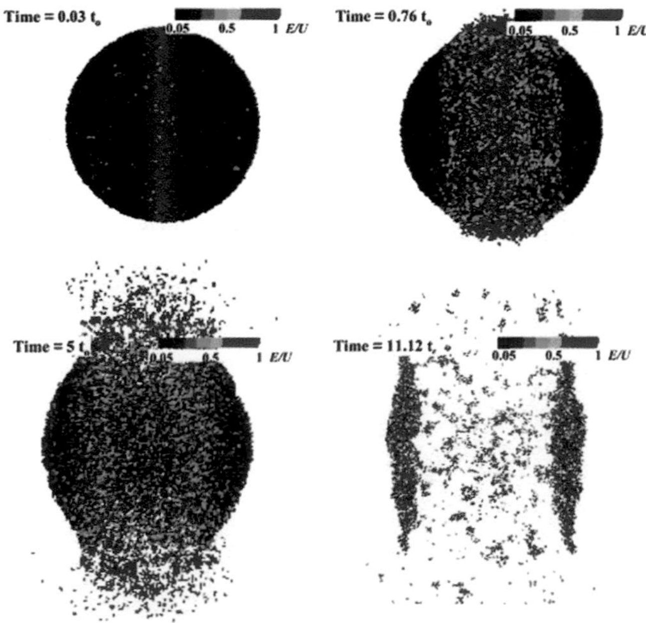

Figure 7.13 Molecular dynamics study of the interaction of a heavy cosmic ray with a small CO-ice grain. The total energy deposited is 8 keV. Each panel is a snapshot at a given timestep. The different colors correspond to energy in units of the binding energy, U_0. Figure taken from [12].

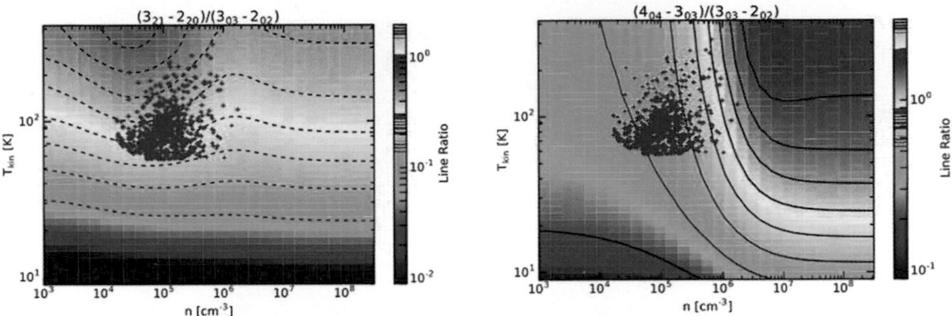

Figure 10.11 Line ratio's of para H_2CO lines. The line ratio in the left panel is mainly sensitive to the temperature. The right panel shows a line ratio that is sensitive to temperature and density (for $n < 10^7$ cm^{-3}). Combination of these two line ratios allows a measurement of density and temperature. The red dots represent observations of both of these line ratios to determine the best fit in temperature and density. Color scales show a range of 2 orders of magnitude in the ratio. The dashed contours range from 0.05 to 0.5. The solid contours range from 0.2 to 2. Figure taken from [115].

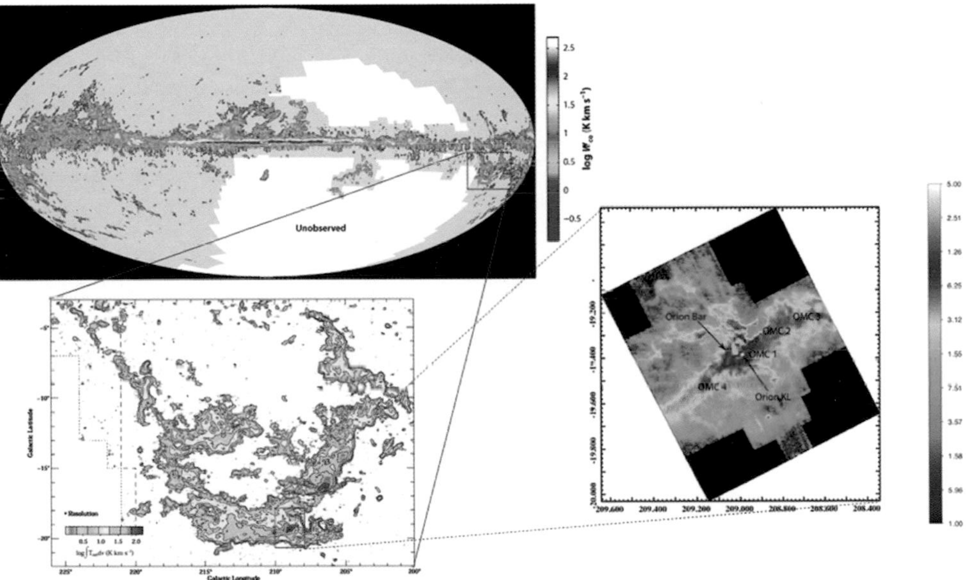

Figure 10.12 Top: Composite of CO $J = 1-0$ surveys of the Milky Way. This image is dominated by local cloud structures. From left to right, we can recognize the Taurus & Perseus molecular clouds (just below the plane), the Cepheus cloud just above the plane, the Ophiuchus Cloud above the (galactic) center with the Lupus cloud left of it, and at the right edge, the Orion-Monoceros cloud complexes. Figure taken from [104]. Bottom left: Blow up of the Orion-Monoceros cloud complex with counter clock-wise: Monoceros with the southern filament sticking up, the Orion A molecular cloud, the Orion B molecular cloud, Orion East, and the Southern filament. Figure taken from [248]. Bottom right: Blow up of the Orion A molecular cloud in CO $J = 2-1$ with in the center the bright BN/KL region, the Orion Bar PDR, the Orion Veil, all superimposed on the integral-shaped filament containing the dense cores OMC1 through OMC4. Figure adapted from [22].

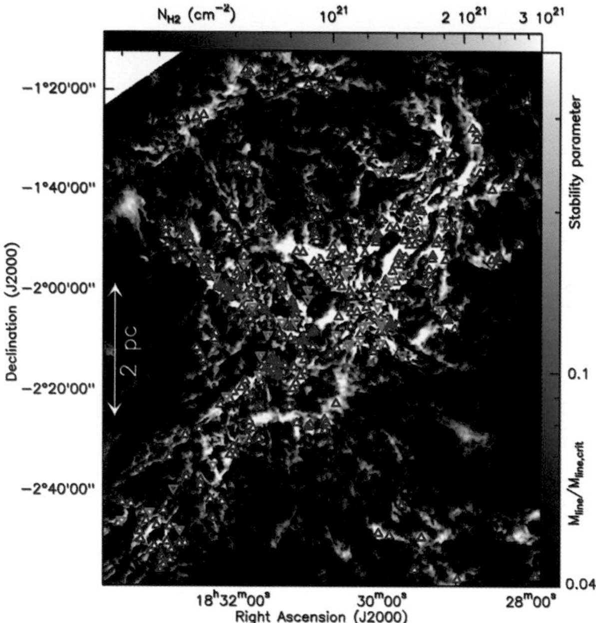

Figure 10.14 Column density map of Aquila derived from Herschel SPIRE/PACS data. The contrast of the filaments with respect to the non-filamentary background has been enhanced using a curvelet transform. The white areas highlight regions where filaments have a mass per unit length larger than half the critical value and are likely to be gravitationally unstable. Locations of candidate prestellar cores and protostellar cores are overlaid as blue and green triangles, respectively. Figure adapted from [124].

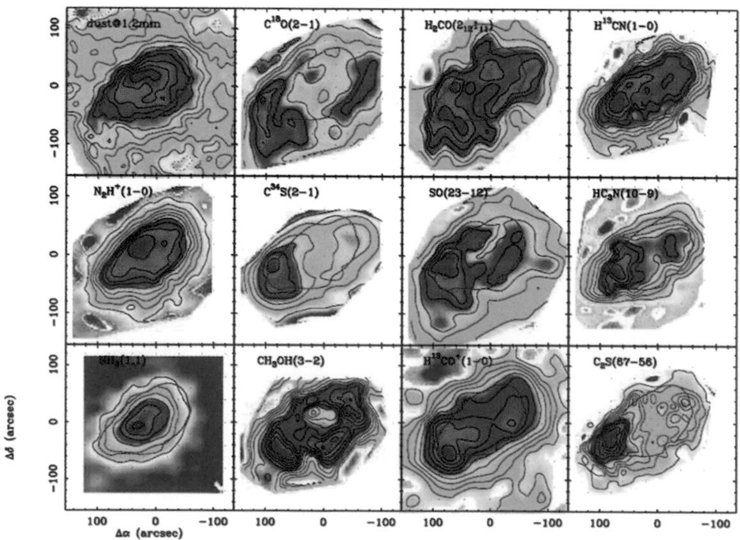

Figure 10.34 Observations of the L1489 core. The dust continuum and the N_2H^+ and NH_3 reveal a centrally condensed core. The other species show a ring-like or centrally depressed spatial distribution relative to the dust continuum indicative of depletion of the emitting species in the densest parts. For ease of comparison, the 75% emission contour of N_2H^+ is superimposed in red on each map. Figure taken from [223].

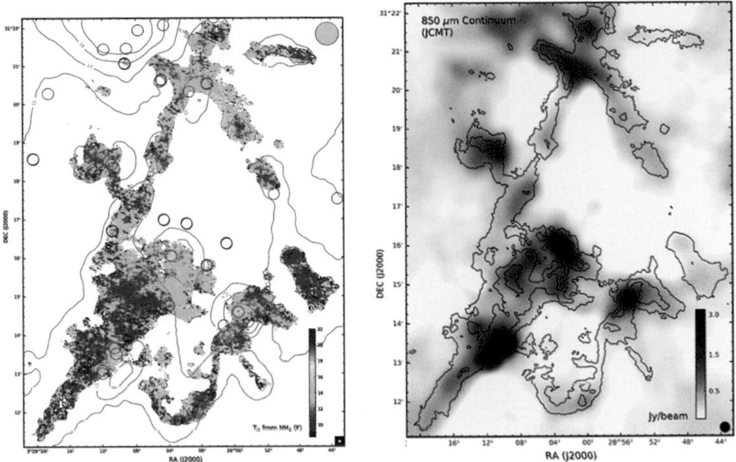

Figure 10.6 Left: The kinetic gas temperature of molecular clouds in the NGC 1333 region in Perseus as derived from ammonia observations using the VLA supplemented by the GBT. The locations of known Class 0/I and Flat YSOs from Spitzer observations are marked as red and black circles (size $\simeq$5000 AU), respectively. Right: The dust emission at 850 μm measured by SCUBA. Contours are the NH_3 (1,1) integrated intensity (0.01, 0.02, 0.04, and 0.08 Jy/beam km s^{-1}). Beam sizes are shown in the bottom left of each panel. Figure kindly provided by Dr Arnab Dhabal, [60].

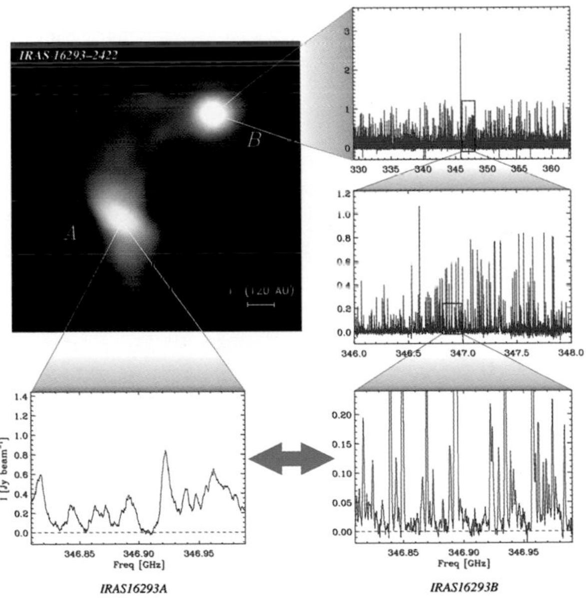

Figure 11.9 ALMA observations of the prototypical protostar, IRAS 16293-2422. The three-color image on the left is a composite of 3mm (red), 1.3 mm (green) and 0.87 mm (blue) continuum emission. The ALMA Protostellar Interferometric Line Survey (PILS) of this low mass solar-type binary reveals the strikingly different spectra of components, A and B. Component B is a face-on disk with very narrow line, which facilitates line identification. Component A (actually also a binary) is edge on and has much broader lines, resulting in line blending. Note how component B shows spectral lines down to the confusion limit. The emission from component A is, however, more extended and dominates single disk spectra. Figure kindly provided by J. Jørgensson; see [158].

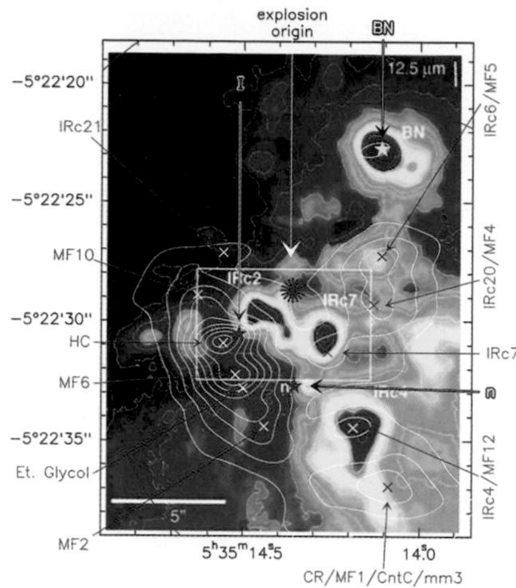

Figure 11.7 The BN/KL region in Orion is host to an explosion caused by a dynamical rearrangement of a mutiple star system. The two runaway stars (BN and I) plus the quasi-static object, n, are indicated by a multipointed star. The other runaway star (source x) is off the map. The presumed position of the origin of the explosion is denoted by a multi-pointed red and blue star. The 12.5 μm dust emission is shown in color with the 1.2 mm continuum emission superimposed as white contours. Mid-IR sources are labeled as IRC*n*. The BN/KL region contains the prototypical hot core sources, the Hot Core (HC) and the Compact Ridge (CR), which break up in mutiple components when observed in different molecular tracers. Dimethyl ether peaks are denoted by crosses (MF*n*) and methyl-formate or ethyl cyanide peaks are labeled as CR*n*. The ethylene glycol peak is also indicated. Figure taken from [227] with data from [96, 113, 131].

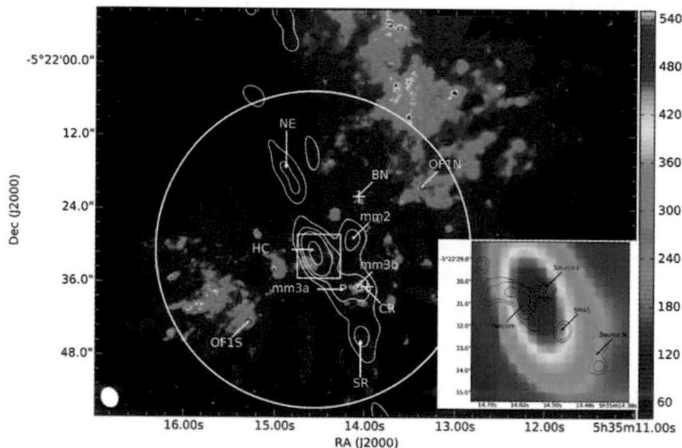

Figure 11.8 The large-scale morphology of the Orion BN/KL region is dominated by shocked H_2 emission associated with the fast shocks (color, brightness in counts per 400 s; [216]). Substructures resolved by SMA-only observations are superimposed (spatial resolution of $\sim$1200 AU). White contours show the 1.3 mm continuum emission at 5, 5, 15, 25, 60 rms levels with σ = 0.04 Jy/beam. The peaks of hot core (HC), mm2, mm3a, mm3b, the southern region (SR), north-east clump (NE), the NW and SE parts of the high-velocity outflow (OF1N, OF1S) are labelled. White crosses denote the BN object and the compact ridge (CR). The beam in the bottom left corner is from the SMA-only continuum at 1.3 mm. The large yellow circle represents the SMA primary beam. Inset: The HC continuum from the SMA at 1.3 mm (color), with black contours of the 865 μm continuum emission at a spatial resolution of $\sim$300 AU from [28] overlaid. The beam in the bottom left corner is from the 865 μm data. The white cross corresponds to the explosion center determined from a trace-back of the H_2 proper motions [19]. Figure adapted from [100] and kindly provided by Siyi Feng.

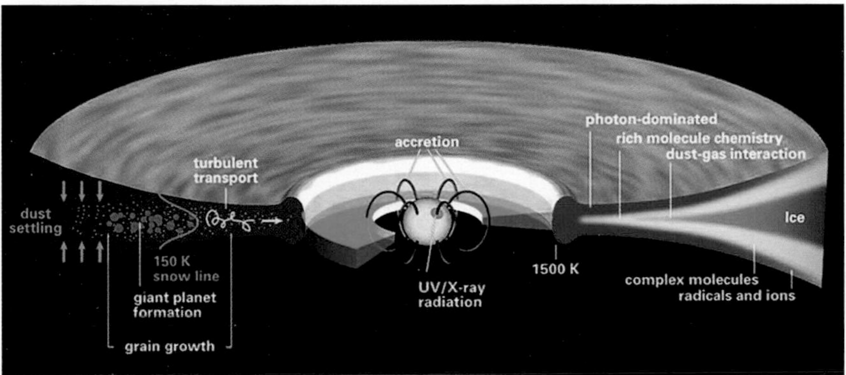

Figure 11.18 Schematic representation of the structure of a protoplanetary disk. The left-hand side summarizes processes that are important in the evolution of the disk and the formation of planets. The right-hand side illustrates the different chemical zones in the disk. Figure taken from [140].

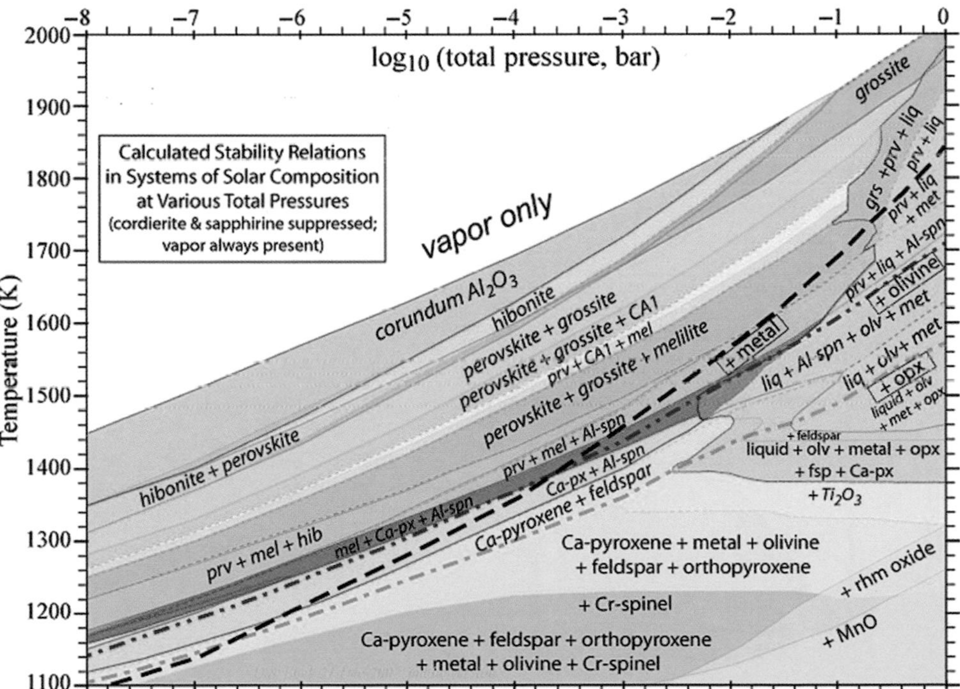

Figure 11.19 Equilibrium stability regime of compounds in a nebula of solar elemental composition calculated using chemical thermodynamics. As the (inner) solar nebula cooled down from high temperatures, different compounds will condens out sequentially. The major condensates of metal, olivine, and orthopyroxene are indicated by dashed, dashed-dot-dot, and dashed-dot lines, respectively. Figure taken from [88].

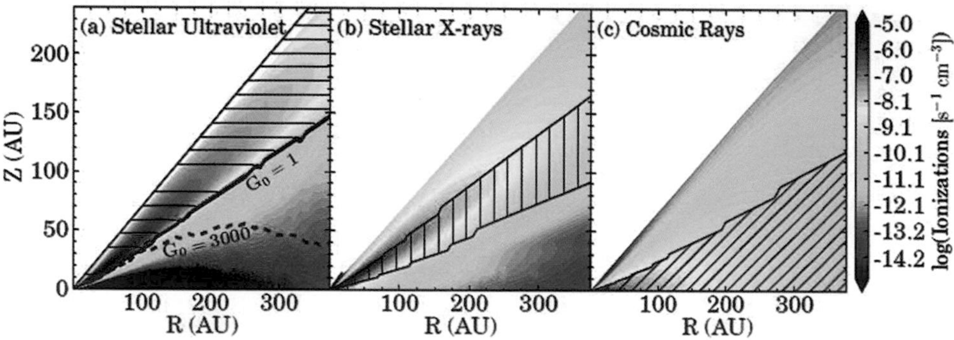

Figure 11.22 The contribution of stellar UV, X-rays, and cosmic rays to the total ionization rate in protoplanetary disks. Colored contours show the volumetric ionization rate due to each source on the same scale. Hatched areas illustrate where each ionizing agent contribute at least 30% of the ionization. Figure taken from [67].

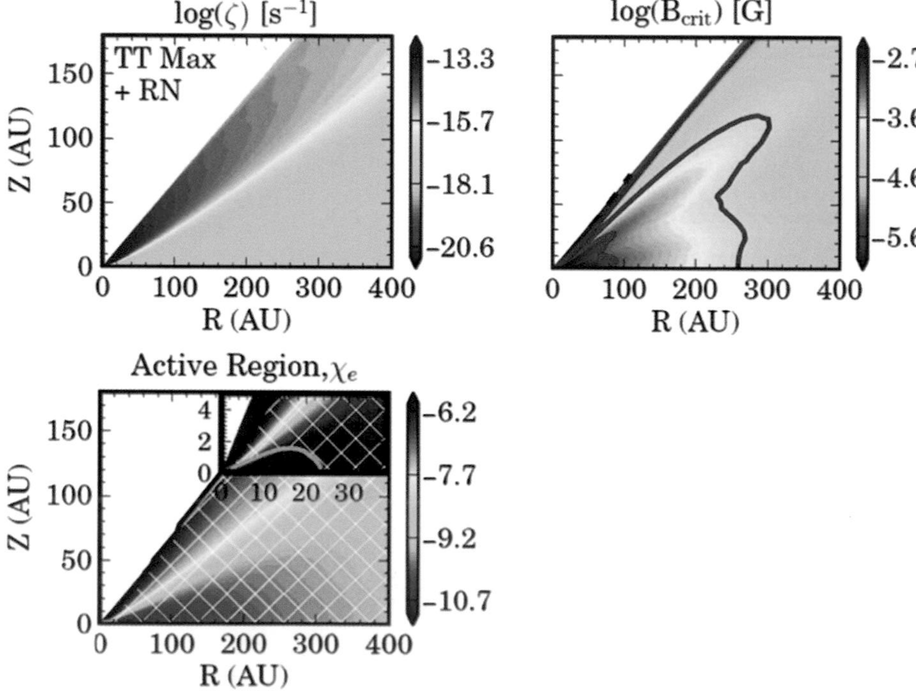

Figure 11.26 Top left: The ionization rate including cosmic rays, X-rays, and radionuclides. Top right: The critical magnetic field involved in the A_m criteria. Magenta, blue, and red lines denote 1 mG, 100 μG, and 10 μG, respectively. Lower panel: MRI active regions are cross-hatched. The inset shows the inner disk with the region that is inactive to MRI (i.e. satisfies R_e but not A_m criteria) outlined in orange. Figure adapted from [67] and kindly provided by Ilse Cleeves.

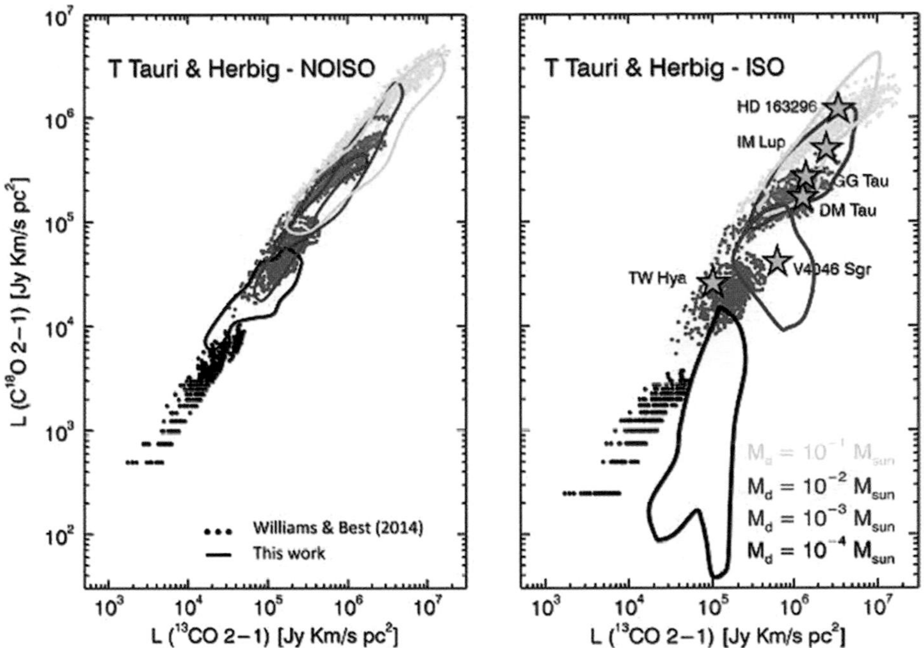

Figure 11.27 Calculated line luminosities of $C^{18}O$ versus ^{13}CO for a large grid of models. Different gray scales represent models with different disk gas masses. Small dots show parameterized models and the solid lines outline model results with a more extensive chemistry. There are also differences in the adopted disk structure between these models. Left: models where isotope-selective processes are not considered (NOISO). Right: models where isotope-selective processes are implemented (ISO). The parametrized models have a $C^{18}O$ abundance reduced by a factor of three to "mimick" the effects of isotope-selective photodissociation. Observations of six well-studied disks are indicated by stars. Figure taken from [199].

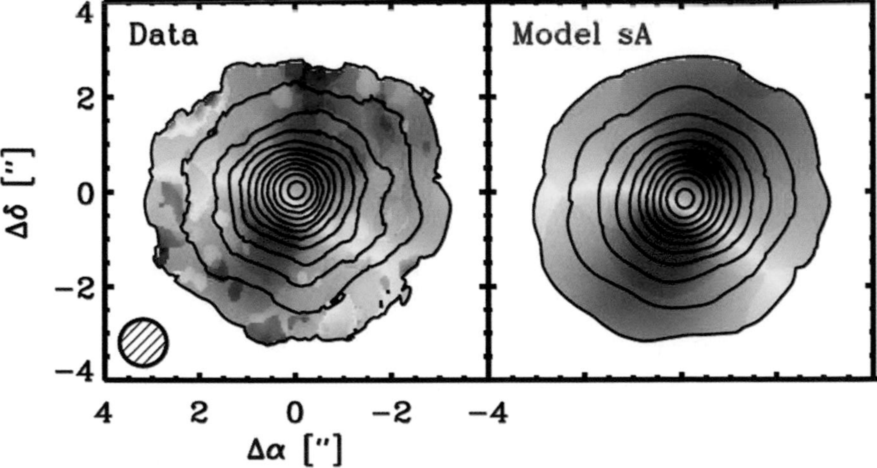

Figure 11.29 Velocity moment map of the observed (left) and model (right) CO $J = 3 - 2$ emission of the TW Hya disk. Contours are the velocity integrated CO intensity (at 0.4 Jy km/s, corresponding to 3σ). Color scale represents the intensity-weighted velocity. Figure taken from [12].

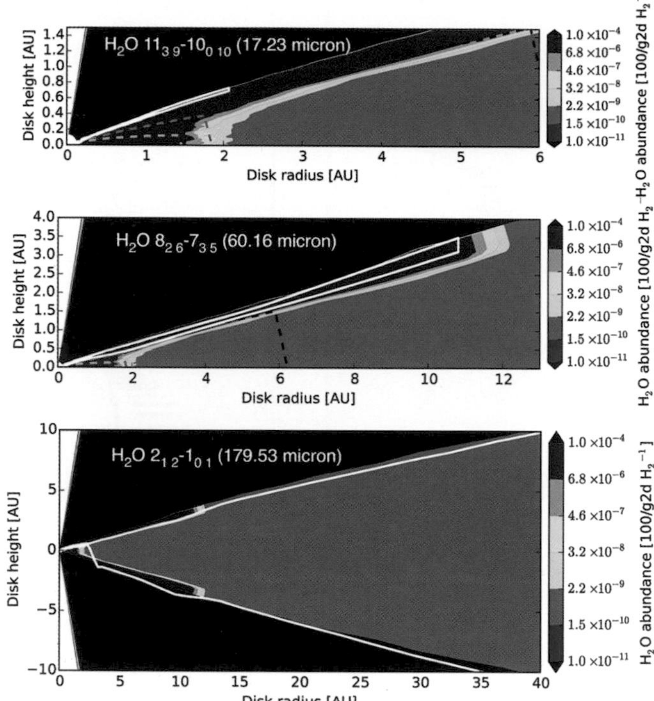

Figure 11.30 Line-emitting regions for three water lines superimposed on the fractional water abundance (color scale). The solid lines indicate the disk region responsible for 10 and 90% of the calculated intensity at line center. Dashed curves indicate water column densities of 10^{18}, 10^{19}, and 10^{20} cm^{-2}, respectively. Observer faces the disk from the top. Figure taken from [32] and kindly provided by Klaus Pontoppidan and Sandra Blevins.

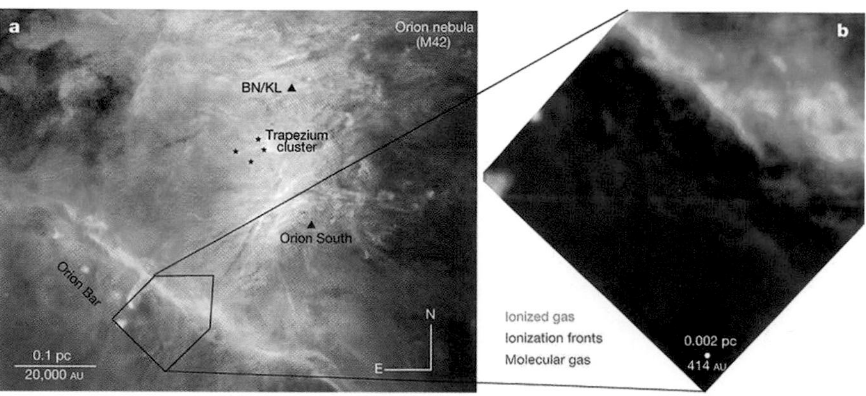

Figure 11.45 Left: The Orion Bar: The star, θ^1 Ori C in the Trapezium star cluster has created an ionized gas region, here traced by the [SII] λ6731 Å line. To the southwest, the ionized gas is separated from the molecular cloud by the Orion Bar PDR. The ionization front – traced by the [OI] λ 6300 Å line – is the boundary of the PDR. The molecular gas in the PDR is traced by the HCO^+ $J = 3-2$ line. Right: Blow up of a small portion of the Orion Bar. Here red is the HCO^+ $J = 4-3$ transition. Figure taken from [124].

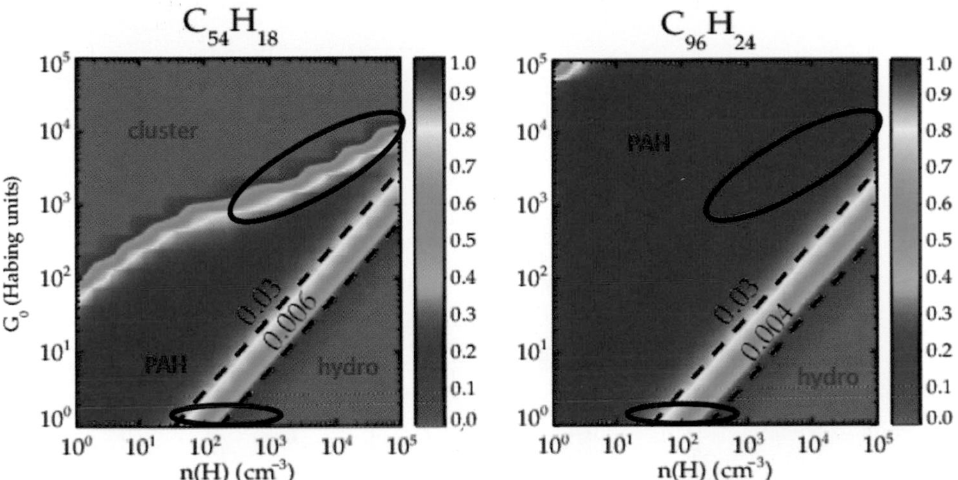

Figure 11.47 The characteristics of circumcoronene (left) and circumcircumcoronene (right) in PDRs. For part of the $n_H - G_0$ parameter space (indicated by PAH), PAHs will be fully hydrogenated; for high (atomic) H densities, PAHs will be "superhydrogenated" (labeled hydro); for high FUV fields PAHs will be rapidly, completely dehydrogenated (labeled clusters). The species in this cluster regime will quickly start to lose C_2 units and be destroyed. The transition between these different regimes is very sharp in the $n - G_0$ plane and the precise location depends on the molecule under consideration. H_2 formation from PAHs will only occur in the transition region between pure C-clusters and normal PAHs. H_2 formation from "superhydrogenated" PAHs can occur throughout the full hydro regime. The ovals indicate conditions typical for PDRs and for the diffuse ISM. Figure adapted from [14].

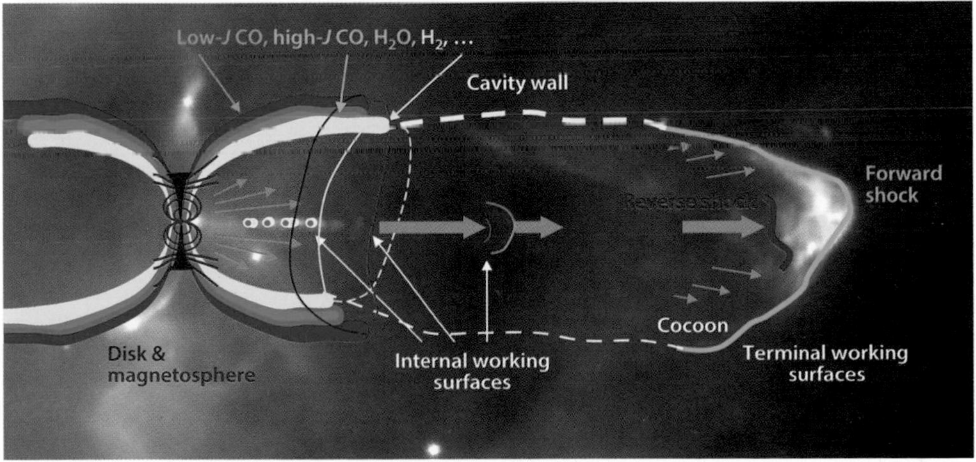

Figure 11.48 Schematic figure illustrating characteristic features of protostellar outflows. The outflow originates from a magnetized (red contours) disk (purple) (both exaggerated in size). Forward shocks (green) sweep up the surrounding material while the reverse shock (magenta) slows down the outflow/jet. Molecules are confined to the outflow walls but can also trace the jet in the youngest flows. Once a flow breaks out of the cloud, it becomes mainly atomic. Stellar photons can freely travel through the cavity and illuminate its walls, creating a PDR. Typical molecular emission regions are indicated. Figure taken from [18].

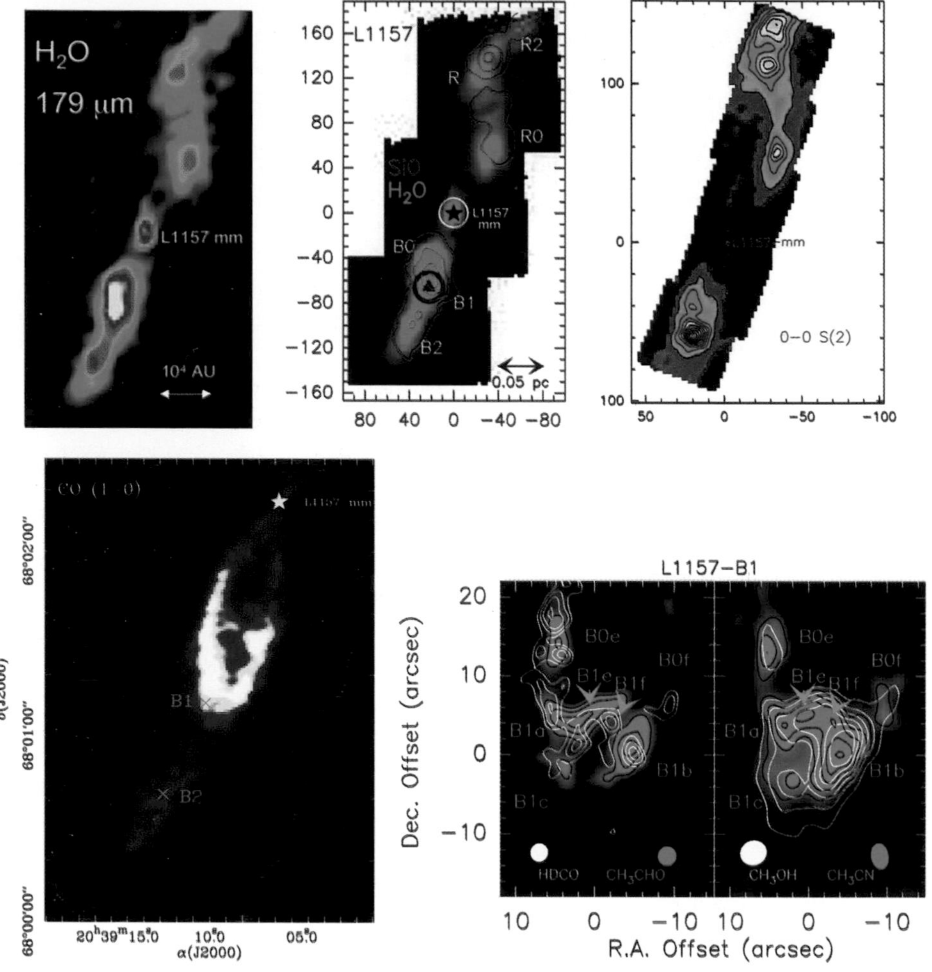

Figure 11.52 The interaction of the outflow from the low mass protostar, L1157mm, with its environment. Top row: The red and blue lobes. The 179 μm H_2O $2_{12}-1_{01}$ transition (left), the SiO contours on top of the H_2O map (center), and the pure rotational H_2 0–0 S(2) emission (right). Bottom row: Blow up of the blue lobe in CO $J = 1-0$ (left). Blow up of just the B1 shock in acetaldehyde with superimposed HDCO contours (middle) and methyl cyanide with superimposed methanol contours (right). CO figure kindly provided by Gemma Busquet. Blow up of the B1 shock region kindly provided by Claudio Codella. Taken from [43, 70, 214, 215].

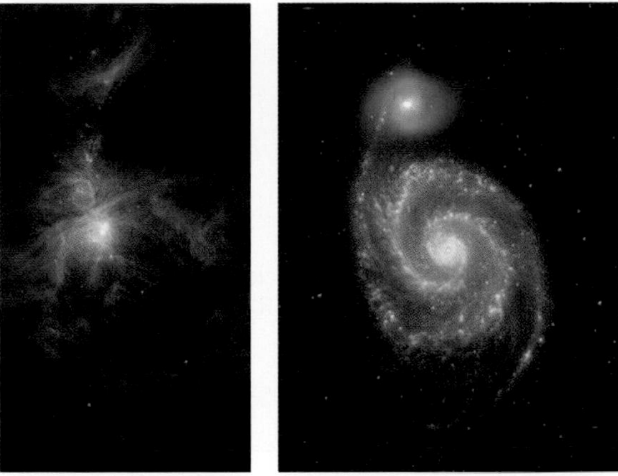

Figure 12.2 The mid-IR emission of (almost) all objects is dominated by fluorescence by large PAH molecules. Composite mid-IR images where the 8 μm (red) image traces primarily emission by PAH molecules in the 7.7 μm feature (see Figure 12.1), while the 4.5 μm image (green) shows ionized gas and dust, and stars show up in 3.6 μm (blue). Left: The Orion star-forming region where massive stars at the center of a star cluster, the Trapezium ionize the gas on a scale of $\simeq$0.2 pc. Far-UV photons from these stars pump PAH molecules on much larger scale sizes ($\simeq$2 pc). Right: The whirlpool galaxy, M51, where the whole ISM on a scale size of 10 kpc is set aglow in the light of the PAHs. The spiral arms are dotted by individual regions of massive star formation akin to the Orion region (left). Images obtained with IRAC on the Spitzer Space Telescope and courtesy of NASA, JPL-Caltech.

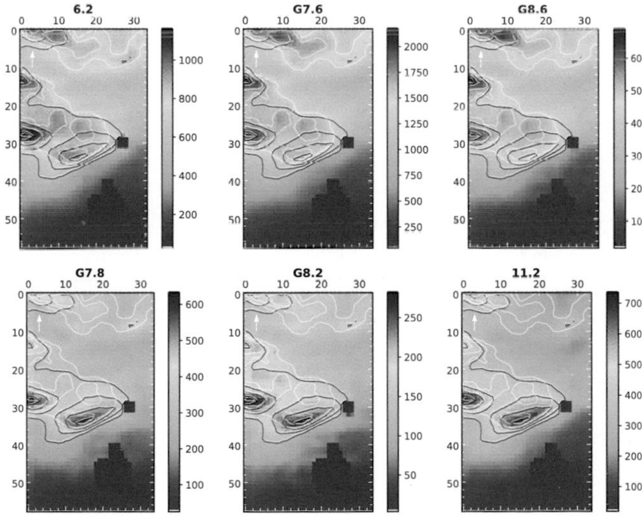

Figure 12.6 Spatial distribution of the emission at selected wavelengths in the 5–15 μm region in the Southern PDR of NGC2023. The AIB emission has been continuum subtracted. Band intensities are in units of 10^{-8} W m^{-2} sr^{-1}. As a reference, the intensity profiles of the 11.2 and 7.7 μm emission features are shown as contours in, respectively, black and white. The label for each panel indicates the emission component. The "G" in front of a label refers to a decomposition of the feature in Gaussian components. Note that the spatial distribution reveals two classes of features with different spatial distribution: 6.2, G7.6, and G8,6 versus G7.8, G8.2, and 11.2. The dark purple rectangle indicates an area removed because of the presence of a confusing protostar. The other dark purple area indicates a region of limited signal-to-noise. Figure adapted from [96].

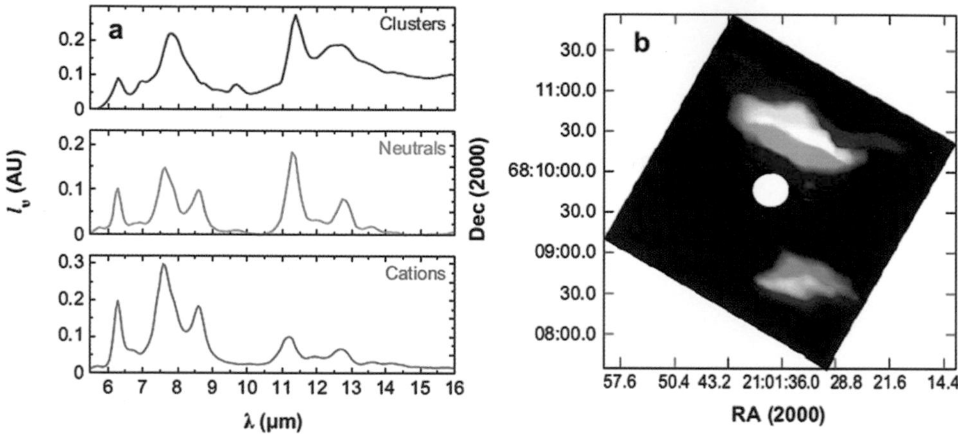

Figure 12.7 (a) Three identified components in the decomposition analysis of ISOCAM spectral-spatial maps of NGC 7023. The components are normalized by their integrated intensity. (b) These components show different spatial distributions in this source and have been attributed to ionic (blue) and neutral (green) polycyclic aromatic hydrocarbon molecules and clusters (red), respectively. In the color scheme used, red and green combine to yellow. Figure adapted from [12, 104].

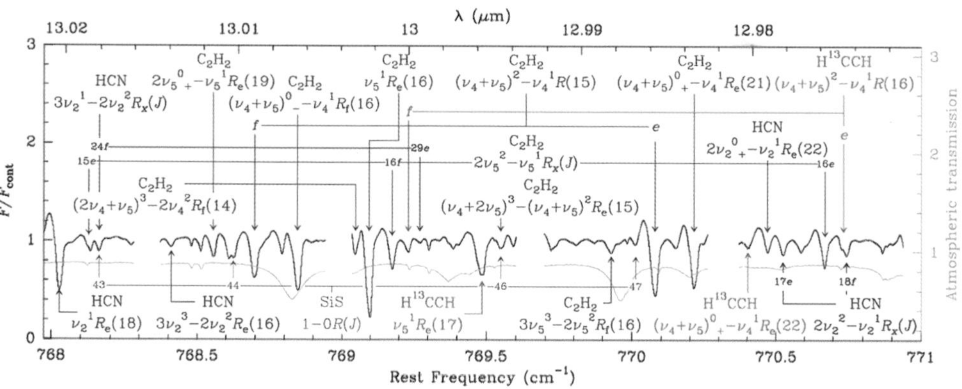

Figure 12.26 A small portion of the 11–14 μm spectrum of the AGB star, IRC+10216, obtained at a resolution of 10^5 using the TEXES instrument at the IRTF, resolving individual transitions of C_2H_2, HCN, SiS, and their isotopologues. The rotational-vibrational states involved are indicated. The atmospheric transmission is shown as a light gray curve. Figures adapted from [42] and kindly provided by Dr. Pablo Fonfria.

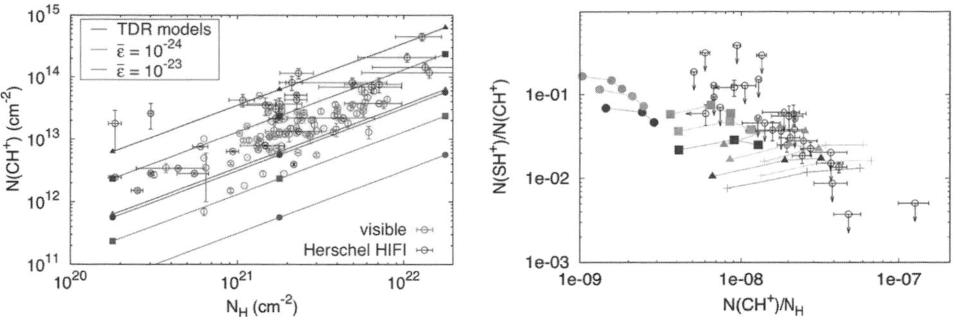

Figure 9.11 Left: Observed CH^+ column densities are compared to TDR models computed for two different average dissipation rates ϵ and three different densities: 10 (crosses), 30 (triangles), 50 (squares), and 10^2 (circles) cm^{-3}. Right: Observed SH^+/CH^+ column density ratios as a function of the average CH^+ abundance. Model predictions (filled symbols) computed for $A_v = 0.1$, 0.2, and 0.4 mag and for several densities: 20 (crosses), 30 (triangles), 50 (squares), and 100 cm^{-3} (circles). Left: The top and bottom curves are for $\epsilon_{tur} = 10^{-23}$ and $\epsilon_{tur} = 10^{-24}$ erg cm^{-3} s^{-1}, respectively. Along each curve, the rate-of-strain varies between 10^{-11} and 10^{-10} s^{-1} from right to left. Reproduced by permission from [37]

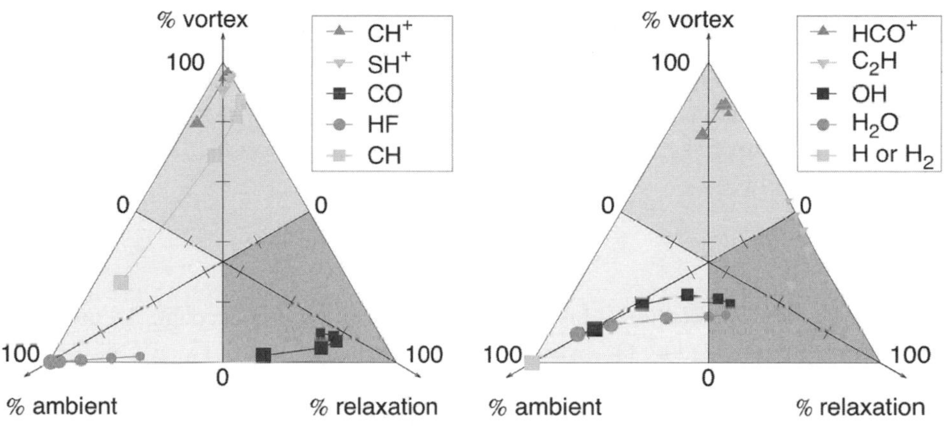

Figure 9.12 Ternary phase diagrams illustrating the contribution of the three different "phases," vortex (top), ambient (left), relaxation (right). The line connecting each corner to its base ranges from 100% to 0% contribution. Models are computed for a total $A_v = 1$ magn and $n = 20$, 30, 50, 100, and 300 cm^{-3} (increasing size of the symbol). Reproduced by permission from [37]

Particularly for lines-of-sight dominated by low densities, the contribution from TDRs can be quite important for a wide variety of species. This is not surprising as abundances in ambient gas typically scale with n_o/G_0 and gas phase chemistry models have great difficulty explaining observations at low densities. Looking at it from the opposite perspective, abundances of many species scale well with each other, reflecting a common dependence on the physical conditions (i.e. n_o/G_0). However, there is quite some scatter and that scatter may reveal the general importance of TDRs in opening up new, high temperature reaction routes. Finally, TDRs may also be the source of rotational excited H_2 in the diffuse ISM. In summary, chemistry provides a memory of the warm dissipation phase, which persists for some 2000 yr and, in order to explain the observations, these (memories of) turbulent dissipation regions fill a few percent of the sightlines in the diffuse ISM.

9.5 The Cosmic Ray Ionization Rate

Cosmic ray ionization is the cornerstone of many gas phase chemistry routes and conversely, molecular observations can be used to determine the cosmic ray ionization rate. In view of their simple chemistry, this is particularly straightforward for diffuse molecular clouds. There are a number of methods that have been used in the past to derive the ionization rate from the observations. In each case, the formation rate of a species is related back to the cosmic ray ionization of H or H_2. Using estimated destruction rates of the molecule under consideration, the observed abundances result then in an estimate of the cosmic ray ionization rate, ζ_{CR}. Here, the focus is on the well-studied diffuse cloud toward ζ Oph summarized in Table 9.1.

9.5.1 *Protonated Molecular Hydrogen, H_3^+*

Chemically, the simplest determination is through the observations of H_3^+, which starts with the cosmic ray ionization of H_2 to H_2^+. The latter reacts quickly with H_2 to protonated H_2, which is then destroyed by dissociative recombination. The primary cosmic ray ionization rate is now given by,

$$\zeta_{CR} = \frac{k_e \, n\,(e)\; n\left(H_3^+\right)}{2.3\, n\,(H_2)}, \tag{9.25}$$

where the numerical factor (2.3) takes secondary ionization into account. Assuming that the electrons are provided by C^+, we can write this as,

$$\frac{\zeta_{-16}}{n} \simeq 10^5 \, \frac{N\left(H_3^+\right)}{N\,(H_2)}, \tag{9.26}$$

where ζ_{-16} is the primary cosmic ray ionization rate in units of 10^{-16} s^{-1}, an electron recombination coefficient of $k_e = 10^{-6}\, T^{-0.45}$ has been adopted, and we have assumed that C^+ provides the electrons with an abundance of 1.4×10^{-4}. Thus, H_3^+ observations provide the cosmic ray ionization rate per H-nucleus relative to the density.

Turning to the observations of ζ Per with $N\left(H_3^+\right) = 8 \times 10^{13}$ cm^{-2} and $N\,(H_2) = 5 \times 10^{20}$ cm^{-2}, we have $\zeta_{-16}/n \simeq 1.8 \times 10^{-2}$ cm^3 s^{-1} and, for a density of 250 cm^{-3} along this line of sight, this results in $\zeta_{-16} \simeq 4.5$ s^{-1}. Actual space densities – derived from C_2 observations – are available for a limited number of lines of sight with H_3^+ and H_2 data. These are compared to this simple model in Figure 9.13. The data indicate that ζ_{CR} is in the range $1{-}50 \times 10^{-17}$ s^{-1}, depending on sightline. This large variation in ζ_{CR} is not understood.

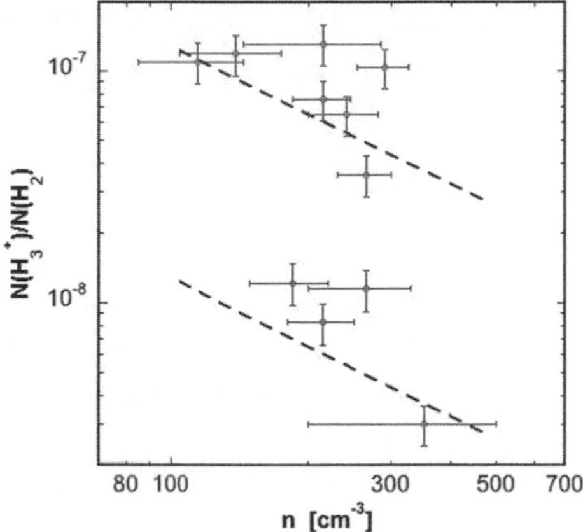

Figure 9.13 Observed column density ratio of H_3^+ to H_2 [1, 50] as a function of the density (derived from C_2 observations [106]. The curves correspond to cosmic ray ionization rates of 10^{-17} and 10^{-16} s^{-1}. Adapted from the discussion in [82]

9.5.2 Argonium, ArH^+

The putative importance of cosmic ray ionization to start nitrogen chemistry in diffuse clouds has been discussed in Section 9.3.3. Here, we mention the role of cosmic rays in the chemistry of argon. Argon has an ionization potential of 15.8 eV and cannot be photoionized in diffuse clouds. Furthermore, as the proton affinity of Ar is less than that of H_2 by 6400K, this reaction channel is also closed in diffuse clouds. Therefore, ionization by cosmic rays is the only way to start Ar chemistry at a rate of $\simeq$10 times that of H. Ar^+ reacts quickly with H_2 to form argonium, but ArH^+ likes to transfer its proton to H_2 ($k = 8 \times 10^{-10}$ cm^3 s^{-1}). We then have,

$$\frac{\zeta_{CR}}{n} = \frac{\zeta_{CR}}{\zeta_{CR}\,(\mathrm{Ar})} \frac{X\left(\mathrm{ArH}^+\right) X\left(\mathrm{H}_2\right) k}{\mathcal{A}\,(\mathrm{Ar})}. \tag{9.27}$$

This can be writen as,

$$\frac{\zeta_{-16}}{n} \simeq 3 \times 10^{-3} \left(\frac{X\left(\mathrm{ArH}^+\right)}{10^{-10}}\right) \left(\frac{X\left(\mathrm{H}_2\right)}{10^{-4}}\right), \tag{9.28}$$

where we have adopted an Ar abundance of $\mathcal{A}\,(\mathrm{Ar}) = 3 \times 10^{-6}$. With the observed ArH^+ abundance (3×10^{-10}; Table 9.1) and a ζ_{-16}/n ratio inferred from H_3^+ observations ($3 \times 10^{-4} - 10^{-2}$), we conclude that the H_2 abundance is very low $< 10^{-4}$ in the layer where ArH^+ is observed. This is only the outer skin of diffuse clouds ($N_{total} \lesssim 10^{20}$ cm^{-2}, Figure 9.3). As the ArH^+ abundance is so sensitive to the H_2 fraction, this species does not provide a good measure of the cosmic ray ionization rate. Instead, it traces well the presence of pure atomic gas.

9.5.3 HD and OH^+: The Role of Large Molecules

There are a number of species whose abundances have been linked in the past to the cosmic ray ionization of H; notably OH^+, OH, and HD. In this, cosmic ray ionization of H is followed by near-resonance charge transfer of H^+ with D or O. This leads to the relevant hydrides, HD, OH, and OH^+ (see Figure 9.14). In this way, molecular abundances are directly tied back to cosmic ray ionization. However, the abundance of the key ion involved, H^+, depends directly on mutual neutralization reactions with large molecular anions and rather than the cosmic ray ionization rate, HD, OH^+, and OH can be better used to study the role of large molecules in the chemistry of diffuse interstellar clouds.

Consider OH^+, whose formation is initiated by the ionization of H (Figure 9.14). This ionization is passed on to O^+, and this can lead to OH^+, which can be observed through its rotational transitions in the sub-millimeter and its electronic transitions in the near-UV. In the absence of large molecules, every ionization of H leads eventually to OH^+, which is then destroyed through dissociative recombination. The cosmic ray ionization rate is then given by,

$$\frac{\zeta_{CR}}{n} = \left(X\left(\mathrm{e}\right)k_6 + X\left(\mathrm{H_2}\right)k_7\right)\frac{N\left(\mathrm{OH^+}\right)}{N\left(\mathrm{H}\right)}. \tag{9.29}$$

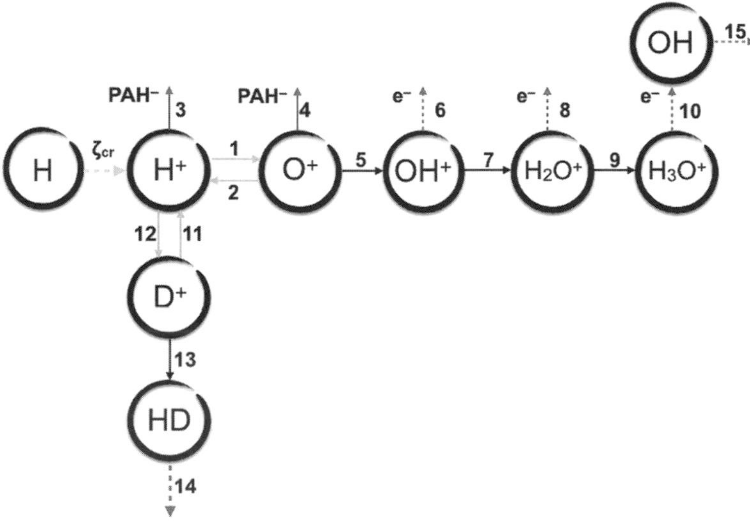

Figure 9.14 Key reactions in the chemistry of atomic ions. Chemistry is initiated by cosmic ray ionization of H. The back and forward arrows indicate charge exchange reactions that set up a rapid steady state between the ions involved. The rightward-pointing solid arrows are reactions with H_2. Upward pointing arrows are neutralization reactions involving either e^- or PAH anions. The end products OH and HD are destroyed through photodissociation. Reactions are identified by number for ease of use in the text in this section.

Recalling the low H_2 abundance implied by the OH^+, H_2O^+, and H_3O^+ and inserting the value for the dissociative recombination rate coefficient, $k_6 = 9.2 \times 10^{-8}$ cm^3 s^{-1} at 50 K, we have,

$$\frac{\zeta_{CR}}{n} = 1.3 \times 10^{-11} \frac{N\left(\mathrm{OH}^+\right)}{N\left(\mathrm{H}\right)}. \tag{9.30}$$

UV and sub-millimeter derived abundances of OH^+ range from 6×10^{-9} to 3×10^{-8} and,

$$\zeta_{CR}/n \simeq 2 \times 10^{-19}\ \mathrm{cm}^3\ \mathrm{s}^{-1} \tag{9.31}$$

This is an order of magnitude less than derived from observations of H_3^+ (Section 9.5.1). In regions where H_2 is abundant – perhaps more relevant in the comparison with H_3^+ – OH^+ will quickly react on with H_2 to ultimately yield H_3O^+, which dissociatively recombines at a rate that is an order of magnitude faster then the dissociative recombination rate coefficient for OH^+. The branching ratio for OH formation in this is 0.18. OH is destroyed by photodissociation. We have then,

$$\frac{\zeta_{CR}}{n} = \frac{k_{15}}{0.18n} \frac{N\left(\mathrm{OH}\right)}{N\left(\mathrm{H}\right)}. \tag{9.32}$$

With $k_{15} = 3.9 \times 10^{-10} \exp\left[-2.2A_v\right] G_0\ \mathrm{s}^{-1}$, we arrive at,

$$\frac{\zeta_{CR}}{n} \simeq 2.5 \times 10^{-17} \frac{G_0}{n} \simeq 2.5 \times 10^{-19}\ \mathrm{s}^{-1}, \tag{9.33}$$

where we adopted the column densities ($N\left(\mathrm{OH}\right)/N\left(\mathrm{H}\right) = 6.5 \times 10^{-8}$), and G_0/n ($\simeq 10^{-2}$) for ζ Per for direct comparison to the H_3^+ observations. This is about an order of magnitude lower than the H_3^+ measurement. Thus, both OH^+ and OH observations imply that only a small fraction of the cosmic ray ionizations flow down through the O^+ channel.

We can also address the cosmic ray ionization rate with deuterium chemistry (Figure 9.14). Again, ignoring mutual neutralization of H^+ with PAH anions for now, the fast charge exchange of H^+ with D followed by reaction of D^+ with H_2 forms HD.[9] HD is then destroyed through photodissociation. Realizing that the charge exchange reactions between H^+ with O and D and D^+ and O^+ with H are very rapid and set up a steady state that connects the relative abundances of the ions through the exothermicity of the reactions, we have,

$$\frac{X\left(\mathrm{D}^+\right)}{X\left(\mathrm{D}\right)} = \frac{X\left(\mathrm{O}^+\right)}{X\left(\mathrm{O}\right)} \exp\left[186/T\right]. \tag{9.34}$$

This allows us to connect the cosmic ray ionization rate of H with the photodestruction of HD,

$$\frac{\zeta_{CR}}{n} = \frac{N\left(\mathrm{HD}\right)}{N\left(\mathrm{H}\right)} \frac{X\left(\mathrm{O}\right)}{X\left(\mathrm{D}\right)} \frac{k_5}{k_{13}} \frac{k_{14}}{n} \exp\left[-186/T\right]. \tag{9.35}$$

[9] This reforms H^+ and does not influence the ionization flow discussed above.

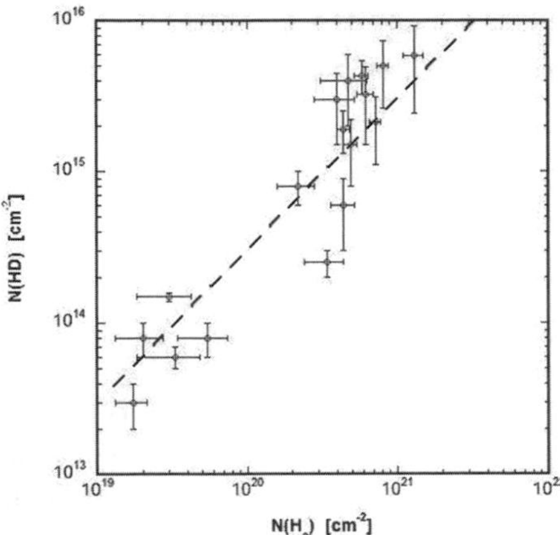

Figure 9.15 The observed relation between the HD and H_2 column densities in the diffuse ISM. Adapted from [56]

We will adopt the unshielded UV photodissociation rate of HD ($k_{14} = 2.6 \times 10^{-11} G_0$ s^{-1} and k_5 and k_{13} are 1.7×10^{-9} and 2.1×10^{-9} cm^3 s^{-1}, respectively). The observed HD abundance toward ζ Oph (2.2×10^{-6}; Table 9.1) is very typical for diffuse clouds (Figure 9.15) and we find,

$$\frac{\zeta_{CR}}{n} = 5.8 \times 10^{-16} \frac{G_0}{n} \exp\left[-186/T\right] \simeq 5.4 \times 10^{-19} \text{ cm}^3 \text{ s}^{-1}, \qquad (9.36)$$

where the right-hand side adopts the physical conditions appropriate for ζ Oph ($n/G_0 = 90$ cm^{-3}, $T = 75$ K). This is very sensitive to the temperature of the gas and ignores dust extinction. Still, we can conclude that the observations seemingly imply a much lower cosmic ray ionization rate then the H_3^+ observations give ($1{-}30 \times 10^{-17}$ s^{-1}).

From this analysis, it is clear that there is a recombination channel for H^+ – mutual neutralization with PAH anions – that rapidly siphons off the H^+ rather than initiating the formation of HD or oxygen-bearing hydrides. Let's analyze the ionization flow from this point of view. In this flow, the reactions between H^+ and O and O^+ and H are very rapid and will set up an equilibrium that determines the abundance ratio of these species. Thus,

$$\frac{X\left(\mathrm{O}^+\right)}{X\left(\mathrm{H}^+\right)} = \frac{X\left(\mathrm{O}\right)}{X\left(\mathrm{H}\right)} \frac{k_1}{k_2} = 3 \times 10^{-4} \exp\left[-227/T\right], \qquad (9.37)$$

where the rate identifiers refer to Figure 9.14, an O abundance of 3×10^{-4} has been adopted, and the reaction rate coefficients are linked through the exothermicity of the forward and backward reaction. At 50K, the O^+ over H^+ ratio is then $\simeq 3.3 \times 10^{-6}$. This might be slightly higher if H_2 is accounted for in this evaluation. We can now compare the ionization flow in Figure 9.14: H^+ with PAH anions and the reaction of O^+ with H_2 that will lead eventually to a dissociative recombination. Comparing these two channels, we have,

$$R_{mn} = \frac{X\left(\mathrm{O}^{+}\right)}{X\left(\mathrm{H}^{+}\right)} \frac{X\left(\mathrm{H}_2\right)}{X\left(\mathrm{PAH}^{-}\right)} \frac{k_5}{k_3} \simeq 5.2 \times 10^{-2} \frac{X\left(\mathrm{H}_2\right)}{f\left(\mathrm{PAH}^{-}\right)}, \tag{9.38}$$

where we have used $k_5 = 1.7 \times 10^{-9}$ cm^3 s^{-1}, $k_3 = 3.7 \times 10^{-7}$ cm^3 s^{-1}, and we have adopted a PAH abundance of 3×10^{-7}. The PAH anion fraction, $f\left(\mathrm{PAH}^{-}\right)$ is set by the balance between electron attachment and photodissociation (cf. Section 12.8.1),

$$f\left(\mathrm{PAH}^{-}\right) = \left(1 + 300/n_o\right)^{-1}, \tag{9.39}$$

where we have used standard ionization (2.2×10^{-8} s^{-1}) and electron attachment (4.4×10^{-7} cm^3 s^{-1}) rates for PAHs of 50 C-atoms (cf. Chapter 8). This results in,

$$R_{mn} \simeq \frac{15\,\mathrm{cm}^{-3}}{n} X\left(\mathrm{H}_2\right), \tag{9.40}$$

and it is clear that the mutual neutralization reaction with PAH anions dominates. We can now define an efficiency, ϵ_{OH^+},

$$\epsilon_{OH^+} = \left(1 + R_{mn}\right)^{-1}, \tag{9.41}$$

which gives the fraction of the H^+ ionizations that flow down to O^+. The cosmic ray ionization rates derived from the OH^+, OH and HD observations (Eq. (9.31), Eq. (9.33), and Eq. (9.36)) have then to be corrected for this efficiency. Conversely, by adopting the ζ_{CR}/n value from the H_3^+ observations, $\simeq 10^{-16}$, we conclude that,

$$R_{mn} \simeq 2 \times 10^{-3}. \tag{9.42}$$

Hence, mutual neutralization of H^+ with PAH anions is very important and only a small fraction of all cosmic ray ionization flows to oxygen-hydrides. This mutual neutralization reaction will be very important for all atomic ions that do not readily make hydride cations (e.g. $\mathrm{H}^+, \mathrm{C}^+, \mathrm{S}^+, \ldots$).

9.6 Diffuse Interstellar Bands and the Organic Inventory of the ISM

The Diffuse Interstellar Bands (DIBs)[10] are a set of several hundred absorption features due to molecular species (Figure 9.16). They were recognized as due to interstellar material in the earliest spectra. By now some 500 DIBs have been catalogued and new ones are reported every year. The strongest and best-studied bands occur in the visible at 4430, 5778, 5780, 5796, 6177, 6196, 6284, and 6614 Å. But as new technologies (e.g. CCDs) developed and new observing windows have opened up (e.g. far-red and near-infrared), our knowledge of the DIB spectrum has slowly but steadily expanded over the last 20 years. DIBs are now known to be present from the near-UV to the far-red and near-IR. Surveys are incomplete for very broad DIBs. Broad DIBs – such as the 4430 Å albeit much weaker – can be expected to be omnipresent. DIBs have also been detected in interstellar

[10] The name diffuse interstellar bands was dubbed early on because the bands are broader than atomic lines and the absorption arises from interstellar gas a long the line of sight.

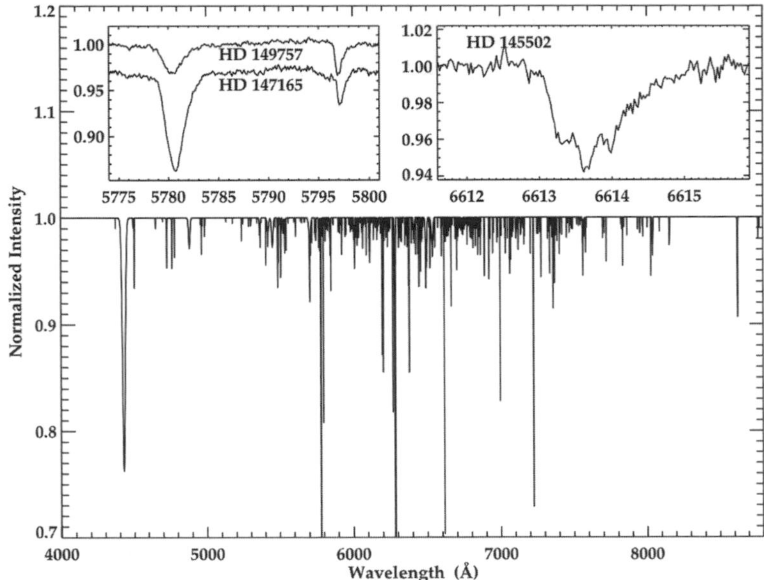

Figure 9.16 Some 500 DIBs are present in visible absorption spectra of the interstellar medium. The synthesized spectrum of Diffuse Interstellar Bands within the 4300–6800 Å wavelength range observed toward BD +63 1964 [21] illustrates the great variety in relative strength and width of the DIBs. The right insert shows the detailed profile of the 6614 Å DIB and its associated substructure observed toward the star HD 145502. The left insert illustrates the large variations in the strength of the DIB bands relative to each other for two well-studied DIBs (5780 and 5797 Å) on the basis of observations of the stars, HD 149757 and HD 147165. Figure courtesy of J. Cami

absorption spectra from extragalactic objects and have been used as quantitative tracers of interstellar gas.

9.6.1 DIB Characteristics

The observed width of the DIBs varies from ∼0.5−30 Å but is typically 0.7 Å. Their strength relative to the continuum is expressed in terms of the equivalent width,

$$W_\lambda = \int \frac{(F_c - F_o)}{F_c} d\lambda \tag{9.43}$$

with F_c and F_o the continuum and observed flux. Most of the DIBs are weak to very weak ($W_\lambda/N_H \simeq 10^{-24}$ Å H-atom^{-1}) but a subset of 30 have appreciable equivalent width ($W_\lambda/N_H \simeq 10^{-23}$ Å H-atom^{-1}) and a few are strong ($W_\lambda/N_H \simeq 10^{-22}$ Å H-atom^{-1}; Figure 9.17). Surveys are likely incomplete at the weak end and have not yet reached the spectral "confusion" limit but it is conceivable that there is a (very weak) DIB at every wavelength. Typically, equal ranges in W contribute equally to the total integrated equivalent width of all DIBs.

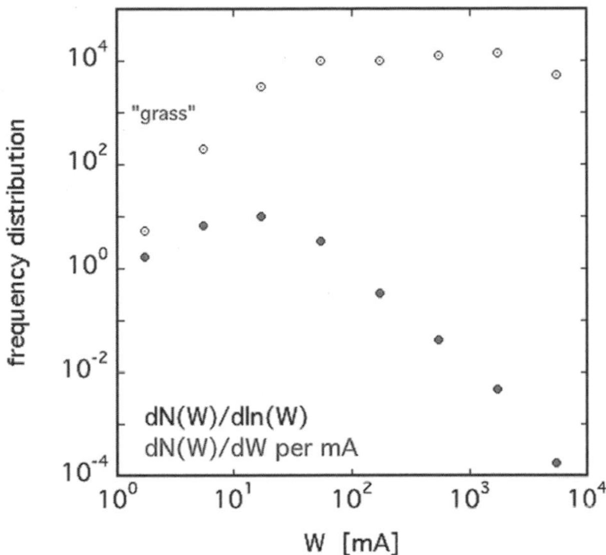

Figure 9.17 The frequency distribution function of DIBs with a given equivalent width (W) in the wavelength range 3900–8100 Å as measured toward the prototypical DIB-survey star, HD 183143, with $N_H = 7.2 \times 10^{21}$ cm^{-2} [45]. Closed and open symbols refer to dN/dW and WdN/dW, respectively. The label "grass" indicates where spectral confusion becomes important. Incompleteness sets in around 30 mÅ.

The DIBs correlate quite well with dust extinction and with atomic H. The correlation with H_2 is much poorer and, in relative terms, DIB strength clearly decreases inside molecular clouds. So, the DIB carriers prefer the conditions of diffuse atomic clouds. The DIBs also correlate reasonably well with each other. However, this correlation is not perfect and small but real variations in relative strength are present for all DIB pairs. Of course, these studies are only meaningful for the stronger DIBs. While, in the past, the DIBs have been divided into a few families, it is now clear that these are, at best, dysfunctional families: DIBs can be divided into groups, and DIBs within each group correlate better with each other than they correlate with outsiders. Nevertheless, in the end it is each (strong) DIB for itself (cf. Figure 9.16). This divison into families could be linked to the interaction of their carriers with UV radiation (e.g. related to the σ Sco (λ5797) or ζ Oph (λ5780) far-UV extinction curves) or the presence of molecules (e.g. the C_2 family, λ4964, 5513). While the actual underlying process(es) are unclear, DIB carriers are (to some extent) sensitive to the local physical conditions through, for example, the ionization balance or the presence of peripheral functional groups.

Following early suggestions – actually ging back to the 1930s – the DIBs are now generally considered to be carried by molecules rather than dust grains. This assessment is based upon the constancy of the peak wavelength, the general invariance of their profiles, and the presence of substructure reminiscent of rotational structure in some DIBs

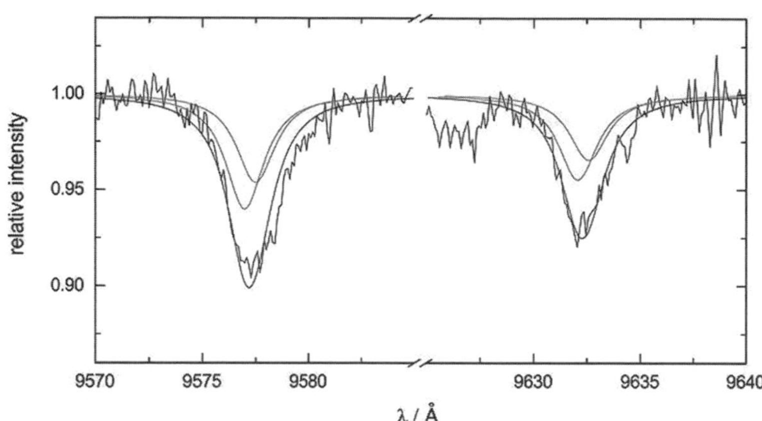

Figure 9.18 Fit to the two far-red DIBs observed toward HD 183143 [30] with the gas phase spectrum of C_{60}^{+}. Figure taken from [6]. The 9632 Å band has been corrected for a weak stellar MgII line.

(cf. Figure 9.16). Furthermore, the enigmatic object, the Red Rectangle, shows *emission* bands at wavelengths that are near to, but not exactly at, the position of some (absorption) DIBs. With distance to the central star, the bands shift closer to the absorption position and their profiles become narrower. The molecules responsible for these bands seem to fluoresce – when the conditions are right – and the observed variations in peak position and width likely reflect a cooling down of the species with distance to the pumping star. The assignment of two far-red DIBs at 9577 and 9632 Å to the molecule C_{60}^{+} really clinches the molecular origin of the DIBs (Figure 9.18). By now five absorption bands have been assigned to C_{60}^{+} and this identification can be considered secure.

9.6.2 Abundance of the DIB Carriers

Using Eq. (4.88), we can derive abundances of the carriers, and these are summarized for the well-studied star, HD 183143, in Table 9.2. Typical measured oscillator strengths for the first transition ($S_0 \rightarrow S_1$) of small PAHs and hydrocarbon chains are quite small, 0.01–0.1. As the size of these species increases, the effective number of electrons involved in the transition is expected to increase and the oscillator strength per C atom is roughly constant at $\sim$0.01. Thus, the carriers of the strong DIBs have to be very abundant (Table 9.2). Higher transitions (e.g. $S_0 \rightarrow S_2$) have larger oscillator strengths. However, by necessity, this would imply small energy gaps ($S_1 - S_2$) and this would imply fast non-radiative decay (e.g. Fermi's golden rule; Figure 8.8; Section 8.2.2) and thus very broad bands. Higher transitions may thus be responsible for very broad bands such as the 4430 Å band (see below) but not for the DIBs in general. The implied abundances of carriers of these strong DIBs is comparable to that of simple diatomics measured in the optical (cf. Table 9.1; Figure 9.19). As the DIB carriers are likely fairly complex species, the fraction

Table 9.2 *Abundance of DIB carriers*[1]

λ Å	EW mÅ	Abundance[2] relative to H	$f_C^{3,4}$	Identification
4430	5700	$4 \times 10^{-9}/f$	1.3×10^{-3}	
5780	779.3	$4 \times 10^{-10}/f$	1.3×10^{-4}	
6284	1884.2	$7 \times 10^{-10}/f$	2.2×10^{-4}	
9577	300	3×10^{-9}	5.2×10^{-4}	C_{60}^{+}
"all"	$10^{-8}/f$	3×10^{-3}		

Notes: [1] For HD 183143. The total H-column density is 7.2×10^{21} cm^{-2}. A C abundance of 3.2×10^{-4} has been adopted. [2] f is the oscillator strength of the transition. [3] Fraction of the elemental carbon locked up in the carrier. [4] Adopting an oscillator strength of $f = 10^{-2}$ per C-atom. For C_{60}^{+} the measured cross section 5×10^{-15} cm^2 results in a column density of 2×10^{13} cm^{-2}.

of the carbon locked up in them is much higher than that of the simple diatomics. The buckyball cation, C_{60}^{+}, is a case in point, locking up some 0.05% of the elemental carbon. Conversely, small hydrocarbon chains are not good candidates for the DIBs. Abundances measured in the very shielded environment of the dark cloud core, TMC 1, are in the range of the DIB carriers. However, given the strong radiation field, these chains will have much lower abundances in diffuse clouds. HC_5N presents a case in point as the upper limit for its abundance in diffuse clouds is two orders of magnitude less then that measured for TMC 1 (Tables 9.1 and 10.1). In the gas phase, these chains are built up from acetylenic species (Section 10.5.1) and their abundance decreases rapidly with chain length (Figure 9.19). It is clear that the carriers of the DIBs have to be formed through different chemical routes (Section 12.7 and, e.g., 12.7.5).

Abundance-wise, interstellar PAHs – abundance of all PAHs corresponds to $\simeq$10% of the elemental C – could be relevant candidates. However, for the small PAHs studied in the laboratory, no match with a strong DIB has (yet) been made. Perusing Figure 9.19, we conclude that if PAHs are responsible for the strong DIBs, these PAHs must be abundant, each locking up $\simeq$1% of the elemental C (e.g. GrandPAHs) and the transitions involved must be strong ($f > 0.1$). Conversely, if the interstellar PAH family is very diverse with many species at very low abundances, they would contribute to the very weak forest of DIB transitions observed but not dominate the spectrum.

Irrespective of the identification, the handful of strong DIBs demonstrate that there is a population of very abundant and, hence, very stable molecules in the ISM. This is indicated in Figure 9.19 by the dashed box. The top of this box represents the abundances required to explain the strongest DIBs (with $f = 0.1$). The bottom of this box corresponds to abundances required to explain the weakest observed DIBs at $f = 0.1$ but, as this limit is set by sensitivity and line-to-continuum issues, this is not a real limit. There could be a myriad of species at abundances too small to be detectable. On the small molecule side, the box is limited by kinetics. Small molecules are not stable in the diffuse ISM. The precise

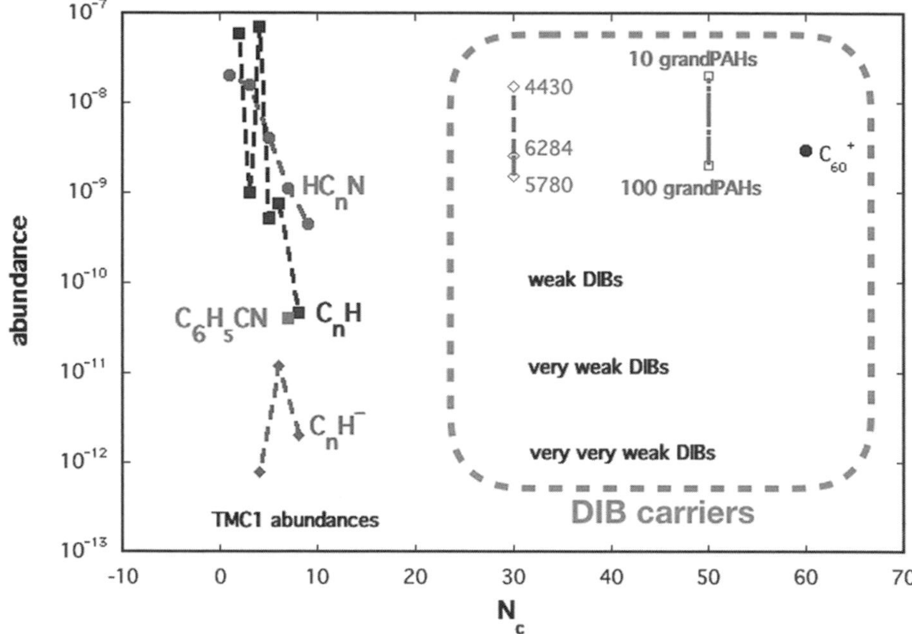

Figure 9.19 Abundances of (potential) DIB carriers. Solid symbols are observed abundances. Open symbols are inferred abundances. Linear carbon chain and benzonitrile abundances are appropriate for the highly shielded environment of the dark cloud core, TMC1 (Table 10.2). Abundances for the carriers of the three strong DIBs have been calculated assuming an oscillator strength of 0.1 (Table 9.2). The same oscillator strength has been adopted for the weak, very weak, and very very weak DIBs. Of course, these could also refer to transitions in more abundant species but with intrinsically small oscillator strengths. The C_{60}^{+} abundance has been measured toward HD 183143. GrandPAH abundances have been marked assuming that 10% of the elemental C is locked up in 10 or 100 equally abundant PAHs in the ISM.

location of this limit is uncertain. In PDRs, the smallest PAHs that could survive are $\simeq$30 C-atoms. For carbon chains or rings, that limit is probably larger as they are more prone to photodissociation. For small species, photodissociation will be key also in the diffuse ISM as is attested to by the observed limit on HC_5N in diffuse clouds. For larger species, including PAHs, the lifetime is set by processing by supernova shock waves ($v_s > 100$ km/s) and there is only a weak dependence on size. The large molecule size limit is also not well defined. There is an abundance limit, of course, but also kinetics must be involved as formation of large species will be kinetically inhibited.

9.6.3 Identification

Electronic transitions provide unique fingerprints of their carriers but this also hampers their identification, as the species has to be known before a supporting laboratory study can be made. Nevertheless, systematic trends exist that can be exploited for

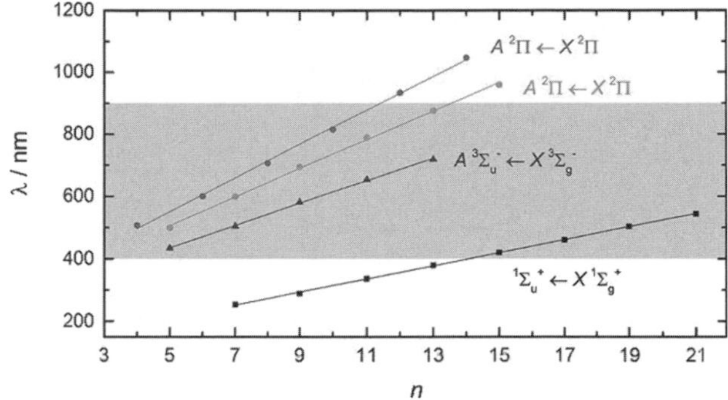

Figure 9.20 Systematic shift with chain length, n, of allowed electronic transitions of neutral (triangles) polyacetylenes HC_nH, their cations (circles), and carbon chains C_n (squares). Most of the DIBs occur in the gray-shaded area. Figure courtesy of [6]

identification purposes. These types of studies have focused on species that, for other reasons, are thought to be abundant in the ISM. Specifically, IR studies have revealed the ubiquitous and abundant presence of large PAH molecules and their derivatives in the ISM. Because of their high stability reasons, fullerenes – in particular C_{60}^+ – have been considered good candidates ever since their first synthesis in the laboratory. Finally, moderately large hydrocarbon chains are known to be present inside dense clouds. Over the years, extensive and systematic studies have been made on the spectroscopic characteristics of these species and, as an example, Figure 9.20 summarizes illustrative results for various carbon chains. The difference between odd and even carbon chains reflects the competition between cumulenic and acetylenic structures and the different chain length implied. A similar figure could also be made for small PAHs (cf. Section 8.2.3). Except for C_{60}^+, none of these species have been linked to strong DIBs and upper limits on their abundance are quite stringent. This is not too surprising as small hydrocarbon chains and small PAHs are not very stable in the harsh conditions of the diffuse ISM.

The DIBs at 9348, 9365, 9428, and 9577 Å have been ascribed to the electronic transitions in C_{60}^+ in view of the good match in peak position, width, and relative intensity with experimental data (to within the experimental uncertainties) and this assignment is now well accepted. Spectroscopic assignment of the experimental bands has been unclear because of various interactions and distortions possible. Figure 9.21 shows the relevant electronic states of this species. Any nonlinear molecule such as C_{60}^+ with degenerate electronic states is unstable to the Jahn–Teller effect and will undergo a geometric distortion to a system of lower symmetry and lower energy thereby lifting the degeneracy. Loss of an electron from the h_u state of C_{60} leads through the Jahn–Teller effect to C_{60}^+ with D_{5d} symmetry. The only symmetry allowed transitions are from the $^2A_{1u}$ ground state to the 3rd and 6th $^2E_{1g}$ doubly degenerate excited electronic states but only the lower one is relevant here (Figure 9.21). Jahn–Teller distortion splits the excited $^2E_{1g}$ into the ^2A_g and ^2B_g states. In these quantum chemical calculations, the two strong transitions at 10382 cm^1

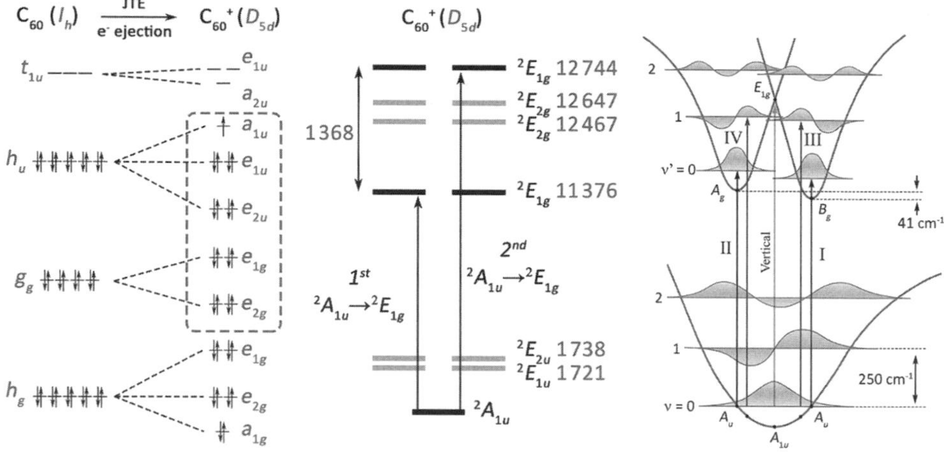

Figure 9.21 Electronic states involved in the C_{60}^+ DIBs. Left: Molecular orbital diagram of C_{60} and C_{60}^+. The Jahn–Teller effect lifts the degeneracy of the electronic states by lowering the symmetry of the molecule through a small geometric distortion. The dashed rectangle denotes the relevant electronic states. Middle: Blow-up of the relevant electronic states with theoretically calculated energies. Dark states are indicated in gray. Right: Vibronic transitions in C_{60}^+. The lower potential is the $^2A_{1u}$ ground state and its vibrational states. The $^2E_{1g}$ state has been split into two states due to the Jahn–Teller effect, the ^2A_g and ^2B_g states. The wave functions of the vibrational states are indicated. Relevant electronic transitions from the ground vibrational state in the ground electronic state to the ground and first excited vibrational state in the two excited electronic states are indicated and correspond to four of the observed interstellar DIBs ascribed to C_{60}^+. Figure courtesy of [70]

and 10442 cm[1] (I and II in Figure 9.21) correspond to the $0-0\,{}^2A_u-{}^2B_g$ and ${}^2A_u-{}^2A_g$ transitions and are responsible for the two strong interstellar C_{60}^+ DIBs. The two weaker ones (III and IV in Figure 9.21) are the first vibronic transitions and are responsible for the two weaker interstellar DIBs.

9.6.4 Profile

Non-radiative Processes

Initially, the diffuse character[11] of the DIBs was interpreted as the result of predissociation but, in the late 1970s, it was realized that non-radiative relaxation processes might cause the observed band broadening. The oscillator strength of the first electronic transition is typically very weak (on a per C-atom basis) and hence high abundances are required to explain the strong DIBs. Higher transitions are stronger but they are inherently very broad. And this linewidth issue is the second – and important – aspect that should be highlighted. When the density of states becomes appreciable, the rate for internal conversion is

[11] A width that was much larger than the Doppler width of atomic transitions in the same interstellar spectra.

very rapid and absorption bands are very broad. A transition dominated by intramolecular relaxation processes will have a Lorentz profile with a FWHM expressed in wavelength space, $\Delta\lambda$, given by,

$$\Delta\lambda \simeq 20 \left(\frac{\lambda}{6000\,\text{Å}}\right)^2 \left(\frac{k_{nr}}{10^{13}\,\text{s}^{-1}}\right) \text{Å}, \tag{9.44}$$

where the non-radiative relaxation rate, k_{nr}, is controlled by an energy-gap law (cf. Figure 8.8; Section 8.2.2). Thus, for visible transitions where the energy gap is 0.5−1 eV (e.g. typically one or more lower lying, excited electronic states), the linewidth will be some 20 Å. Hence, except for possibly the 4430 Å band, none of the strong DIBs involves a transition from the ground-state to $n = 2$ or higher. Conversely, most of the relevant transitions of carbon chains, PAHs, and fullerenes that we expect to be present are much broader than the typical DIB observed but interstellar spectra should be "riddled" with them. The "hunt" for such broad DIBs in interstellar spectra is, however, challenging.

Rotational Structure

When non-radiative processes are slow, the band width is dominated by vibronic structure and this may give rise to P and R or P, Q, and R band structure depending on the rotational constants involved, the rotational excitation temperature, turbulent line width, and spectral resolution (cf. Chapter 3). The profile of the 6614 Å band provides a potential example of P, Q, and R structure (Figure 9.16) and the spectrum of Her 36 (Figure 9.22) drives the importance of rotational broadening home. Ignoring turbulence and spectral resolution issues, the rotational excitation temperature can be related to the observed width by,

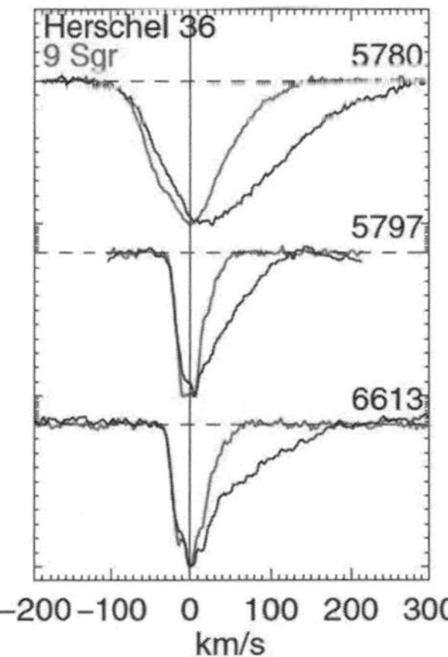

Figure 9.22 Comparison of the extreme gray shading in the profiles of several DIBs observed toward Herschel 36 (black) and the typical profiles (9 Sgr, gray) of these DIBs scaled to the central depth. The velocity zero points has been set by the interstellar KI 7698.965 line. The deep broad absorption near 6379.3 Å in 9 Sgr is a stellar absorption line. Figure courtesy of [83]

$$\Delta\nu \simeq 2.5\frac{kT_{rot}}{hc}\frac{\Delta B}{B} + 4B\sqrt{\frac{kT_{rot}}{hcB}}, \tag{9.45}$$

with B the rotational constant and ΔB the difference in rotational constant between the states involved (cf. Chapter 3). The first term described the width of the P or R branch while the latter term describes the P–R separation. If we focus on this band separation, adopt $B \simeq 2 \times 10^{-3}\,(50/N_c)^2$ appropriate for PAHs, and consider the 6614 Å band with $\Delta\nu \simeq 1.4$ cm^{-1} (Figure 9.16), we arrive at a rotational temperature of ∼100 K. For DIBs where the P,Q,R rotational structure is not resolved, the first term in Eq. (9.45) dominates. The typical width of 0.7 Å for DIBs corresponds then to a rotational temperature of ≃300 K with $\Delta B/B \simeq 3 \times 10^{-3}$ appropriate for PAHs. Clearly, if the line width is dominated by rotational broadening, the absorbing molecules have to be highly excited.

The rotational excitation temperature is a balance between UV photon excitation followed by vibrational relaxation, collisional (de)excitation, and radiative relaxation. We can illustrate the issues well using PAHs as an example. In the diffuse ISM, the collisional deexcitation rate, $n\gamma_{col} \simeq 10^{-9}n \simeq 5 \times 10^{-8}$ s^{-1} is quite slow; comparable to the UV pumping rate, $k_{uv} \simeq 4 \times 10^{-8}(N_C/50)$ s^{-1}. As discussed in Section 12.6 in connection to the Anomalous Microwave Emission, typical J_{max} for PAHs are ≃300. Rotational transitions occur then in the 10 GHz range and have typical Einstein A's $\simeq 3 \times 10^{-8}$ $(\mu_e/1\,\mathrm{D})^2$ s^{-1}. As each absorbed UV photon leads to an IR cascade emitting ≃50−100 photons, angular momentum is mainly built up by emission of ro-vibrational photons with $\Delta J = 0, \pm 1$. As discussed in Section 12.6, the IR cascade gives rise to a Gaussian population distribution centered on $J_{ir} = (hc\overline{\nu}/6hcB)^{1/2} \simeq 290\ (N_c/50)$, where $\overline{\nu}$ is the average IR photon energy and the factor 6 results from the summation over K states. Collisional deexcitation and rotational emission slightly modify this distribution, and reducing the most probable J to the range ≃150. As an example, for circumcoronene, $B = 2 \times 10^{-3}$ cm^{-1}, and with $\overline{\nu} = 10^3$ cm^{-1}, J_{ir} corresponds to a rotational excitation temperature of, $T_{rot} \simeq 250$ K and, allowing for the other processes, $T_{rot} \simeq 100$K (Section 12.6).

One obvious consequence is that differences in physical conditions will result in variations in the rotational excitation and, hence, in the line profile of the DIBs. In that respect, the much wider profile for the DIBs observed toward Her 36 (Figure 9.22) must imply a much higher level of excitation for the carriers along this sight-line than in the general ISM (cf. Eq. (9.45)). UV absorption studies of H_2 place the material seen in absorption toward this star very close (≃0.1 pc) to an O7.5V star, corresponding to a UV field of $G_0 \simeq 10^5$ and likely a concommitant high density (∼10^4 cm^{-3}). The physical conditions are thus clearly different and the calculated most probable J is ≃300; about 2−3 times higher than in the general ISM.

Finally, note that, if the DIB carrier is a small molecule, UV excitation will be followed by electronic relaxation rather than IC and IVR. In that case, the radiative pumping rate of the rotational levels is much smaller. The rotational excitation temperature of the species is then set by the balance between collisional (de)excitation and spontaneous emission. Unless the transition is forbidden, the density is well below the critical density and the excitation temperature in the ISM is very low. Molecular species tend to be in their rotational ground state in the diffuse ISM unless there is an active "pump" such as a UV-driven, IR cascade.

Dark States

A further aspect of UV photo-pumping to consider is the excitation of IR-dark states. After UV excitation, the IR cascade, enabled by rapid IVR, will quickly deexcite the molecule. However, when only $\simeq$1000 cm^{-1} of internal energy is left in the molecule, the density of states will have dropped so low that the states will decouple (Section 8.5.3). The energy will then be trapped in those (low-lying) modes that are excited at decoupling. This excitation can be lost radiatively or collisionally. Consider, as an example, again PAHs: The lowest modes are drumhead modes at $\simeq 59\ (50/N_C)$ cm^{-1} with an integrated strength of $\simeq 2.4 \times 10^{-19}(50/N_C)^{4/5}$ cm/molecule, which corresponds to an Einstein A of $9.5 \times 10^{-4}(50/N_c)^{14/5}$ s^{-1}. For magnetic-dipole or electric-quadrupole transitions, the decay rates will be a factor $\sim\alpha^2 = (1/137)^2 \simeq 5 \times 10^{-5}$ (α is the fine-structure constant) smaller. For IR dark states, radiative decay and collisional decay will have similar rates, 5×10^{-8} s^{-1} and these rates are comparable to the UV excitation rate, $k_{uv} \simeq 4 \times 10^{-8}(N_C/50)$ s^{-1}. While this discussion centered on PAHs, similar considerations apply to C-chains as their lowest vibrational modes shift through the 100 to 10 cm^{-1} range when the chain length increases from 4 to 10 C-atoms. Hence, we conclude that IR dark states can be appreciably excited in diffuse clouds with a population that will depend on the local physical conditions (G_0 and n_H).

Thus, besides the vibronic sequence generated by absorption from the ground vibrational state in the ground electronic state to excited vibrational states in the excited electronic state, we can also expect a vibronic sequence due to absorption from the IR-dark vibrational state in the ground electronic state to ground and excited vibrational states in the excited electronic state. The relative strength of these bands will depend on the Franck–Condon factors connecting the states involved and the excitation of the IR-dark state. If the ground and excited electronic states have similar molecular geometries, then shifts may be quite small but would likely still affect the band profile, causing, perhaps, spectral substructure (cf. Figure 9.16). In addition, while both absorptions originate from the same species, the relative strength of the bands would depend on the local physical conditions. Thus, even if absorptions arise in the same molecules, small intensity variations are possible if dark states are involved. In addition, the rotational Franck-Condon factors can also introduce relative intensity variations.

9.7 Further Reading and Resources

Following a predictive study by [110], molecular absorption lines in the optical were first identified by [18, 75]. Over the years, comprehensive studies have been performed by for example [15, 25, 39, 90]. Pure rotational line studies of molecules in diffuse and molecular clouds have been performed by Liszt and Lucas and their coworkers (see for example [65, 66, 67, 68, 69]). Ultraviolet studies of electronic transitions have been done with Copernicus, HST, and FUSE and some relevant studies are [23, 47, 91, 97, 101]. The Herschel Space Observatory has opened up the sub-millimeter wavelength region to systematic studies of pure rotational transitions of, e.g., hydrides [32, 34]. The use of C_2 as a probe of the physical conditions of diffuse clouds is described in [115]. A comprehensive study is provided by [106].

The characteristics of the ISM in thermal and pressure equilibrium was first studied by [27]. Comprehensive later studies are provided in [121, 122]. The importance of the mechanical energy input by massive stars for the structure of the ISM is reviewed in [74]. Numerical simulations elucidating the structure of the ISM with feedback is presented in [54]. A detailed discussion of the physics underlying the multiphase structure of the ISM can be found in chapter 7 of the monograph [112].

Molecular hydrogen formation through radiative association is discussed in [59]. The various aspects of H_2 formation on grain surfaces are well discussed in [46]. The HI–H_2 transition is discussed in great detail by [4, 109]. The analysis of the data in Figure 9.3 with the model in Section 9.2.1 relies on that study. The early study, [97], on H_2 abundances in the diffuse clouds using Copernicus provided the first determination of the H_2 formation rate on interstellar grain surfaces. Experimental studies have demonstrated that H_2 formation on crystalline surfaces can only occur efficiently over a limited temperature range [88, 89]. Reference [42] reports the much wider temperature range for efficient H_2 formation on amorphous surfaces. Models ([14]) have shown that H_2 formation should, indeed, be efficient in diffuse clouds. A critical evaluation of the observational, experimental, and astronomical issues associated with H_2 formation in the ISM is presented in [119].

The monograph [58] provides an overview of dissociative recombination reactions for astrophysically relevant molecular cations.

Observational studies of N-hydrides are discussed in [86, 87]. The issues with nitrogen chemistry in diffuse clouds have been highlighted in [3]. The importance of the ortho/para ratio of H_2 for N-chemistry has been discussed in [41, 60]. Nitrogen chemistry in dark clouds has been modeled by [61]

The chemistry of halogen in the ISM has been extensively discussed by [81]. The important reactions of H_2 with Cl and F have been studied at low temperatures by [113].

It has been long known from optical ground-based studies that the CH column density correlates well with that of H_2 [24, 66, 102] and indeed the CH column density has often been used as a proxy for the H_2 column density. HIFI has extended these observations from nearby stars to absorption line studies along sightlines throughout the galaxy [32].

Models for the abundance of CH^+ in shocks were first developed by [22] and were critically evaluated against observations by [38]. The importance of turbulent dissipation regions for the CH^+ "problem" was first addressed by [53] and detailed models were developed by [35, 37]. The use of an effective temperature to account for the ion–neutral drift velocity was introduced by [29].

Reference [73] presents a seminal study of H_3^+ and the cosmic ray ionization rate of the diffuse clouds including both new measurements of the H_3^+ electron recombination rate (which was controversial up to that time) as well as direct observations of the H_3^+ ro-vibrational transitions in the near-IR toward background sources. A larger data set was analyzed by [50]. The OH^+ data obtained with HIFI/Herschel was analyzed by [49]. The importance of large molecules for the ionization fraction of diffuse clouds was first addressed by [62] and subsequently stressed by [65]. A comprehensive study of the interplay of cosmic ray ionization and mutual neutralization by large molecules in diffuse clouds is presented in [82]. A thorough analysis for the cosmic ray ionization rate for the ζ Per sightline is provided by [63].

Diffuse Interstellar Bands were first recognized in the 1930s [76] but the field was largely dormant until George Herbig's seminal paper [43]. Some 500 DIBs have now been catalogued [45]. Relevant reviews include [44] and the proceedings of the two DIB conferences [7, 111] provide a good cross cut. The earliest studies linked the DIBs to molecular absorption but for a long time, carriers associated with interstellar dust were deemed plausible. In the late 1970s, the link to molecules was revived with the detection of relatively complex, long chain molecules in space [19]. This has lead to systematic (and sustained) laboratory efforts to measure the electronic transitions of small hydrocarbon chains [52]. The assignment of the IR emission features to large Polycyclic Aromatic Hydrocarbon molecules really brought home the message that space is host to a wide array of large complex molecules that could well be the carriers of the DIBs [40, 94, 95]. The assignment of five DIBs to C_{60}^+ [5, 118] has settled the issue of molecular versus dust carriers in favor of the former. Reference [96] studied the behavior of the DIBs in emission in the Red Rectangle and pointed out its fluorescence origin in large molecules. Systematic studies of the electronic absorption spectra of relevant candidates have been performed by [6, 40, 94, 95]. A model for the rotational excitation of PAHs has been developed by [92]. Rotational contours for PAHs and fullerenes have been calculated by [10, 20]. The extreme profile of Her 36 has been analyzed by [83] and ascribed to vibrational excitation and collisional deexcitation in relatively small molecules. However, that is at variance with the modest linewidths in the general ISM and, rather, excitation variations are involved. The discussion on the rotational excitation of PAHs in space was taken from [92, 114] and applied to DIB profiles as well as the expected rotational emission and the connection to the observed Anomalous Microwave Emission. Effective techniques for searching for rotational emission of interstellar PAHs – based upon "comb" filtering techniques – have been developed by [2]. The importance of vibrational excitation for the substructure in DIB profiles has been examined in the context of the Red Rectangle by [100].

9.8 Exercises

9.1 Limiting abundance of molecules that can be measured in diffuse clouds (Go back to Chapter 3 for relevant equations):

- Electronic transitions. Assume that an absorption feature of 3×10^{-2} depth can be measured at the 3σ level for a system with a spectral resolution of 10,000. Adopt an oscillator strength, $f = 5 \times 10^{-3}$ for the transition at optical (4300 Å) wavelengths.

- Rotational transitions. Assume that an absorption line of 0.05 can be measured at the 3σ level for a line at 533 GHz with a width of 2 km/s. Assume an Einstein $A_{ul} = 5 \times 10^{-4}$, $E_{ul} = 25$ K, and that the excitation is dominated by the galactic background radiation field with an excitation temperature of 3.3 K.

9.2 C_2 absorption lines:

- The equivalent with of the R(0) absorption line of C_2 at 8757.7 Å is measured to be 0.9 mÅ toward the star ζ Oph. With an oscillator strength of 1.7×10^{-3}, calculate the C_2 column density in this state.
- The equivalent width of the Q(10) line at 8780.1 Å toward this star is 0.65 m. The oscillator strength is 8.5×10^{-4}. This state is 200 K above ground. Assuming thermodynamic equilibrium, calculate the temperature of the absorbing gas.
- Explain why pure rotational radiative transitions are not expected to affect the level populations of this molecule under conditions appropriate for diffuse interstellar clouds.
- The rotational level populations in the ground vibrational state can be affected by electronic fluorescence. Electronic excitation after absorption of a photon followed by radiative decay to the ground state may leave the species in a different rotational state than from which it started. With a typical photon excitation rate of $6 \times 10^{-9}\ G_0$ s^{-1}, a collisional cross section of 5×10^{-16} cm^2, and a kinetic temperature of 100 K typical for diffuse clouds, estimate the range of density and interstellar radiation field intensity for which the level populations will probe the physical conditions in diffuse clouds. Discuss your results.

9.3 DIBs:

- With an equivalent width of 72 mÅ measured toward ζ Oph, calculate the column density of the carrier of the 5780 Å DIB, assuming an oscillator strength of unity.
- Explain why this substantial column density for such a large molecule does not surprise you.
- Compare the abundance of this carrier with those of the simple molecules listed in Table 9.1.

9.4 The relationship between n and G_0:

- The size of an HII region, R_s, is set by the ionizing photon luminosity of a star, $\mathcal{N}_{ion}$, and the density of the ionized gas, n_e. Derive the Strömgren relation that relates these quantities by equating the number of recombinations to the number of ionizations in the volume ($\mathcal{N}_{ion} = 4\pi\ \alpha\ R_s^3 n_e^2$ with α the electron recombination coefficient of H^+ (2.6×10^{-13} cm^3 s^{-1}).
- Calculate the radius of the HII region for an O9 star ($\mathcal{N}_{ion} = 3.6 \times 10^{48}$ photons s^{-1}) and a density of 0.25 cm^{-3}.
- Calculate G_0 if the stellar luminosity is 1.2×10^5 L$_\odot$.

- Evaluate n/G_0 in the surrounding neutral shell, assuming pressure equilibrium between the ionized gas and the neutral gas. Adopt $T_i = 8000$ K and $T_n = 80$ K.

9.5 Show that the overall rate for H_2 formation through the H^- route in a diffuse cloud is given by,

$$R_{H^-} \simeq 7.5 \times 10^{-23} \left(\frac{T}{100\,\mathrm{K}}\right) \left(\frac{X(e)}{1.4 \times 10^{-4}}\right) nn(\mathrm{H}) \quad \mathrm{cm}^{-3}\,\mathrm{s}^{-1} \tag{9.46}$$

9.6 Molecular hydrogen formation in the early universe:

- In the early universe ($z > 100$), H_2 is formed through the H^- channel (cf. Section 9.2.2). At this point in time, there is a small amount of residual hydrogen ionization ($X(e) \simeq 3 \times 10^{-4}$) left after recombination. Because of the expansion, the radiation temperature is given by $T_R = T_o(1+z)$ with $T_o = 2.73$ and photoionization of H^- is unimportant. Adopting a temperature of 300 K, calculate the abundance of H^-.
- Here, we will adopt a Hubble constant of $H_o = 100h = 67$ km s^{-1} and a ratio of the density to the critical density of $\Omega_b = 4 \times 10^{-2}$ with the critical density given by $n_{cr} \simeq 10^{-5}h^2$ cm^{-3}. The density is then given by $n = \Omega_b n_{cr}\,(1+z)^3$. For a closure parameter of unity, the relationship between time and z is given by $dt/dz = -H_o^{-1}\,(1+z)^{-5/2}$. Estimate the molecular hydrogen abundance around $z = 100$.

9.7 Oxygen chemistry in diffuse clouds:

- Show that, at 100 K, the neutralization channel through charge transfer of H^+ with O – followed by dissociative recombination of the molecular ions formed – is more important than radiative recombination of H^+ whenever the molecular hydrogen abundance is larger than $\sim 10^{-2}$. Adopt 1.7×10^{-9} cm^{-3} s^{-1} for the rate of O^+ with H_2, an H^+ radiative recombination rate of 8×10^{-12} cm^{-3} s^{-1}, and a forward and backward rate of the charge exchange reaction of 4×10^{-11} and 3×10^{-10} cm^{-3} s^{-1}, respectively.
- Show that CR ionization of H dominates over CR ionization of H_2 for O-hydride formation in diffuse clouds when $H/H_2 > 0.1$.
- Derive Equation (9.11), adopting 10^{-9} and 6.4×10^{-10} cm^3 s^{-1} for the reactions of OH^+ and H_2O^+ with H_2 and a total dissociative recombination rate of $4.2 \times (300/T)^{1/2}$ cm^3 s^{-1} for the latter cation. What molecular hydrogen fraction is implied by the observations toward G10.6-0.4 [31]?

9.8 Nitrogen chemistry in diffuse clouds:

- Using the observations summarized in Table 9.1, demonstrate that the reaction of CH + N dominates over C_2 + N and that C^+ + NH can give a substantial contribution. Adopt for the rates of these reactions, 1.7×10^{-10}, 5×10^{-10}, and 1.4×10^{-9} cm^3 s^{-1}, respectively.

9.9 Discuss the formation and destruction of H_2 in the diffuse ISM.

9.10 Compare and contrast the chemistries of O and C of the diffuse ISM.

9.11 Discuss how the cosmic ray ionization rate can be determined from absorption line measurements of H_3^+.

9.12 Turbulent dissipation regions:

- Compare and contrast the chemistry of diffuse clouds and turbulent dissipation regions.
- Demonstrate quantitatively that Reactions (9.21) and (9.22) in diffuse clouds cannot explain the observed column densities of CH^+, CH^+, and OH.

9.13 Discuss the observed characteristics of the DIBs and the inferred properties of its carriers.

Bibliography

[1] Albertsson, T., et al., 2014, *ApJ*, 787, 44
[2] Ali-Haimoud, Y., 2014, *MNRAS*, 437, 2728
[3] Awad, Z., Viti, S., Williams, D. A., 2016, *ApJ*, 826, 207
[4] Bialy, S., Sternberg, A., 2016, *ApJ*, 822, 83
[5] Campbell E. K., Holz M., Gerlich D., Maier J. P., 2015, *Nature*, 523, 322
[6] Campbell E. K., Holz M., Gerlich D., Maier J. P., 2017, *J Chem Phys*, 146, 160901
[7] Cami, J., and Cox, N. L. J., 2014, IAU symposium 297, *The Diffuse Interstellar Bands* (Cambridge: Cambridge University Press)
[8] Cartledge, S. I. B., Lauroesch, J. T., Meyer, D. M., Sofia, U. J., 2004, *ApJ*, 613, 1037
[9] Cardelli, J. A., Mathis, J. S., Ebbets, D. C., Savage, B. D., 1993, *ApJ*, 402, L17
[10] Cossart-Magos, C., Leach, S., 1990, *A & A*, 233, 559
[11] Crawford, I. A., Barlow, M. J., Diego, F., Spyromilio, J., 1994, *MNRAS*, 266, 903
[12] Crawford, I. A., Williams, D. A., 1997, *MNRAS*, 291, L53
[13] Crutcher, R. M., Watson, W. D., 1981, *ApJ*, 244, 855
[14] Cuppen, H., Herbst, E., 2005, *MNRAS*, 361, 565
[15] Danks, A. C., Federman, S. R., Lambert, D. L., 1984, *A & A*, 130, 62
[16] de Boer, K. S., Morton, D. C., 1979, *A & A*, 71, 141
[17] De Luca, M., Gupta, H., Neufeld, D., et al., 2012, *ApJ*, 751, L37
[18] Douglas, A. E., Herzberg, G., 1941, *ApJ*, 94, 381
[19] Douglas, A. E., 1977, *Nature*, 269, 130
[20] Edwards, S. A., Leach, S., 1993, *A & A*, 272, 533
[21] Ehrenfreund, P., Cami, J., Dartois, E., Foing, B. H., 1997, *A & A*, 317, L28
[22] Elitzur, M., Watson, W. D., 1978, *ApJ*, 222, L141
[23] Federman, S. R., Glassgold, A. E., Jenkins, E. B., Shaya, E. J., 1980, *ApJ*, 242, 545
[24] Federman, S. R., 1982, *ApJ*, 257, 125
[25] Federman, S. R., Danks, A. C., Lambert, D. L., 1984, *ApJ*, 287, 219
[26] Federman, S. R., Cardelli, J. A., van Dishoeck, E. F., Lambert, D.L., Black, J. H., 1995, *ApJ*, 445, 325
[27] Field, G. B., Goldsmith, Habing, H. J., 1969, *ApJ*, 155, L149
[28] Flagey, N., Goldsmith, P. F., Lis, D. C., et al., 2013, *ApJ*, 762, 11

[29] Flower, D., Pineau des Forets, G., Hartquist, T. W., 1985, *MNRAS*, 216, 775
[30] Foing B. H., Ehrenfreund, P., *A & A*, 1997, 317, L59
[31] Gerin, M., de Luca, M., Black, J., et al. 2010, *A & A*, 518, L110
[32] Gerin, M., de Luca, M., Goicoechea, J. R., et al., 2010, *A & A*, 521, L16
[33] Gerin, M., Levrier, F., Falgarone, E., et al., 2012, *Philosophical Transactions of the Royal Society of London Series A*, 370, 5174
[34] Gerin, M., Neufeld, D. A., Goicoechea, J. R., 2016, *Annu Rev Astron Astrophys*, 54, 181
[35] Godard, B., Falgarone, E., Pineau Des Forêts, G., 2009, *A & A*, 495, 847
[36] Godard, B., Falgarone, E., Gerin, M., et al., 2012, *A & A*, 540, A87
[37] Godard, B., Falgarone, E., Pineau Des Forêts, G., 2014, *A & A*, 570, A27
[38] Gredel, R., van Dishoeck, E. F., Black, J. H., 1993, *A & A*, 269, 477
[39] Gredel, R., Pineau des Forêts, G., Federman, S. R., 2002, *A & A*, 389, 993
[40] Gredel, R., Carpentier, Y., Rouillé, G., et al., 2011, *A & A*, 530, A26
[41] Grozdanov, T. P., McCarroll, R., Roueff, E., 2016, *A & A*, 589, A105
[42] He, J., Frank, P., Vidali, G., 2011, *Phys Chem Chem Phys*, 13, 15803
[43] Herbig, G. H., 1975, *ApJ*, 196, 129
[44] Herbig, G. H., 1995, *Annu Rev Astron Astrophys*, 33, 19
[45] Hobbs, L. M., York, D. G., Thorburn, J. A., et al., 2009, *ApJ*, 705, 32
[46] Hollenbach, D., Salpeter, E. E., 1970, *ApJ*, 163, 155
[47] Hupe, R. C., Sheffer, Y., Federman, S. R., 2012, *ApJ*, 761, 38
[48] Indriolo, N., Oka, T., Geballe, T. R., McCall, B. J., 2010, *ApJ*, 711, 1338
[49] Indriolo, N., Neufeld, D. A., Gerin, M., et al., 2015, *ApJ*, 800, 40
[50] Indriolo, N., McCall, B. J., 2012, *ApJ*, 745, 91
[51] Jochims, H. W., Baümgartel, H., Leach, S., 1996, *A & A*, 314, 1003
[52] Jochnowitz, J. B., Maier, J. P., 2008, *Annu Rev Phys Chem*, 59, 519
[53] Joulain, K., Falgarone, E., Pineau des Forets, G., Flower, D., 1998, *A & A*, 340, 241
[54] Kim, C. G., Ostriker, E. C., Kim, W. T., 2013, *ApJ*, 776, 1
[55] Knauth, D. C., Andersson, B.-G., McCandliss, S. R., Warren Moos, H., 2004, *Nature*, 429, 636
[56] Lacour, S., André, M. K., Sonnentrucker, P., et al. 2005, *A & A*, 430, 967
[57] Lambert, D. L., Sheffer, Y., Gilliland, R. L., Federman, S. R., 1994, *ApJ*, 420, 756
[58] Larsson, M., Orel, A. E., 2008, *Dissociative Recombination of Molecular Ions* (Cambridge: Cambridge University Press)
[59] Latter, W. B., Black, J. H., 1991, *ApJ*, 372, 161
[60] Le Bourlot, J., 1991, *A & A*, 242, 235
[61] Le Gal, R., Hily-Blant, P., Faure, A., et al., 2014, *A & A*, 562, A83
[62] Lepp, S., Dalgarno, A., van Dishoeck, E. F., Black, J. H., 1988, *ApJ*, 329, 418
[63] Le Petit, F., Roueff, E., Herbst, E., 2004, *A & A*, 417, 993
[64] Lis, D. C., Schilke, P., Bergin, E. A., et al., 2014, *ApJ*, 785, 135
[65] Liszt, H. 2003, *A & A*, 398, 621
[66] Liszt, H., Lucas, R., 2002, *A & A*, 391, 693
[67] Liszt, H. S., Petit, J., Gerin, M., Lucas, R., 2014, *A & A*, 564, A64
[68] Liszt, H. S., Lucas, R., Black, J. H., 2004, *A & A*, 428, 117
[69] Liszt, H. S., Pety, J., Lucas, R., 2008, *A & A*, 486, 493
[70] Lyhkin, A. O., Ahmadvand, S., Varganov, S. A., *J. Phys Chem Lett*, 10, 115
[71] Lyu, C. H., Smith, A. M., Bruhweiler, F. C., 2001, *ApJ*, 560, 865
[72] Maier, J. P., Lakin, N. M., Walker, G. A. H., Bohlender, D. A., 2001, *ApJ*, 553, 267
[73] McCall, B. J., Huneycutt, A. J., Saykally, R. J., et al. 2003, *Nature*, 422, 500

[74] McCray, R., Snow, T. P., 1979, *Annu Rev Astron Astrophys*, 17, 213
[75] McKellar, A., 1940, *PASP*, 52, 187
[76] Merrill, P. W., 1934, *PASP*, 46, 206
[77] Meyer, D. M., Cardelli, J. A., Sofia, U. J., 1997, *ApJ*, 490, L103
[78] Morton, D. C. 1975, *ApJ*, 197, 85
[79] Neufeld, D. A., Godard, B., Gerin, M., et al., 2015, *A & A*, 577, A49
[80] Neufeld, D. A., Black, J. H., Gerin, M., et al., 2015, *ApJ*, 807, 54
[81] Neufeld, D., Wolfire, M., 2009, *ApJ*, 706, 1594
[82] Neufeld, D., Wolfire, M., 2017, *ApJ*, 845, 163
[83] Oka, T., Welty, D. E., Johnson, S., et al., 2013, *ApJ*, 773, 42
[84] Pabst, C., et al., 2017, *A & A*, in press
[85] Pan, L., Padoan, P., 2009, *ApJ*, 692, 594
[86] Persson, C. M., Black, J. H., Cernicharo, J., et al., 2010, *A & A*, 521, L45
[87] Persson, C. M., De Luca, M., Mookerjea, B., et al., 2012, *A & A*, 543, A145
[88] Pironello, V., Biham, O., Liu, C., Shen, L., Vidali, G., 1997, *ApJ*, 483, L131
[89] Pironello, V., Liu, C., Roser, J., Vidali, G., 1999, *A & A*, 344, 681
[90] Porras, A. J., Federman, S. R., Welty, D. E., Ritchey, A. M., 2014, *ApJL*, 781, L8
[91] Rachford, B., et al., 2009, *ApJS*, 180, 125
[92] Rouan, D., Léger, A., Omont, A., Giard, M., 1992, *A & A*, 253, 498
[93] Roueff, E., 1996, *MNRAS*, 279, L37
[94] Salama, F., Bakes, E. L. O., Allamandola, L. J., Tielens, A. G. G. M., 1996, *ApJ*, 458, 621
[95] Salama, F., Galazutdinov, G. A., Krełowski, J., et al., 2011, *ApJ*, 728, 154
[96] Sarre1, P. R. Miles, J. R. Scarrott, S. M., 1995, *Science*, 269, 674
[97] Savage, B. D.,Drake, J. F., Budich, W., Bohlin, R. C., 1977, *ApJ*, 216, 291
[98] Savage, B. D., Bohlin, R. C., Drake, J. F., Budich, W., 1977, *ApJ*, 216, 291
[99] Schilke, P., Neufeld, D. A., Müller, H. S. P., et al., 2014, *A & A*, 566, A29
[100] Sharp, R. G., Reilly, N. J., Kable, S. H., Schmidt, T. W., 2006, *ApJ*, 639, 194
[101] Sheffer, Y., Federman, S. R., 2007, *ApJ*, 659, 1352
[102] Sheffer, Y., Rogers, M., Federman, S. R., et al., 2008, *ApJ*, 687, 1075
[103] Snow, T. P., 2005, IAU Symposium 231 *Astrochemistry: Recent Successes and Current Challenges* (Cambridge: Cambridge University Press), 175
[104] Snow, T. P., Jr., Smith, W. H., 1981, *ApJ*, 250, 163
[105] Snow, T. P., McCall, B. J., 2006, *Annu Rev Astron Astrophys*, 44, 367
[106] Sonnentrucker, P., Welty, D. E., Thornburn, J. A., York, D. G., 2007, *ApJ*, 168, 58
[107] Sonnentrucker, P., Welty, D. E., Thornburn, J. A., 2007, *ApJS*, 168, 58
[108] Sonnentrucker, P., Neufeld, D. A., Phillips, T. G., et al., 2010, *A & A*, 521, L12
[109] Sternberg, A., Le Petit, F., Roueff, E., Le Bourlot, J., 2014, *ApJ*, 790, 10
[110] Swings, P., Rosenfeld, L., 1937, ApJ, 86, 483
[111] Tielens, A. G. G. M., Snow, T. P., 1995, *The Diffuse Interstellar Bands, Astrophysics Space Sciences Library* (Dordrecht: Kluwer), 202
[112] Tielens, A. G. G. M., 2005, *Physics and Chemistry of the Interstellar Medium* (Cambridge: Cambridge University Press)
[113] Tizniti, M., et al., 2014, *Nature Chemistry*, 6,141
[114] Ysard, N., Verstraete, L., 2010, *A & A*, 509, A12
[115] van Dishoeck, E. F., de Zeeuw, T., 1984, *MNRAS*, 206,383
[116] van Dishoeck, E. F., Black, J. H., 1988, *ApJ*, 334, 771
[117] van Dishoeck, E. F., Black, J. H., 1989, *ApJ*, 340, 273

[118] Walker, G. A. H., Campbell, E. K., Maier, J. P., Bohlender, D., Malo, L., 2016, *ApJ*, 831, 130

[119] Wakelam, V., Bron, E., Cazaux, S., et al., 2017, *Molecular Astrophysics*, 9, 1

[120] Welty, D. E., Xue, R., Wong, T., 2012, *ApJ*, 745, 173

[121] Wolfire, M. G., Hollenbach, D., McKee, C. F., Tielens, A. G. G. M., Bakes, E. L. O., 1995, *ApJ*, 443, 152

[122] Wolfire, M. G., McKee, C. F., Hollenbach, D., Tielens, A. G. G. M., 2003, *ApJ*, 587, 278

10

Molecular Clouds

Molecular hydrogen is the most abundant molecule in the Universe. In the Milky Way, about 80% of the H_2 is in the form of giant molecular clouds with H_2 column densities of $\simeq 8 \times 10^{21}$ cm^{-2}. With typical masses of $10^5 - 5 \times 10^6$ M$_\odot$, giant molecular clouds are the most massive objects in the galaxy apart from the central supermassive black hole. In contrast to diffuse clouds, giant molecular clouds are self-gravitating and, in the Milky Way, they are the exclusive sites of star formation. Observations of other galaxies show that giant molecular clouds are formed in spiral arms where gas is compressed by the spiral density shock wave. Giant molecular cloud formation may result from either coagulation of smaller molecular clouds – as in the molecular ring of our own galaxy – or by the large scale conversion of atomic gas into molecular gas promoted by the higher densities and assisted by the shielding against dissociating FUV radiation afforded by the resulting larger column densities. The turbulence generated by the (continuous) accretion process provides a counterforce against the self-gravity of the gas and prevents the wholesale collapse of the cloud into new stars and the typical star formation efficiency of a giant molecular cloud is only $\sim$5%. While turbulence only slows down the star formation process, the mechanical action of massive stars formed in giant molecular clouds leads to the destruction of the star formation cores either through their stellar winds – sweeping up gas into shells – or through their radiation by creating ionized and photodissociated gas flows from their surfaces. Either way, cloud material is dispersed into the general ISM where nearby supernovae sweep it up in superbubble walls. So giant molecular clouds play an important role in the ecology of spiral galaxies.

This chapter starts off by discussing the analysis of observations of molecular clouds to derive temperatures, densities, and column densities. This is followed by a discussion of the characteristics of molecular clouds in terms of masses, sizes, densities, and their scaling relationships, as well as the (infamous) X_{CO} factor. The processes involved in the energy balance of molecular clouds are discussed in Section 10.3. Gas phase molecular abundances are discussed in Section 10.4 while relevant gas phase chemistry processes are presented in Section 10.5. Gas–grain interaction is discussed in Section 10.6, starting with interstellar ices and their characteristics, and the grain surface chemistry routes involved, and followed by a discussion of the balance between accretion and ejection and the effect on gas phase abundances.

As a guide for courses on molecular astrophysics, Section 10.1 and Section 10.2 focus on the analysis of observations of molecular clouds. Section 10.3 discusses the energy balance of molecular gas. A molecular astrophysics course that is focused more on the chemical aspects would center on Sections 10.5 and 10.6. The resources and further reading Section (10.7) provides the student with entries into the literature and could serve as the basis for student essays on specific aspects of this chapter.

10.0.1 Diffuse Versus Molecular Clouds

Molecular clouds differ in a number of important aspects from the diffuse clouds discussed in Chapter 9. Figure 9.1 illustrates the commonly used nomenclature in the classification of clouds, and molecular clouds have a much higher column density ($N_H \simeq 1.5 \times 10^{22}$ H nuclei/cm^2 or an $A_V \simeq 8$ magn) than diffuse clouds. Nevertheless, the surface layers of molecular clouds are still dominated by penetrating FUV photons and they resemble in many ways photodissociation regions. PDRs are discussed in depth in Section 11.6. Here, we just emphasize that the extent of the region dominated by FUV photons depends on the characteristics examined. The HI/H_2 transition will occur at a typical depth of $N_H \simeq 2 \times 10^{19}$ H-nuclei/cm^2 (Figure 6.11 and Eq. 6.38 with $G_0/n_H \simeq 3 \times 10^{-3}$). The gas temperature will typically drop from $\simeq$60 K to 20 K between a column of 1 $\times$ and 4 $\times$ 10^{21} H-nuclei/cm^2 while the dust temperature drops from $\simeq$16 K to 10 K over that same column. Carbon will transform from C^+ to C and then CO at a depth of 8 $\times$ 10^{20} and 2 $\times$ 10^{21} H-nuclei/cm^2, respectively. Oxygen chemistry is affected to a similar depth (2 $\times$ 10^{21} H-nuclei/cm^2), mainly through the balance of accretion on grains and photodesorption (cf. Figure 10.36, Section 10.6.3). As a result of the decreasing importance of UV photochemistry, gas phase chemistry in molecular clouds is primarily driven by cosmic ray ionization. Indeed, protonated molecular hydrogen, H_3^+, which regulates dense cloud, gas phase chemistry, rises by a factor of $\simeq$30 between 10^{20} H-nuclei/cm^2 and 2 $\times$ 10^{21} H-nuclei/cm^2.

The other key difference with diffuse clouds is the importance of grains. Photo desorption, driven by the high radiation field in diffuse clouds, keeps dust surfaces "clean." In contrast, inside molecular clouds, ice mantles rapidly build up and these become an important repository of species. As mentioned above, this is of major influence for the oxygen chemistry, making H_2O ice the important reservoir of oxygen inside dense clouds for $N_H > 2 \times 10^{21}$ H-nuclei/cm^2. CO will start to condense out at the same depth as the dust temperature will have dropped below the sublimation temperature of pure CO ice as well as for CO mixed into H_2O ice. The rate at which CO depletes will be slightly slower than for atomic O, though.

The decrease in the radiation field also has major influence on the energy balance of the gas. First, the photo-electric effect is no longer the dominant heating agent inside dense clouds. Cosmic ray heating takes over in a global sense, while locally intermittent turbulent energy dissipation can dominate. Second, molecules dominate the gas phase composition and rotational level separations are much smaller than atomic fine-structure line separations. Molecular clouds are thus much less heated and cool more efficiently than atomic clouds, and this leads to the temperature gradient noted above. Finally, molecular clouds are self-gravitating. While gravitational collapse is outside the scope of this book, one consequence is much higher gas pressures, $\simeq 3 \times 10^4$ K cm^{-3} in molecular cloud cores than in the diffuse ISM.

This discussion assumed a homogeneous cloud. However, in reality, molecular clouds are highly structured, reflecting the high turbulence (cf. Section 10.2.2 and 10.2.3). This has large consequences for the penetration of photons from the ambient interstellar radiation field as part of the sky seen by a parcel of gas may be highly obscured while other parts may be quite "transparent" (cf. Figure 6.8 and Section 6.3.2). So the column densities quoted above should be scaled using the effective opacity of clumpy media.

10.1 Analysis of Observations of Molecular Clouds

The analysis of molecular observations has been discussed in Chapter 4. Figure 4.8 is very illustrative. Here, we will focus on those aspects relevant to molecular clouds and illustrate various methods that are commonly used. Each of these methods has its advantages and disadvantages and this should be kept in mind when applying these methods to the analysis of data.

10.1.1 Gas Temperature

Rotation Diagram

There are various ways to determine the gas temperature. The gas temperature can be rather directly determined from excitation (e.g. rotational diagram) analysis as long as the density exceeds the critical density and the lines are optically thin (cf. Section 4.2.2). The low lying, pure rotational transitions of H_2 are a case in point. However, excitation of these transitions requires high temperatures and this is not relevant for dark clouds. The ro-vibrational transitions of CO are often used as with the proper Doppler shift they can be readily observed in absorption against bright background sources in the M band with high resolution spectrographs. Figure 10.1 shows the results for the massive protostar, W33A. Besides the main isotope, $^{12}C^{16}O$, transitions of $^{13}C^{16}O$ and $^{12}C^{18}O$ are also present in the spectrum. Two temperature components are evident; a cold component associated with the outer envelope or surrounding molecular cloud and a warm component reflecting the presence of warm gas in the disk (e.g. Hot Core, Section 11.3) surrounding the YSO. The difference in temperature derived for the cold component from the different isotopes may well reflect gas at a density below the critical density coupled with line trapping in the pure rotational lines of the main isotope (see Section 4.2.2 and below).

This method does require a bright mid-IR background source, preferably with a largely featureless continuum. High spectral resolution is required to shift the molecular cloud CO lines out of the telluric lines. Bright background K to M giants have strong and broad photospheric CO lines and would require helpful Doppler shifts. Also, studies of the warm

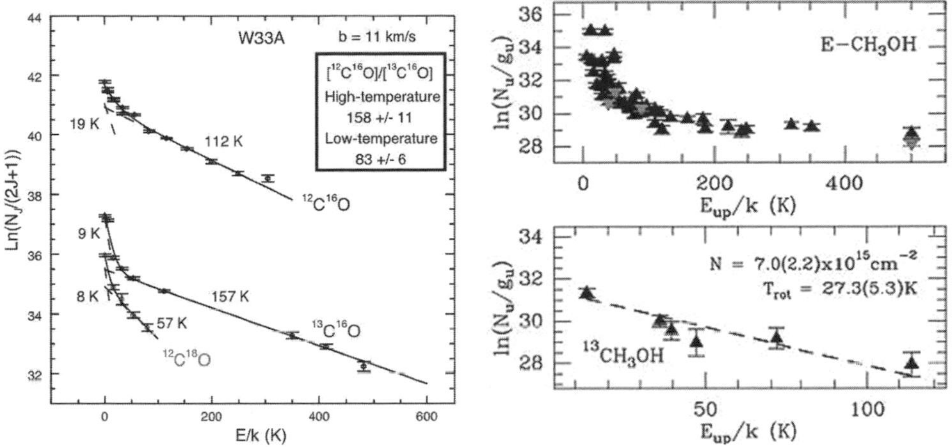

Figure 10.1 Rotational diagram analysis (see Chapter 3). Left: Analysis of the mid-IR, ro-vibrational spectrum of CO toward the massive protostar, W33A. The Boltzmann distribution is well fit by two components, representing the cold gas in the envelope or foreground and warm gas in the hot core associated with the protostar. Figure taken from [215]. Right: Rotational diagram analysis of the sub-mm transitions of E-CH_3OH and its ^{13}C isotope toward the intermediate mass protostar, Cep E-mm. Figure taken from [166]

gas often reveal the presence of multiple velocity components that are likely associated with the acceleration zone of the disk wind and this may further complicate the analysis. Finally, other species (e.g. CS, H_2O, NH_3, C_2H_2, HCN, CH_4, SO_2) are accessible in the 5−8 μm region from airborne altitudes (e.g. SOFIA) and in the L-, N- and Q-band windows from the ground. An example is shown in Figure 4.10.

Rotation diagram analysis can also be done using pure rotational lines in the sub-millimeter. Figure 10.1 shows the results of such a study using methanol lines toward the intermediate mass protostar, Cep E-mm, obtained in two settings (1.3 and 3 mm). The main isotope[1] shows pronounced curvature in the Boltzmann diagram. Possible origins of such curvature are discussed in Section 4.2.2. Here, the low excitation lines are highly optically thick and unreliable temperature indicators through the Boltzmann diagram analysis. In contrast, the ^{13}C isotope is optically thin and reveals gas with an excitation temperature of 27 K. The high excitation (E/k = 150−500 K) lines of the main isotope are actually optically thin and originate in the warm (∼400 K) gas of a hot core.

Optically Thick Rotational Transitions

Millimeter observations of the low lying, pure rotational transitions of the main isotope of CO also provide a straight forward way to derive the gas temperature. These lines are highly optically thick and hence the peak intensity is given by the Planck function

[1] Methanol A and E refer to the spin orientation of the H-atoms in the CH_3 group. In methanol A, they are parallel.

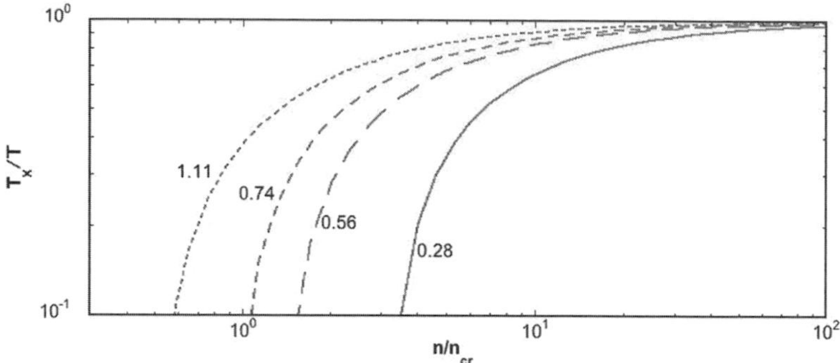

Figure 10.2 The ratio of the excitation temperature to the kinetic temperature as a function of the ratio of the density to the critical density for a two level system (see Chapter 4 for details). The curves are labeled with the ratio of the value for kT_X/E_{ul}.

(cf. Eq. (4.59)), which depends solely on the excitation temperature. For a two-level system, the excitation temperature is equal to the kinetic temperature if the density well exceeds the critical density, n_{cr}, of the transition; viz.,

$$\frac{T}{T_X} = \left(1 - \frac{kT_X}{h\nu}\ln\left[1 + \frac{n_{cr}}{n}\right]\right)^{-1}. \tag{10.1}$$

This relationship is illustrated in Figure 10.3. When the density is much less than the critical density, the level falls out of thermodynamic equilibrium and the kinetic temperature of the gas much exceeds the observed excitation temperature. This difference between kinetic temperature and excitation temperature also depends on the energy level separation (i.e. $kT/h\nu$) as non-LTE effects become more pronounced for low values of this ratio. Typically, when $n \gg 10n_{cr}$, the measured excitation temperature is a good measure for the kinetic temperature (Figure 10.2).

For an optically thick line, the critical density is given by (cf. Section 4.1.6),

$$n_{cr} = \frac{\beta\left(\tau_p\right) A_{ul}}{\gamma_{ul}}, \tag{10.2}$$

where A_{ul} and γ_{ul} are the Einstein A and the collisional deexcitation rate of the upper level and the escape probability is,

$$\beta\left(\tau_p\right) = \left(\tau_p\sqrt{\pi\ln\left[\tau_p\right]}\right)^{-1} \tag{10.3}$$

with τ_p the peak optical depth given by,

$$\tau_p = \frac{\lambda^2}{8\sqrt{\pi}} A_{ul}\frac{N_u}{b}\left(\frac{g_u N_l}{g_l N_u} - 1\right), \tag{10.4}$$

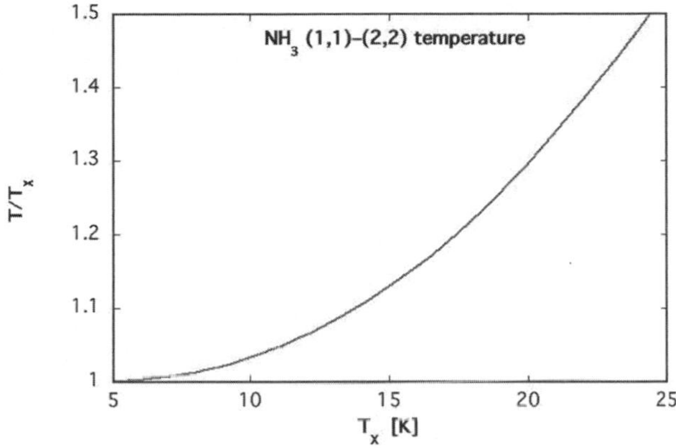

Figure 10.3 Ammonia as a temperature indicator. The ratio of the kinetic temperature, T, to the observed excitation temperature as a function of the excitation temperature, T_x. The excitation temperature is derived from the observed level populations of the two lowest metastable levels of NH_3 (levels (1,1) and (2,2)). Figure adapted from [38, 240]

where the upper level population is approximately given by (cf. Section 3.2.8),

$$N_u = \frac{hcB}{kT_X} \exp\left[-J\,(J+1)\,hcB/kT_X\right] N_{\rm tot}. \tag{10.5}$$

Inserting numerical values for the $1-0$ transition of CO, we have at 10 K,

$$\tau_p \simeq 35 \left(\frac{X_{\rm CO}}{10^{-4}}\right) \left(\frac{3\,{\rm km/s}}{b}\right) \left(\frac{N_{\rm H}}{10^{22}}\right). \tag{10.6}$$

Thus, for a typical molecular cloud column density ($N_H \simeq 1.5 \times 10^{22}$ cm^{-2}), this line will be highly optically thick and the critical density will be reduced by line trapping effects; e.g. $\beta \simeq 5 \times 10^{-3}$ and the critical density is only n_{cr} (CO J $= 1-0$) $\simeq 5$ cm^{-3}. The transitions of isotopologues are not as optically thick and hence in the low density gas of the main molecular cloud, their excitation temperature does not provide a good measure of the kinetic temperature.

Metastable Levels

The metastable backbone levels of prolate or oblate symmetric top molecules are also particularly suited for temperature determinations as transitions between different K-ladders are forbidden and have, therefore, very small Einstein A's and low critical densities. Ammonia is often used for this as the levels are split into inversion doublets – associated with movement of the N atom through the plane of the H's – and transitions between these states occur close together and at low frequencies that can be readily observed from the ground in one setting (which simplifies relative calibration). These transitions are further split because of hyperfine interaction (cf. Section 3.2.5) into 18 components but they separate out into 5 groups. If the excitation of these hyperfine levels is the same, then their relative strength can be determined theoretically. Therefore, these hyperfine transitions are a sensitive measure of the optical depth. We can write for the ratio of the antenna temperature for the satellite to the main component,

$$\frac{T_{mb}(J,K,m)}{T_{mb}(J,K,s)} = \frac{1-\exp\left[-\tau\left(J,K,m\right)\right]}{1-\exp\left[-a\tau\left(J,K,m\right)\right]}, \tag{10.7}$$

where m and s refer to the main and satellite components and the constant a is the relative intensity for the satellite compared to the main component ($a = 0.28$ and 0.22 for the (1,1) NH_3 hyperfine transitions). While the main and satellite lines of the (1,1) hyperfine components can be observed, for the (2,2) line only the central line is typically strong enough to be detectable. In the Rayleigh–Jeans approximation, the excitation temperature for the hyperfine-structure lines is given by,

$$T_{hf} = \frac{T_{mb}}{1-\exp\left[-\tau\left(J,K,m\right)\right]} + T_{BG}. \tag{10.8}$$

The populations of different metastable levels are related by a rotational excitation temperature. For the (2,2) to (1,1) levels this is,

$$\frac{T_0}{T_X} = \ln\left[\frac{-0.28}{\tau\left(1,1,m\right)}\ln\left[1-\frac{T_{mb}\left(2,2,m\right)}{T_{mb}\left(1,1,m\right)}\left(1-\exp\left[-\tau\left(1,1,m\right)\right]\right)\right]\right] \tag{10.9}$$

with T_0 the (2,2)–(1,1) energy separation in degrees Kelvin (41.2 K).

The kinetic temperature can be derived from an excitation analysis of the metastable levels. Consider a three-level system with levels (1,1), (2,2), and (2,1), where (1,1) and (2,2) are metastable. We will assume that collisional excitations of level (2,1), on the other hand, will quickly radiatively decay to the (1,1) state. So, we can ignore the radiative terms in this three level system, and we can write for the excitation temperature of levels (1,1) and (2,2),

$$T_X = T\left(1+\left(\frac{T}{T_0}\right)\ln\left[1+\frac{\gamma_a}{\gamma_b}\right]\right)^{-1}, \tag{10.10}$$

where the γ_i refer to the collision rates between (2,2) and (2,1) ($i = a$) and (2,2) and (1,1) ($i = b$). This relation is appropriate as long as the density is below the critical density of the (2,1) level ($n < 10^5$ cm^{-3}) and the temperature is less than 40 K. Assuming that para H_2 is the dominant collision partner in cold dense cores and taking the temperature dependence of the collisional rates into account, we can invert this relationship to yield,

$$T = T_X\left(1-\frac{T_X}{T_0}\ln\left[1-1.608\exp\left[-25.25/T_X\right]\right]\right)^{-1}. \tag{10.11}$$

This relationship is illustrated in Figure 10.3. Thus, as long as the temperature is low (and $n \lesssim n_{cr}$), the populations provide a good measure of the kinetic temperature of the gas. For high temperatures, the excitation temperature appproaches the limit,

$$T_x \simeq 41\left(\ln\left[1+\frac{\gamma_a}{\gamma_b}\right]\right)^{-1}\ \mathrm{K}. \tag{10.12}$$

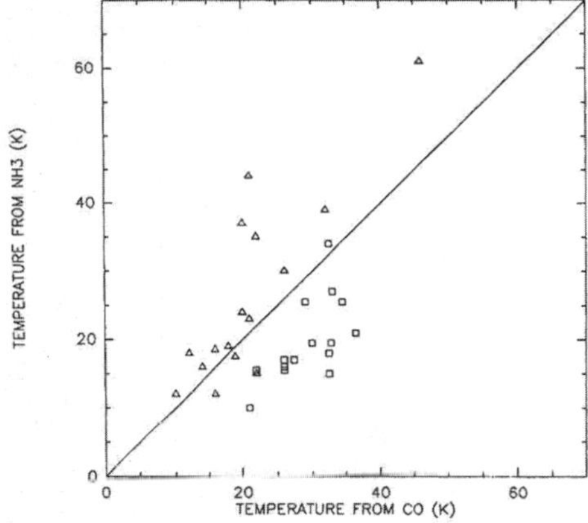

Figure 10.4 Comparison of kinetic temperatures of dark clouds derived from the pure rotational transitions of CO and from the inversion levels of NH_3. Figure adapted from [59]

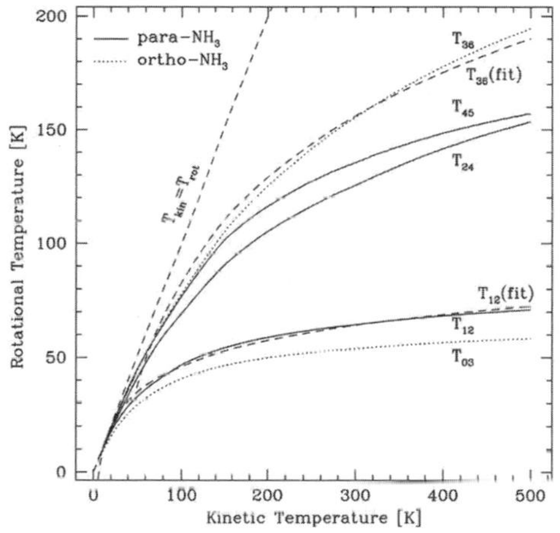

Figure 10.5 The high metastable levels of ammonia as a temperature indicator. The relationship between measured rotational excitation temperatures and the kinetic gas temperatures. Subscripts denote the metastable levels involved in the transitions. The dashed straight line indicates where both temperatures would be identical. Figure taken from [168], which also provided simple analytical fits (dashed lines)

Transitions between higher level metastable levels will show similar behavior and can be fitted with similar expressions. As illustrated in Figure 10.5, these ratios provide reliable temperature estimates from observed excitation temperatures over a larger temperature range as the energy level separation is larger. Of course, they need much higher temperature gas to be appreciably excited (cf. Figure 4.8; Chapter 4).

The use of the metastable lines of NH_3 as thermometers is widespread. Figure 10.6 shows the results of a large scale study of the NGC 1333 region in Perseus. The dust

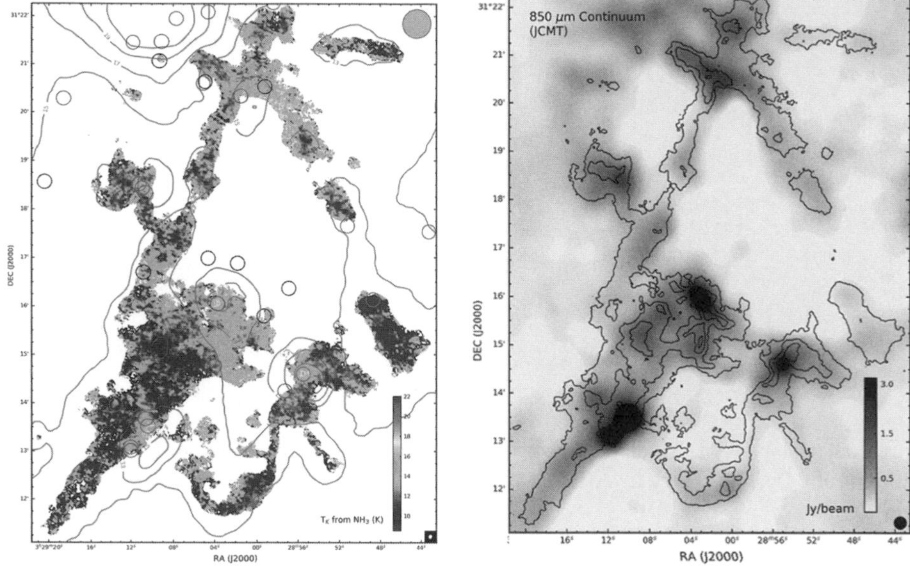

Figure 10.6 Left: The kinetic gas temperature of molecular clouds in the NGC 1333 region in Perseus as derived from ammonia observations using the VLA supplemented by the GBT. The locations of known Class 0/I and Flat YSOs from Spitzer observations are marked as red and black circles (size $\simeq$5000 AU), respectively. Right: The dust emission at 850 μm measured by SCUBA. Contours are the NH_3 (1,1) integrated intensity (0.01, 0.02, 0.04, and 0.08 Jy/beam km s^{-1}). Beam sizes are shown in the bottom left of each panel. Figure kindly provided by Dr Arnab Dhabal, [60]. (A black and white version of this figure will appear in some formats. For the color version, please refer to the plate section.)

emission study reveals a very filamentary structure with several dense cores. Star formation is closely associated with these dense cores. Over much of the cloud, the kinetic gas temperature is 10−12 K. Higher temperature regions are associated with outflow activity such as HH 7−11 and HH 12.

Gas temperatures derived from the pure rotational transitions of CO and from the inversion levels of NH_3 are compared in Figure 10.4, illustrating reasonable agreement. Other symmetric top molecules (e.g. CH_3CN) can be used in a similar fashion.

Molecular Excitation Analysis: Temperature

The temperature (and density, see below) can also be derived from analysis of well-chosen molecular transitions (cf. Figure 4.9; Section 4.2.3). The transitions involved should span a range in excitation energy and critical density and they have to be (preferably) optically thin. Formaldehyde and methanol are often used for this purpose. Again, the relative populations of the different K-ladders are largely controlled by collisions and transitions involving different K-ladders provide good temperature tracers. For para formaldehyde, the transition ratios $3_{22}-2_{21}/3_{03}-2_{02}$ and $3_{21}-2_{20}/3_{03}-2_{02}$ are often used as they are readily accessible in a single setting ($\nu \simeq$218 GHz within a 1 GHz bandwidth).

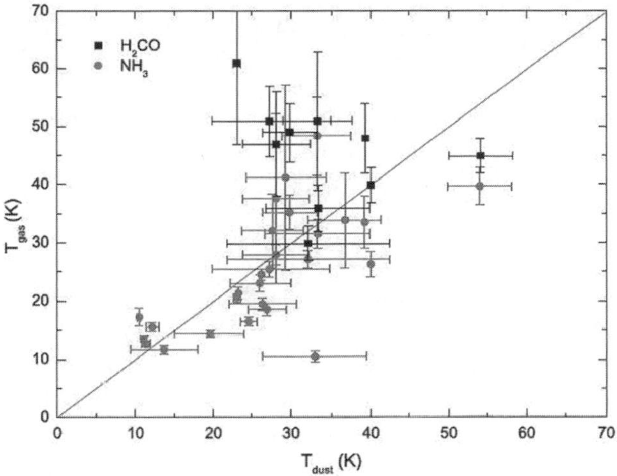

Figure 10.7 Temperatures derived from H_2CO and NH_3 analysis as a function of the dust temperatures. All formaldehyde results shown also have an ammonia result. Figure taken from [224]

Figure 10.7 compares H_2CO temperatures with those derived from NH_3 inversion studies for the same sources. Typically, formaldehyde indicates higher gas temperatures than ammonia in these sources. These differences probably reflect the importance of temperature gradients as these two species may probe different regions in a source. The importance of a temperature gradient is also indicated by the elevated dust temperatures for these sources.

Dust Emission

Collisions couple the dust and gas temperature and at high densities ($n \gtrsim 10^5$ cm^{-3}) expected differences in temperature are $\simeq$1 K (Section 10.3). The dust temperature can be determined from fits of the Spectral Energy Distribution (SED) measured over the far-IR to sub-mm wavelength range. We will come back to this in Section 10.1.2. Figure 10.9 shows the dust temperature derived this way for the dark cloud core, B68. The dust temperature is $\simeq$9 K in the deep interior and increases slightly toward the surface (e.g. $T_d \simeq 13$ K at $A_v = 1$ magn).

10.1.2 Density

CO Column Density

Densities can be derived from measured molecular column densities assuming a molecular abundance and a scale size. Hydrogen column densities of molecular clouds are determined through studies of CO isotopologues. Consider Equation (10.6) but now for ^{13}CO, which has an abundance $\simeq 1/65$ compared to the main isotope. So, for a column less than 10^{22} H-nuclei cm^{-2}, the optical depth is < 0.5 and we can derive the column density from,[2]

[2] See Section 4.2.1 for a discussion of the approximations involved.

$$N\,(\mathrm{H_2}) = 1.7 \times 10^{20} \left(\frac{X(^{13}\mathrm{CO})/X(^{12}\mathrm{CO})}{65} \right) \left(\frac{10^{-4}}{^{12}\mathrm{CO}} \right) \frac{\int T_A dv}{1 - \exp\left[-5.3/T_X\right]}, \quad (10.13)$$

where the denominator in the last fraction represents the rotational partition function. Besides the CO abundance and the isotope ratio, this also depends on the excitation temperature. Often, the excitation temperature is derived from ^{12}CO measurements, but note that, while the ^{12}CO transition may be thermalized because of line trapping, this is not the case for the ^{13}CO transition. For large column densities, the ^{13}CO line also becomes optically thick and a rarer isotope has to be used, e.g., $\mathrm{C^{16}O/C^{18}O} \simeq 500$ and $\mathrm{C^{18}O/C^{17}O} \simeq 4$ and column densities can be scaled from Eq. (10.13). Densities then follow from an adopted scale size estimated, for example, from projected sizes assuming a cloud (spherical) symmetry. This method has to assume an abundance which has to be calibrated against other means (e.g. dust, see below). Besides the drawbacks mentioned above, this is not a reliable method in dense cores as CO is known to freeze out. Furthermore, galactic isotopic elemental abundance gradients introduce another uncertainty when comparing clouds in different parts of the galaxy.

Dust Column Densities

Dust provides another means to assess the density and density distribution of molecular clouds. The reddening of stars behind a dark cloud (Figure 10.8) depends on the column density; viz.,

$$m\,(\lambda_1) - m\,(\lambda_2) = M\,(\lambda_1) - M\,(\lambda_2) + A\,(\lambda_1) - A\,(\lambda_2)\,, \quad (10.14)$$

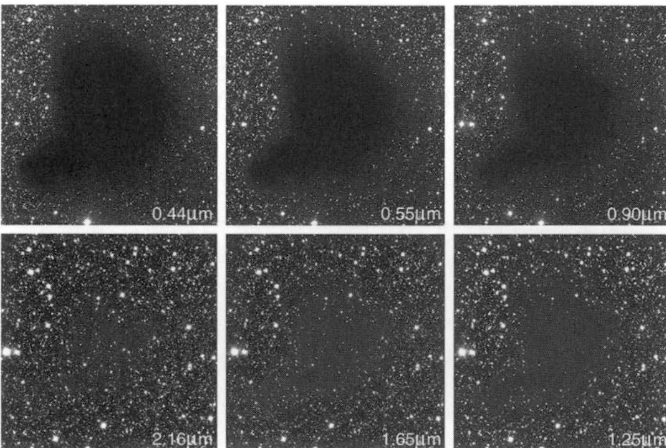

Figure 10.8 While Barnard 68 is very opaque in the visible and nary a star shines through except in the outskirts, in the infrared, this dark cloud becomes quite transparent. This reflects the strong reddening associated with dust extinction. The many stars present in the IR allow an accurate estimate of the column density distribution from star counts. Extinction decreases from left to right in the top row and then again from right to left in the bottom row. Figure taken from [5]

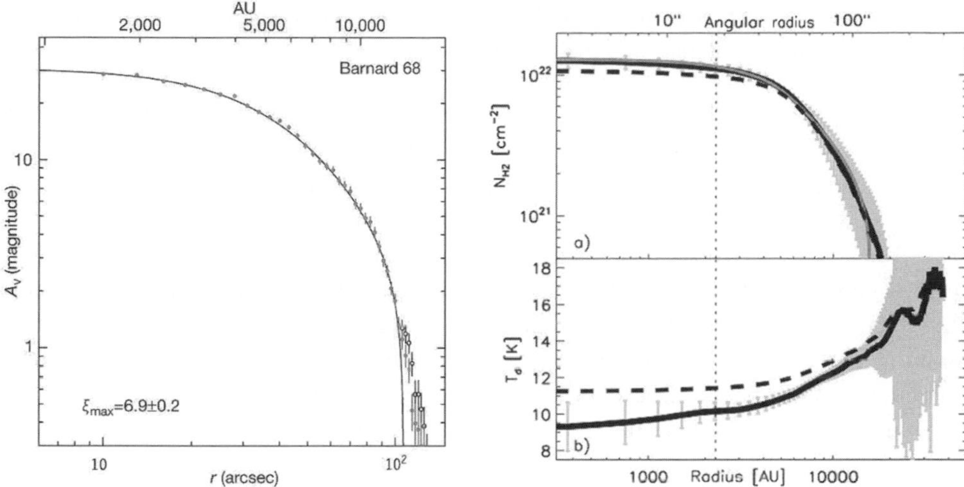

Figure 10.9 Left: Color excess of background stars allows the determination of (azimuthally averaged) column density distribution – in units of visual magnitude – for B68 (gray dots; black dots include the protrusion toward the South-East; see Figure 10.8). The curve represents the best fit using a Bonnor–Eberts sphere. The ξ_{max} ($\equiv$ (R/a) $\sqrt{4\pi G\rho_c}$, with R the size of the cloud, a the sound speed, and ρ_c the central density) value of the fit is indicated. Figure taken from [5]. Right: Column density (a) and dust temperature (b) profiles of B68 obtained from a pixel-by-pixel analysis of the far-IR/sub-mm SED using two different methods (black solid and dashed lines). The gray curve represents the best-fit Bonnor–Ebert model to the solid curve. The vertical dotted line indicates the effective resolution. Note that $N_{H_2} = 10^{21}$ cm^{-2} corresponds to $A_v = 1$ magn. Figure taken from [201]

where m_i, M_i, and A_i are the observed and intrinsic magnitudes of the star and the extinction at wavelength, i. There are about 3000 stars visible in the B68 image (Figure 10.8) of which $\sim$1000 heavily extincted (c.g. not present at visible wavelengths). The color excess measurements toward individual stars can then be appropriately smoothed and, by adopting an extinction law, these color excesses can be related to the total visual extinction. Extinction over the 0.35–3.5 μm range is often represented by a powerlaw $A(\lambda) \propto \lambda^{-1.8}$. Conversion from optical extinction to hydrogen nuclei density is then done using the relationship derived from UV measurements of stars behind diffuse clouds ($N_H = 1.9 \times 10^{21} A_V$ cm^{-2}). The results of this analysis for B68 are shown in Figure 10.9. The azimuthal column density distribution is well represented by a Bonnor–Ebert sphere (Section 11.2) appropriate for a self-gravitating, pressure-confined, isothermal sphere with a center to boundary density contrast of $\simeq$17. As an example, the derived radius and total mass for B68 are 0.06 pc and 2.1 $M_\odot$.

Alternatively, the molecular cloud dust column density distribution can be derived from the observed Spectral Energy Distribution (SED) in the far-IR and sub-mm. For optically thin, dust emission, the surface brightness is given by,

$$S(\nu) = B(\nu, T_d)\,(1 - \exp[-\tau(\nu)]) \simeq B(\nu, T_d)\,\tau(\nu), \tag{10.15}$$

with T_d the dust temperature and τ the dust optical depth at frequency ν. The optical depth is often represented by a modified black body with,

$$\tau(\nu) = \tau(\nu_0) \left(\frac{\nu}{\nu_0} \right)^{\beta}, \tag{10.16}$$

with ν_0 a reference frequency. Modern analysis relies on Planck and Herschel data and uses $\nu_0 = 353$ GHz (850 μm). There are three free parameters (T_d, β, and τ_0), which are fitted for each pixel. A dust model is then needed to relate derived τ_0's to A_V's and N_H's. Dust models for the far-IR are notoriously unreliable as, until Herschel/Planck, there was little data to compare them to and dust properties in dense clouds are likely very different from that in the diffuse ISM (the traditional arena where dust models are validated against extinction measurements). The Planck collaboration has made an all-sky analysis of the dust emission and derived a value for the dust cross section at 353 GHz,

$$\kappa_{353} \simeq 1.2 \times 10^{-26} \text{ cm}^2 \text{ H-atom}^{-1} \tag{10.17}$$

valid over the range $10^{21} < N_H < 3 \times 10^{22}$ H-atom cm^{-2}. For molecular clouds, the far-IR SED analysis can be calibrated against the near-IR dust extinction analysis described above. Again, these column densities can then be converted into densities using a model for the cloud geometry.

Molecular Excitation Analysis: Density

The density can also be determined from suitably chosen line pairs and an excitation analysis (cf. Section 4.2.3). This is similar to the discussion for the temperature (Section 10.1.1). Here we illustrate this method for the density with NH_3, which has been widely used in dense core studies. Figure 10.10 shows the results of an excitation calculation for ammonia. The ratio of inversion transitions measures the kinetic temperature, largely independent of density (cf. Eq. (10.10) and Figure 10.3). In practise, only the excitation

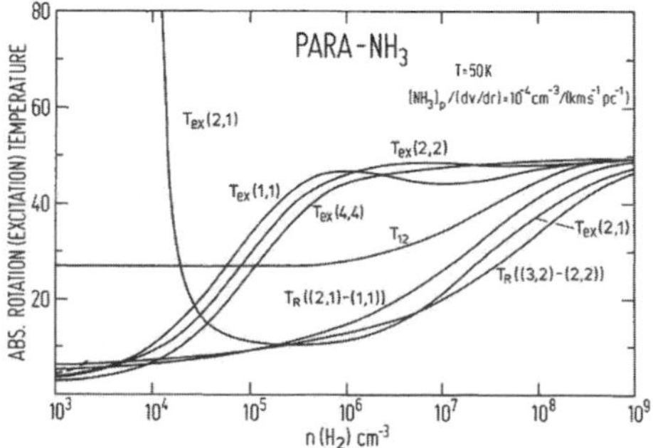

Figure 10.10 Calculated rotational excitation temperatures for different NH_3 levels. The lines marked T_{ex} refer to excitation temperature derived for the inversion transitions (T_{hf} in Eq. (10.7)). The curves marked T_{12}, $T_R((2,1)-(1,1))$, $T_R((3,2)-(2,2))$ measure the excitation (Eq. (10.10)) of levels (1,1)–(2,2), (2,1)–(1,1), (3,2)–(2,2), respectively. Parameters involved in the calculation are indicated. Figure adapted from [240]

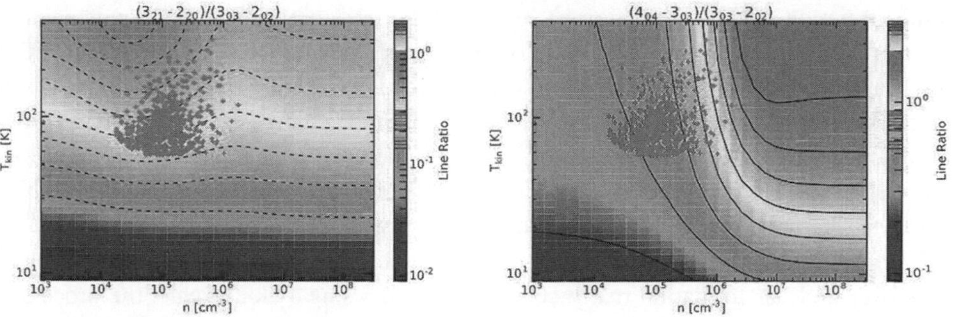

Figure 10.11 Line ratio's of para H_2CO lines. The line ratio in the left panel is mainly sensitive to the temperature. The right panel shows a line ratio that is sensitive to temperature and density (for $n < 10^7$ cm^{-3}). Combination of these two line ratios allows a measurement of density and temperature. The red dots represent observations of both of these line ratios to determine the best fit in temperature and density. Color scales show a range of 2 orders of magnitude in the ratio. The dashed contours range from 0.05 to 0.5. The solid contours range from 0.2 to 2. Figure taken from [115] (A black and white version of this figure will appear in some formats. For the color version, please refer to the plate section.)

temperature of the (1,1) inversion transition can be measured for molecular clouds. Once the temperature is known, the relative population of the nonmetastable levels provides a measure of the density.

Other molecules can also provide good measures of the density and temperature. Figure 10.11 illustrates the use of this method for two para H_2CO line ratios. The combination of these two line ratios allows simultaneous determination of density and temperature for the emitting gas. Line combinations from many other species can be used as well, as long as the density (or density range) of the object falls within the critical density range of the transitions involved. Other species that are often used are CS and HCN, which trace a range of densities as well. In all these analyses, care should be taken as the difference in transition frequencies implies very different beam sizes.

Rotation Diagrams

Subthermal excitation of rotational levels can have a pronounced effect on the rotation diagram (Section 4.2.2). Levels at higher excitation energies generally have larger Einstein A's and hence they fall out of LTE well before lower energy levels (Figure 4.4). If the temperature is known, this behavior can be used to determine the density of the gas.

10.2 Characteristics of Molecular Clouds

Molecular clouds are self-gravitating where (enhanced) thermal pressures as well as magnetic fields and turbulence support the gas against gravity. Massive stars form in dense molecular cloud cores and their interaction with the environment is central to the evolution of the ISM. In recent years, a general paradigm for the intertwined processes of molecular cloud formation and star formation has emerged. Formation of molecular clouds involves multiple steps starting with the conversion of warm diffuse gas into HI superclouds, CO-dark translucent clouds, and Giant Molecular Cloud associations through the effects of compression by spiral arms and self-gravity or through coalescence of smaller CO-dark clouds or molecular clouds in spiral arms. Shocks due to converging and colliding flows – driven by stellar feedback such as expanding HII regions, stellar winds, and SNe blast waves and by turbulence – produce high density sheet-like structures and convert atomic gas into molecular gas. These structures fragment into filaments and cores, where gravity takes over and star formation really commences. On a single cloud scale, the process of gravitational collapse is counteracted by the turbulent pressure support generated by the cloud accretion process itself as well as by the feedback from (massive) YSOs. At the same time, internal turbulence continuously forms and dissolves denser filamentary structures – dominating the observed structure of molecular clouds – and self-gravity selects from the ensuing core mass distribution function those that will collapse to form stars. As a result, the initial mass function resembles closely the core mass spectrum. Thus, the overall distribution of molecular gas, the properties of individual clouds, and the star formation process are connected through turbulence at a deep level.

This discussion starts by considering the ensemble of molecular clouds in the Milky Way, then focuses on the internal structure of molecular clouds, summarizes observed scaling relationships for the physical properties of clouds and their relationship to turbulence, and briefly mentions the evolution of molecular clouds before turning to the so-called X_{CO} factor linking observed CO intensities to cloud masses.

10.2.1 The Distribution of Molecular Clouds

The main constituent of molecular clouds, H_2, is difficult to observe directly and CO is often used as a proxy. Over the years, a number of CO surveys have charted the Milky Way and studied the characteristics and statistical properties of molecular clouds (Figure 10.12). Molecular clouds span a wide range in physical properties. They run from less than a parsec and a mass of ~ 1 M$_\odot$ to tens of parsecs with a mass of 10^6 M$_\odot$. Barnard clouds (aka: Bok Globules) are small (<1 pc), isolated, nearby ($\lesssim 200$ pc) dark clouds that are typically spherical and have densities of $\sim 10^4$ cm^{-3} and masses of $10-50$ M$_\odot$. More often than not, they harbor low mass protostars. B68 is an example of such a globule (Figure 10.8) without an embedded object. Most of the molecular gas in the Milky Way is, however, in much more massive clouds with sizes of $10-100$ pc and masses of $10^4 - 6 \times 10^6$ M$_\odot$. The average density is quite low ($\sim 10^2$ cm^{-3}) but such clouds are very inhomogeneous with structures on all scales and densities. Molecular clouds have very sharp boundaries in CO but are embedded in extended halos in which hydrogen is in the form of HI or H_2. Over the years it has become clear that large amounts of interstellar gas are not traced either by the HI 21 cm line or by CO rotational transitions. Infrared surveys – from IRAS to COBE to Planck – as well as γ-ray surveys have revealed excess emission related to gas not accounted for by HI and CO. Likewise, OH observations reveal the presence of widespread

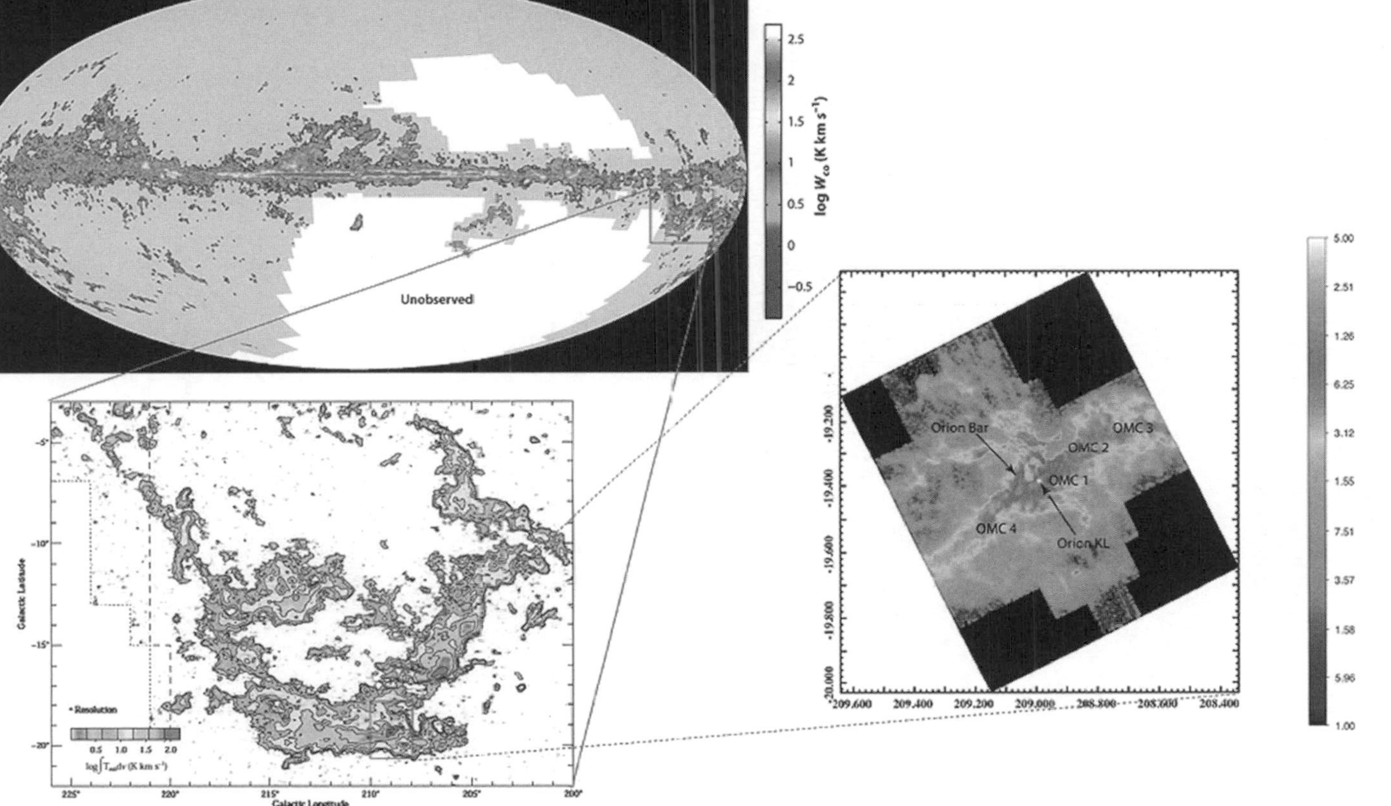

Figure 10.12 Top: Composite of CO $J = 1 - 0$ surveys of the Milky Way. This image is dominated by local cloud structures. From left to right, we can recognize the Taurus & Perseus molecular clouds (just below the plane), the Cepheus cloud just above the plane, the Ophiuchus Cloud above the (galactic) center with the Lupus cloud left of it, and at the right edge, the Orion-Monoceros cloud complexes. Figure taken from [104]. Bottom left: Blow up of the Orion-Monoceros cloud complex with counter clock-wise: Monoceros with the southern filament sticking up, the Orion A molecular cloud, the Orion B molecular cloud, Orion East, and the Southern filament. Figure taken from [248]. Bottom right: Blow up of the Orion A molecular cloud in CO $J = 2 - 1$ with in the center the bright BN/KL region, the Orion Bar PDR, the Orion Veil, all superimposed on the integral-shaped filament containing the dense cores OMC1 through OMC4. Figure adapted from [22] (A black and white version of this figure will appear in some formats. For the color version, please refer to the plate section.)

gas halo's surrounding molecular clouds not traced by HI or CO. In this so-called CO-dark molecular gas, hydrogen is in molecular form but carbon is in C^+ rather than CO.

Various techniques have been developed to recognize and identify clouds (and structures in these clouds) in CO surveys and these have been validated against numerical hydrodynamic simulations. For molecular gas, a convenient representation of the cloud distribution function is,

$$\frac{dN}{d\ln M} \simeq N_u \left(\frac{M_u}{M}\right)^{\alpha} \qquad M < M_u, \tag{10.18}$$

with $N(M)$ the number of clouds in the mass interval, $[M, M + dM]$, and where M is in units of $M_\odot$. Here, N_u is the number of clouds at the upper mass cut off, M_u. For the cumulative distribution, we have,

$$N(> M) = \frac{N_u}{\alpha}\left(\left(\frac{M_u}{M}\right)^{\alpha} - 1\right) \qquad M < M_u. \tag{10.19}$$

The upper mass cut off is well determined to be $= 6 \times 10^6$ $M_\odot$. Fitting the data in the surveys and extrapolating to the full Milky Way disk yields $N_u = 25$ and $\alpha = 0.6$. However, the total mass in molecular clouds is then only 4×10^8 $M_\odot$, which falls short of the total (CO-derived) molecular mass of the Milky Way of 10^9 $M_\odot$. This is related to the detection limit of the CO surveys used in this analysis. So, either N_u is larger ($\simeq$65), the powerlaw is steeper ($\alpha \simeq$0.85), or a combination of both. In any case, most clouds are small but most of

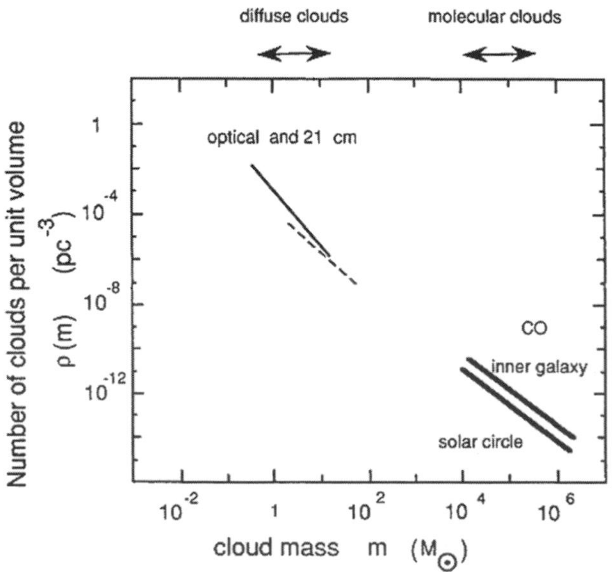

Figure 10.13 The cloud mass spectrum spans a wide range – from diffuse atomic clouds to giant molecular clouds – in a more or less continuous fashion. Figure taken from [62]

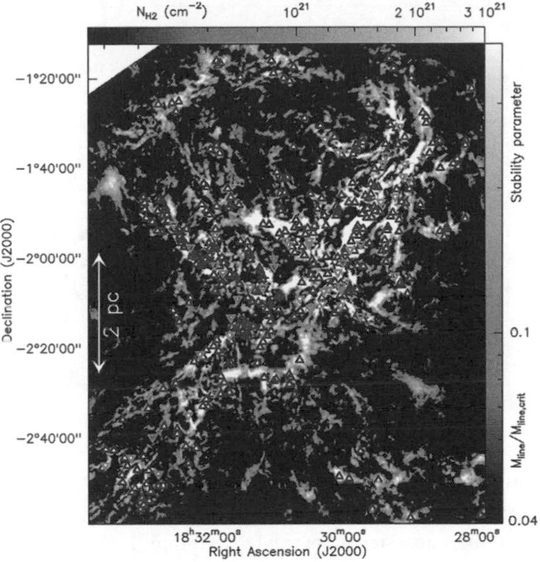

Figure 10.14 Column density map of Aquila derived from Herschel SPIRE/PACS data. The contrast of the filaments with respect to the non-filamentary background has been enhanced using a curvelet transform. The white areas highlight regions where filaments have a mass per unit length larger than half the critical value and are likely to be gravitationally unstable. Locations of candidate prestellar cores and protostellar cores are overlaid as blue and green triangles, respectively. Figure adapted from [124] (A black and white version of this figure will appear in some formats. For the color version, please refer to the plate section.)

the CO-derived molecular mass is in large clouds. There is a clear cut-off at high mass of the cloud distribution function, which is likely related to the physical processes involved in the formation and destruction of clouds. Finally, surveys show that giant molecular clouds are limited to spiral arms but smaller molecular clouds are present in the interarm region.

Internally, molecular clouds show a range of properties as well, with dense, filamentary structures embedded in low density, interclump material. Herschel sub-millimeter dust emission studies have been very instrumental in elucidating their internal structure as their large dynamic range allows a quantitative assessment. Using various techniques, the filamentary structure of molecular clouds can be greatly enhanced (Figure 10.14). Measured over the whole cloud, about 15% of the mass is in the form of these filaments. But for high column densities ($A_V > 7 - 10$ mag), 50 to 75% of the gas mass is in filaments and some 15% is in dense prestellar cores. The derived core mass distribution resembles the initial mass function of stars but shifted to higher masses by some 0.3 $M_\odot$ (Figure 10.15). Isolated dark clouds are well described by isothermal Bonnor–Ebert spheres (Figure 10.9), which may evolve through quasistatic contraction as, for example, supporting magnetic field slowly leaks out throughout the process of ambipolar diffusion (Section 11.2). Eventually, these cores form protostars surrounded by protoplanetary disks. Analysis of the filametary

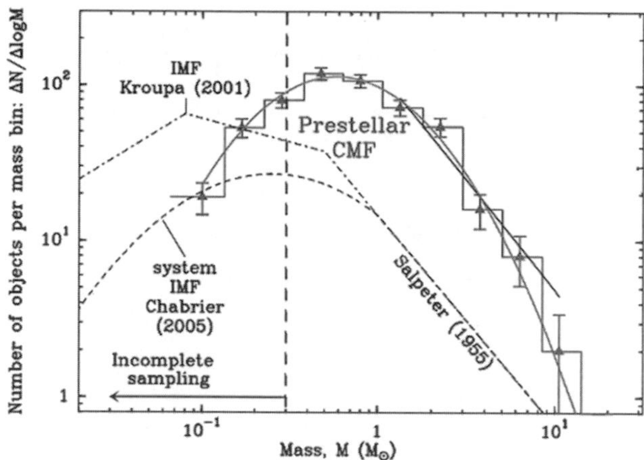

Figure 10.15 The core mass distribution function as derived from IR emission maps of the Aquila molecular cloud. For comparison, various determinations of the initial mass function of stars are also shown. The dashed vertical line indicates incomplete sampling. Figure adapted from [8, 124]

structures that dominate the Herschel images of molecular clouds reveals a radial density profile for the filaments that can be described by,

$$\rho\,(r) = \frac{\rho_c}{\left(1 + \left(r/R_{flat}\right)^2\right)^{p/2}}, \tag{10.20}$$

with ρ_c the central density, R_{flat} ($\simeq$0.05 pc for local clouds) the radius of the inner flat region, and p ($\simeq$ 2) the power law dependence at large radii. These filaments might represent the dense gas resulting from shock compression in converging flows of interstellar turbulence. If these filaments exceed a critical mass per unit length (whose value depends on the means of internal support), they will fragment into dense prestellar cores (Figure 10.15). These dense cores can again be described as isothermal Bonnor–Ebert spheres.

10.2.2 Scaling Relations

Ananlysis of molecular observations has revealed a number of global scaling relations, linking the velocity dispersion, σ, to the size, R, density, n, and mass of the cloud or cloud structure. The precise coefficients depend to some extent on the definition of size and are also somewhat uncertain as some of the observed velocity width and density estimates have issues.

$$\sigma = 0.7 \left(\frac{R}{\text{pc}}\right)^{0.5} \quad \text{km/s} \tag{10.21}$$

$$\sigma = 0.42 \left(\frac{M}{\text{M}_\odot}\right)^{0.20} \quad \text{km/s} \tag{10.22}$$

$$N_H = 1.5 \times 10^{22} \qquad \text{cm}^{-2} \tag{10.23}$$

$$n = 5 \times 10^3 \left(\frac{1\text{ pc}}{R}\right) \qquad \text{cm}^{-3} \tag{10.24}$$

These relations for the physical characteristics of clouds hold over about three orders of magnitude from prestellar cores (0.1 pc) to the largest cloud sizes (100 pc). The constant column density corresponds to an A_V of 8 or a surface density of 160 $M_\odot$ pc^{-2}.

As these relations show, $\frac{M}{R} \propto \sigma^2$, which is the relationship for a virialized cloud.[3] The virial theorem for a gravitationally bound, spherical cloud with a R^{-1} density profile and supported by (turbulent) pressure reads,

$$\sigma^2 = \frac{2GM}{9R} \qquad \text{or} \tag{10.25}$$

$$\left(\frac{\sigma}{\text{km/s}}\right)^2 = 1040\frac{(M/1\,M_\odot)}{(R/1\text{ pc})}. \tag{10.26}$$

While the observational validity of this relation cannot be rigorously established given the uncertainties, these observed scaling relations have nevertheless led to the notion that molecular clouds are in approximate virial equilibrium.

These CO surveys have provided us with an overview of the properties of typical molecular clouds in the Milky Way. As these properties are related through powerlaws, the term "typical cloud" is inherently connected to the property under consideration: e.g. most molecular clouds are small but most of the mass is in large molecular clouds. The total molecular gas mass in the Milky Way is $\simeq 10^9$ $M_\odot$ and there are some 5×10^6 molecular clouds but most of those are tiny. The upper mass cut off is 6×10^6 $M_\odot$ and there are about 1500 giant molecular clouds (GMCs) more massive than 10^5 $M_\odot$. Focusing on these GMCs, we arrive at an average size of 60 pc and an average projected surface area of 1700 pc^2 or a total surface area of molecular clouds of 2.6×10^6 pc^2. The average surface density is 1.5×10^{22} H-nuclei/cm^2 and the average density is 100 cm^{-3}. As mentioned, these values greatly depend on how the calculation is done and different values can be found in the literature. Perhaps, the key point to take away is that, in a powerlaw distribution, there really is no such thing as a typical cloud.

10.2.3 The Probability Density Distribution

Compressible, supersonic turbulence creates strong density fluctuations and the molecular cloud Probability Density Function (PDF) is often approximated by a log normal density distribution. Heuristically, this can be understood as the cumulative effect of turbulent energy dissipation through strong shocks in localized regions. After the initial compression, the warm, shocked gas cools down quickly further compressing the gas and, on timescales longer than the cooling timescale, the system can be treated as largely isothermal. Rarefaction waves will counteract this densification process. Over time, each parcel of gas will be

[3] In a virialized cloud, the potential energy equals twice the kinetic energy of the gas.

shocked/rarefied, randomly, by a large number of shocks/rarefaction waves each of which can produce a density contrast, δ_i. The resulting density is then the product of these,

$$\rho(t) = \rho(t_0) \prod_i \delta_i. \tag{10.27}$$

Equivalently,

$$\ln[\rho(t)] = \ln[\rho(t_0)] + \sum_i \ln[\delta_i]. \tag{10.28}$$

If the density fluctuations are uncorrelated, this will lead to a gaussian distribution in the log of the density by the central limit theorem,

$$PDF(s) = \frac{1}{\sqrt{2\pi}\sigma_s} \exp\left[-(s - s_0)^2 / 2\sigma_s^2\right], \tag{10.29}$$

where

$$s = \ln\left[\rho/\rho_0\right], \tag{10.30}$$

with ρ_0 the volume averaged density, and σ_s the dispersion. Numerical simulations give for the dispersion,

$$\sigma_s^2 \simeq \ln\left[1 + \left(\frac{\beta}{1+\beta}\right) b^2 \mathcal{M}^2\right]. \tag{10.31}$$

Here, β is equal to the ratio of the thermal to the magnetic energy density and $\mathcal{M}$ is the rms Mach number of the turbulence. The parameter, b, depends on the forcing of the turbulence and is, typically, $b \simeq 0.4$. Hence, if β is smaller, shocks will be magnetically cushioned and the density contrast is less. Smaller Mach numbers for the turbulence will also lead to weaker shocks and smaller density fluctuations. Self-gravity adds then a powerlaw density distribution at the high density end.

While this discussion focused on the density distribution, any physical variable that is controlled by random processes creating unrelated fluctuations will lead to log-normal distributions. That holds of course also for, e.g., sizes and masses and the relationship between these is then an expression of the underlying energy source, the gravitational energy of the cloud.

10.2.4 Evolution of Molecular Clouds

In terms of the galactic distribution, molecular gas is largely confined to the inner galaxy (Figure 10.16) and to the galactic plane with a scale height of 90 pc. Giant molecular clouds are confined to the spiral arms and most molecular gas can be found in the spiral arms. However, smaller clouds are present in the inter arm region as well. The lifetime of Giant Molecular Clouds is estimated to be some 30 Myr but smaller clouds may live much longer. If giant molecular clouds are assembled from smaller molecular clouds in spiral arms and then fall apart again after passage through the arm, the "lifetime" of molecular gas may actually be much longer than this 30 Myr.

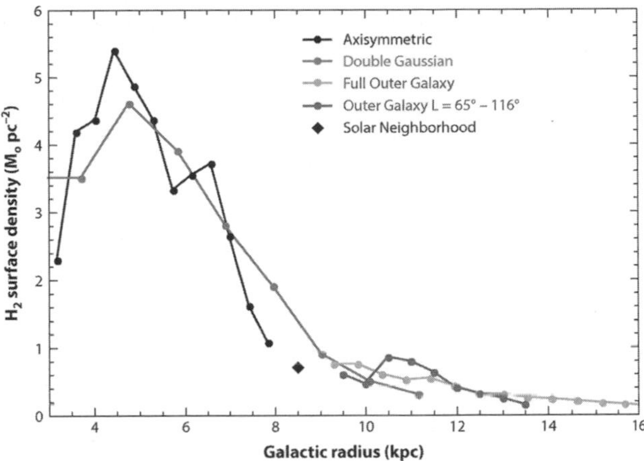

Figure 10.16 The molecular mass surface density as a function of galactocentric radius. Figure adapted from [104]

One class of giant molecular cloud formation mechanisms relies on converging flows to bring material together and this process naturally injects turbulence into the cloud. On a local scale, these converging flows may be driven by stellar feedback – e.g. thermally expanding HII regions, stellar winds, supernova bubbles – and this may lead to the formation of modest-sized molecular clouds, $\lesssim 10^4$ $M_\odot$. The RCW 120 and λ Ori bubbles are examples of such local scale processes where sometimes new star formation has already been triggered. The concerted action of many supernovae in a rich OB association can lead to the formation of superbubbles and those can transport much more material. Giant molecular clouds can be built up this way and new star formation initiated. The giant molecular cloud associated with the active region of massive star formation in W3 is a prime example, of the product of the mechanical action by the rich cluster, IC 1805, in W4. Giant molecular clouds can also form through collisions in spiral arms where clouds are concentrated and the collision timescale allows the build up of giant molecular clouds. The mini starburst, W43 – located at the meeting point of the Scutum-Centaurus spiral arm and the central bar – offers an example of such activity. In the Cygnus X region, converging flows likely played a role in the formation of the dense molecular ridge and the DR 21 star formation activity and this was probably driven by the mechanical action of the $\sim$100 O stars in the Cyg OB 2 association. Giant molecular clouds could also be produced by large scale, self-gravitating instabilities; i.e. the magneto-Jeans-instability, a gravitational instability in a magnetized, thin disk. In this case, the turbulence is inherited from the ISM. Again, spiral arms are the preferred sites for GMCs due to both the increased surface density and the surpression of the effects of velocity shear. Analytical theory and simulations indicate an upper mass for GMCs through this mechanism of 10^7 $M_\odot$. In contrast, the Parker instability – concentrating gas by sliding it down the field lines produced by buckling of a horizontal magnetic field threading a gravitating, thin disk – seems to have issues in producing the characteristics of GMCs.

Star formation is a very inefficient process, globally converting some 5% of the molecular cloud's mass into stars. Molecular clouds must be disrupted, therefore, before star formation can go to completion. Protostellar jets can have some local effects – on the pc-scale-size – but do not affect clouds on a global scale. Photoionization can play a role in molecular cloud disruption as a champagne flow from a fully developed HII region can "launch" gas flows at a few times the sound speed in the ionized gas and that exceeds the escape velocity of the cloud. Taking M42 in Orion as an example, the ionized gas mass loss rate from the Orion core is some 10^{-4} $M_\odot$/yr and over a million years, some 100 $M_\odot$ will be lost. This corresponds to some 3% of the mass of the OMC 1 core. The Trapezium stars will move away with time, limiting further disruption. Chance encounters of massive stars with molecular clouds may add to this as these are common in the crowded environment of OB associations. As examples of such activity, σ Ori will remove some 500 $M_\odot$ from the Orion B molecular cloud over its interaction period ($\simeq$1 Myr) and the star ξ Per will remove a similar amount from the California molecular cloud. The kinetic energy input by stellar winds from massive stars is of more importance. Again, looking at Orion, the stellar wind bubble blown by θ^1 Ori C has swept a shell (the veil) containing some 2600 $M_\odot$, comparable to the remaining mass in the OMC 1 core. This shell expands at 13 km/s, which exceeds the escape velocity of the cloud, and this material will disperse into the surroundings. Mechanical energy input by supernova, on the other hand, is much less important. SN explosions occur after some 5 Myr and the random motion of stars will have created enough distance that the direct impact on the cloud will be limited. The Orion belt stars are a case in point. Rather, the SN energy will be expended in rejuvenating hot and tenuous gas, filling the bubbles created by earlier SN explosions. In the ecology of the galaxy, SNe are the streetcleaners of the Milky Way, sweeping up molecular cloud material ablated by stellar winds and ionized champagne flows into large shells. From that perspective, the total mass of swept up gas in the walls of the Orion–Eridanus superbubble amounts to 2.5×10^5 $M_\odot$ and hence the concerted "pecking" of all the OB stars in the Orion association has removed the equivalent of one (modest) GMC from this region.

10.2.5 The X_{CO} Factor

Given the relationships derived in Section 10.2.2, the cloud mass can be derived from CO observations. The observed integrated surface brightness of the CO line is related to the column density and the CO luminosity to the mass of the cloud. Defining the so-called $X_{\rm CO}$ factor,

$$N\left(\mathrm{H_2}\right) = X_{\rm CO} \int T_B dv, \tag{10.32}$$

we have that the mass of a cloud and the luminosity measured in ^{12}CO are directly related: $M = \pi R^2 N(\mathrm{H_2})\, \mu m_H = \pi R^2 \mu m_H X_{\mathrm{CO}} \int T_B dv = \mu m_H X_{\mathrm{CO}} L_{\mathrm{CO}}$. Various methods exist to determine the X_{CO} factor. Replacing the integral by $T_B \delta v_{FWHM} = 2.35 T_K \sigma$, and using the virial theorem and the relationships between column density, mass, and radius, we find,

$$X_{\mathrm{CO}} \simeq 3 \times 10^{20} \left(\frac{10\,\mathrm{K}}{T_K}\right) \left(\frac{n}{10^3\,\mathrm{cm}^{-3}}\right)^{1/2} \quad \mathrm{cm}^{-2}/\mathrm{K\ km\ s}^{-1} \tag{10.33}$$

This relationship does depend on the temperature and weakly on the density. The relation between the observed CO intensity can also be calibrated against other means of measuring the column density or mass of a cloud, such as the near IR extinction,

$$X_{\mathrm{CO}} \simeq 2 \times 10^{20} \quad \mathrm{cm}^{-2}/\mathrm{K\ km\ s}^{-1}, \tag{10.34}$$

the far-IR emission as determined from the IRAS data,

$$X_{\mathrm{CO}} \simeq 1.8 \times 10^{20} \quad \mathrm{cm}^{-2}/\mathrm{K\ km\ s}^{-1}, \tag{10.35}$$

the far-IR emission as determined from the Planck data,

$$X_{\mathrm{CO}} \simeq 2.5 \times 10^{20} \quad \mathrm{cm}^{-2}/\mathrm{K\ km\ s}^{-1}, \tag{10.36}$$

^{13}CO calibrated against A_V and the local dust-to-gas ratio,

$$X_{\mathrm{CO}} \simeq 1.8 \times 10^{20} \quad \mathrm{cm}^{-2}/\mathrm{K\ km\ s}^{-1}, \tag{10.37}$$

and γ-rays,

$$X_{\mathrm{CO}} \simeq 1.7 \times 10^{20} \quad \mathrm{cm}^{-2}/\mathrm{K\ km\ s}^{-1}. \tag{10.38}$$

Each of these determinations of the X_{CO} factor has its own issues and accompanying uncertainties but it is gratifying to see that they converge to a value of $\simeq 2 \times 10^{20}$ cm^{-2}/K km s^{-1} or 4.3 M$_\odot$ pc^{-2} $\left(\mathrm{K\ km\ s}^{-1}\right)^{-1}$. As a word of caution, the X_{CO} factor should never be used to determine column densities of individual lines of sight. It provides estimates for ensembles of clouds. Also, in order to be meaningful, data derived from these different calibrations should not be mixed. Finally, as emphasized by Eq. (10.33), the X_{CO} factor will depend on the temperature and density of the gas and this should be kept in mind when analyzing, for example, regions of massive star formation. At low column densities, CO is not a good tracer of gas anymore as it will be dissociated by penetrating FUV photons from the interstellar radiation field. In the solar neighborhood, this occurs for a column density of $\simeq 2 \times 10^{20}$ H-atoms/cm^2 (cf. Figure 9.5). This value will depend on the dust properties and will be very different in, for example, low metalicity environments or in regions where coagulation has affected the FUV properties of the dust.

10.3 The Energy Balance

10.3.1 Heating of Molecular Clouds

Cosmic Ray Heating

Low energy cosmic rays (∼1–10 MeV) are most efficient in ionizing and heating the gas. A high energy proton can ionize a gas atom. The substantial kinetic energy (∼35 eV) of the resulting primary electron can be lost through (elastic) collisions with other electrons or through ionization or excitation of gas atoms or molecules. The total ionization rate, ξ_{CR}, including secondary ionizations, depends then on the energy of the primary and the electron fraction, x_e,

$$n\,\xi_{CR} = n\,\zeta_{CR}\left[1+\phi^{H}(E,x_e)+\phi^{He}(E,x_e)\right], \tag{10.39}$$

where ζ_{CR} is the primary ionization rate, and the $\phi^i(E,x_e)$'s are the average number of secondary ionizations of H and He per primary ionization. When the electron abundance is low, the number of secondaries is about 0.8.

Because their propagation is also most affected by the interstellar and solar system magnetic field, the flux of low energy cosmic rays is not well constrained by direct observations but can be inferred from observations of molecular ions, in particular H_3^+ (Section 9.5). With a primary ionization rate of 2×10^{-16} s^{-1}, the total cosmic ray ionization rate is $\xi_{CR} \simeq 3\times10^{-16}$ s^{-1}. The heating rate is then given by,

$$n\,\Gamma_{CR} = n\,\zeta_{CR}\,E_h(E,x_e), \tag{10.40}$$

with $E_h(E,x_e)$ the average heat deposited per primary ionization. For low degrees of ionization, $E_h(E,x_e)\simeq 7$ eV. The cosmic ray heating rate is then,

$$n\,\Gamma_{CR} = 3\times10^{-27}\,n\left[\frac{\zeta_{CR}}{2\times10^{-16}}\right] \quad \text{erg cm}^{-3}\,\text{s}^{-1}. \tag{10.41}$$

Inside dense molecular clouds, however, low energy cosmic rays – which dominate the ionization rate in the diffuse ISM – may be attenuated and typical adopted ionization rates are an order of magnitude lower ($\zeta_{CR} \simeq 2\times10^{-17}$ s^{-1}; Sections 10.5.3 and 10.5.4) and this has been assumed in Figure 10.17.

Dust–Gas Heating

Dust and gas are not in thermodynamical equilibrium as their heating and cooling processes are very different. As a result, they acquire (slightly) different temperatures. Typically, except for the edges of molecular clouds and PDRs, the dust is warmer than the gas. Gas atoms (H, H_2, He) will collide with a grain, possibly stick and thermally accommodate and then be released again into the gas phase. This collisional heating process can be an important gas heating source. The heating rate is then given by,

$$n\Gamma_{g-d} = n\,n_d\sigma_d\left(\frac{8kT}{\pi m}\right)^{1/2}(2kT_d - 2kT)\,\alpha_a, \tag{10.42}$$

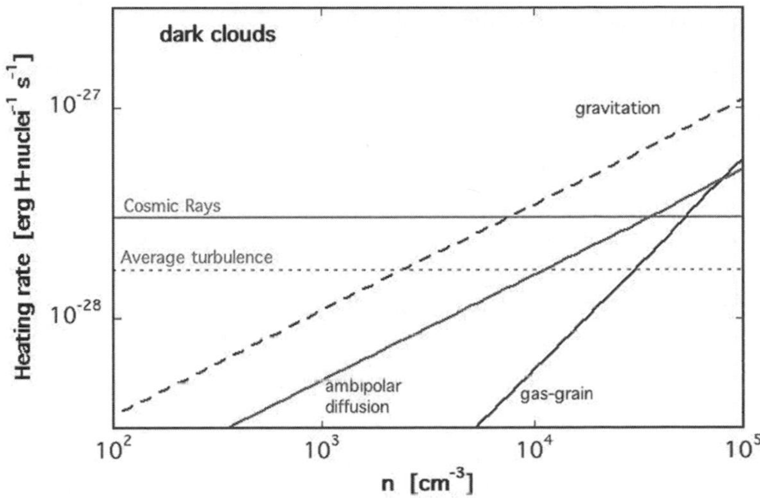

Figure 10.17 The contribution of different heating processes (plotted per H-nucleus) in dark cores as a function of the density.

with $n_d\sigma_d$ the dust density and geometric cross section ($10^{-21}\,n$ cm^{-1}), and $2kT$ the average kinetic energy of a gas atom striking the grain surface. The accommodation coefficient, α_a, measures how well the gas atom accommodates to the grain; i.e. an α_a of unity corresponds to a a bouncing particle which has completely thermalized and leaves with $2kT_d$. Experimental and theoretical studies suggest, $\alpha_a \simeq 0.15$. The heating rate is then,

$$n\Gamma_{g-d} \simeq 10^{-33}\,n^2\,T^{1/2}\,(T_d - T) \qquad \text{erg cm}^{-3}\ \text{s}^{-1}. \tag{10.43}$$

This relation is shown in Figure 10.17 for $T = 8$ K and $\Delta T = 2$ K.

The temperature of a dust grain is set by the radiative energy balance. The energy absorbed by a dust grain is given by,

$$\Gamma_{abs} = 4\pi\sigma_d \int_0^\infty Q\,(\lambda)\;J\,(\lambda)\;d\lambda, \tag{10.44}$$

with $J\,(\lambda)$ the mean intensity of the radiation field at wavelength λ. The energy emitted by a dust grain is,

$$\Gamma_{em} = 4\pi\sigma_d \int_0^\infty Q\,(\lambda)\;B\,(T_d,\lambda)\;d\lambda, \tag{10.45}$$

with T_d the radiative equilibrium dust temperature and B the Planck function. Consider a grain at the surface of a molecular cloud (e.g. exposed to the Habing field over 2π steradians), and using the absorptive properties of silicate material, we have,

$$T_{sil} = 14\left(\frac{1000\ \text{Å}}{a}\right)^{0.06} \qquad \text{K}. \tag{10.46}$$

Graphite grains will reach very similar temperatures,

$$T_{gra} = 16 \left(\frac{1000\ \text{Å}}{a}\right)^{0.06} \quad \text{K}. \tag{10.47}$$

Deeper in the cloud, the interstellar radiation field is attenuated and the dust temperature drops. Theory predicts that the dust temperature drops by about a factor 2 for an externally heated cloud. As an example, Herschel observations of B68 reveal a drop from about 16 at the surface to 8 K in the deep interior (Figure 10.9).

Heating Due to Ambipolar Diffusion

In a partially ionized gas, such as a molecular cloud core, ions and neutrals may develop a small (drift) velocity difference, for example, through the counterplay of magnetic fields and gravity. Ambipolar diffusion plays a role in the slow leakage of magnetic fields from pre-stellar molecular cloud cores and the accompanying settling into a isothermal sphere structure. The friction between the ions and the neutrals heats the gas at a rate,

$$n\Gamma_{ad} = n_e n_n m < \sigma v > v_d^2, \tag{10.48}$$

with n_e the ion (electron) density, n_n the density of neutrals, $< \sigma v >$ the average (Langevin) collision rate (2×10^{-9} cm^3 s^{-1}), and v_d the neutral–ion drift velocity. In molecular cloud cores, calculated drift velocities are $\simeq 5 \times 10^3$ cm s^{-1}, which corresponds to a drift of 0.5 pc on a typical ambipolar diffusion timescale of 10^7 yr. The degree of ionization is $\simeq 10^{-5}/n^{1/2}$ (cf. Section 10.5.3). The heating rate (Figure 10.17) is then,

$$n\Gamma_{ad} \simeq 1.6 \times 10^{-30}\, n^{3/2} \quad \text{erg cm}^{-3}\ \text{s}^{-1}. \tag{10.49}$$

Gravitational Heating

Gravitational collapse will heat the gas by compression. Adopting the free fall solution, we find,

$$n\Gamma_{gravity} = \frac{5}{2} kT\, v \left|\frac{dn}{dr}\right| = \frac{15}{4} kT \frac{n}{\tau_{ff}}, \tag{10.50}$$

with the free fall timescale given by,

$$\tau_{ff} = \left(\frac{3\pi}{32 G n m_H}\right)^{1/2} \simeq \frac{4 \times 10^7}{n^{1/2}} \quad \text{yr}. \tag{10.51}$$

In a dense cloud core, the gravitational heating rate (Figure 10.17) is given by,

$$n\Gamma_{gravity} \simeq 4.3 \times 10^{-31}\, n^{3/2}\, T \quad \text{erg cm}^{-3}\ \text{s}^{-1}. \tag{10.52}$$

Turbulence

Molecular clouds are in approximate virial equilibrium where self gravity is balanced by turbulence (Section 10.2.2) with the turbulent energy per unit mass given by $1/2\sigma^2 = GM/9R$. Numerical simulations show that turbulence is typically dissipated on a crossing time scale and the average energy dissipation rate per unit volume can then be written as,

$$n\Gamma_{turb} = \frac{4\pi^2}{27} G\sigma \left(\mu m_H\right)^2 N_H n_H \tag{10.53}$$

$$\simeq 1.7 \times 10^{-28} \left(\frac{R}{\text{pc}}\right)^{1/2} n_H \qquad \text{erg cm}^{-3}\ \text{s}^{-1}, \tag{10.54}$$

with the scaling relations describing molecular clouds (Section 10.2.2).

In incompressible turbulence, energy is fed into the system at the largest scales and then cascades down to the molecular dissipation scale where viscous interaction releases the energy as heat. The velocity field of incompressible turbulence is described as eddies in eddies, in eddies, in eddies ... In the Kolmogorov description of turbulence, the energy flow,

$$\dot{\epsilon} = v^2/\left(\ell/v\right), \tag{10.55}$$

with ℓ the size scale, through the system is constant. The turbulent kinetic energy density at scale k $(= 2\pi/\ell)$ is then,

$$E(k)dk = (2/3)(2\pi\dot{\epsilon})^{2/3}k^{-5/3}dk. \tag{10.56}$$

With a constant energy flow through the system, the turbulent heating rate is $(1/2)n_H m_H v^3/\ell$, which can be estimated from observations of the turbulent characteristics at one scale size. The observations described in Section 10.2.2 lead then to the rate as derived above and this heating rate permeates the medium.

Interstellar turbulence is, however, supersonic and compressible. Dissipation is then intermittent and concentrated in limited regions of space and time where velocity gradients reach large values. Both compressible and incompressible turbulence is dissipated in a sound crossing timescale and, hence, both lead to the same average heating rate. However, for compressible turbulence, dissipation cannot be described by an average rate. Rather, energy is dissipated in shocks that compress and heat the gas instantaneously and cooling occurs on a timescale of $\sim 10^3$ yr. Dissipation will scale with $\ell^{-\beta}$ with β between 3 and 5 depending on the geometry of the dissipative structures (filaments or sheets). Thus, while on average, the heating by turbulence is as important as e.g., cosmic rays, at any one time, turbulent dissipation is limited to very small regions in space that get temporarily heated to elevated temperatures ($\simeq$100–1000 K) and then cool very quickly through emission in high level CO and H_2 (as well as gas–grain collisions).

The rapid dissipation of turbulent energy in dense molecular clouds requires a rapid injection rate. The turbulent dissipation time scale is $\simeq 3\ R^{1/2}$ (pc) Myr (cf. Eq. (10.21)), which is shorter than the cloud lifetime. Protostellar outflows are insufficient to resupply the turbulence. Typically, the star formation efficiency is only 5% and only a small fraction – estimates range from 0.1 to 0.4 – of the gravitational energy is converted into mechanical energy of the outflow and much of that may be radiated away in strong shocks at the working surfaces of jets rather than lead to turbulent energy input into the general cloud. Thus, stellar outflows can be important sources of tubulence on a small scale, but they cannot maintain a turbulent energy comparable to the gravitational energy of the cloud. Moreover, clouds without star formation show similar turbulence as clouds with star formation. Likely, turbulent energy injection is related to gas accretion onto clouds in converging flows. The turbulence of molecular clouds is then linked to the general turbulence in the ISM. Of course, that just displaces the issue to the larger scale of the global ISM. Turbulence in the ISM is generated by the mechanical energy injection by massive stars in their stellar winds and when they explode as supernovae. With a mechanical energy of 10^{51} erg per SNe, a SN rate of 1 per 100 yrs, and an ISM volume of $\simeq$200 kpc^3, the total energy input is 5×10^{-26} erg cm^{-3} s^{-1}. Much of this energy may set up a disk-halo circulation flow through superbubble break out and the kinetic energy of these flows contains some 15% of SN energy. That would be sufficient to maintain the turbulent energy of the ISM. A final source of interstellar turbulence is provided by the inflow of material from the intergalactic medium. The current mass flow from the intergalactic medium is only 150 $M_\odot$/yr and this is not an important source of turbulent energy input. However, in the early Universe, this was more important.

Heating of Molecular Cloud Cores

Figure 10.17 compares the different heating processes for conditions relevant for the dark cores in molecular clouds: no FUV radiation, a gas temperature of 8 K, a gas–grain temperature difference of 2 K, and a cosmic ray ionization rate of 2×10^{-17} s^{-1} per H-nucleus. At low densities, cosmic rays dominate but at high densities, gas–grain collisions take over. If there is a YSO heating dust in its surroundings, gas–grain collisions might be more important. Ambipolar diffusion is never really important as a heating source. During the collapse phase of the cloud, compressional heating can be quite important. On average, turbulence is not important, but then this is not an average process. If at any one time, the release of energy is localized in a volume containing 1% of the mass, then – in that volume – turbulent energy input overwhelms all other heat sources. But the resulting increase in temperature leads to a quick cooling (Section 10.3.2) and return to the steady state. Finally, we note that the thermal energy input timescale is some 4×10^4 yr, much shorter than other relevant timescales.

10.3.2 Cooling of Molecular Clouds

Excitation of molecules is discussed in Chapter 4 and the calculation of emergent intensities is described in Section 4.1.7. While this has been used to analyze observations in Section 10.1, these lines also cool the gas. The calculation of the contribution of a species to the cooling rate of the gas is given by a summation over all relevant levels. When a line is optically thick, the energy balance of different regions in the cloud may be coupled through

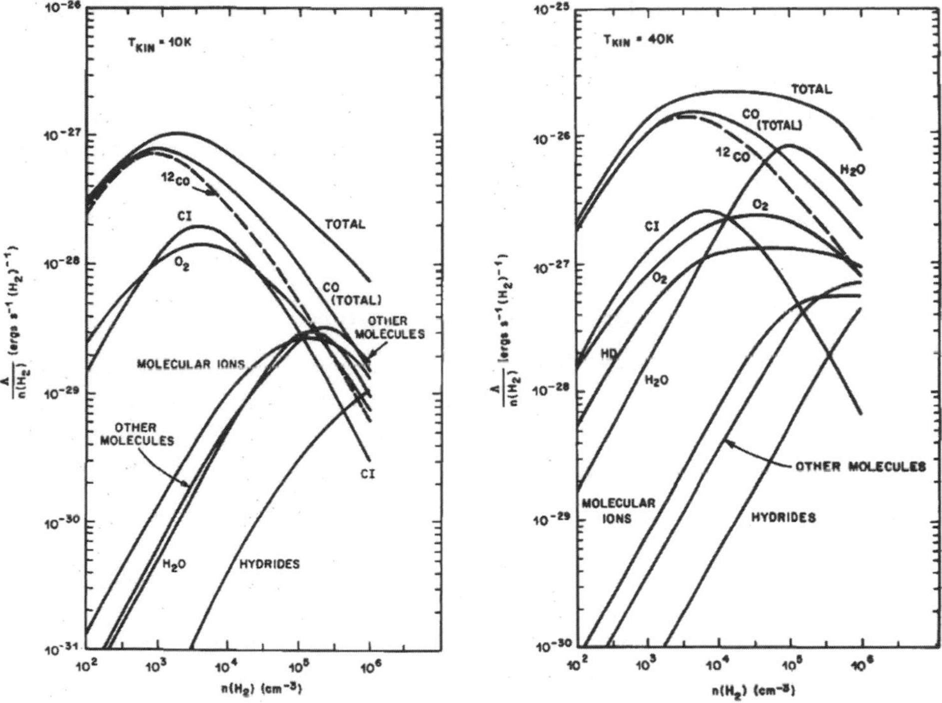

Figure 10.18 The calculated cooling rate per H_2 molecule for 10 and 40 K molecular gas as a function of density. Figure taken from [84]

the radiation field. This turns gas cooling into a nonlocal problem that is nontrivial to solve. Calculation of cooling laws generally makes use of the escape probability formalism to evaluate the effect of optical depth (cf. Section 4.1.6). Here, we will adopt that stance as well.

The many levels available and the small energy separation between levels make rotational transitions of molecules a very effective way of cooling gas. As a result, molecular clouds are generally cold. Figure 10.18 summarizes calculated cooling rates for interstellar molecular clouds. These cooling rates depend on assumed abundances, or more precisely, on the ratio of the abundance to the velocity gradient that sets the optical depth of the transitions involved (cf. Eq. 4.50). For the results shown in Figure 10.18, assumptions have been made that were reasonable at that time. At low densities, CO dominates the cooling because of its high abundance in molecular clouds. Around a density of 10^3 cm^{-3}, ^{13}CO starts to contribute to the cooling. Despite the low isotope ratio of ^{13}C/^{12}C (=65 in the ISM), ^{13}CO cooling can become comparable to the ^{12}CO cooling as the latter is optically thick. Because of the low critical density of the low-J levels of CO, its importance diminishes at higher densities where other species – first C and O_2 and then H_2O – start to contribute or even take over.

Following theoretical studies at that time, a very high gas phase abundance was adopted for O_2 ($X(O_2)/X(CO) = 0.25$). Likewise, H_2O had a high assumed abundance ($X(H_2O)/X(CO) = 0.1$). These high abundances are in conflict with current observations for dark clouds (Table 10.2) and this should be kept in mind when evaluating the importance of these cooling agents. Water and other hydrides have strong transitions at higher frequencies and become progressively more important at higher temperatures. As a result of their large dipole moments, molecular ions are also important coolants despite their low abundance.

10.4 Molecular Abundances

With the opening up of the sub-millimeter wavelength range (Figure 10.19), it has become increasingly clear that molecular clouds are teeming with a wide variety of molecules. Table 10.1 provides a summary of identified molecules as of June 2019. New species are added at a rate of a few per year. In many objects, the exquisite sensitivity of ALMA reveals the presence of lines essentially at every frequency. Digging "new flowers" from among the "weeds" of well-known and abundant molecules presents a challenge as multiple lines will have to be unambiguously assigned before a species can be considered solidly detected, particularly at the spectral confusion limit. ALMA's high spatial resolution is of great use here as different species often have different spatial distributions, reflecting their sensitivity to the rapidly changing environment, and this spatial segregation can be used to unravel the spectral complexity (Figure 11.9).

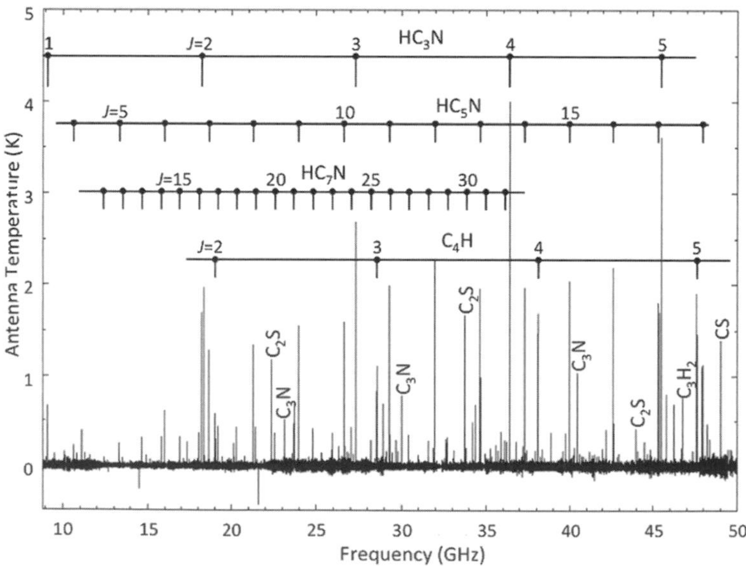

Figure 10.19 Spectral survey of the dark cloud core TMC-1 in the 8.8–50 GHz region reveal a complex array of hydrocarbon species. Figure kindly provided by Masatoshi Ohishi, see [119]

Table 10.1 *Identified interstellar and circumstellar molecules*[1]

Homonuclear Molecules, Simple Hydrides, Oxides, Sulfides, Acids, Salts						
H_2	H_2O	CO	CS	HCl	NaCl*	NaCN*
O_2	NH_3	CO_2	OCS	HF	KCl*	KCN*
N_2	CH_4	SiO	SiS	HNO	AlCl*	MgCN*
	SiH_4^*	SO	PN	HSCN	AlF*	SiCN*
	H_2S	SO_2	N_2O	HNO_2	CH_3CN	FeCN*
	H_2O_2	PO				MgNC*
	CH_3SiH_3					AlNC*
						SiNC*
Alkanes, Alkenes, Alkynes, their Derivatives, and related Compounds						
$C_2H_4^*$	CH_3C_2H	C_2	HCN	$(CN)_2$	CH_2CHCN	HNC
C_2H_2	CH_3C_4H	C_3	HC_3N	CH_3CN	CH_2C_2HCN	CH_3NC
C_6H_2	CH_3C_6H	C_5^*	HC_5N	CH_3C_3N	CH_3CH_2CN	HCCNC
C_3H_6		C_3O	HC_7N	CH_3C_5N	C_3H_7CN	HNCCC
		C_3S	HC_9N	SiH_3CN		
		C_4Si*	HCP			
Aldehydes, Alcohols, Ethers, Esters, Ketones, Amides						
H_2CO	CH_3OH	HCOOH	H_2CCO	CH_3NH_2	H_2CS	NH_2CN
CH_3CHO	CH_3CH_2OH	$HCOOCH_3$	CH_3CONH_2	CH_2NH	CH_3SH	NH_2CH_2CN
CH_3CH_2CHO	CH_2CHOH	CH_3COOH	HCOCN	CH_3CHNH	CH_3CH_2SH	HNCHCN
NH_2CHO	$(CH_2OH)_2$	$(CH_3)_2O$	CH_3CH_2CN	CH_2CNH		
CH_2OHCHO	CH_2CCHOH	$CH_3CH_2OCH_3$	$HOCH_2CN$			
C_2H_5OCHO	CH_3OCH_2OH	$(CH_3)_2CO$				
CH_3NHCHO		CH_3OCOCH_3				
HC_2CHO						
Molecular Cations						
H_3^+	CO^+	$HCNH^+$	OH^+	NO^+	NH_4^+	HCl^+
ArH^+	HCO^+	HC_3NH^+	H_2O^+	HCS^+	N_2H^+	H_2Cl^+
CH^+	HOC^+		H_3O^+		NC_2NH^+	CF^+
CH_3^+	$HOCO^+$		SO^+			
HS^+	H_2COH^+					
	H_2NCO^+					
Molecular Anions						
C_4H^-	C_6H^-	C_8H^-	CN^-	C_3N^-	C_5N^-	
Cyanic acid and related compounds						
HONC	HNCO	CH_3NCO	HNCS	NCO		
Cyclic Molecules						
C_3H_2	SiC_2	c-C_3H	CH_2OCH_2	c-SiC_3	H_2C_3O	C_2H_4O
Radicals						
OH	C_2H	C_3H_2	CN	HCO	C_2S	SiH ?
HO_2	C_3H	C_4H_2	C_3N	CH_3O	NS	SiC*
CH	C_4H	C_6H_2	C_5N*	C_2O	CP*	Si_2C*
CH_2	C_5H		CH_2CN	HC_2O	CCP	SiN*
NH	C_6H		CH_2N	C_5O	HCS	C_4Si
NH_2	C_7H		HC_2N*	HC_5O	HSC	
SH	C_8H		H_2CN	HC_7O	AlO	
HS_2			HNC_3		TiO	
NO				TiO_2		
Aromatics and Fullerenes						
C_6H_5CN	$C_6H_6^*$?	C_{60}	C_{70}	C_{60}^+		

[1]Taken from the CDMS data base (https://cdms.astro.uni-koeln.de/classic/molecules june 2019. Species denoted with $*$ have only been detected in the circumstellar envelopes of carbon-rich stars.

Perusing the list of interstellar molecules, we recognize simple, saturated hydrides such as H_2, H_2O, NH_3, and CH_4, which we expect to be present as hydrogen is so abundant. Many molecules are, however, unsaturated, radicals, or ions. The presence of anions is a relatively recent discovery. These species are very reactive and their presence attests to the importance of kinetics in space. We recognize long chain species (e.g. C_n, C_nH, $HC_{2n+1}N$), running up to 11 atoms, which have an acetylenic structure. In addition, aldehydes, alcohols, and ethers reveal a rich organic inventory, particularly in regions of star formation. Of course, large polycyclic aromatic hydrocarbon species as a class are an important component of the molecular inventory of space but the only ones securely identified are C_{60} (Section 12.4) and its ion (Sections 9.6.1 and 9.6.3). The presence of the simple aromatic species C_6H_6 in the protoplanetry nebula AFGL 618 still has to be confirmed with high spectral resolution studies; JWST may help but clear confirmation may have to await SPICA. Despite a putative sighting in California in 2017, the presence of "GrandPAHs" in space (Section 12.3.9) still awaits confirmation. It should be noted that the small aromatic species benzonitrile (C_6H_5CN) has been unambiguously detected in the dark cloud core TMC1, but its presence is not directly related to the inventory of large PAHs (Section 12.7.3).

This list of interstellar molecules is biased by selection effects. First, linear molecules will have much smaller partition functions (kT/hcB versus $(kT)^{3/2}/(hcABC)^{1/2}$) and will have therefore comparatively stronger transitions. Second, the strong electronegativity of, e.g., the cyanide group results in a large dipole moment and therefore strong rotational transitions. A similar point applies to the molecular anions. Third, really large molecules, such as PAHs, have small rotational constants and their rotational transitions fall in the relatively little explored 10–50 GHz range.[4]

10.5 Gas Phase Chemistry

Dark cloud abundances of some of the more relevant species are presented in Table 10.2. These have typically been determined from a few observed rotational transitions and a measured or adopted excitation temperature, while accounting for optical depth effects through isotope measurements. There is no direct measurement of H_2 in dark clouds and the total column density of H-nuclei is estimated from CO and its isotopologues calibrated against dust extinction or emission adopting dust properties measured in the diffuse ISM (cf. Section 10.1.2). Abundances are thus really measured against dust columns. The listed abundances should be compared to molecular abundances in diffuse clouds and in Hot Cores/Corinos (Tables 9.1 and 11.1) to appreciate the importance of environment for the molecular inventory of space. Table 10.2 has entries for the prototypical dark cloud L134N and for a source rich in carbon chain molecules, TMC-1. Deep inside dense cores – B68 (Figure 10.8) is a good example – molecular abundances are observed to sharply decrease (Section 10.6.3).

[4] They contribute to the anomalous microwave emission (Section 12.6).

Table 10.2 *The composition of dark clouds*[a]

Species	TMC1	L134N[b]	Species	TMC1	L134N[b]
H_2	(1)[b]	(1)[b]			
CO	8 (−5)	8.7 (−5)	CS	4.0 (−9)	1.7 (−9)
HCO^+	8.0 (−9)	1.2 (−8)	HCS^+	4.0 (−10)	
HOC^+		< 7.0 (−13)	H_2CS	7.0 (−10)	
HCO			OCS	2.0 (−9)	
H_2CO	5.0 (−8)	2.0 (−8)	C_2S	8.0 (−9)	
CH_3OH	3.0 (−9)	3.0 (−9)	C_3S	1.0 (−9)	
HCOOH	2.0 (−10)	3.0 (−10)	H_2S	7.0 (−10)	3.0 (−9)
CH3CHO	6.0 (−10)	6.0 (−10)	SO	2.0 (−9)	
H_2CCO	6.0 (−10)		SO_2	1.0 (−9)	
C_2O	6.0 (−11)		NH	3 (−10)	
C_3O	1.0 (−10)	< 5.0 (−11)	NH_2	6 (−11)	
O_2	< 7.7 (−8)	< 1.7 (−7)	NH_3	2.0 (−8)	2.0 (−7)
OH	3.0 (−7)	7.5 (−8)	N_2		3 (−5)
H_2O	< 7.0 (−8)	5.0 (−9)[d]	N_2H^+	4.0 (−10)	5.0 (−10)
H_2O_2	2.5 (−11)		NO	2.7 (−8)	2.0 (−8)
CH	1.6 (−8)	1.0 (−8)	CH_2CN	5.0 (−9)	
C_2	5.0 (−8)		CH_3CN	6.0 (−10)	
C_2H	6.0 (−8)	2.3 (−9)	CH_3C_2H	6.0 (−9)	
c-C_3H	1.0 (−9)	4.3 (−10)	CH_3C_4H	4.0 (−10)	
l-C_3H	8.4 (−11)	1.3 (−10)	CH_3C_3N	8.0 (−11)	
C_4H	7.1 (−8)	1.8 (−9)	C_2CHCN	4.0 (−9)	
C_5H	5.1 (−10)	< 5.0 (−11)	HC_2NC	5.0 (−10)	
C_6H	7.5 (−10)	< 4.3 (−10)	HNC_3	6.0 (−11)	
C_8H	4.6 (−11)		C_4H^-	8.0 (−13)	< 4.0 (−12)
c-C_3H_2	5.8 (−9)	2.1 (−9)	C_6H^-	1.2 (−11)	
l-C_3H_2	2.1 (−10)	4.2 (−11)	C_8H^-	2.1 (−12)	
CN	3 (−10)	4.8 (−10)	C_3N^-	< 7.0 (−11)	
C_3N	6.0 (−10)	< 2.0 (−10)	C_6H_5CN	4.0 (−11)	
C_5N	3.1 (−11)				
HCN	2 (−8)	4.0 (−9)			
HNC	2 (−8)	6.0 (−9)			
$HCNH^+$	2.0 (−9)				
HC_3NH^+	1.0 (−10)				
HC_3N	1.6 (−8)	4.3 (−10)			
HC_5N	4.0 (−9)	1.0 (−10)			
HC_7N	1.1 (−9)	< 2.0 (−11)			
HC_9N	4.5 (−10)				

[a]Abundance relative to H_2, adopting $N\,(H_2) = 10^{22}$ cm^{-2}. Taken from the compilations of [134, 213], which should be consulted for the original references, supplemented by the studies of [61, 88, 120, 144, 164, 179]. [b]Also known as L183. [c]Abundance of H_2 is not measured directly. [d]Abundance measured toward the outer regions of the dark cloud, L1544, and originating from the cloud surface [47].

The species listed in Table 10.2 can be broken up in various classes. Carbon monoxide and the simple ions and radicals (OH, HCO^+, N_2H^+) have very similar abundances in these two dark clouds. These two clouds are also alike in the oxygen-bearing species, H_2CO, CH_3OH, HCOOH, and CH_3CHO. Likely, these two classes of species have very similar production routes and rates in these two dark cloud cores. The first group are undoubtedly gas phase products but the latter group might originate from grains (Section 10.6). The abundances of the hydrocarbon chains, on the other hand, are very different in these two clouds, and this becomes more pronounced when going to longer chains in the various series (C_nH, HC_nN).

Recapping the discussion in Section 6.13, gas phase chemistry is kinitically controlled. Cosmic ray ionization produces a low level of ionization ($X_e \sim 10^{-7}$) that can help drive the chemistry through ion–molecule reactions, which enhances the reaction cross section by almost two orders of magnitude (Tables 4.1 and 6.1) and the Coulomb energy can assist in bond breaking or rearrangement. Given its high abundance, reactions with H_2 dominate for ions unless these are endothermic or an activation barrier is involved. In this case, dissociative electron recombination (generally) takes over. For neutral species, reactions with protonated species are important while reactions with cosmic-ray-produced He^+ are a loss channel. If proton transfer is inhibited, neutral–neutral reactions with small radicals have to be considered. If all else fails, the "'scoundrel rule" applies and grain-surface chemistry may be invoked. Ignoring the surfaces of molecular clouds where penetrating FUV photons are important, there are two differences with the chemistry of diffuse clouds (Section 9.3). First, hydrogen will be predominantly molecular and cosmic ray ionization will now predominantly flow through H_2^+ to H_3^+. Second, there are no photons around to ionize carbon (or photodissociate CO).

Oxygen chemistry is started by reaction of atomic O with H_3^+ to form OH^+ but the next steps are the same as in diffuse clouds (Section 9.3.1): reactions with H_2 and dissociative recombination to OH and H_2O. However, in actuality, oxygen chemistry largely takes place on grain surfaces where atomic O is hydrogenated (Chapter 10.6). For carbon, reactions of H_3^+ with atomic C bypass the slow radiative asscociation rates that hamper the initiation of C-chemistry in diffuse clouds (Section 9.3.2). Hydrogenation stops again at CH_3^+ – because of endothermicity – and methane is not the end product. Rather, the hydrocarbon radicals and their ions formed react with O or OH to form CO. CO is very stable and will lock up "all" the carbon in dark clouds. Reaction of CO with protonated molecular hydrogen will lead to the formyl radical cation, which will dissociatively recombine with electrons forming CO back again and so this is not a route to molecular complexity. Instead, carbon will be slowly broken out of CO by cosmic-ray-produced UV photons (cf. Section 6.3.5) leading to C and by reactions with cosmic-ray-produced He^+, forming C^+. These species can then be used to form, for example, carbon chains (Section 10.5.1). The proton affinity of atomic N is less than that of H_2 and hence nitrogen chemistry does not profit directly from cosmic ray ionization but has to be initiated by neutral radical chemistry (Figure 9.6). Reaction of N with OH forms NO, which reacts rapidly with N to form N_2. Molecular nitrogen is very stable and is the dominant nitrogen reservoir in dark clouds, which can be broken up by reactions with cosmic-ray-produced He^+ (into N and N^+) and photodissociated by cosmic-ray-produced UV photons (into N+N). Now, HCN is chemically related to atomic N while NH_3 is part of the N^+ initiated cycle (Figure 9.6). In ammonia synthesis, the reaction of N^+ with H_2 has a small energy barrier ($\Delta E/k = 214$ K) but this is much less important for reactions with ortho H_2, which is 173 K above the ground (para) state. Ammonia synthesis in the gas phase is therefore very sensitive to the o/p ratio of H_2, which is not directly measurable. H_2 is formed in the high temperature limit (o/p=3) on grain surfaces but when released into the gas phase, the o/p ratio is reset by reactions with H_3^+, which scramble the H's, and by accretion onto grain surfaces where "collisions" with paramagnetic impurities (e.g. the lone electron pair on O) can flip the nuclear spin. Note that the reaction of NH_3^+ with H_2 has a small energy barrier ($\simeq$1000K). However, this reaction is assisted by tunneling and the reaction rate coefficient is still $\simeq 10^{-12}$ cm^3 s^{-1}. This makes this reaction very competitive with dissociative recombination. Finally, proton transfer from H_3^+ to N_2 forms N_2H^+, which cycles back to N_2 through dissociative recombination. Some early studies suggested that recombination might lead to breaking of the strong N_2 bond but that is unexpected and more recent studies conclude that there is no significant production of NH with an upper limit of $0.07^{+0.02}_{-0.04}$ for the branching ratio. Cosmic-ray-produced photons can break the strong N_2 bond (cf. Section 6.3.5) and this rate is about as large as that for CO photodissociation.

Two time-dependent effects enter into these considerations. First, gas phase chemistry of molecular clouds requires the presence of molecular hydrogen, which is formed on grain surfaces on a time scale of,

$$\tau_{chem}\left(H_2\right) \simeq \frac{7 \times 10^8}{n} \quad \text{yr} \tag{10.57}$$

or about 10^6 yr for a typical molecular cloud. Once H_2 is abundant, chemistry can cycle the gas to molecules. The second rate limiting step is the ionization of H_2, which initiates gas phase chemical routes. CO formation requires that for each C atom one H_2 molecule is ionized. Thus, the timescale to cycle the gas to molecules is then,

$$\tau_{chem} = \frac{\mathcal{A}_C}{\xi_{CR}} \simeq 1.5 \times 10^5 \quad \text{yr,} \tag{10.58}$$

with $\mathcal{A}_C$ the gaseous carbon abundance. Cosmic rays also ionize helium, and He^+ is generally a molecular destruction agent, which slows down these atom-to-molecule conversion processes somewhat. In general, the fraction of the ionizations that are effective in driving the chemical conversion has to be taken into account in these timescale estimates and, once the gas has a large molecular fraction, ionization may merely cycle CO to HCO^+ with H_3^+ and back again to CO with an electron and the conversion timescale will increase.

Summarizing this discussion, kinetics rules the chemistry of molecular clouds. Oxygen is largely atomic (or locked up as H_2O in ice mantles; Section 10.6.1), carbon is predominantly in CO, and nitrogen is in the form of N_2. Gaseous hydrides and complex organic molecules have low abundances as their formation routes are inhibited by activation barriers or not readily accessible from the main reservoir. At early times, carbon is atomic, leading to an early peak in the abundance of acetylenic-like species (cf. Section 10.5.1). So-called early time chemical model abundances refer to this early peak in the abundance of these species. Early time should be understood here as referring back to a start as atomic gas, actually with all carbon as C^+ and the chemical clock refers to H_2 formation. Recall, though, that while the lifetime of Giant Molecular Clouds may only be some 30 Myr – tied to their sojourn time in spiral arms – gas may be molecular for some 100 Myr as molecular gas is rarely converted back into atomic gas (cf. Section 10.2.4). Conversely, one of the goals of astrochemistry is to identify relevant chemical routes toward interstellar species, providing a chemical clock to time astronomical processes.

10.5.1 Carbon Chains

The presence of carbon chains is a defining, observed, characteristic of dark cloud chemistry. We recognize pure carbon chains (C_n), as well as carbon chain radicals terminated by a H, N, or O atom (C_nH, C_nH_2, C_nN, C_nO), the cyanopolyyenes, $HC_{2n+1}N$, and further variations on this theme. We note that simple alkanes, alkenes, and alkynes have no permanent dipole moment and hence can only be detected through their ro-vibrational transitions in the mid-IR in absorption against a strong background source such as a massive protostar or asymptotic giant branch star (where some of these species have been detected). Conversely, hydrocarbon chain radicals are relatively easily detected in space due to their large dipole moment, coupled with their linear structure, which limits the partition function. This just illustrates that our knowledge of the molecular complexity of space is very biased by detection techniques.

Inside dense molecular clouds, most of the carbon is locked up in CO. Complex molecule formation has to start with breaking the carbon out of CO (or activating the CO, cf. Section 10.6.1). In the gas phase, cosmic-ray-produced FUV photons photodissociate CO (Section 6.3.5), making atomic C available for the build-up of hydrocarbon chains. Equating the CO dissociation rate with the chemical build-up timescale (Eq. (10.58)), we find,

$$\frac{\mathrm{C}}{\mathrm{CO}} = \frac{1-\omega}{p_{CO}\mathcal{A}_O} \simeq 6 \times 10^{-3}. \qquad (10.59)$$

This C/CO ratio is some two orders of magnitude larger than without cosmic-ray-produced photons. Cosmic rays also ionize helium, and the He^+ ion can destroy molecules. In particular, He^+ breaks C^+ out of CO and makes it available for the formation of more complex hydrocarbon species. However, this is much less efficient than the cosmic-ray-induced photons. Carbon is also very abundant early on in the evolution of a cloud from a diffuse atomic – with carbon mainly as C^+ – to the molecular phase. The presence of reactive carbon atoms and ions may be a natural consequence of the evolution of molecular clouds as they constantly grow through the addition of material from an extensive, largely atomic halo in converging colliding flows and this material may be mixed throughout the molecular cloud through turbulence before chemical equilibration occurs. Astronomically speaking, this turns hydrocarbon chemistry into a "tool" to study accretion and turbulent mixing in molecular clouds.

Referring back to Figure 9.4, hydrocarbon chain formation in dark cloud cores starts off with the formation of small hydrocarbon radicals,

$$H_3^+ + C \longrightarrow CH^+ + H_2, \tag{10.60}$$

which then react with H_2 to form the hydrocarbon ions CH_3^+ and CH_5^+ and after dissociative electron recombination or proton transfer to CH_3 and CH_4 (Section 9.3.2). More complex hydrocarbons can now be formed through an insertion reaction with C^+,

$$C^+ + CH_4 \longrightarrow C_2H_3^+ + H \tag{10.61}$$

$$\longrightarrow C_2H_2^+ + H_2 \tag{10.62}$$

The latter reacts (somewhat slowly) with H_2 (10^{-11} cm^3 s^{-1}) to form $C_2H_3^+$ and dissociative recombination leads to the formation of acetylene, C_2H_2. Condensation reactions are also of importance in the build up of molecular complexity,

$$C_2H_2^+ + C_2H_2 \longrightarrow C_4H_3^+ + H \tag{10.63}$$

$$\longrightarrow C_4H_2^+ + H_2 \tag{10.64}$$

Finally, neutral–neutral reactions are of interest for the build-up of carbon chains,

$$C + C_2H_2 \longrightarrow C_3H + H \tag{10.65}$$

and

$$C_2H + C_2H_2 \longrightarrow C_4H_2 + H. \tag{10.66}$$

Ion–molecule reaction schemes toward HC_3N go through dissociative recombination of its protonated form HC_3NH^+. The latter is formed through a number of channels, including $C_2H_2^+$ + HCN. But this route is effectively excluded as an important source for HC_3N as the protonated cyanoacetylene recombines equally to both HNC_3 and to HC_3N and this is at odds with the observations. Instead, the reaction of the cyanogen radical, CN, with acetylene forms an efficient route toward cyanoacetylene,

$$CN + C_2H_2 \longrightarrow HC_3N + H, \tag{10.67}$$

where CN itself is also a product of neutral radical–radical reactions (Figure 9.6 and Section 9.3.3). Similar reactions with diacetylene and triacetylene can form longer cyanopolyynes. HCN is formed from N + CH. Its isomer, HNC, forms through proton transfer to HCN, forming $HCNH^+$. This ion recombines dissociatively to both HCN and HNC (and CN). Proton transfer to HNC also leads to $HCNH^+$ but of interest here is that the reaction of HNC with C_2H leads to HC_3N. In an environment rich in atomic N, the CN channel to cyanoacetylene will be preferred and in an environment with hydrocarbon radicals, the HCN/HNC channel may become important. Again, longer cyanopolyynes can be formed analogously with relevant acetylenic species. The cyanopolyynes illustrate the importance of neutral–neutral radical reactions in molecular clouds. When these have no barriers, they can be as fast as ion–molecule reactions at low temperatures but radicals have much higher abundances than comparable ions in interstellar clouds.

The reaction network for hydrocarbon formation sketched here will lead to a preponderance of hydrocarbon radicals and acetylene derivatives. Recapitulating, it all starts with breaking the C out of CO and it is clear that these hydrocarbon species will be much less abundant than CO and, as multiple steps are involved, that their abundance will drop rapidly with chain length.

10.5.2 Deuterium Chemistry

Originally, observational studies of interstellar deuterated molecules were inspired by the constraints this might provide on cosmology and on stellar nucleosynthesis. However, observations quickly revealed that isotopologues of molecules with a hydrogen replaced by a deuterium atom typically have abundances (Table 10.3) far in excess of the cosmological D/H ratio (1.5×10^{-5}). Hence, it was realized that, instead of probing cosmology, deuterated molecules provide direct insight into chemical routes in the ISM. There is a small zero-point energy difference between deuterated species and their hydrogenated counterparts that affects the reaction exo/endothermicity. The observed, high deuteration in the ISM reflects this slightly higher stability of deuterated species.

Table 10.3 *Deuteration of molecules in molecular clouds*[a]

Species	TMC1	L134N[b]	Species	TMC1	L134N[b]
D_2H^{+d}	0.7		NH_2D	0.01	0.1
DCO^+	0.02	0.08	ND_2H		0.005
N_2D^+	0.08	0.33	ND_3		0.005[c]
DCN	0.023		HDCO	0.005–0.11	
DNC	0.15		D_2CO^e	0.04	
C_2D	0.048		HDCS	0.02	
C_4D	0.04		c-C_3HD	0.09	0.1
DC_3N	0.03–0.1		c-C_3D_2	0.006	
DC_5N	0.08–0.16		CH_2DC_2H	0.06	
ND	> 0.02		HDS	0.1	
			D_2S	0.01	

[a] Abundance relative to the (fully) hydrogenated species. Taken from the compilations of [130], which should be consulted for original references. [b] Also known as L183. [c] ND_3 abundance observed toward the dark core, HMM1, for which the NH_2D/NH_3 and NHD_2/NH_3 ratios are 0.39 and 0.09, respectively [93]. [d] D_2H^+/H_2D^+ ratio. Measured in the ambient cloud of IRAS 16293-2422. Similar value for the p-D_2H^+/o-H_2D^+ ratio in the prestellar core, L16293E [46, 94]. [e] Observed in L1544 [12].

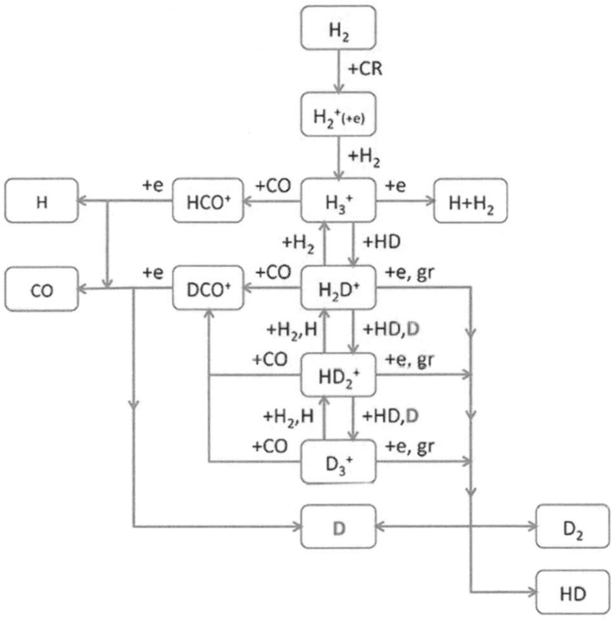

Figure 10.20 Deuterium chemistry of protonated carbon monoxide and the degree of ionization. Figure taken from [226]

The reactions of interest are of the form,

$$XH^+ + HD \leftrightarrow XD^+ + H_2 + \Delta E. \tag{10.68}$$

The energy difference, $\Delta E/k$, is of the order of a few hundred degrees and these reactions therefore only play a role at low temperatures. The route involving H_3^+ is of particular interest (Figure 10.20),

$$H_3^+ + HD \leftrightarrow H_2D^+ + H_2 + \Delta E. \tag{10.69}$$

The endothermicity of this reaction is 232 K when all species are in their ground state.[5] Ignoring other reactions, we find in LTE,

$$\frac{H_2D^+}{H_3^+} = \frac{HD}{H_2} \exp\left[\Delta E/kT\right]. \tag{10.70}$$

As HD and H_2 are the main reservoirs of H and D, at a temperature of 10 K, the H_2D^+/H_3^+ ratio is greatly enhanced over the elemental abundance ratio. Including kinetics, deuterium fractionation is limited by electron recombination and proton/deuteron transfer to other

[5] Note that H_2, H_3^+, and H_2D^+ all have ortho and para states that should be treated as distinct chemical species. This complicates the analysis of this reaction system but does not change the general principle: The zero-point-energy difference in this reaction drives the system to the right and this can lead to large deuterium fractionations.

species. Typically, in one third of the collisions, the enhanced fractionation of H_2D^+ is passed on to neutral molecules with deuteron affinities larger than H_2 such as CO,

$$H_2D^+ + CO \rightarrow DCO^+ + H_2. \tag{10.71}$$

We find then for the deuterium fractionation,

$$\frac{X\left(H_2D^+\right)}{X\left(H_3^+\right)} = \frac{k_f\, X\,(HD)}{k_b\, X\,(H_2) + k_e\, X\left(e^-\right) + k_{CO}\, X\,(CO) + \cdots}, \tag{10.72}$$

and

$$\begin{aligned} \frac{X\left(DCO^+\right)}{X\left(HCO^+\right)} &= \frac{X\left(H_2D^+\right)}{3\,X\left(H_3^+\right)} \\ &= \frac{2}{3}\,\frac{\mathcal{A}_D}{\exp[-\Delta E/kT] + 193\,X\left(e^-\right) + 1.2\,X(CO) + \cdots}, \end{aligned} \tag{10.73}$$

where $\mathcal{A}_\mathcal{D}$ is the elemental deuterium abundance and we have used the reaction rates, $k_f = 1.4 \times 10^{-9}$ cm^3 s^{-1}, $k_e = 6 \times 10^{-8}\,\sqrt{300/T}$ cm^3 s^{-1}, $k_{CO} = 1.7 \times 10^{-9}$ cm^3 s^{-1}. For typical electron abundances ($X\left(e^-\right) = 10^{-8} - 10^{-6}$) and CO abundances ($X\,(CO) = 10^{-5} - 10^{-4}$), the fractionation of deuterated molecules can greatly exceed ($X(DCO^+)/X(HCO^+) \sim 3 \times 10^{-2}$) the elemental deuterium abundance (Table 10.3).

When most of the CO and other gas phase species are depleted out in an ice mantle, the proton/deuteron transfer channel to CO and other species is closed and instead H_2D^+ will react on with HD to form HD_2^+ and even D_3^+ (Figure 10.20). As these species are then the dominant cations, the deuterium balance can be combined with the ionization balance,

$$\frac{X\left(H_2D^+\right)}{X\left(H_3^+\right)} = k_f\, X\,(HD)\left(\frac{n}{\zeta_{CR}\, k_e}\right)^{1/2} = 7 \times 10^{-1}\left(\frac{n}{10^4\ \mathrm{cm}^{-3}}\right)^{1/2}. \tag{10.74}$$

In this case, sequential reactions of the isotopologues of H_3^+ with HD can drive the deuteration of H_2 all the way to D_3^+. Hence, at high densities, high depletions, and high deuterium fractionation of H_3^+ go hand-in-hand but there will be very few heavy species around to hand this enhanced fractionation to and the effects of this deuteration – and high gas phase depletion – may be best studied through observations of H_3^+ and its isotopologues.

Finally, it should be noted that, given the important role of H_3^+ in interstellar ion–molecule chemistry, reaction (10.69) dominates deuterium fractionation at low temperatures. However, as the temperature increases, this reaction "shuts" off and other reactions take over; notably,

$$\mathrm{CH_3^+ + HD \leftrightarrow CH_2D^+ + H_2 + \Delta E} \tag{10.75}$$

and

$$\mathrm{C_2H_2^+ + HD \leftrightarrow C_2HD^+ + H_2 + \Delta E} \tag{10.76}$$

with exothermicities of 370 and 550 K, respectively.

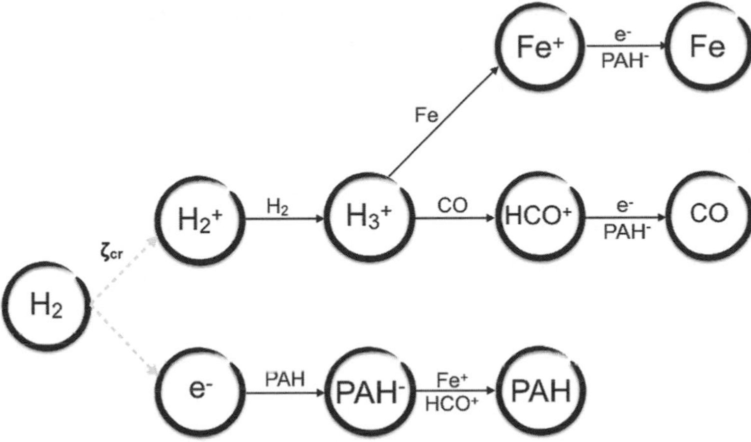

Figure 10.21 The chain of ionization in molecular cloud cores starts with cosmic rays ionizing H_2. The positive charge is transferred from one molecular ion to the next as well as to atoms. The electron can also attach to large molecules. Eventually recombination/mutual neutralization of these negatively and positively charged species closes the ionization loop.

10.5.3 The Degree of Ionization

Ions play a central role in gas phase chemistry of molecular clouds but also regulate the coupling between magnetic fields and the gas and thus control important dynamical processes such as ambipolar diffusion – the slow leakage of supporting magnetic fields out of cloud cores as ions slip past the neutrals – and C-shocks (Section 11.7). The chain of ionization starts with cosmic ray ionization of molecular hydrogen forming H_2, which is quickly transformed into H_3^+ (Figure 10.21). As dissociative recombination of H_3^+ is relatively slow, inside dense molecular clouds where the electron fraction is typically less than 5×10^{-7}, the proton is preferentially passed on to other molecular species with higher proton affinities (e.g. CO; Table 6.7) or H_3^+ charge exchanges with neutral atomic species (Mg, Fe). Atomic ions will recombine only very slowly (radiatively) with electrons and, if these metals are present at gas phase abundances commensurate with those measured in diffuse clouds, this channel will dominate the ionization balance. The slow radiative recombination rate of atoms results in a high degree of ionization in models and this leads to serious issues in explaining observed molecular abundances. Much better agreement is achieved when the metal abundances are artificially lowered – so-called low metal abundance models – as this effectively closes this recombination channel. In this case, the ionization stays with molecular ions and the ionization balance is set by the much faster dissociative recombination of these species.

However, electrons can attach to large neutral PAH molecules and this changes the analysis. If PAH anions are the dominant negative charge carriers (e.g. whenever the electron fraction is less than $2\times10^{-7}\left(X\left(PAH\right)/10^{-7}\right)$), mutual neutralization of these anions with atomic or molecular cations will control the ionization balance and the actual abundance of metal atoms is no longer relevant. With the mutual neutralization rate coefficient given in Section 8.6.5 and a photodetachment rate of $k_{pd} \simeq 4 \times 10^{-10}\, N_c \exp\left[-0.25 A_v\right]\ \mathrm{s}^{-1}$, the PAH anion route will dominate for $A_V \gtrsim 12$; e.g. in dense cores. Finally, when depletion sets in, proton transfer reactions become of little importance and H_3^+ (and its isotopologues) is the prime cation. While PAHs will freeze out on a longer timescale than species such as CO, likely, at that point, PAH anions are not important either and the ionization balance is controlled by dissociative recombination of H_3^+ and e^-.

For dense cores, the degree of ionization is given by,

$$X_e = \sqrt{\frac{2.3\zeta_{CR}}{n\left(\mathrm{H_2}\right) k}} \simeq 2 \times 10^{-7} \left(\frac{\zeta_{CR}}{3 \times 10^{-17}\ \mathrm{s}^{-1}}\right)^{1/2} \left(\frac{10^4\ \mathrm{cm}^{-3}}{n\left(\mathrm{H_2}\right)}\right)^{1/2}, \qquad (10.77)$$

where charge equality has been assumed and X_e stands for the degree of ionization, irrespective if these are electrons or PAH anions. Here, ζ_{CR} is the primary cosmic ray ionization rate, the factor 2.3 takes secondaries into account, and k refers to the relevant recombination rate coefficient. If multiple charge carriers are relevant, then the neutralization rate coefficient has to be appropriately weighted. As the rate coefficients involved are very similar, the degree of ionization is not very sensitive to the details (but the abundances of the ions involved are). We conclude that the degree of ionization drops with increasing density. Figure 10.22 shows the results of a calculation appropriate for a dense core. Depletion is ignored and a constant cosmic ray ionization rate of $3 \times 10^{-17}\ \mathrm{s}^{-1}$ has been adopted. At low densities, molecular ions and electrons dominate but already at a density of $3 \times 10^3\ \mathrm{cm}^{-3}$ the electron abundance has dropped enough that PAH anions take over. At very high densities, charge transfer from molecular ions to metal atoms becomes important and metals become the counter ions. Finally, we note that the presence of PAHs

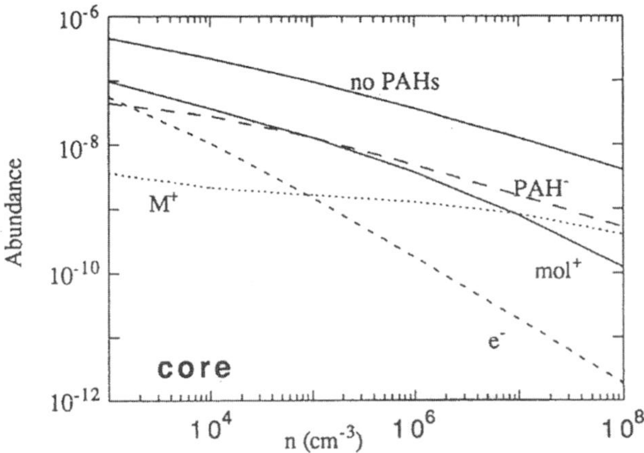

Figure 10.22 The charge distribution in dense cores of molecular clouds calculated including PAHs, metal ions (M+) and molecular ions (mol+). The adopted PAH and metal abundances are 10^{-7} and 10^{-6}, respectively. The degree of ionization in the absence of PAHs is shown for comparison.

decreases the degree of ionization by about an order of magnitude and as metals are not important, no artificial depletion of them has to be included in models.

The cosmic ray ionization rate can thus be determined from observations of molecular ions. Protonated H_2 is observed through its near-IR ro-vibrational transitions and hence requires an IR-bright background source. This species has been observed toward four massive protostars. Simplistically, we can write this as,

$$\zeta_{CR}/n\,(H_2) = X\left(H_3^+\right)\,X\,(CO)\,k, \tag{10.78}$$

with the $X(i)$'s the observed abundances and k the relevant reaction rate coefficient. Observed H_3^+ column densities are $1-5\times10^{14}$ cm^{-2}, corresponding to abundances $1-3\times10^{-9}$ cm^{-3} and $\zeta_{CR}/n\,(H_2) \simeq 1-3\times10^{-22}$ s^{-1} cm^3. These sources have a complex structure with clear gradients in the physical conditions. Adopting the source structure derived from sub-millimeter observations, cosmic ray ionization rates of $1-2\times10^{-16}$ s^{-1} are implied. These derived rates are similar to those derived for diffuse clouds (Section 9.5.1).

However, within this scheme, this high cosmic ray ionization rate would lead to an order of magnitude higher HCO^+ abundances than observed. This discrepancy between the H_3^+ and HCO^+ derived cosmic ray ionization rates has been ascribed to the presence of unrelated diffuse clouds along the sightline contributing to the H_3^+ absorption. Alternatively, CO could be almost fully depleted in a small region ($N \simeq 10^{21}$ cm^{-2}) in the cloud, leaving H_3^+ as the dominant ion. Cosmic ray ionization is then balanced (Eq. (10.78)) by dissociative electron recombination of this species.

As H_3^+ is difficult to observe directly, studies on the degree of ionization in molecular clouds rely on abundances of species further along the chain of ionization. The DCO^+/HCO^+ ratio, R_D, is often used for this. Referring back to Figure 10.20, we can rearrange Equation (10.73),

$$\frac{X_-}{10^{-7}} = \frac{0.52}{R_D} - \frac{5.0}{f_d}, \tag{10.79}$$

where we have ignored the backward reaction (Eq. (10.69)), only included CO as deuteron receptor, and inserted numerical values. The depletion factor, f_d (≥ 1), takes into account that, in dense cores, CO can freeze out. Here, X_- stands for the negative charge carrier and the numerical values refer to electrons. If PAH anions dominate the charge balance, then the numerical factor (0.52) changes slightly and X_- refers to the anion abundance. The derived electron fraction (Eq. (10.79)) can be related back to the cosmic ray ionization rate (Eq. (10.77)). Specifically, HCO^+ formation is connected to H_3^+, which is related to the ionization rate, through reactions with CO and is destroyed through reactions with the main anion reservoir. We can write,

$$R_H \equiv \frac{X\left(HCO^+\right)}{X\,(CO)} = \frac{\zeta_{CR}/n\,(H_2)}{k_- X_-\left(\left(k_-/k_{CO}\right) X_- + X\,(CO)\right)}, \tag{10.80}$$

where X_- and k_- refer to the abundance of the main anion and the rate at which recombination occurs. R_H and R_D depend on three parameters, the degree of ionization X_-, the

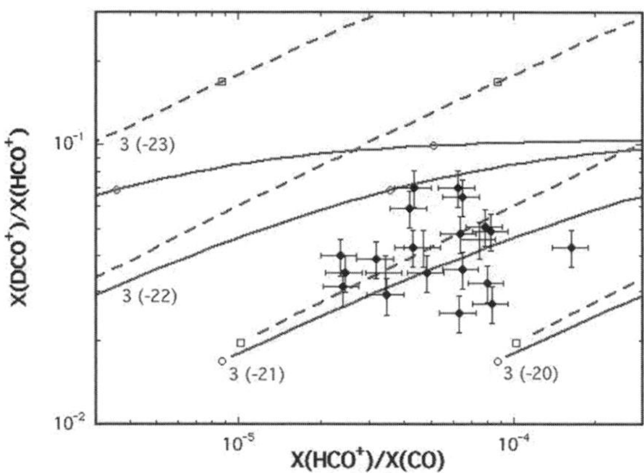

Figure 10.23 The HCO^+/CO abundance ratio (R_H) is shown as a function of the DCO^+/HCO^+ abundance ratio (R_D). Along the curves, the electron abundance varies from 10^{-6} (left) to 10^{-9} (right). The curves are labeled by the adopted values for $\zeta_{CR}/n\,(H_2)$. Solid/dashed curves refer to an adopted CO abundance of 8×10^{-5} and 8×10^{-6} (a depletion by a factor 10), respectively. Lines of constant degree of ionization run vertically and symbols (circles/squares) on the curves, from left to right, refer to values of 10^{-6}, 10^{-7}, and 10^{-8}, respectively. PAH anions are assumed to be the main negative charge. The data points are taken from [41]. Figure adapted from [43]

cosmic ray ionization rate divided by the density, $\zeta_{CR}/n\,(H_2)$, and the depletion factor of CO, f_d, and this can be solved to derive their relationship (Figure 10.23). R_D and R_H both smoothly increase with X_- until R_D reaches the maximum allowed value for the adopted CO abundance. The relation shifts to lower R_H for lower adopted $\zeta_{CR}/n\,(H_2)$. We note that, for high degrees of ionization, the relation does not depend on the degree of CO depletion.

Comparing this simple analytical result with the observations, we conclude that the degree of ionization ranges from 10^{-7} to 10^{-6} and $\zeta_{CR}/n\,(H_2) \sim 3 \times 10^{-21}$ s^{-1} cm^3 (Figure 10.23). With densities ranging from 5×10^3 to 5×10^4 cm^{-3}, we arrive at a cosmic ray ionization rate of 3×10^{-17} s^{-1}. Over the range probed by these observations, the results are not very sensitive to the adopted depletion factor. But that doesn't mean that depletion is, in general, unimportant; only that HCO^+ and DCO^+ observations tend to be biased against regions where CO is strongly depleted (cf. Section 10.6.3). A somewhat hidden result from this analysis is that HCO^+ is not the main cation in dense clouds and neither is H_3^+. Rather, the dominant cations must be metals such as Mg^+ and Fe^+. Finally, it should be noted that H_2, H_3^+ and H_2D^+ have ortho and para forms and the spin statistics has influence on the outcome of the reactions relevant for this analysis as the rotational energy level separation is comparable to the zero-point energy difference driving the deuteration.

10.5.4 Penetration of Cosmic Rays

Low energy cosmic rays are the dominant source of ionization and photodissociation inside dense clouds (Sections 6.3.5 and 10.5.3). They are also the dominant heating source of the gas (Section 10.3.1) and the resulting ionization drives ion–molecule chemistry of molecular clouds (Section 10.5). Observations of H_3^+ in diffuse clouds reveal high cosmic ray ionization rates, $\zeta_{CR} \simeq 2 \times 10^{-16}$ s^{-1} (Section 9.5), but the cosmic ray ionization rate in dense molecular gas is much less ($\zeta_{CR} \simeq 3 \times 10^{-17}$ s^{-1}; Section 10.5.3). This decrease in the ionization rate reflects the energy loss of cosmic rays as they penetrate ever larger column densities of gas.

This energy loss and its effects on the cosmic ray flux can be modeled. Define the energy loss function, $\mathcal{L}_{CR}$, as,

$$\mathcal{L}_{CR}(E) = -\frac{dE}{dN(H_2)}. \tag{10.81}$$

These energy loss functions have been measured and theoretically calculated. In the thick target approximation, we can define the range as,

$$R(E) = \frac{1}{n(H_2)} \int \frac{dE}{\mathcal{L}_{CR}}, \tag{10.82}$$

and a cosmic ray with initial energy, E_0, will reach energy, E, after penetrating a column density of,

$$N(H_2) = n(H_2)\,(R(E_0) - R(E)). \tag{10.83}$$

With conservation of the number of cosmic ray particles, we find for the spectrum, j,

$$j(E,N)\,dE = j(E_0,0)\,dE_0, \tag{10.84}$$

and

$$\frac{dE}{dE_0} = \frac{\mathcal{L}_{CR}(E)}{\mathcal{L}_{CR}(E_0)}. \tag{10.85}$$

These sets of equations can be solved for the spectrum of CRs as a function of the penetration depth.

Figure 10.24 compares models for the penetration of comic rays into molecular clouds with existing observations. In this comparison, densities have to be adopted (see Section 10.5.3) and these are quite uncertain. Also, the column density have been estimated from measured $C^{18}O$ column densities, assuming a CO depletion factor, and these are quite uncertain. Then, the observations are somewhat ill-defined averages over the line of sight. In all, this comparison should be considered indicative of the behavior rather than provide quantitative estimates.

10.5.5 Carbon and Nitrogen Isotopologues

In diffuse clouds and in the surfaces of molecular clouds and protoplanetary disks, CO photodissociation can affect the carbon isotope distribution. CO photodissociation occurs through line absorption into predissociative, excited states. As a result, the CO photodissociation rate is affected by self-shielding. For CO column densities of $\sim 10^{15}$ cm^{-2}, these

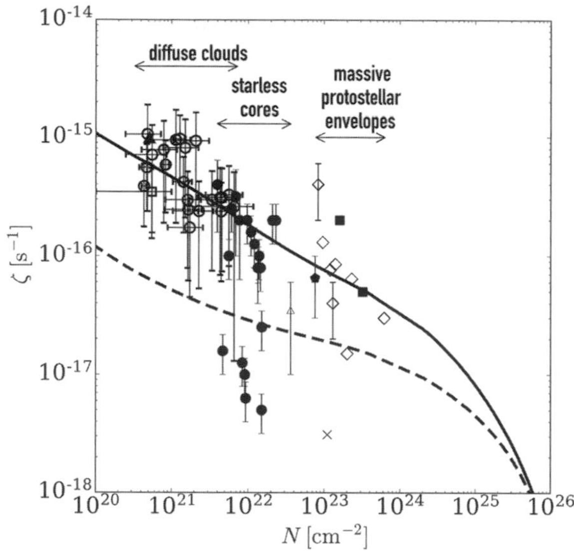

Figure 10.24 Total cosmic ray ionization rate as a function of column density. Filled circles and upper limits are the H_3^+ observations of diffuse clouds. Open circles are the data from dense cores. The triangle is B68. The diamonds are protostellar envelopes. Reference [170] should be consulted for original references, except for the empty square [78] and the solid pentagon [74]. The solid curves represent the model results for the proton ionization rate, assuming different incident spectra, taken from figure F.1 in [171]. Figure kindly provided by Marco Padovani, adapted from [170, 171]

lines become saturated and the photodissociation rate decreases rapidly. As ^{13}CO (and other isotopologues) are much less abundant than the main isotope, the depth in the cloud where self-shielding might become important will be isotope sensitive. This isotope selective effect will lead to fractionation and the abundance of ^{13}CO (and other isotopologues) will be decreased in the surface layer of a cloud relative to that of ^{12}CO. A CO column density of $\sim 10^{15}$ cm^{-2} corresponds typically to a H-nuclei column density of 3×10^{20} cm^{-2} (Figure 9.5). For the diffuse cloud toward ζ Oph, the difference in photodissociation rate is calculated to amount to a factor of 3–6. The precise value depends on the details: the excitation temperatures of H_2 and CO, as there is some overlap in lines, the Doppler broadening of the lines, and the dust properties. The "excess" of ^{12}CO is also passed on to daughter products such as HCO^+. The photodissociation of CO will preferentially produce ^{13}C (and ^{17}O and ^{18}O) for depth in excess of this column density of $\sim 10^{15}$ cm^{-2} and this fractionation can be passed on to C daughter products, including, for example, acetylenic products formed through C-insertion reactions (Section 10.5.1). However, in this surface layer, C will be mostly ionized by penetrating UV photons and C^+ is the main (e.g. unfractionated) carbon reservoir.

As for hydrogen, isotope exchange reactions can fractionate carbon-bearing species because of the zero-point energy difference. The main reaction of interest is,

$$^{13}C^{+} + ^{12}CO \leftrightarrow ^{12}C^{+} + ^{13}CO + \Delta E = 35 \text{ K}. \tag{10.86}$$

The reaction,

$$^{13}CO + H^{12}CO^{+} \leftrightarrow ^{12}CO + H^{13}CO^{+} + \Delta E = 9 \text{ K}, \tag{10.87}$$

can also be of interest. Because the zero-point energy difference is much smaller than for hydrogen-bearing species, fractionation effects will be much more modest. When C^+ is the main carbon reservoir, process (10.86) by itself will lead to a fractionation, $R(\mathrm{CO})$, of ^{13}CO relative to ^{12}CO, given by,

$$R(\mathrm{CO}) = K(T). \tag{10.88}$$

The equilibrium constant, $K(T)$, is approximately given by $\exp\left[\Delta E/kT\right]$, which for a typical temperature of a diffuse cloud (50 K) amounts to $R \simeq 2$. This will counteract the effect of photo-fractionation described above.

In dense clouds, CO is the main carbon reservoir and C^+ is generated through reactions of cosmic-ray-produced He^+ with CO ($\rightarrow C^+ + O$), which breaks out both isotopes equally facile. If CO is the main reaction partner for C^+, reaction (10.86) will preferentially deplete $^{13}C^+$ and a fractionation factor, $R(C^+)$, given by,

$$R(C^{+}) = K(T), \tag{10.89}$$

which is about 30 at 10 K. Thus, as $^{13}C^+$ is rapidly sequestered back into CO, species derived from C^+ will be isotopically light and carbon-isotope fractionation can be much more important inside dense clouds. The resulting C^+/CO ratio is $< 10^{-3}$ and the CO isotope ratio is not affected. Species derived from CO can be fractionated in a similar fashion through ion–molecule reactions (cf. Eq. (10.87)) and this amounts to a factor 2 fractionation. This reaction can also fractionate in diffuse clouds but then the effect is at the 10–20% level as $T \simeq 50$ K. Grain surface products from CO (e.g. H_2CO, CH_3OH, CO_2; Section 10.6.2) will not be fractionated.

Cosmic rays also produce UV photons (through excitation/fluorescence of the Lyman–Werner bands of H_2) and these can photodissociate CO into C and O (Section 6.3.5). The isotopic effects of this depend on the details of the line overlap of the CO-isotopes with the H_2 lines as well as with the main CO isotope (through the shielding factor) and this has not yet been calculated but one might expect a factor of a few. As neutral carbon does not flow directly back to CO, carbon-bearing species deriving from C may show a different behavior than those deriving from C^+. Through reactions with H_3^+, neutral C atoms will flow toward small hydrocarbon radicals (Section 9.3.2) and are converted on grain surfaces to CH_4 and CH_3CH_2OH and similar molecules (Sections 7.4 and 10.6.2). Note that He^+ is some 50 times more important than cosmic ray photons in the destruction of CO.

Observationally, species such as CN, HCN, and CS do not show fractionation in dense clouds relative to CO. Perhaps they are formed early on when C^+ is the main (and unfractionated) reservoir or C rather than C^+ is involved in their formation (and cosmic ray induced photons do not fractionate). It remains puzzling though that the effects of reaction (10.86) are not obvious and possibly this suggests the importance of other loss routes for C^+ in dense clouds such as mutual neutralization with PAH anions that can be important if PAHs are the dominant negative charge carriers and CO is mildly depleted (Section 10.5.3).

Finally, we can consider scrambling of the C isotopes within carbon chain molecules. In contrast to CN, CS, and HCN, observations have revealed that CCH is isotopically light but that the abundance ratio of $C^{13}CH/^{13}CCH$ is larger than unity ($\simeq$1.6). Here the reaction,

$$\mathrm{H} +^{13}\mathrm{CCH} \leftrightarrow \mathrm{H} + \mathrm{HC}^{13}\mathrm{C} + \Delta\mathrm{E} = 8\ \mathrm{K}, \tag{10.90}$$

may play a key role. The other radical that can play a role in the formation of cyanopolyynes, CN, can also be fractionated,

$$^{13}\mathrm{C}^+ +^{12}\mathrm{CN} \leftrightarrow\ ^{12}\mathrm{C}^+ +^{13}\mathrm{CN} + \Delta\mathrm{E} = 31\ \mathrm{K}. \tag{10.91}$$

Similar fractionation patterns of C_2S, C_3S, and C_4H have also been observed. Thus, the isotope fractionation pattern may contain a record of the formation route(s) involved and this may well vary from region to region following the local conditions and the history of the region.

Nitrogen isotope fractionation has been more more difficult to determine observationally as the main isotopes are optically thick. Also, the elemental $^{14}N/^{15}N$ ratio is high ($\simeq$440) and isotope lines are very weak. Isotope ratios are then often determined from mixed isotopic species (e.g. $H^{13}CN/HC^{15}N$) with added uncertainty. As for C-isotope fractionation, N-isotope fractionation is driven by small zero point energy differences, typically 20–30 K, and again this is only of importance at the very low temperatures of cold cores. Gas phase formation of nitrogen hydrides is driven by the reaction of N^+ with H_2 (Section 9.3.3) and the small endothermicity in this reaction is mostly compensated by the excitation energy of ortho H_2. The reaction pathway becomes then very sensitive to the exact o/p ratio of H_2 in dark clouds, which is not well known, but expected to be very small. The reaction of $^{15}N^+$ with $^{14}N_2$ will slightly enrich molecular nitrogen and its daughter products (N_2H^+). Atomic N is slightly isotopically light. There are some subtle points: The abundance of $^{15}NNH^+$ is less than that of $N^{15}NH^+$ due to a small difference in zero point energy. Atomic N^+ will be isotopically light but its daughter products (NH_2 and NH_3) can still be enriched in ^{15}N due to the small difference in endothermicity ($\simeq$5 K) for the reaction of $^{15}N^+$ with o-H_2 as compared to the reaction of the main isotope. The nitriles derive from the reaction of atomic N with small hydrocarbon radicals and are therefore (slightly) isotopically light.

This discussion has focused on molecular clouds as very low temperatures are required for C and N isotope fractionation. However, the resulting fractionated species may be sequestered away in interstellar ice grains that will eventually build comets and planetesimals. In this way, low-temperature fractionation may also be relevant for the isotope fractionation patterns of the different molecular reservoirs in solar system bodies. The diverse fractionation pattern of meteorites and comets (Figure 11.36) attests to the presence of multiple chemical origins for isotope fractionation in space (Section 11.5).

10.5.6 Anions

In recent years, a number of carbon chain anions have been detected in space (Table 10.1). Observed abundance ratios between the anion and its parent species can be as high as 0.1 for the largest species (Table 10.2). Molecular anions are formed through radiative attachment reactions (Section 6.10),

$$A + e^- \leftrightarrow \left(A^-\right)^* \rightarrow A^- \tag{10.92}$$

where $\left(A^-\right)^*$ is an excited molecular anion that can autoionize or radiatively stabilize. For the larger hydrocarbon chains considered here, reaction rate coefficients of $\simeq 2 \times 10^{-7}$ cm^3 s^{-1} at 10 K are indicated. Destruction goes through mutual neutralization ($k_{mn} \simeq 4 \times 10^{-7}$ cm^3 s^{-1} with metal ions) and through "burning" reactions with atomic O ($k_{\rm O} \simeq 6 \times 10^{-10}$ cm^3 s^{-1}). In steady state, we have for C_6H^-,

$$\frac{X\left(C_6H^-\right)}{X\left(C_6H\right)} = \left[\frac{k_{mn}}{k_{ra}} + 3 \times 10^{-3}\frac{X\left(O\right)}{X\left(e\right)}\right]^{-1} \simeq 0.15, \tag{10.93}$$

for an adopted O abundance of 1.5×10^{-4} and a degree of ionization of 10^{-7}, in reasonable agreement with the observations (Table 10.2). The presence of carbon chain anions promotes growth of larger chains through, for example, associative electron detachment reactions with carbon and hydrogen atoms (Section 6.11). Experimentally, for small carbon chain anions, carbon is lost in the reaction with atomic oxygen and nitrogen to CO and CN^-. For larger chains, associative detachment is also observed. Overall, the effect of carbon chain anions in models is quite modest.

PAH anions have a much more pronounced effect in models of interstellar chemistry. For astrophysically relevant PAHs (e.g. with $N_c \sim 50$), radiative stabilization should dominate over autoionization and the rate is then approximately the Langevin rate. In this case, PAH anions are the dominant negative charge carriers in dense ($n > 10^4$ cm^{-3}) cloud cores (Figure 10.22; Section 10.5.3). The degree of ionization is then set by the balance of cosmic ray ionization with mutual neutralization reactions of cations and PAH-anions. As the latter are much faster than the other available recombination channel for metal cations (radiative recombination), this has the effect of greatly surpressing the degree of ionization (Figure 10.22). In this way, PAH anions resolve one of the outstanding issues of astrochemical modeling; e.g. models with metal abundances typical for diffuse clouds but without PAHs predict too high degree of ionization and hence such pre-PAH models had to assume that metals were highly depleted while models with PAHs fare well in comparison to observations.

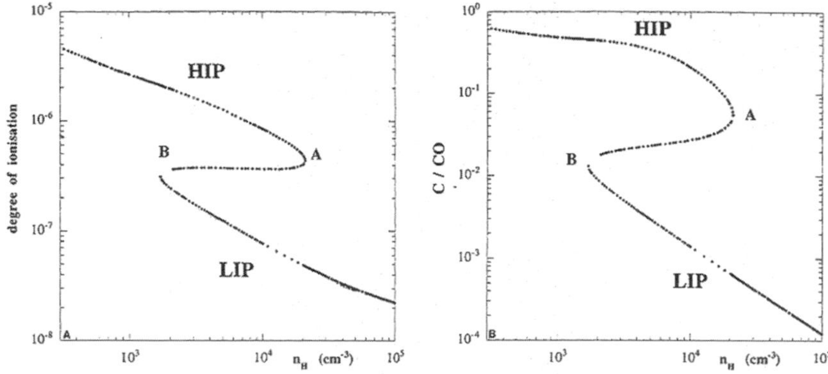

Figure 10.25 Steady state solutions for the electron fraction (left) and the C/CO ratio (right) as a function of density. Over a range of densities (between points A and B), two stable solutions – the high ionization phase (HIP) and the low ionization phase (HIP) – separated by an unstable solution are apparent. Figure taken from [128]

10.5.7 Bifurcation in Interstellar Chemistry

As discussed in Section 6.13.4, interstellar chemistry can show complex bifurcation patterns; e.g. for certain sets of parameters, multiple solutions are possible for the steady state abundances of chemical species. This is illustrated in Figure 10.25 for the ionization fraction and the C/CO ratio. High Ionization Phases (HIP) are characterized by a high degree of ionization carried by atomic ions (S^+, C^+, H^+), low molecular ion abundances (e.g. HCO^+), high abundances of atomic C and hydrocarbon daughter products, and low abundances of the oxygen-bearing species, O_2, OH, and H_2O. Low Ionization Phases (LIP) are just the opposite with low electron abundances carried by molecular ions, low C/CO ratios, and high O_2 abundances. Essentially, in the high ionization stage, atomic ions recombine radiatively and hence very slowly. The high electron abundance surpresses the molecular ion abundance; in particular H_3^+ and that surpresses molecule build-up. In addition, atomic ions (H^+, S^+) open up other destruction channels for neutral molecules (e.g. O_2).

There is not one simple set of reactions that can be isolated that drives the bistability. In a general sense, chemical bifurcations in space reflect the inherent nonthermodynamic equilibrium introduced by cosmic ray ionization. Hence, the precise location of the bistability regime is affected by the sulfur and metal abundances, recombination with PAH anions, the cosmic ray ionization rate, and the depletion of molecules on grains. In addition, uncertainties in the reaction rates also influence the location of the bistability regime. Figure 10.26 shows four trajectories in the $T-$C/CO phase space. In these particular calculations there are two attractors, corresponding to the HIP and LIP solutions, separated by a saddle point. The separatrices divide the phase space into the different regimes and, as illustrated, a small change in the initial conditions may drive the solutions very far apart.

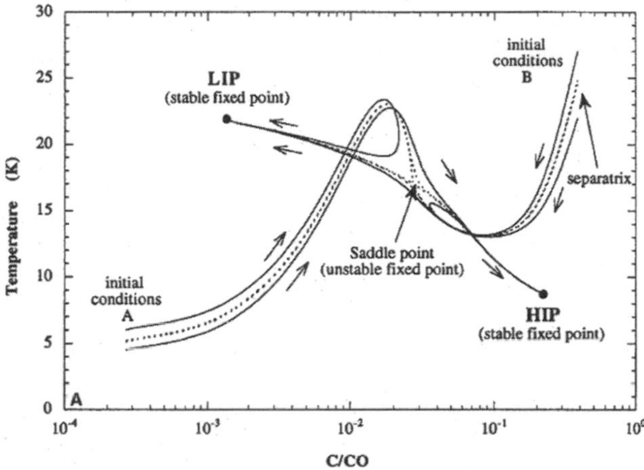

Figure 10.26 Four trajectories of the time-dependent solution in the $T-$C/CO phase space calculated for a density of 10^4 cm^{-3}. The solutions start with different initial conditions in either regime A or B The two stable solutions, HIP and LIP, are separated by the saddle point located on the intersection of the two separatrices. The crossing of the trajectories is a projection effect. Figure taken from [128]

10.5.8 Summary

Gas phase models of interstellar clouds have been very succesful. Among the key successes are explaining the high abundance of CO, the presence of ions such as H_3^+ and HCO^+, the presence of radicals such as OH and CH, the presence of isomers such as HNC/HCN, and the high observed deuteration in species such as DCO^+. Species that have notoriously been difficult to explain include CH_3OH, H_2CO, and NH_3. But otherwise, at this qualitative level, gas phase models agree well with the observations. Quantitively, however, the agreement is less impressive. The most extensive reaction networks contain some 4000 reactions among 400 species, and reach an order of magnitude agreement with observed abundances for some 80% of the species. Part of this is because our chemical knowledge is imperfect: less than 20% of the reactions included have actually been studied in the laboratory or by quantum chemical methods. The remainder are educated guesses. It is not even known whether all the important chemical routes are included. Conversely, some routes may have been included but are inhibited at low temperature. Moreover, the term "agreement" is misleading because often observations (of particularly the complex hydrocarbons) are compared to models that relate to an early time ($\sim 10^5$ yr) in a time-dependent calculation. Gas phase models really have difficulty explaining observations of complex hydrocarbons at late time because almost all the carbon tends to be locked up in the unreactive CO. At early times, this issue is negated as these models start with all the carbon initially available in the form of the highly reactive species, C^+. Once some H_2 is formed, hydrocarbon radical chemistry really takes off but in less than 10^6 yr all of these hydrocarbons are converted into CO. Hence, molecular abundances are then tied to the history of the cloud and the processes that play a role in the conversion from diffuse atomic clouds to dense molecular clouds. Molecular clouds may form through large scale colliding flows producing cold atomic gas that is converted into molecular gas (Section 10.2.4). This may grow further through mass transfer across cloud boundaries from surrounding, large scale, atomic halos. The implications of such a dynamic picture for the composition and evolution of molecular gas is only now being pursued.

There are alternative solutions to the issue of the high abundance of hydrocarbon species. One class of models has C/O elemental abundance ratios that are larger than 1. In this case, the formation of CO does not consume all of the C and hydrocarbon chemistry will be very efficient. Such a large C/O ratio is sometimes attributed to the formation of ice mantles, because, observationally, O seems to be locked more in interstellar ices than C. Some gas phase models use the ices as repositories, where (early time) molecules are kept to be judiciously injected into the gas phase when needed; an approach that has been very effective in explaining Hot Cores (see Section 10.1). Other models invoke the effects of shocks or turbulence to drive the chemistry away from CO toward hydrocarbons. Be that as it may, at present, it seems fair to say that these various models are just a stand-in for an efficient way of breaking the carbon out of CO and it seems clear that, if such a key ingredient of the models is not well understood, a good agreement between models and observations might just be fortuitous. Finally, grain surface chemistry may have contributed to the abundance of some species and we will examine this in the next section. In all, this summary should be read as a positive statement: Qualitatively, we do understand gas phase chemistry, and the ion–molecule and neutral–neutral reaction networks outlined in the previous sections are key. Understanding the quantitative aspects of astrochemistry may bring new insights into the physics of molecular clouds.

10.6 Gas–Grain Interactions

There are various indications that gas–grain interaction is important in molecular clouds. IR observations show the presence of interstellar ices in such environments and we will discuss relevant observations in Section 10.6.1 and the chemistry of interstellar ice in Section 10.6.2. Observations have also shown that the abundance of gas phase species decreases dramatically inside dense cores, presumably because they freeze in ice mantles (Section 10.6.3). Finally, there is observational evidence for the presence of species – at low abundance – that are difficult to form in the gas phase but readily form on grain surfaces, indicating that species can also be released back into the gas phase from interstellar ices.

10.6.1 Interstellar Ices

Composition

Infrared spectra of protostars embedded in or of background stars located behind molecular clouds reveal broad absorption bands due to simple molecular ices (Figure 10.27). Early ground-based studies were confined to the L and M bands and revealed that these bands were omnipresent in molecular clouds. The high resolution and broad wavelength coverage

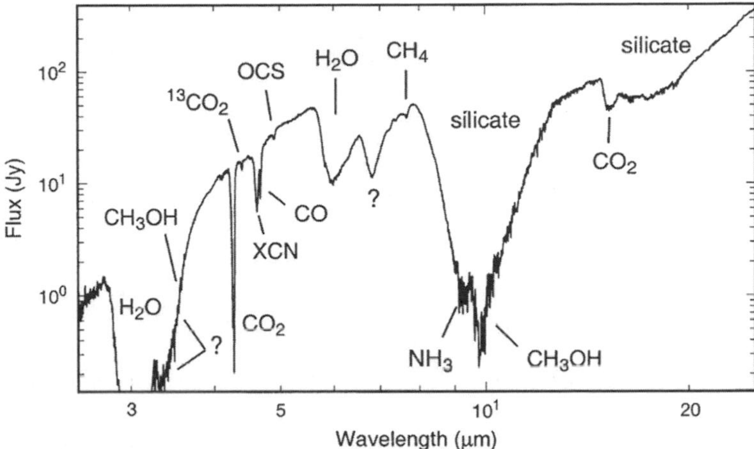

Figure 10.27 The mid-infrared spectrum of the protostar W33A observed with the Short Wavelength Spectrometer on board the Infrared Space Observatory shows a variety of absorption features. Except for the 10 μm and 18 μm silicate bands, these features are due to simple molecules in an ice mantle. Figure taken from [80]

of ISO/SWS, brought the richness of these spectra really home, while the sensitivity of Spitzer/IRS demonstrated the widespread nature of interstellar ices.

Vibrational spectroscopy has been discussed in Section 3.3. While that discussion focused on the characteristics of gas phase species, much of the discussion on identification aspects is relevant for ices as well. Comparison of these spectra with the spectra of laboratory analogs provides direct identification of the solid compounds present in space (Section 3.3 and Table 3.5). Molecules that are unambiguously identified through their fundamental modes are: H_2O (with bands at 3.07, 6.0, and 45/60 μm), CO (both ^{12}CO & ^{13}CO at 4.67 μm), CO_2 (with bands at 4.27 (both $^{12}CO_2$ and $^{13}CO_2$), at 15.2, and weak combination bands at 2.70 and 2.78 μm), CH_3OH (bands at 3.53, 3.85, and 9.74 μm), OCS (band at 4.90 μm), OCN^- (band at 4.62 μm; labeled XCN in Figure 10.27), NH_3 (bands at 2.96 and 9.0 μm), and CH_4 (bands at 3.33 and 7.7 μm). Several other weak features are present and have been tentatively assigned to modes of H_2CO, HCOOH, and SO_2, but those assignments are less certain. In addition, there are a number of broad features consisting of blends of multiple components, whose identification could not be ascertained from the low resolution Spitzer IRS spectra. The most striking aspect of the infrared spectrum in Figure 10.27 is the apparent simplicity of the composition of interstellar ices. Many species are highly hydrogenated (H_2O, NH_3, CH_4) but the spectra also show mixed oxidation/reduction patterns (CO, CO_2, CH_3OH, and possibly H_2CO).

Band Profile

Ice band profiles are sensitive to the conditions of the ice. In particular, H_2O ice directly deposited from water vapor at low temperatures has an amorphous structure with no long

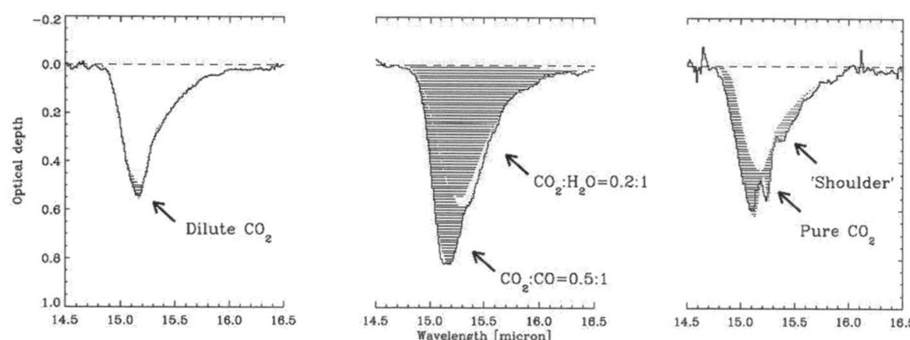

Figure 10.28 The interstellar CO_2 band observed toward the low mass protostars, IRS 51, SVS 4–5, and RNO 91. The hatched areas indicate the five different components that can be identified by comparing these and other spectra. These components are labeled by likely identifications. Figure taken from [185]

range order and is very porous. Upon warm up, the 6 μm ice bending mode loses it pronounced long wavelength wing and the 3 μm stretching mode shifts to longer wavelengths. This is due to compaction of the ice and loss of micropores. Upon futher warm up, to 140 K in the laboratory, amorphous ice anneals to cubic ice, I_c, and shortly thereafter to hexagonal ice, I_h. At this point the 3 μm ice band sharpens and shows clear substructure. These temperature changes are related to structural changes and are irreversible. Similar spectral changes due to amorphous to crystalline transformations occur for other pure ices, but are generally less pronounced. In ice mixtures, interaction with neighboring molecules will shift bands. For van der Waals interactions, these shifts are subtle but if H-bond formation is involved, these shifts can be pronounced. In an ice mixture, bands are also broadened as a result of a distribution of binding sites. Upon warm up, ice mixtures can segregate out, particular if the compounds are immiscible. This behavior becomes quite pronounced near the crystallization temperature as the latent heat of crystallization propagates the crystallization front through the ice, expressing foreign substances in front of it. If this heat-up is rapid, traces of other species can remain behind, encapsulated in chlathrate structures. We encountered this behavior in Section 7.6.1 where the sublimation of mixed ices was discussed. These structural changes in the ice leave their imprint in the detailed profiles of the vibrational modes.

Turning now to interstellar ice, subtle profile details present in the observed ice bands reveal that the interstellar ice is more complex than a simple mixture of the identified compounds. Of particular interest are the CO fundamental at 4.67 μm and the CO_2 bending mode at 15.2 μm. Careful comparison of the detailed profiles in a large sample of interstellar spectra reveals that each of these two features consists of a blend of multiple components (Figure 10.28). For CO_2, the band contains contributions from a polar[6]

[6] Polar and nonpolar refer here to the ability to form hydrogen bonds not the polarity of the mixture.

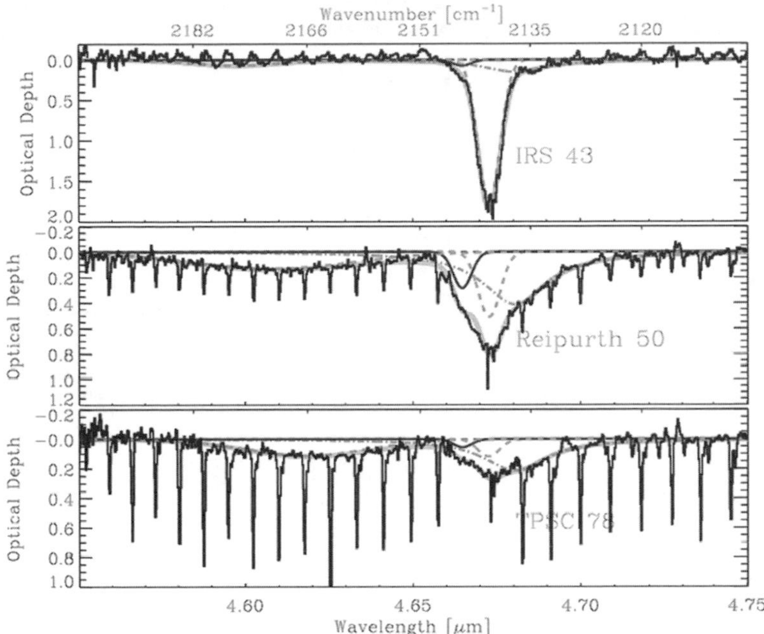

Figure 10.29 The interstellar CO band observed toward the low mass protostars, IRS 43, Reipurth 50, and TPSC 78. Comparison of this band in a large sample of stars reveals the presence of three independent components, the central narrow component due to solid CO in a non-polar environment, the redshifted component due to CO in a polar environment, and a blue shifted component, likely associated with a mixed CO_2:CO component. Figure taken from [184]

(H_2O-rich) component, a nonpolar (CO_2:CO$\sim$1 : 1) component, a very dilute CO_2 in CO component, a pure CO_2 component, and a CO_2:$CH_3OH$$\sim$1 : 1 component. The first two of these CO_2 components dominate the profiles of almost all observed bands. The strength of the polar component correlates well with the H_2O ice band while the nonpolar component correlates well with the nonpolar CO band. For CO we recognize a narrow component due to pure CO, a broad component due to CO in polar environment, and a blue component, likely due to CO:CO_2 mixture. There is somewhat of an issue with the assignment of the CO polar component to a H_2O-rich mixture as H_2O deposited at low temperatures forms a porous structure with many dangling OH bonds. In a mixture with CO, the latter give rise to a redshifted CO component in laboratory studies that has never been observed in space. One possibility is that – in contrast to laboratory experiments where H_2O ice is deposited from H_2O vapor – interstellar H_2O ice is formed through chemical reactions and the release of reaction energy immediately compacts the ice. Alternatively, porous H_2O ice can be "compacted" by UV irradiation and energetic particle bombardment. As a third possibility, the interstellar polar CO component may be due to CO:CH_3OH mixtures (Figure 10.30). Besides the presence of multiple ice components contributing to a band

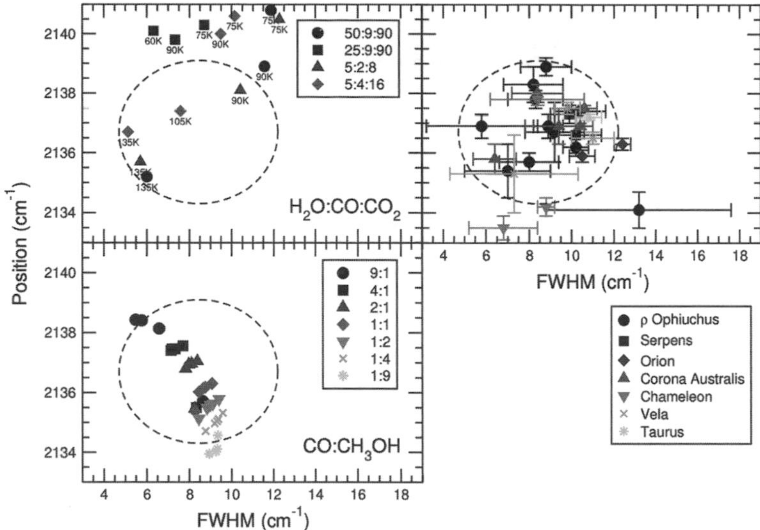

Figure 10.30 The origin of the interstellar polar-CO ice component. Right: The bandwidth versus peak position of the observed interstellar polar CO component. The dashed line has been drawn to guide the eye in the other panels. Left: Measured bandwidth versus peak position for various CO:H_2O ice and CO:CH_3OH ice mixtures. Figure taken from [184]

profile, particle shape can also affect the absorption profile of ices for sub-micron-sized grains if the modes are intrinsically very strong (e.g. pure CO, CO_2). Finally, the effect of ice temperature is evident in the observed profile of the 3 μm ice band observed toward a few sources (notably BN) as well as the long wavelength pure phonon modes (observed in a few sources). The narrow components of the CO_2 bending mode are also indicative of heat-up and segregation.

Disentangling the many factors influencing the observed band profiles is not easy. Degeneracies can be assessed by considering the full ice band spectrum from 2.5 to 100 μm. Also, by analyzing a large sample of sources, the relationship between different components can become clearer and evolutionary scenarios can be developed. Analysis of isotope bands has also been helpful in some cases. Figure 10.31 illustrates the results of such an analysis.

Column Densities

Column densities of interstellar ice compounds can be derived from the observed absorption features coupled with the intrinsic strength, A_i, of these bands measured in the laboratory (Table 3.6),

$$N_i = \frac{\tau_i \, \Delta\nu_i}{A_i}, \tag{10.94}$$

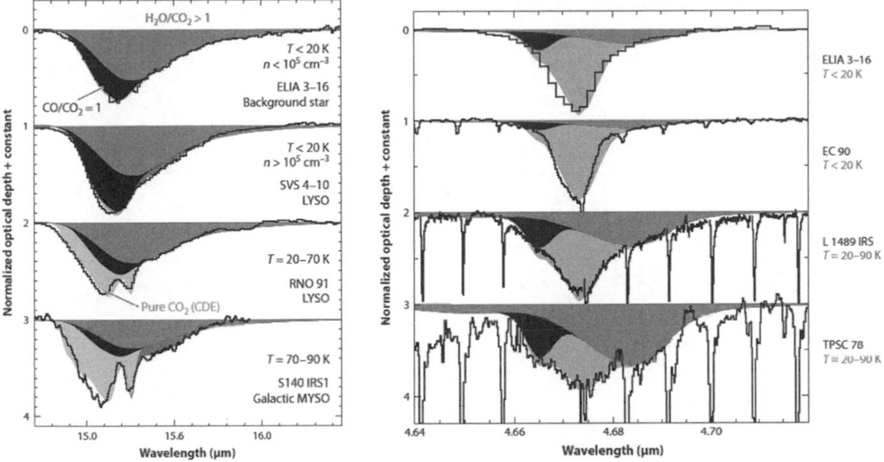

Figure 10.31 Illustrative profiles of the 15.2 μm CO_2 bending mode observed toward a sample of sources arranged in a physical scenario. The different components contributing to the detailed profiles are indicated in gray. The spectra have been arranged to illustrate the effects of density (e.g. chemical formation history) and temperature (segregation and sublimation). Figure taken from [37] and references therein

where $\tau_i \Delta\nu_i$ is a shorthand notation for the integrated strength of the observed band. For strong bands, particle shape effects can influence the intrinsic strength of the transition. These column densities can be translated into abundances through a measurement of the optical depth of the 10μm silicate feature – corrected for underlying emission – and using the conversion factors, $A_V/\tau_{sil} = 18$ and $N_H/A_V = 1.9 \times 20^{21}$ H-atom cm^{-2}. These H-nuclei column densities are quite uncertain, because radiative transfer is dominated by (poorly constrained) geometry, scattering at the shorter wavelengths, and differences in grain properties due to, e.g., grain growth. Generally, the composition is therefore normalized to that of H_2O. Table 10.4 summarizes the chemical make-up of interstellar ices along the lines-of-sight through several molecular clouds.

A number of points can be made: 1) It is striking that the composition of interstellar ice is very different from that in the gas phase (Tables 10.2 and 10.4) and a very different chemistry must be at work. 2) Water ice is the dominant ice component while carbon is divided between CO, CO_2, and CH_3OH. The abundances of CO and CO_2 are very stable relative to H_2O while CH_3OH is highly variable from one sightline to the next. Clearly, there are multiple factors involved in the abundance of the latter. 3) The simplicity is somewhat misleading as there are multiple CO and CO_2 components. Abundances for these components are listed separately in Table 10.4. 4) Absolute abundances of H_2O ice are $\simeq 5 \times 10^{-5}$ relative to H_2 and, hence, an appreciable fraction of the available elements are locked up in ices inside molecular clouds. This is also apparent from gas phase studies as the abundance of many gaseous molecules drops precipitously in the densest cores (Section 10.6.3).

Table 10.4 *The composition of interstellar ice*[a]

Species	Quiescent cloud[b]	Low mass protostar[c]	High mass protostar[d]	Comets[e]
H_2O	100	100	100	100
CO (total)	25	5	13	23
CO (H_2O-ice)	3	10	6	–
CO (pure CO)	22	4	3	–
CO_2 (total)	21	19	13	6
CO_2 (H_2O-ice)	18	(10)[f]	9	–
CO_2 (CO-mix)	3	(10)[f]	2	–
CO_2-CH_3OH complex	–	–	1.2	–
CO_2 (pure crystalline)	–	–	1.0	–
CH_4	< 3	< 1.4	1.5	0.6
CH_3OH	10	30	18	2.4
H_2CO	–	–	6	1.1
HCOOH	< 1	< 1	7	0.09
OCS	< 0.2	< 0.1	0.2	0.4
NH_3	< 8	< 11	15	0.7
OCN^-	< 0.5	< 0.2	3.5	0.1[g]

[a]Interstellar ice abundances taken from the comprehensive study by [81] supplemented by [20, 34, 36, 123, 184, 185, 185]. Cometary abundances refer to comet Hale–Bopp at 1 AU from the Sun [29]. [b]Observations toward the star, Elias 16, behind the Taurus cloud. With a visual extinction of 23.5 magn., $X\,(H_2O) \simeq 5.6 \times 10^{-5}$. [c]Observations toward the low mass protostar, Elias 29, in the ρ Oph cloud. With a visual extinction of 48 magn., $X\,(H_2O) \simeq 3.7 \times 10^{-5}$. [d]Observations toward the luminous protostar, W33A. With a visual extinction of 50 – 100 magn., $X\,(H_2O) \sim 6 \times 10^{-5}$. [e]Methanol abundances are highly variable [245]. [f]Scaled from the averages ratios for these components observed toward low mass YSO [185]. [g]Taken as HNCO.

Chemically speaking, hydrogenation is clearly important for the bulk composition of interstellar ice but oxidation also occurs. The high abundance of CO – a highly volatile species – in quiescent sightlines as compared to YSO sightlines likely reflects thermal processing of dust in the envelopes by the newly formed star. The high abundance of CO_2 in quiescent sightlines implies that energetic processing of ices by UV photons likely plays little role. As UV photolysis is truly the "last refuge of a scoundrel," we will focus the discussion of the bulk composition of interstellar ices on grain surface chemistry routes. We will revisit the importance of energetic processing of interstellar ices when we discuss abundances of trace species and the composition of Hot Cores and Hot Corinos in Section 11.1.1.

10.6.2 Chemistry of Interstellar Ice

Recapping the "rules of engagement" (Section 7.5), interstellar grain surface chemistry is in the diffusion limit where transport to the surface is the limiting factor. Species that have accreted are either very mobile, searching for a reaction partner, or are stuck near the site where they were accreted, awaiting a reaction partner. On an H_2O ice surface at 10 K, the former include H, D, and N. On an O_2 or CO ice surface, C and O could diffuse at 10 K as well. At higher temperatures, other small radicals will start to diffuse (but quite generally their abundance in the accreting gas is too low to be relevant for the bulk composition of the ice). Besides these radicals the main coreactants are CO and N_2 but the latter is unreactive. Other species are at the trace level.

In the diffusion limit, the diffusing species can search the surface and "collide" often with potential coreactants before a new accreting species arrives. In order of preference, reactions to be considered are radical–radical without reaction barriers or radical–molecule reactions with submerged barriers along the reaction pathway. If no such coreactant is present, atomic H (or D) can react with species with appreciable activation barriers. For each collision, the reaction probability is small but many, many, many collisions will occur before sublimation occurs or the next migrating radical adsorbs. The limiting barrier is then $\simeq$3800 K for a single coreactant and, if the surface concentration of the coreactant is high, the limiting barrier can be as high as 8000 K. To conclude these rules, at low temperatures, there will be an appreciable concentration of H_2 on the surface and reactions involving H_2 are important for barriers less than $\simeq$5000 K. Grain surface chemistry predominantly involves hydrogenation and oxidation reactions and leads to the formation of simple species; e.g. H_2O, CO, CO_2, CH_3OH, H_2CO, CH_4, and NH_3. Relevant reactions are summarized in Tables 7.3–7.5.

H_2O Formation

Chemical reactions involved in H_2O formation have been discussed in Section 7.4.4 and are summarized in Figure 10.32. The formation of water can be initiated by the reaction of atomic O with H if the next acreted (reactive) species is an H-atom and no other coreactant is accessible to O on the surface. The hydroxyl radial will react with H_2 – with an activation barrier of $\simeq$1400 K – if the surface coverage of the latter is high. If another O accretes first, O_2 will be formed instead (Figure 10.32). Molecular oxygen will react on to water as well. The initial step ($H + O_2 \rightarrow HO_2$) has been studied extensively experimentally and theoretically in the gas phase and, while there is a submerged barrier – possibly associated with an avoided crossing – this reaction has been shown to occur activationless. Studies in and on an O_2 surface indicate that this reaction is activationless on ices as well. In the gas phase, the $H + HO_2$ reaction has three reaction channels: 2OH, $H_2O + O$, and $H_2 + O_2$ (branching ratio = 0.9:0.07:0.03 at 300 K). In O_2 ices, H_2O_2 is the main product as the excess energy can be transferred to the matrix and the cage effect can also hamper the separation into (reactive) products. There is experimental evidence for a (minor) OH channel, likely because some of the excited OH radicals (there is 2 eV of excess energy for

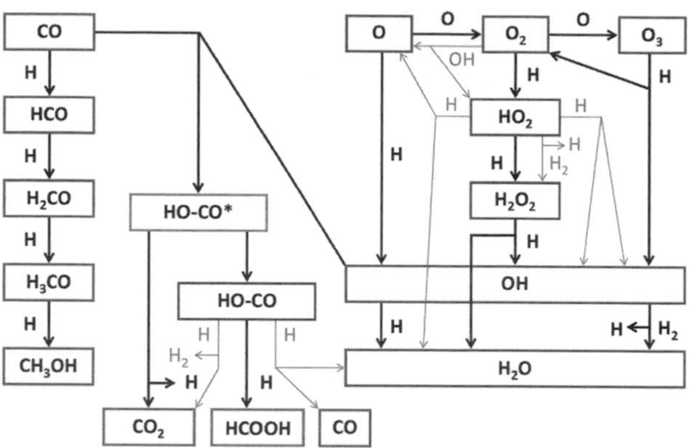

Figure 10.32 Grain surface chemistry network. Reactions involved in the hydrogenation of CO (left branch) leading to the formation of H_2CO and CH_3OH and the formation of H_2O (right branch) on grain surfaces. The two reaction networks intersect in the formation of CO_2 (middle). Black arrows indicate reactions that are central to the network but laboratory studies reveal a number of minor channels (gray arrows). Figure kindly provided by Sergio Ioppolo and updated from [229] with the latest experimental studies, see [117]

the OH radicals) manage to "escape" their cage. Peroxide will be attacked by H (H_2O_2 + H → H_2O + OH) – simultaneously breaking the O–O bond and forming an O–H bond – and the barrier is around 2500 K. The OH formed this way will rapidly react with H_2 (see above). When molecular oxygen is present on the grain surface, further reactions with atomic oxygen will lead to ozone formation. The barrier for this reaction is less than 140 K, submerged, and proceeds in low temperature ice matrices. The reaction H + O_3 → OH + O_2 has a measured Arrhenius energy of 460 K in the gas phase. However, this apparent activation energy results from steric effects. The reaction proceeds by the formation of a van der Waals complex. Along the minimum energy path, where H attacks a terminal O in an out-of-plane approach, the van der Waals complex has a small submerged barrier and, at very low energies, the long range potential steers and reorients the reactants along this pathway. When the temperature increases, the "cone-of-acceptance," and hence the reaction rate, increases. Hence, if there are no steric factors associated with the surface structure, this reaction can also take place on ice surfaces. Laboratory studies show that ozone is hydrogenated to OH + O_2 at low temperatures. On interstellar ice, the OH will form H_2O with H_2. Finally, note that all routes involving OH are catalytic pathways for the reaction of O + H_2 → H_2O. And in that sense, H_2O formation on grain surfaces is not a "sink" for atomic H at low temperatures.

Hydrogenation of CO

As discussed in Section 7.4.4, in comparison to water formation, the chemistry of CO is simple (Figure 10.32). Laboratory experiments have shown that hydrogenation of CO is very efficient even at low temperatures. The initial step, the formation of the formyl radical from carbon monoxide, and the third step, the formation of the methoxy radical from formaldehyde, have appreciable activation barriers. Gas phase kinetic and quantum chemistry studies (Section 7.4.4) place these barriers at 2700 K and 2100 K, respectively. Table 7.4 provides the barrier properties of relevant reactions derived from experimental studies on ice surfaces. These barriers are little affected by surrounding H_2O or CO molecules. The two intermediate steps involve reactions of H atoms with radicals and will be activationless. One additional complication is that the reaction of atomic hydrogen with formaldehyde can also lead to H abstraction and reformation of the formyl radical. In the gas phase at 300 K, theory shows that abstraction is faster than addition by about a factor of 5, which grows to a factor 30 at 20 K. The presence of H_2O changes these branching ratios through dipolar interaction and now addition is greatly favored. Studies on (mixed) isotopically labeled (formaldehyde) ices (see Section 10.6.2) show that both processes also occur on ice. Extrapolation of these studies to the interaction of H with H_2CO is, however, not straightforward as the inferred reaction rates still include the unknown and unequal surface concentrations of H and D (Section 7.4.4). All the H is in principle available for CO hydrogenation as no atomic H is consumed in H_2O formation. However, species such as H_2S and N_2H_2 can act as H-atom scavengers forming H_2 through abstraction reactions. Abundances of methanol and formaldehyde are then controlled by how "fast" such scavengers are kept at a low abundance by the burying process in the ice mantle and this depends on the grain sizes; e.g. on a grain with 10^6 sites, an H_2S will act as a scavenger for many more H accretion events than on a grain with 10^4 sites. Hence, beside gas phase composition (relative accretion rates) and temperature (mobility), ice mantle composition may also depend on the grain size.

CO_2 Formation

The direct reaction of atomic O with CO to form CO_2 at low temperatures is controversial. Matrix studies show that atomic oxygen can diffuse through inert matrices and react with CO to form CO_2. However, in mixed H_2O:CO and O:O_2:CO ices, the reaction is very slow as most of the O is lost to O_2 or O_3 formation and very little CO_2 is formed ($\sim 10^{-3}$). Of course, competition may be irrelevant on interstellar ice surfaces as accretion timescales are very long. Quantum chemical studies show that this reaction is strongly inhibited as there is a large activation barrier on the triplet surface. The discrepancy between the experiments and between (some) experiments and theory is not well understood. It may reside in a spin-forbidden crossing from the excited triplet ($^3B_2$) to the singlet ground state ($X^1\Sigma^+$). In any case, the reaction is slow on experimental timescales and this has led to a search for other surface chemistry routes toward CO_2. In particular, the reaction of OH with CO will lead to CO_2. This reaction proceeds through the HOCO intermediary where the H has to tunnel out. The OH radical is immobile on an ice surface at 10 K, but OH may form near a trapped CO molecule. Alternatively, a newly formed OH – in the H_2O formation scheme – could be translational hot (Section 7.4.7) and this excited OH could hop around on the surface and react with a nearby CO before losing its translational energy (this would be in competition with reactions with H_2). Any vibrational excitation of the newly formed OH is known to assist in the tunneling process. Experiments show that the OH plus CO reaction is fast on low temperature ices. The reaction of O with HCO is also of interest as a possible pathway to CO_2. In the gas phase, the two product channels (CO_2 + H and CO plus OH) have a branching ratio of about unity.

Formation of CH_4 and NH_3

Methane is formed from atomic C through sequential reactions of H atoms. As most of the accreted carbon is in the form of CO in dense cores (Eq. (10.59)), the abundance of CH_4 in ices is small. Note that the sticking coefficient of C^+ is expected to be very small and the cation plays no role in surface chemistry. Ammonia is formed from sequential reactions of accreted atomic N with atomic H. The main nitrogen reservoir is N_2 and reaction with cosmic-ray-produced He^+ or with cosmic-ray-produced photons forms atomic N. In the gas phase, atomic N is lost through neutral–neutral reactions; e.g. with OH, which itself is produced in the chain following cosmic ray ionization of H_2 (and destroyed in the reaction with O). The atomic N abundance is then,

$$\frac{X\,(\mathrm{N})}{X\,(\mathrm{N_2})} \simeq \frac{1}{2}\frac{1}{2}\frac{1}{10} \simeq 0.02, \tag{10.95}$$

where the successive factors $1/2$ reflect that two N's are needed to make an N_2 and the ratio of the neutral–neutral and the ion–molecule reaction rates. The factor 10 is the He over H cosmic ray ionization rate. So, grain surface reactions produce both these species at the few percent level. While that makes them traces in the ice, they are much more abundant than direct accretion of the parent from the gas phase would indicate (Table 10.2).

Deuterium Chemistry

Deuterium fractionation on grain surfaces is in the first instance driven by the relative accretion probability of atomic H and D from the gas. The atomic H and D abundances are set by dissociative electron recombination of molecular ions (Figure 10.20), and their abundance ratio is linked to the abundance ratio of the molecular ions involved (Section 10.5.2). This leads to,

$$X(D)/X(H) \simeq X(DCO^+)/X(HCO^+) \simeq 2\times 10^{-2}, \tag{10.96}$$

where the value is derived from observations (Table 10.3). The atomic D/H ratio in the gas phase is thus very much enhanced above the elemental abundance ratio ($\mathcal{A}_D \simeq 1.5 \times 10^{-5}$). After accretion, this enhanced atomic D/H ratio can be passed on to grain surface chemistry products. The grain surface chemistry routes can then be expanded to include deuterium (Figure 10.33). Direct substitution of D for H in the reaction schemes yields abundance ratios of deuterated species to their hydrogenated counterparts given by the relative accretion rates of D and H. Take ammonia as an example, and considering that only reactions with D (probability p_D) and with H (probability $1 - p_D$) are relevant, we have: $NH_2D/NH_3 = 3p_D/(1-p_D)$, $NHD_2/NH_3 = 3p_D^2/(1-p_D)^2$ and $ND_3//NH_3 = p_D^3/(1-p_D)^3$. Given the difficulty to form NH_3 in the gas phase (Section 9.3.3), gaseous

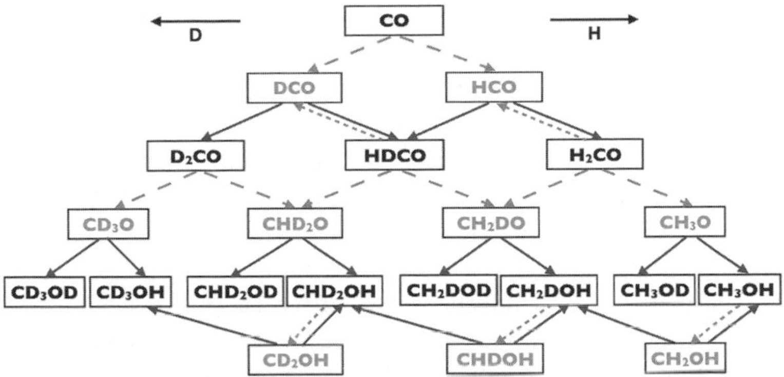

Figure 10.33 Reactions involved in the hydrogenation and deuteration of CO on grain surfaces leading to the formation of H_2CO and CH_3OH and their isotopologues. Solid arrows indicate H (right arrows) and D (left arrows) addition reactions with radicals (e.g. without activaton barriers). Long dashed arrows indicate H and D addition reactions with activation barriers. Short dashed lines indicate reactions abstracting H. To avoid cluttering the diagram, reactions abstracting H from isotopically mixed methyl groups on methanol are not shown but these will equally well occur. Figure adapted from [48, 106]

ammonia has been attributed to grain surface chemistry followed by ejection. The predicted deuterium fractionation ratios are in good agreement with the observed ratios in the gas phase of dark clouds (Table 10.3), lending some support for this suggestion. However, it should be noted that proton/deuteron transfer from the isotopologues of H_3^+ followed by dissociative electron recombination (mainly) back to ammonia will link the ammonia isotopic ratios to those of H_3^+ and the D/H in accreting gas is also set by these ratios.

Further deuteration can occur because of H (or D) abstraction reactions as the resulting radical has another chance to react with D (or H). As these abstraction reactions possess activation barriers, there will be a kinetic isotope effect,[7] which tends to concentrate D in the species. Consider the route from carbon monoxide to formaldehyde and methanol. Reaction with H forms HCO, which then reacts on to H_2CO. There is a similar sequence with D that leads to D_2CO and HDCO is formed in mixed reaction systems (Figure 10.33). Formaldehyde will react on with atomic H and D and, with addition reactions, this is the route toward methoxy and methanol, which can then also be formed in the various isotopologues. However, the abstraction channel will lead back to the formyl radical and, because of the zero-point energy difference, H abstraction is more likely than D abstraction. This kinetic isotope effect can drive the formaldehyde deuteration to much higher levels than the D/H ratio in the accreted gas. Experiments show similar kinetic isotope effects in the exchange reactions on the methyl group of methanol (Figure 10.33) but not on the hydroxyl group. The extreme deuteration of formaldehyde and methanol – as compared to

[7] D barriers are generally higher and the larger mass inhibits tunneling of D-bearing species relative to their H-breaing counterparts.

other species – observed in Hot Corinos (Section 11.3.2; Table 11.2) attest to the importance of these H-abstraction reactions and the highly reducing conditions when ices formed in the preceding, cold, prestellar core phase. Rapid exchange on the hydroxyl group is observed in laboratory studies on mixed HDO and CD_3OH ices at temperatures above 120 K before crystallization of H_2O ice. Such H/D exchange has been studied for other hydrogen bonding systems (H_2O/D_2O and NH_3/D_2O) and is related to proton transfer and ion defects.

The multiple routes involved in water formation make evaluating deuteration somewhat involved. The reactions of H and D with either atomic or molecular oxygen or ozone occurs activationless and there will be no kinetic isotope effect. There is no intermediate complex formed in the hydrogenation of peroxide and hence no scrambling of the isotopes. Hence, H_2O_2 with D will form HDO and OH, and D_2O_2 with H will form HDO and OD. The reaction of the mixed isotope with H (or D) can lead to HDO + OH (HDO + OD) and H_2O + OD (D_2O + OH). Any OH or OD formed can react with H_2, HD, or HD on the surface. Again, there is no intermediate complex and no scrambling of the isotopes. Relative to H transfer from H_2 (or HD), transfer of D from HD or D_2 occurs at a rate of 0.1. Because of the difference in zero-point energy, these molecular hydrogen isotopologues have different sublimation rates and their surface concentrations reflect this; e.g. HD/H_2 is $\simeq 4 \times 10^{-3}$ at 10 K. Hence, even if the D/H ratio in the accreting gas is very high, the abundance of D_2O produced through the hydroxyl route will be quite low. In environments where thermal evaporation plays a role, proton-exchange reactions in a hydrogen bonding network can quickly equilibrate the fractionation between H_2O, HDO, and D_2O (and also CH_3OH/CH_3OD and their isotopologues) but that is not the case for ice mantle sputtering or cosmic-ray-driven evaporation.

Summary

Grain surface chemistry models of interstellar clouds have been quite succesful. The key success is the identification of efficient chemical routes toward all the major species: H_2O, CO, CO_2, CH_3OH, H_2CO, CH_4, and NH_3. Dedicated laboratory studies have been very instrumental in this and the routes discussed in the previous sections have all been confirmed by such studies (Section 7.4). The complexity of multiple component ices is also relatively easy to understand as a combination of accretion in low density gas (3×10^3 – 10^4 cm^{-3}) producing H_2O-rich ices as atomic H is very abundant (H/(CO+O) $\gg$ 1) and accretion at high densities ($>3 \times 10^4$ cm^{-3}), resulting in CO-rich ices as atomic H has a low abundance (H/(CO+O) $\ll$ 1). This mix of ices from low and high density environments along a sightline may be the consequence of accretion in a turbulent environment where parcels of gas are cycled through denser phases or the natural outcome of accretion in a dynamically evolving molecular cloud where accretion starts at densities when the accretion timescale becomes faster than the dynamical timescale but completes when the density is much higher. The high observed deuterations in Hot Cores (Table 11.2, Section 11.3.2) – in particular, the high abundance of multiply deuterated species (D_2CO, CHD_2OH, CD_3OH) is also a natural outcome of grain surface chemistry in a highly reducing environment.

Even more so than for gas phase models, reproducing abundances are an issue. This reflects partly the issues with accounting for the statistical nature of grain surface chemistry in models (Section 7.7). Moreover, laboratory studies cannot really directly determine reaction rate coefficients as the surface concentration of, e.g., H (or O) are convolved into the derived rates (Section 7.4.4). Those surface concentrations depend then on the mobility of the species involved, which are very uncertain. Furthermore, for many reactions, tunneling is of key importance and it is often treated in an approximate way (or even ignored in a thermal hopping-over-the-barrier approach). While the exact treatment is always an issue, laboratory experiments can quantify the competition between different reactions and that is all that is needed, but quantifying all rates on a relative scale has not been fully completed. Finally, while originally the detection of CO_2 was claimed as evidence for the importance of energetic processing of ices, that is no longer considered necessary as experiments show that surface chemistry leads to efficient CO_2 production. Hence, the importance of energetic processing is still unclear (c.f. Section 11.3.2).

As studies of grain mantle chemistry are still searching for accurate and yet efficient methods to evaluate gas and grain surface chemistry simultaneously (Section 7.7) and given the rapidly expanding set of specific reaction systems studied in the laboratory, model studies are lagging in this field. Here a selective set of studies will be used to highlight the different approaches used (for references see the Further Reading section).

The earliest studies evaluated the composition of interstellar ices using a Monte Carlo approach for grain surface chemistry. In this study, the grain surface and gas phase chemistry are decoupled by adopting (relative) accretion rates of the (major) gas phase species. The results showed that at low densities (e.g. high H/CO ratios in the accreting gas; $n < 10^4$ cm^{-3}), water and methanol are abundantly formed. At intermediate densities (H/CO$\sim$1; $n \simeq 10^5$ cm^{-3}), water is still formed efficiently as OH formation through hydrogenation of H_2O_2 or O_3 is followed by reaction with H_2 present on the surface, releasing a new H for reaction. At high densities ($n > 10^5$ cm^{-3}), the abundance of water starts to drop and the mantle evolves toward a CO-dominated composition. CO_2 was mainly formed through the reaction OH + CO rather than CO + O.

More recent studies use an updated grain surface reaction network (based on recent laboratory and quantum chemical studies) and include the effects of grain porosity. These studies adopt a rate equation approach to follow the chemistry – during the collapse of a cloud – validated against a Monte Carlo study for a more limited reaction set. This approach allows a very extensive parameter study. These studies conclude (similarly) that CH_3OH ice is formed at low densities, relatively early on in the collapse. Furthermore, adsorbed CO shifts from a H_2O-rich to a CH_3OH-rich environment during the evolution but there is no obvious CO-rich ice phase. In later studies, the rate equation approach is modified to better account for the competition between tunneling and migration. This study does not allow newly formed OH to move to adjacent (CO-containing) sites. While overall observed H_2O, CO, and CH_3OH abundances are reasonably reproduced at 10 K, too little CO_2 is formed in the ice. Better agreement with the observations is obtained at elevated dust temperatures (>12 K) when the coreactants can migrate.

Another recent study focuses on microscopic aspects of ice formation in the CO-system but limited to hydrogenation of CO (Figure 10.32). Again, this study is based on recent laboratory studies and (thermally driven) rates derived from experiments, using a Monte Carlo approach on a lattice that follows the detailed morphology of the growing ice mantle; e.g. including the effect of kinks, surface structure, and morphology of the growing ice layer. Gas accretion rates are initially set to $X(H)/X(CO) = 10^4/n$ but evolve with time as CO depletes onto grains. The results show efficient methanol formation and, at high densities (10^5 cm^{-3}), a thin layer of "pure-CO" is formed covered by thick layers dominated by CH_3OH.

Finally, the hybrid method (see Section 7.7) has been used to study ice chemistry. This method allows a large reaction set (3075 reactions among 284 gas phase species and 56 surface species with 151 reactions). The grain surface reaction set is based upon laboratory studies. Because this study addresses observations toward ρ Oph, adopted densities are quite high, resulting in a low CH_3OH-to-CO ratio, but in these mantles, the ratio of CO and CO_2 to H_2O is in good agreement with observations.

10.6.3 Depletion in Dense Cores

Dense prestellar cores show low molecular abundances in their central regions. Table 10.5 summarizes abundances derived for the dense prestellar core, B68, whose physical structure – density and temperature profiles – has been well studied (see Figures 10.8 and 10.9). These abundances have been derived toward the central position by adopting the physical structure derived from dust observations and (naively) assuming constant abundances. Comparison with the results for TMC1 and L134N reveal that many of these species are highly depleted compared to typical dark cloud abundances. Indeed, while the sub-mm dust emission reveals a dense, centrally condensed core in these objects, the spatial distribution of these species shows a ring-like structure. This is illustrated in Figure 10.34 for the dense core, L1489; e.g. compare the CH_3OH distribution with the

Table 10.5 *Depletion in dense cores*[a]

Species	B68[a]	TMC1[b]	L134N[b,c]
$C^{18}O$	3.0 (−8)	1.6 (−7)	1.6 (−7)
CS	4.0 (−10)	4.0 (−9)	1.7 (−9)
HCO^+	$<$ 1.4 (−10)	8.0 (−9)	1.2 (−9)
N_2H^+	2.0 (−11)	4.0 (−10)	5.0 (−10)
NH_3	7.0 (−10)	2.0 (−8)	2.0 (−7)
H_2CO	2.0 (−10)	5.0 (−8)	2.0 (−8)
C_3H_2	1.2 (−11)	5.9 (−9)	2.1 (−9)

[a]Derived assuming a constant abundance throughout the cloud. [b]Typical dark cloud abundances (see Table 10.2). [c]Also known as L183.

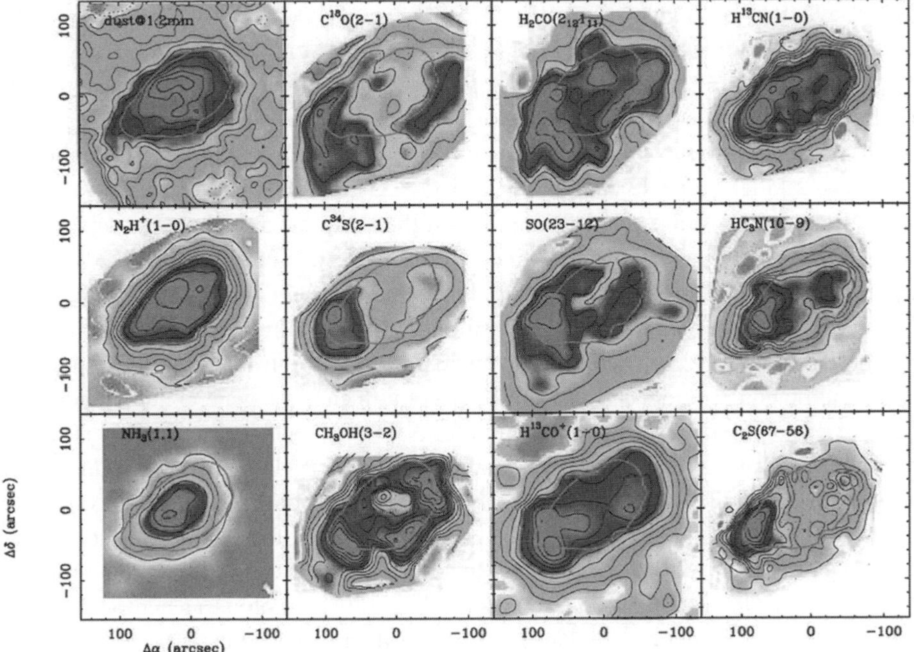

Figure 10.34 Observations of the L1489 core. The dust continuum and the N_2H^+ and NH_3 reveal a centrally condensed core. The other species show a ring-like or centrally depressed spatial distribution relative to the dust continuum indicative of depletion of the emitting species in the densest parts. For ease of comparison, the 75% emission contour of N_2H^+ is superimposed in red on each map. Figure taken from [223] (A black and white version of this figure will appear in some formats. For the color version, please refer to the plate section.)

sub-mm dust continuum map. Analysis of this type of data is often done by assuming a constant abundance envelope surrounding a core with a negligible abundance. The radius of this core and the abundance in the envelope are then left as free parameter in the fitting procedure. These studies lead to the general conclusion that molecules are heavily depleted in the densest parts of the core. Nitrogen-bearing species are an exception to this general behavior. Like the dust emission, N_2H^+ and NH_3 also show a centrally condensed peak (Figure 10.34) but this may reflect excitation effects and chemistry in the outer portion of the cloud.

Accretion

The observed, pronounced depletion of molecules in the densest centers of prestellar cores is evidence for the importance of accretion of gas phase molecules onto dust grains, forming a molecular ice mantle. The accretion timescale is given by,

$$\tau_{acc} \simeq 3 \times 10^5 \left(\frac{10^4\,\mathrm{cm^{-3}}}{n}\right)\left(\frac{10\,\mathrm{K}}{T}\right)^{1/2} \quad \mathrm{yr}, \tag{10.97}$$

and, for densities well in excess of 10^4 cm^{-3}, this is much shorter than the fastest, dynamical timescale, the free-fall timescale ($\tau_{ff} \simeq 4 \times 10^5 \left(10^4 \text{ cm}^{-3}/n\right)^{1/2}$ yr). Here, we have tacitly assumed that all species will stick upon collision, which is a reasonable assumption for all neutral species except H_2 and He. The relative depletion rates of gas phase species will be modulated by their thermal velocities and O depletion (e.g. H_2O formation) will be about $\sqrt{2}$ faster than CO depletion (e.g. CO, CO_2, CH_3OH ice formation). PAHs will disappear at an even slower (5−10) rate. It will then take $\simeq$2.5 O-accretion timescales for the atomic O to CO ratio in the gas phase to reach unity. Initially, the CO to accreted O ratio in the ice will be $\simeq$0.75 the gas phase value but this will quickly rise to unity. While this will cause a gradient in the ice composition, the overall effect is limited, but it will lead to ice layering. Finally, ions will have low sticking coefficients but their abundance is low and recombination rates are fast and this does not influence the general depletion process.

The origin of the lower depletion factors for N_2H^+ and NH_3 in dense cores is involved. The main nitrogen reservoir, N_2 – and parent species for N_2H^+ – has a sublimation temperature very similar to CO and should freeze out as readily in dense cores. Chemical models suggest that N_2 and its daughter products are only formed at high densities and relatively late in the evolution of the core. In addition, CO has a higher proton affinity than N_2 (Table 6.6) and at low electron fractions (high densities), proton transfer to CO becomes the main destruction channel for N_2H^+. Hence, when CO (and electrons) is depleted at high densities, the N_2H^+/N_2 ratio will increase and this may also contribute to a centrally peaked emission region. However, its effect is limited under these circumstances as the ratio of proton transfer with CO to the recombination rate is given by,

$$\frac{\text{proton tranfer}}{\text{recombination}} \simeq 0.2 \left(\frac{X\,(\text{CO})}{8 \times 10^{-5}}\right) \left(\frac{10^{-7}}{X\,(\text{e})}\right). \qquad (10.98)$$

Thus, electron fractions of 10^{-8} – densities of 10^6 cm^{-3} – are required before CO depletion will noticably affect the N_2H^+/N_2 ratio.

The depletion increases markedly toward the center of dense cores and it is, therefore, difficult to determine their physical conditions, as emission from the envelope overshines the core emission. At that point, only hydrogen compounds will be left in the gas phase. Specifically, under high density and high depletion factors, continued reactions of H_3^+ and its isotopologues with HD will "push" deuterium fractionation of H_3^+ all the way to D_3^+ (Figure 10.21) and H_2D^+ and HD_2^+ become the best tracers of such cores.

Ejection

Dust temperatures at the edge of a molecular cloud are low enough ($T_d < 15$ K for $A_v >$ 0.2 mag) that essentially all species (except H_2 and He) should freeze out upon contact with a grain surface in the form of an ice mantle. However, photodesorption by penetrating ultraviolet photons from the ambient interstellar radiation field can manage to keep grain surface "clean" at the edges of molecular clouds. Experimental and theoretical photodesorption studies have been discussed in Section 7.6.2 (see Table 7.7). They range from a high of $\simeq 10^{-2}$ for CO to a low of $<3 \times 10^{-5}$ for CH_3OH. The thickness of the molecular cloud surface layer where grains are kept "ice-free" by photodesorption can be estimated by balancing accretion and photodesorption,

$$A_v = 2.1 + \frac{1}{1.8} \ln\left[\left(\frac{Y_{pd}}{10^{-3}}\right)(G_0)\left(\frac{10^3\ \mathrm{cm}^{-3}}{n}\right)\right], \tag{10.99}$$

where the factor 1.8 is an approximation for the average UV over visual dust absorption ratio. This factor is about twice as large for desorption of CO and N_2. Note that these latter two species are very volatile with sublimation temperatures of $\simeq$16 K. The dust temperature at the edge of a dark cloud is lower than this but if there is a star somewhat close to the cloud, heating of the dust by this star may keep CO and N_2 from depleting. Photodesorption will keep a surface layer of the cloud free from ice and at the same time provide a source of gas phase species. In a global sense, taking accretion as the rate limiting step, for densities in excess of $n > 10^3$ cm^{-3}, grain surface chemistry followed by photodesorption in surface layers is faster than gas phase chemistry routes (cf. Eq. 10.58 and 10.97). Ice column densities have been measured as a function of A_v for a number of clouds (cf. Figure 10.35). This type of data is often fitted to an expression linked to Eq. (10.99) using, e.g., the relation

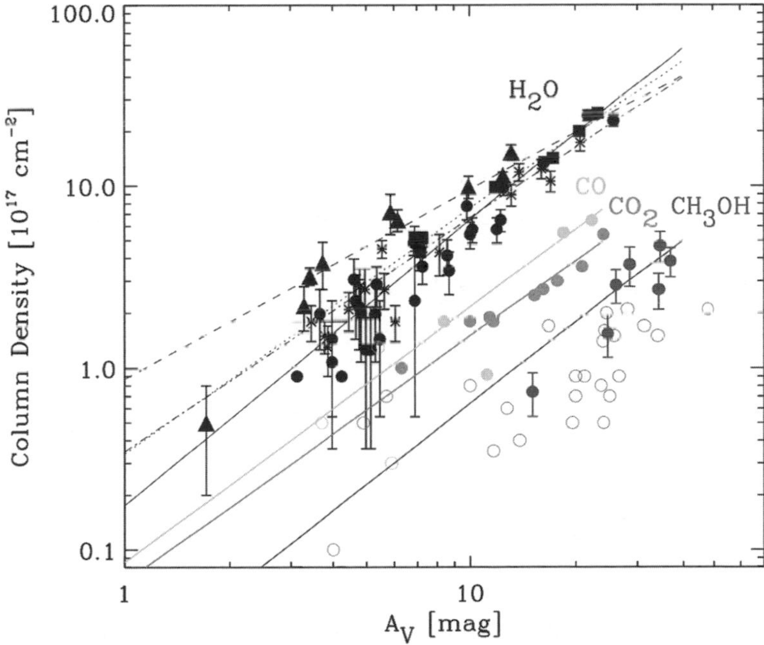

Figure 10.35 H_2O, CO, CO_2, and CH_3OH ice column densities as a function of visual extinction for field stars behind molecular clouds. Filled symbols detections, open symbols upper limits. For H_2O, the different symbols refer to different clouds: circles, Taurus (solid lines); asterisks, Lupus IV (dashed-dotted line); triangles, L183 (dashed line); squares, IC 5146 (dotted line). CO and CO_2 refer to Taurus, while for CH_3OH data toward a variety of clouds has been used. Figure kindly provided by Adwin Boogert [37]

$A_v = aN + A_0$ with a and A_0 fitting parameters. Care should be taken in interpreting these A_0's as limiting magnitudes for ice band formation as sensitivity, band strength, and abundance affect these thresholds as well. Indeed, the data displayed in Figure 10.35 shows no clear evidence for the presence of a minimum threshold in A_v in any of these species. Observations of gaseous water, on the other hand, reveal clearer evidence for the effects of photodesorption in the surface layers (see below).

Inside dense cores, only CR-induced photons are present and the photon flux is down by some four orders of magnitude (Section 6.3.5). Nevertheless, this can maintain a low abundance of species in the gas phase. Equating the accretion rate with the desorption rate driven by cosmic ray produced photons, we find for CO,

$$X\,(\mathrm{CO}) \simeq 10^{-6} \left(\frac{10^4\,\mathrm{cm}^{-3}}{n}\right) \left(\frac{\zeta_{CR}}{3\times 10^{-17}\,\mathrm{s}^{-1}}\right) \left(\frac{Y}{10^{-2}}\right), \tag{10.100}$$

where we have tacitly assumed that the ice consists of CO. For mixed molecular ices, the desorption rate has to be decreased accordingly. For water ice, the expected abundance of photo-ejected species is an order of magnitude less ($X\,(\mathrm{H_2O}) \simeq 10^{-7}\left(10^4\,\mathrm{cm}^{-3}/n\right)$) as $Y_{\mathrm{H_2O}} \simeq 10^{-3}$. Molecular dynamics studies indicate that OH is the main photo-ejected product with H_2O down by a factor of 3. While the observed abundances in dark cloud cores are in reasonable agreement with these expectations, the observed H_2O/OH ratio is less than these results would indicate (Table 10.2). The estimated abundance of methanol, on the other hand, is very low, $<3\times 10^{-9}$; in good agreement with the observations. Photodesorbed products of CH_3OH irradiation would be more abundant, though.

Grain surface chemistry products may also be ejected into the gas phase upon formation (Section 7.6.3). In general, a species will be formed in a highly vibrationally and/or rotationally excited state and some of the excess energy can go into translational motion of the molecule on the surface. If we assume that a fraction, f_{ej}, of the reactions lead to ejection and balance the accretion of H atoms (the main reaction partner) with the accretion rate of gaseous species, we have,

$$X_i \simeq 4\times 10^{-6}\,\frac{10^4\,\mathrm{cm}^{-3}}{n}\,\frac{\zeta_{CR}}{10^{-17}\,\mathrm{s}^{-1}}\,\frac{f_{ej}}{10^{-2}}. \tag{10.101}$$

There are no laboratory studies on f_{ej} and often, rather arbitrarily, a value of 10^{-2} is adopted and the actual value may be much less as vibrational modes are not well coupled to center of mass translational modes. As for photodesorption, mediation may be through an excited electronic state and lead to fairly high ejection fractions. Note that this process can lead to ejection of any of the products; e.g. including the radical intermediaries such as OH, HO_2, HCO, CH_3O, and CH_2OH. But as these react readily with gaseous radicals (e.g. H), their abundance will be quite low ($10^{-11}\left(f_{ej}/10^{-2}\right)$ for a reaction rate of $10^{-12}\,\mathrm{cm}^3\,\mathrm{s}^{-1}$).

Finally, impacts by 10–100 MeV/nucleon heavy cosmic rays will deposit energy at a rate of 4×10^{11} (MeV/E) eV/cm and heat an ice mantle to temperatures of 35–200 K. This can lead to sublimation of volatile ice mantle species (Section 7.6.5). Taking a typical

Fe cosmic ray intensity, $\mathcal{N}_{CR}$ of 10^{-4} particles cm^{-2} s^{-1} sr^{-1}, the timescale for a cosmic ray hit is,

$$\tau_{CR} = \pi a^2 4\pi \mathcal{N}_{CR} \simeq 8 \times 10^4 \left(\frac{1000\ \text{Å}}{a}\right)^2 \quad \text{yr.} \tag{10.102}$$

With this timescale, the ejection rate of volatile species from 1000 Å grains is $\simeq$10 molecules/yr. Equating this again with the accretion rate, cosmic rays can maintain a gas phase abundance of volatile species such as CO of,

$$X_i \simeq 3 \times 10^{-6}\,\theta_i \left(\frac{10^4\ \text{cm}^{-3}}{n}\right), \tag{10.103}$$

where θ_i is the concentration of these species in the grain mantle. The initial energy deposition leads to a thermal spike, which can lead to appreciable ejection of species, including species that are not that very volatile (e.g. H_2O) but also large molecules. Additionally, UV photolysis of ice grains may have produced stored radicals inside the ice. Upon warm up of the ice, these radicals may become mobile and recombine. The released chemical energy may much exceed the deposited cosmic ray energy and in laboratory experiments "explosive" desorption events have been observed.

Observational Evidence

Observationally, there is evidence for the importance of the injection of grain surface products into the gas phase in dense and quiescent molecular cloud cores. Observations – first with Submillimeter Wave Astronomy Satellite, SWAS, and the ODIN satellite and then with the Herschel Space Observatory – demonstrated that the abundance of O_2 is surprisingly low in molecular clouds. Theoretically, steady state models predict that O_2 is the main reservoir of oxygen in the gas phase. Gaseous H_2O, another potential important reservoir of oxygen, is also shown to have a low abundance in molecular clouds. Driven by these observations, detailed models have been developed for gaseous H_2O and O_2 and these place water formation on grain surfaces followed by photodesorption (as H_2O and OH) into the gas phase (and then photodissociation to OH and O) and hence localize gaseous H_2O in molecular cloud surface layers (Figure 10.36). Observationally, it is well known that water ice is a major reservoir of oxygen in molecular clouds (Table 10.4). Adopting realistic photodesorption yields, these model calculations provide a good fit to observations of molecular clouds. Observations of the dense core L1544 reveal a P Cygni profile for the water line. The absorption line corresponds to a line-of-sight averaged H_2O abundance of 5×10^{-9}. Likely, this line arises in this surface layer and reflects an H_2O abundance of $\sim 10^{-7}$. The weak emission feature reflects gaseous water in the dense core with an abundance of $\sim 10^{-9}$.

Hydrogen peroxide (H_2O_2), as well as the hydroperoxyl radical (HO_2), have been detected toward the source, ρ Oph A, with similar abundances of $\simeq 10^{-10}$. There is no obvious gas phase reaction route toward H_2O_2 while it is a main intermediary in the H_2O

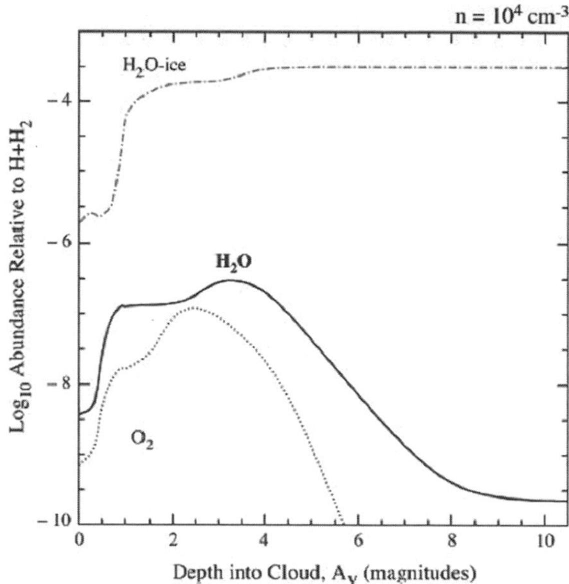

Figure 10.36 Calculated abundances of H_2O and O_2 in molecular clouds. At the surface oxygen is mainly atomic but accretion on grain surfaces produce ice. Photodesorption leads to gaseous H_2O and OH. The OH radical will react with O to form O_2. Figure adapted from [111]

formation routes on grain surfaces (Figure 10.32; Section 10.6.2). Models including gas and grain surface chemistry predict HO_2/H_2O_2 ratios of $\simeq$3 where both species are formed on the grain surface and it is assumed that 1% of all reactions on the grain surface result in ejection. The radical abundance in interstellar ices is expected to be very low. Hence, it is unlikely that HO_2 (and through guild by association, H_2O_2) measures a "bulk" sublimation process such as cosmic-ray-driven desorption. Rather, the presence of these species may reflect a surface ejection process such as molecular formation or photodesorption. While the values adopted for these processes in the model are uncertain, the key point is that these species are difficult to form in the gas phase but are key intermediaries on grain surfaces and this provides a bench mark for models.

Both H_2CO and CH_3OH are difficult to form through direct gas phase routes and readily produced by grain surface chemistry. Hence, their presence in the gas phase is often taken as an indication of a contribution of grains to the gas phase composition of dark clouds. Observations and models of the Horsehead region support this scenario with efficient contribution from grains to gaseous CH_3OH in the dense core. The low photodesorption yield of CH_3OH in pure methanol ices suggests then that this methanol is ejected as a byproduct of knock-on processes of other species present in the ice or result from cosmic ray heating of the ice. In the latter case, the methanol-bearing grains must be small to allow the ice to reach the sublimation temperature of methanol.

10.7 Further Reading and Resources

William and Caroline Herschel discovered the presence of dark regions in the distribution of stars in the Milky Way (recall the famous quote: "Hier ist wahrhaftig ein Loch im Himmel"), which were later shown to be optically opaque interstellar clouds [18]. When in the early days of radio astronomy, an absence of an increased HI was noted for these dark nebulae, it led to speculation ([31]) that this was due to the dominance of molecular gas at high densities. It took, however, CO observations to bring home the point that these clouds were indeed molecular [246]. CO is now widely used as a tracer of molecular gas and systematic surveys of the Milky Way have been performed [58, 102]. For a review see [104].

The difference between diffuse atomic and molecular gas is discussed in [216]. The role of PDRs in molecular cloud surfaces is reviewed by [110]. The chemical properties of the PDR surfaces of molecular clouds can be gleaned from the comprehensive survey of [85]. The penetration of FUV photons in inhomogeneous molecular clouds and its effects on maps in common tracers has been studied by [221]. The importance of turbulence for the chemical structure of molecular clouds and common atomic/molecular tracers has been brought home by postprocessing of numerical simulations [97, 212].

The importance of ammonia as an astronomical tool for molecular clouds has been extensively discussed by [109]. The use and reliability of the inversion transitions as temperature indicators has been investigated by [118, 240]. Updated collisional rates have been published by [38, 59]. Other symmetric top molecules used for this purpose include methyl cyanide and this has been discussed by [136]. The ammonia study of NGC 1333 is reported in [60]. The use of formaldehyde as a thermometer has been discussed by [138].

Isotope abundance gradients in the galaxy have been studied by [148].

Density determination using near-IR extinction has been perfected by [5]. Dust extinction from the optical through the near-IR has been studied by [140]. This has been extended to 8 μm by [116] and to 20 μm by [52]. Density determinations from dust SED fitting has been widely used based upon IRAS all sky maps. The Planck collaboration [182] has calibrated the sub-mm dust cross section per H-nuclei against all sky maps. Its use has been validated against the near-IR extinction method by [135]. This analysis shows that widely used dust models cannot be reliably extrapolated to the far-IR. Better results are obtained by dust models developed for dense cloud that include the effects of coagulation [165, 167].

CO surveys of molecular clouds in the Milky Way have been reviewed by [104]. IR studies revealing the presence of excess dust emission from gas not in HI or CO clouds have been performed by [95, 181, 189, 190]. Relevant γ-ray surveys were performed by [28, 89]. OH studies on gas halos surrounding molecular clouds have been presented in [4, 132, 243, 254]. Modern cloud decomposition algorithms to identify cloud structures as well as (temporary) coherent turbulent structures include dendrograms, principle component analysis, delta variance, and centroid velocity increments [24, 25, 26, 40, 146, 198, 199]. The molecular cloud mass distribution was taken from [247] who analyzed the data of all available CO surveys. Reviews on molecular cloud evolution and the role of star formation therein are [64, 104, 145, 153]. The role of turbulence in molecular clouds is reviewed in [83, 122]. The log-normal probability density distribution is discussed in various places in the literature [97, 98]. Reference [112] provides an improved and more physically motivated PDF. The filamentary structure of molecular clouds is described by [6, 7, 154, 204]. Scaling relations for molecular clouds date back to the seminal study by [126]. Studies for "small" clouds, where NH_3 observations could be used, were very illuminating as well [147]. The CO surveys established their relevance on large scales in galactic and extragalactic environments [58, 102, 206, 217]. For reviews on this see [104]. A comprehensive review on the X_{CO} factor is provided by [32].

Triggered star formation was introduced by [68]. The bubble HII region, RCW 120, and its associated star formation has been studied by [257]. The W3 region is discussed in [192] and the mini starburst in W43 by [16, 159]. The Cyg X and DR 21 regions have been studied by [205]. The discussion on the interaction of massive stars with the Orion molecular clouds leans heavily on [163, 169].

The cosmic ray ionization rate in molecular gas has been reviewed by [170], while [139] provides an accurate determination of the cosmic ray ionization rate in the dense prestellar cire, B68. A general review on turbulence in molecular clouds is provided by [14]. The importance of turbulence for the heating of molecular clouds has been addressed by [107, 175, 183]. The thermal and emission signatures of intermittency in turbulence have been discussed for diffuse clouds by [82]. Accretion onto molecular clouds as an energy source for interstellar turbulence is discussed, among others, by [122].

The dissociative recombination of N_2H^+ has been studied experimentally by [152, 237] to address the branching ratio.

The importance of neutral radical–radical reaction was really brought home by studies on N chemistry, particularly the formation of the cyanopolyynes. Reference [210] presents an early experimental study on the reaction of neutral radicals and the importance of CN in the formation of cyanopolyynes. Astrochemical models quickly bore out the importance of these types of reactions [50, 100, 113]. The possible role of atomic N in cyanopolyyne formation has been considered by [73].

A seminal study of deuterium fractionation driven by ion–molecule gas phase reactions is presented in [244]. A more recent review is provided by [149]. The role of the ortho/para ratio in the chemistry of H_3^+ has been studied by [71, 114, 173]. The importance of doubly and triply deuterated H_3^+ in cores with depleted CO abundances was stressed by [193]. Supported by extensive theoretical models [172, 211, 241], the two isotopomers, H_2D^+ and HD_2^+, have been observed and used to measure the characteristics of very dense molecular cloud cores [45, 46, 94, 177, 194, 235].

Observations of H_3^+ toward massive protostars have been reported by [141] and these data have been analyzed in terms of the implications for the cosmic ray ionization rate by [233]. The use of DCO^+ as an indicator of the electron abundance has a long history. The description here has been taken from [43]. A more recent review is given by [44]. Observations have been published, among others, by [41, 253]. The issues arising from the ortho and para states in H_2, H_3^+, and H_2D^+ have been stressed by [209].

Carbon isotope fractionation through selective photodissociation was originally discussed by [15]. The effect of self-shielding on CO photodissociation rates was quantified by [234] and this has been updated by [238]. Reference [96] has studied the fractionation effects in N_2 photodissociation. Isotope studies have been used by [69, 218, 249] to address the "chemical" relationship of CH_3OH and CO. Fractionation of species such as CN, HCN, and CS has been studied by [133, 148]. The fractionation of carbon chain molecules has been studied by [202, 203, 225, 255]. Nitrogen fractionation in molecular clouds has been studied by [195, 228]. A more recent study by [200] has made a very careful assessment of the reactions involved and the isotope fractionation patterns that evolve from this. Isotopic fractionation of various reservoirs in carbonaceous meteorites has been reviewed by [207]. An origin of these fractionation patterns in various chemical processes in space has been studied by [3, 65, 137, 250].

The chemistry of anions in space has a long history [57, 143]. A key predictive study was [99]. The role of the H^- anion in the formation of H_2 in the early Universe has long been appreciated [180]. Detection of the first anionic molecular species, however, had to await the early twenty-first century when sensitive laboratory spectroscopy studies pinned down their rotational spectra and allowed astronomical searches [142]. Recent reviews provide entries to relevant laboratory and theoretical astrochemistry studies [27, 101]. The role of PAH anions in interstellar chemistry was pioneered by [131]. More recent studies are provided by [42, 79, 239].

The importance of bifurcation in interstellar chemistry was recognized by [127, 128]. Further studies are presented in [30, 49, 242].

A recent review of observations of interstellar ice, including a discussion on the assignments, is presented in [37]. Some relevant studies on the solid CO and CO_2 ice bands are [20, 33, 35, 51, 77, 184, 185, 231]. Methanol has been studied by [245]. References [33, 77, 214] provide evidence for temperature effects in interstellar ice bands. The influence of reactions, UV photolysis, and ion-bombardment on the structure of H_2O ice has been studied by [1, 160, 174, 188]. The importance of particle shape effects on the profiles of the CO and CO_2 ice bands has been discussed by [17, 77, 231].

The basic reaction network, including the importance of hydrogenation reactions and the role of tunneling in grain surface chemistry, was introduced by [229]. At that time, activation barriers of reactions were based on gas phase studies. Over the last 15 years, a number of groups have taken up the challenge to identify key reaction routes and quantify reaction rates experimentally. A pertinent review is [92]. Some relevant experimental studies on H_2O formation are [55, 117, 150] for the peroxide route and [151, 196] for the ozone route. Chemical steps in the specific reactions have been clarified by theoretical studies [155, 232, 256]. The key catalytic reaction OH + H_2 has been studied experimentally on ice surfaces by [161] and theoretically by, among others, [158]. While initial studies on the H–CO system only observed efficient H_2CO formation [108], later studies [75, 105] demonstrated that this reflected the H-flux in these particular experiments and the viability of the hydrogenation of CO all the way to CH_3OH is now well established (see also, [106, 117]). Quantum chemical studies of the barriers involved in the reactions in this route have been performed by [87, 252]. Matrix isolation studies reveal that the reaction O + CO occurs readily [72]. Studies on interstellar ice analogs show little or no CO_2 formation, likely because of competition with other reactions ([90, 187, 197]). Theory reveals a high barrier for this reaction on the triplet surface [86].

Astronomically, the important role of interstellar grain surface chemistry in the deuteration of species such as H_2O, H_2CO, and NH_3 was recognized early on [230], including the effect of H-abstraction/D-addition reactions. Driven by observations of high abundances of deuterated molecules in the warm gas of hot cores (cf. Section 11.1) – thought to reflect evaporating ices accreted in a preceding prestellar core phase – the early CO hydrogenation reaction network was expanded to include deuterium and the astronomical implications assessed [48]. Subsequently, extensive laboratory studies have confirmed the general schemes as outlined and quantified the reactions involved [92, 106, 157]. Quantum chemical studies further support these schemes [87]. Hydrogen/deuterium exchange in ices has a long history [21, 53, 251]. Relevant here, are the studies in mixed H_2O/CD_3OD ices [186, 219, 220].

Reference [229] performed the early Monte Carlo approach for grain surface chemistry and this was followed up by with an updated reaction scheme by [121]. The very recent study by [227] used an updated grain surface reaction network with a rate equation approach that was validated on smaller systems. Reference [76] used a modified rate equation approach. The effects of the microscopic aspects of ice formation in the CO-system was performed by [56]. The hybrid method was pioneered by [67].

The depletion of gaseous species at high density cores is well documented and some of the relevant references are [19, 63, 223]. Reference [2] performed a model study of the (depletion) behavior of species in dense cores. Models for the deuterium shift in H_3^+ with increasing density and depletion have been developed by [193] and this has led to several studies on the properties of dense cores using these isotopologues [46, 94, 176, 235, 236].

Over the years, an increasing number of photodesorption studies on astrochemically relevant ices have been published. Some relevant studies are [9, 11, 23, 54, 70, 156, 162].

The model studies that place H_2O in molecular clouds in the surfaces of molecular clouds were reported by [111]. The water study of L1544 was done by [47]. The peroxide studies in ρ Oph were done by [67, 178, 179]. The importance of cosmic rays for the evaporation of ice mantles has been demonstrated by, in particular, [39]. This has been analyzed in an astronomical context by [208]. The CH_3OH study of dark cloud cores was performed by [91].

10.8 Exercises

10.1 Kinetic temperature of the gas:

(a) Derive equation (10.1) and calculate the excitation temperature as a function of the density for the $J = 1 - 0$ transition of CO. Relevant parameters are given in Chapter 2. Compare your results against Figure 10.2.

(b) Derive Equation (10.10) and calculate the excitation temperature as a function of the density for NH_3. Relevant parameters are given in the text. Compare your results against Figure 10.3.

10.2 Show that the reaction of NH_3^+ with H_2 is important in dark molecular clouds but not in diffuse clouds.

10.3 Both protonated cyanoacetylene and protonated CO are formed through proton transfer to their parent species and (mainly) destroyed through dissociative recombination or mutual neutralization reactions. Compare the abundance ratio of these ions to their parents in TMC-1 (Table 10.2). What could be the origin of these very different ratios?

10.4 Charge balance in molecular clouds:

(a) Describe the different processes involved in the schematic charge flow illustrated in Figure 10.21.
(b) Derive the general expression linking the electron abundance with the cosmic ray ionization rate. Derive simplified expressions for when metals are highly abundant, molecular ions dominate, the dominant charge carrier is PAH anions, and the dominant charge carriers are PAHs. Compare the numerical coefficients that appear in these expressions. Relevant rates can be found in Chapters 5 and 7.
(c) Show that reactions with CO take over as the dominant destruction agent for H_3^+ for electron fractions less than $\simeq 5 \times 10^{-7}$. Adopt a dissociative electron recombination rate for H_3^+ of $10^{-6}/T^{0.45}$ cm^3 s^{-1} and a rate coefficient of $3 \times 10^{-9}/T^{0.14}$ cm^3 s^{-1} for the reaction CO + H_3^+.
(d) Peruse Figure 10.23. Why is there a limit on the value for R_D for a given depletion factor? And why is the depletion factor unimportant for high degrees of ionization?
(e) Calculate the degrees of ionization for TMC-1 and L143N and the cosmic ray ionization rate if their densities are 3×10^4 cm^{-3}.

10.5 Calculate the timescale for ^{13}C fractionation of CO in dense cloud cores.

10.6 Discuss the different ways to measure the physical characteristics (density, temperature, mass, size) of molecular clouds.

10.7 Describe the population of molecular clouds in the galaxy with an emphasize on the scaling relations. Link this to the origins and evolution of molecular clouds in galactic environments.

10.8 Discuss the various heating and cooling mechanisms in the energy balance for molecular clouds. Why are molecular clouds typically $\simeq$10 K?

10.9 Compare and contrast the composition of molecular clouds (Table 10.2) with that of diffuse clouds (Table 9.1) and identify the processes that drive these differences.

10.10 Derive Equation (10.93).

10.11 Sensitivity and ice column density measurements:

(a) Calculate the minimum H_2O ice column density that can be measured in the L band, if $\Delta\tau = 0.05$ is the detection threshold for this broad feature. With an H_2O ice abundance of 10^{-4}, calculate the corresponding H-nuclei column density and the minimum A_v. Compare this to Eq. (10.99).

(b) The much narrower CH_3OH band has a limiting optical depth for detection of 0.03. Calculate the minimum CH_3OH ice column density that can be measured in the L band and compare to that of H_2 derived above. Use a typical interstellar ice abundance to infer the minimum H-nuclei column density and A_v.

(c) Do the same for the CO and CO_2 ice bands.

(d) Consider the implications of this analysis for Figure 10.35.

Relevant parameters can be found in Table 3.6. Use $N_H/A_v = 1.9 \times 10^{21}$ H-nuclei/cm^2.

10.12 Hydrogenation of CO on interstellar ice:

(a) Consider the simplified CO hydrogenation scheme illustrated in Figure 10.32. Derive an expression relating the abundances of H_2CO and CH_3OH with that of accreting H and CO.

(b) Calculate the CH_3OH/CO ratio at a density of 3×10^4 and 10^5 cm^{-3}.

(c) Why are you allowed to ignore loss of atomic H to H_2O formation? What other species could act as sinks for atomic H?

(d) How would these results change if H-abstraction reactions would also be taken into account?

(e) Describe why abstraction/addition reactions are the key to deuterium fractionation of methanol.

10.13 Derive Equation (10.99) and use this to understand Figure 10.36.

10.14 Derive Equation (10.100).

10.15 Compare and contrast the gas phase (Table 10.2) and ice composition (Table 10.4) of molecular clouds.

10.16 Describe the roles of gas phase and grain surface chemistry in the molecular inventory of molecular clouds.

Bibliography

[1] Accolla, M., Congiu, E., Dulieu, F., et al., 2011, *Phys Chem Chem Phys*, 13, 8037

[2] Aikawa, Y., Ohashi, N., Inutsuka, S. I., Herbst, E., Takakuwa, S., 2001, *ApJ*, 552, 639

[3] Aléon, J. 2010, *ApJ*, 722, 1342

[4] Allen, R. J., Ivette R., M., Black, J. H., Booth, R. S., 2012, *AJ*, 143, 97

[5] Alves, J. F., Lada, C. J., Lada, E. A., 2001, *Nature*, 409, 159
[6] Alves, J., Lombardi, M., Lada, C. J., 2007, *A & A*, 462, L17
[7] André, Ph., Men'shchikov, A., Bontemps, S., et al., 2010, *A & A*, 518, L102
[8] André, P., Di Francesco, J., Ward-Thompson, D., et al., 2014, *Protostars and Planets VI*, 27
[9] Andersson, S., Al-Halabi, A., Kroes, G. J., van Dishoeck, E. F., 2006, *J Chem Phys*, 124, 064715
[10] Apponi, A. J., Ziurys, L. M., 1997, *ApJ*, 481, 800
[11] Arasa, C., Andersson, S., Cuppen, H., van Dishoeck, G. J., Kroes, G. J., 2011, *J. Chem. Phys.* 134, 164503
[12] Bacmann, A., Lefloch, B., Ceccarelli, C., et al. 2003, *ApJL*, 585, L55
[13] Bacmann, A., Daniel, F., Caselli, P., et al., 2016, *A & A*, 587, A26
[14] Ballesteros-Paredes, J., Klessen, R. S., Mac Low, M., Vazquez-Semadeni, E., 2007, *Protostars and Planets V*, 63
[15] Bally, J., Langer, W. D., 1982, *ApJ*, 255, 143
[16] Bally, J., Anderson, L. D., Battersby, C., et al., 2010, *A & A*, 518, L90
[17] Baratta, G. A., Palumbo, M. E., Strazzulla, G., 2000, *A & A*, 357, 1045
[18] Barnard, E. E., 1919, *ApJ*, 49, 1
[19] Bergin, E. A., Alves, J., Huard, T., Lada, C. J., 2002, *ApJ*, 570, L101
[20] Bergin, E. A., Melnick, G. J., Gerakines, P. A., et al., 2005, *ApJ*, 627, L33
[21] Bertie J. E., Devlin, J. P., 1983, *J Chem Phys*, 78, 6203
[22] Berné, O., Marcelino, N., Cernicharo, J., 2014, *ApJ*, 795, 13
[23] Bertin, M., Romanzin, C., Doronin, M., et al., 2016, *ApJ*, 817, L12
[24] Bertram, E., Shetty, R., Glover, S. C. O., Klessen, R. S., Roman-Duval, J., Federrath, C., 2014, *MNRAS*, 440, 465
[25] Bertram, E., Konstandin, L., Shetty, R., Glover, S. C. O., Klessen, R. S., 2015, *MNRAS*, 446, 3777
[26] Bertram, E., Klessen, R. S., Glover, S. C. O., 2015, *MNRAS*, 45, 196
[27] Bierbaum, V. M., 2011, *IAU Symposium*, 280, 383
[28] Bloemen, J. B. G. M., Strong, A. W., Blitz, L., et al., 1985, *19th Intern. Cosmic Ray Conf.*, Vol. 1, 329
[29] Bockelée-Morvan, D., et al., 2000, *A & A*, 353, 1101
[30] Boger, G. I., Sternberg, A., 2006, *ApJ*, 645, 314
[31] Bok, B. J., Lawrence, R. S., Menon, T. K., 1955, *PASP*, 67, 108
[32] Bolatto A. D., Wolfire M., Leroy A. K., 2013, *Annu Rev Astron Astrophys*, 51, 207
[33] Boogert, A. C. A., Ehrenfreund, P., Gerakines, P. A., et al., 2000, *A & A*, 353, 349
[34] Boogert, A. C. A., Tielens, A. G. G. M., Ceccarelli, C., et al., 2000, *A & A*, 360, 683
[35] Boogert, A. C. A., Blake, G. A., Tielens, A. G. G. M., 2002, *ApJ*, 577, 271
[36] Boogert, A. C. A., Pontoppidan, K. M., Knez, C., et al., 2008, *ApJ*, 678, 985
[37] Boogert, A. C. A., Gerakines, P. A., Whittet, D. C. B., 2015, *Annu Rev Astron Astrophys*, 53, 541
[38] Bouhafs, N., Rist, C., Daniel, F., et al., 2017, *MNRAS*, 470, 2204
[39] Bringa, E. M., Johnson, R. E., 2004, *ApJ*, 603, 159
[40] Burkhart, B., Falceta-Gonalves, D., Kowal, G., Lazarian, A., 2009, *ApJ*, 693, 250
[41] Butner, H. M., Lada, E. A., Loren, R. B., 1995, *ApJ*, 448, 207
[42] Carelli, F., Satta, M., Grassi, T., Gianturco, F. A., 2013, *ApJ*, 774, 97
[43] Caselli, P., Walmsley C. M., Terzieva, R., Herbst, E., 1998, *ApJ*, 499, 234
[44] Caselli, P., 2002, *Plan Space Sci*, 50, 1133
[45] Caselli, P., van der Tak, F., Ceccarelli, C., Bacmann, A, 2003, *A & A*, 403, L37

[46] Caselli, P., Vastel, C., Ceccarelli, C., et al., 2008, *A & A*, 492, 703
[47] Caselli, P., Keto, E., Bergin, E. A., et al., 2012, *ApJ*, 759, L37
[48] Charnley, S. B., Tielens, A. G. G. M., Rodgers, S. D., 1997, *ApJ*, 482, L203
[49] Charnley, S. B., Marwick, A. J., 2003, *A & A*, 399, 583
[50] Cherchneff, I., Glassgold, A. E., 1993, *ApJL*, 419, L41
[51] Chiar, J. E., Gerakines, P. A., Whittet, D. C. B., et al., 1998, *ApJ*, 498, 716
[52] Chiar, J. E., Tielens, A. G. G. M., 2006, *ApJ*, 637, 774
[53] Collier, W. B., Ritzhaupt, G., Devlin, J. P., 1984, *J Phys Chem*, 88, 363
[54] Cruz-Diaz, G. A., Martín-Doménech, R., Muñoz Caro, G. M., Chen, Y. J., 2016, *A & A*, 592, A68
[55] Cuppen, H. M., Ioppolo, S., Romanzin, C., Linnartz, H., 2010, *Phys Chem Chem Phys*, 12, 12077
[56] Cuppen, H. M., van Dishoeck, E. F., Herbst, E., Tielens, A. G. G. M., 2009, *A & A*, 508, 275
[57] Dalgarno, A., McCray, R. A., 1973, *ApJ*, 181, 95
[58] Dame, T. M., Hartmann, D., Thaddeus, P., 2001, *ApJ*, 547, 792
[59] Danby G., Flower D. R., Valiron P., Schilke P., Walmsley C. M., 1988, *MNRAS*, 235, 229
[60] Dhabal, A. Mundy, L. G., Chen, C.-y., Teuben, P., Storm, S., 2019, *ApJ*, in press
[61] Dickens, J. E., Irvine, W. M., Snell, R. L., et al., 2000, *ApJ*, 542, 870
[62] Dickey, J. M., Lockman, F. J., 1990, *Annu Rev Astron Astrophys*, 28, 215
[63] Di Francesco, J., Hogerheijde, M. R., Welch, W. J., Bergin, E. A., 2002, *AJ*, 124, 2749
[64] Dobbs, C. L., Krumholz, M. R., Ballesteros-Paredes, J., et al., 2014, in *Protostars and Planets VI* Tucson: University of Arizona Press, 3
[65] Dominguez, G., 2010, *ApJ*, 713, L59
[66] dos Santos, S. F., Ngassam, V., Orel, A. E., Larson, Å., 2016, *Phys Rev A*, 94, 022702
[67] Du, F., Parise, B., Bergman, P., 2012a, *A & A*, 538, A91
[68] Elmegreen, B. G., Lada, C. J., 1977, *ApJ*, 214, 725
[69] Favre, C., Carvajal, M., Field, D., et al., 2014, *ApJS*, 215, 25
[70] Fayolle, E. C., Bertin, M., Romanzin, C., et al., 2011, *ApJ*, 739, L36
[71] Flower, D. R., Pineau des Forêts, G., Walmsley, C. M., 2006, *A & A*, 449, 621
[72] Fournier, J., Deson, J., Vermeil, C., Pimentel, G. C., 1979, *J Chem Phys*, 70, 5726
[73] Freeman, A., Millar, T. J., 1983, *Nature*, 301, 402
[74] Fuente, A., Cernicharo, J., Roueff, E., et al. 2016, *A & A*, 593, A94
[75] Fuchs, G. W., Cuppen, H. M., Ioppolo, S., et al., 2009, *A & A*, 505, 629
[76] Garrod, R., T., Pauly, T., 2011, *ApJ*, 735, 15
[77] Gerakines, P. A., Whittet, D. C. B., Ehrenfreund, P., et al., 1999, *ApJ*, 522, 357
[78] Gerin, M., de Luca, M., Black, J., et al. 2010, *A & A*, 518, L110
[79] Gianturco, F. A., T Grassi, T., Wester, R., 2016, *J Phys B: At Mol Opt Phys*, 49, 204003
[80] Gibb, E. L., et al., 2000, *ApJ*, 536, 347
[81] Gibb, E. L., Whittet, D. C. B., Boogert, A. C. A., Tielens, A. G. G. M., 2004, *ApJS*, 151, 35
[82] Godard, B., Falgarone, E., Pineau Des Forêts, G. 2014, *A & A*, 570, A27
[83] Goldbaum, N. J., Krumholz, M. R., Matzner, C. D., et al., 2011, *ApJ*, 738,101
[84] Goldsmith, P. F., Langer, W. D., 1978, *ApJ*, 222, 881
[85] Gong, M., Ostriker, E. C., Wolfire, M. G., 2017, *ApJ*, 843, 38
[86] Goumans, T. P. M., Andersson, S., 2010, *MNRAS*, 406, 2213

[87] Goumans, T. P. M., 2011, *MNRAS*, 413, 2615; errata: 2012, *MNRAS*, 423, 3775
[88] Gratier, P., Majumdar, L., Ohishi, M., et al., 2016, *ApJS*, 225, 25
[89] Grenier, I. A., Casandjian, J.-M., Terrier, R., 2005, *Science*, 307, 1292
[90] Grim, R. J. A., D'Hendecourt, L. B., 1986, *A & A*, 167, 161
[91] Guzmán, V. V., Goicoechea, J. R., Pety, J., et al., 2013, *A & A*, 560, A73
[92] Hama T., Watanabe, N., 2013, *Chem Rev*, 113, 8783
[93] Harju, J., Daniel, F., Sipilä, O., et al., 2017a, *A & A*, 600, A61
[94] Harju, J., Sipilä, O., Brünken, S., et al., 2017b, *ApJ*, 840, 63
[95] Hauser, M. G., Arendt, R. G., Kelsall, T., et al., 1998, *ApJ*, 508, 25
[96] Heays, A. N., Visser, R., Gredel, R., et al., 2014, *A & A*, 562, A61
[97] Hennebelle, P., Falgarone, E., 2012, *A & A Rev*, 20, 55
[98] Hennebelle, P., 2017, *Mem Soc Astron It*, 88, 513
[99] Herbst, E., 1981, *Nature*, 289, 656
[100] Herbst, E., Lee, H. H., Howe, D. A., Millar, T. J., 1994, *MNRAS*, 268, 335
[101] Herbst, E. 2009, Submillimeter Astrophysics and Technology: A Symposium Honoring Thomas G. Phillips, 417, 153
[102] Heyer, M. H., Brunt, C., Snell, R. L., Howe, J. E., Schloerb, F. P., Carpenter, John M., 1998, *ApJS*, 115, 241
[103] Heyer, M., Krawczyk, C., Duval, J., Jackson, J. M., 2009, *ApJ*, 699, 1092
[104] Heyer, M., Dame, T. M., 2015, *Annu Rev Astron Astrophys*, 53, 583
[105] Hidaka, H., Watanabe, N., Shiraki, T., et al., 2004, *ApJ*, 614, 1124
[106] Hidaka, H., Watanabe, M., Kouchi, A., Watanabe, N., 2009, *ApJ*, 702, 291
[107] Hily-Blant, P., Falgarone, E., Pety, J., 2008, *A & A*, 481, 367
[108] Hiraoka, K., Miyagoshi, T., Takayama, T., et al., 1998, *ApJ*, 498, 710
[109] Ho, P. T. P., Townes, C. H., 1983, *Annu Rev Astron Astrophy*, 21, 239
[110] Hollenbach, D. J., Tielens, A. G. G. M., 1999, *Reviews of Modern Physics*, 71, 173
[111] Hollenbach, D., Kaufman, M. J., Bergin, E. A., Melnick, G. J., 2009, *ApJ*, 690, 1497
[112] Hopkins, P., 2013, *MNRAS*, 430, 1880
[113] Howe, D. A., Millar, T. J., 1990, *MNRAS*, 244, 444
[114] Hugo, E., Asvany, O., Schlemmer, S. 2009, *J Chem Phys*, 130, 164302
[115] Immer, K., Kauffmann, J., Pillai, T., Ginsburg, A., Menten, K. M., 2016, *A & A*,
[116] Indebetouw, R., Mathis, J. S., Babler, B. L., et al., 2005, *ApJ*, 619, 931
[117] Ioppolo, S., Cuppen, H. M., Romanzin, C., et al., 2008, *ApJ*, 686, 1474
[118] Juvela, M., Harju, J., Ysard, N., Lunttilla, T., 2012, *A & A*, 538, 133
[119] Kaifu, N., Ohishi, M., Kawaguchi, K., et al., 2004, *Publ Astron Soc Japan*, 56, 69
[120] Kawaguchi, K. Kasai, Y. Ishikawa, S.-I., 1994, *ApJ*, 420, 95
[121] Keane, J., PhD, University of Groningen
[122] Klessen, R. S., Hennebelle, P., 2010, *A & A*, 520, 17
[123] Knez, C., Lacy, J. H., Evans, N. J., II, van Dishoeck, E. F., Richter, M. J., 2009, *ApJ*, 696, 471
[124] Könyves, V., André, P., Men'shchikov, A., et al., 2015, *A & A*, 584, A91
[125] Koussa, H., Bahri, M., Jaïdane, N., Lakhdar, Z. B., 2006, *J Mol Struct*, 770, 149
[126] Larson R. B., 1981, *MNRAS*, 194, 809
[127] Le Bourlot, J., Pineau des Forêts, G., Roueff, E., 1995, *A & A*, 297, 251
[128] Le Bourlot, J., Pineau des Forêts, G., Roueff, E., Schilke, P., 1993, *ApJ*, 416, L87
[129] Le Gal, R., Hily-Blant, P., Pineau des Forêts, G., Rist, C., Maret, S., 2014, *A & A*, 562, A83
[130] Le Petit, F., Roueff, E., 2003, *Dissociative Recombination of Molecular Ions with Electrons*, 373

[131] Lepp, S., Dalgarno, A., 1988, *ApJ*, 324, 553
[132] Liszt, H., Lucas, R., 1996, *A & A*, 314, 917
[133] Liszt, H. S., & Ziurys, L. M., 2012, *ApJ*, 747, 55
[134] Loison, J. C., Wakelam, V., Hickson, K. M., Bergeat, A., Mereau, R., 2014, *MNRAS*, 437, 930
[135] Lombardi, M., Bouy, H., Alves, J., Lada, C. J., 2014, *A & A*, 566, A45
[136] Loren, R. B., Mundy, L. G., 1984, *ApJ*, 286, 232
[137] Lyons, J. R., Young, E. D., 2005, *Nature*, 435, 317
[138] Mangum, J. G., Wootten, A., Plambeck, R. L., 1993, *ApJ*, 409, 282
[139] Maret, S., Bergin, E. A., 2007, *ApJ*, 664, 956
[140] Martin, P. G., Whittet, D. C. B., 1990, *ApJ*, 357, 113
[141] McCall, B. J., Geballe, T. R., Hinkle, K. H., Oka, T., 1999, *ApJ*, 522, 338
[142] McCarthy, M. C., Gottlieb, C. A., Gupta, H., Thaddeus, P., 2006, *ApJ*, 652, L141
[143] McDowell, M. R. C., 1961, *The Observatory*, 81, 240
[144] McGuire, B. A., Burkhardt, A. M., Kalenskii, S., et al., 2018, *Science*, 359, 202
[145] McKee, C. F., Ostriker, E., 2007, *Annu Rev Astron Astrophys*, 45, 565
[146] Meyer, C. D., Balsara, D. S., Burkhart, B., Lazarian, A., 2014, *MNRAS*, 439, 2197
[147] Myers P. C., Dame T. M., Thaddeus P., et al., 1986, *ApJ*, 301, 398
[148] Milam, S. N., Savage, C., Brewster, M. A., Ziurys, L. M., Wyckoff, S., 2005, *ApJ*, 634, 1126
[149] Millar, T. J., 2003, *Space Sci Rev*, 106, 73
[150] Miyauchi, N., Hidaka, H., Chigai, T., et al., 2008, *Chem Phys Letters*, 456, 27
[151] Mokrane, H., Chaabouni, H., Accolla, M., et al., 2009, *ApJ*, 705, L195
[152] Molek, C. D., McLain, J. L., Poterya, V., Adams, N. G., 2007, *Journal of Phys Chem A*, 111, 6760
[153] Molinari, S., Bally, J., Glover, S., et al., 2014, in *Protostars and Planets VI* Tucson: University of Arizona Press, 125
[154] Motte, F., André, P., Neri, R., 1998, *A & A*, 336, 150
[155] Mousavipour, S. H., Saheb, V., 2007, *Bull Chem Soc Jpn*, 80, 1901
[156] Muñoz Caro, G. M., Chen, Y. J., Aparicio, S., et al., 2016, *A & A*, 589, A19
[157] Nagaoka, A., Watanabe, N., Kouchi, A., 2007, *J Phys Chem A*, 111, 3016
[158] Nguyen, T. L., Stanton, J. F., Barker, J. R., 2011, *J Phys Chem A*, 115, 5118
[159] Nguyen Luong, Q., Motte, F., Schuller, F., et al., 2011, *A & A*, 529, A41
[160] Oba, Y., Miyauchi, N., Hidaka, H., Chigai, T., Watanabe N., Kouchi, A., 2009, *ApJ*, 701, 464
[161] Oba, Y., Watanabe, N., Hama, T., Kuwahata, K., Hidaka, H., Kouchi, A., *Ap J*, 2012, 749, 67
[162] Öberg, K. I., van Dishoeck, E. F., Linnartz, H., 2009, *A & A*, 496, 281
[163] Ochsendorf, B. B., Brown, A. G. A., Bally, J., Tielens, A. G. G. M., 2015, *ApJ*, 808, 111
[164] Ohishi, M., Irvine, W. M., Kaifu, N., 1992, in IAU symposium 150 "Astrochemistry of Cosmic Phenomena," ed. P. D. Singh (Kluwer: Dordrecht), 171
[165] Ormel, C. W., Min, M., Tielens, A. G. G. M., Dominik, C., Paszun, D., 2011, *A & A*, 532, A43
[166] Ospina-Zamudio, J., Lefloch, B., Ceccarelli, C., et al., 2018, *A & A*, 618, A145
[167] Ossenkopf, V., Henning, T. 1994, *A & A*, 291, 943
[168] Ött, J., Henkel, C., Braatz, J. A., Weiss, A., 2011, *ApJ*, 742, 95
[169] Pabst, C., Higgins, R., Goicoechea, J. R., et al., 2019, Nature, 565, 618
[170] Padovani, M., Galli, D., Glassgold, A. E., 2009, *A & A*, 501, 619
[171] Padovani, M., Ivlev, A. V., Galli, D., et al., 2018, *A & A*, 614, A111

[172] Pagani, L., Vastel, C., Hugo, E., et al., 2009, *A & A*, 494, 623
[173] Pagani, L., Roueff, E., Lesaffre, P., 2011, *ApJ*, 739, L35
[174] Palumbo, M. E., Baratta, G. A., Leto, G., Strazzulla, G., 2010, *J Molecular Structure*, 972, 64
[175] Pan, L., Padoan, P., 2009, *ApJ*, 692, 594
[176] Parise, B., Belloche, A., Du, F., Güsten, R., Menten, K. M., 2011, *A & A*, 526, A31
[177] Parise, B., Belloche, A., Du, F., et al., 2011, *A & A*, 526, A31
[178] Parise, B., Bergman, P., Du, F. 2012, *A & A*, 541, L11
[179] Parise, B., Bergman, P., Menten, K., 2014, *Faraday Discussions*, 168, 349
[180] Peebles, P. J. E., Dicke, R. H., 1968, *ApJ*, 154, 891
[181] Planck Collaboration, Ade, P. A. R., Aghanim, N., et al., 2011, *A & A*, 536, A19
[182] Planck Collaboration XI, *A & A*, 571, A11
[183] Pon, A., Johnstone, D., Kaufman, M. J., Caselli, P., Plume, R., 2014, *MNRAS*, 445, 1508
[184] Pontoppidan, K. M., et al., 2003, *A & A*, 408, 981
[185] Pontoppidan, K. M., et al., 2008, *ApJ*, 678, 1005
[186] Ratajczak, A., Quirico, E., Faure, A., Schnitt, B., Ceccarelli, C., 2009, *A & A*, 496, L21
[187] Raut, U., Baragiola, R. A., 2011, *ApJ*, 737, L14
[188] Raut, U., Fama, M., Loeffler M. J., Baragiola, R. A., 2008, *ApJ*, 687, 1070
[189] Reach, W. T., Koo, B-C., Heiles, C., 1994, *ApJ*, 429, 672
[190] Reach, W. T., Wall, W. F., Odegard, N., 1998, *ApJ*, 507, 507
[191] Rimola, A., Taquet, V., Ugliengo, P., Balucani, N., Ceccarelli, C., 2014, *A & A*, 572, A70
[192] Rivera-Ingraham, A., Martin, P. G., Polychroni, D., et al., 2013, *ApJ*, 766, 85
[193] Roberts, H., Herbst, E., Millar, T. J., 2003, *ApJ*, 591, L41
[194] Roberts, H., Millar, T. J., 2007, *A & A*, 471, 849
[195] Rodgers, S. D., Charnley, S. B., 2008, *ApJ*, 689, 1448
[196] Romanzin, C., Ioppolo, S., Cuppen, H., et al., 2011, *J Chem Phys*, 134, 084504
[197] Roser, J. E., Vidali, G., Manicò, G., Pirronello, V., 2001, *ApJ*, 555, L61
[198] Rosolowsky, E., Leroy, A., 2006, *PASP*, 118, 590
[199] Rosolowsky, E. W., Pineda, J. E., Kauffmann, J., Goodman, A. A., 2008, *ApJ*, 679, 1338
[200] Roueff, E., Loison, J. C., Hickson, K. M., 2015, *A & A*, 576, A99
[201] Roy, A., André, P., Palmeirim, P., et al., 2014, *A & A*, 562, A138
[202] Sakai, N., Saruwatari, O., Sakai, T., Takano, S. Yamamoto, S., 2010, *A & A*, 512 A31
[203] Sakai, N. J., Takano, S., Sakai, T., et al., 2013, *Phys Chem A*, 117, 9831
[204] Schisano, E., et al., 2014, *ApJ*, 791, 27
[205] Schneider, N., Csengeri, T., Bontemps, S., et al., 2010, *A & A*, 520, A49
[206] Scoville, N. Z., Yun, M. S., Sanders, D. B., Clemens, D. P., Waller, W. H., 1987, *ApJS*, 63, 821
[207] Sephton, M. A., 2002, *Natural Products Reports*, 3, 261
[208] Shen, C. J., Greenberg, J. M., Schutte, W. A., van Dishoeck, E. F., 2004, *A & A*, 415, 203
[209] Shingledecker, C. N., Bergner, J. B., Le Gal, R., et al., 2016, *ApJ*, 830, 151
[210] Sims, I. R., Queffelec, J. L., Travers, D., et al., 1993, *Chem Phys Lett*, 211, 461
[211] Sipilä, O., Hugo, E., Harju, J., et al. 2010, *A & A*, 509, A98
[212] Smith, R. J., Glover, S. C. O., Clark, P. C., Klessen, R. S., Springel, V., 2014, *MNRAS*, 441, 1628

[213] Smith, I. W. M., Herbst, E., Chang, Q., 2004, *MNRAS*, 350, 323
[214] Smith, R. G., Sellgren, K., Tokunaga, A. T., 1989, *ApJ*, 344, 413
[215] Smith, R., Blake, G. E., Boogert, A. C., Ponttopidan, K. M., 2017, *Lun Plan Inst XLVIII*, 2998
[216] Snow, T. P., McCall, B. J., 2006, *Annu Rev Astron Astrophys*, 44, 367
[217] Solomon P. M., Rivolo A. R., Barrett J., Yahil A., 1987, *ApJ*, 319, 730
[218] Soma, T., Sakai, N., Watanabe, Y., Yamamoto, S., 2015, *ApJ*, 802, 74
[219] Souda, R., Kawanowa, H., Kondo M., Gotoh, Y., 2003, *J Chem Phys*, 119, 6194
[220] Souda, R., 2004, *Phys Rev Let*, 93, 235502
[221] Spaans, M., 1996, *A & A*, 307, 271
[222] Spezzano, S., Brünken, S., Schilke, P., et al., 2013, *ApJ*, 769, L19
[223] Tafalla, M., Santiago-García, J., Myers, P. C., et al., 2006, *A & A*, 455, 577
[224] Tang, X. D., Henkel, C., Chen, C. H. R., et al., 2017, *A & A*, 600, A16
[225] Taniguchi, K., Ozeki, H., Saito, M., 2017, *ApJ*, 846, 46
[226] Taquet, V., Ceccarelli, C., Kahane, C., 2012, *ApJ*, 748, L3
[227] Taquet, V., Ceccarelli, C., Kahane, C., 2012, *A & A*, 538, A42
[228] Terzieva, R., Herbst, E., 2000, *MNRAS*, 317, 563
[229] Tielens, A. G. G. M., Hagen, W., 1982, *A & A*, 114, 245
[230] Tielens, A. G. G. M., 1983, *A & A*, 119, 177
[231] Tielens, A. G. G. M., Tokunaga, A. T., Geballe, T. R., Baas, F., 1991, *ApJ*, 381, 181
[232] Troe, J., Ushakov, V. G., 2008, *J Chem Phys*, 128, 204307
[233] van der Tak, F. F. S., van Dishoeck, E. F., 2000, *A & A*, 358, L79
[234] van Dishoeck, E. F., Black, J., 1988, *ApJ*, 334, 771
[235] Vastel, C., Phillips, T. G., Yoshida, H., 2004, *ApJ*, 606, L127
[236] Vastel, C., Caselli, P., Ceccarelli, C., et al., 2006, *ApJ*, 645, 1198
[237] Vigren, E., Zhaunerchyk, V., Hamberg, M., et al., 2012, *ApJ*, 757, 34
[238] Visser, R., van Dishoeck, E. F., Black, J. H., 2009, *A & A*, 503, 323
[239] Wakelam, V., Herbst, E., 2008, *ApJ*, 680, 371
[240] Walmsley, C. M., Ungerechts, H., 1983, *A & A*, 122, 164
[241] Walmsley, C. M., Flower, D. R., Pineau des Forêts, G., 2004, *A & A*, 418, 1035
[242] Wakelam, V., Herbst, E., Selsis, F., Massacrier, G., 2006, *A & A*, 459, 813
[243] Wannier, P. G., Andersson, B.-G., Federman, S. R., Lewis, B. M., Viala, Y. P., Shaya, E., 1993, *ApJ*, 407, 163
[244] Watson, W. D., 1974, 188, 35
[245] Whittet, D. C. B., Cook, A. M., Herbst, E., Chiar, J. E., Shenoy, S. S., 2011, *ApJ*, 742, 28
[246] Wilson, R. W., Jefferts, K. B., Penzias, A. A., 1970, *ApJ*, 161, L43
[247] Williams, J., McKee, C. F., 1997, *ApJ*, 476, 166
[248] Wilson, B. A., Dame, T. M., Masheder, M. R. W., Thaddeus, P., 2005, *A & A*, 430, 523
[249] Wirström, E. S., Geppert, W. D., Hjalmarson, Å., et al., 2011, *A & A*, 533, A24
[250] Wirström, E. S., Charnley, S. B., Cordiner, M. A., et al., 2012, *ApJ*, 757, L11
[251] Wooldridge, P. J., Devlin, J. P., 1988, *J Chem Phys*, 88, 3086
[252] Woon, D. E., 2002, *ApJ*, 569, 541
[253] Wootten, A., Loren, R. B., Snell, R. L., 1982, *ApJ*, 255, 160
[254] Xu, D., Li, D., Yue, N., Goldsmith, P. F., 2016, *ApJ*, 819, 22
[255] Yoshida, K., Sakai, N., Tokudome, T., et al., 2015, *ApJ*, 807, 66
[256] Yu, H. G.; Varandas, A. J. C., 1997, *J Chem Soc Faraday Trans*, 93, 2651
[257] Zavagno, A., Russeil, D., Motte, F., et al., 2010, *A & A*, 518, L81

11
Star Formation

Molecular clouds are the site of star formation. The study of star formation is, therefore, squarely in the realm of molecular astrophysics. Molecular transitions provide good thermometers and barometers. In addition, the high spectral resolution of heterodyne devices turn rotational line observations into probes of the kinematics of the star formation process. Last, but certainly not least, molecular observations provide direct insight into the molecular composition of clouds. Hence, molecular astrophysics can address the key questions of star formation: What are the physical processes that control the formation of stars? When and how does collapse ensue? How did this change over cosmic times? What are the chemical processes that control the organic and volatile inventory of regions of star and planet formation? How did this influence the emergence of life on the Earth? What does this mean for the possibility of life elsewhere in the Universe?

In this chapter, the molecular astrophysics of star formation will be discussed. However, we will not focus on the implications for our understanding of the star formation process and a specialized monograph should be consulted for deeper insight into star and planet formation processes. Molecular clouds and dense cores have been discussed in Chapter 10. The aspects that are relevant for star formation – the formation of the isothermal prestellar core – will be briefly summarized here. We will continue with a discussion of Hot Cores and Hot Corinos (Section 11.3): regions of very dense and warm gas near high and low mass protostars, respectively. We will follow this with a discussion of the physics and chemistry of protoplanetary disks (Section 11.4). After formation of a massive star, an HII region will develop and the nonionizing photons will create a photodissociation region (PDR) that separates the ionized gas from the surrounding molecular cloud. The physical and chemical structure of the PDR is controlled by the penetrating FUV photons and this interaction is described in Section 11.6. The star formation process is accompanied by strong jets and disk outflows that process the surrounding molecular cloud materials as described in Section 11.7. Finally, maser action is a common characteristic of many molecules in a dense and warm environment and is often used to trace active sites of (massive) star formation. The physics of molecular masers is discussed in Section 11.8.1. The chapter ends with a limited summary of relevant literature that can serve as an entry point (Section 11.9) and a number of exercises (Section 11.10).

A course centered on star formation might focus on deriving the physical conditions from observations of regions of star formation, including Sections 11.3.1, 11.4.1, 11.4.3, 11.6.2, 11.7, and 11.8. I also like the concept of chemical timescales that can be used to probe the star formation process that is well illustrated by the composition of Hot Cores and Hot Corinos (Section 11.3.2) but more generally relevant. A course centered on astrochemistry would focus more on sections 11.3.2, 11.4.2, 11.5, 11.6, and 11.7.3.

11.1 Introduction

As the column density of a cloud increases, gravity takes over as a controlling force. At the same time, self-shielding and shielding by dust against penetrating dissociating photons allows molecules to survive and the clouds turn molecular. As a result of the shielding, the heating rate of the gas decreases. In addition, molecular rotational levels are more "finely" spaced than atomic fine-structure levels and, hence, molecular gas can cool to lower temperatures. Molecular clouds are therefore much cooler than diffuse clouds. Gravity is counteracted by turbulence generated by the cloud accretion process as well as by internal sources. Dissipation of this turbulence leads to the formation of dense filaments with a wide range of densities. Gravity takes over for those filaments that are dense enough. Such filaments may fragment into dense cores, which then collapse and form stars.

The paradigm for low mass star formation (Figure 11.1) starts with turbulently generated, filamentary density structures, which constantly form and break up. The densest of these structures will start to contract, fragment into dense cores, and collapse under their own gravity to form a central object, the nascent protostar. Because of conservation of angular momentum, an accretion disk forms around the protostar. Angular momentum transport in the disk due to, e.g., viscous forces will allow further accretion onto the protostar while the disk spreads out. In addition, some of the disk material may be ejected through a gentle disk wind or, if it originates from the stellar accretion zone, in the form of a jet. This, too, will take away some of the angular momentum, promoting accretion. Finally, in turbulent magnetized disks, the magneto-rotational instability can redistribute angular momentum in the disk, allowing matter to fall inward. The density of these disks is very high and their surfaces are irradiated by stellar, FUV, EUV, and X-ray photons, creating a warm, dense PDR. Deeper in the disk, viscous dissipation may lead to additional heating during the initial stages of the collapse. The gas is turbulent and sub-micron-sized dust grains will couple to this turbulence. This will lead to rapid collisional growth into large aggregates. This starts the planet formation process. Skipping over the details (as they are obscure), eventually a planetary system is formed and the disk is dissipated. During the accretion process, the jet and/or disk outflow will drive strong shock waves into the surroundings and the envelope is, eventually, dissipated. The protostar and its surrounding disk become visible as T Tauri stars or Herbig AeBe stars, depending on the stellar mass (less than 2 $M_\odot$ and 2−8 $M_\odot$, respectively). During these phases, the central protostar derives its luminosity from its slow contraction along the Hayashi and Henyey tracks and, apart from a brief phase of deuterium burning, nuclear energy generation does not start until the star reaches the

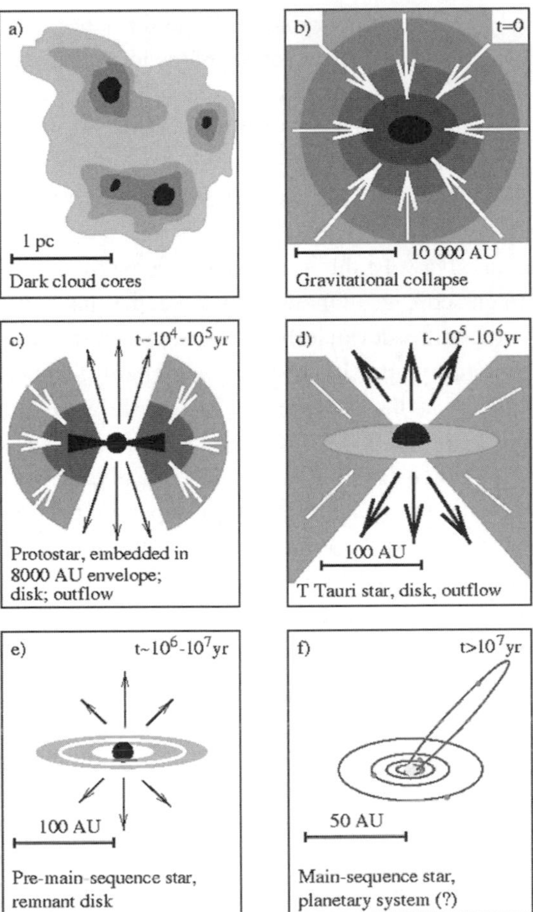

Figure 11.1 The six stages of low mass star formation. After a dense filament fragments and forms a dense core (stage a), gravity can take over forming a protostar (stage b). Material is continually fed into the star through a disk but some of this material is ejected in a jet/wind, clearing out the envelope (stage c) and leaving a T-Tauri star surrounded by a disk in which planets will form (stage d). This planet formation timescale can last for a long time until the dust disk has cleared (stage e). Eventually, the star will settle on the main sequence and a star and associated planetary system remain (stage f). Figure kindly provided by Michiel Hogerheijde and adapted from [258]

main sequence. The timescale to reach the main sequence is given by the Kelvin–Helmholtz timescale, $\tau_{KH} = E_{int}/L = |E_{gr}|/2L = GM^2/2RL \simeq 10^7\ (M_\odot/M)^{5/2}$ yr (where the E's stand for the internal (thermal) and gravitational energies, L is the luminosity, G the gravitational constant, and M and R the mass and radius of the star.) For comparison, observationally, the mass accretion timescale is $\simeq$2.3 Myr while the dust disk dissipation timescale is $\simeq$2.9 Myr.

There is no generally agreed upon scenario for the formation of high mass stars. High mass stars could be formed in a scaled-up variant of the low mass star formation paradigm, starting in InfraRed Dark Cloud cores (IRDC), the high mass equivalent of dark cloud cores. These IRDCs are characterized by such high column densities that they appear as patches of obscuration against the galactic emission at mid-IR wavelengths. One clear difference with low mass stars is that high mass stars will rapidly reach the main sequence, leading to a large increase in stellar luminosity that can stop or even reverse accretion through radiation pressure. Naively, this would lead to a maximum stellar mass well below that observed on the main sequence. One way to overcome this bottleneck in massive star formation is to have accretion occur mainly through an accretion disk that intercepts only a small fraction of the stellar luminosity. Most of the stellar light is then redirected through the outflow poles created by stellar jets, the so-called flashlight effect. When the effective temperature of the star exceeds 25,000 K ($M > 13\ M_\odot$), the star can ionize its environment, creating an HII region in which H is photo-ionized. Initially, the infall will compress this ionized gas region close to the stellar surface (a hypercompact HII region). However, eventually, as the infall thins and the ionizing photon flux increases, the overpressure of the $\simeq$8000 K, ionized gas will lead to a reversion of the infall and an expansion of the ionized gas volume (an ultracompact HII region). Again, a nonuniform density distribution in the environment may still allow additional accretion through, e.g., a circumstellar disk structure. As an alternative to this monolithic collapse of a massive dense cloud core, fragmentation may lead to the formation of a large number of smaller cores, none of which is dense enough that radiation pressure becomes important. In this scenario, massive stars are then built up by collisional coalescence of stars formed in a dense cluster environment, rather than by continued gas accretion. Once these stars are massive and hot enough, they will start to ionize the gas around them and thereby disrupt the cloud cores. These cores can also be destroyed by the strong stellar winds of the more massive O stars (spectral type earlier than O6 (30 $M_\odot$, $T_{eff} > 40,000$ K) with stellar wind, kinetic luminosities exceeding 10^{36} erg/s. So, there are a number of stages that can be discerned: prestellar core formation and fragmentation, accretion, and disruption. But the relative importance of and interrelationship of these phases is poorly understood. In this chapter, the molecular astrophysics aspects of this physical evolution will be discussed. This physical evolution of the region is accompanied by large variation in the chemical inventory as the differences in physical conditions activate new chemical processes. This chapter also discusses these chemical changes and the processes that may drive them.

11.2 Prestellar Cores

Prestellar – or starless – cores are the first stage in star formation[1] where gravitational collapse has produced a centrally peaked density distribution but no star is apparent yet. There are two classes of models for the formation of these cores. First, the supporting

[1] Also sometimes called class -1 protostars (to distinguish them from class 0, I, and II) or pre-protostellar cores.

magnetic fields may leak out slowly through the ambipolar diffusion process and the timescale is set by, $\tau_{AD} \simeq 10\tau_{ff}$, with the free-fall timescale, τ_{ff} (for a homogeneous cloud: $\tau_{ff} = \sqrt{3\pi/32G\rho} = 10^5 \sqrt{10^4/n\,(\mathrm{H_2})}$ yr with G the gravitational constant and ρ and n the mass and number density). In the other scenario, turbulence will create density fluctuations and gravity "selects" those that are dense enough to gravitationally collapse. The timescale is then 1 to 2 free-fall times.

Prestellar cores have been extensively studied in, e.g., nearby dark clouds. Typically, these cores have masses and sizes in the range 0.5–5 $M_\odot$ and 0.02–0.2 pc. Corresponding mean densities are 10^4–10^5 cm^{-3}. The core, B68 (Figure 10.8), is the prototypical example. The density distribution within prestellar cores has been measured through various means (cf. Section 10.1.2) and is, generally, well described by a Bonnor Ebert sphere. For an isothermal sphere, the pressure is given by, $P = \rho C_s^2$, with ρ the density and C_s the sound speed. The equation of hydrostatic equilibrium reduces to,

$$r^2 \frac{d \ln\left[\rho\left(r\right)\right]}{dr} = -\frac{Gm\left(r\right)}{C_s^2} \tag{11.1}$$

with $m(r)$ the mass enclosed within radius r. With the continuity equation,

$$\frac{dm\left(r\right)}{dr} = 4\pi r^2 \rho\left(r\right), \tag{11.2}$$

we find the Lane–Enden equation,

$$\frac{1}{r^2}\frac{d}{dr}\left(\frac{r^2}{\rho}\frac{d\rho}{dr}\right) = -\frac{4\pi G\rho}{C_s^2}. \tag{11.3}$$

The solution of this equation is,

$$\rho\left(r\right) = \frac{C_s^2}{2\pi G r^2}. \tag{11.4}$$

This solution – the singular isothermal sphere – has a singularity at the origin. The isothermal sphere is a special case of a polytropic stellar model; i.e. with $\gamma = 1$ or equivalently polytrope index $n = \infty$. This model has an infinite radius and an infinite mass.

Nonsingular and finite solutions to the Lane–Emden equations can be obtained by imposing a confining pressure, P_m, at the outer radius, R. Introducing the dimensionless variables, $\theta = \ln\left[\rho/\rho_c\right]$ and $\xi = r/R_c$ with ρ_c the central density and $R_c = C_s/\sqrt{4\pi G\rho_c}$ the characteristic radius, the Lane–Emden equation can be rewritten as two coupled differential equations,

$$\frac{d\theta}{d\xi} = \frac{u}{\xi^2} \tag{11.5}$$

$$\frac{du}{d\xi} = -\xi^2 \exp\left[\theta\right]. \tag{11.6}$$

These can be solved numerically – inside-out – with the boundary conditions, $\theta\left(0\right) = 0$ and $d\theta/d\xi = 0$ at $\xi = 0$. There is a family of solutions characterized by one parameter; e.g. ρ_c or equivalently P_m or the total mass (Figure 11.2). For large ξ, the solution approaches

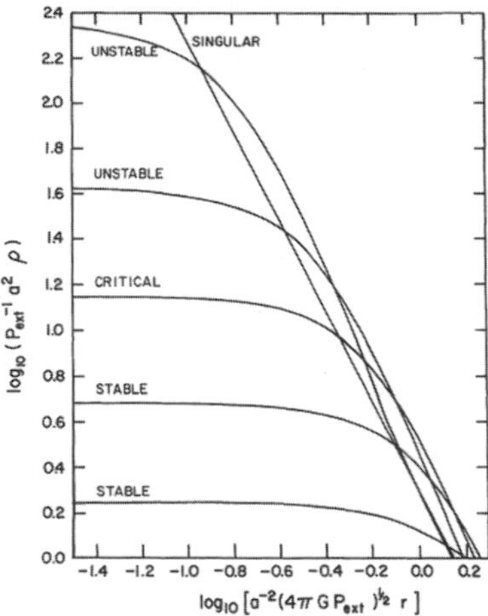

Figure 11.2 Normalized density distributions of isothermal spheres. The outer radius of each sphere is given by the intercept of the corresponding curve with the abscissa. The curve labeled critical corresponds to the critical mass (Eq. (11.7)). Solutions with lower central densities are stable. More centrally condensed solutions are unstable. Figure taken from [257]

the isothermal sphere while, in the inner part ($r \lesssim R_c$, the density flattens to the central density ρ_c. Stable solutions require that $\xi(R) \leq 6.5$ or equivalently that the mass of the cloud is less than a critical mass, M_{cr},

$$M_{cr} = \frac{1.18C_s^4}{P_m^{1/2}G^{3/2}} \simeq 0.66\left(\frac{T}{10\,\mathrm{K}}\right)^2\left(\frac{3\times10^5\,\mathrm{cm}^{-3}\,\mathrm{K}}{P_m/k}\right)^{1/2} \quad \mathrm{M}_\odot \tag{11.7}$$

In terms of the mass of the core, there is a critical pressure that can be applied before collapse ensues,

$$P_{m,cr} = \frac{3.15C_s^8}{G^3M_{cr}^2} \simeq 3.8\times10^5\left(\frac{1\,\mathrm{M}_\odot}{M}\right)^2 \quad \mathrm{cm}^{-3}\,\mathrm{K}. \tag{11.8}$$

Or, equivalently, for a given (turbulent) pressure, masses above a limited value will be gravitationally unstable and collapse.

If the core is "nudged" above the stability criteria through either mass accretion or an external pressure increase, the prestellar core will collapse in an inside-out manner in which material falls inward and a rarefaction wave propagates outward. In the inner region the density distribution is then given by the free-fall solution, viz.,

$$n(r) = \frac{1}{4\pi\mu m_H}\sqrt{\frac{\dot{M}}{2Gt}\frac{1}{r^{3/2}}} \tag{11.9}$$

$$\simeq 2.2\times10^7\left(\left(\frac{\dot{M}}{10^{-5}\,\mathrm{M}_\odot}\right)\left(\frac{10^5\,\mathrm{yr}}{t}\right)\left(\frac{10^{15}\,\mathrm{cm}}{r}\right)^3\right)^{1/2} \quad \mathrm{cm}^{-3}. \tag{11.10}$$

The free fall velocity is given by,

$$v(t) = \sqrt{\frac{2G\dot{M}t}{r}} \simeq 5.15 \left(\left(\frac{\dot{M}}{10^{-5}\,\mathrm{M_\odot}} \right) \left(\frac{10^5\,\mathrm{yr}}{t} \right) \left(\frac{10^{15}\,\mathrm{cm}}{r} \right) \right)^{1/2} \quad \mathrm{km/s} \quad (11.11)$$

with $\dot{M}$ the mass accretion rate,

$$\dot{M} = -0.975\,\frac{a^3}{G} \simeq 10^{-5} \left(\frac{a}{0.35\,\mathrm{km/s}} \right)^3 \quad \mathrm{M_\odot/yr}, \quad (11.12)$$

with a the sound speed in the core. The structure consists then of a central object, surrounded by a circumstellar disk, surrounded by an inner collapsed region with a structure described above and embedded in the prestellar core given by the Bonnor–Ebert sphere solution. The released gravitational energy provides the protostellar luminosity,

$$L_\star = \frac{GM_\star\dot{M}}{R_\star} \simeq 22 \left(\frac{M_\star}{\mathrm{M_\odot}} \right) \left(\frac{\dot{M}}{10^{-5}\,\mathrm{M_\odot}} \right) \left(\frac{10^{12}\,\mathrm{cm}}{R_\star} \right) \quad \mathrm{L_\odot} \quad (11.13)$$

If the envelope rains in on the disk, then the envelope accretion luminosity is set by the disk radius where the accretion occurs. Material is then fed through the disk to the protostar, and the accretion from the disk onto the protostar sets the luminosity of the system. Also, the jet from the protostar and/or the wind from the disk will affect the structure of the collapsing envelope.

As discussed in Section 10.1.2, molecular observations can be used to determine the density in prestellar cores. Figure 11.3 illustrates three methods to derive the density profile of prestellar cores. Each of these methods has its drawbacks, particularly as dust properties inside dense cores are not well known and are prone to differ from those in the diffuse ISM. Typically, these studies reveal a relatively flat interior density distribution and an r^{-2} behavior at larger radii. The density distribution measured for the B68 globule is well represented by a Bonnor–Ebert sphere (Figure 10.9). This model is, however, not unique and other interpretations are possible, including cores in the early stages of their collapse as well as turbulent structures.

This discussion has focussed on well-studied, isolated, small cores in nearby molecular clouds such as Taurus. InfraRed Dark Clouds (IRDCs) are cold. dense, isolated clouds that appear in absorption against the bright mid-infrared emission of (mainly, but not exclusively) the inner galaxy. There are some 10^4 IRDC known with typical optical depths of 1–5 in the mid-IR, corresponding to hydrogen column densities of $0.4\text{–}2 \times 10^{23}$ H-nuclei cm^{-2}. These cores are the larger brethren of the dark cores discussed above with masses in the range $10^2 - 3 \times 10^3$ M$_\odot$, densities of order 10^5 cm^{-3}, and sizes 0.3–0.7 pc. Some 20% of these IRDCs are starless; the remainder contain an embedded protostar or an early phase (hypercompact or ultracompact) of an HII region.

Molecular observations provide a premier tool to study the kinematics of star-forming regions. Molecular clouds are very turbulent with turbulent velocity dispersions, $\sigma = 0.7R_{pc}^{1/2}$ km/s (Section 10.2.2), which corresponds to a FWHM of emission lines

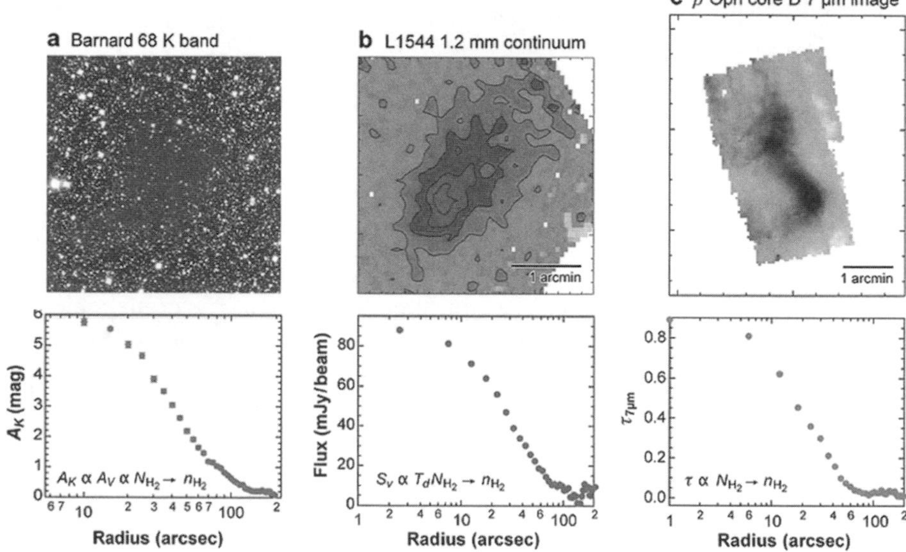

Figure 11.3 Three methods to derive the density distribution of prestellar cores. (a) H-K color excess is transformed into K band and then V band extinction. Standard dust-to-gas ratio provides then the column density. (b) Sub-millimeter emission maps can be inverted into column densities assuming a dust temperature and a dust emissivity "calibrated" on Planck observations of the diffuse ISM. (c) The inner galaxy is bright in the 8 μm PAH emission and foreground clouds appear in absorption against this continuum. Extinction laws convert the measured 8 μm extinction into a visual extinction, which can be converted into a H column density using standard dust-to-gas ratios. Each of these column density maps can be converted into a density map assuming a geometry. Figure taken from [25]

of $1.2 R_{pc}^{1/2}$ km/s. Infall velocities are of similar order only very close to a (low mass) protostar,

$$v_{\rm infal} \simeq 1 \left(\frac{M}{1\,{\rm M}_\odot}\right)^{1/2} \left(\frac{10^3\,{\rm AU}}{r}\right)^{1/2}. \tag{11.14}$$

Very quickly after the formation of the protostar and its surrounding disk, some of the gravitational energy generated by the accretion is channeled into a fast jet or broader outflow at velocities of 10−100 km/s, which can drive strong shocks into the surrounding medium (Section 11.7). Their velocity signature can be recognized from spatially resolved studies as red and blue shifted components. The disk itself is characterized by Keplerian motion with rotational velocities,

$$v_{\rm rot} \simeq 1 \left(\frac{M}{1\,{\rm M}_\odot}\right)^{1/2} \left(\frac{10^3\,{\rm AU}}{r}\right)^{1/2}, \tag{11.15}$$

and this is discussed in Section 11.4.

Detecting the signature of infall in prestellar cores is of special interest. Infall will generate an asymmetric line profile with a brighter blue shifted peak than the red shifted peak,

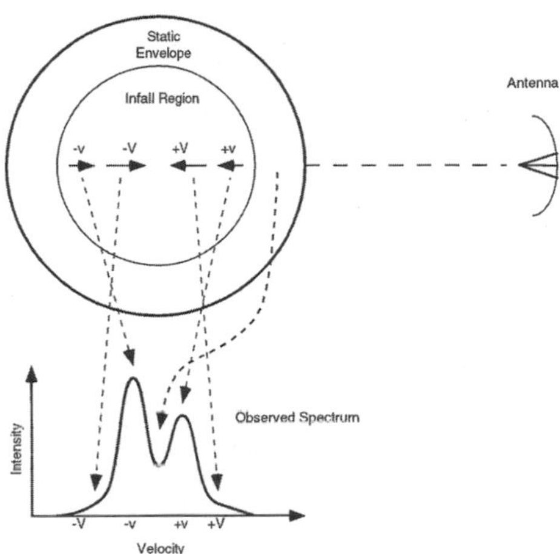

Figure 11.4 Schematic line profile for a prestellar core undergoing inside-out collapse. The static envelope produces the central self-absorption dip, the blue shifted peak comes from the back of the cloud, and the red shifted peak from the front of the cloud. The faster collapse near the center produces line wings, but these are usually confused by outflow wings. Figure taken from [95]

where the static surrounding cloud will produce a central absorption dip (Figure 11.4). The line profile of CS is optically thick and, for L1498, indeed reveals the signature of infall (Figure 11.5). A similar example is shown in Figure 1.11 for B335. For L1517, on the other hand, the asymmetric CS line profile is due to outflow activity. Detailed studies of the line profiles of CS in the prototypical prestellar core, B68, show an organized velocity field indicating small-amplitude, nonradial oscillation or pulsation of the outer layers of the cloud about an equilibrium configuration.

The chemistry of prestellar cores is dominated by the interaction of gas and grains and this has been discussed extensively in Section 10.6.3. The chemical structure is summarized in Figure 11.6. Penetrating FUV photons from the ambient interstellar radiation field dominate the surface regions, keeping grains "clean" and photodissociating and photoionizing gas phase species (Section 10.6.3). Regular ion–molecule chemistry takes over at an $A_v \simeq 2$, but as the density increases rapidly inward, accretion on dust surfaces quickly takes control of the chemical composition of the core. Some ice mantle species are returned to the gas phase upon formation and/or by photodesorption driven by cosmic-ray-produced photons and/or by sputtering due to cosmic ray hits. However, gas phase abundances quickly decrease inward to higher densities. Some species may "profit" from the decrease in gaseous destruction agents; e.g. when CO depletes, N_2H^+ will attain higher abundances. Likewise, protonated molecular hydrogen will become more and more enriched in deuterium (cf. Figure 10.20). Eventually, in the core, only these hydrides will remain in the gas phase. After formation of a protostar, the core will heat up from the inside out and ices condensed during the preceding, cold core phase will sublimate and a Hot Corino/Hot Core will appear (Section 11.3).

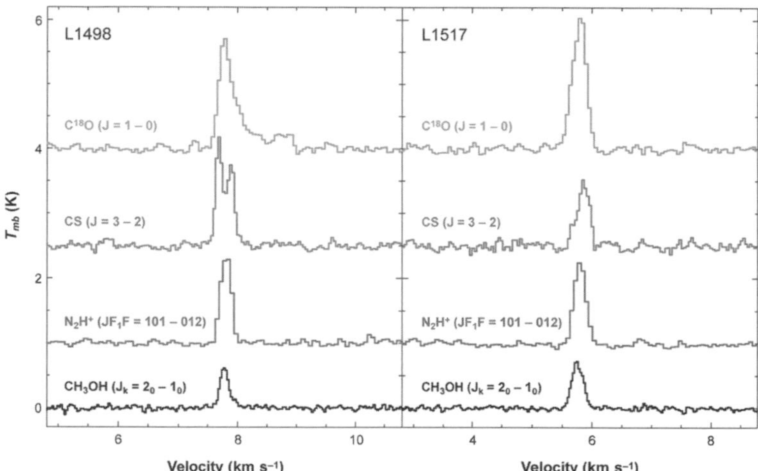

Figure 11.5 Observed line profiles of prestellar cores, L1498 and L1517. The spatial distribution of these species in L1498 is shown in Figure 10.31. The excitation characteristics combined with the chemical history drives the different profiles of these species. CO traces, e.g., low density material. CS is optically thick and reveals the kinematics of infall (L1498) or outflow (L1517). N_2H^+ is more centrally peaked, while CH_3OH traces a zone where dust accretion has not taken its full toll. Figure taken from [25]

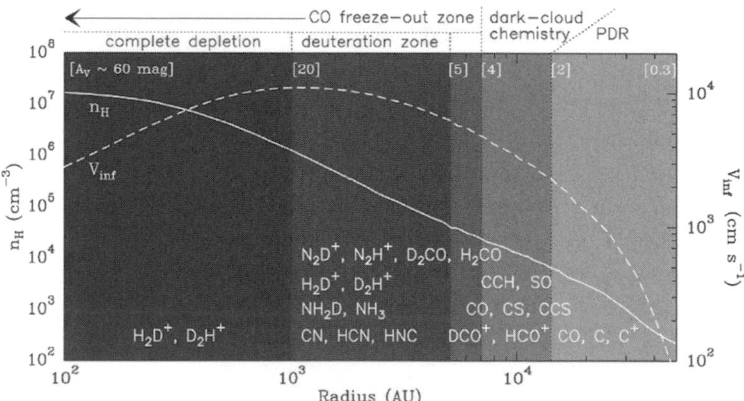

Figure 11.6 Schematic of the physical and chemical structure of prestellar cores. The composition of the PDR surface is dominated by penetrating FUV photons that photodesorb ice mantle species and photodissociate and photoionize gaseous species. Slightly deeper into the core, ion–molecule chemistry dominates. At high densities, gaseous species condense out into an ice mantle. Because of the depletion of, e.g., CO in the next zone, protons are transferred to species with smaller proton affinities and the abundance of, e.g., N_2H^+ increases. In the central core, depletion is "complete" and only the pure hydrides, H_2, H_3^+, HD, H_2D^+, D_2H^+, and D_3^+, and He remain in the gas phase. Figure taken from [51]

11.3 Hot Cores and Hot Corinos

Hot cores are compact (<0.1 pc) regions of warm (>100 K), dense (>10^7 cm^{-3}) gas near massive protostars. Chemically, hot cores are characterized by high abundances of saturated species, such as H_2O, NH_3, H_2CO, CH_3OH, or CH_3CN and CH_3CH_2CN. These species are called "complex organic molecules" (COMs) in astronomical parliance.[2] These hot core regions show strong chemical differentiation with some sources dominated by oxygen-bearing molecules while in others nitrogen-bearing species prevail. This is well illustrated for the Orion BN/KL region (Figures 11.7 and 11.8) at the center of the Orion Molecular Cloud 1, OMC1, where the Hot Core and the Compact Ridge can be separated kinematically as well as spatially using interferometers: e.g. The Hot Core is a dense ($\simeq 10^8$ cm^{-3}), warm ($\simeq$200K) clump at $v_{LSR} = 5-7$ km/s and $\Delta v \simeq 3-4$ km/s characterized by a N-rich inventory. The Compact Ridge is a slightly less dense ($\simeq 5 \times 10^7$ cm^{-3}), slightly cooler ($\simeq$100 K) clump at $v_{LSR} = 7-8$ km/s and $\Delta v \simeq 2-3$ km/s characterized by an O-rich inventory. Both are embedded in the extended ridge, a lower density ($\simeq 10^6$ cm^{-3}) and cooler ($\simeq$60 K), quiescent cloud traced, for example, in HCN and CS at $v_{LSR} = 8-9$ km/s and $\Delta v \simeq 3-4$ km/s. In addition, there is the plateau source representing shocked gas with molecules such as SO and SiO at $v_{LSR} = 5-6$ km/s and $\Delta v \simeq 20$ km/s. There is also a high velocity flow, which is quite apparent in the near-IR H_2 emission with $v_{LSR} = 10$ km/s and $\Delta v \simeq 150$ km/s. These H_2 structures are the fingerprints of an explosive event that took place some 300 yr ago, presumably linked to the dynamical decay of a multiple system of massive stars. Detailed interferometric studies have revealed that the structure of these regions is actually much more complex than this simple set of components would indicate with the bulk of the dust emission arising from many small and compact clumps. Likewise, the dichotomy in oxygen- versus nitrogen-bearing species is more complex than previously thought; e.g. seemingly related species – such as CH_3OH and CH_3CH_2OH – which all concentrate in the Compact Ridge area, have a very different distribution when examined at high spatial resolution. On the other hand, formic acid (HCOOH), ethylene glycol ($(CH_2OH)_2$), dimethyl ether (CH_3OCH_3), methyl formate ($HCOOCH_3$), and ethanol (CH_3CH_2OH) behave more similarly. Abundances for the Orion Hot Core and Compact Ridge are summarized in Table 11.1 as telling examples of an N-rich and a O-rich hot core environment. These are largely taken from high resolution ALMA studies supplemented with some older single dish data. When perusing this data, the caveats on the observed small-scale variations should, however, be kept in mind. In general, Hot Cores are associated with the surfaces of disks surrounding the luminous, massive protostar. However, the Orion Hot Core and Compact Ridge are actually clumps heated from the outside (see below) in a ridge fragmented by the explosion event.

Regions with similar characteristics as Hot Cores – but much smaller (<100 AU) and therefore labeled Hot Corinos – have also been discovered around solar-type protostars. For Hot Corinos as well, care should be taken to disentangle contributions from multiple

[2] The acronym SOM might be more appropriate as these are really simple organic molecules.

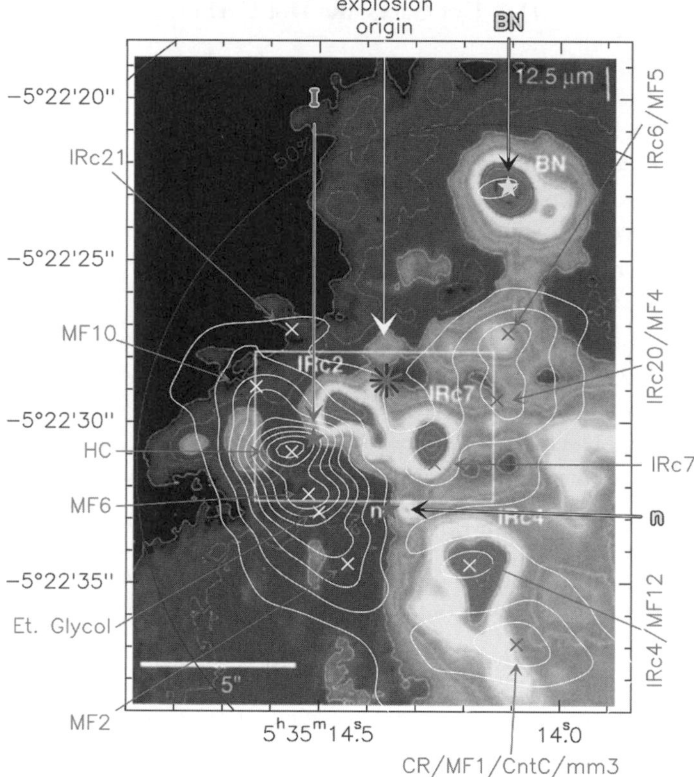

Figure 11.7 The BN/KL region in Orion is host to an explosion caused by a dynamical rearrangement of a mutiple star system. The two runaway stars (BN and I) plus the quasi-static object, n, are indicated by a multipointed star. The other runaway star (source x) is off the map. The presumed position of the origin of the explosion is denoted by a multi-pointed red and blue star. The 12.5 μm dust emission is shown in color with the 1.2 mm continuum emission superimposed as white contours. Mid-IR sources are labeled as IRCn. The BN/KL region contains the prototypical hot core sources, the Hot Core (HC) and the Compact Ridge (CR), which break up in mutiple components when observed in different molecular tracers. Dimethyl ether peaks are denoted by crosses (MFn) and methyl-formate or ethyl cyanide peaks are labeled as CRn. The ethylene glycol peak is also indicated. Figure taken from [227] with data from [96, 113, 131] (A black and white version of this figure will appear in some formats. For the color version, please refer to the plate section.)

components present in single dish beams (Figure 11.9). In addition, as for Hot Cores, the spatial structure may be quite complex with multiple small components contributing to the emission. The composition of the prototypical Hot Corino associated with the low mass protostar, IRAS 16293-2422, is summarized in Table 11.1. As the line profiles attest, the Hot Corino emission is associated with the (surfaces of) disks. While such disks may live some 3 Myr, Hot Corinos are only observed in the earliest, embedded class 0 protostellar phase. This short Hot Corino lifetime may reflect the rapid conversion of organics back to CO in warm gas on a timescale of $\sim 10^5$ yr.

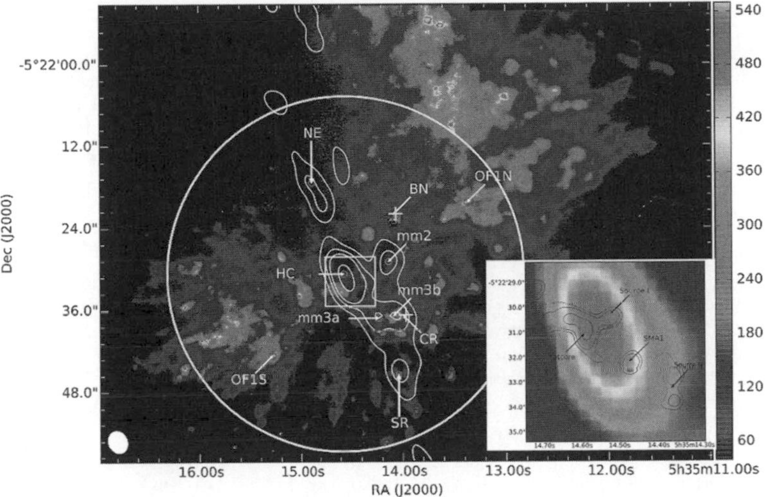

Figure 11.8 The large-scale morphology of the Orion BN/KL region is dominated by shocked H_2 emission associated with the fast shocks (color, brightness in counts per 400 s; [216]). Substructures resolved by SMA-only observations are superimposed (spatial resolution of ~1200 AU). White contours show the 1.3 mm continuum emission at 5, 5, 15, 25, 60 rms levels with $\sigma = 0.04$ Jy/beam. The peaks of hot core (HC), mm2, mm3a, mm3b, the southern region (SR), north-east clump (NE), the NW and SE parts of the high-velocity outflow (OF1N, OF1S) are labeled. White crosses denote the BN object and the compact ridge (CR). The beam in the bottom left corner is from the SMA-only continuum at 1.3 mm. The large yellow circle represents the SMA primary beam. Inset: The HC continuum from the SMA at 1.3 mm (color), with black contours of the 865 μm continuum emission at a spatial resolution of ~300 AU from [28] overlaid. The beam in the bottom left corner is from the 865 μm data. The white cross corresponds to the explosion center determined from a trace-back of the H_2 proper motions [19]. Figure adapted from [100] and kindly provided by Siyi Feng (A black and white version of this figure will appear in some formats. For the color version, please refer to the plate section.)

11.3.1 Physical Conditions

Column Density and Density

The column density and density of Hot Core gas can be estimated analogously to molecular clouds (Section 10.1). Dust or CO measurements are converted into column densities and, with a scale size, the density can be determined. Table 11.1 summarizes the results for the Hot Core and Compact Ridge in Orion. Sub-millimeter dust emission is converted directly into a dust optical depth. Adopting interstellar dust emissivity properties, these optical depths can be converted into hydrogen column densities; viz.,

$$N\left(\mathrm{H}_2\right) = \frac{I\left(\nu\right)}{B\left(T_d,\nu\right)\kappa_d\left(\nu\right)\delta_d\mu m_H} \tag{11.16}$$

$$\simeq 2\times 10^{24}\left(\frac{I_{1.3\,\mathrm{mm}}}{10^4\,\mathrm{MJy/sr}}\right)\left(\frac{10^2\,\mathrm{K}}{T}\right)\left(\frac{1.3\,\mathrm{mm}}{\lambda}\right)^2\left(\frac{0.9\,\mathrm{cm}^2/\mathrm{g}}{\kappa_{1.3\mathrm{mm}}}\right), \tag{11.17}$$

Table 11.1 *Characteristics of "Hot Cores"*

		Orion "N" Hot Core[a]	Orion "O" Hot Core[b]	IRAS 16293-2422 Hot Corino[c]	dark cloud TMC1	B1-b Hot Corino[d1]
T	[K][d]	120–225	90–160	300	10	200
$N\,(H_2)$	$[10^{24}\ cm^{-2}]$[e]	1.1	0.6	17	0.01	14
size	$[10^{16}\ cm]$	2.0	2.0	0.2	30	0.12
$n\,(H_2)$	$[cm^{-3}]$[f]	1.2(8)	5.0(7)	∼1(10)	3.3(4)	1.0(10)
CO	carbon monoxide	2.3 (−4)	4.6 (−4)	2.0 (−5)	8.0 (−5)	
^{13}CO	carbon monoxide	1.9 (−6)	2.5 (−6)			
$C^{18}O$	carbon monoxide	1.8 (−7)	1.8 (−7)			
HCO^+	formyl cation	1.3 (−9)	5.0 (−10)		8.0 (−9)	
CO_2[j]	carbon dioxide	6.0 (−8)				
H_2O[g]	water	4.0 (−4)	4.0 (−6)	5.0 (−6)	< 7.0 (−8)	
H_2CO	formaldehyde	3.6 (−8)	8.0 (−8)	1.6 (−7)	5.0 (−8)	
CH_3OH	methanol	2.1 (−7)	4.2 (−7)	1.7 (−6)	3.0 (−9)	2.4 (−8)
CH_3CH_2OH	ethanol	6.3 (−9)	1.1 (−8)	2.0 (−8)		1.4 (−11)
CH_3OCH_3	dimethylether	1.3 (−7)	2.4 (−7)	2.0 (−8)		7.1 (−10)
CH_3COCH_3	acetone	4.5 (−9)	1.0 (−9)			
CH_3COOH	acetic acid		2.0 (−8)[h]	5.0 (−10)		
HCOOH	formic acid	2.4 (−9)[i]	5.0 (−10)[i]	4.8 (−9)	2.0 (−10)	
$HCOOCH_3$	methylformate	2.3 (−8)	1.2 (−7)	2.2 (−8)		2.10 (−11)
$HCOOCH_3CH_2$	ethylformate		1.5 (−9)			3.3 (−11)
CH_3CHO	acetaldehyde	1.6 (−9)	1.8 (−8)	1.0 (−8)	6.0 (−10)	5.7 (−11)
$HOCH_2CHO$	glycolaldehyde	7.0 (−10)	≤ 6.7 (−10)	5.8 (−9)		3.6 (−11)
$OHCH_2CH_2OH$	ethylene glycol	9.0 (−9)	1.6 (−8)	1.8 (−8)		2.0 (−11)
CH_2CO	ethenone	6.9 (−10)	1.5 (−9)	1.3 (−11)	6.0 (−10)	
$HCONH_2$	formamide	4.0 (−9)		8.5 (−10)		2.8 (−13)
C_2H_2	acetylene	5.0 (−8)[j]				
HCN	hydrogen cyanide	7.0 (−8)[j]			2.0 (−8)	
HC_3N	cyanoacetylene	1.0 (−8)	4.3 (−9)	2.6 (−11)	1.6 (−8)	
CH_3CN	acetonitrile	1.2 (−7)	4.3 (−8)	4.7 (−9)	6.0 (−10)	
CH_2CHCN	vinyl cyanide	2.0 (−9)	≤ 3.0 (−11)	< 1.4 (−11)		2.8 (−12)
CH_3CH_2CN	propanenitrile	8.8 (−9)	2.5 (−9)	5.3 (−10)		
HNCO	isocyanic acid	9.9 (−9)	3.6 (−9)	2.3 (−9)		2.1 (−11)
CH_3NCO	methylisocyanate	7.0 (−9)				
CS	carbon monosulfide	2.1 (−8)	1.3 (−8)		4.0 (−9)	
H_2CS	thioformaldehyde	1.6 (−9)	2.5 (−9)		7.0 (−10)	
OCS	carbonyl sulfide	1.4 (−7)[k]	2.0 (−7)			
C_2S		5.0 (−11)	1.2 (−11)		8.0 (−9)	
C_3S		2.0 (−11)			1.0 (−9)	
SO_2	sulfur dioxide	3.5 (−7)	2.0 (−7)		1.0 (−9)	
SO	sulfur monoxide	6.9 (−8)	5.3 (−8)		2.0 (−9)	
HCS^+	thioformyl cation	5.0 (−11)	1.3 (−10)			

[a] Hot Core in Orion BN/KL as a template for a source rich in N-bearing species. Abundances compiled from [58, 100]. [b] Compact Ridge in Orion BN/KL as a template for a source rich in O-bearing species (Figure 11.7). Abundances compiled from [100]. [c] Hot Corino associated with IRAS 16293-2422B, a disk around the northern protostar in this binary system (see Figure 11.9). [d] Temperature range corresponds to excitation temperature of different molecules. [d1] Incipient hot corino, B1-b [186]. [e] Column density derived from submillimeter continuum. The range corresponds to the different assumed temperatures. [f] Density derived from the column density estimates and the size. [g] Derived from the $H_2^{18}O$ abundance [205]. [h] Ethylene glycol and acetic acid emission originates from a compact source near the hot core [97]. [i] Formic acid emision has an extended component and a compact component aligned with the ethylene glycol and acetic acid emission [179, 226]. [j] Detected in absorption in the mid-IR toward the source IRC 2 [35, 241]. This sightline skirts the hot core, proper. [k] Abundance derived in the mid-IR toward the source IRC 2 is only 3.0(−8) [94].

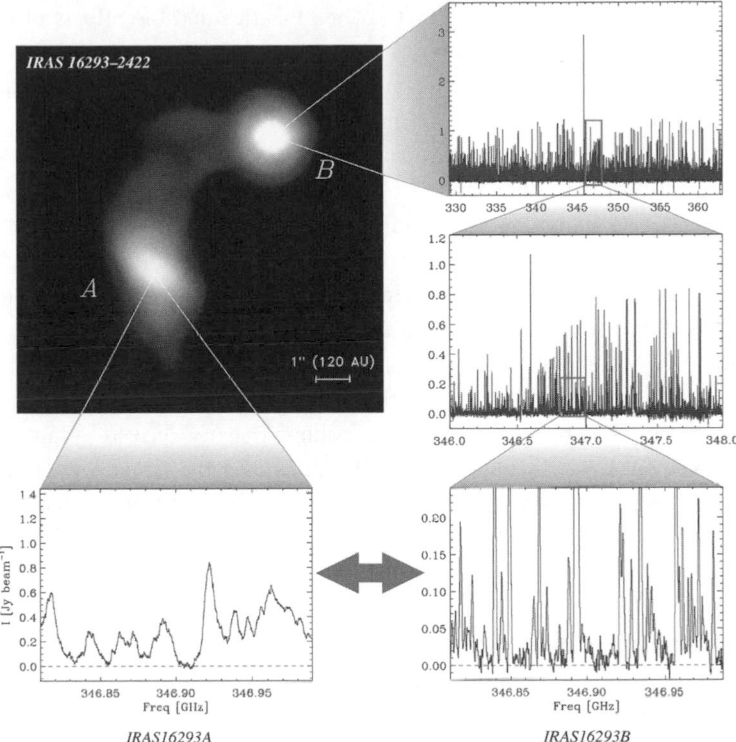

Figure 11.9 ALMA observations of the prototypical protostar, IRAS 16293-2422. The three-color image on the left is a composite of 3mm (red), 1.3 mm (green) and 0.87 mm (blue) continuum emission. The ALMA Protostellar Interferometric Line Survey (PILS) of this low mass solar-type binary reveals the strikingly different spectra of components, A and B. Component B is a face-on disk with very narrow line, which facilitates line identification. Component A (actually also a binary) is edge on and has much broader lines, resulting in line blending. Note how component B shows spectral lines down to the confusion limit. The emission from component A is, however, more extended and dominates single disk spectra. Figure kindly provided by J. Jørgensson; see [158] (A black and white version of this figure will appear in some formats. For the color version, please refer to the plate section.)

with I the observed sub-millimeter surface brightness, κ the dust emissivity per unit dust mass, δ_d the dust-to-gas ratio per unit mass, and m_H the hydrogen mass. The second line pertains to 1.3 mm, assumes a dust opacity of 0.9 cm^2/g appropriate for coagulated dust, a dust-gas ratio by mass of 10^{-2}, and we have taken the Rayleigh–Jeans limit. We note that 10^4 MJy/sr corresponds to 250 mJy in a 1$''$ beam. As it is difficult to determine the spectral energy distribution for a small core embedded in a more complex region – particularly if there is unrelated foreground absorption – dust temperatures cannot be determined directly. Instead, the dust temperature is often set equal to the gas temperature estimated from molecular excitation studies (see below). Even more so than for molecular clouds, the dust properties at millimeter wavelengths in hot cores can be expected to be very different from those in the diffuse ISM due to the growth of ice mantles and the effects of coagulation. The value adopted above is a value calculated for a coagulated size distribution of dust and is about a factor of 3−5 larger than standard models for dust in the ISM would indicate. Note that the dust opacity is wavelength (and temperature) dependent.

The total hydrogen column density can also be estimated from CO isotopologue studies (i.e. $C^{18}O$), adopting a CO abundance (8×10^{-5}) and an isotope ratio ($^{16}O/^{18}O$=550). For the CO $J = 2{-}1$ transition, this leads to column densities of 1.1×10^{24} and 6.5×10^{23} H_2 molecules/cm^2 for the Hot Core and Compact Ridge in Orion in a $5.5'' \times 4.4''$ beam, respectively. As Hot Cores are typically quite warm, much of the population may be in the higher J-states. It can then be advantageous to observe higher J transitions and determine the temperature of the emitting gas directly. The highest transition observable from the ground is the $J = 8{-}7$ transition at 877 GHz. With HIFI/Herschel, a wide range of transitions could be observed and the column density derived by summation. This resulted in somewhat lower average column densities over the larger region (see Exercise 11.2). As for molecular clouds, densities can also be estimated from molecular excitation characteristics (Section 4.2.3 and 10.1).

Temperature

Rotation diagrams are often employed to derive the gas temperature in Hot Cores. Current wideband spectrometers allow a wide range of K transitions associated with a single J transition to be measured in a single setting, allowing accurate relative calibration and probing a wide range in molecular excitation. Densities in these regions are very high and an LTE assumption is often justified. On the other hand, optical depth effects can often be quite important (Section 4.2.2). Figure 11.10 illustrates the results for one such analysis for the methyl cyanide transitions around 220 GHz from the Orion Hot Core. In this case, optical depth effects are quite important. The wide spectral range probed by HIFI/Herschel was particularly suited for this type of analysis. For methanol, torsional

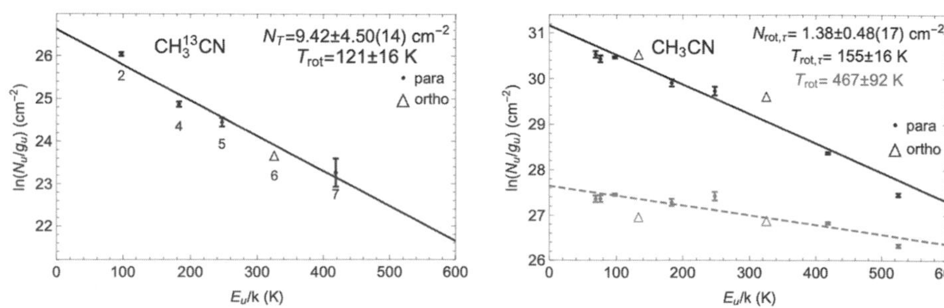

Figure 11.10 Rotation diagram of CH_3CN (right) and $CH_3{}^{13}CN$ (left) $J = 12 \rightarrow 11$ transitions around 220 GHz in the Hot Core region in Orion. Dots and triangles represent para and ortho, respectively. Left: Data for $CH_3^{13}CN$. The line is fit for para lines assuming they are optically thin. Right: Data for CH_3CN.The top line represents the results of an iterative optical depth correction procedure for the main isotope. Figure kindly provided by Siyi Feng, adapted from [100]

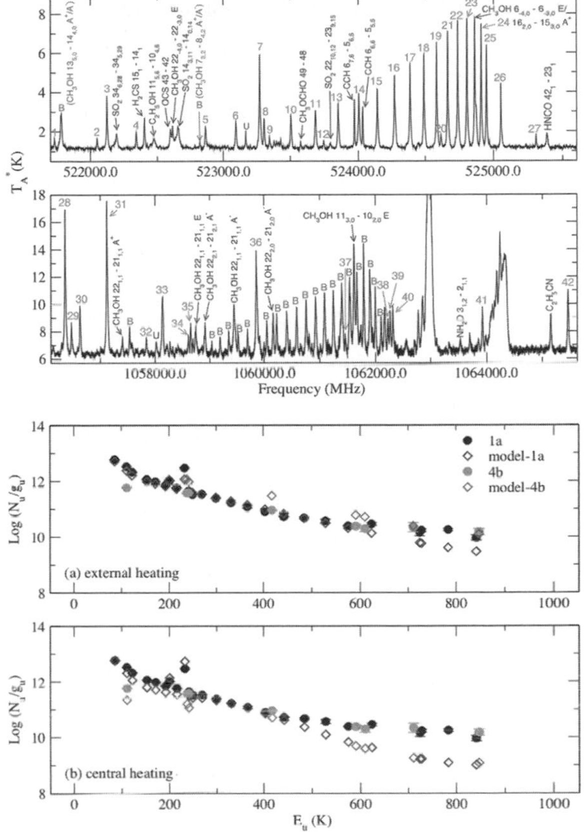

Figure 11.11 Top: HIFI spectra of the Orion KL region around 525 GHz and 1.06 THz showing a forest of CH_3OH transitions. Numbered labels indicate isolated methanol transitions, many of them associated with $20_{K,0} \rightarrow 20_{K-1,0}$ transitions. Blended methanol transitions and transitions from other molecules are indicated. Methanol lines blended with other parity states are labeled B. Bottom: Rotational diagram for isolated CH_3OH transitions compared to detailed model results for a clump heated from the inside (bottom) and from the outside (top). Figure taken from [287]

motion associated with the internal rotation of the methyl group with respect to the OH-bond can couple to other degrees of freedom and, as a result, the abundant methanol molecule has numerous and often very strong rotational transitions spanning the range from the far-infrared to the millimeter range. Figure 11.11 illustrates the richness of the rotational emission spectrum and the results of a study of methanol emission in Orion. These lines originate from the Compact Ridge. The results show a curved relationship as a function of excitation energy, which may reflect opacity effects, non-LTE excitation, temperature gradients, or a combination of these (cf. Section 4.2.2). These observations are compared to statistical equilibrium model calculations, which take non-LTE and optical depth effects into account. In this case, the curvature in the rotational diagram reflects a temperature gradient. There has been a longstanding debate whether the Compact Ridge (and the Hot Core) in Orion is externally heated by a nearby protostar as the very narrow linewidth and high temperature and column density are inconsistent with normal, centrally heated star-forming cores. Analysis of this data shows that a clump heated from the outside is more appropriate for the Orion Compact Ridge than heating by an embedded star. As revealed by interferometric studies, a similar situation pertains for the Orion Hot Core, which also seems to be heated from the outside. For the Hot Core, heating might be due to the shock interaction with the explosive event but that is not applicable for the Compact Ridge.

Cooling Lines

Broadband spectral surveys of protostars allow us to probe the cooling line budget of protostellar regions. Results for the intermediate mass protostar OMC2-FIR 4 are summarized in Figure 11.12. CO is the dominant coolant, emitting 60% of the cooling line flux. H_2O and CH_3OH are distant seconds coming in at 13 and 9%, respectively. Over the same frequency range, continuum emission amounts to 7 $L_\odot$ compared to 0.1 $L_\odot$ in lines. Shorter wavelength CO lines ($J = 14-46$ contribute another $\simeq$0.3 $L_\odot$. Most of the continuum emission, estimated to be in the range of 40–90 $L_\odot$, also escapes at shorter wavelength. A detailed comparison of the cooling budget for other sources has not been made, yet. We do note though that, for the Orion KL region, SO_2 and CH_3OH dominate the emission in the

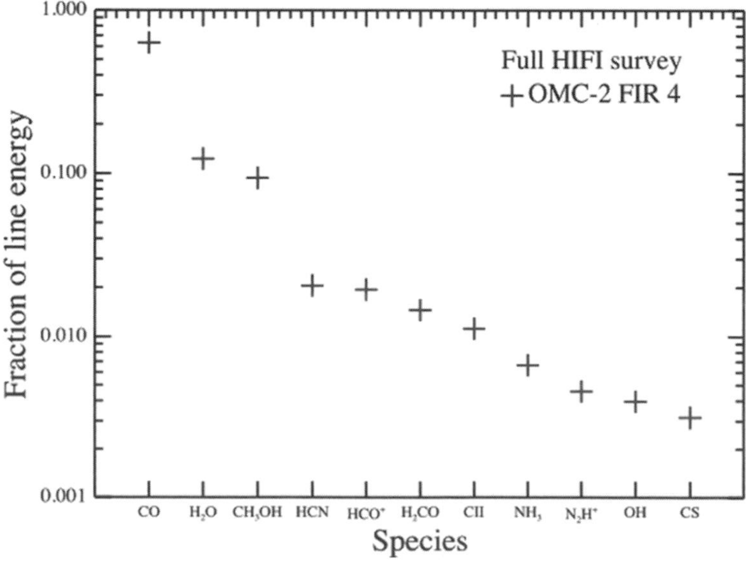

Figure 11.12 Contribution of the line emission by different species to the cooling budget of the intermediate mass protostar, OMC2-FIR4, in the 480–1902 GHz range. Fractions are relative to the total observed line emission. Figure taken from [160]

795–903 GHz range (each at ≃23%) over CO (16%), while for OMC2 FIR4, CO is three times more important than CH_3OH in this wavelength region.

11.3.2 Chemistry of Hot Cores

There are a number of clues that imply a strong involvement of interstellar ices in the composition of Hot Cores. First, interstellar ices in dark clouds show high abundances of H_2O and CH_3OH, much higher than the gas phase, implying an efficient surface chemistry route toward these species. These species are also very abundant in the gas phase of Hot Cores.[3] Second, the organic inventory of Hot Cores is very different from that of dark clouds (Compare Table 11.1 with Table 10.2), indicative of a very different chemistry. In contrast to dark clouds, Hot Cores are dominated by saturated species, indicative of H-tunneling under reducing conditions on grain surfaces (cf. Section 7.4.4). Third, Hot Cores show high deuterium fractionation (Table 11.2). Deuterium fractionation of gas phase molecules is generally understood to reflect chemistry at very low temperatures where small zero-point energy differences ($\Delta E \sim$200–500 K, Section 10.5.2) result in large disparities between the forward and backward reaction rates, which favors high abundances of deuterated species. Yet, high temperatures ($T > 100$ K) are a defining characteristic of Hot Cores and this fractionation mechanism should be of little consequence in this environment. Perusing the deuterium fractionation patterns in the Hot Corino associated with IRAS 16293-2422 provides direct evidence for deuterium chemistry on grain surfaces (Table 11.2). The observed deuterium fraction of HDO, CH_3OD, and DNCO indicates formation from an accreting gas with a D/H fraction of $\simeq 2 \times 10^{-2}$ and such a high atomic D over H ratio reflects the high H_2D^+/H_3^+ ratio at low temperatures in the gas phase (cf. Section 10.5.2). The much higher HDCO, D_2CO, CH_2DOH, CHD_2OH, and CD_3OH fractionations are a direct consequence of the H abstraction reactions that characterize the chemistry of CO (cf. Section 10.6.2). The relatively low D_2O fraction is a result of the rapid reaction of OD/OH with H_2/HD and the rather modest enhancement in the HD/H_2 fraction on an ice surface as compared to the atomic D/H in the accreting gas. All of this implies ice formation in dark cloud cores under reducing conditions when H abstraction and addition reactions coupled with the kinetic isotope effect drive fractionation. As an aside, we note that – in contrast to Hot Corinos – Hot Cores (Table 11.2) seem to have formed under less extreme reducing conditions (e.g. ice mantle formation at somewhat higher densities).

For these reasons, it is now generally accepted that the gaseous organic inventory of Hot Cores reflects the formation of interstellar ices in the preceding dark core phase and their sublimation when the ice reaches a temperature of ~100 K once the region heats up due to the presence of a newly formed star. The ice acts thus as a repository of "cold" chemistry, preserving, e.g., the high level of saturation and the high deuterium fractionation into the Hot Core phase. Conversely, this must also imply a very recent event as the sublimated gas

[3] Note that this is not true for all species as NH_3 and CO_2 are much less abundant in the gas phase of Hot Cores than in ices in dark clouds.

Table 11.2 *Deuterium fractionation in "Hot Cores"*

	"N" Hot core[a]	"O" Hot core[b]	Hot corino[c]
HDO/H_2O	3.0 (−3)	3.8 (−3)	3.4 (−2)
D_2O/H_2O			7.0 (−5)
NH_2D/NH_3	6.8 (−3)	≤1.0 (−2)	
$HDCO/H_2CO$	≤5.0 (−3)	6.6 (−3)	3.3 (−2)
D_2CO/H_2CO			9.2 (−2)
CH_2DOH/CH_3OH	≤4.2 (−3)	5.8 (−3)	3.7 (−1)
CH_3OD/CH_3OH	≤1.8 (−3)	5.0 (−3)	1.8 (−2)
CHD_2OH/CH_3OH			7.4 (−2)
CD_3OH/CH_3OH			1.4 (−2)
$DCOOCH_3/HCOOCH_3$		3.9 (−2)	1.5 (−1)
$HCOOCH_2D/HCOOCH_3$		2.5 (−2)	
CH_2DOOCH/CH_3OOCH			6.1 (−2)
CH_3OOCD/CH_3OOCH			5.9 (−2)
CHD_2OOCH/CH_3OOCH			12 (−2)
NH_2CDO/NH_2CHO			2.0 (−2)[d]
$NHDCHO/NH_2CHO$			2.0 (−2)[d]
DNCO/HNCO			2.0 (−2)[d]
CH_2DCN/CH_3CN	0.01		
HDS/H_2S	<4.9 (−3)		

[a]Hot Core in Orion BN/KL (see Figure 11.7) as a template for a source rich in N-bearing species. Abundances compiled from [206]. [b]Compact Ridge in Orion BN/KL (see Figure 11.7) as a template for a source rich in O-bearing species. Abundances compiled from [206]. [c]Hot Corino associated with IRAS 16293-2422A (see Figure 11.9). Abundances compiled from [53, 74, 75, 83, 181, 230]. [d]Measured toward IRAS 16293-2422B. Spectral confusion limits the detectability of such weak lines from complex species in component A [76].

has not yet had the time to "equilibrate" with the gas (e.g. $\tau_{chem} \sim 10^5$ yr) and "burn" back to CO or lose its deuterium signature. Besides sublimation, sputtering in shocks may also act to release these molecules into the gas phase.

The details of this scenario and in particular the actual chemistry involved in the formation of the complex organic molecules is still highly contentious. Various schemes have been considered, including sublimation of grain surface chemistry products, sublimation of simple species followed by gas phase reactions, and photolysis of ices followed by sublimation. Each of these have their pros and cons and it is instructive to consider these in more detail here.

A successful model has to explain the high abundance of COMs, their deuterium enrichment, the dichotomy in O-rich and N-rich cores, and the chemical specificity. As to the O-rich versus N-rich cores, the difference is in the high abundance of nitriles; the abundance of oxides are quite comparable between the two types of cores (Table 11.1). The chemical specificity issue is well illustrated by comparing the abundances of the three isomers, methyl formate, acetic acid, and glycol aldehyde. The abundance of these three species differ by more than a factor of 100 where we note that the thermodynamically least favored species (methyl formate) is the most abundant one. The same holds for ethanol and dimethyl ether. All three arguments, in general, and these two sets of species, in particular, illustrate well the importance of kinetics and how specific chemical routes must control the organic inventory of Hot Cores.

Grain Surface Chemistry Routes

As H_2O, CH_3OH, H_2CO, CH_4, and NH_3 are thought to result from grain surface reactions, more complex species – such as CH_3CH_2OH, CH_3OCH_3, HCOOH, and $CHOOCH_3$ – may also reflect hydrogenation and oxidation reactions on grain surfaces and putative chemical routes have been identified (Figure 11.13). These routes toward further chemical complexity have not been tested experimentally and, if there are no activation barriers involved, isomerization at intermediate steps may also, or even preferentially, lead to the

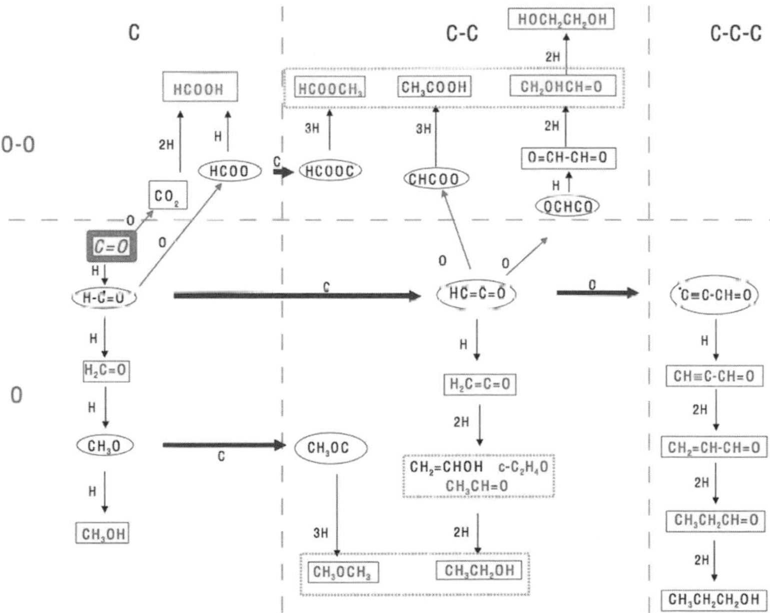

Figure 11.13 Chemical routes toward complex organic species on grain surfaces. The left-hand side summarizes the activation of CO through reaction with H (cf. Figure 10.32). Intermediate radicals such as HCO and CH_3O can react with other atoms than H, leading to larger organic molecules (thick rightward and upward arrows) and possible routes toward, among others, HCOOH, CH_3CH_2OH, and CH_3OCH_3 are indicated. Boxed molecules are often observed to be abundant in Hot Cores. Molecules in large dotted boxes are isomers. Figure taken from [243] as adapted from [61]

most stable products instead. The reaction of N with HCO may be a case in point as this may lead to HNCO rather than NCHO (which would hydrogenate further to $CHONH_2$). Because of the low abundance involved, direct confirmation of the importance of these routes through ice absorption studies is difficult to obtain.

As a general point, almost all of the carbon is locked up in CO in dense clouds (C/CO$\sim 6\times10^{-3}$, Section 10.5.1). As a result, the grain mantle abundance of organic species formed by reaction with accreted atomic C is expected to be very low and complex organic species with multiple C atoms are difficult to form. Atomic O, on the other hand, is very much available (as it is converted into H_2O on grain surfaces). So, oxidation might be rapid and efficient HCOOH formation could well result from HCO + O. The alternative route – OH + CO $\rightarrow$ HOCO* $\rightarrow$ HOCO followed by HOCO + H $\rightarrow$ HCOOH – is also important.

Hydrogenation of CO is an important step in the "activation" of CO on grain surfaces. Reactions of intermediate radicals in this route might form an interesting alternate route toward COMs on grain surfaces. Co-deposition studies of atomic H with $H_2CO/CH_3OH/CO$ mixtures support this as they (Figure 11.14) reveal the presence of small amounts of glycolaldehyde, ethylene glycol, and methyl formate but no evidence for glyoxal. In this scheme, H-addition and abstraction reactions compete and lead to the (repeated) formation of the intermediate, transient radicals, HCO and CH_2OH when the H-flux is high; e.g. H_2CO and CH_3OH cycle back to HCO and CH_2OH, respectively (Section 10.6.2). These radicals react readily with the next (accreted) atomic H that "passes" by to reform the parent species and, at any time, the abundance of these radicals is

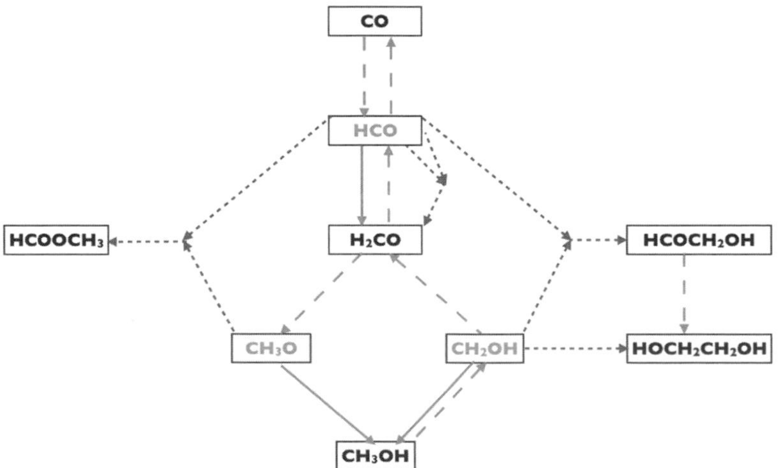

Figure 11.14 Grain surface chemical routes toward glycol aldehyde, ethylene glycol, and dimethyl ether involving radicals. The radicals HCO, CH_3O, and CH_2OH are formed as transient intermediaries in the hydrogenation of CO to CH_3OH. Stable species are bold. The other species are radicals, dashed (solid) arrows involve reactions with(out) barriers. Dotted arrows are radical–radical reactions. The others are H-addition or -abstraction reactions. Figure adapted from [45, 66, 292]

low. Occasionally, a radical may be formed while another radical is present at an adjacent site. These two radicals would then be able to react before a next H atom accretes and reacts with either of them. Given the intermediate radicals, this scheme could lead to the formation of glyoxal (HCO+HCO), glycolaldehyde (HCO+CH_2OH), and ethylene glycol (CH_2OH+CH_2OH). However, glyoxal will fragment to formaldehyde and carbon monoxide as the barrier for this reaction lies below the energy of the formyl entrance channel. So, only the latter two are relevant. Formation of dimethyl ether requires the presence of CH_3O formed by hydrogenation of H_2CO. But the reaction probability of CH_3O would not be (much) enhanced by H-abstraction reactions as the dominant route is through CH_2OH. The high deuterium fractionation observed for formaldehyde and methanol in IRAS 16293-2422 (Table 11.2) attests to the importance of this repeated cycling through the intermediate radicals HCO and CH_2OH in the surface hydrogenation of CO and provides credence to this scheme. However, this seems to have played much less of a role for Hot Cores.

As mentioned, under highly reducing conditions, these radical reaction probabilities are greatly enhanced because of H-abstraction reactions. H reaction with H_2CO can add to form CH_3O or abstract an H to reform HCO (cf. Figure 10.31). In the gas phase, the latter is much preferred over the former, by about a factor of 30 at low temperatures. In contrast, on H_2O ice, addition may be slightly more likely, reflecting dipolar interactions. In contrast, a CO-rich ice would not lead to such drastic changes in activation barriers. Here too, the resulting composition will be sensitive to the physical conditions in the core. The reaction of H with HCO may lead back to CO or to H_2CO. In this cycle, addition is preferred over abstraction but the magnitude of the difference in the rates varies much with the level of quantum theory. Multiple passes through the HCO intermediary are thus possible but not highly likely.

Consider now a CO surface with N-sites. If the probability that this H atom reacts with a CO molecule when it is in the associated site is p (typically p is assumed to be $\sim 10^{-5}$), then this new H-atom can visit $1/p$ sites before a reaction with CO occurs. Assume now that a HCO radical has just formed, and that the next accretion is again an H atom. The probability[4] that this H-atom will not find the HCO after k steps is $\pi / \ln [k]$ and, hence, the probability of the formation of a second HCO rather than a reaction with the HCO already present to form H_2CO can be rather large. Assuming $pN > 1$, we get $\simeq 0.27$ in this example. However, the probability of two adjacent HCOs is small, $4\pi/(-(N \ln[p]))$ $\simeq 10^{-5}$. When $pN < 1$, the evaluation becomes slightly more complex as several widely separated HCOs may be present but the gist is the same: An accreted H-atom can "survey" a large area for a radical coreactant before a reaction with an activation barrier would occur. Note that, in the experiments, N is kept "artificially" small by the high H flux (and H_2 formation rate) and this probability is 7×10^{-3} if $N = 300$. At elevated temperatures, HCO will be mobile and the reaction probability would increase dramatically, but then H may not stick around long enough for HCO to form.

[4] Essentially, Polya's return to the origin probability.

Thus, grain surface chemistry readily forms rather simple species such as H_2CO, CO_2, CH_3OH, and HCOOH, but it is challenging to form abundant amounts of COMs. In particular, a dimethyl ether abundance comparable to methanol – as for the Compact Ridge in Orion – is very difficult to attain. The atom–radical reaction scheme can lead to a wide diversity of species. Radical–radical reations, on the other hand, are more restrictive. Both schemes will profit much from a high abundance of H in the accreting gas, as this cycles the products more often through some of the radical intermediaries. This leads to chemical specificity as it will, for example, prefer CH_2OH "daughters" over CH_3O daughters. That seems to be at odds with the organic inventory of Hot Cores and Hot Corinos. Further comparison of COM abundances relative to CH_3OH between Hot Cores and Hot Corinos – representing very different reducing conditions – may provide good tests for these schemes.

Gas Phase Chemistry Driven by Sublimating Ices

The release of H_2O, CH_3OH, and NH_3 from the ices into the gas phase can drive a rich chemistry in these warm and dense environments (Figure 11.15). In these models, methanol plays a key role in driving molecular complexity because protonated methanol is readily formed and can transfer CH_4^+ groups (instead of H^+). Chemical routes toward the complex (oxide) species observed in hot cores can then be formulated. For example, the following routes have been proposed for dimethyl ether and methyl formate,

$$H_3^+ \xrightarrow{H_2O} H_3O^+ \xrightarrow{CH_3OH} CH_3OH_2^+ \xrightarrow[-H_2O]{CH_3OH} CH_3OCH_4^+ \xrightarrow{e} CH_3OCH_3 \quad (11.18)$$

$$\xrightarrow[-H_2]{H_2CO} HC(OH)OCH_3^+ \xrightarrow{e} HCOOCH_3 \quad (11.19)$$

$$\xrightarrow[-H_2O]{HCOOH} HC(OH)OCH_3^+ \xrightarrow{e} HCOOCH_3 \quad (11.20)$$

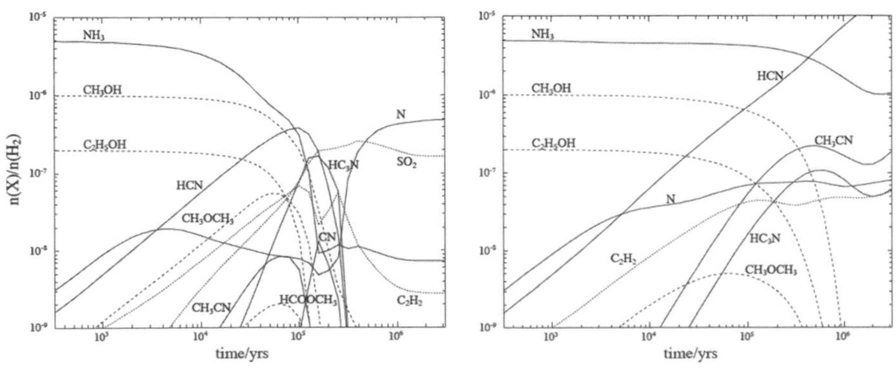

Figure 11.15 The evolution of COMs driven by gas phase ion–molecule chemistry after evaporation of ices containing H_2O (off scale), NH_3, CH_3OH, and CH_3CH_2OH. Left: T = 100 K. Right: T = 300 K. Figure taken from [61]

However, these proposed routes face two issues. The first is the transfer of the methyl group. Consider the dimethyl ether formation pathway (Eq. (11.18)): alkylation of protonated methanol with methanol. Mixed isotope studies show that only in $1/5$ of the collisions, the methyl group is transferred (the remainder is merely proton transfer, which is immaterial) and protonated dimethyl ether is formed. The resulting rate leaves this route still relevant. The route toward methyl formate (Eq (11.19)) has a barrier that hampers formation of protonated methyl formate. This route leads to complex formation (i.e. $[CH_3OH_2 \cdot OCH_2]^+$) but there is a large barrier to isomerization to protonated methylformate from this complex. The third reaction (Eq (11.20)) where protonated methanol reacts with formic acid in the third step also leads to complex formation but in this case, the transition state to protonated methylformate lies below the energy of the reactants by 0.15 eV. Moreover, in this case, proton transfer is energetically not feasible (cf. Table 6.6 in Chapter 6). While experiments show a preference for adduct formation, this may reflect rapid collisional stabilization at these high pressures. So, kinetically, this is a much more favorable route toward methyl formate. Besides these kinetic arguments, the abundance of the coreactants also enters in these evaluations. Methanol and formaldehyde are very abundant in hot cores but formic acid is not. Indeed, with the abundances given in Table 11.1, the third route is excluded as an important pathway toward methyl formate.

The second issue to be considered is the branching ratios in step four, dissociative electron recombination. Experiments show that fragmentation into multiple heavy fragments dominates this process. Protonated dimethyl ether, for example, leads mainly to methanol and methyl as well as two methyl groups and hydroxyl. Dimethyl ether is also formed but only with a branching ratio of 0.07. Likewise, recombination of $CH_3COH_2^+$ does not lead to measurable amounts of acetaldehyde. The predominance of the formation of multiple heavy fragments is a common characteristic of dissociative electron recombination and is a general hindrance to these reaction schemes. As an alternative, rather than electron recombination, the last step might involve proton transfer. Perusing proton affinities of relevant species (Table 6.7 in Chapter 6), only NH_3 is a viable candidate but in order to take over, the ammonia abundance should be $X\,(NH_3) \gtrsim 10^{-5}\left(X\,(e)\,/10^{-8}\right)$. As mentioned above, the NH_3 ice abundance is typically this high but the abundance of gaseous ammonia is much less in Hot Cores.

In these classes of models, the dichotomy in O-rich and N-rich Hot Cores is largely a temperature effect. Again, cosmic ray ionization drives the chemistry. The resulting He^+ reacts with CO to form C^+, which reacts with NH_3 to $HCNH^+$. This ion recombines to HCN and CN. The H_3^+ produced by cosmic rays reacts with CH_3OH to form $CH_3OH_2^+$ (which will go on to drive the complex O-bearing species). Besides proton transfer, this reaction will also yield CH_3^+ (in about half of the collisions). The methyl cation reacts with HCN to form CH_3CNH^+, which will then lead to CH_3CN through proton transfer to NH_3 (or dissociative electron recombination). The CN radical can react with C_2H_2 (presumed to be released from the ice) to form HC_3N. Vinyl cyanide would form from the reaction of CN with C_2H_4 (also presumed to be released from the ice). The N-routes are favored at high temperatures because NH_2 produced by dissociative recombination of NH_4^+ can rapidly react with H_2 and oxygen is locked up in H_2O, keeping nitrogen in NH_3 rather than channeling it to N_2.

The timescale for the formation of these complex organics is some 3×10^4 yr but after $\sim 10^5$ yr gas phase chemistry will have driven all the sublimated organics to CO. So, within this model, the organic inventory provides a chemical clock with which the evolution can be timed and this derived timescale is very similar to the lifetime estimated for Hot Cores from considerations centered on the evolution of regions of massive star formation. Deuterium fractionation provides a very similar chemical clock as proton transfer followed by dissociative recombination will drive the fractionation back to the values consistent with the gas temperature. Hot Corinos around low mass protostars are associated with protoplanetary disks (cf. Figure 11.9) and this lifetime argument implies that only during the short-lived deeply embedded phase, COMs will be abundant. With a typical lifetime of $\simeq 3$ Myr, the abundances of COMs can be expected to be be down by some two orders of magnitude in the later phases of disk evolution. This is a general argument, independent of the formation route of the COMs.

Energetic Processing of Interstellar Ices

UV photolysis and/or ion bombardment of interstellar ices during the prestellar core phase will produce (hot) radicals – specifically H, OH, CH_3, CH_3O, NH_2 – and some of these will react with other molecules while others will rattle around until they lose their excess energy and become trapped in the matrix. Upon warm up, the trapped radicals will diffuse and can recombine with each other, forming more complex species (Figure 11.16). The more volatile products will evaporate with the main ice components, leaving behind

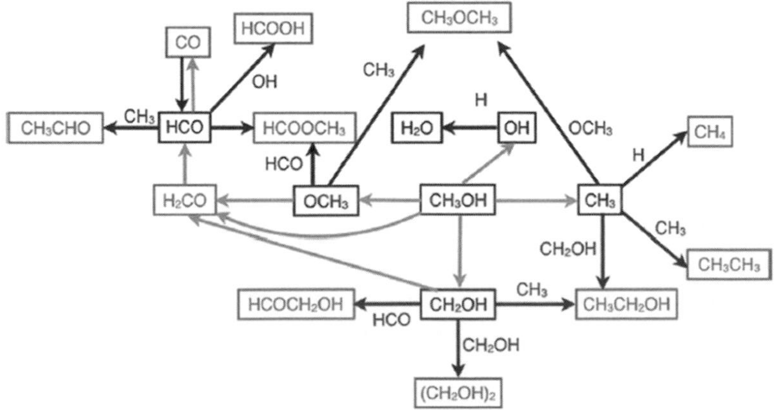

Figure 11.16 UV photolysis produces radicals that can react on to form more complex organic molecules. This figure summarizes the proposed chemical network for the UV photolysis of CH_3OH. Gray arrows indicate UV dissociation reactions. Reactions involving radical recombination (after warm-up) are shown with black arrows. Gray boxed species are COMs typically observed in Hot Cores/Hot Corinos. Figure taken from [221]

a residue. The UV absorption cross sections are comparable to those in the gas phase ($\sigma_{uv} = 5 \times 10^{-18} - 3 \times 10^{-17}$ cm^2; Section 6.3). However, the photodissociation yield is very small. This reflects the cage effect as the products are "locked" in their sites by neighboring species and will likely recombine and those products that are energetic enough to escape their site (e.g. H) have a high probability to find a radical nearby produced by an earlier photodissociation event and recombine there with no net radical production. Measured "radical production" cross sections are therefore only $1{-}5 \times 10^{-19}$ cm^2. The photon energy absorbed goes therefore mainly into heat of the grain. With a cosmic-ray-produced UV radiation field inside a dense cloud of 1.5×10^3 photons cm^{-2} s^{-1}, the rate at which a molecule is photodissociated to a radical is $\simeq 5 \times 10^{-16}$ s^{-1} or $\simeq 1.5 \times 10^{-2}$ Myr^{-1}. So, over the lifetime of a dense core, $\simeq$1Myr, the radical fraction produced is $\simeq$0.01: this 1% radical "limit" is mainly set by the reaction of diffusing radicals.

While this has little influence on the ice composition, this production rate can produce sufficient radicals to account for many (but not all) of the observed COM/CH_3OH ratios in Hot Cores and Corinos. The implied fluence is rather small, 5×10^{16} photons cm^{-2}, and compares to typically 1 minute in laboratory settings. Laboratory studies employ much longer UV exposure times to allow for detection using, e.g., IR spectroscopy but that has the drawback that secondary photolysis products can start to play a role. As an aside, this analysis assumes that all ice molecules see the same (incident) UV flux. This implicitly assumes that the ice is very thin, $\lesssim$300 Å. Adopting the surface area of dust in the diffuse ISM, the calculated thickess of ice mantles is only 175 Å, if all the O condenses out as H_2O. As to first order, the surface area is conserved during coagulation in dense clouds until the aggregates compact collisionally, this is a good assumption. Finally, interstellar ices will experience much higher UV fluences in molecular cloud surface layers. At $A_v = 4$, the fluence will be a factor of 50 higher than in a dense core. As the exposure time may be a factor of 10 higher, UV photolysis may be more pronounced and could conceivably have more of an influence on the global composition of interstellar ices, although there is no observational evidence to support this.

An active community has extensively studied this chemistry over the years in the laboratory. Recent studies have focused on identifying the chemical routes involved and astronomical models have been developed, incorporating ice photochemistry under astrophysically relevant conditions. The photolysis of pure H_2O results in the formation of HO_2, H_2O_2, O_2, and O_3 due to recombination processes involving H, O, and OH. Given the observed composition of interstellar ices and the organic inventory of Hot Cores/Corinos, methanol photolysis is particularly relevant. Figure 11.16 illustrates the chemistry routes possible in methanol ice photolysis. Methanol is broken down in a number of steps to H_2CO and CO. Radicals produced are predominantly H, OH, CH_2OH, CH_3O, CH_3 and the recombination products include formic acid, dimethyl ether, methyl formate, acetaldehyde, glycol aldehyde, ethylene glycol, and ethanol. As the radicals involved as intermediaries in the hydrogenation of CO on grain surfaces are also intermediaries in the energetic processing of methanol, there is some communality in the products of these two schemes. However, different routes are possible in photolysis and the branching ratios are very different. Hence, the resulting product abundances are quite different. In addition, steric factors may inhibit the recombination process. The main product in the photolysis of pure CH_3OH are H_2CO, CH_2OH, and CH_3. Recombination of "hot" radicals leads to formation of CH_3CH_2OH and CH_3OCH_3 formation through recombination of CH_3 with CH_2OH, $(CH_2OH)_2$, and CH_3O.

The prolonged exposure to UV photons (fluences in excess of 10^{17} photons cm^{-2}) in the laboratory experiments gives rise to secondary photolysis products such as CO and HCO. Their recombination with first generation radicals gives rise to CO_2 (CO + OH), H_2CO (HCO + H), and CH_3CHO (HCO + CH_3). Turning back to the more relevant first generation products, the experiments show a high ethanol over methyl formate ratio ($\simeq 5$) and, in contrast to the gas phase, in the solid phase, H loss from the methyl group is apparently much preferred over H loss from the hydroxyl group. That may well reflect the H-abstraction reaction from the methyl group in methanol by photolytically produced hot H atoms. Moreover, upon warm up, ethanol formation occurs at higher temperatures than dimethyl ether formation, presumably because H-bonding of CH_2OH into the methanol network inhibits its reaction with CH_3. In a H_2O ice, the additional H and OH radicals will have little influence on the products. However, diffusion will be more inhibited. The addition of CO on the other hand will first lead to a greatly enhanced production of CO_2 through reaction with OH. In addition, HCO is then, in a way, a primary product. This will greatly enhance – by a factor of 5 in a 10:1 mixture of CO:CH_3OH – formaldehyde, formic acid, acetaldehyde, methyl formate, and glycol aldehyde. Prolonged exposure of CO:CH_3OH mixtures to UV photons reveals the formation of H_2CO_3 through reaction of OH with CO forming HOCO (as a minor channel compared to CO_2) followed by the reaction of HOCO with OH. It should be noted that, for mixtures dominated by CO, experiments show that diffusion of radicals is inhibited and products are largely formed upon warm-up.

The other abundant interstellar ice species to consider is NH_3. Photolysis of pure NH_3 ice produces NH_2 and NH radicals; however, their reaction upon warming leads exclusively to N_2. In the presence of small amounts of H_2O, the reaction of photolytically produced OH radicals with NH_3 leads to a large enhancement of trapped NH_2 radicals, presumably because OH has enough energy to overcome the small activation barrier (1600 K) in this reaction. NH_2 can then react on with OH to form NH_2OH. In a CO matrix, NH_3 is photodissociated into NH_2 and H. The latter reacts with CO to form HCO, and HCO reacts with the amino radical to form formamide, H_2NCHO. The photolysis of complex ice mixtures of ammonia, water, and carbon containing species such as methane, methanol, and carbon monoxide leads to the formation of NH_4^+ and OCN^-.

These laboratory experiments show that prolonged energetic processing of interstellar ice analogues leads to a rich and complex mixture of organics. Typically, a micromole of material is processed, resulting in the formation of some 1000 compounds each at the nanomole level. The kinetics of this chemistry is impossible to unravel but it is clear that pathways are multifold and there is little chemical specificity left after prolonged "cooking." This prolonged exposure to energetic radiation may not be very relevant for molecular cloud conditions where relevant fluences are modest. But there is an interesting question – which will not be further pursued here – whether amino acids and other biologically interesting species could be formed this way. In addition to the rich chemical inventory, UV photolysis of simple ices also results in the formation of a highly refractory residue containing species with masses up to 4000 amu. Hexamethylenetetramine (HMT, $C_6H_{12}N_4$) is a dominant component of these residues (upto 50% by mass). Besides the species mentioned above in the photolysis of CH_3OH mixtures, the chemistry involves the formation of HCOOH, which with NH_3 forms the acid–base pair $HCOO^-$ and NH_4^+. As the ice is warmed, first the ices and then the more volatile compounds in the organic residue sublimate and the characteristic IR bands of HMT appear. Dedicated warm-up experiments of (unphotolyzed) ice mixtures has elucidated the chemical routes involved. They start with the formation of formaldehyde, which reacts with ammonia to form aminomethanol (NH_2CH_2OH). Dehydration of this species in the presence of formic acid leads to methylenimine (CH_2NH), which polymerizes to the protonated cyclic trimer trimethylenetriamine-H^+ (TMT, $C_3H_9N_3$) in a salt with $HCOO^-$ at about 260 K. At 330 K, this salt is converted into HMT. The chemical steps can be summarized as,

$$CH_3OH + NH_3 + H_2O \xrightarrow{\mathrm{uv}} H_2CO + HCOOH + NH_3 \xrightarrow{T>70\,\mathrm{K}} NH_2CH_2OH \quad (11.21)$$

$$NH_2CH_2OH \xrightarrow[\mathrm{HCOOH}]{T>180\,\mathrm{K}} CH_2NH \xrightarrow{T>260\,\mathrm{K}} \mathrm{TMTH^+HCOO^-} \quad (11.22)$$

$$\mathrm{TMTH^+HCOO^-} \xrightarrow{T>330\,\mathrm{K}} \mathrm{HMT} + \mathrm{HCOOH(g)}. \quad (11.23)$$

Formaldehyde and formic acid play key roles in this chemical scheme. With this starting mixture, photolysis is only required to form these species or conversely, if these species are formed through grain surface chemistry, HMT will also be formed. Thermal processing is, however, key as it drives off volatile ices and compounds and assists in overcoming the energy barrier in aminomethanol (500 K), $TMTH^+$ (7300 K), and HMT (11,000 K) formation.

Returning to the formation of COMs, UV photolysis of ices rich in methanol, in small doses, gives rise to a very relevant chemistry as this species donates CH_3 groups. However, in contrast to the gas phase route toward COM discussed above, other radicals will be present (CH_2OH, HCO) as well and a much more diverse chemistry ensues. For small doses, H_2O:CH_3OH ices will primarily produce ethanol with traces of methyl formate. In CO-rich methanol mixtures, aldehydes are a major product. Methyl formate or dimethyl ether are not expected to be abundant products. From this perspective, UV photolysis of interstellar ices does not seem to play a major role in the observed organic inventory of Hot Cores/Corinos. Take the ethanol/methanol ratio as a tracer; the observed ratio CH_3OH/CH_3CH_2OH of 40–100 (Table 11.1) implies only very modest exposure to UV photons, 10^{15} photons s^{-1}. There is a caveat as warm-up occurs on a dynamical timescale in a hot core/corino and small differences in diffusion barriers will control the chemical routes favored. Hence, even more than in the laboratory experiments, diffusion and reaction of H trapped in the ice will come first, followed by CH_3 and then OH. And this aspect has not been fully explored. CO_2 and HCOOH are sometimes considered indicators of the importance of UV photolysis of ices. However, experiments show that there are efficient surface chemistry routes toward these species and they may not be good discriminants. Additional issues to consider are the absence of intermediate radicals such as HCO in cold ices. The dichotomy of N-bearing and O-bearing Hot Cores is also not easily explained in a UV photolysis scenario. Ice photolysis may play much more of a role in a UV-rich environment; e.g. for $A_V \lesssim 4$ magn in a molecular clouds or at the surface of a protoplanetary disk.

Ion–Molecule Chemistry in Interstellar Ices

This is a largely unexplored variant of the chemical schemes where CH_3 is transferred from methanol to other species in ice-loaded, gaseous environments. Accepting that interstellar ices are charged and that these charges are localized, new interesting chemical routes akin to ion–molecule gas phase reactions, but now in the solid state, open up. Specifically, Figure 11.17 illustrates a possible route toward dimethyl ether, based upon molecular beam experiments on methanol droplets where hydrolysis of methanol, facilitated by the Coulomb forces of the charge center, leads to dimethyl ether formation. As for gas phase ion–molecule chemistry, the presence of other species may lead to a diverse organic inventory.

The discussion of the warm up of NH_3 ices with acids such as HCOOH or HNCO already illustrated the importance of ions. In interstellar spectra, the 4.62 μm band is attributed to the OCN^- ion. Also, trapped PAHs will act as electron donor/acceptors and localize charge, preventing electron diffusion and recombination, while trapped atoms such as Na and K can donate electrons. Studies on dipole alignment of molecular films also point toward the importance of electric fields for the evolution of interstellar ices. Laboratory experiments show that, upon warm up to ~80 K, methanol segregates out from the water in ice mixtures. The profile of the 15 μm CO_2 ice band shows that methanol segregation

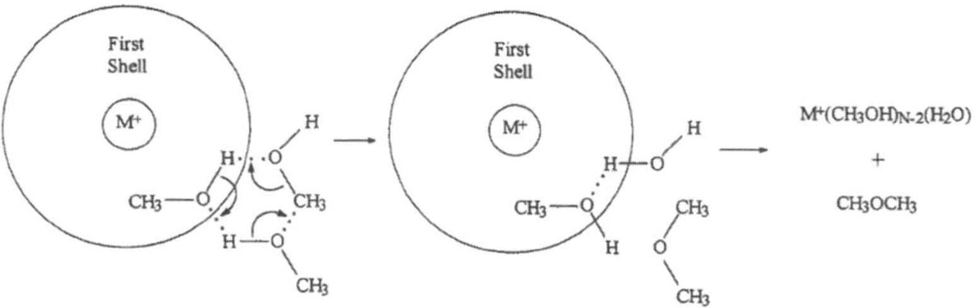

Figure 11.17 Possible chemical route toward molecular complexity in methanol-rich interstellar ices. The circle around the central (alkali) M^+ ion represents the first solvation shell of methanol. Methanol molecules in the second solvation shell and hydrogen bonded to the first are (stereochemically) in position to react. In this way, two methanol species are converted into dimethyl ether and water ($M^+(CH_3OH)_N \rightarrow M^+(CH_3OH)_{N-2}H_2O + CH_3OCH_3$). Figure taken from [249]

can occur in space. Hence, the concept of methanol-rich zones in interstellar ice mantles surrounding charge centers is conceivable (but has not been explored seriously). This variant of ion–molecule chemistry in a droplet-like environment could thus conceivably occur on astronomical timescales ($\sim 10^4$ yr). While there is good experimental support for this reaction scheme in droplets, the importance of methyl transfer without fragmentation in ices is unproven and, at present, this whole scheme is highly speculative. Regardless, this discussion serves as a warning that issues generally ignored may actually be very important: ionization and localization of charges in ices, stereochemistry, thermal segregation of molecular mixtures, molecular solubility in water/methanol mixtures, and solid-state ion–molecule chemistry.

11.4 Protoplanetary Disks

Protoplanetary disks around low mass stars are the site of planet formation in the Universe. We have realized that essentially every star in our galaxy has a planctary system and that there are many planets in the habitable zone of these stars. The wide range of planetary systems and of planetary properties indicates that the processes in protoplanetary disks are rich and varied. It is clear that the origin and evolution of these disks and the planet therein has become one of the key questions in astrophysics. While gravity (and conservation of angular momentum) plays a key role in the formation of protoplanetary disks, their subsequent evolution is largely driven by micro and macro physical and chemical processes involving the gas and dust. The organic inventory of protoplanetary disks, in particular in the habitable zone, is of key interest to the origin of life and the question "Are we alone?" And these questions are very much tied to chemical processes in the gas and on ice grains.

Among the key questions that molecular astrophysics can help address are: "What are the physical conditions in the disk?," "What are the run of density and temperature with radius and with height in the disk?," "What is the mass of the disk?," "What is the velocity field in the disk?," and "What is the turbulent behavior of disks and is it driven by, e.g., the magnetorotational instability?." This section first introduces the physical structure of protoplanetary disks and the key physical processes involved. Next, the chemical processes that play a role in disk formation are discussed. The degree of ionization is at the base of both the turbulence of disks as also ion–molecule chemistry and the organic inventory of the observable molecular zone. Finally, the analysis of molecular observations of protoplanetary disks are summarized.

11.4.1 Physical Processes

Protoplanetary disks are the natural outcome of the gravitational collapse of a rotating molecular cloud core as angular momentum forces material in a disk around the central protostar. Figure 11.18 illustrates the structure of a protoplanetary disk and the different processes that are important. Protoplanetary disks evolve viscously, transporting material inward to the star and angular momentum outward. As molecular viscosity is insufficient,

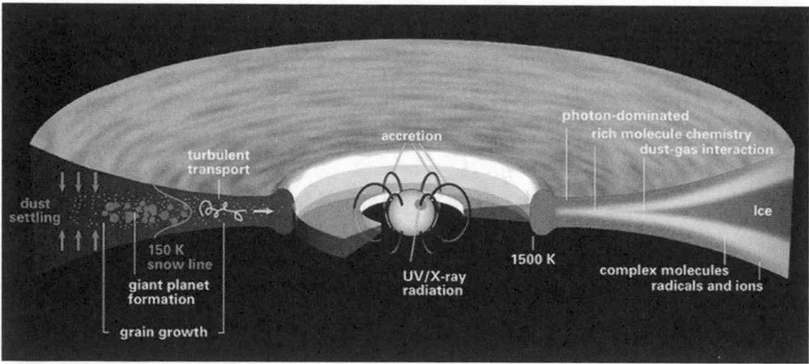

Figure 11.18 Schematic representation of the structure of a protoplanetary disk. The left-hand side summarizes processes that are important in the evolution of the disk and the formation of planets. The right-hand side illustrates the different chemical zones in the disk. Figure taken from [140] (A black and white version of this figure will appear in some formats. For the color version, please refer to the plate section.)

this viscosity is generally presumed to be turbulent in nature and linked to the magnetorotational instability (often abbreviated to MRI). Except for the inner zone where thermal ionization is important ($T > 10^3$ K), the very weakly ionized gas is imperfectly coupled to the magnetic field. As a result, dead zones may develop near the mid-plan where ionization levels are very low. MRI would be dampened and no angular momentum transport would occur accross these dead zones.

The vertical structure of a protoplanetary disk is often approximated by an isothermal (in the vertical direction), thin disk with a negligible mass compared to the star. Hydrostatic equlibrium yields then a gaussian density distribution with,

$$\rho(z) = \rho_0 \exp\left[-z^2/2h^2\right], \tag{11.24}$$

where the scale height is related to the Keplerian orbital velocity, Ω_K, and the sound speed, C_s, through $h = C_s/\Omega_K$ where $\Omega_K = v_K/r$ with $v_K = \sqrt{GM_\star/r}$ the Keplerian rotational velocity. Typically, protoplanetary disks are very thin and the scale height is a small fraction of the radius, $h/r \sim 0.05$.

The radial density distribution is often described by a minimum mass solar nebula,

$$\Sigma_{mmsn}(r) = 1.7 \times 10^3 \left(\frac{\mathrm{AU}}{r}\right)^{3/2}. \tag{11.25}$$

A self similar solution for a viscous disk is provided by,

$$\Sigma(r) = \Sigma_0 \left(\frac{r}{r_c}\right)^{-\gamma} \exp\left[-\left(\frac{r}{r_c}\right)^{2-\gamma}\right], \tag{11.26}$$

where Σ_0 is a normalization parameter, r_c is a radial scaling parameter, and the exponential term describes a cut off to the size of the disk. In some approximations, this surface density roll off is ignored. Resolved dust continuum observations yield γ between 0 and 1 with typically $\gamma = 0.9$ – considerably flatter than the minimum mass solar nebula – and r_c between 5 and 200 AU with typically $r_c = 200$ AU. Molecular observations typically yield gas disks whose size well exceeds that of the dust disk. Typical masses of disks associated with T Tauri stars in nearby regions of star formation are 5×10^{-3} $M_\odot$ and disk-to-star mass ratios of 5×10^{-3}.

The dominant velocity structure of a protoplanetary disk is the Keplerian rotational velocity, v_K. The azimuthal velocity distribution for a static (or slowly evolving disk) is given by,

$$\frac{v_\phi^2}{r} = \frac{GM_\star}{r^2} + \frac{1}{\rho}\frac{dP}{dr}, \tag{11.27}$$

which yields for $\Sigma \propto r^{-1}$ and $T \propto r^{1/2}$,

$$v_\phi = v_K \left(1 - \frac{11}{4}\left(\frac{h}{r}\right)^2\right)^{1/2}. \tag{11.28}$$

So, the small gas pressure gradient produces very small deviations from Keplerian velocities, $\simeq (h/r)^2 = 2.5 \times 10^{-3}$ or 75 m/s at 1 AU. The dust does not experience this pressure gradient and an important consequence of this is that the dust and gas move at different velocities. The aerodynamical drag due to the mismatch in velocity leads then to slow spiraling-in of the dust.

For a disk with viscosity, ν, density fluctuations on a scale Δr are smoothed out on a time scale of $\tau_{vis} \simeq (\Delta r)^2/\nu$. Hence, the viscous timescale of a disk with a characteristic scale size of r is then $\tau_{vis} \simeq r^2/\nu$. The steady-state solution for the disk structure is then (without derivation),

$$\Sigma = \frac{\dot{M}}{3\pi\nu}\left(1 - \left(\frac{R_\star}{r}\right)^{1/2}\right). \tag{11.29}$$

The molecular viscosity of the gas is $\nu = v_{th}\lambda$ with λ the mean free path $((n\sigma)^{-1}$. For a typical disk at 1 AU, we have $n \simeq 10^{15}$ cm^{-3}, $\sigma \simeq 3$ Å^2, and $v_{th} \simeq 10^5$ cm/s, resulting in $\lambda \simeq 3$ cm and $\nu \simeq 3 \times 10^5$ cm^2/s. The molecular viscous timescale is then $\simeq 10^{13}$ yr and not relevant for the evolution of a disk. The principles of a viscous disk can be retained for a disk dominated by MHD turbulence, which is "local" in nature. The viscosity is then typically prescribed by an α parameter,

$$\nu = \alpha C_s h, \tag{11.30}$$

where α is considered to be a more slowly varying function of temperature, density, or radius than ν. In the most simple approximation, α is set to a constant. While this is a useful approach to get insight into the structure of protoplanetary disks, studies of protoplanetary disk evolution studies will have to link the viscosity to the physical condition of the gas through the relevant macro and micro physical and chemical processes. Finally, near the star, the stellar magnetic field shapes the structure. Specifically, accretion on to the star will be along magnetic field lines. For reasonable stellar magnetic field strengths, this region has a size of $\simeq$15 $R_\odot$.

An infinitely thin disk irradiated by a central star with size $R_\star$ and temperature, T_{eff}, which radiates like a black body, will have a temperature distribution, $T = 1.33$ $T_{eff}\,(R_\star/r)^{3/4}$. Such a passively heated, "black body" disk will have a sound speed scaling as $r^{-3/8}$ and a scale height, $h/r \propto r^{1/8}$. While this is a very approximate calculation, this flaring is a common feature of a protoplanetary disk (e.g. T typically falls of more slowly than $1/r$). It implies, of course, that the disk will proportionally intercept more radiation at larger radii and the flare will be even more pronounced. It is somewhat involved to solve simultaneously the equations of thermal balance and hydrostatic equilirium but the results show that near the star ($r/R_\star < 3$) the behavior will be as described above but then changes over to a somewhat flatter dependence $T \propto r^{-0.6}$ for $3 < r/R_\star < 1000$ while for large distances, it goes to the limit $T \propto r^{-1/2}$. A further complication arises because small dust grains are not black bodies and absorb well at short wavelength (optical and UV star light) but radiate inefficiently at longer wavelength (IR). Hence, small dust grains will be hotter than black bodies at a given radius; typically, the dust opacity is parameterized by $\kappa_{dust} \propto \nu^\beta$ where β depends on grain material, porosity, and temperature and ranges between $\simeq$1 and 4. This non–black-body behavior implies for the Planck averaged dust opacity, $<\kappa_{dust}> \propto T^\beta$, and results in hotter dust and a shallower decline of the dust temperature with radius. Finally, for externally irradiated protoplanetary disks, the interior layers will be colder than the photosphere. This is often represented by two separate layers: a top layer heated by the stellar light and a mid-plane heated by the warm dust surface layer, respectively. As an example, a model for a disk around a star with $M_\star = 0.5\ M_\odot$, $T_{eff} = 4000$ K, $R_\star = 2.5\ R_\odot$, and $\Sigma = 10^3\,(r/1\,\mathrm{AU})^{-3/2}$ g/cm^2, has a surface layer with $T \simeq 550\,(r/1\,\mathrm{AU})^{-2/5}$ K and an internal layer with $T \simeq 150\,(r/1\,\mathrm{AU})^{-3/7}$ K.

For a disk with viscous heating, the energy input due to viscous dissipation, $(9/4)\nu\Sigma\Omega^2$, is balanced by radiation, $2\sigma_{sb}T^4_{disk}$, with σ_{sb} the Stefan–Boltzmann constant. This can be rewritten as,

$$T^4_{disk} = \frac{3GM_\star\dot{M}}{8\pi\sigma_{sb}r^3}\left(1-\left(\frac{R_\star}{r}\right)^{1/2}\right), \tag{11.31}$$

which scales as $r^{-3/4}$ far from the inner boundary. This energy release in the inner disk will lead to a vertical temperature gradient as the radiation has to diffuse out. For a highly optically thick disk, the mid-plane temperature is then given by,

$$T^4_{mp} \simeq \frac{3}{4}\tau T^4_{disk} + T^4_{irr}, \tag{11.32}$$

where the optical depth is given by $\tau = \Sigma\kappa_R$, where κ_R the Rosseland opacity, in units of cm^2/g, is normalized on the gas mass and T_{irr} is the surface temperature due solely to irradiation. Observationally, the radial temperature distribution can be represented by a power law with exponent between -0.4 and -0.7; well in the range of these expectations.

The temperature structure has been discussed here in terms of radiative diffusion in a disk whose opacity is dominated by dust and links therefore the dust temperature to the radiation field. The gas temperature will be closely tied to the dust temperature through collisions (cf. Section 10.3 and Figure 10.17). The gas and dust temperature will decouple in the surface layer where direct FUV or X-ray heating of the gas raises its temperature relative to the dust. This is akin to PDRs (Section 11.6). In the inner regions, close to the star, or in the mid-plane when viscous heating is very important, the temperatures may become very high and all dust sublimates. Radiative transfer is then dominated by molecular opacity and cooling has to occur through ro-vibrational transitions in the near-IR.

The thermal structure controls the location of the snow lines.[5] For irradiated disks, the temperature increases upward in the nebula and the snow line is closer to the star in the mid-plane. The H_2O snow line is sketched in Figure 11.18 but there will be a snow line for every major ice component. For H_2O ice, the snow line corresponds typically to a dust temperature of 150 K – or $r \sim 3$ AU for a T Tauri disk – while for CO ice, the sublimation temperature is $\simeq$20 K and hence its snow line is located much further out in the disk ($\simeq$100 AU). For mixed ices, the sublimation behavior is more complex (cf. Section 7.6.1). For trace species mixed into the ice, warm up can lead to clathrate formation and encapsulation into large H_2O ice cages. Phase transitions – such as the high density to low density H_2O ice or amorphous-to-crystalline ice transition – release latent heat that can be used to express inclusions and lead to segregation into homogeneous compounds that will sublimate separately.

This evolution in the physical structure of the disk is accompanied by an (independent) evolution of the properties of the dust. Small dust grains will collide, stick, and grow into aggregates. The outcome of a collision between aggregates will depend on velocity as low velocities will allow hit-and-stick, at intermediate velocities compaction will take over and, at high velocities, collisions will lead to fragmentation of the aggregates into their constituent multimers or even monomers. These critical velocities depend on the structure and material properties. In a general sense, two extremes can be discerned in the aggregation of dust: Particle–Cluster Aggregation (PCA), in which clusters only collide with monomers and Cluster–Cluster aggregation (CCA) where growth takes place through collisions between clusters of similar size and structure. The former will lead to homogeneous and relatively compact aggregates with a filling factor of 0.15. CCA will lead to fractal growth with a fractal dimension of $\simeq$2.

The dynamics of dust aggregates will depend on their coupling to the turbulence in the gas disk and this will scale with the surface-area-to-mass ratio of the aggregate. For compact grains, and PCA growth, the inertia of aggregates will increase with size and these grains will gradually sediment to the mid-plane. In contrast, for CCA growth, the surface-area-to-mass ratio does not really change much and these particles will behave for

[5] The position where sublimation of an ice balances accretion from the gas phase. Its location depends on the dust temperature and the partial pressure of the species in the gas phase.

much of their growth as much smaller grains and are lofted quite high up in the nebula. As particles grow in size and their relative velocities increase, the growth regime will shift from hit-and-stick to compaction and then fragmentation upon collision. When compaction sets in, dust grains will settle quickly to the mid-plane. The outcome of the collisional processes strongly depend on the material properties; Specifically, the larger surface free energy of H_2O ice – linked to H-bonding in the ice network – and the smaller Young's modulus result in a larger deformation zone and an increased contact area. This will lead to build up of much larger aggregates of H_2O ice grains than of silicates (or CO ice). Hence, aggregates can grow to much larger sizes when they are covered by H_2O ice.

As mentioned above, grains will slowly spiral in as they develop a drift velocity relative to the gas. The rate at which this happens depends on the inertia of the grain. The dust stopping time is set by the rate at which it encounters its own mass in gas,

$$t_s = \frac{m_d}{\rho v_{th}} \tag{11.33}$$

Large grains will orbit at close to the Keplerian speed and they will experience a headwind at Δv as the gas is orbiting at this slightly smaller velocity. The rate at which the grain's orbit will decay – relative to the orbital frequency, Ω_K – is given by,

$$\frac{1}{\Omega_K}\frac{dr_d}{dt} \simeq -\frac{2\Delta v}{\tau_s}, \tag{11.34}$$

where we have introduced the Stokes number, $\tau_s = \Omega_K t_s$. Very small dust grains are, of course, coupled to the gas and drift in with the gas at a terminal drift speed,

$$\frac{1}{\Omega_K}\frac{dr_d}{dt} \simeq -2\tau_s\frac{\Delta v}{\Omega_K}. \tag{11.35}$$

Considering a disk where the gas drift is given by $v_\theta = v_K\sqrt{1-\eta}$ (cf. Eq. (11.28)). A general expression for the decay of a particles orbit is then given by,

$$\frac{1}{\Omega_K}\frac{dr_d}{dt} \simeq \frac{1}{\Omega_K}\frac{v_r/\tau_s - \eta v_K}{\tau_s + \tau_s^{-1}}. \tag{11.36}$$

The peak radial drift will occur for particles with $\tau_s \simeq 1$, and such particles will typically drift into the star in some 10^3 orbits. For homogeneous particles, this size is of the order of 1 m; this is the so-called meter-sized barrier as it is difficult to grow particles beyond this size in a dust disk. Much of astrophysics of planet formation is concerned with getting submicron dust particles grow "safely" beyond this meter-sized barrier either by having growth take place in regions of high dust density linked to eddies, (streaming) instabilities, or pressure traps, or the dust–gas separation associated with large scale radial dust drift, or in the cold trap associated with gas cycling across the H_2O snow line, or by keeping particles porous until they have grown well beyond the meter-sized barrier before compacting them and dropping them in the mid-plane. These large boulders can then grow in successive steps; first to km-sized planetesimals, which start to interact gravitationally toward 1000 km-sized embryos and then planets. Planets can have a major (gravitational) feedback on the surrounding gas disk by opening up gaps and creating density concentrations.

11.4.2 Chemical Processes

Armed with the understanding of the structure of protoplanetary disks and the physical processes involved in the previous subsection, we can go back to Figure 11.18 and recognize the importance of turbulent transport through the disk and accretion on the star as well as dust sublimation and snow lines, grain growth, dust settling, and planet formation. This short survey suffices to identify the key questions in the evolution of protoplanetary disks and concommittant planet formation. These questions include: What is the nature and driving force of turbulence in protoplanetary disks and how does that influence dust dynamics and growth? What is the mass, density, and thermal structure of the disk and what are the processes regulating them? Of much interest is also the organic inventory of the disk, particularly in the habitable zone as that is directly tied to the question of the origin of life in the Universe, in general, and on Earth, in particular. Molecular observations can shed much light on these key questions as they directly probe the characteristics of protoplanetary disks and the processes involved.

Much of this is tied to micro-chemical processes occuring in the gas and on grains (Figure 11.18). Close to the star, densities and temperatures will be very high, all dust will sublimate,[6] and chemical equilibrium will ensue. Further out in the disk, we can recognize different layers, each dominated by their own chemical processes leading to distinct molecular inventories. The surface layer will be dominated by penetrating UV and X-ray photons from the central star, in a PDR. Next is the molecular zone where ionization by X-rays and cosmic rays dominate the chemistry. This molecular layer is bounded on one side by the PDR and on the other side by freeze out on cold dust grains. The mid-plane is icy in nature and controlled by gas–dust interaction. Turbulent diffusion will rapidly mix material in the vertical direction but radial transport on a longer timescale is also important.

Equilibrium Chemistry

Studies of the chemistry of the inner solar nebula have long relied on thermodynamic calculations of molecular equilibria. During the earliest phases of accretion, the inner nebula will be very hot, and even the most resilient compounds will sublimate close to the star. As this material subsequently cools down, compounds will start to condense out sequentially. As densities are also very high, the chemistry is well described by chemical thermodynamics (Chapter 5). Condensation will start with the rare-earth elements such as Os, Zr, and Re (Figure 11.19) followed by aluminum and titanium oxides such as corundum (Al_2O_3) and perovskite ($CaTiO_3$) at about 1700 K.[7] These react on with the gases to form spinel ($MgAl_2O_4$), melilite ($Ca(AlMg)(SiAl)_2O_7$), and diopside ($CaMg(SiO_3)_2$) at somewhat lower temperatures (1450–1650 K). Major condensation – around 1400–1500 K – involves condensation of Fe, Mg, and Si and, depending on pressure, this may start with metallic iron condensation followed by olivine (forsterite, Mg_2SiO_4) or the other way around. Forsterite will react on with SiO in the gas phase around 1300 K to form

[6] Even the most refractory material will sublimate around $\simeq$1500 K.
[7] The exact temperature depends on the (partial) pressures of the elements involved (Chapter 5).

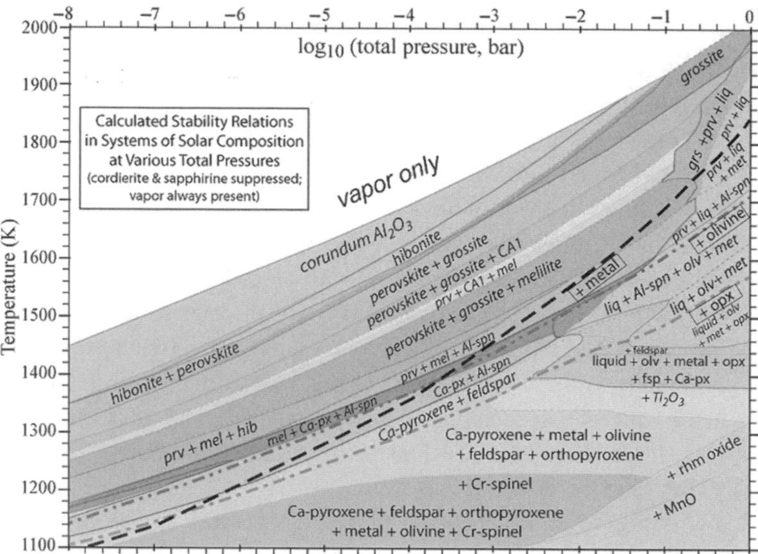

Figure 11.19 Equilibrium stability regime of compounds in a nebula of solar elemental composition calculated using chemical thermodynamics. As the (inner) solar nebula cooled down from high temperatures, different compounds will condens out sequentially. The major condensates of metal, olivine, and orthopyroxene are indicated by dashed, dashed-dot-dot, and dashed-dot lines, respectively. Figure taken from [88] (A black and white version of this figure will appear in some formats. For the color version, please refer to the plate section.)

orthopyroxene (enstatite, $MgSiO_3$). Diopside is converted into Ca-Al-Na silicates such as plagioclases (e.g. anorthite, $CaAl_2Si_2O_8$) and feldspar (e.g. albite $NaAlSi_3O_8$) around 1000–1200 K, while the Mg-rich end members of olivine and pyroxene are converted into mixed-Mg-Fe compounds around 700 K. Upon further cooling, these models predict that metallic iron is converted into sulfides ($\sim$700 K) and oxides, while the silicates react on with water to form phyllosilicates.

This condensation sequence applies to thermodynamic equilibrium. However, not all compounds in the condensation sequence may be accessible as they may involve reactions between solid phases, which are not necessarily in contact or they involve reactions with activation barriers that become insurmountable at low temperatures. The latter is particularly important if nucleation is involved as the first chemical steps in the formation of particles are generally dominated by kinetics rather than thermodynamics.

The primitive components in ordinary chondritic meteorites record the first steps of formation of solids in our Solar System. This calculated condensation sequence bears a striking resemblance to the major mineral assemblages and depletion patterns observed in these meteorites. Chemical and textural evidence attests to the importance of gas–grain interaction in the early chemical history of the inner solar system. However, the various minerals present in meteorites are not in equilibrium with each other and require isolation of high temperature condensates before cooling is complete. Hence, as condensation proceeded and grains grew in size, a fraction ($\sim$1%) of the compounds formed were steadily sequestered away presumably in a dust-rich, gas-poor, mid-plane layer where further reactions were inhibited. This storage was likely associated with a dead zone in the disk to stop their diffusion into the Sun. Eventually, these grains aggregated into planetesimals and then larger bodies. A substantial amount of chondritic meteorite material, including the Ca and Al rich inclusions (CAI) and the chondrules, has seen temperatures in excess of $\sim$1500 K. The mid-plane temperature depends on the accretion rate and the viscosity. Models for the solar nebula attain such high temperatures for the first $\simeq 10^4$ yr out to some 5 AU and for $\simeq 10^6$ yr out to $\simeq$1 AU. Radioactive aluminum-26 (half life of 720,000 yrs) has been used to date the formation of (precursors of) the most primitive CAIs to within a time span of $\simeq 40,000$ yr but reprocessing may have occured over 0.2 Myr. Chondrule precursors are formed typically later, $1-3$ Myr, than the CAIs. The melting events that transformed these precursor minerals into chondrules took place in a dust-rich, colder (< 650 K) environment where transport was unimportant and likely these are associated with disk dead-zones. In contrast, the matrix material in chondritic meteorites contains much material that has never equilibrated with the nebula at high temperatures and – in viscous evolution models – has to represent material that was brought in later.

At the densities (in excess of 10^{12} cm^{-3}) and temperatures (in excess of 1000 K) characteristic for the inner nebula, carbonaceous materials including PAHs will be destroyed through attack by H, OH, and O on timescales of 10^5-10^6 yr, producing, e.g., C_2H_2. While the acetylene can be used to temporarily build up more complex species, eventually, all will converted to CO, CO_2, and CH_4. In this regime, three body reactions become important and the chemistry is more akin to terrestrial environment except that timescales are much longer. At somewhat lower temperatures ($\sim$500 K), Fischer–Tropsch reactions on iron and silicate surface can efficiently convert CO into CH_4. The methane produced is then available for further hydrocarbon chemistry in the gas phase.

The PDR Surface

The surface layers of protoplanetary disks around T Tauri stars are irradiated by strong stellar EUV, FUV, and X-ray radiation fields from the star. At 1 AU from a classical T Tauri star, the FUV radiation field corresponds to $\simeq 10^7$ Habings (Figure 11.20). The FUV flux is dominated by Lyα photons with a minor contribution from continuum ($\sim$10%). Densities in the disk PDR are $\simeq 2 \times 10^9$ (AU$/r$). These are very extreme conditions for a PDR (Section 11.6) and photochemistry dominates the energy balance and molecular composition. The differences in the SED of the incident radiation field will selectively influence the resulting photo-rates with much larger impact on, e.g., H_2O and HCN than H_2 and CO. The high X-ray flux is another difference with galactic PDRs as this will lead to a much higher ionization rate in the deeper, molecular, PDR zones. Large PAHs can still survive under these conditions and heat the gas well above the dust temperature. This leads to a warm molecular atmosphere. The prevalence of radiation as well as He^+ ions that can break C out of CO and N out of N_2 combined with the high temperatures and densities opens up molecular pathways that are normally closed (Figure 11.21). The C broken out

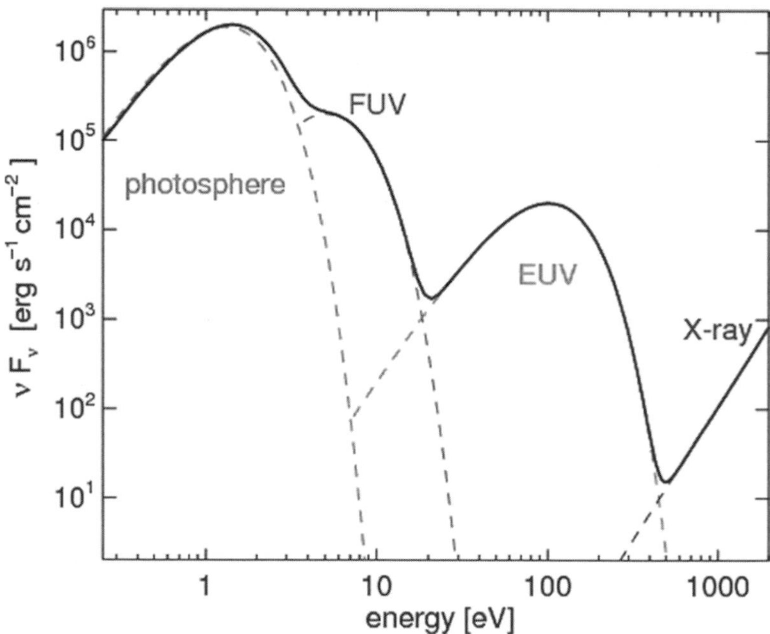

Figure 11.20 Schematic of the radiation field incident on a protoplanetary disk surrounding a classical T Tauri star at 1 AU. The actual radiation field consists mostly of line radiation. For comparison, one Habing field corresponds to 1.6×10^{-3} erg s^{-1} cm^{-2}. Figure taken from [259]

of CO can react with H_2 all the way to CH_4. Intermediate hydrocarbon radicals can react with C en route toward C_2H_2. The reactions with H_2 also open the formation route toward NH_3 and to HCN. These layers are then abundant in C_2H_2, HCN, CH_4, and NH_3 besides, of course, H_2O and CO (Figure 11.21).

Further out in the disk, the temperature and density have dropped and these reaction channels close down again. At that point, regular PDR chemistry takes over (Section 11.6). As dust in the deeper layers is very cold, gas phase species freeze out in ice mantles. When these grains are turbulently transported to the disk photosphere, photodesorption and sublimation will return species to the gas phase. Photo chemistry will then convert these species to atoms and ions and form a surface layer dominated by O and C^+. Models then result in the typical layered PDR structure with an atomic surface layer, which transforms into a radical zone and then a molecular zone. Specifically, C^+ makes place for C and then CO with depth into the disk. In the radical zone, species such as CN are (relatively) abundant. Penetrating X-ray photons can keep up a high ionization rate. Ion–molecule reactions result in high abundances of HCO^+ and N_2H^+.

Two wrinkles have been introduced in these types of models. When the dust temperature is between 20 and 150 K, CO may sublimate but H_2O ice would stay on the grains.

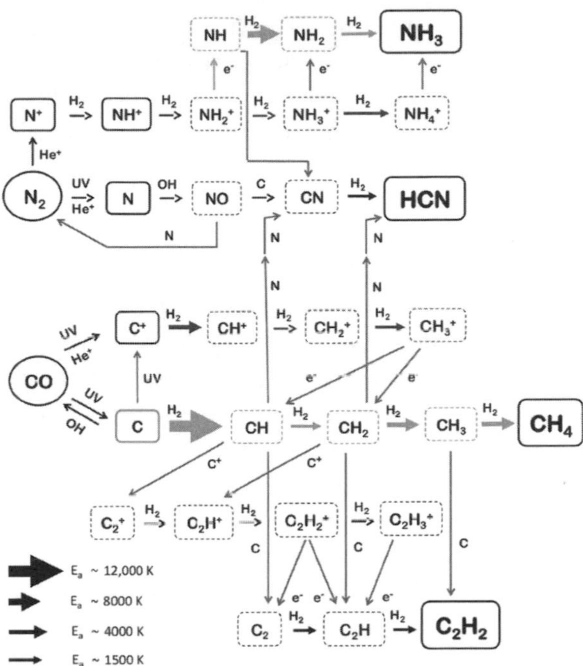

Figure 11.21 Main reaction routes to HCN, C_2H_2, CH_4, and NH_3 in warm gas in protoplanetary disks. Chemistry starts with breaking C out of CO and N out of N_2 through FUV photons or He^+ ions. At high temperatures, the chemical routes starting with the reaction of C with H_2 to CH and NH to NH_2 dominate over lower temperature chemical routes of C^+ going to CH^+ and NH_2^+ going to NH_3^+ and NH_4^+. The different activation energy (E_a) barriers for the chemical reactions are represented by different types of arrows; the thicker the arrow, the higher the activation energy. Figure taken from [23] after [3]

The CO can then be photodissociated and the resulting atomic O converted into H_2O on the grain surface and sequestered in the ice. The C on the other hand would be converted into CH_4, which – like CO – would sublimate back into the gas phase (and be photodissociated there). This would lead to an increase in the gas phase C/O ratio, which might reach values >1 and start a C-rich chemistry. The second aspect is related to the isotopic composition of minerals condensing from the gas phase in the inner part of the disk (see Section 11.4.2). One model for isotopic variations in, e.g., O attributes them to self-shielding of CO. This results in isotope selective photodissociation rates, which are greatly enhanced for the $^{12}C^{17}O$ and $^{12}C^{18}O$ isotopologues as compared to the main species, $^{12}C^{16}O$. Then, if the liberated O is converted into H_2O and locked up in ices, ices will be enriched in the minor O isotopes while the gas phase will be enriched in the main isotope. This process will only work in the outer, temperate zones of the protoplanetary disk but if this enriched gas could then be transported inward, this could lead to an enrichment in ^{16}O and hence in the minerals condensing out from the gas. While isotope selective photodissociation is amenable to precise modeling and accurate photodissociation rates can be calculated, many other processes are involved that are only partly understood, including accretion and reaction of species on grain surfaces, photodesorption and sublimation of (mixed) ices, aggregation and fragmentation of ice grains, and transport of gas versus solids toward the inner solar nebula.

The Icy Mid-plane

Out in the nebula, dust temperatures in the mid-plane will be very low and molecules will freeze out on the dust. Grain surface chemistry has been discussed extensively in Chapter 7 and applied to the formation of interstellar ice inside molecular clouds in Sections 10.6.2 and 10.6.3. The abundance of atomic H is particularly relevant as hydrogenation reactions, in particular CO to H_2CO and CH_3OH, is key to the ice composition. In molecular clouds, the atomic H abundance is tied to cosmic ray ionization of H_2. In protoplanetary disks, X-ray ionization takes over. For a typical T Tauri star at 10 AU, X-rays produce an ionization rate in excess of 10^{-17} s^{-1} for columns of 10^{24} H nuclei cm^{-2}. This produces enough H atoms to hydrogenate all CO in 10^6 yr, provided that CO and H reside on a grain long enough for reactions to occur and that the CO is not sequestered away inside an H_2O ice mantle.

Deeply embedded, low mass protostars have associated Hot Corinos that show the sub-millimeter signatures of complex organic species (Section 11.3.2). These species are thought to originate from sublimating ices. ALMA observations associate these Hot Corino's with the surfaces of protoplanetary disks (Figure 11.9). Hot Corinos are common in class 0 protostars and imply that this phase is rather short-lived ($<10^5$ yr) as these species – once released into the gas phase – quickly burn back to CO. Complex organic species are not very abundant during the later stages of the evolution of young stellar objects. Emission from H_2CO and CH_3OH has been detected toward the nearby YSO, TW Hya, but at low to moderate abundances. Presumably, this reflects that, in the disk photosphere of the outer nebula, photodesorption controls their release into the gas phase but as fragments. Moreover, the strong UV flux will quickly photodissociate these species in the photosphere. Inside the snow line, thermal sublimation may release these species "intact" into the gas phase and drive the chemistry. Curiously, CO_2 should be released from the ices as well but there is no evidence for its presence in IR spectra. The absence of this species in Hot Cores, Hot Corinos, and protoplanetary disks is not understood. Herschel detected H_2O in the spectrum of TW Hya. This emission originates from the cold, outer disk surface layers. As for studies of H_2O emission from molecular clouds, these observations probe the efficiency of photodesorption and reaccretion and grain surface chemistry in protoplanetary disk surfaces rather than the main reservoir of water in protoplanetary disks. Essentially, all of the water that contributes to the formation of planetary oceans is locked up in mm- to cm-sized icy dust grains in the mid-plane and then sequestered away in cometesimals or planetesimals and is therefore not amenable to direct observations.

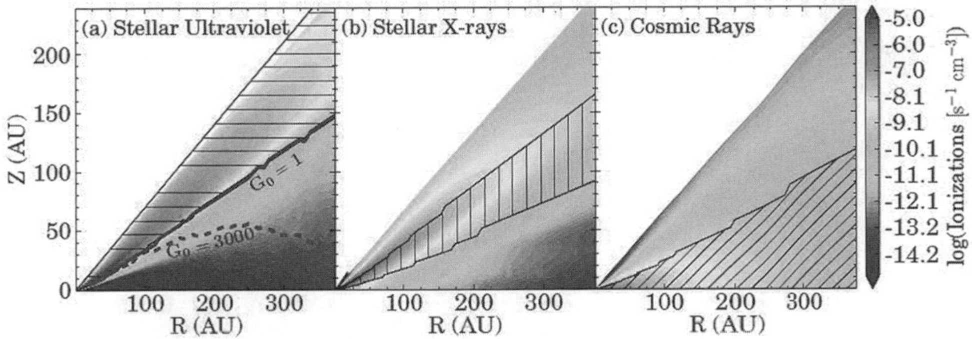

Figure 11.22 The contribution of stellar UV, X-rays, and cosmic rays to the total ionization rate in protoplanetary disks. Colored contours show the volumetric ionization rate due to each source on the same scale. Hatched areas illustrate where each ionizing agent contribute at least 30% of the ionization. Figure taken from [67] (A black and white version of this figure will appear in some formats. For the color version, please refer to the plate section.)

11.4.3 The Degree of Ionization

Ionization is a key aspect of many physical and chemical processes in protoplanetary disks, including turbulence driven by the magnetorotational instability and ion–molecule driven chemistry. Ionization can be due to FUV photons ionizing minor species such as C and S in the disk PDR to a depth of 10^{-2} g/cm^2, to X-rays penetrating to depth of $\simeq$0.5 g/cm^2, to Cosmic Rays penetrating to depth of $\simeq$50 g/cm^2, and to short-lived radionuclides (e.g. ^{26}Al) decay throughout the disk. Figure 11.22 illustrates the regions where these different processes dominate the degree of ionization.

Photo-ionization by FUV photons is discussed in Section 11.6. As the FUV photons couple with a high cross section to the outer electrons of abundant species the degree of ionization is limited only by the abundance of these species and are $\simeq 10^{-4}$ for C in the highest layers and $\simeq 10^{-5}$ for S in slightly deeper layers. Sodium with its low ionization potential and a gas phase abundance of 2×10^{-7} can be ionized even somewhat deeper in. Except perhaps for C, dust attentuation of ionizing photons is the limiting factor and FUV ionization is limited to a thin surface layer, $A_V \simeq 5$. For ISM dust abundances, this would correspond to a column density of H-nuclei of $\simeq 10^{22}$ cm^{-2} but due to dust settling this may be larger in protoplanetary disks. FUV ionization is important as it controls the PDR signatures of protoplanetary disks.

T Tauri stars have enhanced X-ray activity with a typical luminosity of 2×10^{30} erg/s. X-ray photons interact with core electrons, which have a much lower cross section and hence penetrate much deeper than FUV photons. Figure 11.23 shows the results of a radiative transfer calculation for the penetration of X-rays into a disk at different distances from a T Tauri star. Typically, X-rays penetrate to a depth of $3 - 10 \times 10^{24}$ cm^{-2}. For the minimum mass solar nebula, the depth dependence of the ionization rate can be well represented by,

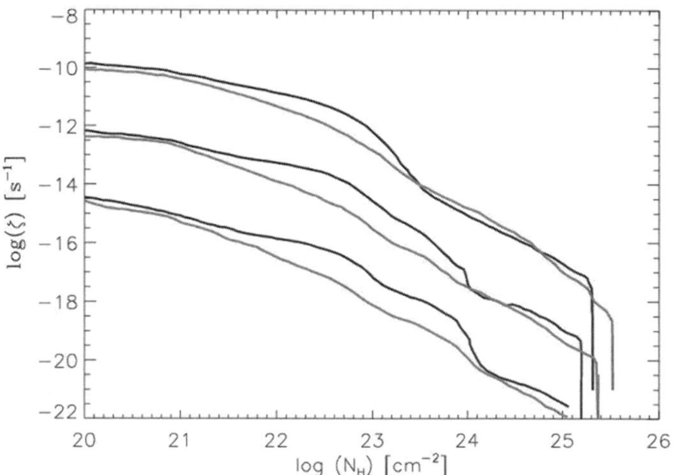

Figure 11.23 X-ray ionization rates plotted versus vertical column density adopting a spectrum appropriate for T Tauri stars at 1, 5, and 10 AU for solar abundances (black) and depleted interstellar abundances (gray). The curves for 5 and 10 AU have been shifted down by factors of 10 and 1000, respectively. Figure taken from [93]

$$\frac{\zeta_{XR}}{n\,(\mathrm{H})} \simeq 10^{-29} \left(\frac{3 \times 10^{24}\,\mathrm{cm}^{-2}}{N_H} \right)^{8/3}. \tag{11.37}$$

Galactic Cosmic Rays are the dominant source of ionization of molecular clouds. Cosmic Ray ionization rates of $\simeq 2 \times 10^{-16}$ s^{-1} per H-nuclei have been measured in the diffuse ISM through molecular absorption and emission lines (Section 9.5). Much lower values, $\simeq 3 \times 10^{-17}$ s^{-1}, have been derived from observations of CO, HCO^+, and DCO^+ in dense cores (Section 10.5.3). This difference in ionization rate likely reflects the attenuation of the lowest energy Cosmic Rays by the surface layers of molecular clouds (Section 10.5.4). As shown in Figure 11.24, for column densities up to 10^{25} cm^{-2} (14 g/cm^2), the attenuation leads to a power law behavior of the ionization rate with column density as the main ionizing agents, low energy cosmic rays, are constantly replenished from the reservoir of high energy Cosmic Rays as they lose energy through interaction with the gas. For higher column densities, the attenuation is best described by an exponential with a critical column density of $\simeq$200 g/cm^2. The Solar wind excludes low energy cosmic rays from the heliosphere. The effects of this on the Cosmic Ray ionization rate are shown in Figure 11.24 for solar maximum and minimum conditions. T Tauri stars have strong stellar winds and their effects are also shown in Figure 11.24 for the range of relevant conditions.

The decay of short-lived radionuclides, such as ^{26}Al with a half live of 700,000 yrs for decay to ^{26}Mg, can potentially provide an important source of ionization in the mid-plane. Studies of Calcium–Aluminum-rich Inclusions (CAIs) in meteorites have revealed that ^{26}Al was live in the early solar nebula with an abundance of 5×10^{-5}. The 1.8 MeV

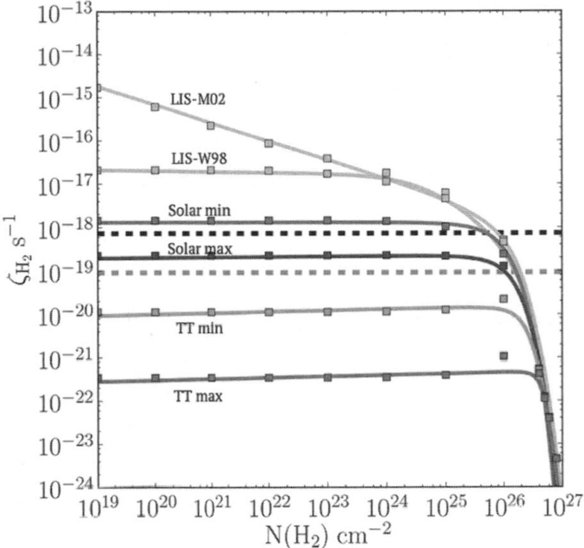

Figure 11.24 Cosmic Ray ionization rates as a function of attenuating column density. The curves labeled LIS-M02 and -W98 are calculated for two extreme cases of the local cosmic ray spectra (cf. [225]). The other curves illustrate the effects of modulation of the penetrating cosmic ray spectrum due to stellar winds. They correspond to the effects of the Solar wind at minimum and maximum as well as estimates of the T Tauri stellar winds at minimum and maximum. For comparison, the heavy dashed line represents the ionization rate due to decay of live ^{26}Al with an abundance inferred from studies of CAIs in the solar nebula while the light dashed line is derived from the mean interstellar ^{26}Al abundance. Figure taken from [67]

γ-ray line produced by the decay of ^{26}Al has been observed to be widespread in the galaxy, implying an average abundance of ^{26}Al of 6×10^{-6} in the ISM. The high abundance of this short-lived radionuclide is maintained by the winds of WR stars and SNe. The much higher abundance of live ^{26}Al in the solar nebula must reflect either the very recent addition and thorough mixing from a nearby SN just before collapse of the parental cloud or spallation by energetic particles near the stellar surface followed by rapid mixing. Figure 11.24 shows the ionization rate derived for these two abundances of live ^{26}Al in a protoplanetary disk. The importance of ionization by the radioactive decay of ^{26}Al will drop exponentially with time after accretion stops.

The ionization balance equates the ionization rate with the recombination rate and this can be solved for the degree of ionization. Ionization scales with the density while recombination scales with density squared. Hence, the key parameter controlling the degree of ionization is ζ_i/n_H with ζ_i the ionization rate per H-nuclei for the relevant process (i.e. cosmic rays, X-rays, or radionuclides). For dense cloud cores (Section 10.5.3), recombination involves a reaction between the electron/PAH anion with the cation (metal cation, molecular ion). As either recombination process scales with the degree of ionization squared, we have,

$$X_e = \sqrt{\frac{2.3\zeta_{CR}}{n\,(\mathrm{H}_2)\,k_{rec}}} \simeq 2 \times 10^{-7} \sqrt{\frac{\zeta_{CR}/n\,(\mathrm{H}_2)}{3 \times 10^{-21}\,\mathrm{cm}^3/\mathrm{s}}}. \tag{11.38}$$

However, in protoplanetary disks, the degree of ionization will drop to much lower values and if there are no PAHs around, gas phase recombination is slow, driving up the dust charge. In that case, the ionization balance for dust grains will set the average charge on grains to be,

$$\langle Z_d \rangle \simeq \tau\psi, \tag{11.39}$$

where $1 + \psi$ is the Coulomb enhancement factor for the grain recombination cross section ($\psi \simeq$3.8 for recombination of heavy atomic cations), and $\langle Z_d \rangle/\tau$ is the grain potential in units of kT ($\tau = akT/e^2$ with a the grain size). Recombination of cations with negatively charged dust grains can then become important in the ionization balance of the gas. The key point is that recombinations scales with the degree of ionization (for the cations) times the density of dust grains (as all grains will be charged). The degree of ionization is then given by,

$$X_e \simeq \frac{\zeta_i/n\,(\mathrm{H})}{k_d x_d} \simeq 6 \times 10^{-11} \left(\frac{\zeta_i/n\,(\mathrm{H})}{10^{-29}\,\mathrm{cm}^3\,\mathrm{s}^{-1}}\right) \left(\frac{1.7 \times 10^{-19}\,\mathrm{cm}^3/\mathrm{s}}{k_d x_d}\right), \tag{11.40}$$

with $k_d x_d$ the recombination rate with dust (per H-nuclei) and n_H the density of H-nuclei,

$$k_d x_d = \pi a^2 \langle v_i \rangle\,(1 + \psi)\; n_d/n_H \simeq 1.7 \times 10^{-19} \left(\frac{\rho_d/\rho}{0.01}\right) \left(\frac{10\,\mu\mathrm{m}}{a}\right) \left(\frac{T}{100\,\mathrm{K}}\right)^{1/2}, \tag{11.41}$$

This holds as long as recombination with dust dominates over radiative recombination of metal ions,

$$\frac{k_d x_d}{\alpha X_e} < 1 \tag{11.42}$$

or

$$X_e < 3.6 \times 10^{-8} \left(\frac{\rho_d/\rho}{0.01}\right) \left(\frac{10\,\mu\mathrm{m}}{a}\right) \left(\frac{T}{100\,\mathrm{K}}\right)^{1/2}. \tag{11.43}$$

Figure 11.25 compares the results for the ionization balance in a PAH and dust-dominated environment. The ionization rate is described by Eq. (11.37) appropriate for X-ray ionization. FUV photo ionization in the PDR only plays a role for column densities much less than 10^{23} cm^{-2} and it is assumed that cosmic ray ionization is negligible because of the T Tauriosphere. For column densities in excess of 3×10^{25} cm^{-2}, the X-ray ionization rate drops precipitously and ionization by radionuclides will take over. The two curves describe the degree of ionization when PAHs are abundant and dominate the recombination process and when large dust grain recombination controls the ionization balance. Observations reveal the signatures of PAHs and (sub)micron-sized dust grains in the spectra of Herbig and T Tauri stars: e.g. their presence in the disk photospheres. While accretion and coagulation are very rapid in the dense midplane, the presence of PAHs and

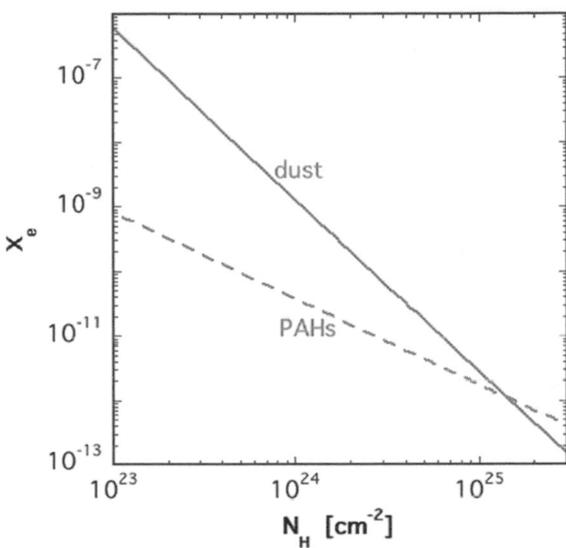

Figure 11.25 Degree of ionization as a function of vertical column density for a protoplanetary disk at $R = 1$ AU. The results for a PAH (dust) dominated ionization balance are shown as dashed (solid) lines. The results for the dust dominated case have been calculated for a minimum mass solar nebula with 10 μm grains, and a heavy atom abundance of 10^{-7}. Figure adapted from [120]

small dust grains in the phtoosphere must imply a balancing process (e.g. "catastrophic" collisions that completely shatter an aggregate) and mixing toward the surface of the turbulent disk. Hence, the PAH/small dust grain scenario may be more relevant.

The degree of ionization is key to ion–molecule reactions in the gas but these reactions are largely confined to the molecular zone that separates the PDR surface from the icy mid-plane. The degree of ionization also controls the magneto rotational instability (MRI), which drives turbulence in the gas. For MRI to operate, the plasma has to couple well to the magnetic field and the neutral fluid has to couple well to the plasma. The first criterion is controlled by the magnetic Reynolds number,

$$Re_m = \frac{C_s h}{\eta} \simeq 1 \left(\frac{X_e}{10^{-13}}\right) \left(\frac{T}{100\,\mathrm{K}}\right)^{1/2} \left(\frac{r}{1\,\mathrm{AU}}\right), \tag{11.44}$$

with η the magnetic resistivity ($C_s^2/4\pi\sigma = C_s^2 m_e \nu_{en}/4\pi e^2 n_e \simeq 234\, T^{1/2}/X_e$ cm^2/s) where σ is the electron conductivity and ν_{en} the electron–neutral collision frequency ($\simeq 8 \times 10^{-10}\, n_n T^{1/2}$ s^{-1}). The critical value to sustain MRI is $R_e \simeq 3000$. The second criterion compares the ion–neutral collision rate with the Keplerian orbital speed (Ω),

$$Am = \frac{k_L X_e n\,(\mathrm{H_2})}{\Omega} = 1 \left(\frac{X_e}{10^{-8}}\right) \left(\frac{n\,(\mathrm{H_2})}{10^{10}\,\mathrm{cm^{-3}}}\right) \left(\frac{r}{1\,\mathrm{AU}}\right), \tag{11.45}$$

where k_L is a typical Langevin collision rate (1.5×10^{-9} cm^3/s). Sufficient coupling requires $Am \simeq 10^2$, which is never attained in disk models. In a weakly ionized plasma,

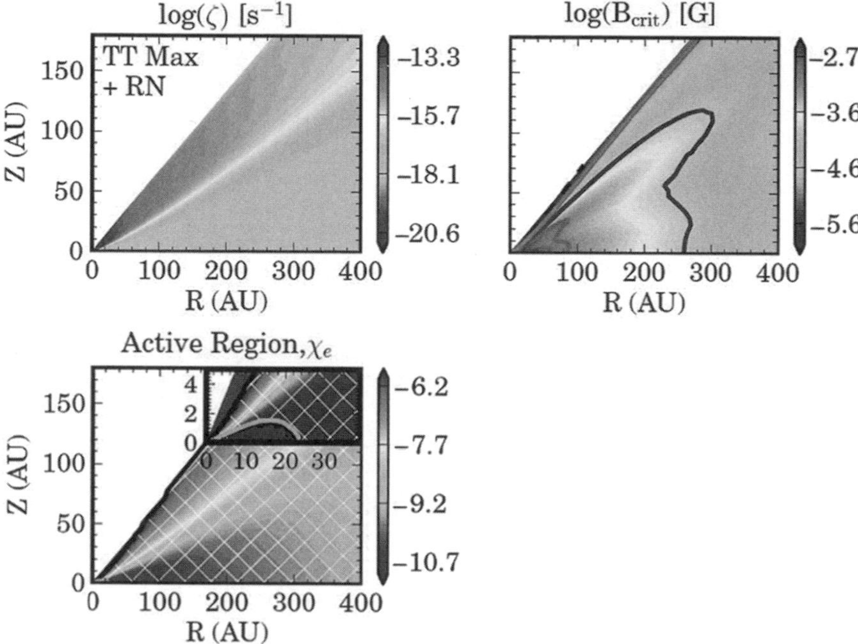

Figure 11.26 Top left: The ionization rate including cosmic rays, X-rays, and radionuclides. Top right: The critical magnetic field involved in the A_m criteria. Magenta, blue, and red lines denote 1 mG, 100 μG, and 10 μG, respectively. Lower panel: MRI active regions are cross-hatched. The inset shows the inner disk with the region that is inactive to MRI (i.e. satisfies R_e but not A_m criteria) outlined in orange. Figure adapted from [67] and kindly provided by Ilse Cleeves (A black and white version of this figure will appear in some formats. For the color version, please refer to the plate section.)

MRI can work (for any Am value) if the ratio of the gas-to-magnetic pressure is larger than a minimum value, β_{min}. Numerical simulations provide the following fit to β_{min},

$$\beta_{min} \simeq \left(\left(\frac{10}{Am^{1.2}} \right)^2 + \left(\frac{8}{Am^{0.3}} + 1 \right)^2 \right)^{1/2}. \qquad (11.46)$$

This results in a maximum value for the magnetic field, which for typical disk models ranges between $B\,(50\text{AU}) = 0.3\text{--}5$ mG. The MRI active and dead zone resulting from these three criteria are illustrated for a typical disk model in Figure 11.26. As this figure illustrates, the mid-plane of the inner disk is a dead zone for MRI. This is a common characteristic for models of protoplanetary disks and directly reflects that ionizing agents do not penetrate that deep.

11.4.4 Observations

Molecules provide powerful probes of the physical and chemical characteristics of protoplanetary disks. In this subsection, observations of protoplanetary disks and their analysis

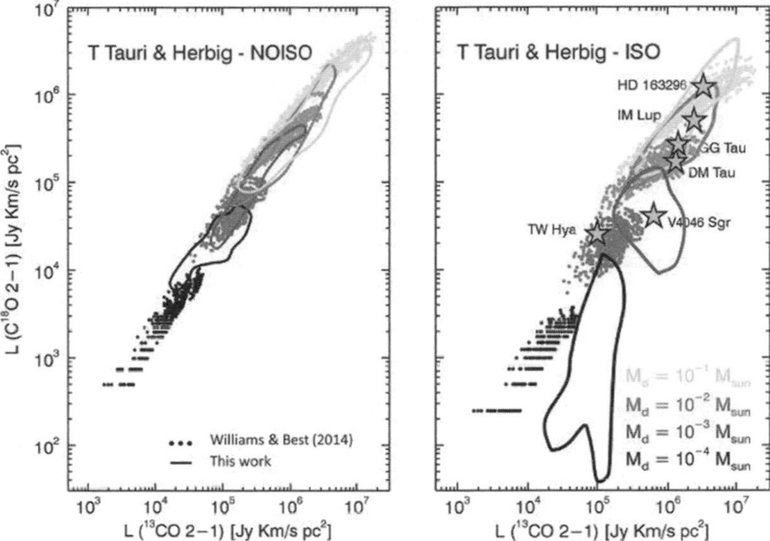

Figure 11.27 Calculated line luminosities of $C^{18}O$ versus ^{13}CO for a large grid of models. Different gray scales represent models with different disk gas masses. Small dots show parameterized models and the solid lines outline model results with a more extensive chemistry. There are also differences in the adopted disk structure between these models. Left: models where isotope-selective processes are not considered (NOISO). Right: models where isotope-selective processes are implemented (ISO). The parametrized models have a $C^{18}O$ abundance reduced by a factor of three to "mimick" the effects of isotope-selective photodissociation. Observations of six well-studied disks are indicated by stars. Figure taken from [199] (A black and white version of this figure will appear in some formats. For the color version, please refer to the plate section.)

are summarized. The emphasis is on deriving key parameters of protoplanetary disks, notably the disk mass, the depletion of molecular abundances related to freeze out – including the snow line – the Keplerian and turbulent velocity fields, and the properties of gas in the inner disk. Throughout this section, it should be kept in mind that the complex thermal, density, and chemical abundance structure of protoplanetary disks greatly complicates the interpretation of this data. As these data are often analyzed by different authors using very different physical and chemical models, comparison of the different results is challenging. Comprehensive models have been developed to assess the importance of these uncertainties in a coherent framework.

Dust Masses

Observations of the thermal emission from dust can be translated into a dust mass. For optically thin emission from dust particles in an isothermal disk, we can write,

$$M_{dust} = \frac{d^2 F_d(\nu)}{\kappa_d(\nu)\, B(\nu, T_d)}, \tag{11.47}$$

where F_d is the observed flux, κ_d is the dust opacity per unit dust mass, B is the Planck function at frequency ν and dust temperature T_d, and d is the distance. For particles that are small compared to the wavelength, the opacity per unit mass is not very sensitive to the size but cm-sized or larger pebbles will be missed by millimeter observations. Assuming a gas-to-dust ratio (typically the ISM value of 100 is adopted), the gas mass is derived. Derived masses are not overly sensitive to the adopted temperature as millimeter emission is on the Rayleigh–Jeans tail of the Planck function. Surveys of nearby regions of star formation in Taurus–Auriga and Ophiuchus yield disk masses of 5×10^{-3} $M_\odot$ (or disk/star mass ratios of 5×10^{-3}) with large variations, likely reflecting differences in dust properties, temperature, and/or disk evolution. Interferometry studies provide information on the radial structure of the disk.

Gas Masses

Gas masses can be traced through CO observations. The main isotope is optically thick but ^{13}CO and C^{18}O can potentially trace gas into the mid-plane. CO emission depends, though, on the temperature and density structure of the disk. Analysis of these observations is then hampered by freeze out of CO in the mid-plane whenever the dust temperature drops below 20 K and by (isotope selective) photodissociation near the surface. Detailed model analysis is then required to assess the importance of density and temperature gradients in the disk, the geometry, and the chemistry involved (Figure 11.27). These models reveal that there is quite some sensitivity to the disk structure and the chemistry involved. Simple fits to these models have been derived that allow conversion of observed fluxes (or flux ratios) into CO masses. Adopting standard CO abundances results then in the gas mass of the disk. Analysis of ALMA CO surveys of nearby star-forming regions reveal surprisingly low CO masses, 0.03 – 1 $M_{♃}$ ($3 \times 10^{-5} - 10^{-3}$ $M_\odot$). This is graphically summarized in Figure 11.28, which shows that CO underestimates the disk mass by typically a factor of 10. This depends somewhat on the model adopted. In principle, the gas mass could be much

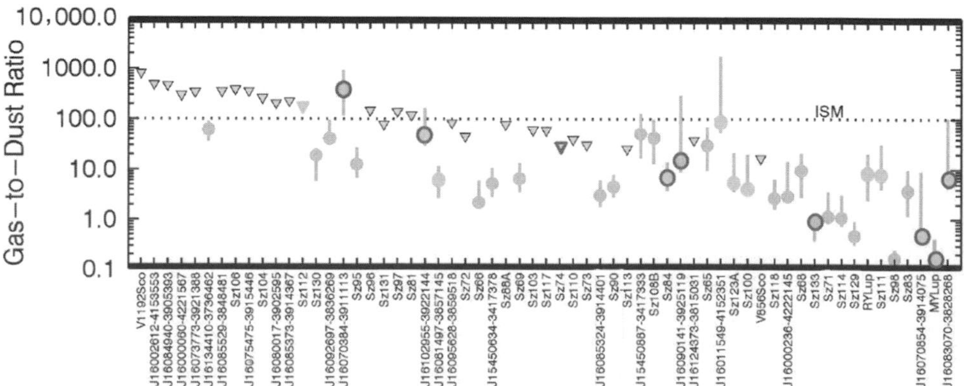

Figure 11.28 Gas-to-dust mass ratio for disks in Lupus. Circled dots are edge on. Upper limits on the CO fluxes are indicated by inverted triangles. The ISM value of 100 for this ratio is indicated. Figure taken from [200]

less than the dust mass if the gas has largely dispersed. However, the implied dispersal time of 10^5 yr (for $\dot{M} \simeq 2 \times 10^{-9}$ $M_\odot$/yr) is implausibly short. Likely, a chemical process is implicated that converts CO into other hydrocarbons, possibly hydrogenation of CO to CH_3OH on grain surfaces and sequestering into ice mantles.

HST/COS UV absorption line can study absorption lines in the Lyman band of H_2 and the $A - X$ bands of CO for heavily inclined disks. For the T Tauri star, RW Aur, which has a disk inclined by about 80° to the line of sight, the derived CO/H_2 abundance ratio is the canonical value of 1.6×10^{-4} and there is no indication of CO depletion in the inner disk. In any case, CO does not seem to be an accurate tracer of the gas mass of protoplanetary disks.

Given the chemical and excitation issues with CO observations, other tracers have been looked for. The ground-state HD line at 112 μm is in principle a good candidate as its abundance should trace that of H_2. In contrast to H_2, the (weak) dipole moment (8.9×10^{-4} D) of HD results in allowed rotational transitions. To derive a gas mass, a temperature has to be assumed: often taken from a disk model sometimes supported by CO rotational line analysis. There is a strong temperature dependence, though, for the derived gas mass as the $J = 1$ level is 128.5 K above ground and really only the warmish gas mass is measured. Analysis of the observations require then detailed thermal models to convert the measured warmish gas mass to the total gas mass and this becomes very model dependent. Also, potentially, some of the D may be locked up in dust or PAHs – as seems to be the case in the local diffuse ISM – and that may introduce another factor of 2 uncertainty. The HD line has been observed with the PACS instrument on Herschel for three T Tauri disks, resulting in disk gas masses about an order of magnitude larger than derived from CO observations.

The Velocity Field

The CO and dust disk emission can be resolved with interferometers. Figure 11.29 shows the SMA data for the well-studied, nearby, relatively massive disk of the T Tauri star, TW Hya. The first moment CO map for this disk reveals the velocity signature of a Keplerian rotating disk. However, the CO emission implies a disk with a radius of $\simeq$200 AU. In contrast, millimeter dust continuum emission reveals a dust disk with a sharp edge at 60 AU. This difference in size may reflect the slow inward migration of mm-sized dust grains under influence of the head wind in the gas disk (cf. Section 11.4.1).

Snow Lines

The location of the H_2O snow line is particularly important for planet formation due to the increase in dust mass as well as "stickiness" of small dust grains associated with H_2O ice. Ground-based, Spitzer, and Herschel observations can trace water emission from protoplanetary disks in the near-, mid-, and far-IR, respectively. Because of freeze out in the deeper layers, the increase in temperature with height, the high incident UV field, and the rapidly increasing dust opacity with depth into the disk, molecular emission lines generally arise from intermediate layers rather than the mid-plane. Figure 11.30 illustrates this for a high energy transition and the ground-state transition of ortho H_2O. As discussed in

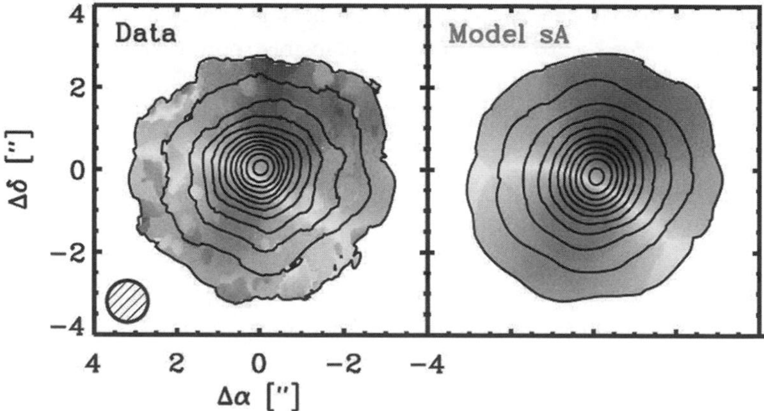

Figure 11.29 Velocity moment map of the observed (left) and model (right) CO $J = 3-2$ emission of the TW Hya disk. Contours are the velocity integrated CO intensity (at 0.4 Jy km/s, corresponding to 3σ). Color scale represents the intensity-weighted velocity. Figure taken from [12] (A black and white version of this figure will appear in some formats. For the color version, please refer to the plate section.)

Sections 10.6.3 and 11.4.2, the measured H_2O abundance is then mainly set by gas–grain interaction; i.e. freeze out, thermal- or photodesorption, and photodissociation as well as gas phase H_2O formation in warm gas and disk processes such as grain coagulation and fragmentation, dust settling, and turbulence.

Carbon monoxide is another major condensate whose snow line may affect planet formation as well as the chemical inventory of protoplanetary disks. Pure CO condenses out at about 20 K and its snow line is therefore located at $\simeq$20 AU in a T Tauri disk. As for H_2O, the bulk of the CO emission originates from the UV-irradiated disk surface layers and does not probe directly the mid-plane snow line. Indirectly, the CO snow line can be traced through N_2H^+ observations. N_2 has a lower proton affinity than CO (Table 6.7) and hence the abundance of N_2H^+ is low in regions where CO is abundant. This has been discussed in connection to depletion in dense cores in Section 10.6.3. Given the low electron density, this correlation of bright N_2H^+ emission with regions of severe CO depletion is even more pronounced in protoplanetary disks. Figure 11.31 shows observations of the protoplanetary disk around TW Hya. Model analysis of these observations places the CO snow line at 30 AU in this disk, somewhat further out than expected from the dust temperature structure derived for this disk. This may reflect photochemistry leading to N_2/CO abundance variations in the surface layers of the disk. The location of the snow line is also affected by diffusion, driven by, e.g., turbulence, mixing CO into the colder outer disk and N_2H^+ into the inner disk on a timescale of $\simeq 10^4$ yr.

Warm Gas in the Inner Disk

Mid-IR spectroscopy with Spitzer has revealed that emission from H_2O, HCN, CO_2, C_2H_2, and OH is widespread among the T Tauri stars but not Herbig Ae/Be stars. The observed

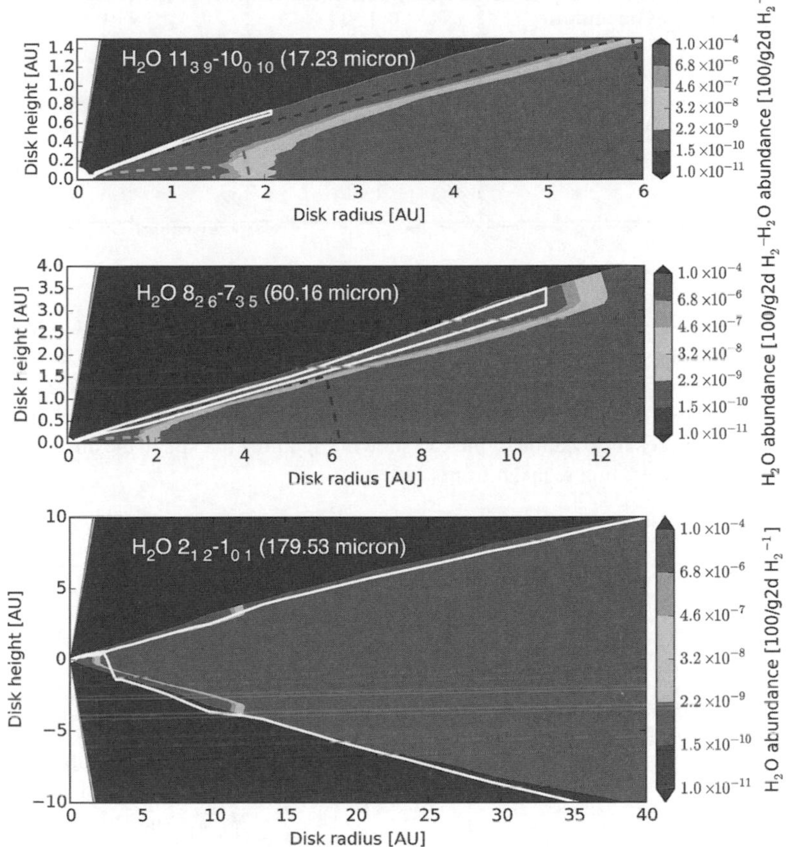

Figure 11.30 Line-emitting regions for three water lines superimposed on the fractional water abundance (color scale). The solid lines indicate the disk region responsible for 10 and 90% of the calculated intensity at line center. Dashed curves indicate water column densities of 10^{18}, 10^{19}, and 10^{20} cm^{-2}, respectively. Observer faces the disk from the top. Figure taken from [32] and kindly provided by Klaus Pontoppidan and Sandra Blevins (A black and white version of this figure will appear in some formats. For the color version, please refer to the plate section.)

emission originates from gas at 500−1000 K. Follow-up, high spectral resolution spectroscopy locates the emission in the inner (0.1−10 AU) region of a Keplerian disk. This will be a prime hunting ground for MIRI/JWST. Ground-based near-IR observations can trace the disk in the ro-vibrational transitions of the OH stretching modes of H_2O and the fundamental and overtone transitions of CO. These features trace very hot ($\sim$1000−1500 K) gas from the inner $\simeq$0.05−10 AU of the disk. While water will be abundant throughout the disk, the emitting zone must be located in the upper regions of the disk as the dust disk is optically thick at these wavelengths. The derived H_2O abundance is less than expected based on the CO abundance, probably reflecting photo-processing of the gas in the disk photosphere. The detailed profiles of the $\Delta v = 2$ CO bandheads reveal the presence of

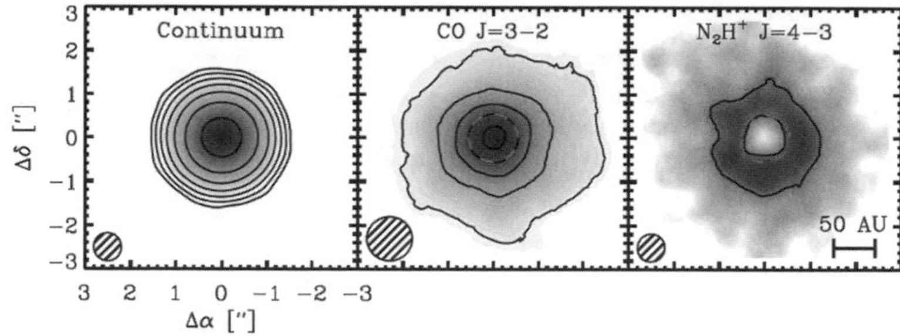

Figure 11.31 Dust, CO, and N_2H^+ emission observed toward TW Hya by ALMA. Contours mark factors of 2 in intensity starting with 5 mJy/beam (dust), linear increase from 1 to 5 Jy km/s beam^{-1} (CO), a single contour of 150 mJy km/s beam^{-1} (N_2H^+). Beam sizes and a physical scale are also indicated. Note the central hole in the N_2H^+ emission, which contrasts with CO and dust emission. The dashed ring indicates the location of the CO snowline required to explain the difference in spatial structure for these species. Figure taken from [240]

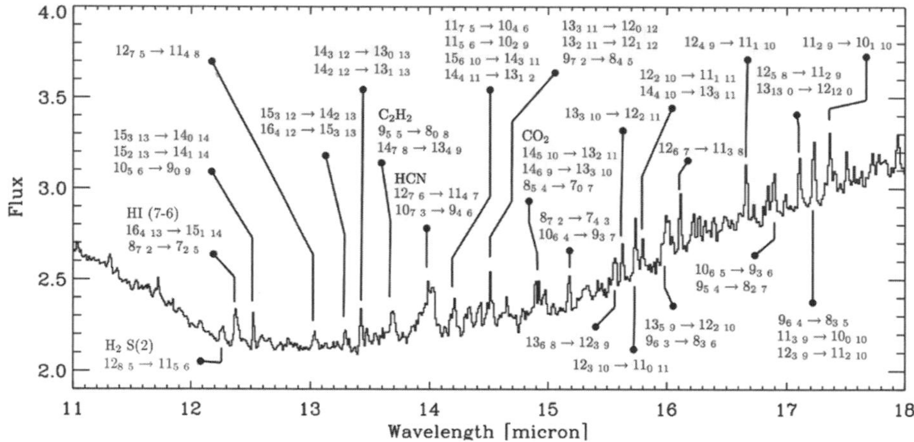

Figure 11.32 The Spitzer mid-IR spectrum of the T Tauri star, RNO 90, illustrate the richness of spectral features due to, in particular, blended lines of H_2O. The Q branches of CO_2 and HCN are also evident. Identifications are as indicated but only strong transitions have been included. The [NeII] line as well as H_2 and HI transitions are also present. Figure taken from [239]

turbulent broadening with derived velocity dispersions in the range 7−15 km/s for the gas in the inner disk. With a judicious choice of species and transitions, pure rotational lines can be used to trace the vertical extent of turbulence in the outer disk. ALMA studies of CO and $C^{18}O$ ($J = 2{-}1$) and DCO^+ ($J = 3{-}2$) have shown that, in contrast to these near-IR lines, turbulent line broadening is of minor consequence with an upper limit of $\simeq 0.05 C_s$, corresponding to an α parameter $<2 \times 10^{-3}$.

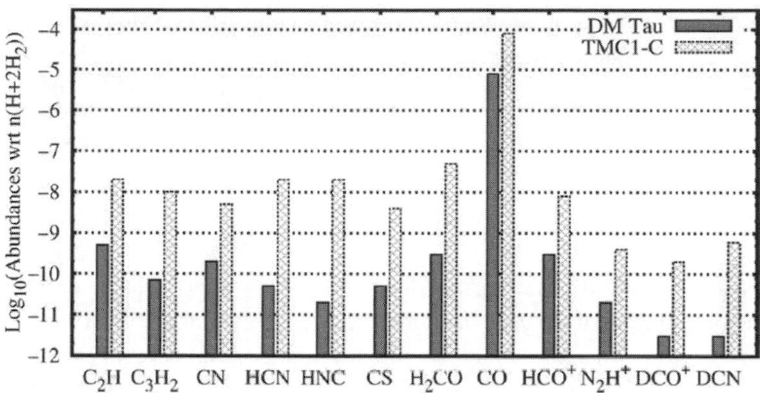

Figure 11.33 Derived abundances in the protoplanetary disk of the T Tauri star, DM Tau, compared to those in the dark cloud TMC 1-C. Figure taken from [140]

Molecular Abundances

A variety of molecules have been detected through their pure rotational transitions in the (sub)millimeter. Analysis of this data requires detailed model studies. As Figure 11.27 illustrates for CO, resulting abundances depend sensitivily on the adopted disk properties and are typically uncertain by a factor $\sim$3. Figure 11.33 summarizes abundances derived for the T Tauri star, DM Tau. Typically derived abundances are down by one to two orders of magnitude compared to dark cloud cores. As for CO, this likely reflects the importance of freeze out in the mid-plane of protoplanetary disks.

The Incident Radiation Field

The intensity and spectral energy distribution of the FUV radiation field has a great influence on the observable characteristics of the PDR surface layer of T Tauri disks. High resolution and high S/N HST observations can probe the characteristics of the FUV radiation field directly, albeit that corrections have to be made for absorption in the HI Lyα line and fluorescence in H_2 and CO lines. The results for 16 classical T Tauri stars show that the average radiation field is 10^7 Habing fields at 1 AU. About 90% comes from the Lyα line while the remainder is due to continuum emission. The latter is mostly ($\simeq$95%) longward of the H_2 limit of 1110 Å. This FUV radiation results from the accretion process onto the star.

11.5 Astrochemistry and the Solar System

Meteorites and comets reveal a diverse organic composition that was likely also sampled by the early Earth. This organic inventory contains a record of the processes relevant for their origin in the solar nebula and clues to the organic inventory available to exoplanetary systems. This was discussed in Section 1.3 (see Figures 1.1 and 1.13). The evidence for an interstellar heritage resides in a similarity of the molecular composition between these solar system objects and star-forming regions in molecular clouds. In addition, their isotopic signature reveals formation in a low temperature setting and the ISM is then an obvious location to seriously consider. This is a rich field of research that is only briefly touched upon here.

11.5.1 The Cometary Organic Inventory

The chemical inventory of comets provides a window on the early phases of the assemblage of the Solar system and can be used to address where interstellar chemistry stops and parent body chemistry starts. In addition, the organic inventory of comets is of interest since some 10% of the volatile budget may have been delivered by comets during the late bombardment and these molecules could also be relevant to jump start life on the early Earth. Comets as they approach the Sun will start to sublimate and the molecular composition of the resulting coma can be studied through the rotational spectra in the sub-millimeter and the fluorescence bands in the visible and near-IR.

The chemical inventory of comets is very rich and diverse and we recognize complex organic molecules that are commonly observed in Hot Corinos around low mass protostars, species that are present in the warm photospheres of (inner) protoplanetary disks, and species that are characteristics for grain surface chemistry and interstellar ices (Figure 11.34). Each of these reservoirs and the chemical processes involved may have contributed to the cometary composition. However, these species are not merely inherited from the parent molecular cloud. Specifically, species such as methanol, formaldehyde, formic acid,

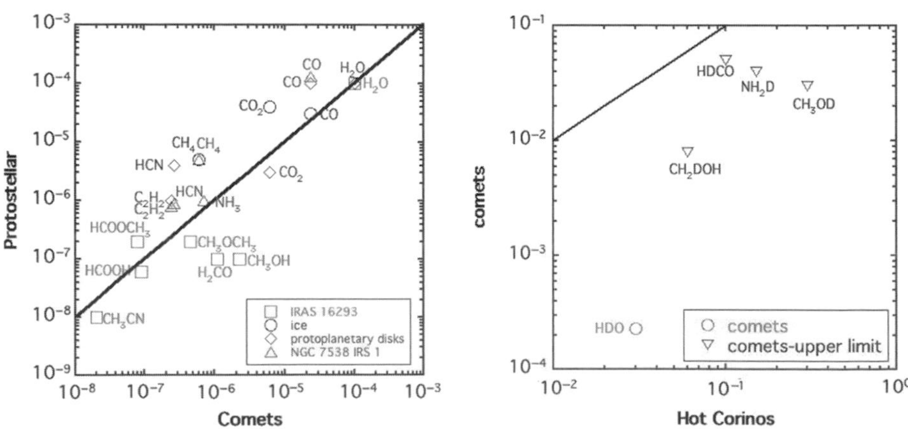

Figure 11.34 Comparison of the chemical composition of solar system comets with that measured in molecular clouds. Left: Molecular abundances. Squares: observations of the Hot Corino in IRAS 16293-2422. Circles: ice measurements in dark clouds. Diamonds: molecular abundances observed in the photospheres of protoplanetary disks. Triangles: mid-infrared absorption line studies toward Hot Cores. Right: Deuterium fractionations in comets compared to those of the Hot Corino in the low mass protostar, IRAS 16293-2422.

methyl formate, and formamide are present in comets at abundances quite similar to those in Hot Corinos. However, the deuterium fractionation of these species is typically much less in comets than in Hot Corinos (Figure 11.34) and this suggests formation (or equilibration) at much higher temperatures, probably in the inner Solar nebula close to the snow line where temperatures are high enough to drive isotope exchange with the main deuterium reservoir kinetically. This could entail a complex sublimation/distillation process whereby interstellar ice grains evaporate when they approach the snow line ($\simeq$4 AU for the Solar system), and the gasses largely/partly equilibrate their deuterium fractionation signature in the warm gas but this then would have to leave their chemical signature in complex organic species intact. Such a model would require extensive circulation of material in the Solar nebula from the warmer inner regions to the cold, outer realm of comets. The high abundance of the hydrocarbon species, acetylene and ethane, in comets is not an indicator of ice chemistry or a Hot Corino signature. Likely, this also indicates the importance of high temperature chemistry in the warm inner nebula (Figure 11.21). Their presence in comets also suggests extensive mixing of material in protoplanetary disks.

11.5.2 Isotopic Signatures

The H, C, and N isotopic composition of the Earth differs from that of the bulk of the solar system and this contains a signature of the reservoir from which this volative inventory was derived. A discussion on the original reservoir is generally started by comparing measured isotopic ratios for different solar system bodies. Figure 1.2 compares the H and N isotopic composition of the Earth with that of different meteorite classes and individual comets. The D/H and $^{15}N/^{14}N$ ratios of the Earth compare well with some asteroidal (meteoritic) values and are distinctly lower than comets. Jupiter family comets do better in the D/H ratio but not for the N isotopes. Of course, the D/H ratio of the Earth is not well known as the equivalent of some two Earth oceans is locked up in the mantle (as hydrated minerals) with a D/H ratio less than the Standard Mean Ocean Water value. The same issue may hold for nitrogen. Here, I mainly want to reiterate that the link to interstellar reservoirs is distant as the measured deuterium fractionation on the Earth is considerably less than that measured in molecular cloud environments and clearly much processing has occured in the solar nebula or planetary body environment.

Figure 11.35 summarizes carbon and hydrogen isotopic composition of different classes of organics in meteorites. This diversity in isotopic composition immediately implies the presence of multiple reservoirs and/or processes involved. Moreover, the C isotope values depend on the chain length, suggesting a kinetically controlled synthesis of higher molecular weight compounds from simpler precursors, presumably in the warm aqueous environment of the early asteroidal history.

11.5.3 Amino Acids

Figure 11.36 illustrates possible chemical processes involved in the synthesis of amino acids in meteorites. In molecular clouds, kinetics drives carbon into CO. In order to

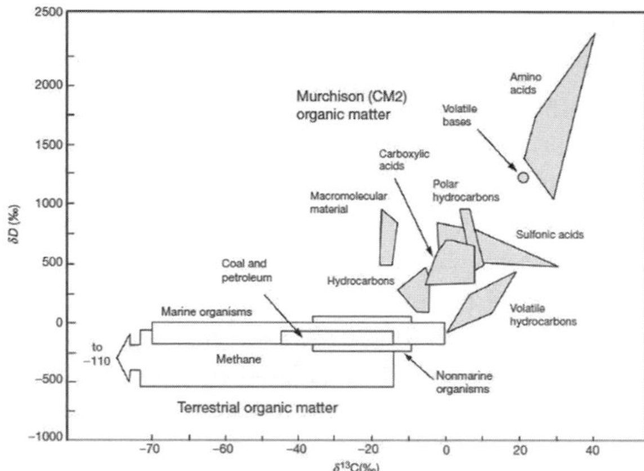

Figure 11.35 Hydrogen and carbon isotopic compositions of classes of organic compounds in meteorites indicate the presence of several initial reservoirs or processes involved in their formation. The comparison to terrestrial compounds illustrates that those are quite distinct from Earth. Figure take from [253]

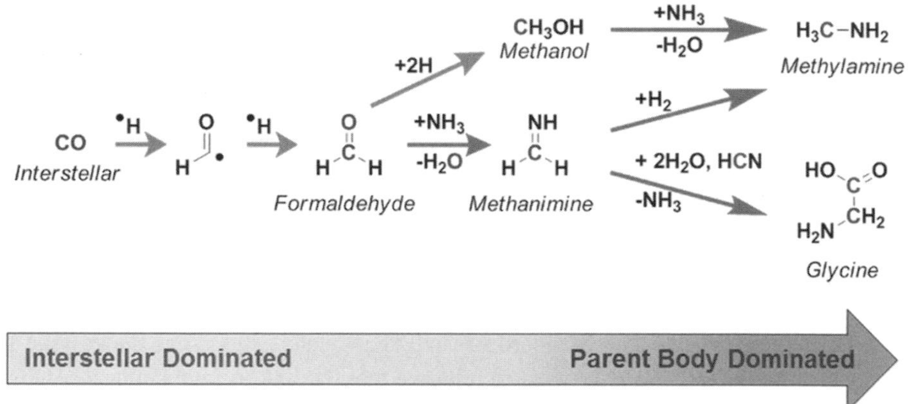

Figure 11.36 Processes relevant to the chemical inventory of meteorites involve a combination of chemistry in molecular clouds and in asteroids. Here, hydrogenation of CO (left) is combined with the Strecker synthesis (right) and amino acids are formed. In this example, the feedstock (H_2CO, NH_3 and HCN) are products of interstellar chemistry – some formed on grain surfaces others in the gas phase – while Strecker synthesis occurs in the aquaous environment of the meteorite parent body. Figure taken from Aponte, J., et al., 2017, *Earth and Space Chemistry*, 1, 3.

start organic chemistry, the carbon in CO has to be activated (cf. Sections 10.5, 10.6.2, and 11.3.2). Cosmic rays produce UV photons and He^+ ions that can both dissociate CO. However, these processes are very inefficient and other carbon-bearing species are less abundant by three orders of magnitude or more. Chemistry on interstellar grain surfaces provides an alternative, and very efficient, way to "jump start" organic chemistry. On grain surfaces, carbon monoxide is readily hydrogenated by atomic hydrogen to form formaldehyde and methanol (Figure 11.36). At the same time, atomic nitrogen is converted into ammonia and these species are quite abundant in interstellar ices (at the few percent level compared to water). Formaldehyde and ammonia are key ingredients for Strecker synthesis of amino acids. This does require hydrogen cyanide, which is a product of gas phase reactions and in interstellar clouds has an abundance less than one part in ten thousand compared to gaseous CO or solid H_2O. Asteroids that are sufficiently distant from the Sun can incorporate these species. The decay of ^{26}Al will heat an asteroid to a few hundred degrees, melting the water, and driving further chemical evolution, in particular, promoting Strecker synthesis. In such an environment, condensation reactions of aldehydes with amonia lead to imine formation. Reaction with a cyanide leads then to amino-nitrile formation, which can then be hydrolysed to an amino acid.

These steps are illustrated in Figure 11.36 for the formation of the simplest amino acid, glycine. Where in this scheme the transition from interstellar to asteroidal chemistry occurs is unclear. In particular, photolysis of interstellar ices consisting of simple molecules can lead to amino acid formation as well – probably in very similar steps – and processing in the aqueous environment might be sidestepped. Indeed, the presence of glycine in the samples returned by the Stardust mission and in comet 67P/Churyumov-Gerasimenko visited by the Rosetta mission suggests that there is a low temperature ice-mediated route toward amino acids. As an alternative, the Urey–Miller experiment illustrates that under reducing conditions, electrical discharges in a CH_4, NH_3, H_2O atmosphere will produce H_2CO. In an aqueous environment, amino acids can then be produced through Strecker synthesis.

Strecker synthesis is not the only option for amino acid formation. Much of the carbon in space is locked up in large Polycyclic Aromatic Hydrocarbon molecules. If this carbon could be activated, other synthesis routes might open up. Photolysis of small PAHs in ices is one such route toward amino acids. Thermodynamic equilibrium calculations suggest that small PAHs will be converted into simple amino acids in a warm aqueous environment. The amino acid glycine has also been recovered from material brought back from comet Wild 2 by NASA's stardust mission. Likewise, ESA's Rosetta mission detected glycine in the coma of comet, 67P/Churyumov-Gerasimenko. Possibly, the presence of glycine in these comets indicates energetic processing of interstellar ice grains. Glycine has been searched for in molecular clouds but these studies were hampered by the large rotational partition function of this molecule and the forest of lines present at low intensity levels. If truly an ice chemistry product, deep searches in Hot Cores/Hot Corinos will be key to establish this link.

11.6 Photodissociation Regions

Traditionally, photodissociation regions (PDRs) are dense regions near bright hot stars, which separate the ionized gas from the surrounding molecular material. The incident stellar FUV photons control the degree of ionization, the energy balance, and the composition of the gas in PDRs. As the key physical and chemical processes are very similar, the atomic phases of the ISM are now also considered PDRs. This is particularly true for diffuse and translucent clouds (Chapter 9). PDRs are bright in atomic cooling lines (e.g. [CII] 158 μm, [CI] 609,270 μm, [OI] 63,145 μm, [SiII] 34 μm), in fluorescent and thermal emission of the ro-vibrational and pure rotational lines of H_2, in mid-J pure rotational lines of CO as well as of molecules such as CN, HCN and HCO^+, in the Aromatic Infrared Bands, and in the dust mid- and far-IR continuum.

11.6.1 PDR Structure

Figure 11.37 summarizes the structure of a PDR. The PDR is separated from the ionized gas by a very thin ionization front in which the degree of ionization of H goes from 1 to 0. The thickness of the ionization front is set by the mean free path,

$$\ell = (n_H \sigma_H)^{-1} = ((1-x)\, n_0 \sigma_H)^{-1} \simeq 2 \times 10^{-4} \left(\frac{10^4 \text{ cm}^{-3}}{n_0} \right) \text{ pc}, \tag{11.48}$$

with n_H the density of neutral H, n_0 the density of H nuclei, σ_H the ionization cross section ($\simeq 5 \times 10^{-18}$ cm^2), and x the degree of ionization. At the surface of the PDR, the strong FUV field has dissociated all molecules, including H_2, and ionized all atoms with ionization potentials less than 13.6 eV. The size of the atomic zone has been discussed in Section 6.3.4 for conditions appropriate for diffuse clouds, $G_0/n_0 \lesssim 4 \times 10^{-2}$, where H_2 self-shielding dominates the penetration of dissociating photons. PDRs are in the opposite limit, where dust dominates extinction. The column density to the depth where half of the H-nuclei are in molecular form is given by (Eq. 9.8, Section 9.2.1),

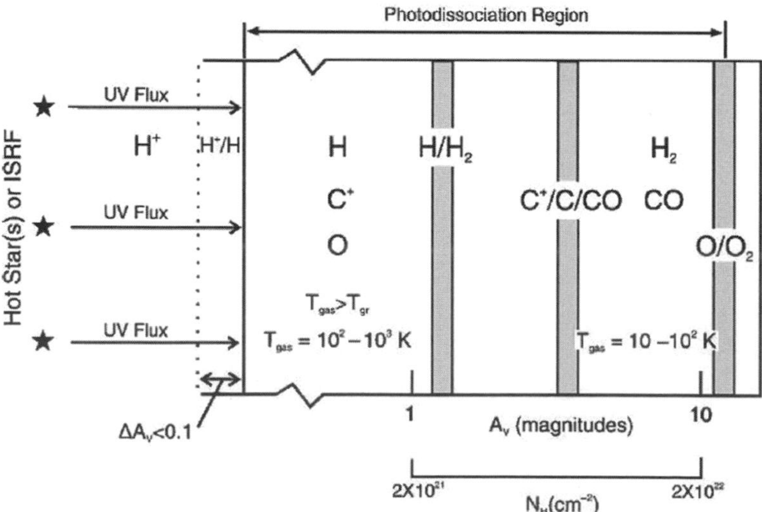

Figure 11.37 Schematic structure of a photodissociation region. The cloud is illuminated from the left by a strong FUV radiation field. The PDR extends from the H ionization front all the way to a dust extinction of some five visual magnitudes. Figure adapted from [269]

$$N_{DF} \simeq 5 \times 10^{20} \ \ln\left[43\frac{G_0}{n}\right]. \tag{11.49}$$

For a PDR with $G_0/n = 1$, we have, $N_{DF} = 1.9 \times 10^{21}$ N-nuclei cm^{-2} or an A_v of 1 mag.

The degree of ionization is dominated by C^+ with small contributions by S^+ and trace elements such as Fe and Na. Dust will attenuate the FUV radiation and that sets the size of the C^+ region,

$$A_v = 4.4 + \frac{1}{3.3}\ln\left[\left(\frac{1.5\times10^{-4}}{\mathcal{A}_C}\right)\left(\frac{10^4\,\text{cm}^{-3}}{n_0}\right)\left(\frac{G_0}{10^4}\right)\right]$$

$$R_{C^+} \simeq 0.27\left(\frac{10^4\,\text{cm}^{-3}}{n_0}\right)\ \text{pc}, \tag{11.50}$$

with $\mathcal{A}_C$ the carbon abundance, G_0 the strength of the incident FUV field in Habings, and the physical size is estimated for the value of the parameters indicated. Actually, deep in the PDR charge exchange reactions of C^+ with, e.g., PAHs can start to play a role in carbon ionization balance as well. The size of the PDR can also be linked to the ionization balance through the competition of ionization of trace species by the UV field and cosmic ray ionization of H_2, which dominates ionization inside molecular clouds (cf. Section 10.5.3). Comparing photoionization of a trace species ($k_i = a_i \exp\left[-bA_v\right]$) with cosmic ray ionization, we have,

$$A_v = \frac{1}{b}\ln\left[\frac{a_i G_0 X_i}{\zeta_{CR}}\right] \simeq 8\ \ln\left[\frac{G_0}{10^4}\frac{3\times10^{-17}\,\text{s}^{-1}}{\zeta_{CR}}\right]$$

$$N_H \simeq 1.5\times10^{22} \quad \text{cm}^{-2}, \tag{11.51}$$

where we have used an ionization rate of 4×10^{-11} s^{-1}, $b = 1.4$, and a gas phase abundance of 3.5×10^{-6}, appropriate for Mg.

Traditionally, the depth of the PDR has been equated with the size of the region where O-chemistry is dominated by FUV photons. The latter was estimated by balancing O_2 formation with photodissociation. It is now recognized that oxygen chemistry is controlled by accretion onto dust grains balanced by photodesorption (Section 10.6.3). For the size of the PDR, this results in,

$$A_v \simeq 5.9 + \frac{1}{1.8}\ln\left[\left(\frac{G_0}{10^4}\right)\left(\frac{10^4\,\text{cm}^{-3}}{n_0}\right)\right] \tag{11.52}$$

$$R_O \simeq 0.36\left(\frac{10^4\,\text{cm}^{-3}}{n_0}\right)\ \text{pc}, \tag{11.53}$$

where again the physical size is estimated for the value of the parameters indicated.

Finally, we can also consider the energy balance and equate the depth of the PDR with the region where the photo-electric effect dominates over cosmic ray heating. Cosmic ray heating is discussed in Section 10.3.1 while photo-electric heating is described in Section 11.8.2. Equating these two rates, yields for the depth,

$$A_v \simeq 7.7 + \frac{1}{1.8} \ln \left[\left(\frac{G_0}{10^4} \right) \left(\frac{3 \times 10^{-17}\,\mathrm{s}^{-1}}{\zeta_{CR}} \right) \right] \tag{11.54}$$

$$R_{heating} \simeq 0.47 \left(\frac{10^4\,\mathrm{cm}^{-3}}{n_0} \right) \mathrm{pc}. \tag{11.55}$$

All these depth estimates are very similar as they are all controlled by the penetration of FUV photons.

In many of these estimates – and, in general, for PDRs – G_0/n is the controlling parameter. This is the equivalent of the ionization parameter in studies of HII regions. G_0 is defined as the 6−13.6 eV FUV field in units of the interstellar radiation field. In Habing fields (1 Habing equals 1.6×10^{-3} erg cm^{-2} s^{-1} in unidirectional field), we have for a star with luminosity, $L_\star$, at a distance, d, from the PDR,

$$G_0 = 625 \frac{L_\star \chi}{4\pi d^2}, \tag{11.56}$$

with χ the fraction of the stellar luminosity between 6 and 13.6 eV. Adopting the Strömgren solution for the size of the HII region and assuming pressure equilibrium between the ionized gas and the PDR, we have,

$$\frac{G_0}{n} \simeq 0.22 \left(\frac{n}{10^4\,\mathrm{cm}^{-3}} \right)^{1/3} \mathrm{cm}^{-3}, \tag{11.57}$$

where a temperature of 8000 K has been assumed for the HII region.

The chemistry of PDRs is very similar to that of diffuse and translucent clouds (Section 9.3). There are a few differences with diffuse clouds. First, the H–H_2 transition occurs somewhat deeper in the cloud (see above). Second, the gas is quite warm ($T \simeq 500$ K) and dense ($n \simeq 10^4$ cm^{-3}) and hence reactions with activation barriers play more of a role. Third, the strong FUV field results in an appreciable population of vibrationally excited H_2. This nonthermal excitation can assist in overcoming activation barriers.

We can recognize four distinct zones in the composition of a PDR (Figure 11.38). The surface layer is atomic consisting of H, C^+, O, and S^+. The H-to-H_2 transition has been discussed above. The second region is the radical zone, which is slightly deeper in the PDR. The UV field is still very strong and – as for diffuse clouds – results in relatively high abundances of small radicals; e.g. radicals such as CN and C_2H, show a broad peak in the spatial abundance distribution. The hydrocarbons result from the reaction chain starting with C^+ and H_2 and, as mentioned, possibly facilitated by H_2 vibrational excitation. The small hydrocarbon cations can recombine and the resulting radicals can react with C^+ to form acetylene-like cations. Further reaction with H_2 and electron recombination leads to, e.g., C_2H. The radicals OH and NH are also more readily formed. The latter reacts with C^+ and winds up (eventually) in HCN. In this radical zone, as for diffuse clouds, the abundance of molecules, radicals, and daughter products is severly limited by photodissociation. This radical zone is followed by a zone in which carbon transits from C^+ into C and then CO. Very deep, we reach the fourth region, the molecular core, where the molecular composition is controlled by cosmic ray ionization. At that point, carbon is mainly in the form of CO

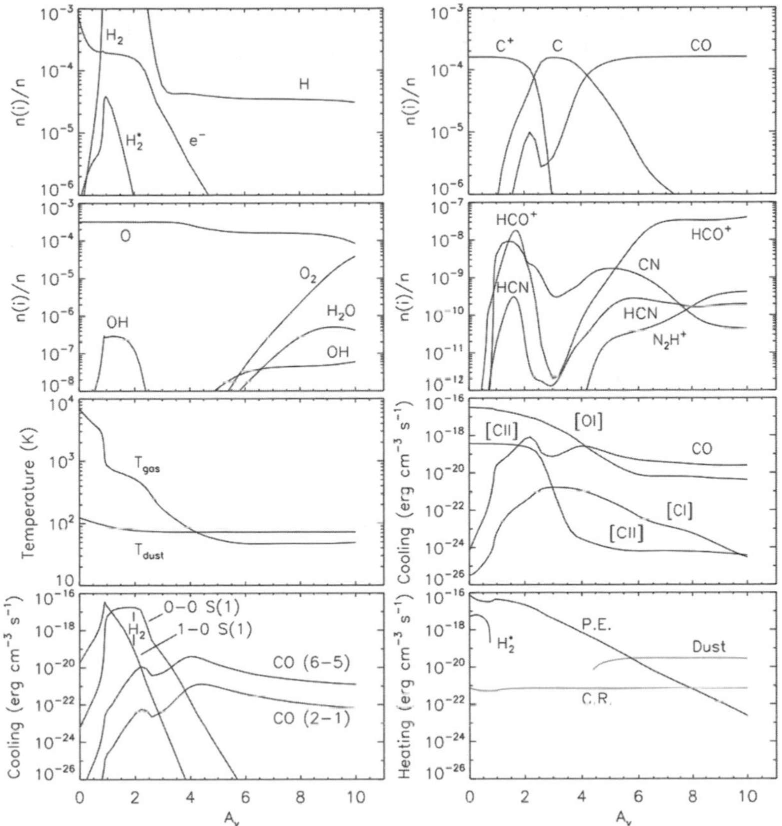

Figure 11.38 The physical and chemical structure calculated for the PDR in Orion ($n = 2 \times 10^5$ cm^3, $G_0 = 10^5$) as a function of visual extinction. The radiation field enters from the left. Top four panels: Abundances of various species. Bottom four panels: Gas and dust temperatures, dominant heating and cooling processes, and distribution of other key emission lines. Figure kindly provided by Mark Wolfire

and nitrogen in the form of N_2. These will react with H_3^+ to form, e.g., HCO^+ and N_2H^+ (see Section 10.5.3). As discussed in Section 10.6.3, freeze out becomes important deep inside cold dense cores. We can evaluate when this becomes important for PDRs. Take oxygen as an example: an accreted O atom has to reside long enough on the grain for an H atom to be accreted, diffuse on the surface, find the O-atom and react to form H_2O. H_2O is bound quite strongly and will not sublimate for temperatures below $\simeq$90 K in the ISM. Near the surface, H_2O photo-desorption may play a role (Section 10.6.3) but we will ignore that here. The critical dust temperature is then,

$$T_{cr} \simeq \left(\frac{E_b\,(\mathrm{O})}{k} - \frac{E_b\,(\mathrm{H})}{4k} \right) \left(\ln \left[\frac{\nu N}{\pi a^2 n\,(\mathrm{H})\, v} \right] \right)^{-1} \simeq 25 \text{ K}, \qquad (11.58)$$

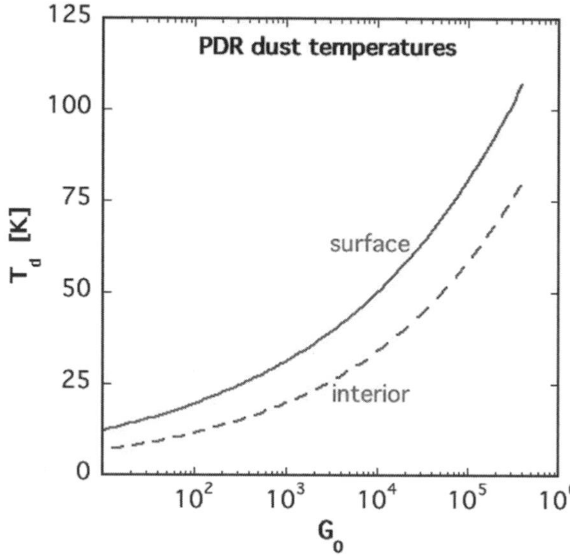

Figure 11.39 Calculated dust temperatures at the surface of the PDR and in the deep interior. Figure adapted from [273]

where the right-hand side adopts binding energies of 1300 and 550 K for O and H, respectively, an atomic H density of 1 cm^{-3}, a pre-exponential frequency factor of 10^{12} s^{-1}, and a gas temperature of 100 K. Comparing this to calculated dust temperatures for PDRs (Figure 11.39), we find that G_0's in excess of $\simeq 2 \times 10^3$ will inhibit H_2O formation and mantle growth. Sublimation of preexisting H_2O ice mantles in the interior of a PDR will require G_0's in excess of $\sim 10^6$. A CO-rich ice mantle, on the other hand, will sublimate for a G_0 as low as 10^3. In the surface layers of a PDR, photodesorption will keep grain surfaces clean (cf. Section 10.6.3).

As discussed in Section 6.13.4, the ISM is generally far from equilibrium. This may allow multiple stable solutions for the structure and composition of a region. For dense regions illuminated by a strong radiation field ($n \gtrsim 10^5$ cm^{-3}, $G_0 \gtrsim 10^5$), the resulting PDRs may show bifurcation. In such an environment, UV pumping of H_2 will result in a high abundance of vibrationally excited H_2. In a warm gas, collisional deexcitation will transfer much of this vibrational energy to heat, keeping the temperature high. However, for low temperatures, collisional deexcitation is not important, H_2 relaxes radiatively, this heat source is not accessible, and the gas stays cold. So, the physical processes involved allow for a high temperature and a low temperature solution. The cloud's history will then control which solution is actually achieved. While this focuses on the thermal behavior of the gas, this bifurcation will also affect the chemical composition.

11.6.2 Observations

PDRs are bright in the atomic fine-structure lines of [OI], [CII], [CI], and [SiII]. Their analysis will not be discussed here. Rather, the focus is on studies of PDRs in the molecular lines; in particular, H_2, low-, and mid-level J CO lines, and lines from simple molecules, including radicals, characteristic for PDRs. Analysis is typically started using simple diagnostic diagrams followed up by detailed model fits. The prototypical PDR diagnostic tool used in these types of analysis employs the dominant, atomic cooling lines of [OI] and [CII]. These lines together measure the total cooling rate of the PDR gas. The ratio to the far-IR dust continuum – the strength of the incident FUV radiation field – is a direct measure of the total heating rate of the medium, which is a strong function of the physical conditions. Together with the [CII] 158 μm/[OI] 63 μm ratio – lines with critical densities and excitation energies spanning the physical conditions of PDRs – this diagnostic diagram has been widely used to analyze PDR gas.

Molecular Hydrogen

Molecular hydrogen can be observed through its electronic transitions in the FUV, its ro-vibrational transitions in the near-IR, and its pure rotational transitions in the mid-IR. The FUV data can be analyzed using the curve of growth technique (Section 4.3.2). Figure 11.40 summarizes the results from the rich H_2 absorption line spectrum observed toward the B1.5V illuminating star of the reflection nebula NGC 2023, HD 37903. The rotational levels in the ground state vibrational level indicate gas with a temperature of 100−200 K. However, the vibrational excitation is much higher than expected for LTE. The high excitation revealed by this spectrum is characteristic for FUV pumping of H_2 by the strong incident FUV radiation field of PDRs. The rotational temperature in the excited vibrational states ranges from about 2000 K for $v = 1$ to 500 K for $v = 12$. The total H_2 column density is $\sim 10^{21}$ cm^{-2}, which can be compared to the total HI column density of 1.5×10^{21} cm^{-2} or a H_2 fraction of 0.57. Most of the population is in the lowest vibrational state – in fact, for 90% in the $v = 0$, $J = 0$, and $J = 1$ states.

PDRs are typically relatively dense and warm and collisions with H^+ will bring the o/p ratio in $v = 0$ to the LTE value, which for $T \gtrsim 200$ K is 3. As Figure 11.40 illustrates, the observed ortho-to-para ratio deviates from the expected equilibrium value of 3. The H_2 ortho-to-para ratio in vibrationally excited states is a complex interplay of collisional processes (driving it to the LTE value at the local temperature) and FUV pumping. If the vibrationally excited level populations are set by FUV pumping of the $v = 0$ population and IR decay, self-shielding will be more important for o-H_2 than for p-H_2 and the observed o/p ratio will be less than the LTE value. Specifically, on the square root portion of the curve of growth, the high temperature o/p (3) would result in an observed o/p ratio of 1.7 in vibrationally excited H_2 levels. For dense ($n > 5 \times 10^4$ cm^{-3}) PDRs illuminated by very strong radiation fields ($G_0 > 10^5$), the gas temperature can exceed 1000 K and collisional

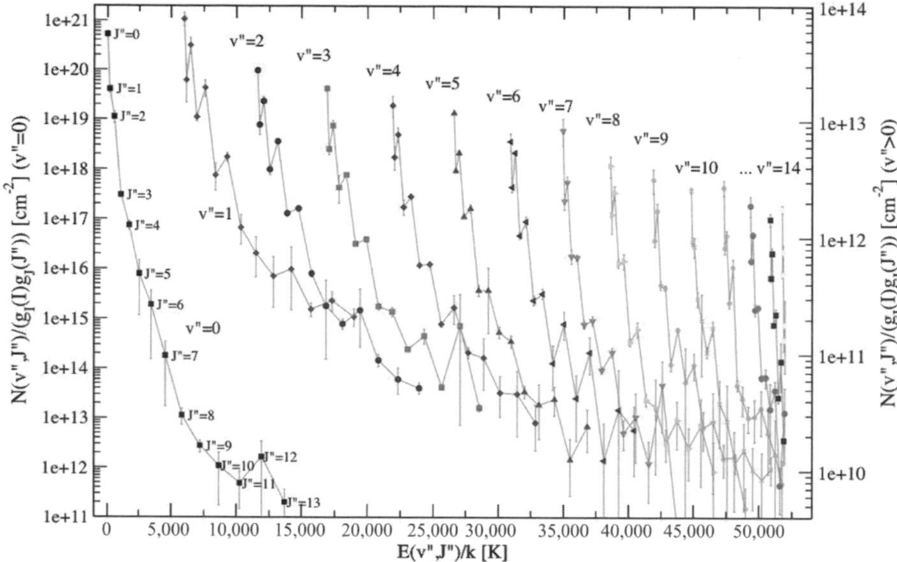

Figure 11.40 The H_2 rotational diagram derived from the Lyman and Werner band transitions in the FUV observed with FUSE and HST in absorption toward the illuminating star of NGC 2023, HD 37903. The level populations in the ground vibrational state refer to the lefthand y-axis. The other vibrational levels refer to the righthand y-axis. The level populations have been normalized by the rotational and the nuclear spin statistical weights (g_J and g_I). Figure taken from [122, 198]

(de)excitation processes can become very important, particularly for $v = 1$. In this case, the observed o/p ratio will go to the LTE value ($o/p = 3$). State-of-the-art PDR models include these aspects, and studies with the Meudon PDR code give a good fit to the data of HD 37903 for a cloud at a distance of 0.45 pc from the star, corresponding to $G_0 \simeq 2 \times 10^3$. In this isobaric model, the density ranges from 500 cm^{-3} in the surface of the PDR to 1900 cm^{-3} on the other side (the side viewed by the observer), while the temperature decreases from 380 to 100 K.

FUV absorption studies require a suitable geometry wth the PDR in front of the illuminating star, which is quite rare. Near- and mid-IR emission studies are more commonly used to study H_2 in PDRs. Figure 11.41 summarizes the near-IR data on the ro-vibrational emission toward the South-Western bar in NGC 2023. In a single grating setting many different vibrational levels can be probed. Given the small Einstein A's, the lines are optically thin and observed intensities can be directly converted into column densities in the excited level (Eq. 4.58, Section 4.1.7). The data on NGC 2023 results in a rich Boltzmann diagram. Again, the data reveal the importance of FUV pumping resulting in a high excitation temperature for the vibrational levels and a much lower rotational excitation temperature within these vibrational levels. Compared to the gas probed in the UV absorption study, for this region, the comparison with a PDR model results in a somewhat denser environment

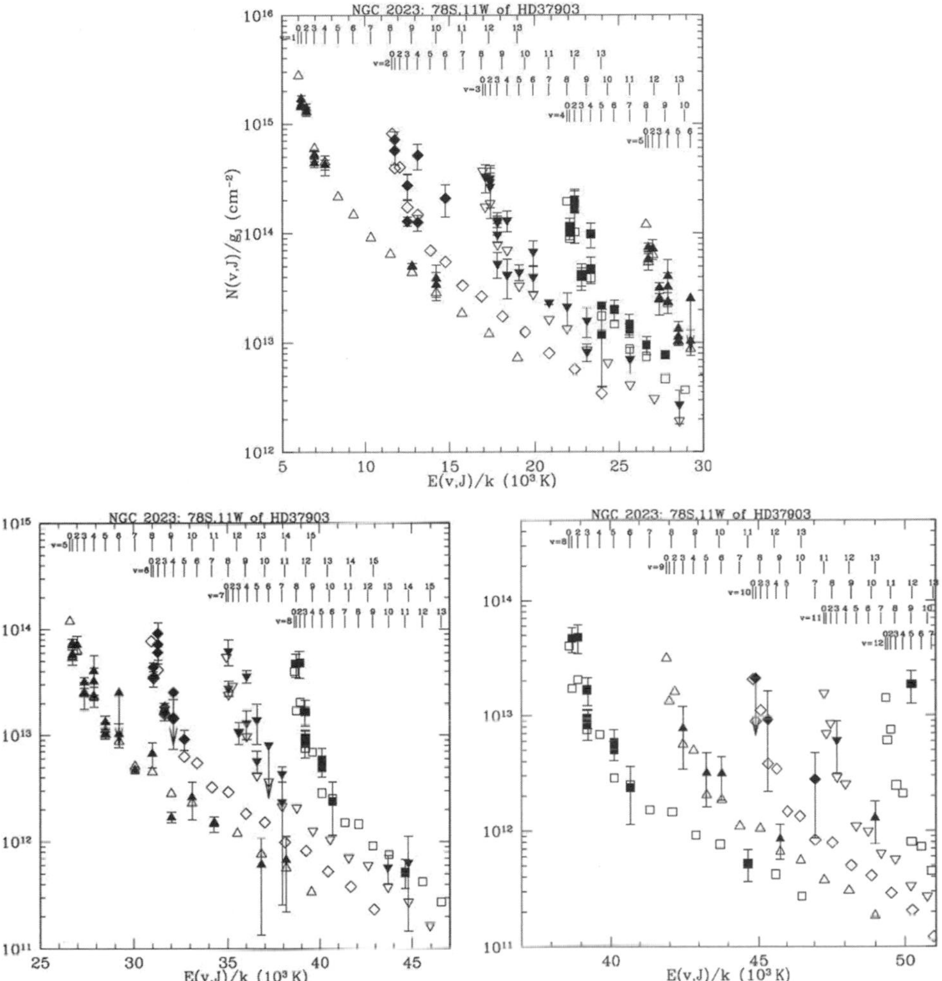

Figure 11.41 H_2 rotational diagram derived from ro-vibrational transitions in the near-IR toward the dense South-Western PDR in NGC 2023 after correction for foreground extinction with A_K = 0.5 mag. Data from [42, 189]. Open symbols: plane-parallel model with $n_0 = 5 \times 10^4$ cm^3 and $G_0 = 8.5 \times 10^3$ observed almost edge-on (a 78° angle) [86]. Figure taken from [85]

illuminated by a somewhat stronger FUV field, placing the PDR at a distance of $\simeq$0.2 pc from the star. Note that these lines do not directly probe the population in the ground state levels ($v = 0$ $J = 0$ and $J = 1$), which dominate the population.

Finally, the mid-IR pure rotational line spectrum of H_2 is accessible from space-based platforms and (partly) from the ground. Spitzer has surveyed a number of PDRs in these lines and JWST is slated to do so at higher spectral and spatial resolution. These quadrupole lines are optically thin and the observed intensity can be directly translated into column

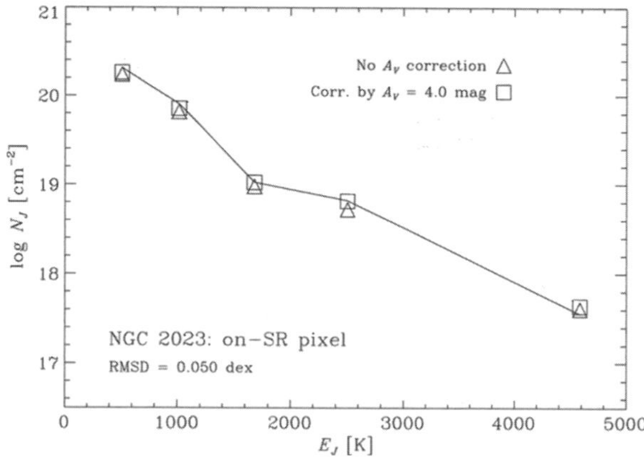

Figure 11.42 H_2 rotational diagram derived from the pure rotational transitions in the mid-IR toward the dense, southern bar in NGC 2023. A small correction for extinction is indicated. Note that the nuclear statistical weight has not been included. The (face-on) model results have been "arbitrarily" shifted by a factor of 0.76 to account for the edge-on geometry and for beam-filling effects. Model parameters [167] are $n_0 = 2 \times 10^5$ cm^3 and $G_0 = 1.7 \times 10^4$. Figure adapted from [255]

densities using the Einstein A of the level (Eq. 4.58, Section 4.1.7). Figure 11.42 summarizes the results for one specific location in the southern bar of NGC 2023. Including an o/p ratio of 1.9, the data can be well fitted by two components at 205 and 685 K, respectively. In a partially atomic/molecular gas, collisional deexcitation of the pure rotational levels is dominated by atomic H collisions. The lowest rotational levels of H_2 span an interesting range in critical density with 6.5×10^1, 1.3×10^3, 1.1×10^4, and 9.0×10^4 cm^{-3} for levels 2–5 at 300 K. The lowest J levels will thus be in LTE in PDRs while higher J levels are controlled by FUV pumping (as is the population of the vibrationally excited levels).

Carbon Monoxide

Warm, dense PDR gas will radiate copiously in the mid-J CO levels. Transitions up to $J = 8-7$ can be observed from the ground. With Herschel, higher transitions up to $J \simeq 40$ were probed. These CO lines span a range in critical densities and excitation energies. At $T = 100$ K, some relevant values for PDR conditions are: for $J = 1$ the critical density is $\simeq 300$ cm^{-3} ($E_1 = 5.5$ K), for $J = 3$ it is $\simeq 10^4$ cm^{-3} ($E_3 = 33$ K), for $J = 7$ it is $\simeq 10^5$ cm^{-3} ($E_7 = 155$ K), and for $J = 16$ it is $\simeq 10^6$ cm^{-3} ($E_{16} = 752$ K). A line ratio will then be roughly sensitive to the range between the two critical densities (Section 4.2.3). Figure 11.43 illustrates this for $J = 6 - 5$ over the $J = 2 - 1$ line. The critical densities are readily recognized. As the depth where CO becomes abundant is set by the penetration of FUV photons and CO is the dominant coolant at that point, the temperature at that depth is not very sensitive to the incident FUV field over much of the parameter space relevant for PDRs. Hence, the CO diagnostic diagram has to be supplemented by a line ratio that is mainly sensitive to the radiation field. The ([OI]63 μm + $[CII]$158 μm/ FIR is often used

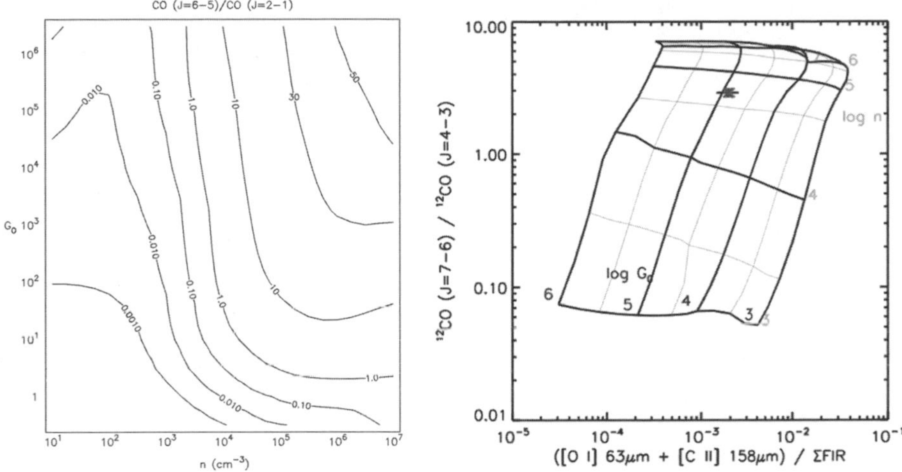

Figure 11.43 Diagnostic diagrams involving CO. Left: The calculated intensity ratio for the CO $J = 6{-}5$ over the $J = 2 - 1$ lines as a function of the incident FUV field, G_0, and the density of hydrogen nuclei, n_H. Right: The calculated intensity ratio for the CO $J = 6{-}5$ over the $J = 2{-}1$ lines versus the ratio of the atomic cooling lines over the far-IR dust continuum. The lines are models with the constant density or FUV field. Labels are the log of these quantities. The two data points refer to S106 and to IRAS23133+6050 (see Figure 11.44). Figure taken from [262]

for this. This ratio measures the total gas cooling rate over the total incident FUV field. Figure 11.43 illustrates how these ratios can be used.

Figure 11.44 summarizes CO observations of the PDRs associated with the Orion Bar and the northwest PDR in NGC 7023. Model studies reveal that the CO emission originates from warm ($\simeq$200 K), dense ($\simeq 10^6$ cm^{-3}), molecular gas. Many PDRs reveal a break in the CO ladder at high J ($\gtrsim$20) due to an emission from traces of CO in the warm atomic surface layer. In very dense clumps, the increased importance of H_2 self-shielding pulls the H_2/H transition to the surface and the formation routes of CO are greatly speeded up. This leads to traces of CO in the warm surface gas. The resulting CO emission is optically thin and only becomes prominent for high J.

The Orion Bar PDR

The Orion Bar is the prototype of a dense PDR associated with a region of massive star formation. The Orion Bar is located to the southeast of the prominent HII region, M42 (Figure 11.45), and is seen almost edge-on. The O6.5 star, θ^1 Ori C is surrounded by a dense ionized gas region ($n \simeq 3000$ cm^{-3}, $T_e \simeq 9000$ K, $P_e \simeq 3 \times 10^7$ K cm^{-3} in the ionized bar) with a size of $\sim$0.5 pc. This star has a strong stellar wind ($\dot{M} \simeq 4 \times 10^{-7}$ M$_\odot$/yr; $v_{\rm wind} \simeq 1200$ km/s), which has created a 2 pc-sized bubble – the Orion veil – filled with hot gas ($n = 0.1 - 0.5$ cm^{-3}; $T_e \simeq 2 \times 10^6$ K; $P \simeq 10^6$ K cm^{-3}). The ionized gas flows from the ionization front into this bubble at $\simeq$10 km/s, adding about 10^{-2} M$_\odot$/yr to

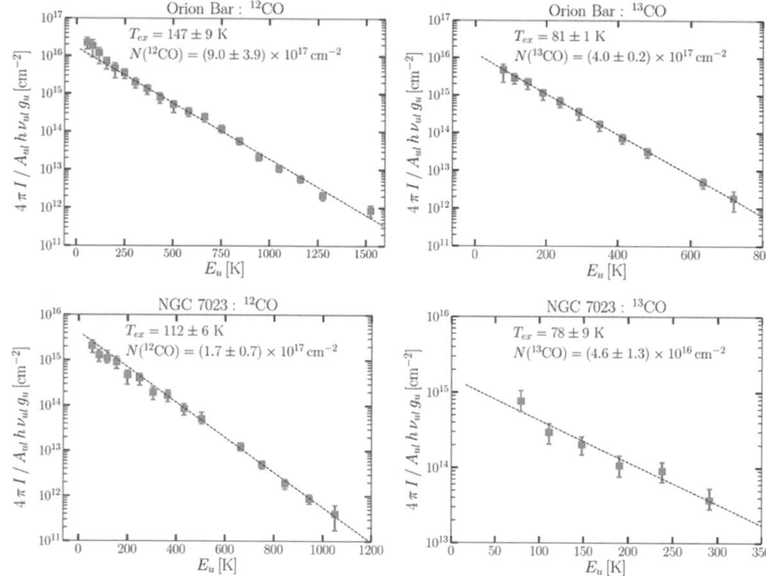

Figure 11.44 CO rotational diagram derived from Herschel PACS, SPIRE, and HIFI observations toward the PDR in the Orion Bar and in NGC 7023. While the results show Boltzmann behavior, optical depth effects are clear. Figure taken from [156]

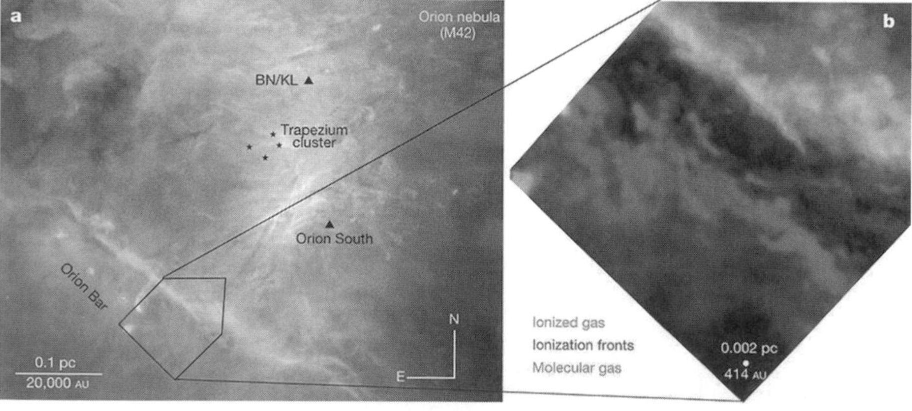

Figure 11.45 Left: The Orion Bar: The star, θ^1 Ori C in the Trapezium star cluster has created an ionized gas region, here traced by the [SII] λ6731 Å line. To the southwest, the ionized gas is separated from the molecular cloud by the Orion Bar PDR. The ionization front – traced by the [OI] λ 6300 Å line – is the boundary of the PDR. The molecular gas in the PDR is traced by the HCO$^+$ $J = 3 - 2$ line. Right: Blow up of a small portion of the Orion Bar. Here red is the HCO$^+$ $J = 4 - 3$ transition. Figure taken from [124] (A black and white version of this figure will appear in some formats. For the color version, please refer to the plate section.)

the interior of the bubble. On the reverse, the bubble presses against the dense starforming core, OMC-1. On the front side, the Veil bubble expands by about 13 km/s into the lower density environment. Eventually, this bubble will burst open and unload the hot X-ray emitting plasma and the photoionized gas into the Orion–Eridanus superbubble. On the molecular core side, a $\simeq$3 km/s shock wave driven by the stellar wind and thermal pressure of the ionized gas has carved out a bowl-shaped structure, sweeping up a dense shell of gas, the prominent PDR. To the southeast, the bowl's curvature leads to limb-brightening: the Orion Bar. The width of the ionization front as measured in the [OI]λ 6300 Å line implies an almost perfect edge-on view ($\simeq 7^\circ$ tilt with respect to the line of sight). The images of the Orion Bar in different atomic and molecular tracers reveal the layered structure of a PDR where the ionization front separates the ionized gas from the PDR (Figure 11.45). The surface of the PDR and the penetration of FUV photons is traced by the PAH emission features. The H_2 dissociation front is evident in the pure rotational ro-vibrational lines of H_2. This is also the zone where reactive molecular radicals are abundant and high level CO is emitted. Deeper in, molecules such as (low$-J$) CO and HCO^+ are present.

Table 11.3 summarizes the conditions of the different regions in Orion. These have been gleaned from different studies, as indicated. On a global scale, studies of the warm dust

Table 11.3 *Characteristics of the Orion Bar PDR*

Physical conditions			
Ionization bar			
n	3×10^3	cm^{-3}	[237]
T	9000	K	[237]
P	5.5×10^7	cm^{-3} K	
Atomic PDR			
G_0	2.6×10^4		[187]
n	5×10^4	cm^{-3}	[270]
N	6.7×10^{22}	cm^{-2}	[223]
T	260	K	[223]
P	1.3×10^7	cm^{-3} K	
T_d	75	K	[250]
H_2 peak			
T	500	K	[9]
N (H_2)	9×10^{20}	cm^{-2}	[9]
Molecular zone			
n (H_2)	2×10^5	cm^{-3}	[260]
T	85	K	[137]
P	1.7×10^7	cm^{-3} K	
N_H	1.3×10^{23}	cm^{-2}	[137, 266]

associated with the Orion Bar PDR reveal a H-nuclei density of $\simeq 5 \times 10^4$ cm^{-3}, column density of 4×10^{22} cm^{-2}, and a mass of $\simeq 100$ M$_\odot$. Sub-millimeter studies show that the swept-up shell has a thickness of $\simeq 0.1$ pc and a line-of-side length of $\simeq 0.28$ pc. These parameters agree well with the studies of the atomic surface layer derived from the atomic fine-structure lines. The pure rotational lines of H_2 reveal a higher temperature but smaller column density, perhaps because this emission is dominated by denser, warmer clumps, or because the temperature increases inward in the PDR as the dominant cooling lines become optically thick. The density in the molecular zone is estimated from excitation studies of CN and CS. Analysis of the CO ladder (Figure 11.44) reveals that most of the CO emission originates from moderately warm, dense molecular gas. While not as apparent in the CO rotational diagram as in other PDRs associated with regions of massive star formation, the high level CO ($J \gtrsim 20$) does originate from very dense warm gas. Likely, this warmer gas resides in dense clumps. The presence of such dense clumps is also aparent in the ALMA study of HCO^+ emission (Figure 11.45).

There are clear pressure variations between these dense molecular clumps ($\simeq 3 \times 10^8$ cm^{-3} K), the molecular and atomic PDR zones ($\simeq 1.5 \times 10^7$ cm^{-3} K), and the ionized gas ($\simeq 8 \times 10^7$ cm^{-3} K). For comparison, the magnetic pressure is estimated to be $\simeq 3 \times 10^7$ cm^{-3} K. These pressures exceed the pressure in the hot plasma, the extended ionized gas nebula, and the Veil nebula and this pressure gradient has likely caused the large scale expansion of the Veil nebula toward us and is at the root of the 1 km/s PDR flow and the 10 km/s ionized gas flow.

11.6.3 Chemistry of PDRs

Molecular abundances are summarized in Table 11.4. These can be compared to the abundances in diffuse clouds and molecular cores (Tables 9.1 and 10.3). The chemistry in the surface layers is much like that in diffuse clouds with simple atom–radical reactions balanced by photodissociation, while the chemistry deeper in resembles that of dense cores. As for diffuse clouds, because the chemistry in the surface layers involves only a few steps, PDRs provide an excellent "test" for our understanding of the initial steps in chemical routes in the ISM. The Orion Bar has been extensively studied at many different wavelength regimes and the key species in this chemistry have been observed, which allows these tests to be performed through simple analysis. Here we will focus on three issues: "What is the role of (vibrationally) excited H_2 in the chemistry in the radical zone?," "What drives the chemistry in the radical zone?," specifically, "Do PAHs contribute to the abundance of small hydrocarbon radicals in that zone?," and "Do PAHs contribute to the formation of H_2 in the PDR surface layers?." We will base our analysis on observations of the Orion Bar PDR as those are the most complete set.

In Chapter 6, we identified the "rules of engagement" for gas phase chemistry: If a species can react rapidly with H_2, then that reaction takes precedence. For ions, the next step to consider (if reaction with H_2 is inhibited) would be dissociative recombination. For neutrals, we have to consider which ion(s) could be precursor(s) and consider the

Table 11.4 *Molecular abundances in PDRs*

Species	Orion Bar[a]	Horsehead[a]	Reference[b]
CO	8.5 (−5)	9.5 (−5)	[266]
HCO	5 (−10)	8.4 (−10)	[80]
HCO^+	5 (−8)	9.0 (−10)	[229]
HOC^+	5 (−12)	4.0 (−12)	[114]
DCO^+	5 (−12)		[229]
H_2CO	6 (−9)	2.0 (−10)	[229]
CH_3OH	1.5 (−9)	1.2 (−10)	[80]
CH_3CHO	2.3 (−10)	6.8 (−11)	[80]
HCOOH	1.5 (−10)	5.2 (−11)	[80]
H_2C_2O	2.5 (−10)	1.5 (−10)	[80]
HNCO	5 (−11)		[80]
CN	5 (−9)		[260]
HCN	1.3 (−8)		[229]
DCN	1.5 (−10)		[229]
HC_3N	3 (−11)	6.3 (−12)	[80]
CH_3CN	6 (−11)	2.5 (−10)	[80]
CH_2NH	3 (−11)		[80]
OH	≳ (−7)		[125]
H_2O	1.0 (−10)		[65]
CH	2.1 (−8)		[202]
C_2H	1.7 (−8)	1.4 (−8)	[79]
c-C_3H	1.0 (−10)	2.7 (−10)	[79]
l-C_3H	3.2 (−11)	1.4 (−10)	[79]
l-C_3H^+	2 (−11)	3.1 (−11)	[79]
c-C_3H_2	6 (10)	1.1 (−9)	[79]
l-C_3H_2	2 (−11)	<4.6 (−11)	[79]
C_4H	2 (−9)	1.0 (−9)	[79]
CH^+	2 (−8)		[202]
OH^+	3 (−9)		[279]
CO^+	1.0 (−11)	<5.0 (−13)	[114]
SH^+	3 (−10)		[202]
SO^+	1.0 (−10)	1.7 (−11)	[114]
CF^+	5 (−11)	5.7 (−10)	[202]
CS	1.3 (−8)	2.0 (−9)	[260]
H_2CS	2 (−10)	3.7 (−11)	[80]
H_2Cl^+	1.5 (−10)		[209]

[a] Abundances vary through these edge-on PDRs (see text) and quoted abundances refer to the peak abundance for each species.
[b] References for Orion Bar data. Horsehead abundances taken from the compilation in [134].

formation and destruction of those. Here, we have to amend that for PDRs by inserting the possibility that a species can react slowly with H_2 in its ground state but rapidly with vibrationally or rotationally excited H_2. Those reactions would depend, of course, on the population of highly excited H_2. For neutrals, we also need to account for the possibility of an insertion reaction with C^+. The FUV field is an important aspect of PDR chemistry: for one, it keeps carbon ionized. However, here we will assume that the radical zone is deep enough in the PDR that photo-destruction of molecular species is not important, where we note that the C^+ abundance stays high because recombination is very slow. We will use these rules to examine the gas phase routes toward many of the species in the radical zone of the PDR. As the chemistry is quite limited, most species are mainly in atomic form and ionization is produced by carbon, this is a good testbed for the initial steps in chemistry in space. In Chapter 7, we considered the microphysics of grain surface chemistry and in Chapter 9, we concluded that grain surface chemistry is key to understanding the molecular inventory of dark clouds, including the surface layers. For PDRs, though, dust temperatures are high and, for example, a physisorbed O-atom will sublimate on a timescale of $\simeq$30 μs, well before another atom can accrete ($\simeq$0.3 s). Grain surface chemistry requires therefore chemisorption and we will examine the implications for H_2 formation. Finally, PAHs undergo a strong photochemical evolution at the surface of a PDR and this may be a source of hydrocarbon species.

Chemistry of Excited H_2

CH^+ has been seen through its pure rotational transitions in the far-IR and sub-millimeter with the PACS and HIFI spectrometer on Herschel. The six transitions reveal curvature in their rotational diagram, indicative of suprathermal excitation as optical depth effects should be unimportant. This may reflect the effects of FUV/visible pumping and/or formation pumping. The observed column density is 6×10^{12} cm^{-2}. Including the ground state would up this to $\simeq 1.2 \times 10^{13}$ cm^{-2}.

The reaction of C^+ with ground-state H_2 has a large activation barrier and is not relevant in PDR conditions. We discussed this issue in connection to the observed high abundance of CH^+ in diffuse clouds in Section 9.4 and concluded that high temperatures were required for this reaction to proceed. Models for Turbulent Dissipation Regions are in good agreement with the observations of diffuse clouds. Turbulent dissipation is not expected to be very relevant in a high UV environment such as the Orion Bar. Instead, in PDRs, the reaction is promoted by rotational and/or vibrational excitation of H_2. The reaction rate coefficient increases rapidly with J and proceeds at a rate of $k_r\ (J = 7) = 10^{-11}$ cm^3 s^{-1} for H_2 in state $v = 0,\ J \geq 7$. Likewise, for H_2 in state $v = 1$, the rate is $k_v = 1.6 \times 10^{-9}$ cm^3 s^{-1}. CH^+ is destroyed through reactions with H and H_2 at rates $k_H = 7.5 \times 10^{-10}$ and $k_{H_2} = 1.2 \times 10^{-9}$, respectively. In steady-state abundance, the CH^+ column density is then,

$$N\left(\mathrm{CH}^+\right) = \frac{\left(N\left(\mathrm{H}_2, J=7\right) k_r + N\left(\mathrm{H}_2, v=1\right) k_v\right) X\left(\mathrm{C}^+\right)}{f_H k_H + f_{H_2} k_{H_2}}, \tag{11.59}$$

where the f_i are the fraction of H-nuclei in that species. The column densities of vibrationally excited H_2 is directly measured in the near-IR and is $N(H_2, v = 1) \simeq 7 \times 10^{16}$ cm^{-2} in the Orion Bar. The column density of the rotational population of H_2 has been measured through 0 – 0 S(1), S(2) and S(4) transitions. Extrapolating from this to $J = 7$ yields, $\simeq 3 \times 10^{18}$ cm^{-2}. With a C^+ abundance of 1.5×10^{-4} and $f_H = 0.5$, we arrive at a column density of 4.4×10^{13} cm^{-2}, somewhat larger than the observed column density. This simple estimate illustrates that vibrational excitation is key to the formation of CH^+ in a PDR setting.

SH^+ has been seen through its pure rotational transitions in the far-IR with Herschel and from the ground with ALMA. The derived column density is $\simeq 1.5 \times 10^{13}$ cm^{-2}. The barrier for reaction of S^+ with H_2 is higher than for C^+. In diffuse clouds, SH^+ is also an indicator of the importance of Turbulent Dissipation Regions (Section 9.4). H_2 vibrational excitation can also facilitate this reaction but vibrational states $v \geq 3$ are now implicated and even then the rate is slow. Esentially, while C^+ can accept the two H_2 electrons in its p-orbitals and form a long-lived complex, S^+ has to accept these electrons in d-orbitals or change its spin. For $v = 3$, the rate is calculated to be 2×10^{-10} cm^3 s^{-1}. The destruction of SH^+ with H is also slow with a rate of $\simeq 2.7 \times 10^{-11}$ cm^3 s^{-1} at 1000 K and the reaction with H_2 is strongly endothermic. FUV photodissociation can then become competitive. Ignoring this route, we have a relationship analogous to that for CH^+,

$$N\left(SH^+\right) = \frac{N\left(H_2, v \geq 3\right) k_v X\left(S^+\right)}{f_H k_H}. \tag{11.60}$$

The observed column of $N(H_2, v \geq 3)$ is only 10^{16} cm^{-2} and the sulfur abundance is 1.4×10^{-5}, which results in an SH^+ column of 2×10^{12} cm^{-2}, almost an order of magnitude less than observed. This presents a challenge for astrochemical models.

The hydroxyl radical is the third species that is strongly inhibited at low temperatures. The reaction of atomic O with H_2 has a barrier of 4600 K and has a reaction rate coefficient of $k_{H_2} \simeq 3.8 \times 10^{-15}$ at $T = 570$ K. In the surface layer, OH is mainly destroyed by UV photons at a rate of $k_{uv}(\mathrm{OH}) = 3.9 \times 10^{-10} G_0 \exp\left[-2.2 A_v\right]$. The OH column density is then,

$$N(\mathrm{OH}) = \frac{k_{H_2}}{k_{uv}(\mathrm{OH})} X(O)\, n_H\, N(H_2) \simeq 3 \times 10^{-7} \frac{n_H}{G_0} N(H_2), \tag{11.61}$$

where we have adopted $A_v = 2$ appropriate for the location of the H_2 peak in the Orion Bar. The column density of warm H_2 is measured to be $\simeq 10^{21}$ cm^{-2} and hence, with $N(\mathrm{OH}) \simeq 6 \times 10^{14}$ cm^{-2}, this route falls somewhat short of the observed OH column density ($>10^{15}$ cm^{-2}). The route through vibrationally excited H_2 also contributes. The reaction rate coefficient is much faster for H_2 in $v \geq 2$ ($k \simeq 2 \times 10^{-12}$ cm^3 s^{-1} for $v = 2$), which compensates for the smaller column density ($\simeq 2 \times 10^{16}$ cm^{-2}), resulting in a contribution of $\simeq 3 \times 10^{14}$ cm^{-2} to the OH column density.

Chemistry in the Radical Zone

These simple radicals are key intermediaries in the chemistry in the radical surface layer. The formation of CH^+ initiates hydrocarbon chemistry and leads to columns of CH_2^+ and CH_3^+ which are a factor of 2 and 6 higher than for CH^+. Recombination of these cations lead to, e.g., CH and CH_2. The neutral hydrocarbon radicals react on with C^+ to form, e.g., C_2H^+ at a rate of 5.2×10^{-10} cm^3 s^{-1}. Reaction with H_2 leads to $C_2H_2^+$ but further reaction with H_2 is inhibited by a 2000 K activation barrier. So, $C_2H_2^+$ recombines with an electron to form C_2H. C_2H is destroyed through reaction with C^+ to C_3^+. The C_2H and CH column densities are then related,

$$\frac{N\left(C_2H\right)}{N\left(CH\right)} = 0.5. \tag{11.62}$$

These species have been observed by HIFI in the same spectral survey and the observed column density ratio, 0.8, is in good agreement with this expectation. The reaction chain is continued as C_3^+ reacts with H_2, at a rate of 2.4×10^{-10} cm^3 s^{-1}. C_3H^+ reacts slowly with H_2 and so its abundance is set by dissociative recombination,

$$\frac{N\left(C_3H^+\right)}{N\left(C_2H\right)} = \frac{k_{H_2}}{k_e} \simeq 3\times10^{-3}. \tag{11.63}$$

The observed ratio of these two species in the ALMA survey is $1.5\simeq10^{-3}$. The slow reaction of C_3H^+ with H_2 forms both $c-C_3H_2^+$ and $l-C_3H_2^+$ (bimolecular; in a ratio of about 1 : 4) as well as $c-C_3H_3^+$ and $l-C_3H_2^+$ (radiative association; in a ratio of about 1.3 : 1.3). These ions preferentially recombine, producing many fragments. The upshot is that $l-C_3H_2$, $c-C_3H_2$, $l-C_3H$, and $c-C_3H$ are formed in a ratio of 0.2:0.2:0.3:0.3. Observations, on the other hand, show a ratio of 1 : 26 : 1.5 : 4.3 and hence there has to be an additional and very efficient route toward cyclopropenylidene. This is the reaction with atomic H that converts $l-C_3H_2$ into $c-C_3H_2$. As the latter is about 55 kJ/mol lower in energy (about 7000 K) and there is no activation barrier, in a region with a high atomic H abundance, this conversion is important. While this explains the preference of $c-C_3H_2$ over $l-C_3H_2$, it does not explain the high $c-C_3H_2$ over $l,c-C_3H$ ratio. Possibly, this involves the reaction of H_2 with $l-C_3H$ in the warm H/H_2 dissociation front, although that reaction involves a transition state about 40 kJ/mol higher in energy ($\simeq$6000 K).

We can examine the chemistry involving C_4H. It is formed through an insertion reaction of C_3H_2 with C^+ followed by recombination and (slowly) destroyed by reactions with H_2 (with a barrier of $\simeq$950 K, so a rate of $\simeq2\times10^{-12}$ cm^3 s^{-1}). We have then,

$$\frac{N\left(C_4H\right)}{N\left(C_3H_2\right)} = \frac{k_{H_2}\left(C_3H_2\right)}{k_{H_2}\left(C_4H\right)} \simeq 6. \tag{11.64}$$

Again, this is in reasonable agreement with the observations.

The hydroxyl radical leads to the formation of CO^+ through the reaction with C^+ ($k\simeq7.5\times10^{-10}$ cm^3 s^{-1}), which is lost through reactions with H_2 ($k\simeq7.7\times10^{-10}$ cm^3 s^{-1}) forming HCO^+ (See Figure 8.3). We have,

$$\frac{N\left(\mathrm{CO}^{+}\right)}{N\left(\mathrm{OH}\right)}=\frac{X\left(\mathrm{C}^{+}\right)}{f_{H_2}}\frac{k_{C+}}{k_{H_2}}\simeq 3\times 10^{-4}. \tag{11.65}$$

While this is in reasonable agreement with the observations, this route is not the origin of the observed HCO^+. Likewise, the equivalent reaction of H_2O plus C^+ to form HCO^+ (See Figure 8.3) is not very relevant in this sense as the abundance of H_2O is low (Table 11.4). The route toward HCO^+ involving the reaction of atomic oxygen with small hydrocarbon radicals and ions is also not very relevant in the radical PDR zone. Rather, the observed HCO^+ must originate from deeper in the cloud through the reaction H_3^+ with CO. The spatial distribution of HCO^+ already indicates that this species is not a radical zone tracer. HOC^+, on the other hand, will be formed through the reaction of CO^+ plus H_2 but the observed HOC^+/CO^+ (1/3) is somewhat less than predicted ($7.5\times 10^{\ 10}/2\times 10^{-10}\simeq 4$) and perhaps the branching ratio between HOC^+ and HCO^+ is less than assumed. As for HCO^+, the reaction of H_2O with C^+ is not important for the formation of HOC^+.

Finally, the reaction of OH with S^+ forms SO^+ at a rate of $\simeq 5\times 10^{-10}$ cm^3 s^{-1}. SO^+ does not react with H_2 and – as SO^+ is located at a depth where photochemistry is no longer important – is destroyed through dissociative electron recombination. We have then,

$$\frac{N\left(\mathrm{SO}^{+}\right)}{N\left(\mathrm{OH}\right)}\simeq 3\times 10^{-3}\,\frac{X\left(\mathrm{S}^{+}\right)}{X\left(\mathrm{e}\right)}\simeq 3\times 10^{-4}, \tag{11.66}$$

which is an order of magnitude less than the observations and sulfur chemistry does not seem to be fully understood.

This discussion centered on taking the observed abundances to identify the key species and their reactions, using the simple "rules of engagement." Except for sulfur, the results are very encouraging and we seem to understand the first steps in the road toward molecular complexity in space well. This bodes well for analysis of other regions in space. Many models have been published for chemical structure of the PDR and have used the Orion Bar as testbed. The best fit models sometimes differ considerably in the adopted conditions. As this discussion implies: to get the chemistry right in the zone that is characteristic for PDRs – radical zone – requires, e.g., a proper treatment of the reactions involving vibrationally – and to a lesser extent rotationally – excited H_2 and this depends very much on the details of the radiative transfer in the FUV pumping lines, the extinction of these photons by dust, and the formation process of H_2 in PDRs. In the analysis, we circumvented these issues by relying on observed column densities but, as discussed in the next subsection, our understanding of H_2 formation in a PDR setting is limited.

H_2 Formation in PDR Gas

Chemistry in the ISM is predicated on the formation of H_2. In the previous subsection, we saw how the chemistry in the PDR radical zone is linked to the presence and characteristics of H_2. While studies on H_2 formation have mostly focused on diffuse clouds (Section 9.2), the rate of H_2 formation has been examined in a few well-studied PDRs through analysis of near- and mid-IR observations of vibrationally and rotationally excited H_2. This analysis shows that the $0-0$ S(3) to $1-0$ S(1) line ratio is a good measure of the H_2 formation rate.

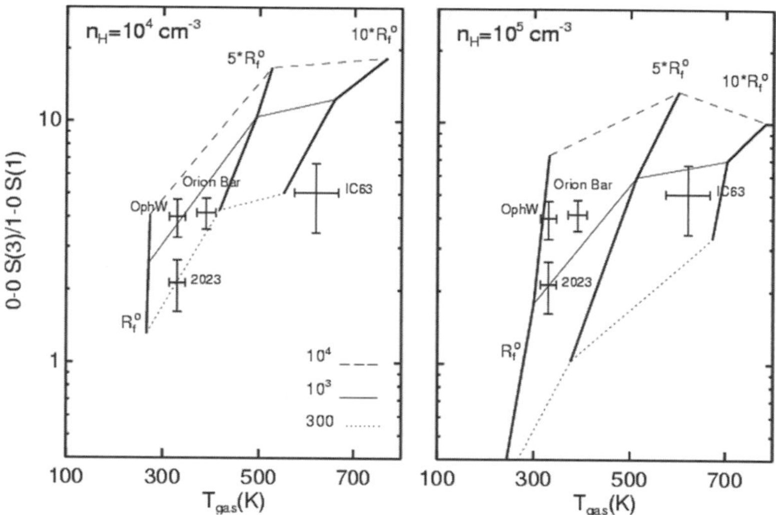

Figure 11.46 The ratio of the pure rotational 0 − 0 S(3) line to the UV pumped 1 − 0 S(1) line is a good measure for the H_2 formation rate. Data points refer to observations of individual PDRs while curves refer to model studies with different H_2 formation rates for different incident FUV fields and densities. Left: $n = 10^4$ cm^{-3}. Right: $n = 10^5$ cm^{-3}. The values of the adopted H_2 formation rate have been normalized to that measured for the diffuse ISM, R_f^0. Figure taken from [136]

This is illustrated in Figure 11.46. For the densities relevant for PDRs, the pure rotational lines measure the kinetic temperature of the gas, while the ro-vibrational line measures the pumping rate, and hence the H_2 destruction rate, which, in steady state, is linked to the H_2 formation rate. The behavior of the models in this diagram can be understood by realizing that, for higher H_2 formation rates, the H_2 dissociation front – where this emission originates – is pulled closer to the surface where the gas is warmer and allows H_2 to compete better with the dust for pumping photons. This analysis shows that, in PDRs, the H_2 formation rate is typically a few times that in the diffuse ISM.

We discussed the formation of H_2 in the chapter on diffuse clouds (Section 9.2) and compared theory with the observations in Figure 9.3. The gas as well as the dust in PDRs are warmer than in the diffuse ISM and hence the sticking coefficient is expected to be low ($\simeq$0.1 at 500 K). Moreover, physisorbed H atoms will sublimate before the next H atom adsorbs and a reaction can occur. For PDRs, chemisorption is, therefore, required for H_2 formation. The role of chemisorption has been studied in detail for graphitic surfaces. For a graphitic surface, an H atom will have to pucker out a C-atom in order to enter the chemisorbed site. This has an activation barrier of $\simeq$2000 K. A chemisorbed H atom is bound by about 0.9 eV but has a barrier against diffusion of about 1.1 eV. Hence, the chemisorbed H atom is immobile. Further adsorption will be preferentially in specific neigbour sites, which show enhanced binding (2 eV) and little or no activation barrier to adsorption. While adsorption of the first H atom will be difficult – with a sticking coefficient of $\simeq$0.04, further H adsorption will be rapid. This will lead to high H coverages in the form of clusters on graphitic surfaces and this will facilitate H_2 formation on interstellar graphite grains through, e.g., the Eley–Rideal mechanism. These (dimer) structures can mediate H_2 formation and greatly increase the Eley–Rideal cross section. However, at best, this will lead to a similar H_2 formation rate in PDRs as in diffuse clouds. In reality, more likely, the H_2 formation rate on dust grains in PDRs is factor of a few less than for diffuse clouds, given that the sticking coefficient may be less and the Eley–Ridal cross section may be less than for the Langmuir–Hinshelwood mechanism.

The observations suggest that additional "dust" surface area is activated for H_2 formation in PDRs compared to diffuse clouds. The total extra surface area involved must be comparable to that of classical dust grains. PAHs and very small grains may provide this extra surface area. Based upon extensive experimental studies, two specific mechanisms have been considered: 1) "superhydrogenation" of PAHs. 2) Photolysis of PAHs. In the first mechanism, extra H atoms are adsorbed at the edge C-atoms, forming a CH_2 group in an sp^3 structure. Adsorption of the first extra H has a small activation barrier (10−30 meV for the cation and 60 meV for the neutral). H_2 formation occurs then through Eley–Rideal abstraction. A cross section for this process has been measured for coronene molecules adsorbed on a graphitic substrate and is very small, 0.06 Å^2. For such a small cross section, this process is unimportant. In the second process, a PAH molecule absorbs one or more FUV photons and becomes highly excited. As a result, an H atom will start to roam the periphery of the molecule forming temporarily an CH_2 group before going back to the regular aromatic structure. There is a barrier for the formation process of the aliphatic group in the range of 3.2−3.6 eV while the barrier to go back is much less (0.7−1.3 eV, depending on the species involved). This back-and forth roaming of the H-atom will continue until the molecule either relaxes through IR emission or through the loss of an H-atom or an H_2 molecule. When the internal excitation of the PAHs is high enough that fragmentation becomes important, experiments show that, for small PAHs, H-loss dominates but for PAHs larger than $\gtrsim$ 32 C-atoms, H_2 loss dominates. The resulting PAH radical can then be hydrogenated again with accreting H atoms, which will be an efficient process. For very high FUV fluxes, a PAH will lose all of its H and these bare C-clusters will then start to lose C_2 units. This destruction process will be very rapid. H_2 formation from normal PAHs will therefore only occur in the transition region between pure C-clusters and normal PAHs where photolysis is important but enough H are present to protect the species against rapid C_2 loss. Figure 11.47 summarizes the fractional abundance of PAHs in different hydrogenation states for different atomic H densities and incident FUV fields. As these results show, H_2 formation through photolysis of PAHs is important in PDR environments but not in the diffuse ISM. As this process is very size-dependent, in a broad PAH family, only a limited subset of PAHs will contribute. So, for this process to be important on an interstellar scale, the interstellar PAH family must have a rather narrow size distribution (e.g. $50 \lesssim N_c \lesssim 75$), but then this is very much the size range of interstellar PAHs.

Photolysis of Hydrogenated Aromatic Carbon grains has also been suggested as a route toward H_2 formation in high FUV field environments. However, this is not an efficient

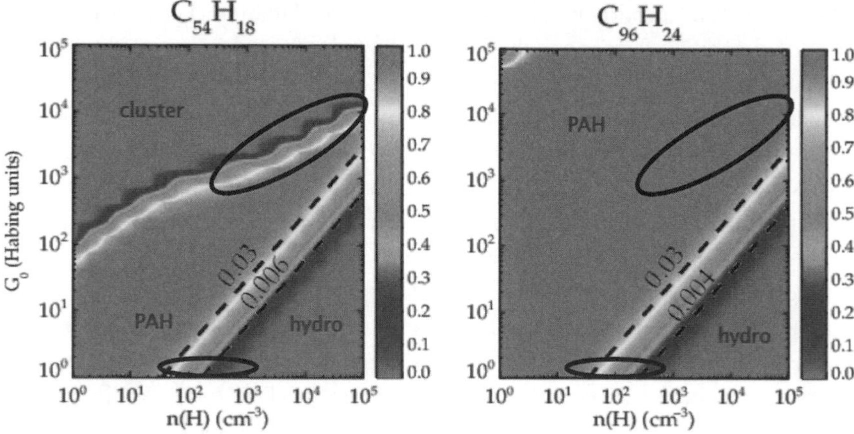

Figure 11.47 The characteristics of circumcoronene (left) and circumcircumcoronene (right) in PDRs. For part of the $n_H - G_0$ parameter space (indicated by PAH), PAHs will be fully hydrogenated; for high (atomic) H densities, PAHs will be "superhydrogenated" (labeled hydro); for high FUV fields PAHs will be rapidly, completely dehydrogenated (labeled clusters). The species in this cluster regime will quickly start to lose C_2 units and be destroyed. The transition between these different regimes is very sharp in the $n - G_0$ plane and the precise location depends on the molecule under consideration. H_2 formation from PAHs will only occur in the transition region between pure C-clusters and normal PAHs. H_2 formation from "superhydrogenated" PAHs can occur throughout the full hydro regime. The ovals indicate conditions typical for PDRs and for the diffuse ISM. Figure adapted from [14] (A black and white version of this figure will appear in some formats. For the color version, please refer to the plate section.)

process as only direct FUV excitation of the aliphatic CH_2 groups may lead to H_2 loss and this electronic excitation will quickly "leak" away to the phonon modes. Moreover, rehydrogenation of photolyzed HAC grains will be hampered by the limited range of thermal H atoms into HAC material.

We have focused here exclusively on the role of PAHs in H_2 formation. As argued before, "regular" grain surface chemistry is not important in the Orion Bar PDR (but may be important in lower radiation field PDRs such as the Horsehead PDR). However, sublimation of ices accreted in a previous dark cloud phase in the deeper zones and photodesorption could be a source of molecular species. The high abundance of CH_3OH and H_2CO in the Orion Bar (Table 11.4) suggests that ices are indeed important.

Finally, chemisorption of other species than H on PAHs has been largely unexplored and this might be interesting avenue to follow.

PAHs and the Organic Inventory of PDRs

In Section 11.6.3, we examined the chemistry of small radicals around the H_2 peak in the Orion Bar PDR and we concluded that this chemistry is driven by the formation of small radicals (OH, CH^+, ...), which drive a rich gas phase chemistry involving, e.g., C^+ insertion reactions to form $C_nH_m^{0,+}$ species. The Orion Bar is a high UV field, high density environment ($G_0 = 2.6 \times 10^4$, $n = 5 \times 10^4$ cm^{-3}), resulting in a high abundance of excited H_2, which is key in the first chemical steps. High abundances of these small hydrocarbon radicals and ions have also been observed in the Horsehead PDR, which is characterized by a similar high density but much lower FUV field as the illuminating star, σ Ori (an O9.5 star) is about 3pc away from the cloud surface ($G_0 = 10^2, n = 5\times10^4$ cm^{-3}). Models have some difficulty explaining the observations of these simple hydrocarbons near the PDR surface. It has been suggested that photolysis of PAHs may contribute to the inventory of small hydrocarbons near the PDR surface and we will explore this here.

We will start by considering lifetime arguments: A small hydrocarbon ion dissociatively recombines on a timescale of $\tau_e \simeq 500/n = 10^{-2}$ yr. Photodissociation of neutral species occurs on a timescale of $\tau_{uv} \simeq 30/G_0 = 0.3$ yr. So, for PAHs to be relevant, they have to "inject" small hydrocarbon species at this rate into the gas. For a typical PAH UV absorption cross section (5×10^{-18} cm^2/C-atom), a PAH size of 50 C-atoms, with an abundance of 10^{-7}, we have an injection rate of $\tau_{PAH} \simeq 10^7/n/G_0 f = 2/f$ yr, where f is the fraction of the UV photons absorbed that lead to hydrocarbon injection. In principle, f is very small and the PAH contribution will be small.

The actual timescale to consider is the dynamical timescale over which "unprocessed" PAHs can be brought to the PDR surface. With a length scale of 0.01 pc and a typical velocity of 1 km/s, we have $\tau_{dyn} = \ell/v = 10^4$ yr. From this perspective, PAH could only be important if they act as catalyst, bringing the reactants together but not being consumed by the reaction. In that case, the limit on the timescale is set by how fast species can be brought to the "PAH-surface." We can balance that rate with the UV dissociation rate of, say, C_3H_2, to evaluate the potential contribution,

$$\frac{N\left(C_3H_2\right)}{n\left(C^+\right)} = \frac{n\left(\mathrm{PAH}\right) k_{pah}}{3k_{uv}} \simeq 10^{-7}\frac{n}{G_0} \simeq 6 \times 10^{-4}, \tag{11.67}$$

where a reaction rate coefficient of $k_{pah} = 3 \times 10^{-9}$ cm^3 s^{-1} has been adopted. With an observed abundance of small hydrocarbons at the PDR surface of $\simeq 10^{-8}$, it is clear that a major reservoir of the C has to be involved (i.e. C^+) and that these reactions have to be very efficient. For small PAHs, C^+ inserts into the π electron system at the adopted rate and this results in a cyclopropa-PAH functional group. The chemistry of large PAHs has not been investigated and the product of the reaction of the next C^+ with such a species (after neutralization) is not known. In terms of UV-driven fragmentation, studies have shown that the weakest bonded functional groups will be lost first and that for large PAHs, the carbon skeleton is unaffected except possibly for isomerization. So, it is conceivable that large PAHs do contribute to the abundance of small hydrocarbons in PDR surfaces but further experimental and theoretical studies will be needed to identify the relevant reaction networks.

It is also relevant to consider the effects of this chemistry on the PAH population. The appropriate timescale is now the dynamical timescale, which we should compare to the reaction timescale of C^+ with a PAH, $\tau \simeq 7 \times 10^5/n \simeq 10$ yr. So, if such reaction routes exist, PAHs could grow in a PDR. Reactions with small hydrocarbon radicals could also be of interest as many of these can react with no or submerged barriers. But then the abundance of these species is much less than for C^+; e.g. to be relevant, abundances $\gtrsim 10^{-7}$ are needed.

11.7 Stellar Jets, Disk Winds, and Outflows

Many accreting, protostellar systems exhibit outflow phenomenae as some of the accretion energy is converted into kinetic energy for a collimated jet or wind. Young stellar objects are indeed a powerful source of mechanical energy. If the wind originates from the disk, then this may also be an efficient way of removing angular momentum and thereby allowing accretion to occur. At the same time, outflows stir up the protostellar environment and are a source of turbulence in the cloud. Moreover, by clearing out the envelope, outflows may also limit the mass of the newly formed star. Interaction of jets with clumps may trigger star formation. Furthermore, outflows drive shocks that can chemically process the surrounding gas and sputter ices. During the embedded phase, stellar outflows are very common and there are hundreds known in nearby star-forming regions. The nearby Perseus molecular cloud is riddled by many jets which show up well in for example Spitzer/IRAC 4.5 μm map – tracing emission from very warm rotationally excited H_2 – as well as in molecular lines of SiO, SO, and CO.

Figure 11.48 illustrates the structure associated with a protostellar outflow/jet. The disk and magnetic field structure channel the ejecta in narrow jets, creating two opposing lobes

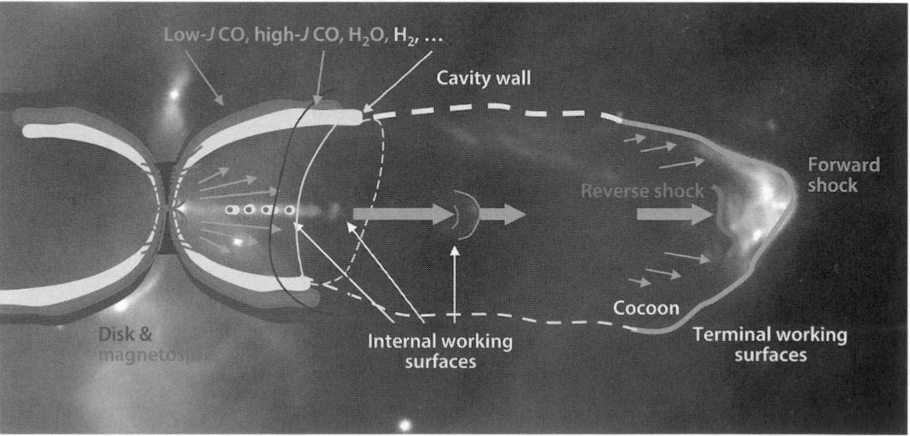

Figure 11.48 Schematic figure illustrating characteristic features of protostellar outflows. The outflow originates from a magnetized (red contours) disk (purple) (both exaggerated in size). Forward shocks (green) sweep up the surrounding material while the reverse shock (magenta) slows down the outflow/jet. Molecules are confined to the outflow walls but can also trace the jet in the youngest flows. Once a flow breaks out of the cloud, it becomes mainly atomic. Stellar photons can freely travel through the cavity and illuminate its walls, creating a PDR. Typical molecular emission regions are indicated. Figure taken from [18] (A black and white version of this figure will appear in some formats. For the color version, please refer to the plate section.)

in a bi-polar outflow. The outflow/jet will drive a shock wave, sweeping up the surrounding gas while compressing and heating it. The ejecta will be slowed down by a reverse shock. This double shock structure is the working surface of the jet. As the jet interacts only over a small area with the environment, the shocked high pressure material is forced sideways. Due to velocity variations in the ejecta, there may be internal working surfaces as well, when faster moving material overtakes previously ejected slower material. Shear-induced Kelvin–Helmholtz instabilities along the cavity wall will lead to mixing of surrounding gas into the jet and momentum transfer from the outflow to the surrounding gas. The cavity that is created will be illuminated by the stellar FUV and X-ray radiation field, creating a PDR-like environment.

In the next subsection, the structure of shocks will be briefly discussed. There are authorative textbooks and review articles on these aspects, which should be consulted for deeper insight. Molecular observations can be used to study the kinematic structure in the outflow and its interaction with the environment as well as determine the physical conditions in the surrounding gas. A few examples will have to suffice to illustrate these aspects of jets and outflows. The influence of shocks on the chemical composition of regions of star formation will be discussed in Section 11.7.3.

11.7.1 Shock Structure

When the velocity of oncoming material exceeds the speed of sound, the medium cannot respond dynamically until this material arrives. This results in a shock, in which the medium is compressed, heated, and accelerated. The shocked gas will then cool through line emission. Generally speaking, there are two classses of shocks: J and C shocks.

Fast shocks in which the gas is so suddenly stopped and heated that no relaxation takes place and the shock heating layer is much thinner than the postshock relaxation layer. These are the so-called J shocks (J stands for Jump). The preshock and postshock conditions can be related through the mass, momentum, and energy conservation equations across the shock front. These jump conditions are known as the Rankine–Hugoniot equations. Solving for the cooling in the gas and its influence on the temperature and density completes the calculation of the shock structure.

For low velocity shocks in a weakly ionized, magnetized plasma, the work is done by the (trace) ions drifting through the neutrals. These heat up the neutral gas over a much longer length scale and the gas can simultaneously cool. The temperature is then set by the balance of heating and cooling. These are the so-called C shocks (C stands for continuous).

Figure 11.49 shows the results of model calculations for J- and C-shocks at different velocities. Whether a shock is a C or a J shock depends on the density, composition, degree of ionization, and the magnetic field. Low shock velocities, low degree of ionization, high density, and high magnetic field strength favor C shocks. Finally, this dichotomy assumes that the shock has attained steady-state structure but when the dynamical age of the system is short, the shock may have developed a magnetic precursor characteristics for a C-shock but still retain a J-type discontinuity.

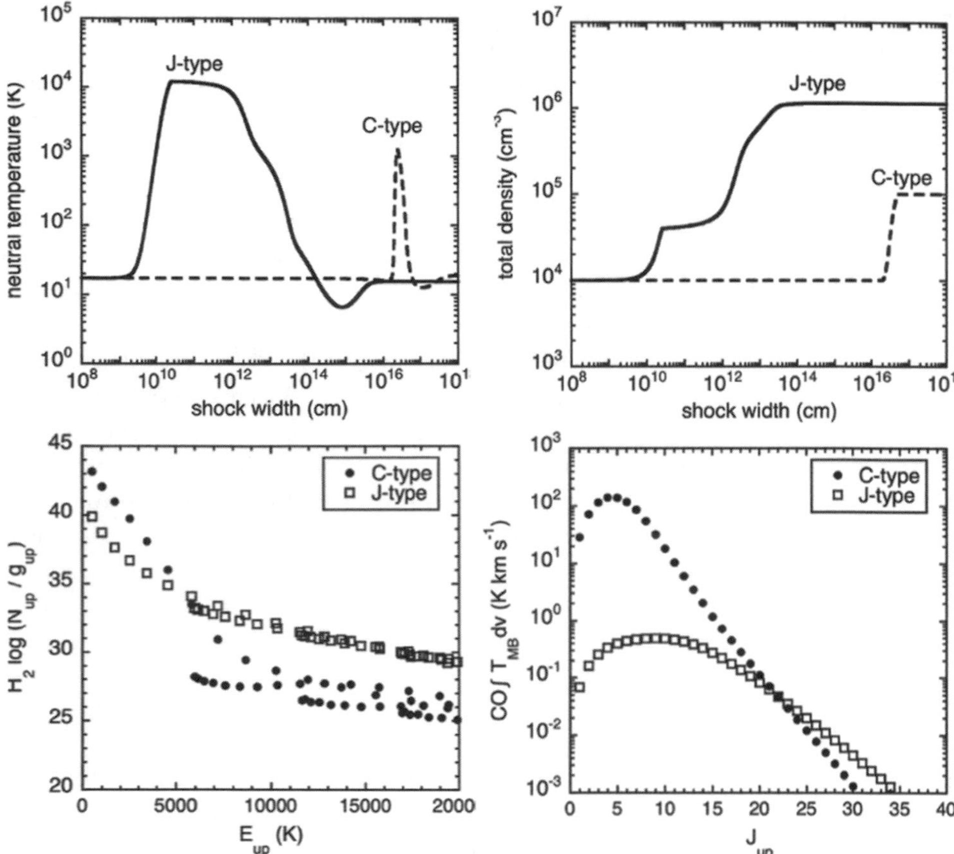

Figure 11.49 The structure and emission characteristics of C and J Shocks. Physical conditions are a shock velocity of 15 km/s, a preshock density of 10^4 cm^{-3} and a perpendicular magnetic field strength of 100 μG (C shock) and 10 μG (J shock). Top row: Temperature and density profiles as a function of position in the shock. Note that the position is plotted on a log scale and in reality the J shock is much thinner than the C shock and is characterized by a very rapid rise and decline. For the J shock, the density increases by a factor 4 in the shock front and then by an additional factor 25 as the postshock gas cools. For the C shock, heating (and cooling) occur on a much longer timescale. Bottom row: Rotation diagram for H_2 (left). Integrated intensities ($\int T_b dV$) of the pure rotational transitions of CO as a function of J (right). Figure adapted from [123, 133]. Figure kindly provided by Dr Antoine Gusdorf.

There are a number of only partially solved issues with shock models. First, if PAHs are present, they are the predominant negative charge carriers and the specifics of this will influence the structure of C shocks. Second, for high velocity J shocks ($\gtrsim$30 km/s), molecular hydrogen will be dissociated and can reform in the cooling, compressed post shock gas. The newly formed H_2 may be released into the gas phase with excess internal energy, which can be transformed to heat by collisions with gas atoms. It is not well understood how much of the formation energy is retained as internal excitation of H_2 rather than transferred to the grain: either the phonon bath or to neighboring molecules through vibrational coupling. In principle, this could lead to a large gas column density maintained at a relatively warmish temperature. Third, as discussed in Section 11.6.3, the H_2 formation rate depends on the temperature of the (cooling) gas and the temperature of the dust (possibly heated by the nearby protostar) and this introduces additional uncertainty in the shock structure of fast J-shocks. Fourth, sputtering of grains will influence the structure and molecular composition of a shock and the details are important but yet uncertain. Much of the molecular gas will be frozen out in ice mantles on silicate or carbonaceous cores. These physisorbed species are bonded by between 0.05–0.5 eV, depending on the species and the surrounding matrix. In a J-shock, the gas is stopped suddenly, but the inertia of grains will lead to gas–grain drift velocities of $0.75 v_s$. "Incoming" gas atoms can then sputter ice mantle molecules. For a C-shock in a molecular cloud, grains may be charged (at least part of the time) and coupled to the magnetic field. The grains may then develop a drift speed relative to the neutral gas comparable to the ion–neutral drift velocity. And again this can lead to sputtering of ice mantles. Calculations show that sputtering becomes important for shock velocities of $\simeq$25 km/s for H_2O but can occur at much lower velocity for CO adsorbed on a CO mantle. Grain core material is more strongly bound ($\simeq$4.5 eV) and hence requires higher interaction energies. Sputtering of silicate grains sets in at $\simeq$30 km/s and a shock of 50 km/s may lead to about 0.05 of refractory grain material sputtered into the gas phase. In the diffuse ISM, some of the Si seems to be in the form of a somewhat less strongly bound material. In addition, some Si may be part of the ice mantle. These would be released at lower velocities.

The structures of C and J shocks are compared in Figure 11.49. As expected, for the same velocity, J shocks reach a higher temperature and compress the gas more than C shocks. The warm postshock gas can drive a rich chemistry. We have seen this in a different setting when the chemical effects of turbulent dissipation zones were discussed in Section 9.4 and when we discussed the chemistry in the warm inner regions of protostellar disks (Section 11.4.2). Any atomic oxygen present in the gas phase will be quickly converted into H_2O in warm H_2 gas but if the shock is dissociative, atomic H will drive the chemistry the other way. Any Si released into the gas phase will react with OH (resulting from O + H_2) to form SiO. Atomic S is converted with H_2 to H_2S in sequential steps and as for H_2O, the presence of H drives this chemistry back. S can react with OH to SO and SO_2. It is clear that the issues mentioned above directly reflect back on the chemistry and emission characteristics of interstellar shocks.

Summarizing this, H_2 and CO are often used as shock tracers as they are important coolants, and H_2O, SiO, SO, and SO_2 are often used as shock indicators in molecular clouds. They do trace shocks well both in morphology and velocity profile. But these species do require OH and hence gas phase O and reformation of H_2 in high velocity shocks. In addition, Si also requires sputtering of (refractory) grain material. The difference in structure of J and C shocks can be seen back in the emission characteristics. Figure 11.49 illustrates the expected emission for H_2 and CO from a C and a J shock.

11.7.2 Physical Conditions in the Orion BN/KL Outflow

The Orion BN-KL region shows multiple outflows including a low velocity ($\simeq$18 km/s) north-east south-west flow (the plateau; Figure 11.7) driven by source I and a spectacular, wide angle, high velocity ($\simeq$30$-$100 km/s) outflow roughly perpendicular to this. The latter figures prominently in H_2 maps of this region in the form of bright H_2 "fingers," glowing in, e.g., the 1$-$0 S(1) line, with [FeII] "fingernails." Some 500 years ago, an exposion – possibly linked to a dynamical encounter in the OMC1 cluster – ejected some 8 $M_\odot$ of material with a total energy of some $\simeq 10^{47}$ erg and a mean velocity of 20 km/s. This violent explosion is shredding the molecular core and the H_2, and other tracers represent the shocks where the ejected bullets interact with the core. The total luminosity in the H_2 lines amounts to $\simeq$120 $L_\odot$.

As we have seen in other environments, H_2 provides a powerful diagnostic of the physical conditions in emitting gas. Figure 11.50 shows the rotation diagram obtained from the 2.5$-$45 μm spectrum of the Orion molecular shock. Comparing and contrasting this diagram with that derived from the observed emisison of a PDR (Figure 11.40) drives home the difference between a FUV-pumped and a thermal population. As optical depth effects are not important for these quadrupole lines, the curvature in the distribution is indicative of suprathermal excitation and "critical density effects" (Section 4.2.2).

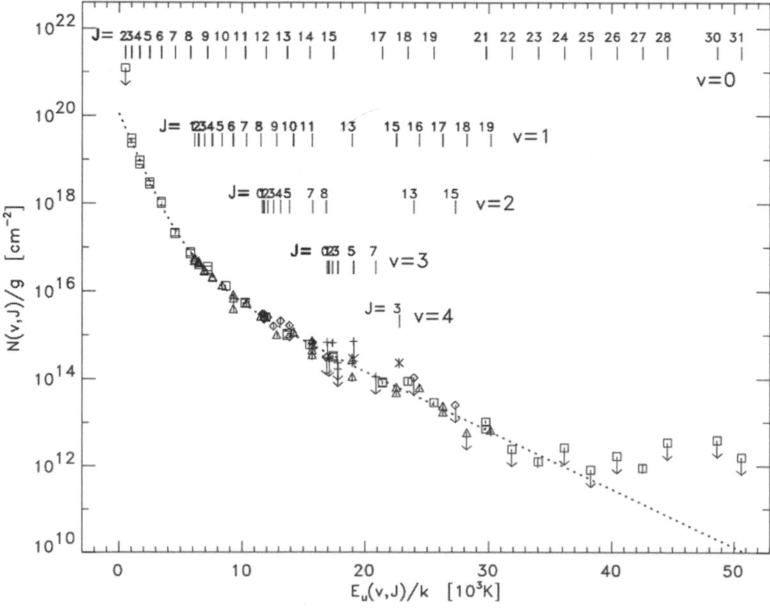

Figure 11.50 H_2 rotation diagram derived from the SWS/ISO 2.5$-$45 μm spectrum. Symbols: $v = 0$ squares, $v = 1$ triangles, $v = 2$ diamonds, $v = 3$ plus, $v = 4$ cross. The data have been extinction corrected. Figure taken from [247]

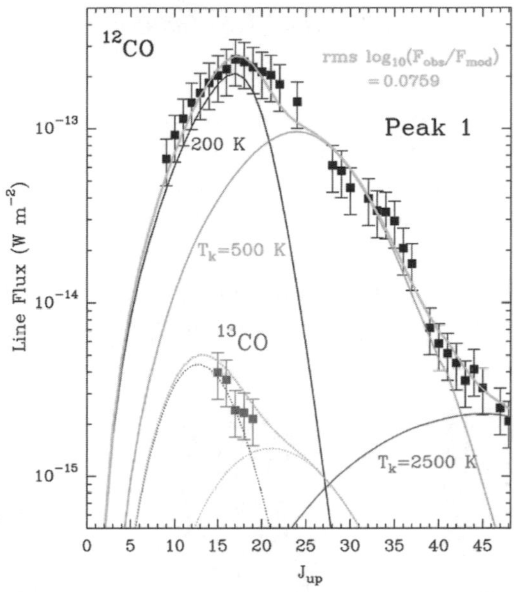

Figure 11.51 Herschel/PACS and SPIRE observations of CO line fluxes in the BN/KL region, Orion Peak 1, are compared to non-LTE excitation and radiative transfer model for ^{12}CO (solid) and ^{13}CO (dashed). The fit reveals three components: the cool ($\simeq$200 K) plateau, the warm ($\simeq$500 K) shock, and the hot ($\simeq$2500 K) shock. The sum is also shown. Figure taken from [126]

The far-IR spectrum of this region shows bright emission in high excitation lines of CO (up to $J = 48{-}47$, $E_u = 6458$ K), H_2O (up to $8_{26}{-}7_{35}$ with $E_u = 1414$ K), and OH (up to $^2\Pi_{1/2}$ $J = 9/2{-}7/2$ with $E_u = 876$ K). The total luminosity in these lines is appreciable: CO 84, H_2O 32, and OH 13 $L_\odot$, respectively. In addition, the gas cools through the ro-vibrational transitions of CO and H_2O with some 20 $L_\odot$. The observed emission reveals a range in excitation conditions reflecting that the bullets propagate into a region with considerable density structure as well as that the shock velocity will vary over the face of the bullet from very fast to much slower away from the stagnation point. Analysis of the data (cf. Figure 11.51) implies that most of the warm CO column density (N (CO) $\simeq 2 \times 10^{19}$ cm^{-2}) arises from $T_{kin} \simeq 200{-}500$ K, dense ($n \simeq 2 \times 10^7$ cm^{-3}) gas associated with low-velocity ($\upsilon_s < 25$ km/s) shocks that do not sputter grain ice mantles and are characterized by a low gas-phase H_2O/CO ($\lesssim 10^{-2}$) abundance ratio. Highly excited CO transitions (N (CO) $\simeq 1.5 \times 10^{16}$ cm^{-2}) as well as the H_2O emission originate from much higher velocity shocks that have sputtered the ice mantles, increasing the H_2O/CO abundance ratio to $\gtrsim 1$, and have resulted in temperatures of $\sim$2500 K in this dense ($n \simeq 2 \times 10^7$ cm^{-3}) gas.

11.7.3 The L1157 Outflow

Physical Structure

The outflow from the low mass, class 0 protostar, L1157mm ($L_\star = 4$ $L_\odot$) in Cepheus at a distance of 250 pc has created two limb-brightened cavities in its environment (Figure 11.52). Because of the precessing of the (intermittent) jet, there are multiple, distinct shocks evident: The blue-shifted and red-shifted shocks are generally known as B0, B1, and B2, and R0, R1, and R2. These shocks represent the working surfaces where the jet interacts with the cavity with B2 and R2 the oldest interaction ($\simeq$4000 yr ago) and B1 and R1 are about 2000 yr old. In interferometric maps, the B1 shock as well as the cavity walls break up

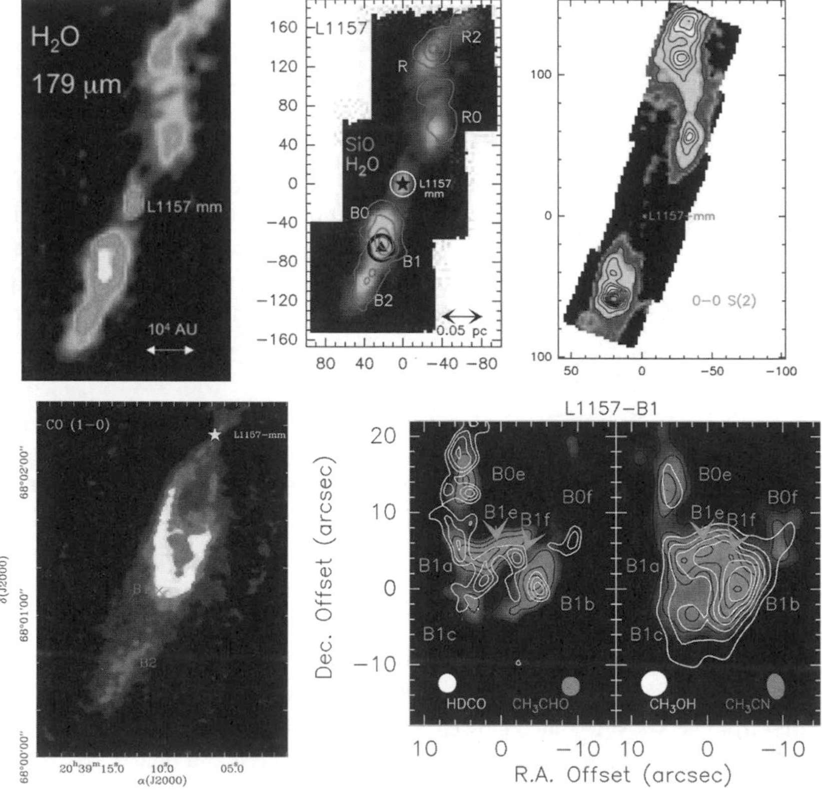

Figure 11.52 The interaction of the outflow from the low mass protostar, L1157mm, with its environment. Top row: The red and blue lobes. The 179 μm H_2O $2_{12}-1_{01}$ transition (left), the SiO contours on top of the H_2O map (center), and the pure rotational H_2 0–0 S(2) emission (right). Bottom row: Blow up of the blue lobe in CO $J = 1-0$ (left). Blow up of just the B1 shock in acetaldehyde with superimposed HDCO contours (middle) and methyl cyanide with superimposed methanol contours (right). CO figure kindly provided by Gemma Busquet. Blow up of the B1 shock region kindly provided by Claudio Codella. Taken from [43, 70, 214, 215] (A black and white version of this figure will appear in some formats. For the color version, please refer to the plate section.)

Table 11.5 *Physical characteristics of the L1157 outflow*

Component	n cm^{-3}	T K	L_{H_2} $L_\odot$	L_{H_2O} $L_\odot$	L_{CO} $L_\odot$
Warm	$1-3 \times 10^6$	250–300	0.06^b	0.002	0.004
Hot	$0.8-2 \times 10^4$	900–1400		0.03	0.01

[a] Taken from [43, 215]. A distance of 250 pc has been assumed. [b] Combined luminosity of both components.

in a number of distinct clumps. The outflow is well separated from the driving source, there is no confusion with other sources, and its size is well matched to the beam size and field of view of single dish/interferometers. This has made this source a welcome target to study the chemical consequences of the interaction of protostellar outflows with the surrounding molecular cloud core. Indeed, position B1 has been dubbed the prototype of a chemically active outflow.

Analysis of the molecular data reveals the presence of multiple components. A high density ($1-3 \times 10^6$ cm^{-3}), warm (250–300 K) component with a low H_2O abundance (10^{-6}) and a low density ($0.8-2 \times 10^4$ cm^{-3}), hot (900–1400 K) component with a high H_2O abundance ($1.2-3.6 \times 10^{-4}$). These results are suggestive of two different shocks propagating into the environment, a high velocity shock ($v_s > 25$ km/s) moving into a low density medium, which has sputtered the ice mantles, and a low velocity shock ($v_s < 25$ km/s) moving into a higher density medium, where most of the oxygen remains frozen out in ice mantles. The walls of the cavity can be probed in various species (e.g. CO, CS) and has a density of 10^5-10^6 cm^{-3}. Molecular hydrogen emission dominates the shock luminosity. In addition to the molecular luminosities summarized in Table 11.5, there is a small contribution by OI (0.002 $L_\odot$) and OH (0.0004 $L_\odot$). The total luminosity of the shock is dominated by the hot component and amounts to about 0.13 $L_\odot$ or about 3% of the stellar luminosity. CO $J = 2-1$ reveals an equivalent H-nuclei mass of about 0.12 $M_\odot$ in the southern lobe, which is a factor of 10 more than the total mass of emitting H_2 ($\simeq$0.016 $M_\odot$). The total kinetic energy, assuming an angle to the line of sight of 81°, is 2.4×10^{45} erg.

Chemistry

The chemical composition in the L1157 B1 shock region is summarized in Table 11.6. This summary is somewhat misleading as, at high resolution, this region breaks up in a number of components, which show compositional variations in some species (Figure 11.52). No attempt was made to separate out these clumps. The L1157 B1 region is rich in the same oxygen-bearing species as Hot Corinos (cf. Table 11.1). It is therefore generally accepted that the organic inventory of shocked regions is due to the release of ice mantle species into the gas phase. A direct comparison of the abundances is somewhat involved as only part of the shock may be strong enough to sputter molecules of the ice. Furthermore, some species (e.g. CO) may come off more easily than others. Perhaps the best way to make a

Table 11.6 *Molecular abundances in the L1157-B1 shock*

Species	Abundance[a]	Reference	Species	Abundance	Reference
CO	1.0 (−4)	assumed	CS	5.0 (−9)	[128]
H_2CO	2.0 (−7)	[193]	H_2CS	2.2 (−8)	[144]
CH_3OH	1.0 (−6)	[41, 176]	C_2S	1.9 (−9)	[144]
CH_3CH_2OH	2.3 (−8)	[176]	SiO	4.5 (−7)	[236]
CH_3OCH_3	2.5 (−8)	[176]	SiS	1.0 (−8)	[236]
CH_3CHO	1.3 (−9)	[70, 176]	H_2S	1.5 (−7)	[143]
$HCOOCH_3$	2.7 (−8)	[176]	SO	1.2 (−7)	[144]
$HCOCH_2OH$	1.7 (−8)	[176]	SO_2	6.2 (−8)	[144]
CH_3COOH	<5.0 (−8)	[176]	OCS	7.8 (−9)	[144]
HCOOH	8.5 (−9)	[176]	NO	2.5 (−6)	[71]
CH_2CO	8.5 (−9)	[176]	PN	4.5 (−10)	[175]
H_2O	2–5 (−5)	[43]	PO	1.0 (−9)	[175]
HCN	5.0 (−8)	[41]	PH_3	<5.0 (−10)	[175]
HC_3N	2.0 (−9)	[194]	HCO^+	4.0 (−8)	[235]
HC_5N	6.0 (−10)	[194]	N_2H^+	1–4 (−9)	[70]
CH_3CN	5.0 (−9)	[68]	HCS^+	3.0 (−10)	[235]
HNCO	1.6 (−8)	[41, 193]	$HOCO^+$	5.0 (−10)	[235]
NH_2OH	7.0 (−9)	[191]	SO^+	3.5 (−10)	[235]
NH_2CHO	1.8 (−9)	[176, 193]	HDO/H_2O	0.001	[69]
			CH_2DOH/CH_3OH	0.03	[110]
			$HDCO/H_2CO$	0.07	[110]
			HDS/H_2S	2.5 (−2)	[143]
			DCN/HCN	0.004	[44]

[a]Relative to H-nuclei ($N_H = 2 \times 10^{21}$ cm^{-2}).

quantitative comparison is by normalizing the measured abudances to those of CH_3OH, a species that is generally considered a parent species made by hydrogenation of CO on grain surfaces. When we make that comparison, we find that the organic composition of the B1 shock in L1157 is within a factor of 2 of that measured for the Hot Corino in IRAS 16293-2422 for H_2O, H_2CO, CH_3CH_2OH, CH_3OCH_3, and $HCOOCH_3$, lending support to a grain mantle origin for these species. Moreover, it suggests that these species are chemically related to methanol, and branching ratios in their formation are insensitive to the conditions. Formic acid, acetaldehyde, isocyanic acid, fomamide, glycol aldehyde, and ketene on the other hand show somewhat larger differences (relative to methanol) between these two sources (factor 3–12). The difference in spatial distribution also suggests that the chemistry is more involved (Figure 11.52). These species have in common that they are "sidechains" in the formation route of CH_3OH and this may be the deciding factor. In grain surface chemistry, HNCO and NH_2CHO result from the reaction of atomic N with HCO (the first step in the hydrogenation of CO toward CH_3OH). In this reaction, the branching ratio between the NCHO and HNCO isomers is not known and may be influenced by both energetics (favoring HNCO) and steric factors (possibly favoring NCHO). Likewise, HCOOH results from the reaction of atomic O with HCO. Possibly, the "parent" region of L1157 B1 had more atomic O and N (relative to H) available during ice mantle formation than in IRAS 16293-2422. Glycolaldehyde may be formed through reaction between the formyl and the hydroxymethyl radicals – intermediaries in the hydrogenation of CO – and again this would be a sidechain in CH_3OH formation. Alternatively, these species may reflect gas phase formation routes toward complex organic species. This route would have to be jumpstarted by seeding the gas phase through injection of parent species from the ice, followed by breakdown to a radical (e.g. NH_3 to NH_2) and then reaction with an abundant parent ice species (e.g. H_2CO). Finally, we note also that the comparison with the composition of the Orion Hot Core (Table 11.1)) is not as favorable, indicating that the composition of ices is sensitive to the difference in the physical evolution between high mass and low mass star-forming regions.

There are some more general inferences to be made from this data. The deuterium enrichment in the "ice-species," water and methanol (but not formaldehyde), observed toward L1157 B1 is more modest than the fractionation observed toward IRAS 16293-2422 (compare Table 11.6 with Table 11.1). Particularly, the modest enhancement in CH_2DOH suggests that ice mantles were formed under less extreme reducing conditions in L1157 B1 than those in IRAS 16293-2422. The other general point to be made about the origin of complex organic species in these environments is that the ices in L1157 B1 were never exposed to strong FUV fields as they are far away from the powering source, L1157mm, and this source is very low luminosity anyway. We already noted that the cosmic-ray-induced radiation field is not strong enough to produce much processing (Section 11.3.2). Likewise, while thermal processing of the mantles resulting in diffusion of radicals may be relevant in a Hot Corino setting, this is not important in sputtering environments. Cosmic ray spikes during the dark cloud phase may still drive temperature fluctuations. A 100 MeV Fe cosmic ray will hit a 1000 Å grain every 80,000 yr and raise its temperature to 35 K, enough to get rapid diffusion of CH_3 but more limited diffusion for HCO and OH. A 10 MeV Fe cosmic ray will raise the temperature to 70 K and much chemistry might result. However, such a hit only occurs every 800,000 yr and is likely not very relevant for the evolution of this L1157 B1 region.

11.8 Masers

Because of their brightness, maser lines are easily detected and can be used to trace star-forming regions. In addition, they are often used to study the kinematic structure of the region, tracing rotation or outflow. Astronomical masers can give insight in the physical conditions, but, as the strength of the maser lines is affected by a number of factors, including physical conditions – velocity gradients, amplification path, and radiation field – in a highly nonlinear way, derivation of physical conditions from the observations is involved.

The Zeeman splitting of OH hyperfine lines provides unique information on the magnetic field strength in regions of massive star formation.

11.8.1 The Physics of Masers

Normally, the population per magnetic sublevel decreases with increasing energy. Under special conditions, population inversion can occur and maser action can ensue. The opacity, and hence the optical depth, scales with $\kappa_{lu} \propto n_l B_{lu} - n_u B_{ul}$ where the second term is the correction for stimulated emission. Using the relationship between the Einstein B coefficients, we can write this as,

$$\kappa_{lu} \propto B_{lu} \left(1 - \frac{g_l}{g_u} \frac{n_u}{n_l} \right). \tag{11.68}$$

With a population inversion, κ becomes negative and stimulated emission can drive maser action. In Section 4.1.2 we examined the two-level system and concluded that the relative level population is given by,

$$\frac{n_u}{n_l} = \frac{\frac{g_u}{g_l} \exp\left[-E_{ul}/kT\right]}{1 + n_{cr}/n}, \tag{11.69}$$

with n_{cr} the critical density, $n_{cr} = A_{ul}/\gamma_{ul}$. This expression shows that, per magnetic sublevel, the upper level population is always less than the lower level population. Essentially, this is the effect of time reversibility in collisional processes[8] as microscopic reversibility demands that upward and downward transitions are related by the Boltzmann factor, e.g., $\gamma_{lu} = \gamma_{ul}\,(g_u/g_l) \exp\left[-E_{ul}/kT\right]$. Hence, a two-level system can never maser. Masering requires the interaction with other levels, either rapid, preferential pumping of population into the upper level or rapid draining of the lower level to other levels. Masering, of course also requires that the density is below the critical density of the masering levels as high densities drive the system to thermodynamic equilibrium.

Molecules are "natural" masers and we can illustrate that here by looking at the rotational energy level system. Consider first a diatomic molecule where, to first order, the energy levels are separated by $2B\,(J+1)$ (Section 3.2.2). The Einstein $A_{J,J-1}$ coefficient scales with $\nu^3 \propto (J+1)^3$ and hence higher levels drain more quickly than lower levels.[9] We have seen that this gives rise to suprathermal excitation and lower levels become overpopulated compared to upper levels (Sections 4.1.8 and 4.2.2). This gives rise to curvature in rotation diagrams and we have seen in previous chapters that this behavior is common in space. So, collisional excitation in a linear molecule leads to pile up of population in lower levels but this is just the opposite of population inversion and no masering will occur. A linear molecule can only maser if there is a radiative pump, which preferentially excites upper levels and this is the case, for example, in SiO masers.

[8] For a gas with a Maxwellian velocity distribution.

[9] Actually, $\propto (J+1)^4\,/\,(2J+3)$ when the dependence on the dipole moment is taken into account as well (Section 3.2.9).

This situation changes for a symmetric top molecule where the energy levels are described by two quantum numbers, the total angular momentum, J, and its projection on the symmetry axis, K. For a given K, the rotational energy level diagram is essentially that of a linear rotor, where energy increases with J as $J\,(J+1)$, except that they start with $J = K$ rather than $J = 0$ (Section 3.2.3). There is no dipole moment along the symmetry axis, resulting in the selection rule, $\Delta K = 0$. Radiative relaxation will thus occur along the K-ladders and the $J = K$ level is metastable. Collisions will then establish LTE among the $J = K$ levels as the critical densities are low but, within each K-ladder, levels can be subthermally excited. Levels within one K-ladder can then easily become overpopulated relative to levels within the adjacent $K = J - 1$ ladder (but radiative transitions are of course forbidden).

11.8.2 Water Masers

Consider now H_2O, which is an asymmetric top. The H_2O energy level diagram is shown in Figure 11.53. The levels are labeled by the total angular momentum, J, and two indices, K_a and K_c, which refer to corresponding prolate and oblate symmetric tops. Transitions between K ladders are now allowed (the selection rules include ΔK_a or K_c is ± 1). The lowest level of each J-ladder – the so-called back bone – have a high absorption cross section and as a result they become readily optically thick. Because of radiation trapping,

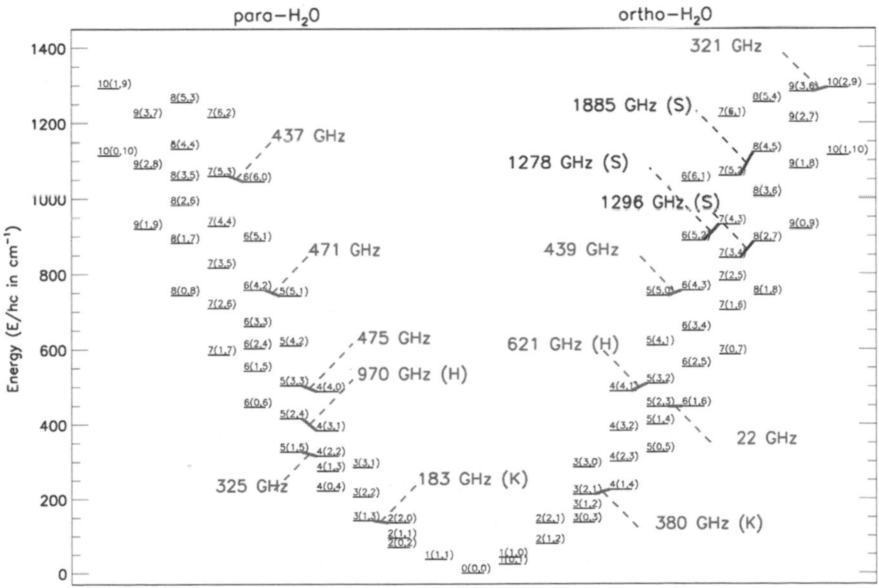

Figure 11.53 Rotational energy level diagram for H_2O. Para levels to the left, ortho levels to the right. J ladders have been shifted to the left/right. States are labeled by $J\,(K_a, K_c)$. Lines indicate maser transitions observed from the ground, the Kuiper Airborne Observatory (K), the Herschel Space Observatory (H), or SOFIA (S). Adapted from [213]

collisions between the backbone levels establishes LTE and most of the population is in these levels. In contrast, the non-backbone levels are predominantly populated by radiative decay into them from backbone levels and their population drains quickly away. H_2O is thus a case where population inversion occurs because the lower level drains more quickly radiatively than the upper level. This reflects that either the lower state has more decay channels and/or its radiative transitions are at higher frequencies and therefore faster, and/or it connects to levels that are substantially subthermally excited. The first H_2O maser discovered in space (22 GHZ; $6_{1,6} - 5_{2,3}$) is a case in point as the transition $5_{2,3} - 4_{1,4}$ is the main drain of the lower level of the masing transition.

We can get some further insight by considering the excitation as a modified two-level system where, besides the maser process connecting the two levels directly, we include pump rates, P_i, into and loss rates, Γ_i, out of the levels. These pump and loss rate are per unit volume and include all the collisional and radiative rates except for the emission in the masering transition. Ignoring the maser action, the level populations are,

$$f_i^0 = \frac{P_i}{\Gamma_i n\,(H_2O)} \qquad i = u, l \tag{11.70}$$

and maser action requires that $f_u^0/g_u > f_l^0/g_l$. Now, maser action will reduce the inversion. If we assume that the maser action does not affect the population in the other levels – a good assumption for H_2O mases in space – and introduce the rate of maser emission, Φ, the actual steady-state level populations can be described by,

$$f_u = \frac{P_u - \Phi}{\Gamma_u n\,(H_2O)} \tag{11.71}$$

$$f_l = \frac{P_l + \Phi}{\Gamma_l n\,(H_2O)}, \tag{11.72}$$

which yields,

$$\frac{f_u}{g_u} - \frac{f_l}{g_l} = \left(\frac{f_u^0}{g_u} - \frac{f_l^0}{g_l}\right)\left(1 - \frac{\Phi}{\Phi_s}\right), \tag{11.73}$$

where the saturated maser intensity,[10] Φ_s, is given by,

$$\Phi_s = \left(\frac{f_u^0}{g_u} - \frac{f_l^0}{g_l}\right) n\,(H_2O) \left[\frac{1}{g_u\Gamma_u} - \frac{1}{g_l\Gamma_l}\right]^{-1}. \tag{11.74}$$

It is clear that the maser intensity has to be less than Φ_s for maser action to occur. We can take Φ_s as a measure for the maser action and we introduce the maser rate coefficient, Q, by $\Phi_s = Qnn\,(H_2O)$. Significant maser action requires a large enough column density. We can define an unsaturated gain length by, $N_s/n\,(H_2O)$, where N_s is given by the value for which $\tau = -1$,

[10] In the literature, sometimes, the intensity of a saturated maser is defined as the maser intensity that reduces the inversion by a factor of 2, which is, of course, simply related to Φ_s.

$$N_s = \frac{4\pi\,\Delta v}{hc\left(f_u^0 B_{ul} - f_l^0 B_{lu}\right)}. \tag{11.75}$$

Saturated maser action occurs when the column density exceeds N_s by a logarithmic factor $\ln[A]$, where A is the maser gain. The maser intensity will drop quickly when the column density drops below $N_s \ln[A]$. The maser gain can be related to the maser rate coefficient through, $A \simeq 4\pi n Q/f_u A_{ul}\Omega$ with Ω the maser beam solid angle. Inserting typical values for interstellar H_2O masers ($n \sim 10^9$ cm^{-3}, $Q \sim 10^{-14}$ cm^3 s^{-1}, $f_u \sim 10^{-2}$, $A_{ul} \sim 10^{-6}$ s^{-1}, and $\Omega \sim 10^{-2}$) we have $\ln A \sim 15$.

In order to determine whether maser action will occur in a transition and calculate the maser intensity, we have to solve the rate equations governing the level populations. The solution to these equations will depend on the ratio, $n\gamma_{ij}/A_{ij}\beta\left(\tau_{ij}\right)$ (cf. Section 4.1.6). Now, the non-maser H_2O transitions are optically thick and $\beta\left(\tau_{ij}\right) \propto 1/\tau_{ij} \propto 1/B_{ij} \propto 1/A_{ij}$. Thus, we can identify ξ as the parameter controlling level populations,

$$n\gamma_{ij}/A_{ij}\beta\left(\tau_{ij}\right) \propto \xi \equiv \frac{n^2 X\,(H_2O)}{\Delta v/\ell}, \tag{11.76}$$

where ℓ is the length over which velocity coherence is maintained and Δv is the line width. As this discussion makes plausible, the maser intensity will be set by the rate coefficient for saturated maser action, Q, and the unsaturated maser column density, N_s. Both of these quantities are controlled by ξ and by the temperature of the gas. The brightness temperature is then given by,

$$T_b = \frac{\lambda^3}{2k}\frac{\Phi_s h\nu}{\Omega}\frac{\ell}{\Delta v}. \tag{11.77}$$

Stimulated emission depends directly on the radiation field and hence the direction with the highest optical depth will "win" exponentially and will dominate the maser emission. Conversely, observers will only see the maser if the beam is directed at them. As the beaming angle is not known, observers specify the maser luminosity assuming it is isotropic (which it never is for masers). So, this procedure will overestimate the luminosity of a single maser spot but, for a group of masers, if they are randomly oriented, this would provide a reasonable estimate of the total maser luminosity. The apparent isotropic photon luminosity of a saturated maser is,

$$L_{ul} = \Phi_{sat} V \frac{4\pi}{\Omega} = \xi Q \frac{\Delta v}{\ell} V \frac{4\pi}{\Omega}, \tag{11.78}$$

with V the volume of the masering region, where the factor $4\pi/\Omega$ is typically very large. The parameter space for interstellar water masers is best described by a scaled ξ parameter,

$$\xi' = \left(\frac{n}{10^9\ \mathrm{cm}^{-3}}\right)^2 \left(\frac{X\,(H_2O)}{10^{-4}}\right)\left(\frac{10^{-8}\ \mathrm{s}^{-1}}{\Delta v/\ell}\right), \tag{11.79}$$

with $\xi = 10^{22}\xi'$.

The level populations for H_2O have been solved for relevant conditions and maser lines have been identified. The results for maser lines at 400 K are shown in Figure 11.54.

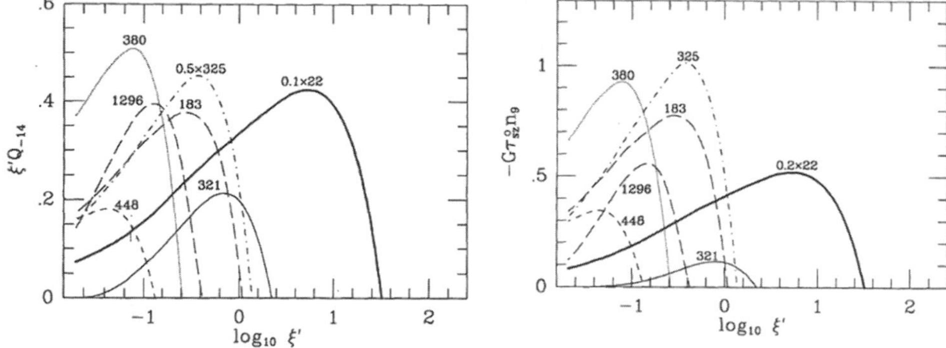

Figure 11.54 Left: Emissivities ($\xi' Q$, with Q in units of 10^{-14} cm^3 s^{-1}) for different H_2O masers as a function of the scaled ξ' parameter and a temperature of 400 K. Curves are labeled by the transitions frequency in GHz (cf. Figure 11.53 except for the 448 GHz line ($4_{2,3} - 3_{3,0}$)). Right: Maser optical depth as a function of the scaled ξ' parameter. n_9 is the density in unit of 10^9 cm^{-3} and G is a geometric factor that enters in the escape formalism (G is 1 for spherical geometry and 3 for a plane parallel slab (cf. Section 4.1.6). Here, $-G\tau^0_{sz} n_9 \propto \xi' \Delta v / N_s$. Adapted from [211]

For each transition, the maser action increases with ξ' until thermal equilibrium is reached and the maser disappears. At higher temperatures, higher levels become appreciably populated and more maser lines appear. Moreover, the intensity of maser lines increases, approximately linearly with T.

Bright H_2O masers require filamentary structures with aspect ratios exceeding $\sim$5, high densities ($n \simeq 10^9$ cm^{-3}), but not too high as the levels thermalize, a high water abundance, and temperatures of a few hundred to a few thousand degrees. Models show that dissociative J shocks can provide these conditions under certain conditions (cf. Figure 11.55 for the 22 GHz maser). J shocks will compress, dissociate, ionize, and heat the gas in the shock front. The hot, dense gas will radiatively cool and compress further. Magnetic cushioning will limit the compression to,

$$n_{ps} \simeq 8 \times 10^8 \left(\frac{n_0}{10^7\,\text{cm}^{-3}}\right) \left(\frac{v_s}{10^2\,\text{km s}^{-1}}\right) \left(\frac{1}{b}\right) \text{cm}^{-3}, \tag{11.80}$$

with n_0 the preshock density, v_s the shock velocity, and where b is related to the magnetic field strength parallel to the shock front by, $B = b n_0^{1/2}$ μG. Once the gas cools down to 10^4 K, the ions recombine, and at 3000 K, newly formed H_2 can "survive." Atomic oxygen will then be transformed into H_2O (and CO) in the warm gas. In these models, H_2 is assumed to be formed efficiently on grain surfaces despite the high gas and grain temperature. Moreover, it is assumed that H_2 molecules are released from the grain with a large fraction of the heat of formation as internal (vibrational) energy. Collisional deexcitation heats the gas. The column density of this warm molecular gas, N_{wmg}, is set by the reformation timescale of H_2, $\tau_{chem}\,(H_2) = \left(k_d n_{ps}\right)^{-1}$ s (cf. Section 10.5),

$$N_{wmg} \simeq n_0 v_s \tau_{chem}\,(H_2) \simeq 1.3 \times 10^{22} \left(\frac{10^{-17}\,\text{cm}^{-3}\,\text{s}^{-1}}{k_d}\right) \text{cm}^{-2}. \tag{11.81}$$

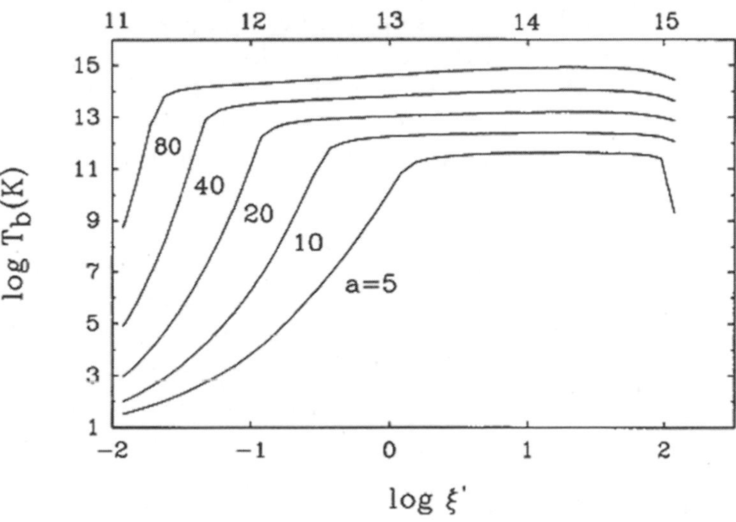

Figure 11.55 Brightness temperature of the 22 GHz H_2O maser as a function of the scaled ξ' parameter at a temperature of 400 K. The different curves are calculated for different aspect ratios, $a \equiv \ell/r_s$ with ℓ the length of the maser path along the line of sight and r_s the size of the maser spot. Taken from [89]

The temperature of the warm molecular gas is set by the balance of the formation heating and cooling by (optically thick) H_2O lines. This yields,

$$T_{wmg} \simeq 400 \left(\frac{n_0}{10^7\,\text{cm}^{-3}}\right) \left(\frac{v_s}{10^2\,\text{km s}^{-1}}\right) \left(\frac{\text{km s}^{-1}}{\Delta v}\right) \text{K}, \tag{11.82}$$

with Δv the width of the cooling lines. This temperature is not very sensitive to k_d or b as both heating and cooling depend in a similar way on these parameters. As the cooling lines are optically thick, this estimate also does not depend on the H_2O abundance. The width of the maser spot is then set by, N_{wmg}/n_{ps}

$$d \simeq 1.7 \times 10^{13}\, b^2 \left(\frac{10^7\,\text{cm}^{-3}}{n_0}\right) \left(\frac{10^{-17}\,\text{cm}^{-3}\,\text{s}^{-1}}{k_d}\right) \left(\frac{10^2\,\text{km s}^{-1}}{v_s}\right) \text{cm}. \tag{11.83}$$

In terms of the ξ' parameter, dissociative shocks have,

$$\xi' \simeq \left(\frac{X\,(\text{H}_2\text{O})}{10^{-4}}\right) \left(\frac{v_s}{10^2\,\text{km s}^{-1}}\right) \left(\frac{n_0}{10^7\,\text{cm}^{-3}}\right) \left(\frac{\text{km s}^{-1}}{\Delta v}\right) \left(\frac{10^{-17}\,\text{cm}^{-3}\,\text{s}^{-1}}{k_d}\right), \tag{11.84}$$

As so often in this chapter, the microphysics controls the characteristics of the region. Specifically, maser action in dissociative J shocks depend on the detailed physics of H_2 formation both in terms of the rate at high gas and dust temperatures as well as in terms of the energy carried off by ejected H_2.

Slow non-dissociative shocks can also produce conditions amenable to water maser action. C shocks naturally lead to large columns of dense, warm molecular gas. Again preshock densities have to be some 10^7 cm^{-3} in order to have postshock densities relevant for maser action. Shock velocities have to be in a relatively narrow range with minimum shock velocities of $\simeq$15 km/s in order to sputter the H_2O ice mantles and no more than $\simeq$45 km/s, as for higher velocities the shocks become dissociative. The density of the postshock warm gas will be less (factor of $\sim$2), as cooling does not compress the gas further. Hence, higher preshock densities are needed to reach masering conditions. In C shocks, the temperature of the postshock, masering gas is not tied to the H_2 reformation heating (cf. Section 11.7.1) and much higher temperatures than 400 K can be reached. Indeed, for velocities in excess of 25 km/s, the warm gas is $1-2 \times 10^3$ K. This difference provides the best discriminant for the question of C versus J shocks, as at high temperatures higher levels can also become appreciably populated and start to maser. In C shocks, as for J shocks, the size of the maser spots is set by the geometry (curvature) of the shock and the need to reach a high enough column density to have appreciable gain (see above). For maser sizes of $1-3 \times 10^{13}$ cm, again, this implies higher preshock densities ($3-10 \times 10^8$ cm^{-3}) than for J shocks. In terms of the H_2O masers in the Orion BN/KL region, this debate on J versus C shocks centers on whether the masers are associated with slow shocks propagating into very dense bullets ejected from the protostar (shocks that are also responsible for the observed H_2 emission) or fast shocks driven into the surrounding, lower density molecular cloud by these bullets. Both of these types of shocks will be present and hence may contribute to the observed maser action. The next generation of large telescopes will reach the size scales at which the location of H_2 emission can be compared to that of maser spots.

This discussion has centered on shocks as the location of H_2O masers, as they produce in a natural way warm, dense molecular gas with coherent velocity structures. The expanding shells of AGB ejecta provide another notable example of conditions where maser action can occur. In regions of star formation, the inner regions of protoplanetary disks associated with protostars are another example of large column densities of warm molecular gas (Figure 11.56). Water masers have also been detected associated with the warm dense gas in subparsec-diameter accretion disks around supermassive black holes. In both cases, maser velocity measurements can be used to trace the dynamics and hence the mass of the central object, the distance of the object, as well as the structure of the disk.

11.8.3 Hydroxyl Masers

In regions of massive star formation, OH maser emission is present during the early evolution of the central object. Early on, the OH maser emission is located in a rotating, circumstellar disk, but after the formation of an ultracompact HII region, a dense toroidal structure remains, which is outlined by the OH masers. These phases of evolution are characterized by bright mid- to far-IR dust emission and that is clearly a requirement. OH maser emission is dominated by the ground state main lines – 1665, 1667 MHz

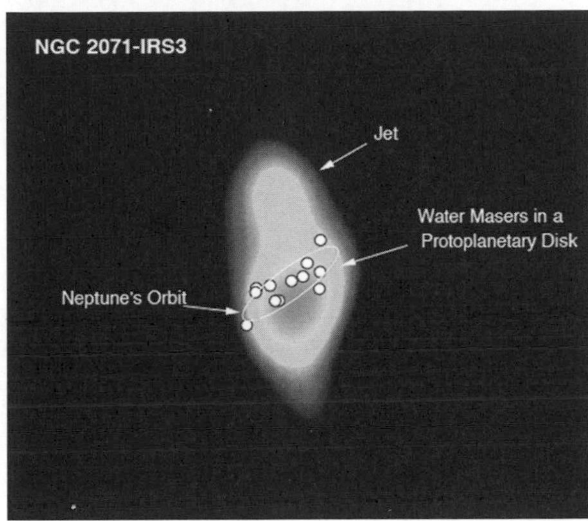

Figure 11.56 VLA continuum emission of the jet associated with the low mass ($\simeq$1 $M_{\odot}$) protostar, NGC 2071 IRS 3. White dots: H_2O maser spots in the protoplanetary disk. The ellipse indicates the size of the orbit of Neptune. Taken from [274]

(cf. Figure 3.5) – where the former is generally the brighter. Higher maser lines can originate from the same region but are less intense. The focus here is on regions of massive star formation but OH masers are also strong in long period variables with strong IR radiation fields such as Miras and OH/IR stars.

The rotational energy levels of OH have been discussed in Section 3.2.6 (Figure 3.6). Weak coupling between the electronic angular momentum and molecular angular momentum leads to lambda doubling of the rotational levels. These Λ-doubles are then further split by coupling between the spin of the unpaired electron and the proton. These hyperfine levels have either parallel or antiparallel spins and are described by the total angular momentum quantum number, $F = J \pm I$. Each Λ-doublet can thus generate four maser lines; the transitions in which F does not change are called main lines. The other two are satellite lines. Allowed transitions require a parity change and $\Delta F = 0, \pm 1$ but $F = 0 \rightarrow 0$ is forbidden. Maser transitions are indicated in Figure 3.6 and correspond to 18, 6, 5, and 2.3 cm.

The inversion of OH masers is the result of preferential pumping of the upper level rather than a depletion of the lower level. As for H_2O, the relevant lines connecting the rotational states are optically thick and the rate, which scales with $\eta_{ij} A_{ij}$, is independent of the intrinsic strength of the transition. Overpopulation reflects the selection rules. Consider here the main line maser at 1665 MHz. Many pumping routes contribute but they all have the same characteristic. Here, we will detail one of the pumping routes to illustrate the principle. Consider pumping by the 119 μm line from $^2\Pi_{3/2}\,(J = 3/2)\; e, F = 1$ to $^2\Pi_{3/2}\,(J = 5/2)\; f, F = 2$ where the e and f stand for $-$ and $+$ parity states. This is followed by collisional deexcitation from $^2\Pi_{3/2}\,(J = 5/2)\; f, F = 2$ to $^2\Pi_{3/2}\,(J = 3/2)\; f, F = 1$

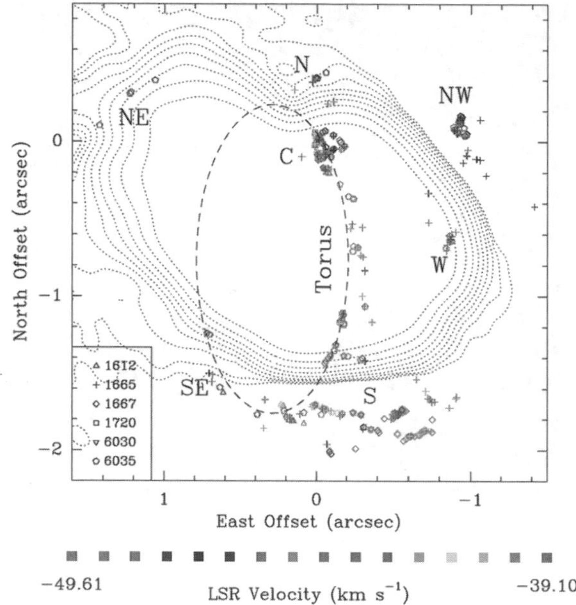

Figure 11.57 The locations of 6 and 1.6 GHz OH masers in W3(OH). Velocities are not corrected for Zeeman splitting, which can be large (several kilometers per second) and unknown (due to unpaired Zeeman components) at 1665 and 1667 MHz. Contours indicate the 8.4 GHz continuum emission and are shown at 4, 8, 16, ... times the rms noise of 11.6 Jy/beam. Key maser groups and the toroidal structure are indicated. Taken from [102]

where radiative decay between these two states is forbidden. Labeling the rotational $^2\Pi_{3/2}$ $(J = 3/2)$ states sequentially starting at 1, we can write for the net result of this pump,

$$k_{1,5}k_{5,3} - k_{3,5}k_{5,1} \simeq B_{1,5}J_{5,1}C_{5,3} - A_{5,1}C_{3,5}, \tag{11.85}$$

where we have assumed that densities are low, so radiative rates dominate over collisional rates, and spontaneous decay dominates over the radiative pump rate and its reverse. Using the relationship between the Einstein coefficients and assuming a low kinetic temperature, we can write,

$$k_{1,5}k_{5,3} - k_{3,5}k_{5,1} \propto \frac{J_{5,1}}{B_{5,1}(T_k)} - 1. \tag{11.86}$$

Inversion requires that this expression is positive. Hence, maser action requires that the (dust) radiation field must exceed the Planck function at the (gas) kinetic temperature and that downward transitions exceed upward collisions. Of course, densities must also be low enough that collisional equilibrium is not established. The restriction on the radiation field translates to constraints on the temperature, emissivity characteristics, filling factor, and optical depth of the dust emission and, in general, OH maser action requires strong mid- to

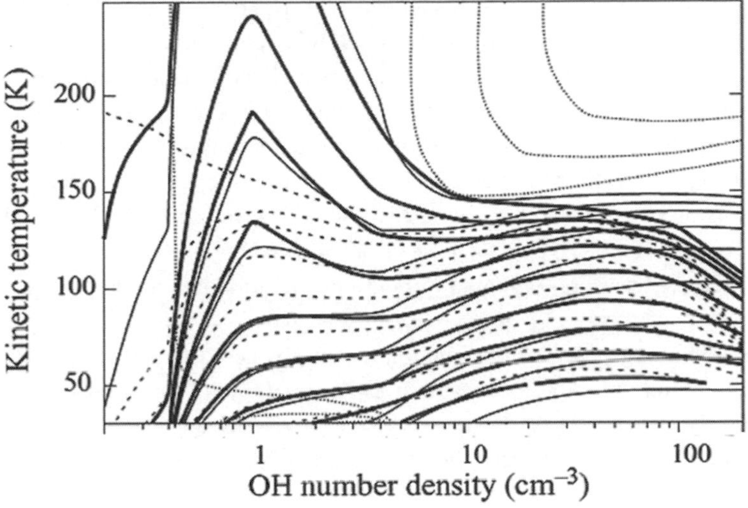

Figure 11.58 Contour plots of maser optical depths. The adopted dust temperature is 155 K and the OH abundance is fixed at 2×10^{-7}. A velocity shift of 0 km s^{-1} is adopted. 1665 MHZ (heavy solid), 1667 MHz (dashed), 1720 MHz (light solid), and 1612 MHz (dotted). Contour levels are 0, 0.5, 1, 2, 4, 8, 16, 32, 64 and increase generally to the bottom right for all transitions except 1612 MHz, where levels increase to the top right. All depths above a value of 16 would result in strongly saturating masers. Taken from [129]

far-IR radiation fields. The real situation is more complicated as several, similar pumping routes contribute. In addition, the far-IR frequencies coupling the hyperfine components in the different rotational states differ by only a few MHz, comparable to the thermal linewidth. Hence, line overlap is an issue and this line locking complicates model studies further.

Typical results for the ground-state maser lines are presented in Figure 11.58. Maser action requires densities in the range 10^4 to 2×10^8 cm^{-3} except for the 1612 MHz line, which requires slightly higher densities. Kinetic temperatures typically have to be less than the dust temperature and the specific column density, N (OH) / (dv/dr), which controls the optical depth, has to be $\sim 10^{11}$ cm^{-2} (km/s)$^{-1}$. The higher frequency maser lines require somewhat higher densities for substantial maser action.

11.8.4 Other Masers

Many other molecules have been detected to maser. Specifically, methanol is rather rich in masing transitions. Observed methanol masers have been divided into two classes. Like OH masers, class II CH_3OH masers – the characteristic maser transition is at 6.67 GHz ($5_1 - 6_0$ A$^+$) – are associated with ultracompact HII regions and W3(OH) provides a prototypical example. Class I CH_3OH masers – strong maser transitions at 36.1 GHz ($4_{-1} - 3_0$ E), 44.0 GHz ($7_0 - 6_1$ A$^+$), and 84.5 GHz ($5_{-1} - 4_0$ E) – are associated with outflow sources but while H_2O masers are associated with the head of the shock, the methanol masers outline the broader interface of the shock with the molecular cloud. Class I masers are favored when the radiation temperature is less than the kinetic temperature of the gas. The radiation field can then be neglected and pumping is due to collisional pumping and radiative decay. Consider the 84.5 GHz maser, three strong downward transitions from 6_{-2}, 5_{-2}, and 6_0 feed into the upper level while only one, the 5_1, level strongly connects to the lower level and no allowed transitions feed into this level from the $K = -1$ ladder. The discussion follows that of H_2O in Section 11.8.2 and comparable equations can be derived. For class II CH_3OH masers, pumping through the OH torsional mode ($\nu = 270$ cm^{-1}) is important.

Maser action in regions of star formation have also been discovered in NH_3 and H_2CO transitions. For ammonia, the metastable transitions ($J = K$) are similar to the H_2O and class I CH_3OH masers. They are associated with shocks and probably have a similar pumping scheme as those masers. For the non-metastable ammonia masers, the large Einstein A values indicate that collisional pumping would require very high densities. Rather, some form of radiative pumping is likely involved. For formaldehyde, IR pumping is indicated. SiO masers have been detected in a few sources, of which the Orion KL region is the best studied. The maser action originates from the dense ($10^9 - 10^{11}$ cm^{-3}), warm ($\sim$1500–2000 K), expanding disk for the $v = 2$ and $v = 1$ transitions. For the $v = 0$ transitions, maser action is ascribed to slightly cooler and lower density outer regions of this disk.

11.9 Further Reading and Resources

There is a long history on the structure of prestellar cores and the onset of gravitational collapse. An early study recognizing prestellar cores is [288]. Observations of prestellar cores have been reviewed by [25]. Hi-GAL observations of infrared dark clouds can be found in [289]. A catalogue of IRDCs is provided by [261]. Reference [257] presents the seminal study on isothermal spheres, the inside-out collapse, and the structure of prestellar cores. The search for infall signature in the line profiles observed toward class -1 and 0 sources was pioneered by [95]. The oscillatory motion of the prestellar core, B68, is reported in [173]. The physical and chemical structure of prestellar cores has been reviewed by [51].

Hot Cores and their characteristics have been reviewed by [59, 172]. For a review of this key phase in the chemical evolution of regions of star formation, see [55]. Early observations of Orion revealed the presence of two distinct regions with very different molecular composition but distinguished by their kinematic signature and distribution: the compact ridge with high abundances of oxygen-bearing organic species and the Hot Core dominated by N-bearing species. This dichotomy is a common characteristic of Hot Cores ([8]). Hot Corinos with very similar characteristics as Hot Cores but around low mass protostars were first described in [38, 52]. An early review is presented by [54].

The analysis of sub-millimeter dust continuum emission from the Orion Hot Core is discussed in [141]. The $C^{18}O$ $J = 2 - 1$ analysis is taken from [100]. The HIFI/Herschel analysis of the $C^{18}O$ ladder in Orion is described in [234]. The discussion on the internally versus externally heated Hot Cores dates back to the seminal study by [31]. For a further discussion based upon radiative transfer studies, see [166]. The analysis of the HIFI/Herschel CH_3OH transitions and the importance of external heating for the Orion Compact Ridge is described in [287]. The insightful Hot Core study is reported in [296]. The cooling budget of OMC2 FUR4 has been investigated by [160]. The abundance of gaseous CO_2 in Hot Cores was studied by [36] and its chemistry was studied by [60].

Hot Core chemistry driven by evaporating ices has been studied by [50, 62]. These were based on experimental studies of alkylization reactions involving alcohols [163, 192]. Experimental studies highlighting the isomerization issues in transfer of CH_4^+ groups from protonated methanol were performed by [150]. The reaction of protonated methanol with formic acid was studied by [72]. The fragmentation issues in the dissociative electron recombination of complex organic species were well illustrated by the study of the protonated dimethyl ether [116, 138]. The importance of NH_3 as a proton scavenger was emphasized by [244]. Experimental studies on the radical surface routes involving HCO and CH_2OH are reported in [45, 66, 99, 292].

Some relevant laboratory studies on the energetic processing of ices by UV photons and energetic ions are [6, 27, 45, 117, 118, 228, 248, 264]. References [153, 219] have focused on identifying the chemical routes involved and [115] have included these in gas–grain models for interstellar chemistry. Reviews can be found in [87, 221, 272]. Photodestruction rates are measured by [73]. The photochemistry of methanol has been discussed by [219] and methanol:water mixtures in [220]. Ammonia photochemistry is described in [159] and the formation of formamide in ammonia:carbon monoxide mixtures is discussed in [157]. Four component mixtures of H_2O, CH_3OH, CO, and NH_3 are described in [6]. The importance of hexamethylenetetramine was emphasized by [27] and its chemistry studied by [281, 282]. The OCN^- assignment for the 4.62 μm band was established by [82]. The importance of PAHs as electron acceptors and donors was demonstrated by [39]. The dipole alignment of ice films has been studied by [22]. The methanol droplet studies have been performed by [249]. Experimental and astronomical evidence for methanol segregation in ices is presented in [34, 81].

The classical paper on the structure and evolution of viscous accretion disks is [184]. A comprehensive overview of the chemistry relevant for protoplanetary disks is presented in [140] while physical processes are reviewed by [16]. Reference [13] provides a review of observations of protoplanetary disks. The temperature structure of flaring, passive disks has been studied by [168] and the two layer approach has been discussed by [64].

There has been a long history of condensation calculations relevant to solar nebula conditions starting with [132]. More recent noteworthy studies are [88, 180]. The "burning" of PAHs in the hot gas of the inner nebula has been investigated by [11, 169]. The role of Fischer–Tropsch in converting CO into hydrocarbons has a checkered past. Recent studies [170, 217, 218, 251] focus on the conversion of CO into CH_4 on metallic and silicate surface as a first step in the formation of hydrocarbons in the gas phase.

The photochemistry of PDR surfaces of protoplanetary disks has been developed by [2, 109]. The role of neutral chemistry in warm PDR gas in the inner disk has been examined by [3]. There is a rich literature on models for the chemistry of irradiated disks with different levels of sophistication for the disk physics and the chemical processes involved [5, 149, 161, 252, 285, 291]. Isotope selective photodissociation of CO and N_2 has been examined by [139, 178, 256, 283]. Possible implications for the isotopic composition of minerals and meteorites have been pointed out by [185]. An entry into the literature for Hot Corino observations has been provided above. Observations of H_2O, H_2CO, and CH_3OH in TW Hya have been reported in [142, 222, 286].

X-ray ionization of protoplanetary disks has been examined in detail by [120, 121, 154], while cosmic ray ionization of protoplanetary disks has been discussed by [67]. The text leans heavily on these papers. Radionuclide ionization is discussed in [275, 277]. The importance of dust in the ionization balance (of molecular clouds) at densities in excess of 10^{10} cm^{-3} was first discussed by [276]. An in-depth discussion of MRI and the degree of ionization is presented in the comprehensive review by [16]. The discussion here derives from [67].

Observations and analysis of dust disk masses have been reviewed by [13]. Extensive model studies of the CO emission from protoplanetary disks with a wide range of properties have been developed by [199, 290]. ALMA observations drive home that CO observations severely underestimate the gas mass of protoplanetary disk [15, 183, 200]. The importance of chemical depletion as well as of the temperature structure for the determination of CO masses of protoplanetary disks has been stressed by [293]. The FUV absorption line study of RW Aur and the derived CO/H_2 ratio is reported in [112]. The FUV spectral energy distribution of classical T Tauri stars has been studied by [111]. Disk masses derived from HD have been reported by [26, 190].

Ground-based and space-based, near- and mid-IR spectroscopy of molecules in protoplanetary disks were pioneered by [46, 48]. Noteworthy contributions are reported in [203, 204, 238, 239, 267]. Herschel PACS and HIFI far-IR and sub-millimeter studies have been reported in [32, 98, 142]. The near-IR study of turbulence in the inner disk was reported by [47] while the ALMA study is discussed in [103]. The study on the CO snowline in TW Hya is reported in [240]. The influence of photochemistry on the differences in the N_2H^+/CO spatial structure has been stressed in [280]. The CO snowline has also been studied in disks around deeply embedded protostars [10]. The effect of turbulently driven diffusion on the location of snowlines has been addressed by [224]. Reference [37] discusses the use of CO_2 as a tracer of disk processes.

Comprehensive models for the structure of protoplanetary disks, the expected variations therein, and the observational consequences have been developed by [161]. The website (www.astro.rug.nl/~prodimo/) of ProDiMo (PROtoplanetary DIsk MOdel) provides a scientific software package that models protoplanetary disks. Similarly, the physical/chemical model DALI (Dust And LInes) has been developed to model molecular line observations of protoplanetary disks (www.mpe.mpg.de/~simonbr/research_dali/index.html).

The RADMC-3D program – accessible at www.ita.uni-heidelberg.de/~dullemond/software/radmc-3d/index.html – is a highly flexible code for "postprocessing" models to compute predictions for gas and dust emission from among other protoplanetary disks. Likewise, LIME (Line Modeling Engine https://lime.readthedocs.io/en/v1.6.1/usermanual.html) is an excitation and radiative transfer code that can be used to predict line and continuum radiation from an astronomical source model. The code, MC-Max (www.exoclouds.com/Software/), models the physics and radiative transfer of dusty environments, including protoplanetary disks.

Compilations and references to studies of the composition of comets can be found in [33, 77, 201]. The importance of the Strecker synthesis for amino acid formation in asteroidal bodies is discussed in [232]. The volatile inventory of the Earth and its isotopic signature is summarized in [188]. The deuterium "trail" from molecular clouds to the Earth is reviewed by [56]. The carbon isotopic composition of meteoritic organics has been summarized by [253, 254].

The first comprehensive models for PDRs were presented in [269] and these have been used in the analysis of much of the early Kuiper Airborne Astronomy observations of the atomic fine-structure lines from PDRs. Over the years, a number of 1-D models have been developed and these have become very sophisticated [101, 167, 177, 246]. The physics and chemistry of PDR and their application to the general ISM has been reviewed by [147, 148]. Chapter 9 in the textbook [273] presents an overview of PDR physics. The chemistry of PDRs has been discussed at length in [263]. Clumpy 3-D models have been developed by [30, 78]. The discussion in Section 11.6.2 and the references in Table 11.3 provide an entrance into the literature on the analysis of molecular data for PDRs. For CO, the reader should also consult [137, 156, 262], while H_2 data is analyzed by [122, 162, 255]. The importance of magnetic fields for pressure support in PDRs has been stressed by [1, 231]. The Cloudy code calculates the structure of the ionized and PDR gas and can be accessed through its home page: www.nublado.org, the Meudon PDR code is available at https://ism.obspm.fr, and the Cologne KOSMA-tau code can be found through www.astro.uni-koeln.de/kosma-tau. The PDRToolbox accessible at http://dustem.astro.umd.edu/pdrt/ provides many diagnostic plots as well as a line ratio fitting program that finds the G_0 and n values that best fit a data set. Different PDR codes have been benchmarked in [245]. The CH^+ observations are reported in [202]. The vibrational population of H_2 was studied by [162]. The rotational population was measured by [9]. Detailed models for the H_2 ro-vibrational population have been presented by [122, 198]. The reaction of C^+, O, and S^+ with rotationally and vibrationally excited H_2 was studied by [119, 265, 294, 295]. The chemistry of the radical molecular cations in the PDR have been studied most recently by [4, 114, 127]. The discussion on $c,l{-}C_3H_2$ and $c,l{-}C_3H$ isomers leans heavily on the studies by [79, 182] for dark cloud cores. Reference [284] presents an overview of astronomical, quantum chemical, and surface chemistry studies of the formation of H_2 under interstellar conditions. The H_2 formation rate in PDRs was measured by [136]. The role PAHs in the formation of H_2 in the ISM has been investigated in the laboratory and in silico by [151, 152, 242, 268]. These results have been used to assess the role of PAHs in the formation of H_2 in the ISM by [14, 40]. The interaction of H atoms and FUV radiation with carbon surfaces has been investigated by [7, 195, 196]. Reference [233] has stressed the possible importance of PAHs for the inventory of small hydrocarbons. Ion–molecule chemistry involving PAHs has been reviewed by [29]. Photochemistry of PAHs and the importance of top-down photochemistry has been investigated by [297]. A pioneering study on PAH photochemistry, [174], was concerned with photolysis of PAH dications, which led to a variety of $C_nH_m^+$ fragments, but this may not be so relevant for large PAHs, where functional group loss is key in the photolysis [63, 298].

There is an extensive literature on protostellar outflows and jets. A recent review is provided by [18]. Models for J- and C-shocks in dense environments are described in [84, 145, 146]. A detailed discussion on the water emission from C-shocks can be found in [165]. Other noteworthy early papers are [105, 207, 208]. A more recent relevant paper is [107]. Nonsteady-state aspects are introduced in [104]. The chemistry of J-shocks and C shocks is discussed in [146, 165]. The chemistry of SiO in shocks has recently been revisited by [133]. Sputtering of ice mantles is discussed in [84]. A recent paper is [155]. The influence of (2D) geometry of a bullet interacting with a medium on the shock spectrum and line profiles is emphasized in [171]. A shock code simulating both continuous (C) type shock waves and jump (J) type shock waves is available at http://vizier.cfa.harvard.edu/viz-bin/VizieR?-source=J/A+A/578/A63 and described in [108] and a set of relevant output files can be obtained at this url as well. The Orion explosion has been discussed by [20, 21]. The SWS/ISO spectrograph has studied the near- and mid-IR emission spectrum of H_2 in the Orion BN/KL region [247]. The far-IR PACS and SPIRE spectrum of the Orion shock has been presented and analyzed in [126]. Another illustration of the power of these types of observations and analysis is presented in [210]. The L1157 outflow has been studied in low J CO, CS, and other molecular species by [17, 24, 128]. Pure rotational H_2 emission was studied with the IRS/Spitzer by [215]. PACS/Herschel studied the excited H_2O and CO emission [43]. References to the organic inventory of L1157 B1 are given in Table 11.6. References relevant to grain surface chemistry routes toward complex organic molecules are provided above in the context of Hot Cores/Hot Corinos. The importance of gas phase synthesis of some of these species has been stressed by [70].

There are two excellent textbooks on interstellar masers [90, 129]. The discussion on the physics of water masers in space is derived from [211, 213]. Dissociative shock models for interstellar H_2O masers were developed by [89]. C-shock model studies were performed by [164]. Searches for H_2O masers associated with high level transitions have been performed by [197, 212, 213] using ground-based, airborne, and space-based platforms. The physics of OH masers – in particular, Eq. (11.85) and (11.86) – is discussed in [129]. Other very relevant studies are [91, 130]. The monograph by Gray [129] provides also a good entry point for the other masers summarized in this section.

11.10 Exercises

11.1 Bonnor–Ebert spheres and prestellar cores:

- Describe the different ways the density distribution of prestellar cores can be determined and their pros and cons.
- Write Eq. (11.3) in dimensionless form and derive Eq. (11.5).
- Integrate this set of equations numerically inside-out with the appropriate boundary conditions for a few values of $\xi_{max} = (R/C_s)\sqrt{4\pi G\rho_c}$. Pick ξ_{max}'s around the value appropriate for B68 ($\xi_{max} = 6.9 \pm 0.2$).
- Compare the derived column density distribution with that observed for B68 (Figure 10.9).
- Show that B68 is gravitationally stable.
- Describe the line profile signature of a collapsing core.

11.2 Hot Core mass:

- Dust continuum
  - At 850 μm, the Hot Core in Orion has a flux of 640 mJy from a region with a size of $0.7 \times 0.6''$. Calculate the H_2 column density. Adopt temperature of 200 K and a dust emissivity law given by $k(\lambda) = 2.3 \times 10^{-25}$ $(160\ \mu m/\lambda)^2$ cm^2/H-atom.
  - Calculate the average density along the sight-line. Adopt a distance of 414 pc.
  - What is the mass of the Hot Core?
- CO emission
  - Go to the Radex online website (http://var.sron.nl/radex/radex.php) and derive the column density for which the $C^{18}O$ $J = 2 - 1$ becomes optically thick. Adopt parameters appropriate for the Hot Core in Orion.
  - What H_2 column density does this correspond to?
  - At what J is the peak in the CO column density?
  - Analysis of the CO ladder implies an average $C^{18}O$ column density of 6×10^{16} cm^{-2} in a region of $10''$ in size. Calculate the total mass of the Orion Hot Core.

11.3 The chemical composition of Hot Cores and Hot Corinos:

- Compare and contrast the chemical inventory of Hot Cores/Hot Corinos and dark clouds.
- Why is it generally considered that grain surface chemistry has been involved in setting the chemical inventory of Hot Cores/Corinos?

11.4 Vertical density distribution for a protoplanetary disk:

- Derive Equation (11.24), starting with the vertical hydrostatic equilibrium equation in cylindrical coordinates. Assume that $z \ll r$ and introduce the sound speed.

- Relate the central density to the surface density.
- What is the mid-plane density for a surface density of 10^3 g/cm^2?

11.5 Show that an infinitely thin disk irradiated by a central star with size $R_\star$ and temperature, T_{eff}, which radiates like a black body has a temperature distribution given by $T = 1.33 T_{eff} \left(R_\star/r\right)^{3/4}$ when $r \gg R_\star$. Hint: Evaluate the stellar flux passing through a surface at a distance r, $F = \int I_\star sin\theta cos\phi d\Omega$ and expand the resulting algebraic expression.

11.6 Describe the different chemical zones in protoplanetary disks and the various processes relevant for their chemical composition.

11.7 Describe the different ionization processes involved in the charge balance of protoplanetary disks.

11.8 Discuss the various observations to measure the mass of protoplanetary disks and their pros and cons.

11.9 Describe how the isotopic composition can be used to trace the chemical history of comets and asteroids.

11.10 Size of the PDR:

- Derive the equation for the column density of the H_2 photodissociation front. Reread Sections 5.2.3 and 9.2.1 and start with the H_2 formation–destruction balance. Approximate $n(H)$ by n in this equation. Use $dN(H_2)/dz = n(H_2)$ to link the H_2 column density to the dust optical depth and substitute this back in to the expression for the H_2 density. The depth of dissociation front is given by $n(H_2)/n = 1/4$.
- Evaluate the column density for various G_0/n values and compare to the results given in Figure 5.14.
- Derive the size of the C^+ region (Eq. (11.50)) by balancing photoionization with radiative recombination. Use the photoionization rate given in Table 6.4, a radiative recombination rate of 2.2×10^{-12} cm^3/s appropriate for 300 K gas, a gas phase carbon abundance of 1.5×10^{-4}, and $N_H/A_v = 1.9 \times 10^{21}$ cm^{-2}.
- Derive the size of the PDR based upon balancing O accretion and photodesorption (eqn. (11.52)). Use a gas phase oxygen abundance of 3×10^{-4}.
- Derive the size of the PDR based upon a comparison of the photo-electric heating rate (Section 12.8.2) with the cosmic ray heating rate (Section 9.2.1).
- Compare these different sizes for a PDR with $G_0 = 10^4$ and $n = 10^4$ cm^{-3}.

11.11 The strength of the incident FUV field.

- Derive Equation (11.57) assuming a luminosity of 7.6×10^5 $L_\odot$. an ionizing photon luminosity of 5×10^{49} photons s^{-1}, and $\chi = 1$.
- The stellar parameters used are appropriate for an O4 star. Determine the sensitivity of G_0/n to spectral type, using the stellar characteristics given in [278].

11.12 CO intensities of PDRs:

- The observed intensities of the $J = 13 - 12$ and $J = 6 - 5$ CO lines in the Orion Bar are 2.5×10^{-8} and 5.3×10^{-9} W/m^2/ster. Go to the PDR toolbox website and download the appropriate figure with the intensity ratio for these lines. Derive the density of the emitting gas, assuming a G_0 of 2.6×10^4 Habings.
- Why are you not surprised that this line ratio implies a high density?
- The use of rotational diagrams was discussed in Section 4.2.2. The rotational population of ^{13}CO in the Orion Bar is characterized by an excitation temperature of 81 ± 1 K [156]. Assume that the cloud is homogeneous and estimate the kinetic temperature of the gas.
- Estimate the critical density for the $J = 13$ level and estimate (Section 4.1.3) the density of the emitting gas.

11.13 The chemistry of PDRs:

- Compare and contrast the chemical composition of PDRs with that of diffuse clouds and dark cloud cores.
- Describe the different chemical zones in PDRs and relate their composition to the dominant chemical processes.
- Show that the reaction of H_2O with C^+ cannot be the source of HCO^+ observed in the Orion Bar.
- Discuss the chemical role of excited H_2.
- Discuss the potential roles of PAHs in the chemistry of PDRs.

11.14 Shocks:

- Describe the differences between J and C shocks and relate these to the relevant processes.
- Compare and contrast the composition of the shocked gas in L1157 B1 with that of the Hot Corino around IRAS 16293-2422A and with the dark cloud core TMC 1. What does this comparison tell us about the chemistry in these different regions?

11.15 Interstellar masers:

- Maser activity requires population inversion. Discuss the different processes that could cause such an inversion.
- Describe the different types of masers that are prevalent in regions of star formation.

Bibliography

[1] Abel, N. P., Ferland, G. J., 2006, *ApJ*, 647, 367
[2] Ádámkovics, M., Najita, J. R., Glassgold, A. E., 2016, *ApJ*, 817, 82
[3] Agúndez, M., Cernicharo, J., Goicoechea, J. R., 2008, *A & A*, 483, 831
[4] Agúndez, M., Goicoechea, J. R., Cernicharo, J., et al., 2008, *A & A*, 713, 662

[5] Aikawa, Y., Herbst, E., 1997, *A & A*, 351, 233
[6] Allamandola, L. J., Sandford, S. A., Valero, G. J., 1988, *Icarus*, 76, 225
[7] Alata, I., Cruz-Diaz, G. A., Muñoz Caro, G. M., Dartois, E., 2014, *A & A*, 569, A119
[8] Allen, V., van der Tak, F. F. S., Sánchez-Monge, Á., et al., 2017, *A & A*, 603, A133
[9] Allers, K. N., Jaffe, D. T., Lacy, J. H., et al., 2005, *ApJ*, 630, 368
[10] Anderl, S., Maret, S., Cabrit, S., et al. 2016, *A & A*, 591, A3
[11] Anderson, D. E., Bergin, E. A., Blake, G. A., et al., 2017, *ApJ*, 845, 13
[12] Andrews, S. M., Willner, D. J., Hughes, A. M., et al., 2012, *ApJ*, 744, 162
[13] Andrews, S. M., 2015, *PASP*, 127, 961
[14] Andrews, H., Candian, A., Tielens, A. G. G. M., 2016, *A & A*, 595, A23
[15] Ansdell, M., Williams, J. P., van der Marel, N., et al., 2016, *ApJ*, 828, 46
[16] Armitage, P. J., 2015, arXiv:1509.06382
[17] Bachiller, R., Pérez Gutiérrez, M., Kumar, M. S. N., Tafalla, M., 2001, *A & A*, 372, 899
[18] Bally, J., 2016, *Annu Rev Astron Astrophys*, 54, 491
[19] Bally, J., Cunningham, N. J., Moeckel, N., et al., 2011, *ApJ*, 727, 113
[20] Bally, J.Ginsburg, A., Silvia, D., Youngblood, A., 2015, *A & A*, 579, A130
[21] Bally, J., Zinnecker, H., 2005, *AJ*, 129, 2281
[22] Balog, R., Cicman, P, Jones, N. C., Field, D., 2009, *Phys Rev Let*, 201, 73003
[23] Bast, J. E., Lahuis, F., van Dishoeck, E. F., Tielens, A. G. G. M., 2013, *A & A*, 551, A118
[24] Benedettini, M., Viti, S., Codella, C., et al., 2013, *MNRAS*, 436, 179
[25] Bergin, E. A., Tafalla, M., 2007, *Annu Rev Astron Astrophys*, 45, 339
[26] Bergin, E. A., Cleeves, L. I., Gorti, U., et al., 2013, *Nature*, 493, 644
[27] Bernstein, M. P., Sandford, S. A., Allamandola, L. J., et al., 1995, *ApJ*, 454, 327
[28] Beuther, H., Zhang, Q., Greenhill, L. J., et al., 2004, *ApJ*, 616, L31
[29] Bierbaum, V. M., Le Page, V., Snow, T. P., 2011, *EAS Publications Series*, 46, 427
[30] Bisbas, T. G., Bell, T. A., Viti, S., Yates, J., Barlow, M. J., 2012, *MNRAS*, 427, 2100
[31] Blake, G. A., Sutton, E. C., Masson, C. R., Phillips, T. G., 1987, *ApJ*, 315, 621
[32] Blevins, S. M., Pontoppidan, K. M., Banzatti, A., et al., 2016, *ApJ*, 818, 22
[33] Bockelée-Morvan, D., et al., 2000, *A & A*, 353, 1101
[34] Boogert, A. C. A., Ehrenfreund, P., Gerakines, P. A., et al., 2000, *A & A*, 353, 349
[35] Boonman, A. M. S., van Dishoeck, E. F., Lahuis, F., et al., 2003, *A & A*, 399, 1047
[36] Boonman, A. M., van Dishoeck, E. F., Lahuis, F., Doty, S. D., 2003, *A & A*, 399, 1063
[37] Bosman, A. D., Tielens, A. G. G. M., van Dishoeck, E. F., 2018, *A & A*, 611, A80
[38] Bottinelli, S., et al., 2004, *ApJ*, 617, L69
[39] Bouwman, J., Cuppen, H. M., Steglich, M., et al., 2011, *A & A*, 529, A46
[40] Bron, E., Le Bourlot, J., Le Petit, F., 2014, *A & A*, 569, A100
[41] Burkhardt, A. M., Dollhopf, N. M., Corby, J. F., et al., 2016, *ApJ*, 827, 21
[42] Burton, M. G., Howe, J. E., Geballe, T. R., Brand, P. W. J. L., 1998, *Publ Astr Soc Austral*, 15, 194
[43] Busquet, G., Lefloch, B., Benedettini, M., et al., 2014, *A & A*, 561, A120
[44] Busquet, G., Fontani, F., Viti, S., et al., 2017, *A & A*, 604, A20
[45] Butscher, T., Duvernay, F., Rimola, A., et al., 2017, *Phys Chem Chem Phys*, 19, 2857
[46] Carr, J. S., Tokunaga, A. T., Najita, J., Shu, F. H., Glassgold, A. E., 1993, *ApJ*, 411, L37
[47] Carr, J. S., Tokunaga, A. T., Najita, J., 2004, *ApJ*, 603, 213
[48] Carr, J. S., Najita, J. R., 2008, *Science*, 319, 1504

[49] Carr, J. S., Najita, J. R., 2011, *ApJ*, 733, 102
[50] Caselli, P., Hasegawa, T. I., Herbst, E., 1993, *ApJ*, 408, 548
[51] Caselli, P., 2011, *IAU Symposium*, 280, 19
[52] Cazaux, S., Tielens, A. G. G. M., Ceccarelli, C., et al., 2003, *ApJ*, 593, L51
[53] Ceccarelli, C., Castets, A., Loinard, L., Caux, E., Tielens, A. G. G. M., 1998, *A & A*, 338, L43
[54] Ceccarelli, C., Caselli, P., Herbst, E., et al., 2007, *Protostars and Planets V*, (Tucson: University of Arizona Press), 47
[55] Ceccarelli, C., 2008, IAU Symposium, 251, 79
[56] Ceccarelli, C., Caselli P., Bockele-Morvan D., et al., 2014, in *Protostars and Planets VI*, eds H. Beuther et al., (Tucson: University of Arizona Press), 859
[57] Cernicharo, J., Marcelino, N., Roueff, E., et al., 2012, *ApJ*, 759, L43
[58] Cernicharo, J., Kisiel, Z., Tercero, B., et al., 2016, *A & A*, 587, L4
[59] Cesaroni, R., 2005, in IAU Symposium, 227, *Massive Star Birth: A Crossroads of Astrophysics*, 227, 59
[60] Charnley, S. B., Kaufman, M. J., 2000, *ApJ*, 529, L111
[61] Charnley, S. B., Rodgers, S. D., 2005, IAU Symposium 231, *Astrochemistry: Recent Successes and Current Challenges*, 237
[62] Charnley, S. B., Tielens, A. G. G. M., Millar, T. J., 1992, *ApJ*, 399, L71
[63] Chen, T., Zhen, J., Wang, Y., et al., 2018, *Chem Phys Lett*, 692, 298
[64] Chiang, E. I., Goldreich, P., 1997, *ApJ*, 490, 368
[65] Choi, Y., van der Tak, F. F. S., Bergin, E. A., Plume, R., 2014, *A & A*, 572, L10
[66] Chuang, K.-J., Fedoseev, G., Qasim, D., et al., 2017, *MNRAS*, 467, 2552
[67] Cleeves, L. I., Adams, F. C., Bergin, E. A., 2013, *ApJ*, 772, 5
[68] Codella, C., Benedettini, M., Beltrán, M. T., et al., 2009, *A & A*, 507, L25
[69] Codella, C., Ceccarelli, C., Lefloch, B., et al., 2012, *ApJ*, 757, L9
[70] Codella, C., Fontani, F., Ceccarelli, C., et al., 2015, *MNRAS*, 449, L11
[71] Codella, C., Viti, S., Lefloch, B., et al., 2018, *MNRAS*, 474, 5694
[72] Cole, C. A., Wehres, N., Yang, Z., et al., 2012, *ApJ*, 754, L5
[73] Cottin, H., Moore, M. H., Bénilan, Y., 2003, *ApJ*, 590, 874
[74] Coutens, A., Vastel, C., Caux, E., et al., 2012, *A & A*, 539, A132
[75] Coutens, A., Vastel, C., Cazaux, S., et al., 2013, *A & A*, 553, A75
[76] Coutens, A., Jørgensen, J. K., van der Wiel, M. H. D., et al., 2016, *A & A*, 590, L6
[77] Crovisier, J., et al., 2004, *A & A*, 418, 1141
[78] Cubick, M., Stutzki, J., Ossenkopf, V., Kramer, C., Röllig, M., 2008, *A & A*, 488, 623
[79] Cuadrado, S., Goicoechea, J. R., Pilleri, P., et al., 2015, *A & A*, 575, A82
[80] Cuadrado, S., Goicoechea, J. R., Cernicharo, J., et al., 2015, *A & A*, 603, A124
[81] Dartois, E., Demyk, K., d'Hendecourt, L., Ehrenfreund, P., 1999, *A & A*, 351, 1066
[82] Demyk, K., Dartois, E., d'Hendecourt, L., et al., 1998, *A & A*, 339, 553
[83] Demyk, K., Bottinelli, S., Caux, E., et al., 2010, *A & A*, 517, A17
[84] Draine, B., Roberge, W. G., Dalgarno, A., 1983, *ApJ*, 264, 485
[85] Draine, B., Bertoldi, F., 2000, *Molecular Hydrogen in Space*, eds. F. Combes and G. Pineau des Forêts, (Cambridge: Cambridge University Press), 131
[86] Draine, B. T., Bertoldi, F., 1996, *ApJ*, 468, 269
[87] Dworkin, J. P., Seb Gillette, J., Bernstein, M. P., et al., 2004, *Advances in Space Research*, 33, 67
[88] Ebel, D. S., 2006, in *Meteorites and the early solar system II* (Tucson: University of Arizona Press), 253

[89] Elitzur, M., Hollenbach, D. J., McKee, C. F., 1989, *ApJ*, 346, 983
[90] Elitzur, M., 1991, *Astronomical Masers, Astrophysics and Space Science Library 170* (Berlin: Springer Verlag)
[91] Elitzur, M., de Jong, T., 1978, *A & A*, 67, 323
[92] Elmegreen, B. G., Lada, C. J., 1977, *ApJ*, 214, 725
[93] Ercolano, B., Glassgold, A. E., 2013, *MNRAS*, 436, 3446
[94] Evans, N. J., II, Lacy, J. H., Carr, J. S., 1991, *ApJ*, 383, 674
[95] Evans, N. J., 1999, *Annu Rev Astron Astrophys*, 37, 311
[96] Favre, C., Despois, D., Brouillet, N., et al., 2011, *A & A*, 532, A32
[97] Favre, C., Pagani, L., Goldsmith, P. F., et al., 2017, *A & A*, 604, L2
[98] Fedele, D., Bruderer, S., van Dishoeck, E. F., et al., 2013, *A & A*, 559, A77
[99] Fedoseev, G., Cuppen H. M., Ioppolo, S., et al., 2015, *MNRAS*, 448, 1288
[100] Feng, S., Beuther, H., Henning, T., et al., 2015, *A & A*, 581, A71
[101] Ferland, G. J., Chatzikos, M., Guzmán, F., et al., 2017, *Rev Mex A & A*, 53, 385
[102] Fish, V. L., Sjouwerman, L. O., 2007, *ApJ*, 668, 331
[103] Flaherty, K. M., Hughes, A. M., Rose, S. C., et al., 2017, *ApJ*, 843, 150
[104] Flower, D. R., Pineau des Forêts, G., 1999, *MNRAS*, 308, 271
[105] Flower, D. R., Le Bourlot, J., Pineau des Forêts, G., Cabrit, S., 2003, *MNRAS*, 341, 70
[106] Flower, D. R., Pineau des Forêts, G., 2010, *MNRAS*, 406, 1745
[107] Flower, D. R., Pineau des Forêts, G., 2013, *MNRAS*, 436, 2143
[108] Flower, D. R., Pineau des Forêts, G., 2015, *A & A*, 578, A63
[109] Fogel, J. K. J., Bethell, T. J., Bergin, E. A., Calvet, N., Semenov, D., 2011, *ApJ*, 726, 29
[110] Fontani, F., Codella, C., Ceccarelli, C., et al., 2014, *ApJ*, 788, L43
[111] France, K., Schindhelm, E., Bergin, E. A., et al., 2014, *ApJ*, 784, 127
[112] France, K., Herczeg, G. J., McJunkin, M., Penton, S. V., 2014, *ApJ*, 794, 160
[113] Friedel, D. N., Widicus Weaver, S. L., 2012, *ApJS*, 201, 17
[114] Fuente, A., Rodrıguez-Franco, A., Garcıa-Burillo, S., et al., 2003, *A & A*, 406, 899
[115] Garrod, R. T., Weaver, S. L. W., Herbst, E., 2008, *ApJ*, 682, 283
[116] Geppert, W. D., Hamberg, M., Thomas, R. D., et al., 2006, *Faraday Discussions*, 133, 177
[117] Gerakines, P. A., Schutte, W. A., Ehrenfreund, P., 1996, *A & A*, 312, 289
[118] Gerakines, P. A., Moore, M. H., Hudson, R. L., 2001, *J Geophys Res*, 106, 33381
[119] Gerlich, D., Disch, R., Scherbarth, S., 1987, *J Chem Phys*, 87, 350
[120] Glassgold, A. E., Lizano, S., Galli, D., 2017, *MNRAS*, 472, 2447
[121] Glassgold, A. E., Najita, J., Igea, J., 1997, *ApJ*, 480, 344
[122] Gnaciński, P., 2011, *A & A*, 532, A122
[123] Godard et al., 2019, *A & A*, 622, 100
[124] Goicoechea, J., Pety, J., Cuadrado, S., et al., 2016, *Nature*, 537, 207
[125] Goicoechea, J. R., Joblin, C., Contursi, A., et al., 2011, *A & A*, 530 L16
[126] Goicoechea, J. R., Chavarría, L., Cernicharo, J., et al., 2015, *ApJ*, 799, 102
[127] Goicoechea, J. R, Cuadrado, S., Pety, J., et al., 2017, *A & A*, 601, L9
[128] Gómez-Ruiz, A. I., Codella, C., Lefloch, B., et al., 2015, *MNRAS*, 446, 3346
[129] Gray, M., *Maser Sources in Astrophysics* (Cambridge: Cambridge University Press)
[130] Gray, M. D., 2007, *MNRAS*, 375, 477
[131] Greenhill, L. J., Gezari, D. Y., Danchi, W. C., et al., 2004, *ApJ*, 605, L57
[132] Grossman, L., Larimer, J. W., 1974, *Rev Geophys Space Phys*, 12, 71
[133] Gusdorf, A., Pineau des Forêts, G., Flower, D. R., 2008, *A & A*, 490, 695

[134] Guzman, V. V., Pety, J., Gratier, P., et al., 2014, 69th International Symposium on Molecular Spectroscopy, RF05
[135] Guzmán, V. V., Pety, J., Goicoechea, J. R., et al., 2015, *ApJ*, 800, L33
[136] Habart, E., Boulanger, F., Verstraete, L., et al., 2004, *A & A*, 414, 531
[137] Habart, E., Dartois, E., Abergel, A., et al., 2010, *A & A*, 518, L116
[138] Hamberg, M., Österdahl, F., Thomas, R. D., et al., 2010, *A & A*, 514, A83
[139] Heays, A. N., Visser, R., Gredel, R., et al., 2014, *A & A*, 562, A61
[140] Henning, T. Semenov, D., 2013, *Chem Rev*, 113, 9016
[141] Hirota, T., Kim, M. K., Kurono, Y., Honma, M., 2015, *ApJ*, 801, 82
[142] Hogerheijde, M. R., Bergin, E. A., Brinch, C., et al., 2011, *Science*, 334, 338
[143] Holdship, J., Viti, S., Jimenez-Serra, I., et al., 2016, *MNRAS*, 463, 802
[144] Holdship, J., Jimenez-Serra, I., Viti, S., et al., 2019, *ApJ*, 878, 64
[145] Hollenbach, D., McKee, C. F., 1979, *ApJ Suppl*, 41, 555
[146] Hollenbach, D., McKee, C. F., 1989, *ApJ*, 342, 306
[147] Hollenbach, D. J., Tielens, A. G. G. M., 1997, *Annu Rev Astron Astrophys*, 35, 179
[148] Hollenbach, D. J., Tielens, A. G. G. M., 1999, *Reviews of Modern Physics*, 71, 173
[149] Hollenbach, D., Gorti, U., 2009, *ApJ*, 703, 1203
[150] Horn, A., et al., 2004, *ApJ*, 611, 605
[151] Hornekær, L., Šljivančanin, Ž., Xu, W., et al., 2006, *Phys Rev Lett*, 96, 156104
[152] Hornekær, L., Rauls, E., Xu, W., et al., 2006, *Phys Rev Lett*, 97, 186102
[153] Hudson, R. L., Palumbo, M. E., Strazzulla, G., et al., 2008, in *The Solar System Beyond Neptune*, eds, M. A. Barucci, H. Boehnhardt, D. P. Cruikshank, and A. Morbidelli (Tucson: University of Arizona Press), 507
[154] Igea, J., Glassgold, A. E., 1999, *ApJ*, 518, 848
[155] Jiménez-Serra, I., Caselli, P., Martín-Pintado, J., Hartquist, T. W., 2008, *A & A*, 482, 549
[156] Joblin, C., Bron, E., Pinto, C., et al., 2018, *A & A*, 615, A129
[157] Jones, B. M., Bennett, C. J., Kaiser. R. I., 2011, *ApJ*, 734, 78
[158] Jørgensen, J. K., van der Wiel, M. H. D., Coutens, A., et al., 2016, *A & A*, 595, A117
[159] Jonusas, M., Krim, L., 2017, *MNRAS*, 470, 4564
[160] Kama, M., López-Sepulcre, A., Dominik, C., et al., 2013, *A & A*, 556, A57
[161] Kamp, I., Thi, W. F., Woitke, P., et al., 2017, *A & A*, 607, A41
[162] Kaplan, K. F., Dinerstein, H. L., Oh, H., et al., 2017, *ApJ*, 838, 152
[163] Karpas, Z., Meot-Ner, M., 1989, *J Phys Chem*, 93, 1859
[164] Kaufman, M. J., Neufeld, D. A., 1996, *ApJ*, 456, 250
[165] Kaufman M., Neufeld, D. A., 1996, *ApJ*, 456, 611
[166] Kaufman, M. J., Hollenbach, D. J., Tielens, A. G. G. M., 1998, *ApJ*, 497, 276
[167] Kaufman, M. J., Wolfire, M. G., Hollenbach, D. J., 2006, *ApJ*, 644, 283
[168] Kenyon, S. J., Hartmann, L., 1987, *ApJ*, 323, 714
[169] Kress, M. E., Tielens, A. G. G. M., Frenklach, M., 2010, *Advances in Space Research*, 46, 44
[170] Kress, M. E., Tielens, A. G. G. M., 2001, *Meteoritics and Planetary Science*, 36, 75
[171] Kristensen, L. E., Ravkilde, T. L., Pineau Des Forêts, G., et al., 2008, *A & A*, 477, 203
[172] Kurtz, S., Cesaroni, R., Churchwell, E., Hofner, P., Walmsley, C. M., 2000, *Protostars and Planets IV* (Tucson: University of Arizona Press), 299
[173] Lada, C. J., Bergin, E. A., Alves, J. F., et al., 2003, *ApJ*, 586, 286
[174] Leach, S., Eland, J. H. D., Price, S. D., 1989, *J Phys Chem*, 93, 7583
[175] Lefloch, B., Vastel, C., Viti, S., et al., 2016, *MNRAS*, 462, 3937

[176] Lefloch, B., Codella, C., Ceccarelli, C., et al., 2017, *MNRAS*, 469, L73
[177] Le Petit, F., Nehm, C., Le Bourlot, J., Roueff, E., 2006, *ApJS*, 164, 506
[178] Li, X., Heays, A. N., Visser, R., et al., 2013, *A & A*, 555, A14,
[179] Liu, S.-Y., Girart, J. M., Remijan, A., Snyder, L. E., 2002, *ApJ*, 576, 255
[180] Lodders, K., 2003, *ApJ*, 591, 1220
[181] Loinard, L., Castets, A., Ceccarelli, C., et al., 2000, *A & A*, 359, 1169
[182] Loison, J. C., Agúndez, M., Wakelam, V., et al., 2017, *MNRAS*, 470, 4075
[183] Long, F., Herczeg, G. J., Pascucci, I., et al., 2017, *ApJ*, 844, 99
[184] Lynden-Bell, D., Pringle, J. E., 1974, *MNRAS*, 168, 603
[185] Lyons, J. R., Young, E. D., 2005, *Nature*, 435, 317
[186] Marcelino, N., Gerin, M., Cernicharo, J., et al., 2018, *A & A*, 620, A80
[187] Marconi, A., Testi, L., Natta, A., Walmsley, C. M., 1998, *A & A*, 330, 696
[188] Marty, B., 2012, *Earth Planet Sci Lett*, 313, 56
[189] McCartney, M. S. K., Brand, P. W. J. L., Burton, M. G., Chrysostomou, A., 1999, *MNRAS*, 307, 315
[190] McClure, M. K., Bergin, E. A., Cleeves, L. I., et al., 2016, *ApJ*, 831, 167
[191] McGuire, B. A., Carroll, P. B., Dollhopf, N. M., et al., 2015, *ApJ*, 812, 76
[192] McMahon, T. B., Beauchamp, J. L., 1977, *J Phys Chem*, 81, 593
[193] Mendoza, E., Lefloch, B., Lopéz-Sepulcre, A., et al., 2014, *MNRAS*, 445, 151
[194] Mendoza, E., Lefloch, B., Ceccarelli, C., et al., 2018, *MNRAS*, 475, 5501
[195] Mennella, V., 2006, *ApJ*, 647, L49
[196] Mennella, V., 2010, *ApJ*, 718, 867
[197] Menten, K. M., Melnick, G. J., Phillips, T. G., 1990, *ApJL*, 350, L41
[198] Meyer, D. M., Lauroesch, J. T., Sofia, U. J., et al., 2001, *ApJ*, 553, L59
[199] Miotello, A., van Dishoeck, E. F., Kama, M., Bruderer, M., 2016, *A & A*, 594, A85
[200] Miotello, A., van Dishoeck, E. F., Williams, J. P., et al., 2017, *A & A*, 599, A113
[201] Mumma, M. J., Charnley, S. B., 2011, *Annu Rev Astron Astrophys*, 49, 471
[202] Nagy, Z., Van der Tak, F. F. S., Ossenkopf, V., et al., 2013, *A & A*, 550, A96
[203] Najita, J., Carr, J. S., Mathieu, R. D., 2003, *ApJ*, 589, 931
[204] Najita, J., Carr, J. S., Glassgold, A. E., Shu, F. H., Tokunaga, A. T., 1996, *ApJ*, 462, 919
[205] Neill, J. L., Wang, S., Bergin, E. A., et al., 2013, *ApJ*, 770, 142
[206] Neill, J. L., Crockett, N. R., Bergin, E. A., Pearson, J. C., Xu, L. H., 2013, *ApJ*, 777, 85
[207] Neufeld, D. A., Dalgarno, A., 1989, *ApJ*, 340, 869
[208] Neufeld, D. A., Dalgarno, A., 1989, *ApJ*, 344, p. 251
[209] Neufeld, D. A., Roueff, E., Snell, R. L., et al., 2012, *ApJ*, 748, 37
[210] Neufeld, D. A., Gusdorf, A., Güsten, R., et al., 2014, *ApJ*, 781, 102
[211] Neufeld, D. A., Melnick, G. J., 1991, *ApJ*, 368, 215
[212] Neufeld, D. A., Wu, Y., Kraus, A., et al., 2013, *ApJ*, 769, 48
[213] Neufeld, D. A., Melnick, G. J., Kaufman, M. J., et al., 2017, *ApJ*, 843, 94
[214] Nisini, B., Benedettini, M., Codella, C., et al., 2010, *A & A*, 518, L120
[215] Nisini, B., Giannini, T., Neufeld, D. A., et al., 2010, *ApJ*, 724, 69
[216] Nissen, H. D., Gustafsson, M., Lemaire, J. L., et al., 2007, *A & A*, 466, 949
[217] Nuth, J. A., Johnson, N. M., Ferguson, F. T., Carayon, A., 2016, *Meteoritics and Planetary Science*, 51, 1310
[218] Nuth, J. A., III, Johnson, N. M., Manning, S., 2008, *ApJL*, 673, L225
[219] Öberg, K. I., Garrod, R. T., van Dishoeck, E. F., Linnartz, H., 2009, *A & A*, 504, 891
[220] Öberg, K. I., van Dishoeck, E. F., Linnartz, H., Andersson, S., 2010, *ApJ*, 718, 832

[221] Öberg, K. I., 2016, *Chem Rev*, 116, 9631
[222] Öberg, K. I., Guzman, V. V., Merchantz, C. J., et al., 2017, *ApJ*, 839, 43
[223] Ossenkopf, V., Röllig, M., Neufeld, D. A., et al., 2013, *A & A*, 550, A57
[224] Owen, J. E., 2014, *ApJ*, 790, L7
[225] Padovani, M., Galli, D., Glassgold, A. E., 2009, *A & A*, 501, 619
[226] Pagani, L., Favre, C., Goldsmith, P. F., et al., 2017, *A & A*, 604, A32
[227] Pagani, L., Favre, C., Goldsmith, P. F., et al., 2017, *A & A*, 624, L5
[228] Palumbo, M. E., Castorina, A. C., Strazzulla, G., 1999, *A & A*, 342, 551
[229] Parise, B., Leurini, S., Schilke, P., et al., 2009, *A & A*, 508, 737
[230] Parise, B., Ceccarelli, C., Tielens, A. G. G. M., et al., 2006, *A & A*, 453, 949
[231] Pellegrini, E. W., Baldwin, J. A., Brogan, C. L., et al., 2007, *ApJ*, 658, 1119
[232] Peltzer, E. T., Bada, J. L., 1978, *Nature*, 272, 443
[233] Pety, J., Teyssier, D., Fossé, D., et al., 2005, *A & A*, 435, 885
[234] Plume, R., Bergin, E. A., Phillips, T. G., et al., 2012, *ApJ*, 744, 28
[235] Podio, L., Lefloch, B., Ceccarelli, C., et al., 2014, *A & A*, 565, A64
[236] Podio, L., Codella, C., Lefloch, B., et al., 2017, *MNRAS*, 470, L16
[237] Pogge, R. W., Owen, J. M., Atwood, B., 1992, *ApJ*, 399, 147
[238] Pontoppidan, K. M., Salyk, C., Blake, G. A., Käufl, H. U. 2010, *ApJ*, 722, L173
[239] Pontoppidan, K. M., Salyk, C., Blake, G. A., et al., 2010, *ApJ*, 720, 887
[240] Qi, C., Öberg, K. I., Wilner, D. J., et al., 2013, *Science*, 341, 630
[241] Rangwala, N., Colgan, S. W. J., Le Gal, R., et al., 2018, *ApJ*, 856, 9
[242] Rauls, E., Hornekær, L., 2008, *ApJ*, 679, 531
[243] Requena-Torres, M. A., Martín-Pintado, J., Martín, S., Morris, M. R., 2008, *ApJ*, 672, 352
[244] Rodgers, S. D., Charnley, S. B., 2001, *ApJ*, 546, 324
[245] Röllig, M., Ossenkopf, V., Jeyakumar, S., Stutzki, J., Sternberg, A., 2006, *A & A*, 451, 917
[246] Röllig, M., Abel, N. P., Bell, T., et al., 2007, *A & A*, 467, 187
[247] Rosenthal, D., Bertoldi, F., Drapatz, S., 2000, *A & A*, 356, 705
[248] Schutte, W. A., Gerakines, P. A., 1995, *Plan Space Sci*, 43, 1253
[249] Selegue, T. J., List, J. M., 1994, *J Am Chem Soc*, 116, 4874
[250] Salgado, F., Berné, O., Adams, J. D., et al., 2016, *ApJ*, 830, 118
[251] Sekine, Y., Sugita, S., Shido, T., et al., 2006, *Meteoritics and Planetary Science*, 41, 715
[252] Semenov, D., Wiebe, D., 2011, *ApJS*, 196, 25
[253] Sephton, M. A., 2014, *Treatise on Geochemistry* vol. 12, 1.
[254] Sephton, M. A., Watson, J. S., Meredith, W., et al., 2015, *Astrobiology*, 15, 779
[255] Sheffer, Y., Wolfire, M. G., Hollenbach, D. J., et al., 2011, *ApJ*, 741, 45
[256] Shi, X., Yin, Q. Z., Gao, H., et al., 2017, *ApJ*, 850, 48
[257] Shu, F. H., 1977, *ApJ*, 214, 488
[258] Shu, F. H., Adams, F. C., Lizano, S., 1987, *Annu Rev Astron Astrophys*, 25, 23
[259] Siebenmorgen, R., Kruegel, E., 2010, *A & A*, 511, A6
[260] Simon, R., Stutzki, J., Sternberg, A., Winnewisser, G., 1997, *A & A*, 327, L9
[261] Simon, R., Jackson, J., Rathborne, J. Chambers, E., 2006, *ApJ*, 639, 227
[262] Stock, D. J., Wolfire, M. G., Peeters, E., et al., 2015, *A & A*, 579, A67
[263] Sternberg, A., Dalgarno, A., 1989, *ApJ*, 338, 197
[264] Strazzulla, G., Brucato, J. R., Palumbo, M. E., Spinella, F., 2007, *Mem Soc Astron Ita*, 78, 681
[265] Sultanov, R. A., Balakrishnan, N., 2005, *ApJ*, 629, 305

[266] Tauber, J. A., Lis, D. C., Keene, J., Schilke, P., Buettgenbach, T. H., 1995, *A & A*, 297, 567
[267] Teske, J. K., Najita, J. R., Carr, J. S., et al., 2011, *ApJ*, 734, 27
[268] Thrower, J. D., Jørgensen, B., Friis, E. E., et al., 2012, *ApJ*, 752, 3
[269] Tielens, A. G. G. M., Hollenbach, D. J., 1985, *ApJ*, 291, 722
[270] Tielens, A. G. G. M., Hollenbach, D. J., 1985, *ApJ*, 291, 747
[271] Tielens, A. G. G. M., Meixner, M. M., van der Werf, P. P., et al., 1993, *Science*, 262, 86
[272] Tielens, A. G. G. M., Allamandola, L. J., 2011, *Physics and Chemistry at Low Temperatures*, ed. L. Khriachtchev (Singapore: Pan Stanford), 341
[273] Tielens, A. G. G. M., 2005, *Physics and Chemistry of the Interstellar Medium* (Cambridge: Cambridge University Press), chapter 9.4
[274] Torrelles, J. M., Gómez, J. F., Rodriguéz, L. M., et al., 1998, *ApJ*, 505, 756
[275] Umebayashi, T., Nakano, T., 1981, *PASJ*, 33, 617
[276] Umebayashi, T., 1983, *Progress of Theoretical Physics*, 69, 480
[277] Umebayashi, T., Nakano, T., 2009, *ApJ*, 690, 69
[278] Vacca, W. D., Garmany, C. D., Shull, J. M., 1996, *ApJ*, 460, 914
[279] van der Tak, F. F. S., Nagy, Z., Ossenkopf, V., et al., 2013, *A & A*, 560, A95
[280] van't Hoff, M. L. R., Walsh, C., Kama, M., Facchini, S., van Dishoeck, E. F., 2017, *A & A*, 599, A101
[281] Vinogradoff, V., Rimola, A., Duvernay, F., et al., 2012, *Phys Chem Chem Phys*, 14, 12309
[282] Vinogradoff, V., Fray, N., Duvernay, F., et al., 2013, *A & A*, 551, A128
[283] Visser, R., van Dishoeck, E. F., Black, J. H., 2009, *A & A*, 503, 323
[284] Wakelam, V., Bron, E., Cazaux, S., et al., 2017, *Molecular Astrophysics*, 9, 1
[285] Walsh, C., Millar, T. J., Nomura, H., 2010, *ApJ*, 722, 1607
[286] Walsh, C., Loomis, R. A., Öberg, K. I., et al., 2016, *ApJ*, 823, L10
[287] Wang, S., Bergin, E. A., Crockett, N. R., et al., 2011, *A & A*, 527, A95
[288] Ward-Thompson, D., Scott, P. F., Hills, R. E., André, P., 1994, *MNRAS*, 268, 276
[289] Wilcock, L. A., Ward-Thompson, D., Kirk, J. M., 2012, *MNRAS*, 422, 1071
[290] Williams, J. P., Best, W. M. J., 2014, *ApJ*, 788, 59
[291] Woitke, P., Kamp, I., Thi, W. F., 2009, *A & A*, 501, 383
[292] Woods, P. M., Slater, B., Raza, Z., et al., 2013, *ApJ*, 777, 90
[293] Yu, M., Evans, N. J., II, Dodson-Robinson, S. E., et al., 2017, *ApJ*, 841, 39
[294] Zanchet, A., Agúndez, M., Herrero, V. J., Aguado, A., Roncero, O., 2013, *AJ*, 146, 125
[295] Zanchet, A., Godard, B., Bulut, N., et al., 2013, *ApJ*, 766, 80
[296] Zapata, .A., Schmid-Burgk, J., Menten, K. M., 2011, *A & A*, 529, A24
[297] Zhen, J., Castellanos, P., Paardekooper, D. M., et al., 2014, *ApJ*, 797, L30
[298] Zhen, J., Castellanos, P., Linnartz, H., Tielens, A. G. G. M., 2016, *Molecular Astrophysics*, 5, 1

12
The Aromatic Universe

12.1 Introduction

The physics and chemistry of large aromatic molecules has been discussed in Chapter 8. Here, we will apply those results to the ISM of galaxies. We will discuss observations of the Aromatic Infrared Bands (AIBs) and derive the characteristics of their carriers. We will learn that the carriers are polycyclic aromatic hydrocarbon molecules (PAHs) containing some 50 C-atoms, that these species are very abundant, locking up some 10% of the elemental carbon, and that they are widespread throughout the interstellar medium. These species typically absorb one FUV photon every year in the diffuse ISM and relaxation occurs through vibrational transitions on a timescale of $\simeq$1 s. Astronomers use PAHs to trace the penetration of FUV photons into photodissociation regions associated with regions of massive star formation or the surfaces of protoplanetary disks. In a way, the AIBs function as FUV "photon counters" and as such photons are emitted by short-living massive stars (5−100 Myr), their intensity provides a measure of the star formation rate. PAHs also play an active role in the ISM. They heat the gas through the photo-electric effect where the excess photon energy is carried away as kinetic energy of the photo-electron, which is then shared with the electrons and atoms in the gas. For completeness, PAHs may also act as formation sites of H_2 and be the origin of small hydrocarbons in PDRs, but this has already been discussed in Section 11.6.3. PAHs are generally thought to be formed in stellar outflows as building blocks of soot particles and then undergo a rich photochemistry in the ISM where they can be broken down to smaller hydrocarbons. This process may weed out the interstellar PAH family and only the most stable species may be able to withstand the rigors of the extreme environments. This survival of the fittest may favor the formation of so-called GrandPAHs. Besides FUV photons, PAHs will also be destroyed by shock waves driven by supernovae, in hot plasmas, and by cosmic ray interactions. All of these aspects will be discussed in this chapter.

The presence of C_{60} in space has been unambiguously established through visible and IR spectroscopy. This is the largest molecule identified so far in space. While fullerenes are not aromatic, many of the relevant processes are akin to those of large PAHs. Moreover, its origin and evolution in space may be closely linked to that of PAHs. Indeed, in some models, interstellar C_{60} can be considered the “grandson” of large PAHs. Hence, the discussion of interstellar C_{60} is included in this chapter.

A course on molecular astrophysics might focus on the physics of large molecules in space, discussed in Sections 12.5 and 12.8. A spectroscopy course would include Sections 12.2 and 12.3. An astrochemistry course would target Section 12.7 but also include Sections 9.5.3, 10.5.3, and 11.6.3. Fullerenes are discussed in Section 12.4 but also, in connection to the DIBs, in Section 9.6.3. Finally, a discussion of the role of PAHs in the Anomalous Microwave Emission is presented in Section 12.6. The section on further reading and resources provides students with an entry point into the literature and a starting point for a more in-depth essay.

12.2 The Aromatic Infrared Bands

In the 70s and 80s, ground-based and airborne spectra of HII regions and planetary nebula revealed broad emission features at mid-IR wavelengths. As, originally, the carriers of these features were unknown, they were dubbed the Unidentified InfraRed – or UIR – bands. Many carriers were proposed in those early years,[1] but it is now well accepted that these bands are characteristic for vibrational modes of Polycyclic Aromatic Hydrocarbon species and they are now commonly called the Aromatic Infrared Bands (AIBs). One of the big revelations of the IRAS mission was the detection of bright, mid-IR emission from IR cirrus clouds in the general diffuse ISM. As dust in these objects is far too cold to emit at these short wavelengths and triggered by pioneering studies of the mid-IR emission from reflection nebulae, it was quickly realized that the mid-IR originated in small species that because of their limited heat capacity become temporarily very hot upon absorption of a UV photon and lose this energy through emission in mid-IR vibrational modes. Subsequent studies with the Infrared Space Observatory (ISO) and the Spitzer Space Telescope revealed that the infrared spectra of almost all interstellar objects, including regions of massive star formation such as HII regions and reflection nebulae; carbon-rich stars in the last phases of their evolution such as post-asymptotic-giant branch stars and planetary nebulae; surfaces of molecular clouds; planet-forming disks surrounding young stellar objects; the general interstellar medium of the Milky Way and star-forming galaxies; galactic nuclei; galactic scale outflows; (ultra)luminous infrared galaxies; and galaxies out to red-shifts of $\sim$4, are dominated by strong, broad emission features. The AIB spectrum is very rich with main bands at 3.3, 6.2, 7.7, 8.6, 11.2, and 12.7 μm perched on broad emission plateaus and accompanied by weak bands at 3.4, 5.2, 5.7, 6.0, 7.4, 12.0, 13.5, 14.2, 15.8, 16.4, 17.0, and 17.4 μm (Figure 12.1). These features are ubiquitous in the ISM on all scales from planet-forming disks around young stellar objects ($\simeq$200 AU) to regions of massive star formation ($\simeq$5 pc), to the scale of whole galaxies (10’s of kpc; Figure 12.2). It is clear that the carriers represent a ubiquitous, widespread and abundant component of the ISM.

[1] A session at an IAU meeting at the time summarized this well by dubbing them the Overidentified InfraRed bands.

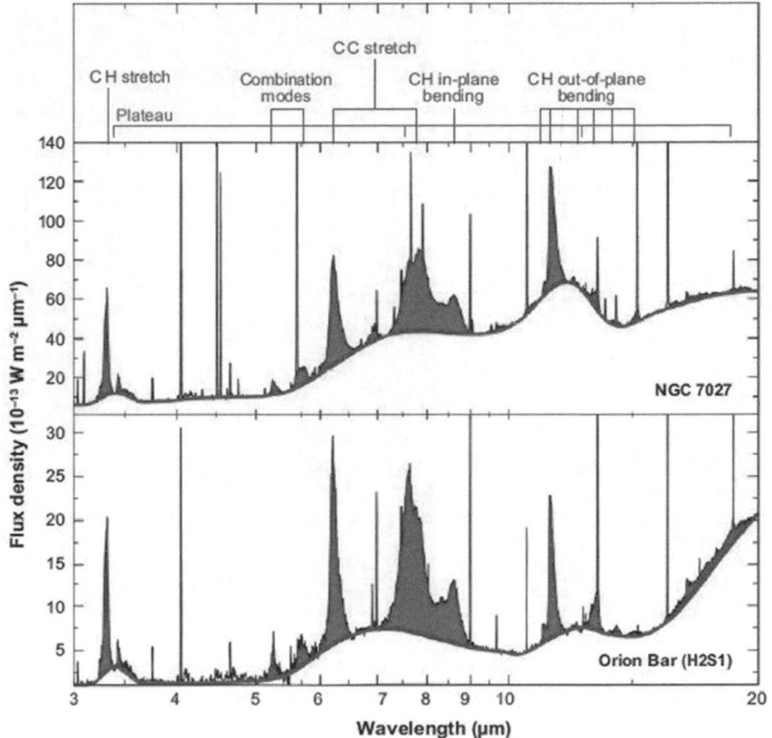

Figure 12.1 ISO/SWS mid-IR spectra of two bright PDRs: the Orion Bar, associated with the M42 HII region and the planetary nebula, NGC 7027. The AIB spectrum is incredibly rich in spectral detail. General identifications of the modes involved are given at the top. Figure taken from [94]

12.2.1 Spectral Variations

Profile Variations

Profile variations, including peak position and width, may reflect the presence of multiple components with a varying relative contribution or differences in excitation conditions. As Figure 12.1 so clearly illustrates, the AIB spectrum is very rich with a multitude of features. At high spectral resolution, some of these features reveal spectral substructure, which pertains to different emission components. The 7.7 and 3.4–3.6 components are cases in point (Figure 12.3). While the pattern of features in the AIR spectrum is quite similar in all objects – with six major bands at 3.3, 6.2, 7.7, 8.6, and 11.2 μm – a small subset of sources show large variation in the position and profile of some of the features. Again, the 7.7 and 3.4–3.6 components are very illustrative (Figure 12.4). Initial studies divided the observed profiles into three class (type A to C). Later, a fourth class (type D) was added. This spectral subclassification mainly pertains to the peak position of the 6.2 and 7.7 μm bands with only minor variations in the 3.3, 11.2, and 12.7 μm band peak positions

Figure 12.2 The mid-IR emission of (almost) all objects is dominated by fluorescence by large PAH molecules. Composite mid-IR images where the 8 μm (red) image traces primarily emission by PAH molecules in the 7.7 μm feature (see Figure 12.1), while the 4.5 μm image (green) shows ionized gas and dust, and stars show up in 3.6 μm (blue). Left: The Orion star-forming region where massive stars at the center of a star cluster, the Trapezium ionize the gas on a scale of $\simeq$0.2 pc. Far-UV photons from these stars pump PAH molecules on much larger scale sizes ($\simeq$2 pc). Right: The whirlpool galaxy, M51, where the whole ISM on a scale size of 10 kpc is set aglow in the light of the PAHs. The spiral arms are dotted by individual regions of massive star formation akin to the Orion region (left). Images obtained with IRAC on the Spitzer Space Telescope and courtesy of NASA, JPL-Caltech (A black and white version of this figure will appear in some formats. For the color version, please refer to the plate section.)

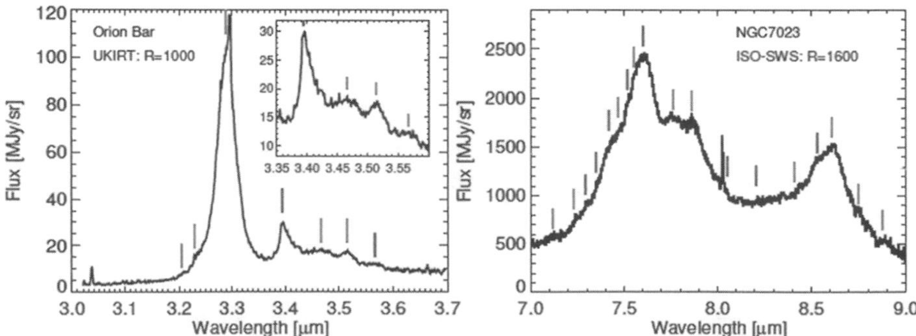

Figure 12.3 The spectral richness of the PAH emission toward two prototypical PDRs shown for the 3 μm (left) and 8 μm (right) region. Sub-components are indicated by vertical bars. Figure taken from [85, 116]

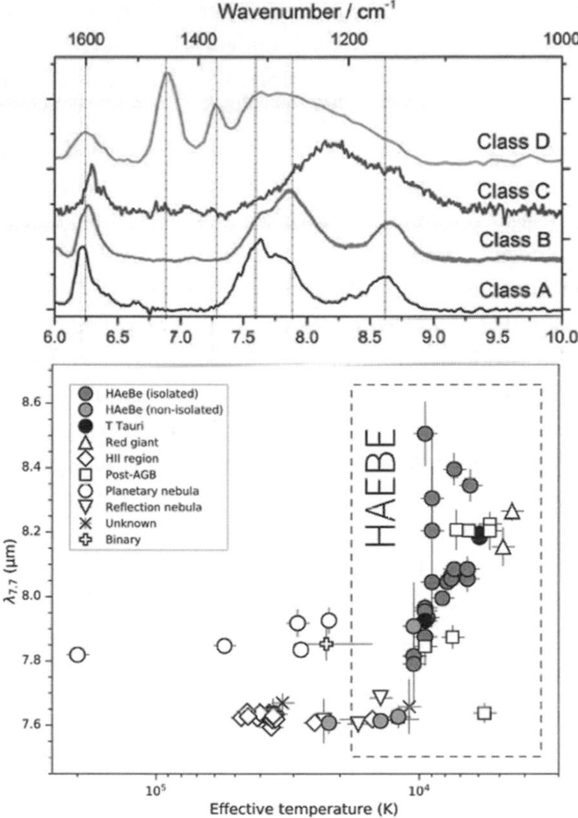

Figure 12.4 Top: Selected IR spectra, illustrating the diversity of AIR emission band profiles in the 6–10 μm range. The profiles have been continuum subtracted and normalized to the maximum intensity of the 7–8 μm complex. Type A: IRAS 23133+6050 (bottom); Type B: HD 44179 (2nd from the bottom); Type C; IRAS 13416-6243 (2nd from the top); Type D: IRAS 05110-6616 (top). Original classification of types A–C was done by [94, 125] based upon ISO/SWS spectra. Type D was added by [78] based upon Spitzer/IRS spectra. Bottom: The peak position of the "7.7" μm band as a function of the effective temperature of the illuminating star. The different symbols indicate different source classes. Data taken from [19, 116]. Figure courtesy of C. Boersma

and profiles. This classification seems to correlate with source properties, as sources with ISM material are dominated by type A emission spectral characteristics, while planetary nebulae, post asymptotic giant branch (post-AGB) objects, and protoplanetary disks associated with Herbig AeBe stars or T Tauri stars are type B. The type C class consists of only a few objects. The best known of these is CRL 2688, a post-AGB descendant from a quite massive progenitor, linking Asymptotic Giant Branch (AGB) stars (e.g. IRC+10216) with PNe (e.g. NGC 7027). Class D also consists of post-AGB objects, but these are low metallicity offshoots of low-mass ($\simeq$1 $M_\odot$) progenitor stars, born 10 Gyr ago when the Milky Way was formed, and not linked to well-known PNe.

The profiles of these different classes are so different that independent emission components might be indicated. However, analysis suggests that there is a continuous variation in peak position, suggesting perhaps an evolutionary link. Specifically, the peak position of the 7.7 μm band seems to correlate with the effective temperature of the illuminating star with a rapid transition from class A to class C within a very narrow effective temperature interval (Figure 12.4). This shift may reflect blending of two or more components coupled with the limited spectral resolution of the observations rather than a genuine, gradual shift of the peak position. Indeed, the presence of a 7.6 and a 7.8 μm component blending with varying contribution into the 7.7 μm band was recognized early on. The 7.7 μm band in the spectrum of NGC 7027, a PN with a 200,000 K central star, is a clear combination of a 7.6 and a 7.8 μm band, which at low resolution would give rise to a peak position clearly shifted from 7.6 μm (and, by the way, would therefore not follow the trend shown in Figure 12.4). In this respect, also consider the peak position of the 7.7 μm feature in the spectra of the three A0/B9, Herbig AeBe stars, Ty Cra, HD 97048, and HD 100546, which shifts from 7.6 to 8.0 μm, and does not follow therefore the trend of Figure 12.4. For these three sources, this spectral variation is accompanied by differences in spatial distribution of the emitting material; e.g., from the surrounding ISM material (Ty Cra) to a combination of emission from ISM and disk material (HD 97048) to emission only from disk material (HD 100546). Hence, the difference between Class A, B, and C may reflect the history of the material; e.g. the prolonged exposure to a strong radiation field in the photosphere of the disk transforms class A into class C.

Strength Variations

Close inspection of the AIB spectrum reveals relative strength variations between the features both when comparing spectra of different nebulae and also when comparing spectra within a given nebula. The earliest example identified consists of the 6.2 and 7.7 μm bands versus the 11.2 μm band where the intensity of the first two is well correlated but varies by a factor of 3 with respect to the latter (Figure 12.5). The 8.6 and 12.7 μ bands similarly correlate with the 6.2 and 7.7 bands. While, globally, the 3.3 and 11.2 μm bands correlate better with each other than with the 6.2 or 7.7 μm bands, variations in the 3.3/11.2 μm intensity ratio are known to exist within sources. These correlation studies were pioneered with ISO/SWS and were obtained in a large aperture. The IRS slit spectrograph on Spitzer opened the way for spatial-spectral correlation studies by isolating emission feature maps at different wavelengths (Figure 12.6). This provides an alternative way to identifying emission "families" in the AIB spectrum. Figure 12.6 shows that components labeled 6.2, G7.6, and G8.6 have similar spatial distributions, which differ from those of the 11.2, G8.2, and G7.8 components. These variations have been attributed to variations in charge state and excitation (e.g. size) of the emitters. In view of the discussion in Section 12.2.1, relative variations in the integrated strength of AIBs may also reflect varying contributions of different emission components in blended profiles. Given Figure 12.3, a full understanding of this spectral-spatial diversity will have to await the superior spatial and spectral resolution of JWST.

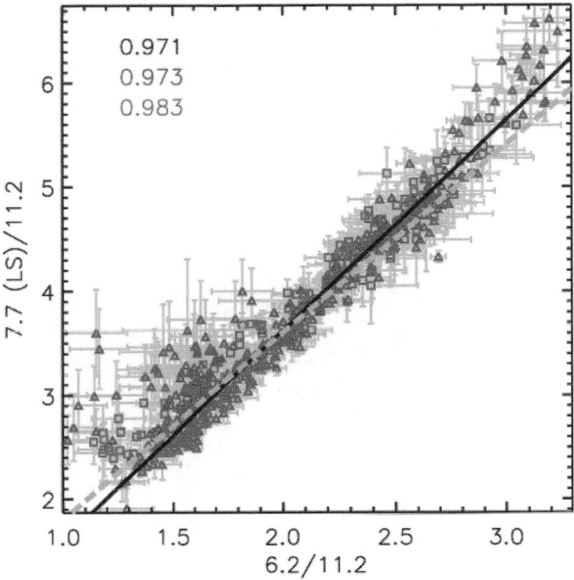

Figure 12.5 There is a good correlation between the ratio of the 6.2/11.2 and 7.7/11.2 μm bands as illustrated here for data of the reflection nebula, NGC 2023. The triangles and squares refer to different regions in this source. Best linear fits are also shown. Values of the correlation coefficient are indicated in the top left-hand corner separately for each data set plus for all the data. Figure taken from [96]

12.2.2 Plateau Emission Components

The AIBs are superimposed on broad emission components whose strength varies relative to that of the main bands. Their profiles are not well characterized and they may consist of several blended components. In addition, it is difficult to disentangle them from the extended red wings of the AIBs. These broad emission components have been attributed to clusters of PAHs or larger carbonaceous moieties containing hundreds of C-atoms.

12.2.3 Mid-IR Continuum Emission

The AIB bands and broad plateaus are perched on a weak continuum, which rises sharply toward longer wavelengths. This continuum varies from source to source. This emission is distinct from the far-IR dust continuum. The latter typically peaks around 100 μm and is due to emission by large dust grains in radiative equilibrium with the ambient FUV-visible radiation field. This shorter wavelength continuum is attributed to very small grains (VSGs) whose temperatures fluctuate somewhat due to their limited heat capacity. They contain some 10^4 C-atoms.

12.2.4 Component Analysis

A very different approach relies on principal component/blind signal decomposition analysis of spatial-spectral ISOCAM maps. As such studies require extended emission, their

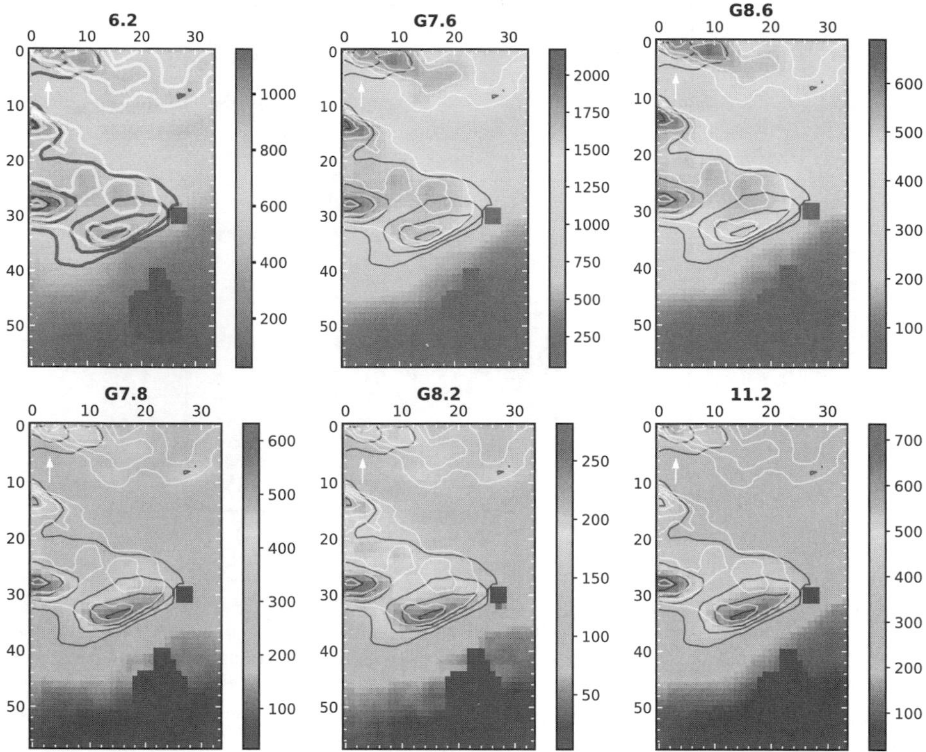

Figure 12.6 Spatial distribution of the emission at selected wavelengths in the 5–15 μm region in the Southern PDR of NGC2023. The AIB emission has been continuum subtracted. Band intensities are in units of 10^{-8} W m^{-2} sr^{-1}. As a reference, the intensity profiles of the 11.2 and 7.7 μm emission features are shown as contours in, respectively, black and white. The label for each panel indicates the emission component. The "G" in front of a label refers to a decomposition of the feature in Gaussian components. Note that the spatial distribution reveals two classes of features with different spatial distribution: 6.2, G7.6, and G8,6 versus G7.8, G8.2, and 11.2. The dark purple rectangle indicates an area removed because of the presence of a confusing protostar. The other dark purple area indicates a region of limited signal-to-noise. Figure adapted from [96] (A black and white version of this figure will appear in some formats. For the color version, please refer to the plate section.)

analysis and conclusions pertain to ISM-like sources. This has resulted in the identification of three distinct spectral components with a very different spatial distribution (Figure 12.7). The first of these is characterized by strong modes in the 6−9 μm range, while the second one has, relatively speaking, much stronger bands in the 11−13 μm range. The third component has broader features that are perched on a continuum. In this latter component, the 7.7 μm band peaks at 7.8 μm while, in the former two, the peak is at 7.6 μm. These three components can be separated in the data as they have very different spatial distributions (Figure 12.7). These three components have been assigned to emission by PAH cations, neutral PAHs, and PAH clusters, respectively. After characterization of the IR spectra of these three emission components in spatially extended nebula, spectra of a wider array of sources have been fit.

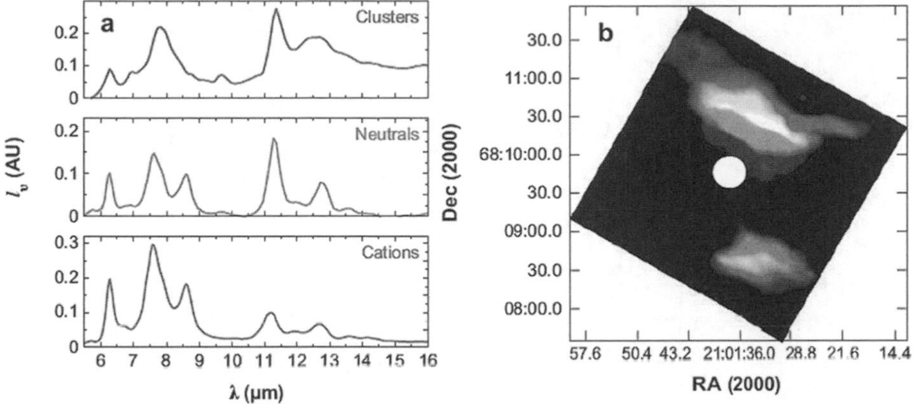

Figure 12.7 (a) Three identified components in the decomposition analysis of ISOCAM spectral-spatial maps of NGC 7023. The components are normalized by their integrated intensity. (b) These components show different spatial distributions in this source and have been attributed to ionic (blue) and neutral (green) polycyclic aromatic hydrocarbon molecules and clusters (red), respectively. In the color scheme used, red and green combine to yellow. Figure adapted from [12, 104] (A black and white version of this figure will appear in some formats. For the color version, please refer to the plate section.)

12.2.5 Spectral Fits

Web-based tools have been developed that fit observed mid-IR spectra returning the integrated strength of gaseous emission lines as well as the AIB bands. This allows a systematic analysis of observed spectra and identification of trends in their relative strength or with other characteristics of the emission region. These tools generally account for the underlying continuum by adopting a number of black-body components modified by dust emissivity properties represented as simple power-laws (e.g. $\kappa \propto \nu^{\beta}$ where β is typically chosen to be 1 or 2). Gaseous emission lines are then included and, as these are unresolved, gaussian line profiles are adopted with a width given by the instrumental resolution. The AIB bands are resolved and a profile has to be specified. Sometimes, a Lorentz profile is used that assumes that the modes can be represented by a harmonic oscillator or that the profile is given by a Gaussian.[2] Atomic or H_2 lines are represented by Gaussian profiles. This analysis is obviously affected by the choice of the components present and their adopted properties. For example, compared to a Gaussian, a Lorentz profile has a much stronger wing and, therefore, much of the underlying plateau component is included in the fit in the latter (Figure 12.8). Thus, in a Lorentz analysis, the derived integrated strength of AIBs is much larger than for Gaussian methods, while, at the same time, the strength of underlying plateaus will be underestimated. Likewise, representing the underlying (VSG) continuum by a limited number of modified black bodies introduces some arbitrariness in the analysis but has the advantage of ease of use. Studies show that these different approaches will differ in the detailed results but identified trends agree well.

[2] The expected profile of PAH emission bands from highly excited PAHs is described in Section 12.5.2.

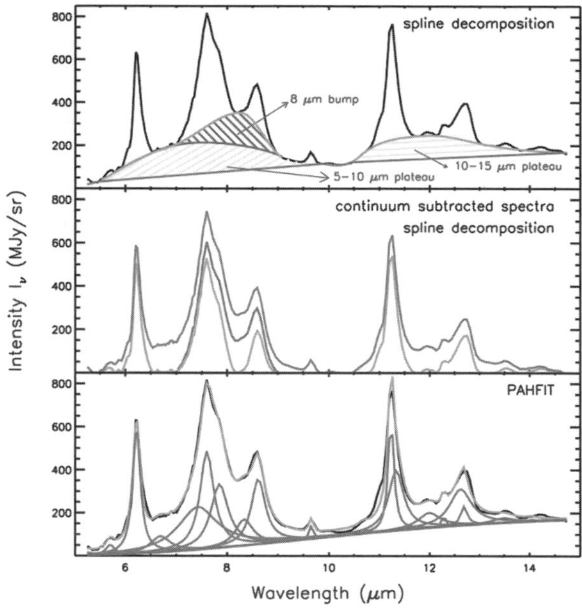

Figure 12.8 Top: a typical AIB spectrum obtained by IRS/Spitzer. The underlying continuum and emission plateaus are identified, which span up a global continuum. In addition, a broad 8 μm bump is indicated. Middle panel: The spectrum derived after subtraction of the global continuum – using a spline – is fitted by Gaussian components. Bottom: The same spectrum fitted by the PAHFIT tool using Lorentzian profiles for the AIBs and Gaussians for the atomic and H_2 emission lines. Figure taken from [96]

12.2.6 Database Fits

Observed spectra can also be analyzed using a database of PAH spectra either measured in the laboratory or calculated with DFT methods. This approach has the advantage that observed trends are directly linked to intrinsic properties of the PAH family (e.g. charge, molecular structure, size). The drawback is that such databases are by necessity incomplete and often have strong biases built in. Moreover, the measured or calculated spectra refer to low temperature molecules and models are required to calculate the expected emission spectra of highly excited molecules.

12.3 Spectroscopy and the Characteristics of the Carriers

12.3.1 General Assignments

Gas phase spectroscopy of large PAHs is very challenging because of the low volatility of these species. Early studies relied, therefore, on matrix isolation spectroscopy. Later, gas phase studies used a messenger technique[3] or used a very sensitive IR detector. But

[3] I.e. attaching a weakly bound noble gas atom to an ion and measuring the ion signal after dissociation of the complex using an IR laser.

as these techniques are very demanding, the number of species studied is very limited. Infrared MultiPhotonDissociation[4] studies of ions are more readily done but suffer from anharmonicity effects. In recent years, ion-dip studies have been used to study the 3 μm region of gas phase PAHs at low temperature and this has now been extended to lower frequency modes.

In contrast, in silico, quantum chemical studies of PAHs are quite routine using DFT in the double harmonic approximation, and (highly symmetric) species as large as ≃400 have been calculated and the results have been collected in a database (containing some 4000 spectra). However, in this approximation, anharmonicity and resonances are not included to the calculated harmonic frequencies. This correction factor is established by comparison with experimental data for a few small PAHs, mostly using matrix isolation spectroscopy. For PAHs, this correction factor is taken to be 0.963, although in recent years, multiple scaling factors have been introduced to account for frequency/mode dependencies. As the matrix typically induces frequency shifts of 0.2%, this is a worthwhile improvement in accuracy. In recent years, Moore's law has allowed calculation up to quartic force field in the vibrational potential and the dipole surface for small PAHs. As there is no simple analytical solution any more, second-order vibrational perturbation theory (VPT2) is required to solve for the vibrational frequencies but then resonance interactions (see Section 12.3.2) have to be explicitly taken into account. This approach has greatly improved the accuracy of theoretical calculations and the agreement with measured spectra is now at the 0.2% level in frequency, removing the need for scaling factors and, in addition, reproducing (relative) intensities much better. Nevertheless, there is still room for further improvement as measurements show that triple combination bands are also important.

The general principles of vibrational spectroscopy have been discussed in Section 3.3. The well-known IR emission features at 3.3, 6.2, 7.7, 8.6, and 11.3 μm are characteristic for the stretching and bending vibrations of aromatic hydrocarbon materials and assignments are summarized in Figure 12.1 and Table 12.1. The 3 μm region is characteristic for CH stretching modes and the 3.3 μm band is due to the CH stretching mode in aromatic species. Pure CC stretching modes generally fall in the range 6.1–6.5 μm, vibrations involving combinations of CC stretching and CH in-plane bending modes lie slightly longward (6.5 − 8.5 μm), and CH in-plane wagging modes give rise to bands in the 8.3−8.9 μm range. The 11−15 μm range is characteristic for CH out-of-plane bending modes. Bands in the 15−18 μm region are due to in-plane and out-of-plane ring bending modes of the C-skeleton. Longer wavelength modes – not yet detected in space – are due to large scale motions of the C skeleton such as the naphthalene butterfly mode (60 μm) and the jumping jack (≃40−120 μm) and drumhead (≃80−1000 μm) modes. These modes are very molecule specific.

These assignments to specific modes of polycyclic aromatic materials are generally agreed upon and noncontroversial. There is more discussion on the assignment of the

[4] An IR-free electron laser tuned to IR absorption bands induces fragmention of ions. This fragmentation is followed with a mass spectrometer while scanning the laser.

Table 12.1 *Common assignments for the Aromatic Infrared Bands*

Band	Assignment	Charge state
3.3 μm	aromatic C–H stretching mode	neutral
3.4 μm	aliphatic CH stretching mode in methyl groups	neutral
	CH stretching mode in hydrogenated PAHs	neutral
5.2 μm	combination mode, CH out of plane bend	neutral
5.65 μm	combination mode, CH out of plane bend	neutral
6.0 μm	CO stretching mode (?)	cation
6.2μm	aromatic CC stretching mode	cations
6.9μm	aliphatic CH bending modes	
7.6μm	CC stretching and CH in-plane bending modes	cation
7.8μm	CC stretching and CH in-plane bending modes	cation
8.6μm	CH in-plane bending modes	cation
11.0μm	CH out-of-plane bending modes, solo	cation
11.2μm	CH out-of-plane bending modes, solo	neutral
12.0μm	CH out-of-plane bending modes, duo	
12.7μm	CH out-of-plane bending modes, duo	cation (?)
13.6μm	CH out-of-plane bending modes, quartet	
14.2μm	CH out-of-plane bending modes, quartet	
15.8μm	in-plane and out-of-plane C–C–C bending modes	neutral
16.4μm	in-plane and out-of-plane C–C–C bending modes (pendant ring ?)	cation (?)
17.4μm	in-plane and out-of-plane C–C–C bending modes	cation
17.8μm	in-plane and out-of-plane C–C–C bending modes	neutral & cation
Plateaus		
3.2–3.6 μm	overtone/combination modes, CC stretch	
6–9μm	blend of many CC stretch/CH in-plane bend modes[a]	
11–14μm	blend of CH out of plane bending modes[a]	
15–19μm	blended in-plane and out-of-plane C–C–C bending modes	neutral

[a] In PAH clusters

weaker features. The 3.4 μm band – which, in interstellar spectra, is only 10% of the 3.3 μm mode but can be as strong as the 3.3 μm band in low mass, low metallicity post-AGB objects – is due to aliphatic CH stretching mode, possibly in methyl groups attached to the PAHs or alternatively to PAHs with extra H bonded to some carbons, destroying the aromatic character and introducing ring puckering. The presence of a feature at 6.9 μm in some spectra may provide some support for the aliphatic interpretation. However, the 6.9 μm band could also be due to a (weak) CC mode as a strong interstellar 3.4 μm band does not always go hand-in-hand with a strong interstellar 6.9 μm band. The alternative, the interpretation in terms of extra-H bonded to the peripheral carbon, is controversial as well because of the limited stability of such hydroPAHs (Sections 8.5 and 11.6.3).

12.3.2 The 3 μm Region

Observationally, this spectral window is dominated by one strong band, at 3.29 μm with weaker bands at 3.40, 3.46, 3.52, 3.55, 3.56, and 3.58 μm (Figure 12.3). The observed peak and profile of the 3.3 μm AIB are very similar for most sources with a very symmetric feature, peaking at 3.29 μm with a width of 0.04 μm. A few sources show small shifts in peak position ($\simeq$0.02 μm). The Red Rectangle, HD 44179, is a good example where the 3.3 μm AIB contains at least two components, at 3.28 and 3.30 μm, with different spatial distributions. In terms of intensity, in general, the 3.3 μm AIB correlates well with the 11.2 μm AIB but not with the 6.2 and 7.7 μm AIBs. There are, however, variations in the relative strength of the 3.3 and 11.2 μm AIBs over individual nebulae. The 3.4–3.6 μm range shows a rich set of features but, except for the 3.4 μm AIB, they have been little studied. Spatial studies reveal variations in the relative strength of the 3.4 μm band within nebulae with larger 3.4/3.3 μm ratios at lower FUV fields. Typically, this ratio is only $\simeq$0.1. Much higher 3.4/3.3 μm ratios have been measured for the so-called 21-μm objects[5] where occasionally the 3.4 μm band is even stronger than the 3.3 μm AIB. Actually, the observed profile of the 3.4 μm band in these objects is also quite different from that observed in the AIB spectrum in the ISM and closely resembles that of the 3.4 μm dust absorption feature.[6]

Spectroscopically, the 3 μm region is dominated by the CH stretching modes and 3.3 μm is very characteristic for the CH stretching mode in PAHs. Harmonic DFT calculations and low resolution experimental studies reveal that the peak position of these modes depends on the molecular structure. Typically, this band falls at 3.26 μm in absorption spectra but for H atoms in armchair edge structures[7] (Figure 12.15), the symmetric CH stretching modes shift slightly to the blue (to 3.21 μm) due to steric hindrance of the two hydrogens facing each other at close distance. At moderate resolution ($R \simeq 3000$), the precise position depends on the number of adjacent H's. Furthermore, the molecular symmetry "controls" the number of allowed transitions and the frequency region over which IR activity is observed (compare pyrene and chrysene in Figure 12.9). There is also a small dependence on charge but the CH stretching modes in cations are very weak compared to the CC modes (see Section 12.3.3) and, likely, cations play no role in the wavelength region in interstellar spectra. Overall, these subtle differences show up in low temperature spectra but will be washed out in emission spectra from highly excited molecules.

The modes in the 3 μm range are very much influenced by resonances that push one band to a lower energy and the other to a higher energy. In addition, intensity can be shared across a resonance and that can result in significant intensity gain for otherwise weak bands. Four different types of resonances between two bands can be discerned: Fermi one resonance when a fundamental is approximately equal to an overtone, Fermi two resonance when a fundamental is approximately equal to a combination band, Darling–Dennison

[5] A class of post-AGB objects with F/G supergiant central stars, descending from 1 $M_\odot$, low metallicity stars formed early on during the assemblage of the Milky Way, and characterized by slowly expanding dust-gas shells.

[6] Generally attributed to methyl groups in hydrogenated amorphous carbon materials.

[7] Also known as bay regions.

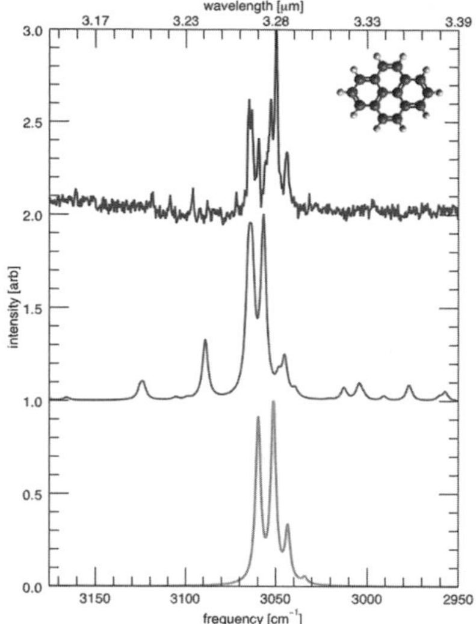

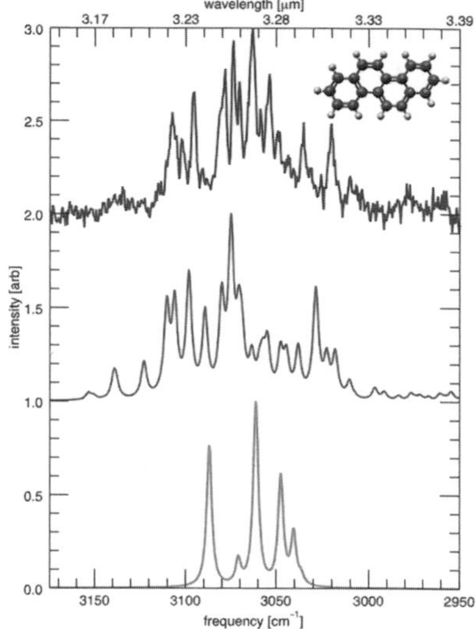

Figure 12.9 Comparison between calculated and measured 3 μm spectra of pyrene (left) and chrysene (right). Top: The gas phase absorption spectrum measured using an ion-dip technique. Middle: Calculated spectrum accounting for anharmonicity and resonances. No scaling factor was applied. Bottom: Calculated spectrum in the double harmonic approximation scaled by a factor of 0.963 to line it up with matrix isolation spectra. Figure adapted from [70, 75]

resonance when two overtones or two combination bands or a mixture of these two are approximately equal, and Coriolis resonance when two fundamentals are approximately equal. Resonances need special care in quantum chemistry studies but can be accounted for by removing related terms from the VPT2 treatment and handled these separately with a method akin to degenerate second-order perturbation theory. Figure 12.9 compares the theoretical results for pyrene and chrysene to low temperature, moderate resolution spectrum measured using an ion-dip technique, illustrating the good agreement in pattern, number of bands, relative intensity, and peak position (to better than 0.15%) possible. Note that no scaling factor has been applied.

The 3.4–3.6 μm region is characteristic for CH stretching modes involving sp^3 carbon. Spectral structure in this region may involve (a)symmetric stretching vibrations in aliphatic functional groups attached to the aromatic skeleton; e.g. methyl (CH_3) and ethyl (CH_2CH_3) groups. Spectroscopically, only the former seems to be relevant in space. The symmetric CH stretch in alkylated PAHs falls near 3.47 μm while the asymmetric stretch occurs close to 3.41 μm. Alternatively, AIBs in this wavelength range may involve so-called superhydrogenated PAHs where an extra H converts an aromatic (sp^2) CH bond into a sp^3 CH_2 group. These species show several well-separated bands in the 3.333–3.636 μm region. The asymmetric aliphatic CH stretch is the strongest and falls near 3.393 μm. The symmetric CH stretch falls close to 3.52 μm. Methylation or hydrogenation also produces changes in the low temperature absorption spectra of the aromatic CH stretching bands. Again, such subtle changes will be washed out in emission spectra.

For completeness, deuteration will shift the CH stretching modes to $\simeq$4.4 μm and 4.6–4.7 μm for the aromatic and aliphatic CD stretching modes, respectively, and weak bands have been detected near these wavelengths in some interstellar sources.

12.3.3 The 5–6 μm Region

The AIB spectrum shows two weak bands, at 5.25 and 5.7 μm, where the latter shows a red-shaded profile. In PAHs, bands in this region are due to combination bands. These combination bands are found to be generated by modes of the same type, i.e., out-of-plane in combination with out-of-plane, and in-plane in combination with in-plane. The out-of-plane out-of-plane combination bands range from 1200–1900 cm^{-1} and the in-plane in-plane combination bands range from 2200–2600 cm^{-1}. The out-of-plane out-of-plane combination bands are found to have more intensity than the in-plane combination bands by approximately an order of magnitude. Bands originating between a combination of one in-plane mode and one out-of-plane mode do not have significant intensity, even in other frequency regions. In the double harmonic approximation, overtone and combination bands do not have intensity but when anharmonicity is accounted for, these bands show weak IR activity. Measured and calculated spectra are compared in Figure 12.10, illustrating the good agreement in pattern, number of bands, relative intensity, and peak position (to better than 0.15%) possible. Note that no scaling factor has been applied.

12.3.4 The 6–9 μm Region

The 6–9 μm region is characteristic for CC stretching modes and CH in plane bending modes. Pure aromatic CC stretching modes fall between 6.1 and 6.5 μm. At slightly longer wavelengths (6.5 μm and 8.5 μm), vibrations involve combinations of CC stretching and CH in-plane bending modes. The CH in-plane wagging vibrations occur in the 8.3 to 8.9 μm range (Figure 12.1). Anharmonic DFT calculations compare very well to experimental spectra (Figure 12.11). Harmonic DFT calculations require a scaling factor (0.963) and do nearly as well. In this wavelength range, the number of bands and their precise position will depend on charge, molecular structure, size, and heterogeneity. Charge has a great influence on the intensity of the bands in this wavelength range: increasing the strength of these bands by an order of magnitude or more upon ionization (Figure 12.12). Given the observed strength of the modes in the 6–9 μm range relative to those in the 11–14 μm range, the former have been assigned to CC modes in PAH cations while the latter (as well as the 3 μm modes) are attributed to CH modes in neutrals.

Neutrals and cation/anions have very similar modes in the 6–9 wavelength range. The change in dipole moment derivative strongly affects the strength but produces only a small shift in peak position (i.e. the main band in the 6.2–6.5 μm range shifts typically by only

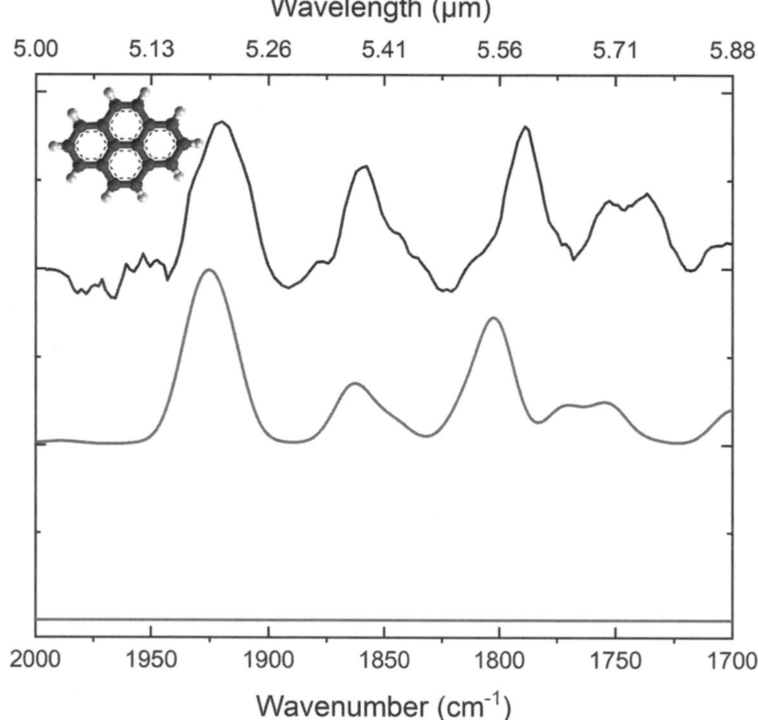

Figure 12.10 Comparison between calculated and measured 5–6 μm spectra of pyrene. Top: The absorption spectrum measured using an ion-dip technique. Middle: Calculated spectrum accounting for anharmonicity and resonances. No scaling factor was applied. Bottom: In the double harmonic approximation, overtone and combination bands have no intensity. Figure adapted from [67, 70]

$\simeq$0.13 μm). Focussing on the cations here, the main band in the highly symmetric coronene and ovalene families occurs at $\simeq$6.36 μm. For PAHs with in excess of 100 C-atoms, more bands appear that blend into a broad feature at longer wavelengths. The main band(s) in the 7.5–8.0 μm range shift systematically with size toward longer wavelength from about 7.6 μm for $C_{96}H_{24}^{+}$ to 7.8 μm for $C_{150}H_{30}^{+}$, and beyond for even larger ones. A strong in-plane CH bending mode near 8.6 μm is a characteristic for highly symmetric, compact PAHs and this band grows in strength with increasing size. Compact, highly symmetric PAHs with in excess of 100 C-atoms have a band near 8.6 μm that rivals in strength the bands in the 7.5–8.0 μm range. On the other hand, the 8.6 μm band is very weak for noncompact and irregular PAHs. Taking this together, the observed 8.6 μm AIB suggests then that highly symmetric, compact PAHs with sizes in the range 50–100 C-atoms dominate the interstellar PAH family.

While overall, the calculated IR spectrum of large, symmetric, compact PAHs is quite similar to the AIB pattern, there are several issues when comparing the observed

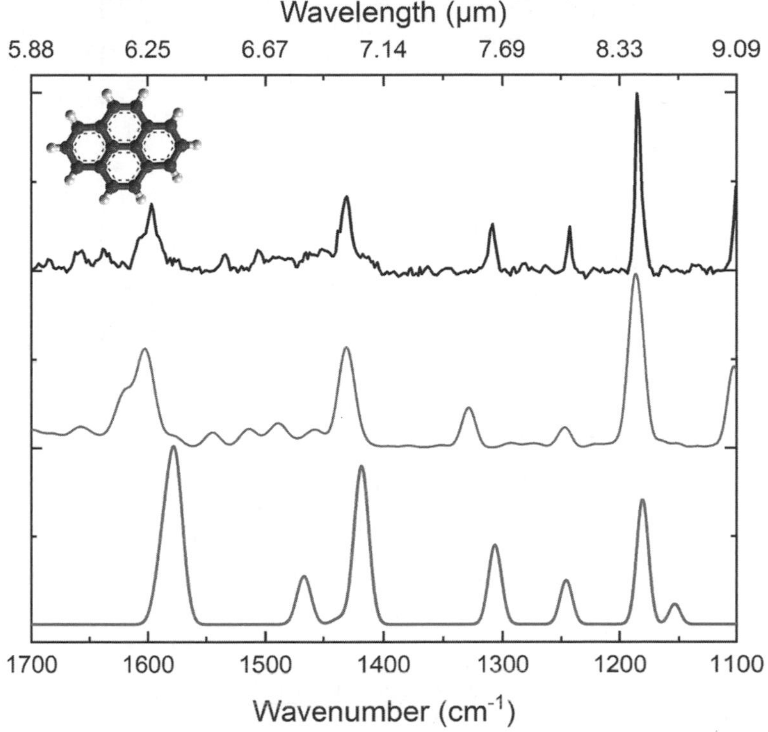

Figure 12.11 Comparison between calculated and measured 6–9 μm spectra of pyrene. Top: The absorption spectrum measured using an ion-dip technique. Middle: Calculated spectrum accounting for anharmonicity and resonances. No scaling factor was applied. Bottom: The double harmonic approximation where the bands have been shifted by a uniform scaling factor (0.963) determined from comparison to matrix isolation spectra. Figure adapted from [67, 70]

interstellar spectra with the calculated spectra. Specifically, the observed peak position at 6.2 μm for ISM sources cannot be explained by large (or small), compact, ionized PAHs as those species have bands at 6.3 μm. Several solutions have been suggested. The first one, N-substitution into the ring, has been best researched. Substitution of N for C shifts the main band in the 6.1–6.5 μm range toward shorter wavelengths and this shift is more pronounced when the N is moved "deeper" into the ring structure. Note that the IR active mode is still a CC mode. The presence of N does not affect the aromaticity of the π−electron system but its high electronegativity does change the electron density distribution in the molecule. The latter produces subtle changes in the CC bond force constants as well as in the dipole derivatives. Together, this produces a blueshift of the aromatic CC stretch. As these effects depend on the magnitude of the change in the electron density distribution, its effect decreases with increasing PAH size. For a PAH of the size of circumcoronene ($C_{54}H_{18}$), substitution with 2 N atoms is required to produce

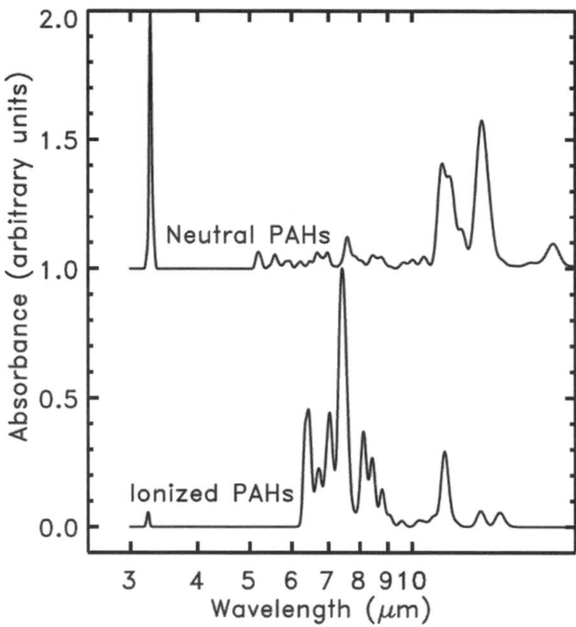

Figure 12.12 The influence of charge on relative intensity of the CC stretching and CH in plane bending modes in the 6−9 μm range relative to the CH stretching modes (3.3 μm) and CH out-of-plane bending modes (11−14 μm). Figure taken from [94]

the same shift as for coronene ($C_{24}H_{12}$). Observed variations in the peak position of the interstellar 6.2μm band are then attributed to variations in the location of the N-atom in the ring system. The chemistry driving these variations is not understood. While N could be incorporated into the PAH skeleton during their formation in stellar outflows where gaseous HCN is available as a growth stock, the spectra of PNe are generally characterized by 6.3 μm bands not 6.2 μm bands. The shift of the band in the 6.1−6.5 μm range from 6.3 to 6.2 μm between PNe and ISM sources requires that, in the ISM, N is deeply inserted into the PAH skeleton after release into the ISM by AGB winds. Alternatively, this requires chemical growth of PAHs in the ISM involving N-bearing compounds.

The second possibility discussed in the literature is the presence of aliphatic groups. This interpretation is driven by the observed characteristics of spectra of the so-called 21 μm sources. The unique AIB pattern in the spectra of these post-AGB sources has strong 3.4 μm bands, which rival and sometimes even exceed the 3.3μm band in strength. At the same time, the 6.2 μm and 7.7 μm AIB shifts toward 6.3 μm and 8.0 μm, respectively (Figure 12.4). The increased importance of the 3.4 μm AIB betrays a high aliphatic character and their detailed profiles are unlike that encountered in other AIB spectra. In effect, the peak position and substructure in the spectra of these peculiar sources are very reminiscent of the (a)symmetric CH stretching modes of methyl (CH_3) or methylene (CH_2) groups.

Studies on HAC films and soot show that as the aliphatic character of these compounds increases the aromatic CC mode shifts from 6.2 to 6.3 μm and the longer wavelength modes shift toward 8 μm; very much in line with the observed differences. These solid compounds are not good analogues for the carriers of the AIBs as the feature-to-continuum contrast is very small and, in the ISM, solids do not attain the high temperatures required to emit at such short wavelengths. There has not been a systematic study of the effect of aliphatics on the IR spectrum of small PAHs. In general, PAHs with aliphatic functional groups have very weak, aliphatic CH bending modes at 6.8 and 7.3 μm. While the ratio of the intrinsic strength of the 6.8 and 3.4 μm modes is only 0.07, after taking excitation into account, the expected intensity ratio of the 6.8/3.4 bands is $\sim$1. Hence, the direct contribution of aliphatic modes to the interstellar emission spectrum in the 6–9 μm range is small. The introduction of aliphatic carbon does not relieve the conundrum that the aromatic CC mode of PAH molecules occurs at 6.3 μm rather than at 6.2 μm as, in HAC-like materials, the 6.3 μm band shifts to even longer wavelengths. Nevertheless, this does point to an interesting avenue of spectroscopic research. Likewise, the introduction of (pentagonal) defects leads to pyramidal carbon, and quantum chemical studies suggest that the increased strain shifts the CC modes to higher frequency and this might be a worthwhile direction to probe as well.

As a third issue to be considered, the importance of PAH clusters has been raised. Blind signal separation analysis of the IR spectral maps of reflection nebulae reveals the presence of a third component whose 6–9 μm bands peak at longer wavelengths than the AIBs (Figure 12.7). This third component has been assigned to PAH clusters. The spectra of stacked van der Waals clusters reveal little effect on the peak position of the CC modes. Possibly, interstellar PAH clusters are linked through aliphatic bridges and this induces spectral shifts, but this has not been studied in detail. Complexing of PAH cations with Fe atoms does shift the CC mode toward shorter wavelengths. However, given the low binding energy of such complexes ($\simeq$2.8 eV), the abundance of Fe-PAH complexes is expected to be small.

12.3.5 The 11–14 μm Region

The 11–14 μm region is characteristic for C–H out-of-plane bending modes (OOPs) and interstellar spectra show two strong bands, at 11.2 and 12.7 μm, and weaker bands at 11.0, 12.0, 13.6, and 14.2 μm (Figure 12.1 and Table 12.1). The relative strength of these bands is known to vary between sources and within sources (Figure 12.13). Anharmonic DFT calculations compare very well to experimental spectra (Figure 12.14). Harmonic DFT calculations require a scaling factor (0.963) but do nearly as well.

These OOP modes provide an excellent tool to study the pattern of peripheral H-atoms. In this pattern, we can recognize rings with solo H atoms (no H atoms attached to adjacent C atoms), duos (two adjacent C atoms, each with an H atom), trios (three adjacent C atoms, each with an H atom), and quartets (four adjacent C atoms, each with an H atom). Due to mode-coupling, the peak position of these modes is very sensitive to the number of adjacent

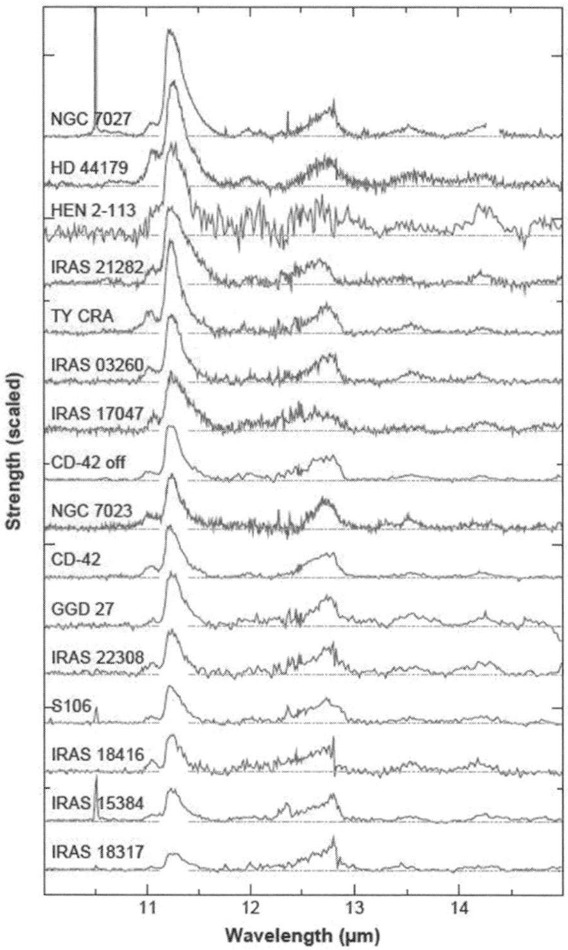

Figure 12.13 An overview of the observed variations in the out-of-plane bending region. The ISO/SWS spectra have been continuum subtracted, scaled to have the same integrated intensity in the 12.7 μm feature, and arranged in order of decreasing 11.2 μm strength (from top to bottom): planetary nebulae, Herbig stars, reflection nebulae, and HII regions. These observed variations are very characteristic for differences in molecular (edge) structure. (Fig. 12.15 illustrates potentially relevant structures). Solo CH modes occur at $\simeq$11.2 μm and are associated with long straight molecular edges, whereas duo and trio modes (at $\simeq$12.7 μm) correspond to corners. Figure adapted from [52]

H-atoms on a ring. Experiments and theory reveal that while the precise wavelengths of these modes depend somewhat on the carrier charge state, the 11.2 μm feature can be safely ascribed to solo CH groups. For neutrals, the 12.7 μm feature is ascribed to trios with possibly a contribution by duos. For ions, this band is due to duos. In terms of molecular structure, solo CH groups are part of long straight edges, whereas duos and

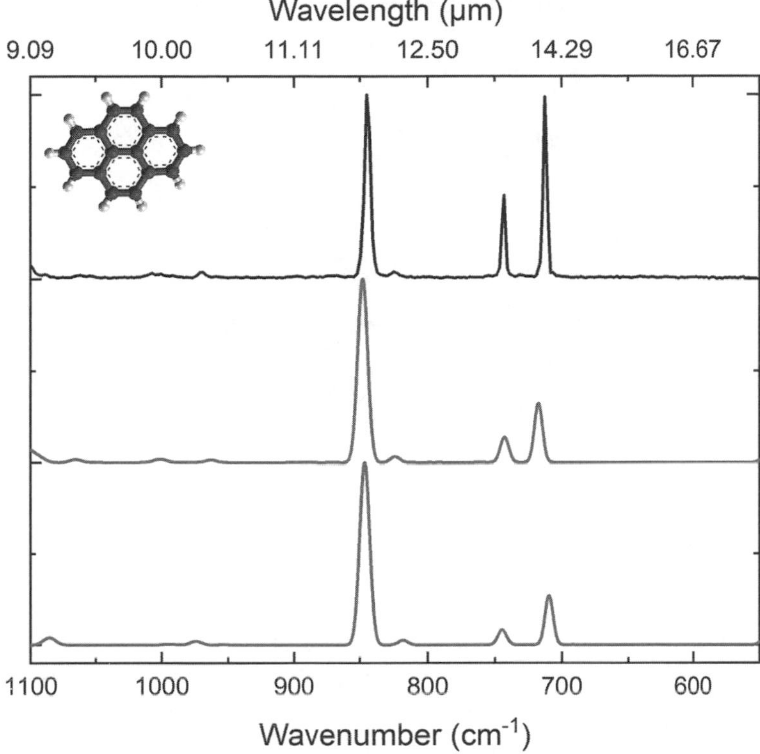

Figure 12.14 Comparison between calculated and measured 9–17 μm spectra of pyrene. Top: The absorption spectrum measured using an ion-dip technique. Middle: Calculated spectrum accounting for anharmonicity and resonances. No scaling factor was applied. Bottom: The double harmonic approximation where the bands have been shifted by a uniform scaling factor (0.963) determined from comparison to matrix isolation spectra. Figure adapted from [67, 70]

trios are characteristic of corners. Focussing on the 11.2 μm to the 12.7 μm features, variations in their relative strength indicate variations in the molecular structure of the emitting PAHs. As illustrated in Figure 12.13, this ratio is indicative of the importance of long straight edge structures relative to "corners" in the PAH molecular structure. The observed variation in the edge structure in, e.g., reflection nebulae is then interpreted in terms of the photochemical evolution of the interstellar PAH family as material flowing through the photodissociation front approaches the illuminating star.

For the 11.2 μm band, the relative intensity (with respect to the CC modes; Figure 12.12) as well as the spatial behavior (Figure 12.7), link this band to the solo H OOPs mode in neutral PAHs. The charge state of the carrier of the 12.7 μm band, on the other hand, is controversial. Observationally, the strength of the 12.7 μm band correlates well with that of the cationic CC modes (the 6.2 and 7.7 μm bands). However, the observed variation in the 12.7/11.2 μm ratio is not caused by variations in the charge state. These

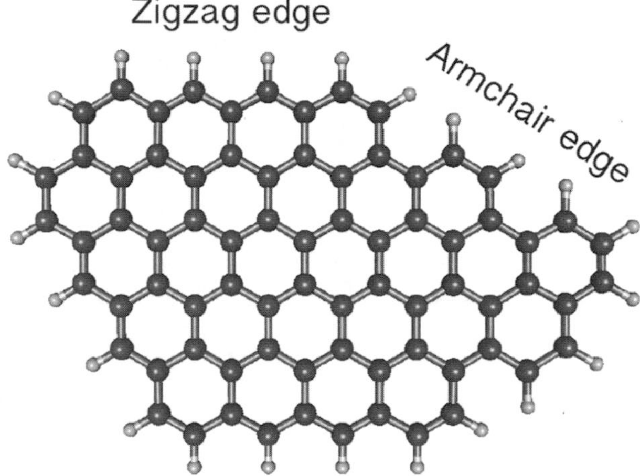

Figure 12.15 Zigzag and armchair molecular edge structures. The former are characterized by solo H OOPs modes while the latter contain duo H OOPs modes.

variations are linked to a change in edge structure of the carriers (Figure 12.15) from zigzag edges (strong 11.2 μm band) to armchair edges (strong 12.7 μm band). Apparently, the change in the physical conditions that drive the variation in the charge state of the emitting PAH family also cause a chemical conversion from "zigzag" to "armchair" PAHs. As PAHs with armchair edges are much more stable than PAHs with zigzag edges, it seems that, in the surface layers of PDRs, UV photons lead to rapid ionization while also reducing the interstellar PAH family to its most stable form, the armchair PAHs. Within this scenario, it is then still possible that the 12.7 μm band is carried by a neutral PAH as the ionization/recombination rates are much faster than photo-chemical rates.

12.3.6 The **14−18** *μm Region*

The 14−18 μm range shows a set of weak absorption bands at 15.8, 16.4, 17.4, and 17.8 μm (Figure 12.16). While the spectra reveal large variations, there is no connection to the classification scheme developed for the 6−9 μm range nor is there a systematic dependence on source type. This spectral region is characteristic for in-plane and out-of-plane C–C–C bending modes of the carbon skeleton and are, therefore, more molecule specific than the shorter wavelength fundamentals. Typically, averaging together a collection of spectra from the experimental or theoretical database results in a broad emission plateau from 16 to 19 μm, with some spectral substructure whose detail depends on the selected mixture. Overall, the presence of a limited number of bands in the observed spectra, coupled with the sensitivity of these bands to the larger molecular structure, implies that the emitting interstellar PAH family is dominated by a few PAHs (see Section 12.3.9).

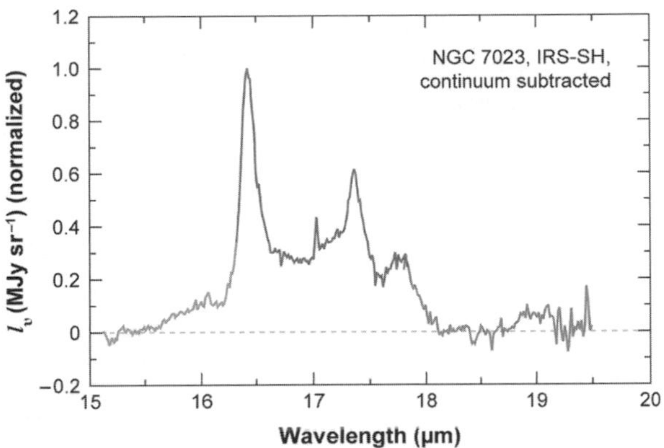

Figure 12.16 The 15−18 μm spectrum of the reflection nebula, NGC 7023, reveals the presence of weak bands at 15.8, 16.4, 17.4, and 17.8 μm. Figure taken from [115]

12.3.7 The Far-IR Spectral Range

No AIBs have been identified at wavelengths longward of 20 μm. Bands in this wavelength range are expected to be much weaker than in the mid-IR as the Einstein A scales with ν^3 and there is no "piling up" of the bands as the peak position of these modes are very molecule specific. In addition, the Herschel/PACS spectrometer only covered a very limited wavelength range per grating setting and that severly hampered the search for $\sim$1% features. Figure 12.17 illustrates the potential richness of this spectral range.

Vibrational modes at these wavelengths involve the whole molecule. This is illustrated for the jumping jack and drumhead modes in Figure 12.18. These bands are very weak with calculated intensities at the 1% level of the strong mid-IR bands (Figures 12.17 and 12.18). Because of the small Einstein A's, emission in these modes will occur when the energy in the molecule is low and the modes have decoupled. This has two effects. First, in contrast to the mid-IR modes, there is no homogeneous IVR line broadening, and these bands will show a P(Q)R rotational profile. For astrophysically relevant PAH sizes, the rotational constant will be $\simeq 10^{-3}$ cm^{-1} and with a typical J_{max} of 200 (Section 12.6), the P(Q)R separation will be $\simeq$0.4 cm^{-1}. Sensitive high resolution studies could then measure the rotational spacing and this would provide a "fingerprint" of the emitting species. Second, IR inactive mode will decay collisionally in a PDR and this provides a heating agent for the gas. As the energy left in these modes is only a few hundred wavenumbers, this is not a very important heating agent compared to, e.g., the photo-electric effect.

12.3.8 Functional Groups

Many of the weaker features in the AIB spectrum may have an origin in sidegroups and possible identifications are summarized in Table 12.1. With the measured or calculated intrinsic strength of the features involved, observed (relative) strengths can be directly

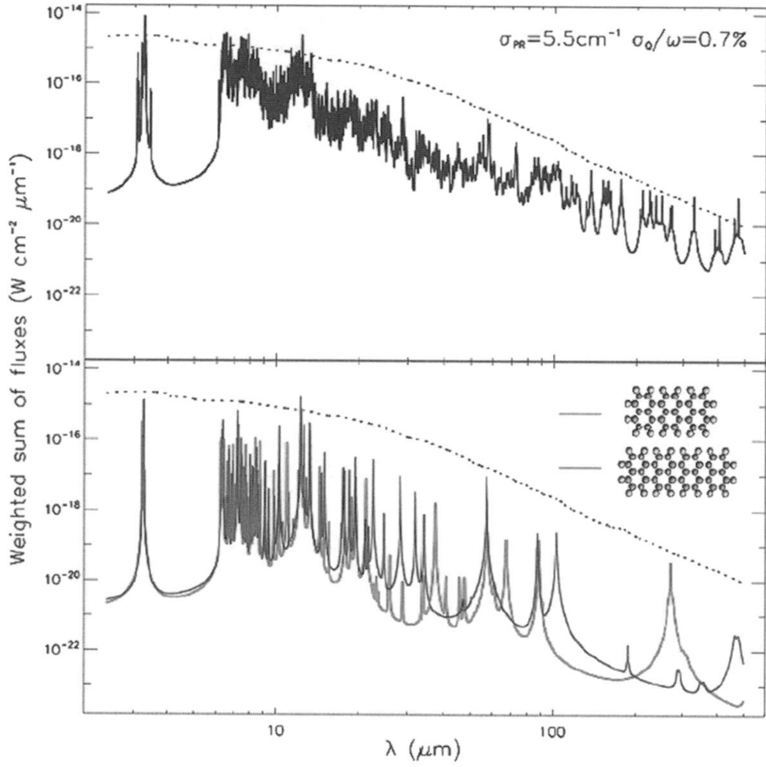

Figure 12.17 Top: Weighted sum of the spectra of 20 PAHs in their neutral and cationic state, calculated for the excitation conditions of the post-AGB object, the Red Rectangle. The dotted line shows, for comparison, the estimated dust continuum in the same source. Bottom: The predicted spectra of two individual PAHs illustrating that far-IR spectroscopy could lead to unique identification of specific species within the interstellar PAH family. Figure adapted from [86]

translated to fractional abundances of these functional groups. If the features are at very different wavelengths, then an excitation dependent correction factor is involved. Consider now the 3.4 μm band due to sp^3 H (either in a methyl group or as an extra H attached to a peripheral C in the aromatic ring), which is close enough to the 3.3 μm band that excitation should not be a major concern. Independent of whether these are alkylated- or hydro-PAHs, on a per H basis, the intrinsic strength ratio of the aliphatic to aromatic CH stretching mode is $\simeq$1.5. The maximum 3.4/3.3 μm ratio is observed to be $\simeq$0.1 and translates into a ratio of methyl groups to aromatic H of $\simeq$0.02 (Table 12.2). In terms of hydro-PAHs, this abundance would be three times higher. Much higher 3.4/3.3 μm ratios have been measured for the so-called 21-μm objects mentioned before. Actually, the observed profile of the 3.4 μm band in these objects is different from that observed in the AIB spectrum in the ISM and resembles that of methyl groups in hydrogenated amorphous carbon materials

Table 12.2 *Functional groups on interstellar PAHs*

Functional group	λ µm	Fraction[a]
aromatic H	3.3	1
H bonded to sp^3 C[b]	3.40	0.02
aliphatic deuterium CH_2D	4.65	0.02
hydroxyl (OH)	2.77	<0.002
amine (NH_2)	2.88–2.95	<0.01
quinone (CO)	5.9	0.006 (?)
nitrile (C≡N)	4.48	<0.01 (?)
N substituted in the ring structure[c]	6.2	0.04
acetylenic (C≡CH)	3.03	<0.003

[a]Fraction relative to aromatic hydrogen., [b]This could be in the form of aliphatic CH_3 or as superhydrogenated PAHs. [c]If the shift of the aromatic CC stretching mode is due to substitution of N in the ring. Other explanations have been offered for this observed shift.

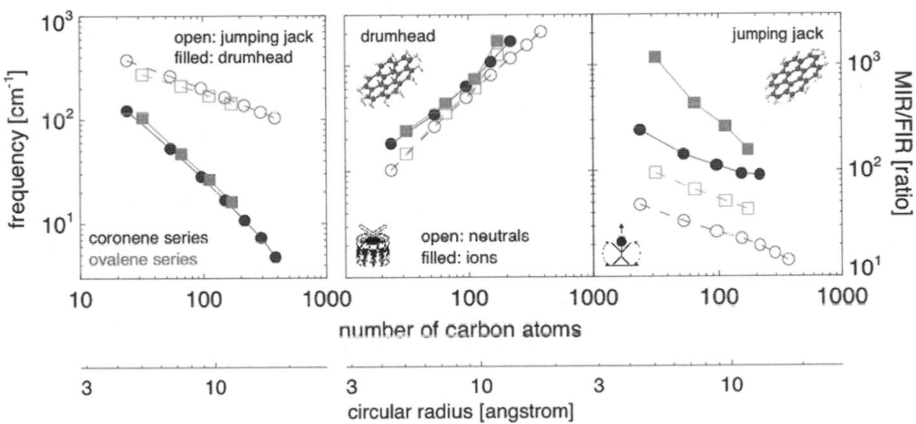

Figure 12.18 The jumping jack and drumhead modes for compact PAHs in the coronene and ovalene series. Left: The frequency of these modes is set by the surface area of the PAHs. Right two panels: The calculated intensity of the far-IR modes relative to the mid-IR modes as a function of PAH size. Figure derived from the NASA AMES PAH IR Spectral Data Base [21] and kindly provided by Christiaan Boersma

(Section 12.3.2).[8] As an example consider IRAS 22272+5435 with a measured 3.4/3.3 µm ratio of 10 at a distance of 0.35″ ($\simeq 3 \times 10^{-3}$ pc), which implies a methyl to aromatic H ratio of $\simeq$2. The observed 3.4/3.3 µm ratio decreases rapidly with distance from

[8] The mid-IR emission features in this object may actually be due to dust rather than molecules.

the central object, presumably as the methyl groups are photolyzed away by penetrating far-UV photons from the interstellar radiation field.

The abundances of other functional groups can be obtained similarly and are summarized in Table 12.2. It is of some interest to consider the abundance of deuterated PAHs. The mass difference shifts the C–D stretching modes to the 4.4–4.7 μm range and weak bands have been observed in this range in some sources. Attributing these to deuterated PAHs results in a aromatic-D/aromatic-H ratio <0.05 but (when detected) the aliphatic-D/aliphatic-H is much larger (0.2–1, depending on the source). As the modes involved in this analysis are somewhat more separated in frequency, these estimates depend more on the excitation of the emitting PAHs than for the 3.4/3.3 μm analysis above and that has to be taken into account. For the sources for which these bands are detected, the derived D-in-PAHs represents some 10 ppm relative to H, which is in line with the observed variations in the D/H abundance in the gas phase in the local ISM. A somewhat different approach has to be taken for N substitution in the ring (rather than as an amine or nitrile). Substitution of N in the aromatic network will shift the aromatic CC stretching mode to higher frequencies (Section 12.3.4). The shift depends on how "deep" the N is substituted in the ring and the size of the PAH. The relative abundance (Table 12.2) is scaled from the shift introduced on the 6.2 μm band as a function of the N/C ratio.

Perusing Table 12.2, it is clear that the overwhelming majority of the peripheral sites of the carrier of the IR emission features contain aromatic hydrogen, with a very small amount of aliphatic hydro or CH_3 groups and possibly some quinone groups (C=O). While in early studies hydrogenated amorphous carbon structures were touted as potential carriers of the AIBs and this idea is every now and then reinvented, the spectroscopy evidence is quite clear: In the ISM, the carriers are polycyclic aromatic hydrocarbon molecules with the possible exception of the substitution of N in the ring. The other exception is presented by this class of special post-AGB objects where, close to the star, aliphatic groups are very abundant. The more massive post-AGB objects and PNe (which are more relevant in terms of the dust and PAH budget of the ISM) show an "interstellar ratio" of the 3.4/3.3 μm ratio.

12.3.9 GrandPAHs

Overall, the consensus in the field is that the interstellar PAH family is very diverse. However, there are signs that, actually, a few large and compact "grand PAHs" dominate the PAH population. First, the 15–20 μm range is dominated by a few AIBs (Section 12.3.6). As this wavelenth range is quite molecule specific, this points toward a presence of a few dominant species. Second, the brightest spots in well-studied PDRs show almost identical AIB spectra despite large spectral variations over each nebula. As PAHs show subtle variations in their spectral characteristics, this similarity suggests that the family of PAHs in these hot spots is nearly identical. Analysis using the PAH IR database suggests that compositional variations are less than 30%. The grandPAH hypothesis suggests that only the most stable PAHs survive the harsh conditions of these hot spots. Such a view of the "survival of the fittest" is in line with observed size and abundance variations of PAHs in

PDRs, pointing toward the importance of photochemical processing as material approaches the illuminating star. In addition, interstellar fullerenes may well be the photochemical progeny of PAHs and C_{60} by itself has an abundance of 3×10^{-9} or 6×10^{-4} of the available C in the ISM. Finally, the DIB spectrum is rich and varied with over 400 bands, but this spectrum is dominated by a few strong bands and that may well point toward the dominance of a few large molecules in the ISM. It should be noted, though, that attempts to fit this characteristic spectrum of PDR hot spots using only the most stable PAHs (e.g. circumcoronene, circumcircumcoronene) have not been successful, possibly because the photochemical route toward grandPAHs is not fully understood.

Identification of any putative grandPAH is of great interest. The feasibility of this has been addressed by creating artificial spectra of PAH families containing 30 PAHs arbitrarily selected from the NASA Ames PAH data base. These were used to create a large set of 100–3300 cm^{-1} "observed" spectra with varying relative abundances. In this exercise, PAHs present at an abundance exceeding 10% could be reliably extracted from this set of spectra. The far-IR modes in themselves provide an excellent probe of grandPAHs in the interstellar PAH family as these modes are intrinsically weak and likely can only be detected if the abundance of the specific species is substantial (Figure 12.18). Both the peak position and the rotational substructure provide direct handles on the identity of the emitting PAH. As there has been no suitable instrument, this has not been further explored.

12.4 Buckminsterfullerene

Because of its symmetry, the fullerene, C_{60}, has a very simple infrared spectrum with bands at 7.4, 8.6, 17.4, and 18.9 μm. As several of these bands coincide with the rich AIB spectrum, identification of C_{60} in the ISM did present a daunting task and did not occur until the unique spectrum of the planetary nebula TC 1 was uncovered. This unusual spectrum is dominated by a few bands that are very different from the PAH bands normally seen and are matched well by the spectra of the fullerenes C_{60} and C_{70} (Figure 12.19). This identification is now widely accepted. Once these bands were recognized as signposts for the presence of fullerenes in space, they were also recognized in spectra dominated by the (regular) AIBs. In general, in the ISM, the presence of C_{60} is recognized from the 18.9 μm band as the 7.04, 8.6, and 17.4 μm bands are more difficult to extract from among the strong PAHs bands. With this caveat, many objects in space show C_{60} in their spectra, including reflection nebulae, HII regions, post-asymptotic-giant-branch (post-AGB) objects, young stellar objects, and the general diffuse ISM. Evidence for the presence of the C_{60} cation has also been uncovered in some sources.

The abundance of C_{60} has been studied in the reflection nebula, NGC 7023. The strength of the 18.9 μm fullerene band increases relative to the AIBs close to the star, attesting to C_{60}/PAH abundance variations in this nebula. Specifically, fullerenes lock up only $\simeq 10^{-5}$ of the elemental carbon in this nebula as compared to $\simeq 7 \times 10^{-2}$ in PAHs. Close to the illuminating star, the C_{60} abundance increases to $\simeq 10^{-4}$ of the elemental C, while the

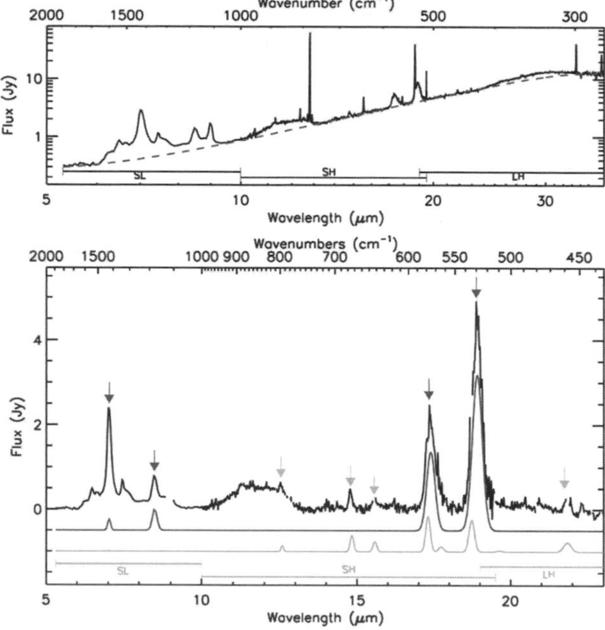

Figure 12.19 Top: The Spitzer/IRS infrared spectrum of the planetary nebula TC 1. The different spectrometer modules are indicated at the bottom. The dashed line indicates the adopted continuum. Bottom: The continuum subtracted spectrum (upper trace) is dominated by a few bands that are well fitted by emission by C_{60} (middle trace) and C_{70} (lower trace) molecules. Figure kindly provided by Jan Cami, adapted from [29]

PAHabundance decreases to $\simeq 2 \times 10^{-2}$. This has been taken as evidence for photochemical processing of PAHs and their conversion to C_{60} (Section 12.7.5).

In the mid 90s, two weak absorption bands in the far red were discovered and ascribed to electronic transitions of C_{60}^{+} based upon matrix isolation studies. It took, however, two decades and very elegant studies to prove that these bands could indeed be confidently ascribed to this cation (Section 9.6.3; Figure 9.18). A further three (weak) far-red DIBS have now been assigned to this species. Using the intrinsic strength of these transitions, the measured equivalent width results in a derived column density of C_{60}^{+} of 2×10^{13} cm^{-2} toward HD 183143 (Table 9.2) and a fraction of the elemental carbon in this one molecule of $\simeq 6 \times 10^{-4}$.

12.5 PAH Emission Models

The excitation of large molecules in space and the resulting IR spectra are discussed in Sections 8.3 and 8.4. Here, we will use these models to analyze interstellar observations to derive the sizes of the emitting species, their abundances and the profiles of the emission bands.

12.5.1 PAH Size

The relative strength of the different vibrational bands of a molecule depend on the excitation of the species and hence its internal excitation during the emission process. As the absorbed UV photon energy does not vary much between different regions in space or from one species to another and, on a per C-atom basis, the heat capacity for PAHs is very similar, this can then be used to derive the size of the species. Indeed, this was the original argument for the connection between the AIBs and ≃50 C-atom species. The 3.3/11.2 μm AIB ratio is the most suitable band ratio for this purpose, because (1) it has the widest frequency span and so is very sensitive to the excitation, (2) both bands are due to CH modes, and (3) both bands are due to neutral species. Combining either of these modes with a CC mode in the 5−9 μm range introduces dependencies on the degree of ionization and the molecular structure. The two strong CC modes – 6.2 and 7.7 μm – on the other hand are so close in frequency that their ratio only becomes dependent on excitation when the emission is on the Wien-side of the curve and for such large species, the emission in these bands will be very weak as most of the energy will come out in low frequency modes. Moreover, in contrast to the CH modes, the integrated strength of these two bands depends strongly on the species.

The calculated 11.2/3.3 μm ratio is compared to observations in Figure 12.20. The calculations include the effects of the energy cascade associated with the emission process. An initial excitation by a 6.5 eV photon is assumed but the results are not very sensitive to

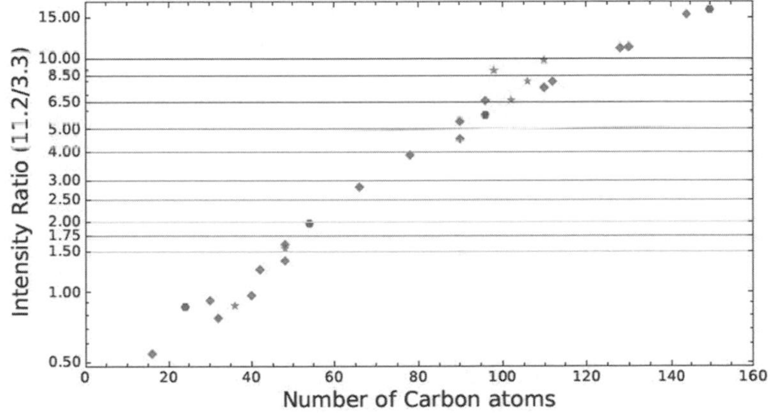

Figure 12.20 Calculated intensity ratio for the the solo out-of-plane bending mode (11.2 μm) over C–H stretching mode (3.3 μm) for 27 different PAHs (coronene family (circles), other compact PAHs (diamonds), and noncompact PAHs (stars)). The excitation is calculated using an average photon energy of 6.5 eV and includes the emission cascade. The intrinsic strength of these modes is taken from the NASA Ames PAH database, but is largely independent of the specific species on a per H basis. The horizontal lines cover the range of this ratio observed in the reflection nebula NGC 7023 and is typical for all regions. Figure taken from [37]

this value. As, per H-atom, the intrinsic strength of these two modes is largely independent of the species involved, the observed ratio is a good measure of the size of the emitter. Observations reveal variations in the 11.2/3.3 μm ratio of a factor of 5 both from one source to another as well as within sources. As Figure 12.20 indicates, sizes in the range of 50–100 C-atoms are implicated.

12.5.2 Anharmonicity and the Band Profile

The influence of anharmonicity on the IR spectra of PAHs has been discussed in Section 12.3. The interaction between the modes will shift the fundamental positions to the red. When other modes are excited by one or more quanta, secondary bands are produced in a pattern that is set by the anharmonic interaction between the two modes (Figure 12.21). A fundamental can couple with many modes (and combination of modes) each characterized by its own anharmonic shift and pattern. For high excitation or a molecule with many, many modes, this substructure will blend into a broad band. As a result, with excitation,

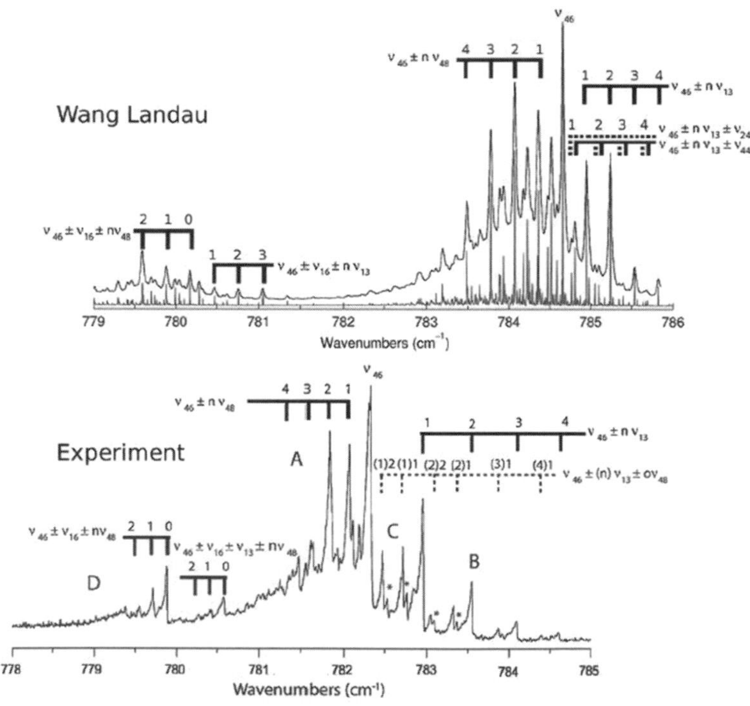

Figure 12.21 Comparison of high-resolution experimental data of thermally excited (300 K) naphthalene (bottom) to a model spectrum that includes the effect of anharmonic coupling between the modes convolved to a resolution of 0.02 cm^{-1} (top). The stick diagram is the 0 K calculated anharmonic spectrum. Experimental data from [100]. Figure adapted from [73]

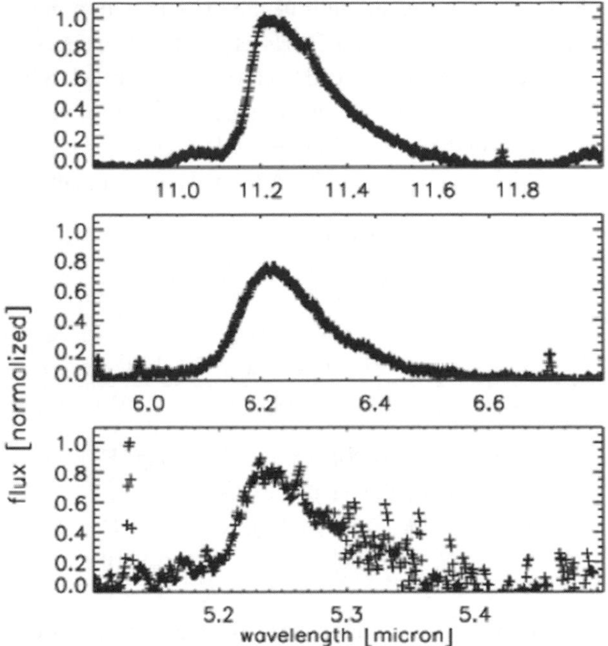

Figure 12.22 Observed profiles of the 11.2, 6.2, and 5.7 μm AIBs. Note the pronounced shaded profiles of these bands. Figure adapted from [18]

the anharmonic shift of the fundamental will become more prominent (Section 8.4.3; Figure 8.15). In the ISM, a molecule excited to a high internal energy will cool through the emission of IR photons and, as a result of this cascade, the anharmonic shift will become smaller and smaller. The observed band profile will then be an integration over this energy cascade, resulting in a profile with a steep blue rise and a gradual red decrease (Figure 8.15). The asymmetry of the profile is then a measure of the internal excitation of the species and hence, with a given (e.g. average) UV photon energy, a measure for the size of the emitting species. The observed profiles of several AIBs show a clear red-shaded profile (Figure 12.22). Quite accurate anharmonic DFT calculations have come within reach for small PAHs and detailed models for the emission profiles of PAHs have been developed. Comparing the observed line profiles with these theoretical studies, a typical size of $\simeq$30 C-atoms is derived.

12.5.3 Abundances

The observed IR intensity in the AIBs is a direct measure of the abundance of excited PAHs in the emission region. However, the "'emission" temperature is not well known and a direct translation of an inferred emission "optical" depth to a PAH abundance is fraught with issues. In addition, the fraction of excited PAHs is very small, $\simeq \tau_{IR}/\tau_{UV} \sim 10^{-5}-10^{-7}$, depending on the PAH size and the strength of the FUV radiation field and the correction factor for the unexcited and unobserved PAHs is large. Fortunately, the fraction of the elemental C, f_c locked up in interstellar PAHs can also be derived, relatively simply, from observations by realizing that PAHs and dust compete for the same photons. As the FUV dust extinction per H-atom in the diffuse ISM is well known, the observed ratio of the AIB flux to that of the dust, f_{IR}, can be related back to the PAH abundance; namely,

$$\frac{\kappa_{\rm PAH}\,(FUV)}{\kappa_{\rm dust}\,(FUV)} = \frac{f_{IR}}{(1-f_{IR})}, \tag{12.1}$$

which can be rewritten as,

$$f_C = \frac{A_v}{N_H}\,\frac{\kappa_{\rm dust}\,(FUV)}{\kappa_{\rm dust}\,(V)}\,\frac{(1-\omega_{\rm dust}\,(FUV))}{\sigma_{\rm PAH}\,(FUV)\,\mathcal{A}_C}\,\frac{f_{IR}}{(1-f_{IR})}, \tag{12.2}$$

with A_v/N_H the dust extinction per H-atom (5.2×10^{-22} mag/(H-atom/cm^2)) $\kappa_{\rm dust}\,(FUV)/\kappa_{\rm dust}\,(V)$ the ratio of the FUV to the visual extinction cross section of the dust ($\simeq$3), $\omega_{\rm dust}\,(FUV)$ the dust albedo in the FUV ($\simeq$0.6), $\sigma_{\rm PAH}$ the FUV absorption cross section of PAHs per C-atom (7×10^{-18} cm^2 (C-atom)$^{-1}$), and $\mathcal{A}_C$ the elemental abundance of carbon (2.7×10^{-4}).

Table 12.3 summarizes the PAH abundance, 14 ppm, as measured toward prominent PDRs such as the Orion Bar, NGC 7023, and NGC 2023 ($f_{IR}\simeq0.13$) and compares it to the fraction of the carbon locked up in other species. The gas phase C-abundance in the ISM

Table 12.3 *Abundances of the carriers of IR emission components*

Carrier	IR emission component	N_c	a Å	$\mathcal{F}_C{}^a$ ppm^c	$f_C{}^b$
PAHs	AIBs	$50-100$	$4-10^d$	14	0.05
$C_{60}{}^e$	18.9 μm feature	60	3.5	5×10^{-2}	2×10^{-4}
$C_{60}^{+\,f}$	9632 & 9577 Å	60	3.5	1.6×10^{-1}	6×10^{-4}
PAH-clusters	plateaus	$100-1000$	$10-20$	8	0.03
very small grains	25 μm cirrus	10^3-10^4	$20-30$	7	0.03
small grains	60 μm cirrus	$\sim10^5$	50	16	0.06
classical grains	$\lambda>100$ μm		>100	85^g	0.31^g

[a] Abundance of C locked up in these species (relative to H). [b] Fraction of the elemental carbon locked up in these species relative to the Solar C abundance. [c] Parts per million. [d] Size corresponds to disk rather than sphere. [e] The highest abundance of C_{60} as measured near the illuminating star of NGC 7023. Deeper in the PDR, the abundance is an order of magnitude smaller. [f] The abundance of C_{60}^+ in the diffuse ISM as measured from the 9632 and 9577 Å DIBs. [g] The abundance of C locked up in dust grains is calculated by subtracting the measured interstellar gas phase atomic C abundance (140 ppm) from the adopted Solar abundance (270 ppm) and correcting for the C in PAHs to small grains range. Some determinations place the Solar C abundance at 320 ppm, while the C-abundance in early type B stars is estimated to be only 210 ppm.

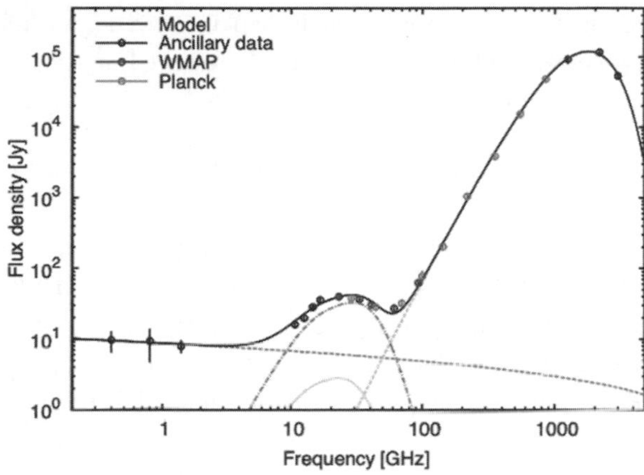

Figure 12.23 Spectrum of a region in the Perseus molecular cloud, combining WMAP, Planck and ground-based and balloon data. The short wavelength region is well described by thermal dust emisison, while at long wavelength the emission is due to bremstrahlung in ionized gas. The presence of an additional emission component – the AME – at intermediate frequencies is quite evident. A complex, multi-component model has been fitted to the data. Figure taken from [101]

is 150 ppm – in the form of C^+ in the diffuse ISM – and the solar C abundance is 270 ppm. We can recognize a number of other IR emission components – PAH clusters, very small grains, and small grains – which together lock up a similar amount of C as determined from the IR emission spectrum and Eq. (12.2). The fullerene, C_{60}, is at the 1% level compared to PAHs. By default, the most important carbon-bearing compound seems to be "large" carbonaceous grains, which in the SED contribute to the far-IR emission. There is some indication that the fraction of the C locked up in PAHs increases with decreasing FUV radiation field. Indeed, a somewhat larger value for the PAH abundance is obtained from an analysis of the mid-IR cirrus emission. The total mid-IR flux is 1.6×10^{-24} erg s^{-1} (H-atom)$^{-1}$ in the diffuse ISM. With an average interstellar UV radiation field of 2.7×10^{-3} erg cm^{-2} s^{-1}, the fraction of the C locked up in PAHs is then $\simeq$0.3.

12.6 Anomalous Microwave Emission

Until the mid-1990s, the emission spectrum of the galaxy in the sub-millimeter to centimeter regime was thought to be dominated by synchrotron and free-free emisson on the low frequency side and thermal dust emission on the high frequency side. Cosmic Microwave Background (CMB) experiments revealed, however, that there is an additional emission component present between $\simeq$ 10 and 100 GHz, which manifested itself in careful decomposition studies. Analysis of CMB data requires a careful correction for foreground emission by the galaxy. This separation is attained using template spectra for the individual components, relying on differences in their spatial morphology. These studies revealed a dust-related component that could not be easily explained by any of the known emission mechanisms, and, hence, was dubbed Anomalous Microwave Emission (AME). At 30 GHz, the typical emission strength is,

$$\frac{T_{AME}}{I_d\ (100\mu\mathrm{m})} \simeq 10 \quad \mu\mathrm{K/\ (MJy/sr)} \tag{12.3}$$

with I_d (100μm) the IR surface brightness at 100 μm. Observed variations in this ratio are about a factor of 2. For the diffuse ISM, a surface brightness of 1 MJy/sr at 100 μm corresponds to a H-nuclei column density of 2×10^{20} cm^{-2}. The AME is generally attributed to the rotational emission of large molecules and/or spinning dust grains. Ascribing all of the observed AME to PAHs and using a PAH abundance in the diffuse ISM of 1.6×10^{-6}, we have,

$$I_{AME} = 2 \times 10^{-25} \quad \mathrm{erg\ s^{-1}\ PAH^{-1}} \tag{12.4}$$

In the diffuse ISM, a PAH absorbs a 10 eV photon every year. Hence, the fraction of the absorbed energy emitted in rotational transitions is small, $\simeq 2 \times 10^{-7}$, and most of the absorbed UV energy is emitted in vibrational transitions (1.6×10^{-24} erg s^{-1} (H-atom)$^{-1}$).

This spectral component is now generally attributed to dipole (rotational) emission from large molecules or small (spinning) dust grains. In either case, the size range implied is less than 10 Å or the equivalent of less than 500 C-atoms. Hence, these species are more properly called large molecules or clusters of molecules. As CMB measurements have to be corrected for galactic AME, detailed models have been developed for this foreground component. Generally, these rely on treating the emission as a "continuum" process by spinning dust grains. While this really concerns molecular emission, which is a quantum process, detailed comparisons have shown that the two descriptions yield very comparable results. Here, we will consider the molecular approach.

The rotational excitation of PAHs has been discussed in Section 9.6.4 with regard to the line profile of the DIBs. The rotational population is a balance between three processes, UV photo excitation followed by the IR emission cascade, collisional excitation, and rotational emission. The first dominates the build up of the rotational population. The other two merely modify the population slightly. The last one gives rise to the AME.

Let's start this discussion by focusing on UV photon excitation followed by vibrational cascade. During the downward cascade, the small difference in Einstein A's slightly favors $\Delta J = +1$ transitions. This is counteracted by the difference in the statistical weight of the rotational states that prefers $\Delta J = -1$. The balance between these two processes leads to a relationship for the average rate of change in J,

$$\Delta J \simeq \frac{1}{J_{ir}} \left(\frac{J_{ir}}{J} - \frac{J}{J_{ir}} \right) \qquad \text{for } J \gg 1. \tag{12.5}$$

This balance between Einstein A's and statistical weight leads to a most probable rotational quantum number ($\Delta J = 0$) given by $J = J_{ir} = (hc\overline{\nu}/6hcB)^{1/2}$, where $\overline{\nu}$ is the average energy of IR photons emitted in the cascade (in wavenumbers) and the factor 6 in this expression accounts for the summation of the energies over all allowed K states for a given J. A more complete analysis shows that the population is indeed well described by a Gaussian distribution,

$$\frac{n\,(J)}{n_0} = (2J+1)^2 \exp\left[-hcBJ\,(J+1)\,/kT_{rot}\right] \simeq 4J^2 \exp\left[-\,(J/J_{ir})^2\right], \qquad (12.6)$$

where $T_{rot} = hc\overline{\nu}/6k$ is an effective rotational excitation temperature ($kT_{rot} = hcBJ_{ir}^2$).

This "radiative" excitation is modified by collisional deexcitation and by pure rotational emission. The collisional deexcitation rate, $n\gamma_{col} \simeq 10^{-9} n \simeq 5 \times 10^{-8}$ s^{-1}, is comparable to the UV pumping rate, $k_{uv} \simeq 4 \times 10^{-8} (N_C/50)$ s^{-1} but the latter induces some $\simeq 80$ ΔJ transitions. As discussed below, the rotational population of PAHs typically peaks for $J \simeq 100-300$. Rotational transition occur then in the $\simeq 10-30$ GHz range and have typical Einstein A's, $A_{rot} \simeq 3 \times 10^{-8}$ $(\mu_e/1\,\mathrm{D})^2$ s^{-1}. We can describe the population flow by the Fokker–Planck equation with systematic term $\Delta J/\tau$. Allowing for collisions and pure rotational transitions, the most probable J is given by,

$$\left(\frac{\Delta J}{\tau}\right)_{ir} + \left(\frac{\Delta J}{\tau}\right)_{col} + \left(\frac{\Delta J}{\tau}\right)_{rot} = 0, \qquad (12.7)$$

where the ΔJ's and τ's represent the average change in J and the characteristic timescale for the processes involved (i.e. IR cascade, collisional deexcitation and rotational emission). This can be rewritten as,

$$10^{-8} G_0 \frac{1}{J_{ir}} \left(\frac{J}{J_{ir}} - \frac{J_{ir}}{J}\right) = 8 \times 10^{-8} \frac{n_H}{50\,\mathrm{cm}^{-3}} \left(\frac{N_C}{50}\right) (J/J_{ir}) + 4 \times 10^{-7} \left(\frac{50}{N_C}\right)^3 \left(\frac{\mu}{1\,\mathrm{D}}\right)^2 (J/J_{ir})^3 \qquad (12.8)$$

where the three terms are in the same order, reasonable estimates have been introduced for the relevant molecular parameters, and the J_{ir} have been introduced in the other terms for ease of comparison.

Balancing these processes results in a simple equation describing the dependence of the most probable rotational quantum number, J, on the physical conditions,

$$\alpha \left(\frac{J}{J_{ir}}\right)^4 + (\beta + 1) \left(\frac{J}{J_{ir}}\right)^2 - 1 = 0, \qquad (12.9)$$

with

$$\alpha = 43 \left(\frac{1}{G_0}\right) \left(\frac{\mu}{1\,\mathrm{D}}\right)^2 \left(\frac{50}{N_C}\right)^3 \qquad (12.10)$$

$$\beta = 8 \left(\frac{N_C}{50}\right) \left(\frac{1}{G_0}\right) \left(\frac{n_H}{50\,\mathrm{cm}^{-3}}\right) \qquad (12.11)$$

where μ is the dipole moment and we have assumed that 80 IR photons are emitted during the cascade. Figure 12.24 shows J as a function of the physical conditions and the

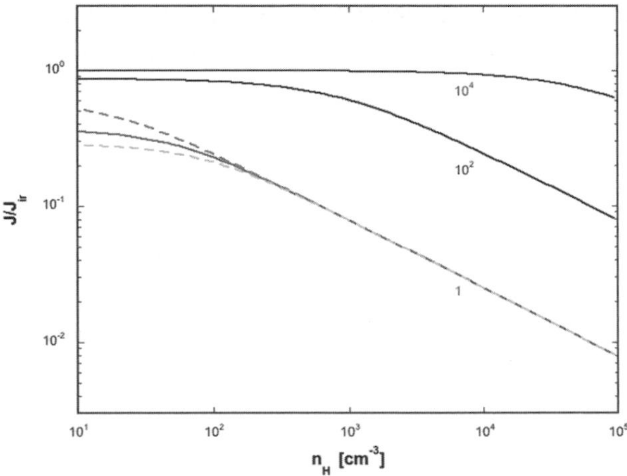

Figure 12.24 The ratio of the most probable rotational quantum number, J, to J_{ir} ($= \sqrt{\overline{\nu}/6B}$, with $\overline{\nu}$ the typical energy of a vibrational photon, and B the rotational constant) as a function of the physical conditions. Solid lines give the relationship for the different G_0 values indicated. The dashed lines illustrate for $G_0 = 1$ the dependence on μ (0.1 and 3). Figure adapted from [109, 131]

characteristics of the PAH/PAH cluster. The results are normalized with respect to most probable rotational quantum number, J_{ir}, when only UV pumping followed by the IR cascade is considered. Not surprisingly, for high UV fields, the rotational population is dominated by the IR cascade, $J \simeq J_{ir}$, and collisional deexcitation only becomes important at very high density (Typically, $G_0/n_H \simeq 1$ in a PDR). Rotational emission is then never really important. In the diffuse ISM with $G_0 = 1$, collisional deexcitation does drive the rotational population down somewhat ($G_0/n_H \simeq 10^{-2}$) and there is a weak dependence on the dipole moment of the species. The most probable rotational quantum number varies between 0.2 and 0.5 times J_{ir}, depending on the precise conditions and PAH properties (Figure 12.24).

For diffuse clouds with $G_0 = 1$ and $n_H = 50$ cm^{-3}, the frequency corresponding to the most probable J is given by,

$$\nu_{rot} = 2BJ \simeq 35 \left(\frac{50}{N_C}\right)\left(\frac{J}{J_{ir}}\right) \simeq 15 \left(\frac{50}{N_C}\right) \quad \text{GHz}. \tag{12.12}$$

The emission spectrum would actually peak at slightly higher frequency, $\simeq$25 GHz. The emission intensity corresponding to the transition $J_{ir} \rightarrow J_{ir} - 1$ is given by,

$$4\pi I_J = A_{J,J-1} h\nu_{J,J-1} \qquad \text{with } J = J_{ir}, \tag{12.13}$$

where the Einstein A of the transition is given by,

$$A_{J,J-1} \simeq \frac{32\pi^4}{3hc^3}\,\nu^3\mu^2, \tag{12.14}$$

where we have adopted $|\mu_{J,J-1}|^2 = \mu^2/2\left(1-(K/J)^2\right)$ appropriate when $J \gg 1$ and we have made the further approximation that $J \gg K$. Here, μ is the electric dipole of the molecule. We have then,

$$A_{J,J-1} \simeq 2.6 \times 10^{-7} \left(\frac{\nu_{rot}}{35\,\mathrm{GHz}}\right)^3 \left(\frac{\mu}{1\,\mathrm{D}}\right)^2. \tag{12.15}$$

The intensity is given by,

$$4\pi I_J \simeq 6 \times 10^{-23} \left(\frac{\nu_{rot}}{35\,\mathrm{GHz}}\right)^4 \left(\frac{\mu}{1\,\mathrm{D}}\right)^2 \quad \mathrm{erg\ s^{-1}}. \tag{12.16}$$

This has to be integrated over the rotational distribution (Eq. (12.6)),

$$4\pi I_{rot} \simeq 2 \times 10^{-22} \left(\frac{\nu_{rot}}{35\,\mathrm{GHz}}\right)^4 \left(\frac{\mu}{1\,\mathrm{D}}\right)^2 \quad \mathrm{erg\ s^{-1}}. \tag{12.17}$$

In the diffuse ISM, this would translate to $4\pi I_{rot} \simeq 5 \times 10^{-23}$ $(\mu/1\ \mathrm{D})^2$ erg s^{-1} PAH^{-1}. Ascribing the AME to PAHs and using Eq. (12.4) for the inferred AME intensity per PAH, the average dipole moment of interstellar PAHs has to be $\overline{\mu} \simeq 0.06$ D. Or, perhaps more realistically, only a few percent of the interstellar PAHs has an appreciable dipole moment. The derived dipole moment scales inversely with the square root of the adopted abundance of PAHs in the diffuse ISM. Also, remember that this ignores rotational emission by large PAH clusters and/or spinning dust grains. Finally, in a PDR, the rotational emission intensity will be slightly larger and shifted to slightly higher frequency as $J \simeq J_{ir}$.

The physical conditions for regions bright in the AME are not well known. If we assume that the emission is dominated by typical diffuse clouds with n_H equals 50 cm^{-3}, then rotational emission by PAHs with sizes of $N_c \simeq 50$ can explain the observations of diffuse clouds well – in peak position and in observed intensity – if the average dipole moment is of the order of 0.06 D. PAH clusters do not contribute, as radiative excitation does not spin up them sufficiently to compensate for the decrease in rotational constant. For grains, the concept of suprathermal excitation has been introduced to account for the high rotational excitation required to explain the observations. This suprathermal excitation is supposed to be driven by the rocket effect associated with H_2 formation. This can lead to high angular momentum, if H_2 formation is limited to a few enhanced binding sites that keep "pushing" the grain in the same direction. For PAHs, this is not as relevant, as many aromatic-H can interact with an incoming H and hence there are many H_2 formation sites distributed along the periphery of the molecule. Moreover, this H_2 formation process is not very efficient. Irrespective of these arguments, assuming the presence of only a few enhanced binding sites, the rocket mechanism has been shown to only play a role for species in excess of 10^4 C-atoms. In actuality, the concept of enhanced binding sites for H_2 formation is not viable for amorphous grains either as binding sites have a broad distribution and, hence, the fraction of the sites involved in H_2 formation on interstellar grains is substantial.

In order to explain the observed AME, emitting PAHs need a dipole moment of $\simeq 0.06$ D. Planar PAHs do not possess a permanent dipole moment. PAHs can become nonplanar and develop a dipole moment through incorporation of a pentagon in its structure;

e.g. corannulene has a dipole of $\simeq$2 D. Alternatively, the peak position of the 6.2 μm band has been interpreted as evidence for the substitution of N atoms in the C-skeleton and such species would have dipole moments of 1 – 10 D, depending on where the N is substituted. Anions and (de)protonated PAHs would also have a permanent dipole moment of $\simeq$1–5 D and protonated PAHs could be important in the diffuse ISM. In PDRs, conditions are such that anions and protonated and deprotonated PAHs will have low abundances. It has been suggested that nitrogenated PAHs could be abundant in PDRs and hence produce rotational emission.

From this discussion it is clear that PAHs could be the carriers of the AME in the diffuse ISM. It is then interesting to consider the rotational emission of grandPAHs. If grandPAHs would dominate the PAH family in the diffuse ISM, then they may well imprint their rotational signature on the AME. A small asymmetry in the molecular structure of the PAH would lead to a "comb-like" structure in the rotational emission that could be pulled out of high resolution data using match-filtering techniques. This would be an interesting avenue to identify the presence of specific PAHs in the ISM.

12.7 Evolution of Interstellar PAHs

Interstellar PAHs are generally thought to be formed in the ouflows of carbon-rich Asymptotic Giant Branch stars (AGB). This is partly based upon the prominence of the AIBs in the spectra of carbon-rich post-AGB objects and planetary nebulae (PNe), the stellar descendants of the AGB phase (Figure 12.1). While AGB objects themselves do not – in general – show the AIBs in their spectra that might just be because of the lack of UV photons rather than the lack of PAHs. Indeed, the AGB star, TU Tau, which has a blue companion, does have very weak AIBs in its spectum. The other reason that outflows of C-rich AGB stars are considered prime sites for PAH formation is that such objects are enshrouded in dusty envelopes. This dust is produced locally by processes akin to those in sooting flames and PAHs are abundant in sooty experiments as well. The first part of this section deals with these processes. Alternatively, during the post-AGB and/or PN phase, AGB ejecta are processed by strong shocks and X-ray/UV photons and this may result in PAH formation. Those same processes may also be important for the subsequent evolution of PAHs in the ISM. Finally, some C-rich PNe show evidence for C_{60} in their IR spectrum. In particular, the spectrum of the PNe, TC1, is dominated by the emission bands from this molecule (Figure 12.19) and thus the chemistry in these environments should also be conducive to producing fullerene structures.

12.7.1 PAH Formation in Stellar Ejecta

Carbon-rich asymptotic giant branch (AGB) stars return a large fraction of their material to the interstellar medium in the form of cool ($T < 1000$ K), gentle ($v < 25$ km/s) stellar wind ($\dot{M} \sim 10^{-7}$–10^{-4} $M_{\odot}$/yr) over a period of $\simeq 10^4 - 10^6$ yr. This material is enriched by the nucleosynthetic products from the stellar core, which are mixed into the surface layers through dredge up events, and carbon and the s-process elements are greatly enhanced in abundance. This also leads to isotope patterns that are nonsolar and very characteristic for these types of stars. A good fraction of the carbon in these outflows is in the form of soot particles, which obscure the star at optical and near-IR wavelengths and emit copiously at mid- to far-IR wavelengths. These stars show a rich molecular inventory through a plethora of vibrational and rotational transitions. The nearby ($d \sim 130$ pc) star IRC+10216 (CW Leo) is the prototypical object (sometimes designated as the smoke stack of the Milky Way). When much of the stellar atmosphere has been returned to the ISM, the white dwarf core will be exposed. During the last phases of this evolution, the star will rapidly transit through the moderate temperature range of post-AGB objects ($4000 < T < 10,000$ K) to the planetary nebulae phase. In the latter phase, the central, hot white dwarf can reach temperatures of 200,000 K and will ionize the detached, slowly expanding, shell of gas ejected during the AGB phase, creating luminous, very intricate nebular objects. The high energy photons can process the molecular envelope. In addition, during the post-AGB and PN phases, jets plow through the ejecta, driving strong shock waves that will also process ejected molecules and dust grains. In terms of PAH formation, we will focus here on the chemical processes near the photosphere of the AGB star that convert the main carbon lifestock, C_2H_2, into larger species and soot particles. The IR spectra shown in Figures 12.25 and 12.26 illustrate the chemical richness of these outflows

The molecular composition of these outflows depend strongly on the C/O abundance ratio. Initially, C/O starts at less than 1 but nucleosynthesis in the deep interiors coupled with mixing can change this ratio to > 1. When C/O < 1, carbon is locked up in CO

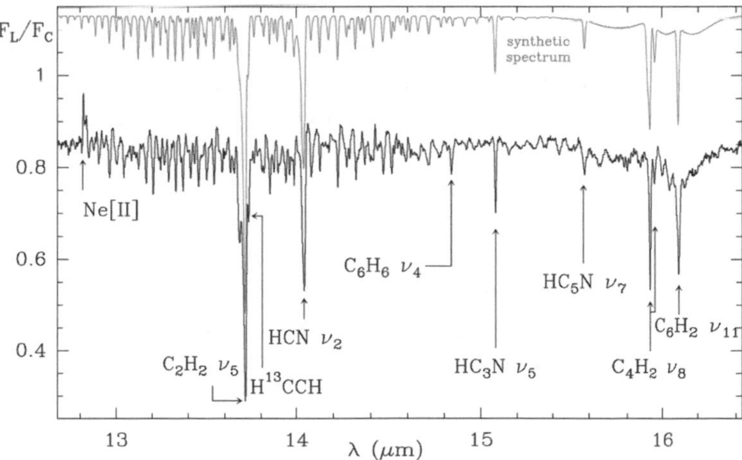

Figure 12.25 Moderate resolution ($\lambda/\Delta\lambda \simeq 3000$) ISO/SWS spectrum of the prototypical C-rich post-AGB object, AFGL 618, is dominated by P, Q, and R transitions of (poly)acetylenes ($C_{2n}H_2$) and cyanopolyynes ($HC_{2n-1}N$). The weak band at 14 μm has been attributed to benzene. For comparison, the top trace gives a calculated spectrum. Figure adapted from [34]

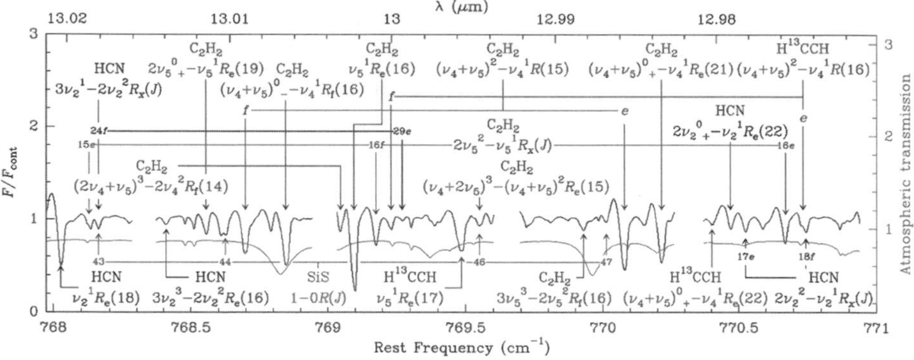

Figure 12.26 A small portion of the 11−14 μm spectrum of the AGB star, IRC+10216, obtained at a resolution of 10^5 using the TEXES instrument at the IRTF, resolving individual transitions of C_2H_2, HCN, SiS, and their isotopologues. The rotational-vibrational states involved are indicated. The atmospheric transmission is shown as a light gray curve. Figures adapted from [42] and kindly provided by Dr. Pablo Fonfria. (A black and white version of this figure will appear in some formats. For the color version, please refer to the plate section.)

and oxygen-bearing species dominate (H_2O, SiO, TiO, ...) in the photosphere and the dust is in the form of simple oxides and/or silicates. When C/O> 1, oxygen is locked up in CO and the excess C forms, e.g., C_2H_2 and HCN in the photosphere, and carbon soot condenses out in the ejecta. The photosphere has typical temperatures and pressures of 2500 K and 10^{-6} atm and a chemistry largely in thermodynamic equilibrium. Stellar pulsations drive shock waves that travel through the photosphere, and lift material to great heights above the stellar surface, while heating the gas and dissociating molecules present. These molecules can then reform in the adiabatically cooling gas behind the shock front. This results in an extended molecular layer where chemical nucleation of nano particles is initiated. These nano particles grow through chemical processes as well as through clustering and coagulation while simultaneously being accelerated by radiation pressure. The gas is momentum coupled to these nano grains.

The astrophysical setting for soot formation is thus a high density zone of cooling gas. In this environment, thermodynamics will want to drive the system to graphitic materials but kinetics determines the pathway. There are many kinetic pathways that can play a role – each favored by specific physical conditions – and that may lead through very different molecular intermediaries to a different soot structure. Models rely on the extensive literature on soot formation for terrestrial environments driven by, in particular, car engine efficiency and polution aspects.

PAHs are an abundant constituent in sooting flames where they may form either as molecular building blocks or intermediaries in the soot condensation process. The formation of the first ring (e.g. benzene) is the bottleneck. Generally, these models rely on neutral–neutral reactions that start with the formation of the first ring, benzene, through polymerization reactions of C_2H_2. One reaction pathway involves the highly stable radical propargyl C_3H_3 reacting with itself to form the phenyl (C_6H_5) radical (+ H). The stable

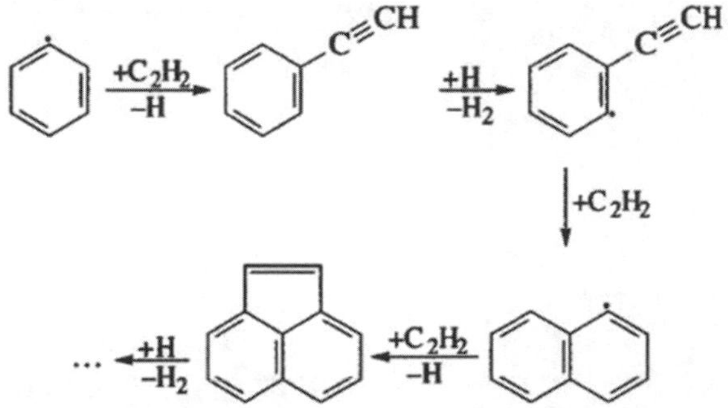

Figure 12.27 The HACA mechanism for the growth of aromatic ring species involves sequential steps of activation through H-atom abstraction reactions followed by C_2H_2 addition. In this example, aromatic ring formation happens after two of these steps, converting benzene into naphthalene. Figure adapted from [69]

propargyl radical may also react with acetylene to the cyclopentadienyl (c-C_5H_5) radical, which then reacts on to benzene. The propargyl radical itself is formed through reactions of acetylene with the CH_2 radical, formed through attack of H atoms on acetylene. An alternative route involves even (rather than odd) C-species; e.g. reaction of the ethynyl radical, C_2H, with acetylene forms 1-buten-3-ynyl (n-C_4H_3) followed by reaction of this species with C_2H_2 to the phenyl radical (C_6H_5). The ethynyl radical is formed through a H-abstraction reaction of H with acetylene. Various other routes are possible too, involving, for example, linear C_6H_n species, condensation of polyacetylenes, or ions.

Once this "nucleation" step has been made, further growth can take place through the so-called HACA mechanism (Hydrogen Abstraction Carbon Addition), involving a repetitive reaction sequence of activation of the "growing" species through abstraction reaction with H, reaction of the resulting radical with acetylene and, when appropriate, cyclization (Figure 12.27). Growth is then favored through the compact PAHs sequence of pyrene, coronene, ... For the pressures in stellar outflows, there is then a window of opportunity – $900\ K < T < 1100\ K$ – in which rapid growth can take place. For temperatures above 1100 K, both types of reactions (HA and CA) are reversible and entropy does then not favor growth. At 1100 K, the CA reaction becomes irreversible (particularly where the most stable PAHs are involved) but HA is still reversible. This will drive rapid PAH formation and soot growth. Below 900 K, the HA reaction freezes out and chemistry stops. PAHs can then still grow through agglomeration – driven by weak van der Waals interaction – and this will lead to large soot particles. Likewise, if the C_2H_2 "stock supply" has been exhausted, further growth will involve coagulation of soot particles themselves. Alternatives to the HACA mechanism have been proposed where, for example, acetylene is activated ($C_2H_2 \rightarrow C_2H$) and the ethynyl radical reacts with PAHs through a submerged barrier.

Sooting experiments support the role of PAHs as the molecular building blocks of carbonaceous soot. Small PAHs are very abundant in sooting flames, typically locking

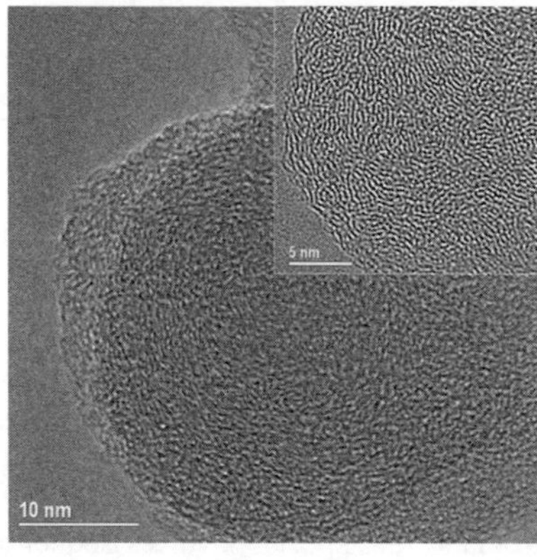

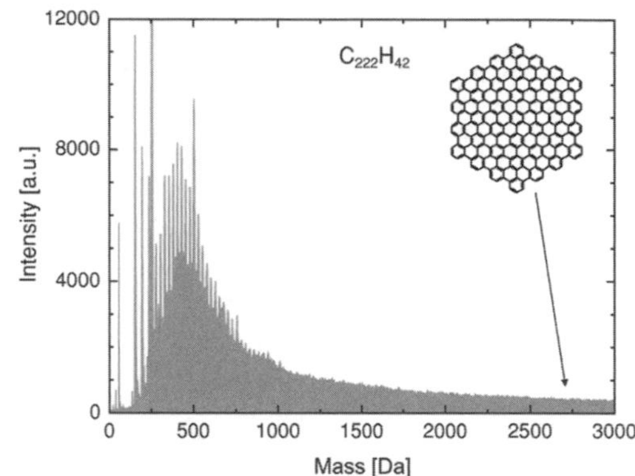

Figure 12.28 Left: High resolution Transmission Electron Microscope image of a typical large carbon nanoparticle produced in a gas phase condensation experiment, revealing its graphene substructure. The inset shows a section of a particle and illustrates the arrangement and mean sizes of the graphene layers. Right: These soot particles have been analyzed using matrix-assisted laser desorption/ionization in combination with time-of-flight mass spectrometry. The resulting mass spectrum reveals the presence of a broad distribution of PAHs with masses up to 3000 Da. One, hypothetical, PAH structure is illustrated. Figures adapted from [58].

up some 5% of the available C. Transmission electron microscope images of soot reveal regions of polyaromatic units with a graphene structure (Figure 12.28). The extent of these regions depends on the conditions of the experiment but can be as large as five stacked layers. Additionally, PAHs can be extracted from the soot. Mass spectrometeric analysis reveals a broad distribution of PAHs centered on $C_{40}H_x$ but which extends to PAHs as large as 200 C-atoms (Figure 12.28). IR spectroscopy of these soot grains supports the largely aromatic nature of this material but also indicates a high degree of disorder, which broadens, in particular, the absorption in the CC stretching region.

This discussion has focused on PAH and soot formation under conditions that are relevant for AGB outflows. It should be mentioned that carbon condensation experiments also find a window for soot formation at much higher temperatures, $\sim$3500 K. The chemical routes involved in this high temperature regime have not been elucidated, yet. It is clear, though, that fullerene-like structures are formed and that this soot contains small, strongly bent graphene layers. This soot is highly disordered and the level of disorder depends on the condensation conditions. When the H concentration is low, the structure is more ordered and shows closed fullerene cages. The order decreases with increasing H content and fullerene cages are often not closed. These cage structures are bonded through van der Waals interaction or are linked by aliphatic-CH_x groups.

In this environment, PAH formation is driven by the presence of atomic H. However, if there were another source of radicals, other, more direct carbon growth processes could take place. In that respect, in the outer layers, penetrating FUV photons can ionize and photodissociate molecules produced at the base of the flow. This gives rise to a rich neutral–neutral and ion–molecule chemistry and, for C-rich outflows, species such as HC_3N, C_4H, ... are routinely observed. If the outflow is very clumpy, then UV would be able to penetrate deeply and this could be a source of radicals in the (inner) regions of the outflow that could promote PAH/soot growth. In this regard, experiments have shown that naphthyl radicals react rapidly with acetylene and vinylacetylene, even at low temperatures, through van der Waals interactions with submerged barriers and these reactions may proceed at lower temperatures than previously thought.

12.7.2 PAH Formation during the Post-AGB Phase

Models have also been developed for the gas phase formation of benzene during the post-AGB phase. This was partly driven by the putative detection of out-of-plane bending mode of benzene in the ISO/SWS spectrum of the prototypical C-rich post-AGB object AFGL 618. These models rely on a warm (250 K) molecular (e.g. with C_2H_2) envelope with an additional source of ionization (e.g. X-rays or shocks). As for molecular clouds, the ionization leads to H_3^+ and this proton is rapidly transferred to C_2H_2 to form $C_2H_3^+$. The latter ion starts a polymerization process through reactions with C_2H_2, leading to $C_4H_3^+$ and $C_6H_5^+$. The radiative association reaction with H_2 followed by dissociative electron recombination leads then to C_6H_6. For a cosmic ray ionization rate of 10^{-15} s^{-1} and a C_2H_2 abundance of 5×10^{-5}, the calculated benzene abundance is 10^{-6}. This reaction scheme also leads to appreciable abundances of the polyynes, 1,3-butadiyne (diacetylene, C_4H_2) and 1,3,5-hexatriyne (triacetylene, C_6H_2) that are observed to be abundant in this source. There has been no attempt to extend these models further to the formation of larger PAHs. However, the phenylium ion ($C_6H_5^+$) reacts rapidly with acetylene to form the adduct $C_8H_7^+$, which isomerizes barrierless to the 2-phenylethenylium ion. Further reaction with C_2H_2, that might lead to protonated naphthalene, $C_{10}H_9^+$, may have activation barriers. In contrast, the benzene cation does not react with acetylene but does react quickly with diacetylene. As mentioned above, neutral–radical reactions may provide a separate pathway for acetylene condensation.

12.7.3 Growth of PAHs in the ISM

In the ISM, PAH formation routes are slightly different from those that are described in Section 12.7.2 for C-rich, ionized ejecta. In this case, acetylene is a minor species with expected abundance of 10^{-10}–10^{-8}, depending on time, through ion–molecule reactions involving small hydrocarbon radicals (cf. Figure 9.5 and Section 9.4.3). Most of the carbon – before CO is formed – is present as C^+. Ion–molecule reaction routes toward benzene in the ISM start then with C^+, which reacts with C_2H_2 to form C_3H^+. This ion reacts with H_2 to $C_3H_3^+$. Reaction with atomic C leads to C_4H^+ and $C_4H_2^+$. With H_2, this leads to $C_4H_3^+$. Protonated diacetylene does not react further with H_2. Rapid reaction with acetylene forms $C_6H_5^+$ which goes on to benzene. This scheme does require abundant C^+ and C and hence, in models for the chemistry of molecular clouds, the abundance of benzene peaks at early time before most of the available carbon is converted to CO but even then the peak abundance is only $\sim 10^{-9}$. Benzonitrile (C_6H_5CN) is currently the only small aromatic molecule identified in molecular clouds (see Table 9.2) at an abundance of 6×10^{-11} and can be formed through a barrierless reaction of the CN radical with benzene. Given the large uncertainties involved, calculated model abundances are in fair agreement with the observations.

12.7.4 From Soot to PAHs

An alternative route toward PAHs in the ISM has been proposed based upon fragmentation of soot particles. As Figure 12.28 well illustrates, PAHs are the building blocks of soot particles. Hence, energetic processing of soot particles may provide a source of PAHs in the ISM. Detailed models have been developed for PAH formation by grain–grain collisions in interstellar shocks. As the layers in graphitic materials are bonded only by van der Waals forces (50 meV/C-atom), the shear modulus for graphitic materials is quite low, $G = 4 \times 10^{10}$ dynes cm^{-2}, and small graphene flakes readily rub even off at low stresses. Hence, the pressures associated with relatively low velocity collisions will create PAH fragments. This shear modulus translates into a critical velocity of only $\sim$1 km s^{-1} for onset of fragmentation and a 50 Å grain will completely fragment a 1000 Å soot particle at an impact velocity of 75 km s^{-1}. Models show that such collisions occur frequently in interstellar shock waves of $\sim$100 km s^{-1}. Such collisions can be expected to lead to a size distribution set by the intrinsic distribution of PAH-building blocks making up the soot particles and, for the conditions probed by the experiments reported in Figure 12.28, this would range from 25 to 60 C-atom with a typical size of 40 C-atoms. Of course, once released into the gas phase (particularly the small), PAHs would be further processed by, e.g., FUV photons (Section 12.7.5).

12.7.5 From PAHs to Fullerenes: Top-Down Chemistry

UV photolysis of PAHs can lead to the formation of smaller species and these include H, H_2, small hydrocarbons, carbon rings, chains, and fullerenes. During the photolysis, fragmentation competes with isomerization (Figure 12.29). As discussed in Section 8.5, fragmentation channels will open up for highly excited PAH species. Experimental and theoretical studies show that the weakest link will go first, leading to the loss of any functional groups, followed by loss of aromatic H (Figure 8.19). For small PAHs, after the loss of a few H-atoms, the C-skeleton itself can be affected through loss of CH and C_2H_2. But PAHs larger than about 32 C-atoms can be almost completely stripped of H-atoms before fragmentation of the C-skeleton occurs (Figure 12.30). It is instructive to compare the fragmentation behavior of large PAHs and fullerenes. After PAHs are fully stripped of H-atoms, their further fragmentation behavior starts to resemble that of fullerenes as the same magic numbers appear in both (Fig. 12.31). Further studies show that the buckyball, C_{60}, does not absorb at 532 nm and hence does not fragment even at high laser fluences. In contrast, C_{70} does and it fragments until it reaches C_{60} where further fragmentation stops. Large PAHs show a similar behavior. After loss of all its H-atoms, the PAH, $C_{66}H_{26}$, starts

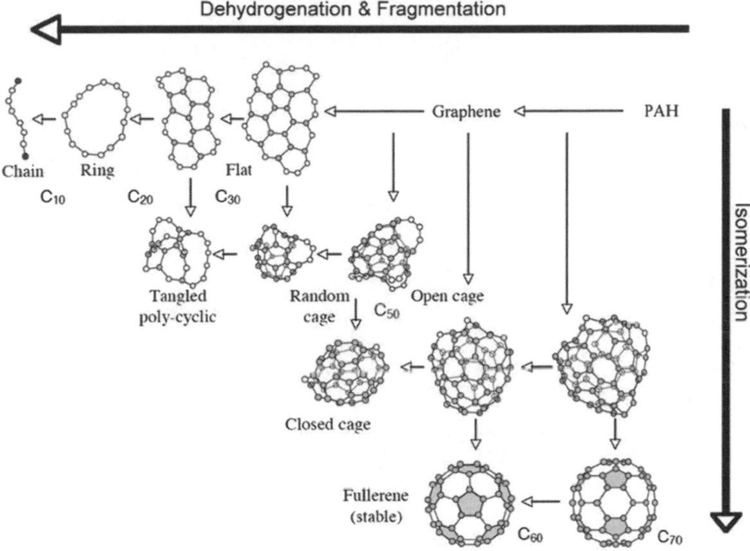

Figure 12.29 Schematic diagram of UV processing of large PAHs. Large PAHs (top right) start by losing all of their H-atoms, which results in the formation of large graphene flakes. Upon further photolysis, C-atoms will be lost, two at a time, breaking the C-skeleton down to smaller and smaller sizes. Eventually, this processing results in a complete breakdown of aromaticity and the formation of carbon chains and rings. Fragmentation competes with isomerization as C-loss from graphene flakes may result in pentagon formation and rolling up of the structure into cages and fullerenes. Figure taken from [13]

to shrink and the high abundance of the C_{60} cluster suggests that some (but not all) of these have isomerized to the buckyball. This top-down scheme for the formation of fullerenes provides an attractive explanation for the increased abundance of C_{60} close to strong UV sources in reflection nebula where the abundance of PAHs has in fact decreased.

12.7.6 PAH Processing by Energetic Ions

Interaction of PAHs with energetic ions can open up new fragmentation channels. At very high energies, interactions will be mediated through the π-electron clouds and this leads to highly excited PAH species, which will fragment through the same statistical process described in Section 8.5 for UV photon excitation. However, at lower energies (10–1000 eV), collisions may lead to direct loss of a C-atom (possibly with an associated H atom) on a femtosecond timescale through a knock-out process. The energy required for this is the displacement energy, which is measured to be 23.3 ± 0.3 eV. Theoretical studies show that the energy transferred by a He atom impacting on a C-atom is well described by $\Delta E_{He} = 2.55 E_{CM}^{0.66}$ and, hence, this corresponds to a threshold energy of about 32.5 eV (Figure 12.32). For fullerenes, this displacement energy is measured to be slightly higher, 24.1 eV.

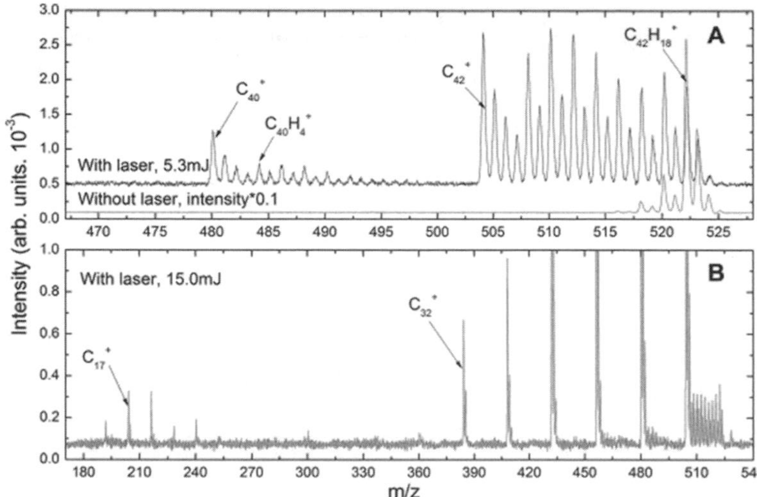

Figure 12.30 Photolysis behavior of large PAHs. Photolysis at 532 nm of the hexa-peri-hexabenzocoronene (HBC, $C_{42}H_{18}^+$) cation isolated in an ion trap. Laser irradiation results in absorption of many photons and highly excited PAHs fragments. Many sequential step of photon absorption followed by fragmentation occur, leading to a range of products that are set by the probability that these sequential steps occur mediated by the internal energy at which these fragmentation channels open up. The products are investigated using a mass spectrometer. For low fluences, only H-atoms are lost and C-loss does not set in until most of the H's are stripped off (lower trace). For high fluences, further fragmentation proceeds and loss of sequential C_2 units becomes apparent. Figure adapted from [135, 136]

This direct interaction may also open up new chemical synthesis routes. This is specifically important when considering (van der Waals) cluster evolution, as the radical formed by the direct impact may rapidly react with a neighboring species creating covalent bonds. Any excess kinetic or chemical energy released into the cluster may be dissipated by evaporation of weakly bonded cluster members. This has been studied for both PAH and fullerene clusters.

In space, energetic collisions can be important in interstellar shocks, in hot plasmas, and by cosmic rays. Supernovae eject typically 1 $M_\odot$ of material at velocities of some 10,000 km/s into the ISM. As these ejecta sweep up surrounding gas, a reverse shock slows down the ejecta heating it to some 10^7 K, while the swept-up gas is shocked by the expanding supernova shock. As the supernova remnant expands adiabatically, sweeping up more and more material, the pressure in the remnant drops, and the strength of the supernova shock decreases. Hence, progressively larger amounts of interstellar material are processed by progressively lower velocity shocks. Evaluating this evolution, processing of interstellar PAHs is dominated by the lowest velocity shock that can still destroy a PAH. In these strong shocks, the gas is stopped in mean-free pathlength while the kinetic energy is converted into heat. But the mean free path of PAHs and dust grains is much longer than for

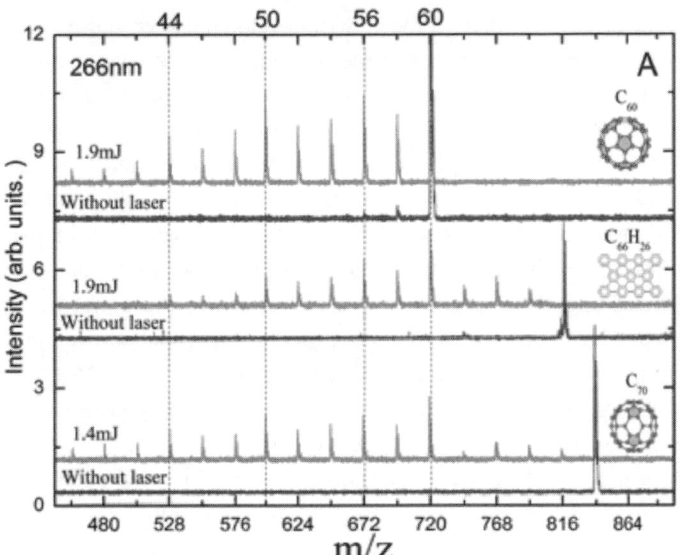

Figure 12.31 Photolysis behavior of large PAHs and fullerenes at 266 nm. The PAHs are first completely stripped of H-atoms. The resulting carbon structures as well as both fullerenes then lose 2C atoms at a time. Note the presence of magic numbers – carbon clusters with higher abundances than neighboring clusters – (labeled at the top) indicating the enhanced stability of specific intermediate fragments. Figure adapted from [135, 136]

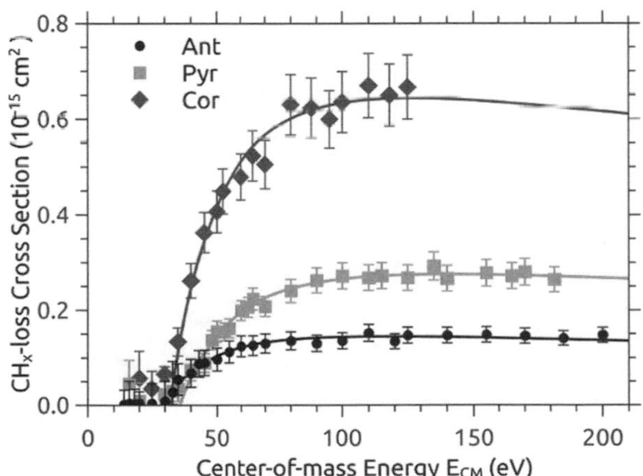

Figure 12.32 Measured cross sections for the loss of CH_x fragments from anthracene, pyrene, and coronene in collisions with He as a function of the center-of-mass energy, E_{CM}. The measured threshold center of mass energy is about 32.5 eV. Figure adapted from [120]

gas atoms and, hence, a large drift velocity will develop between these species and the gas. In this scenario, the destructive effect of the impactor is set by the mass (energy) and the abundance. Near threshold, collisions with helium will dominate as it is still quite abundant (0.1 by number relative to H) but carries four times the energy.

In hot ($>10^6$ K) thermalized plasmas, destruction is dominated by electronic energy transfer. The much higher collision rate of energetic electrons dominates energy transfer and therefore destruction. Calculated lifetimes of PAHs exposed to a plasma are of the order of years and much shorter than the lifetime of supernova remnants. Supernova ejecta are highly clumped and PAHs (and dust grains) formed in these ejecta can survive much longer inside these dense clumps as they are processed by much slower shocks.

PAHs are also destroyed by interaction with high energy (>10 MeV/nucleon) Cosmic Rays particles. Energy transfer is then completely dominated by electronic interaction, and fragmentation is described by the statistical process (Chapter 8.5). The results do depend on how much of the cosmic ray energy remains as vibrational excitation (f) rather than being lost, e.g., in the Auger process. For typical values, calculated lifetimes are $\simeq 10^8$ yr.

Figure 12.33 compares calculated lifetimes of large PAH molecules in the ISM against these different processes. The relative importance of these processes is taken into account by including detailed models for the evolution of supernova remnants and the diffusion of cosmic rays. These lifetimes depend on the displacement energy of a carbon atom for the knock-out process and the C binding energy for the statistical process and in this particular calculation these have been set to 7.5 eV and 4.5 eV, respectively. The choice for the displacement energy is low compared to current estimates but the effects of this have not yet been evaluated. The resulting lifetime in the diffuse ISM is then calculated to

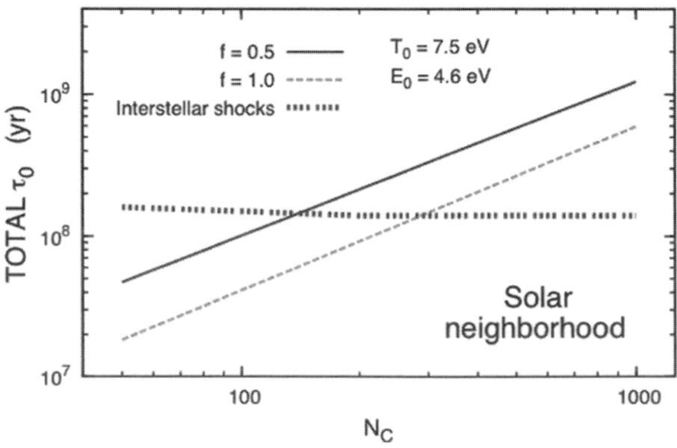

Figure 12.33 Calculated lifetime in the ISM as a function of the size of the PAH. Destruction involves either processing in interstellar shocks or by cosmic rays. These calculations involve models for the evolution of supernova remnants and for the cosmic ray flux in the galaxy. Values adopted for the displacement energy and C binding energy in PAHs are indicated. The parameter f describes the fraction of the energy transferred by the cosmic ray that goes into vibrational excitation. Figure taken from [82]

be between 50–150 Myr, depending on the size of the species involved. At the large size end, destruction is dominated by cosmic rays while at the small size end, supernova-driven shocks are key.

12.7.7 Evolution of PAH Clusters

As discussed in Sections 12.2.2 and 12.2.4, the observed IR spectra show evidence for the presence of emission component(s) that have been attributed to PAH-clusters. Such PAH clusters can form upon collision in a dense cloud core on a timescale of $\simeq 10^6 \left(10^4\,\mathrm{cm}^{-3}/n\right)$ yr. Binding energies of PAH clusters have been discussed in Section 8.7.7 (Figure 8.24). As the van der Waals binding is the weakest link, UV absorption will tend to break up clusters upon photon absorption in a PDR. A coronene dimer is only bound by 1 eV and every photon will lead to fragmentation. A circumcoronene dimer, on the other hand, will have to absorb some 10^5 photons before fragmentation occurs and for circumcoronene multimers, a multiple photon absorption event is required. If we balance the dimerization rate with the dimer photo-fragmentation rate, we find that circumcoronene dimers will grow for $A_V \gtrsim 6$ mag in a PDR ($G_0/n \simeq 1$). Larger clusters can grow even closer to the PDR surface. Experiments show that, when covalent bonds are formed between the monomers of a cluster, they will evolve toward larger PAHs rather than sublimate. This provides a bottom-up photochemical evolution pathway that may counteract the photochemical breakdown of large PAHs (Section 12.7.5).

12.8 PAHs and the Photo-Electric Heating of Interstellar Gas

12.8.1 PAH Charge

In PDRs and in diffuse clouds, the charge is set by a balance between photoionization and electron recombination, viz.,

$$\frac{X_{i+1}}{X_i} = \frac{k_{uv}\,(i)}{X_e n k_e}, \tag{12.18}$$

where X_i is the abundance of the PAH with charge state i, $k_{uv}\,(i)$ and k_e are the photoionization rate of ion, i, and the recombination rate (cations) or electron attachment rate (neutral), and X_e is the electron abundance. These processes are discussed in Sections 8.5.3, 8.6.1, and 8.6.2. Together with the conservation law, this set of equations can be solved. The charge distribution depends strongly on the ionization parameter, $\gamma = G_0 T^{1/2}/n_e$. As γ increases, the level of ionization increases. There is a weak dependence on PAH size as the ionization rate increases faster than the recombination rate (γ increases proportionally to $N_c^{1/2}$). However, more importantly, for larger PAHs more ionization stages become accessible at high γ.

Figure 12.34 shows calculated charge distributions for two PAHs, circumcoronene ($C_{54}H_{18}$) and circumcircumcoronene ($C_{96}H_{24}$), that span the relevant size range for three different physical conditions. As γ increases, the charge distribution shifts from

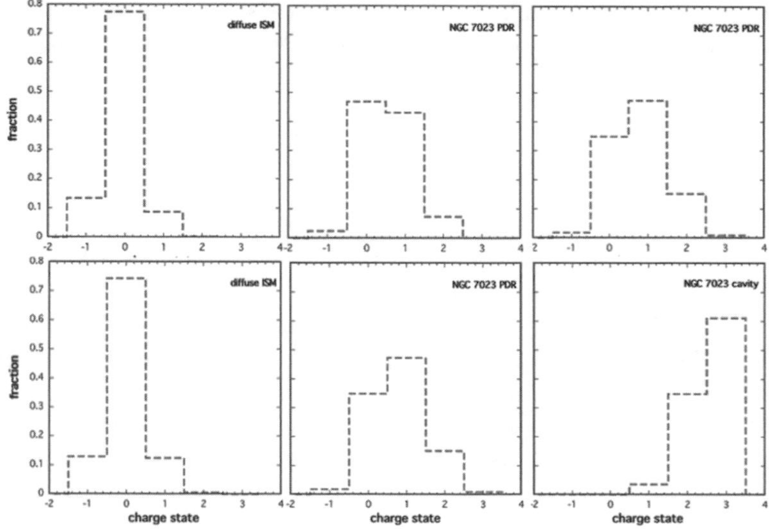

Figure 12.34 Calculated charge distributions for circumcoronene ($C_{54}H_{18}$, top row) and circumcircumcoronene ($C_{96}H_{24}$, bottom row) for three ISM environments: a diffuse cloud ($G_0 = 2$, $n = 50$ cm^{-3}, $T = 80$ K, left column), the bright PDR in NGC 7023 ($G_0 = 2600$, $n = 2 \times 10^4$ cm^{-3}, $T = 500$ K, right column) and the cavity in NGC 7023 at 15$''$ from the illuminating star in NGC 7023 ($G_0 = 2 \times 10^4$, $n = 5 \times 10^3$ cm^{-3}, $T = 500$ K, middle column). An electron abundance of 1.6×10^{-4} has been assumed. The ionization parameter γ increases from 2×10^3 to 6×10^5 K$^{1/2}$ cm^3.

predominantly neutral in diffuse clouds to equal amounts of neutral and singly ionized species at the surface of PDRs to highly ionized close to an illuminating star. As G_0/n_e scales only with $n^{1/3}$ for PDRs, charge distributions are expected to be very similar in all PDR surfaces.

12.8.2 Photo-Electric Heating

The ionization of large molecules in the diffuse ISM is an important source of gas heating in the ISM as the photo-electron can carry away a large fraction of the excess photon energy as kinetic energy, which it will quickly share with other electrons in the gas. Photoionization of PAHs is discussed in Section 8.5.3. The contribution to the photoelectric heating rate, $\Gamma_{pe,i,j}$, of interstellar gas by PAH, j, in charge state, i, is given by,

$$\Gamma_{pe,i,j} = 4\pi \, X(i,j) \int_{\nu_{i,j}}^{\nu_H} \mathcal{N}(\nu) \, \sigma_{i,j}(\nu) \, E_{kin}(i,j) \, d\nu, \tag{12.19}$$

where $X(i,j)$ and σ are the abundance and photoionization cross section of this species, $\mathcal{N}$ is the average photon intensity at frequency ν, and E_{kin} is the kinetic energy of the photo-electron and it is tacitly assumed that the electron quickly shares this energy with other electrons in the gas. The integral is over the relevant frequency range from the ionization potential ($h\nu_{i,j} = IP_{i,j}$) to the hydrogen ionization potential ($h\nu_H = IP_H$).

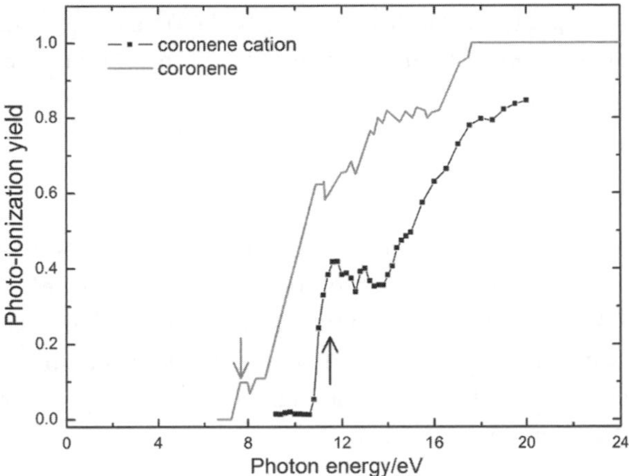

Figure 12.35 Measured photo-electric yields for coronene and the coronene cation. The arrows indicate the calculated ionization potentials (Table 8.3). Figure adapted from [126, 139]

The ionization cross section is the product of the absorption cross section and the ionization yield. UV photoabsorption cross sections have been measured and calculated for small PAHs (Section 8.2.3 and Figure 8.10). Figure 12.35 (see also Figure 8.20) shows the measured photoionization yield for neutral and singly ionized coronene. Apart from the shift in ionization potential, the ionization yields behave very similarly and are well described by the simple formula given in Section 8.5.3. The kinetic energy of the photoelectron can be measured using photoelectron-photoion coincidence techniques but this has not been quantified. Often – rather arbitrarily – a value of half the excess photon energy is assumed. The total photo-electric heating rate is then the sum over all ionization stages and over all PAHs,

$$n\Gamma_{pe} = \sum_j \sum_i \Gamma_{pe,i,j}, \tag{12.20}$$

where the summation is limited to the accessible charge states (e.g. for which $IP_i < 13.6$ eV in the diffuse ISM or a PDR).

Section 8.5.3 and Table 8.3 summarize ionization potentials of some relevant PAHs. The ionization potential of a charged PAH is often approximated by the classical expression appropriate for a circular disk,

$$IP_i = 4.4 + \left(Z + \frac{1}{2}\right)\frac{25.1}{N_C^{1/2}} \quad \text{eV}. \tag{12.21}$$

In the limit of large PAHs, the ionization potential goes to the graphene work function (4.4 eV), but for small PAHs and high ionization stages, IP can exceed 13.6 eV. For example, the second ionization potential of pyrene, $C_{16}H_{10}$, is 16.6 eV, well above the hydrogen ionization limit.

Some insight in the photoelectric heating rate by PAHs can be gained by considering a single PAH with two accessible charge states (neutral and singly ionized). The photoelectric heating efficiency is then,

$$\epsilon = \overline{Y} f(i = 0) \left(\frac{E_{kin}}{h\nu} \right), \tag{12.22}$$

with $f(i = 0)$, the neutral fraction and $\overline{Y}$ the average yield of photo-electrons. This is discussed in Section 8.5.3 and we can approximate $\overline{Y} \simeq 0.3$ for a 10 eV photon and a 6 eV ionization potential. For comparison, the photo-electric yield of bulk materials is $\sim$0.05 as photon absorption occurs rather deep in the material ($\simeq$200 Å) compared to the mean-free path of the electrons ($\simeq$10 Å). If we assume that half of the excess energy remains in the molecule as internal excitation, we have,

$$\epsilon = \frac{\overline{Y}}{2} f(i = 0) \left(1 - \frac{IP}{h\overline{\nu}} \right) \simeq 0.06 f(i = 0), \tag{12.23}$$

and we find that 6% of the photon energy absorbed by neutral PAHs goes into heating the gas. Absorbed energy that is not lost through ionization will result in fluorescence or fragmentation. This efficiency is much higher than for dust grains as the yield is much higher and, in grains, photo-electrons lose energy in migrating to the surface.

The PAH neutral fraction is set by the ionization balance, which for a PAH with two ionization stages is given by,

$$f(Z = 0) = \left(1 + \frac{\overline{Y} k_{uv}}{n_e k_e} \right)^{-1}. \tag{12.24}$$

Adopting a size of 50 C-atoms and inserting values (see above and Sections 8.5.3 and 8.6.1), we have,

$$f(Z = 0) = \left(1 + 7 \times 10^{-5} \gamma \right)^{-1}, \tag{12.25}$$

with $\gamma = G_o T^{1/2}/n_e$ the ionization parameter. The efficiency of the photon energy converted into gas heating by PAHs is then,

$$\epsilon = \frac{0.06}{1 + 7 \times 10^{-5} \gamma}. \tag{12.26}$$

So, in a diffuse cloud ($T = 100$ K, $n_e = 0.008$ cm^{-3}, and $G_0 = 1$), we have $f(Z = 0) \simeq 0.85$ and $\epsilon \simeq 0.05$. In a dense PDR ($\gamma \simeq 10^5$ K$^{1/2}$ cm^3), PAH charging is more important with $f(Z = 0) \simeq 0.13$ (cf. Figure 12.34) and the efficiency is almost an order of magnitude smaller, 0.008. This is a general trend as γ increases, species are charged up, and the photo-electric heating efficiency will drop. Typically, PAHs have several ionization stages with $IP < 13.6$ eV and these are all accessible in the diffuse ISM. Because of the increased Coulomb barrier, the efficiency will, however, drop with increasing IP.

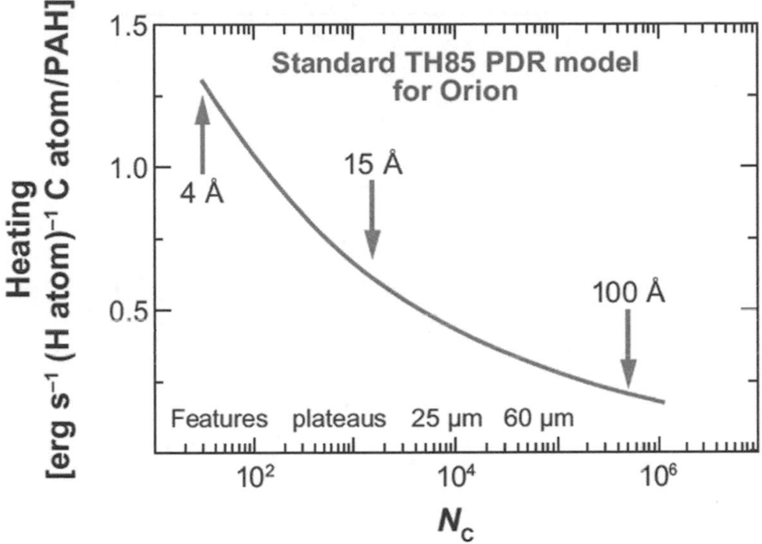

Figure 12.36 The calculated contribution to the photoelectric heating of interstellar gas by species of different number of carbon atoms. Typical sizes are indicated. Equal areas under the curve correspond to equal contributions to the heating. Adopted physical conditions pertain to those of the well-known Orion Bar PDR. Figure adapted from [7]

Theoretical calculations on the heating efficiency, ϵ, by a size distribution of PAHs ranging from $\simeq$50 C-atoms all the way to the classical grain domain (1000 Å) are shown in Figure 12.36 and these have been fitted by a simple analytical formula,

$$\epsilon = \frac{4.87 \times 10^{-2}}{1 + 4 \times 10^{-3}\gamma^{0.73}}. \tag{12.27}$$

This is the equivalent of Eq. (12.26). The dependence on γ is slightly less steep because of the presence of multiple charge states. Here it should be noted that large grains that do not contribute to the photo-electric heating of the gas (see below) also do not absorb much of the FUV photon energy and hence even the numerical values are quite similar.

The photo-electric heating rate by PAHs should then be compared to that by PAH-clusters, very small grains, and "classical" grains. Typically, approximately half of the heating originates from PAHs and PAH clusters ($<10^3$ C atoms). The other half is contributed by very small grains (15 Å $< a <$ 100Å). Classical grains do not contribute noticeably to the heating. The heating rate can then be expressed as a heating efficiency, ϵ, such that the heating rate is given by,

$$\Gamma_{pe} = 10^{-24} \epsilon\, n\, G_0 \qquad \text{erg cm}^{-3}\ \text{s}^{-1}, \tag{12.28}$$

with G_0 the intensity of the UV radiation field in units of the average interstellar radiation field and ϵ is given by Eq. (12.27). This expression has been widely used in astronomical

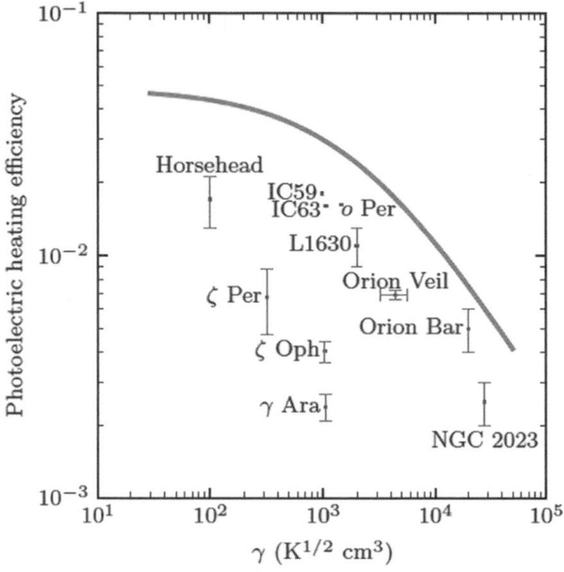

Figure 12.37 The calculated photo-electric heating efficiency (solid curve) as a function of the ionization parameter, $\gamma = G_0 T^{1/2}/n_e$, is compared to observations of PDRs and diffuse clouds. Figure adapted from [90]

models.[9] This efficiency is shown in Figure 12.37. For PDRs, the photo-electric heating rate can be measured by comparing the intensity of the IR cooling lines ([CII] 158 μm and [OI] 63μm) with the intensity of the radiation field (derived from the far-IR dust continuum). For diffuse clouds, the intensity of the dominant cooling line ([CII] 158μm) can be derived from observations of the UV absorption line starting in the excited $^2P_{3/2}$ fine-structure level of C^+. With an estimate of the UV radiation field available for heating, the heating efficiency can be derived. Given the approximate nature of these models, theory and observations show a reasonable agreement (Figure 12.37).

12.9 Further Reading and Resources

Over the years, a number of reviews of the Aromatic Infrared Bands and interstellar PAHs have appeared [4, 103, 123]. The interested reader should also consult the proceedings of a number of workshops and symposia [5, 59, 66]. The spectral assignments discussed in Section 12.3 date back to [4] and have been updated with more recent studies [20, 21, 70, 95, 106, 127]. Matrix isolation studies have been performed by various groups [53, 54, 84, 121], messenger techniques were employed by the Oomens group in Nijmegen [97], the

[9] For high temperatures, a small correction factor has to be included to account for the energy loss by recombining electrons.

single photon studies were done by the Saykally group in Berkeley [112], while the ion-dip studies were done by the Buma group in Amsterdam [75].

Fully reduced Spitzer/IRS spectra are publicly available and can be accessed online at http://sha.ipac.caltech.edu/applications/Spitzer/SHA/. The spectral tool PAHFIT is widely used to analyze mid-IR spectra. It is described in [117] and can be downloaded from http://tir.astro.utoledo.edu/jdsmith/research/pahfit.php. The spectral tool PAHTAT is maintained by Olivier Berné. It is described in [98] and can be downloaded from http://userpages.irap.omp.eu/~cjoblin/PAHTAT/Site/PAHTAT.html. The PAH database maintained by Christiaan Boersma at www.astrochem.org/pahdb/ is described in [11] and regularly updated with new spectra and improved analysis tools. These different tools have been compared to each other in [45, 107].

References [94, 125] have classified AIB spectra obtained by ISO/SWS in terms of peak position and profiles. The 3.3μm AIB has been studied at high resolution by [88]. The presence of two independent components in the 3.3μm AIB was well established by [32]. High resolution studies of the weak AIBs between 3.4 and 3.6 μm have been reported in [88]. Intensity variations have been addressed by [37, 52, 94]. The weak combination bands in the 5–6 μm range have only been studied for a few sources [18]. The AIBs in the 11–14 μm range have been studied systematically in the ISO/SWS spectra by [52] and interpreted in terms of the molecular edge structure. Spatial variations in the reflection nebulae, NGC 7023 and 2023, were analyzed by [22, 96] using Spitzer/IRS spectral-spatial maps. The 15–18 μm range has been studied in the laboratory by [84] and theoretically by [20].

The overwhelmingly aromatic character of the AIB carriers has been clear from the earliest studies [4]. The results in Table 12.2 were taken from [123]. The spectral-spatial study of the behavior of the 3.3 and 3.4 μm AIB are reported in [48]. The deuterated PAH analysis was taken from [39]. The importance of excitation in this analysis was stressed by [89]. The observed variations in the gas phase D/H ratio in the local Solar Neightborhood have been reported by [68]. The fractional abundance of N required to shift the aromatic CC mode to 6.2 μm is discussed in [55].

Quantum chemical studies on the IR spectroscopy of PAHs have been presented by [8, 9, 10, 65, 74, 106]. The CH stretching region has been studied at high spectral resolution by [75, 76, 77] for a limited number of small, gaseous PAHs. The mid-IR wavelength range has been measured in the laboratory by [67]. A theoretical analysis of the anharmonic behavior of the vibrational modes of PAHs is presented in [67, 70, 71, 72, 87]. Detailed DFT studies on the relative strength of the CH aliphatic mode have been reported in [130], while laboratory studies are reported in [77] and analyzed in [72]. The influence of aliphatic carbon on the spectrum of soot particles has been reported by [99]. The possible importance of strained C-skeletons on the position of the 6.2 μm band has been pointed out by [46]. The spectral characteristics of stacked PAH clusters has been investigated by [108]. IR spectra of Fe-PAH complexes have been studied by [122]. The spectra of large regular and irregular PAHs have been studied systematically by [10, 106]. The effect of endoskeletal N on the 6.2μm band has been studied in detail by [55]. Interstellar IR spectra in the 11–15 μm range have been analyzed by [52] and linked to the molecular edge structures of interstellar PAHs. The importance of armchair versus zigzag edges was pointed out by [33]. The spectral-spatial behavior of the 11−14 μm OOPs modes in interstellar spectra has been studied by [23, 95]. Long wavelength skeletal modes of PAHs have been studied by [21, 106].

The far-red DIBs at 9670 and 9632 Å were first discovered by [40]. Reference [30] conclusively demonstrated, using a He-messenger technique, that these bands can be ascribed to C_{60}^{+}. Further laboratory and observational studies have confirmed this identification [31, 64].

The original argument that the AIB bands are carried by large molecules was presented in [114]. The PAH abundance analysis is derived from [3]. Analysis of the 3.3/11.2 μm AIB ratio in terms of the size of the emitting PAHs has been studied by [3, 4, 37, 106]. The PAH abundance is derived from IR cirrus studies in [26] and the dependence of the PAH abundance on the radiation field is reported in [25].

The characteristics of the Anomalous Microwave Emission (AME) have reviewed by [38]. Models for the rotational excitation of PAHs have been developed by [109] and applied to the AME by [131]. The power of the rotational "comb-like" pattern in the rotational spectra for the identification of grandPAHs has been advocated by [2].

The charge distribution of astrophysically relevant PAHs has been addressed by [7, 110]. The photo-electric heating of interstellar gas has been studied by [7, 126]. Photoionization of PAHs has been studied in the laboratory by [27, 60, 124, 139]. Comparison between observed and theoretical photo-electric efficiency has been studied by [6, 90, 111].

The discussion on the chemical routes toward PAHs owes much to the very detailed chemical kinetics studies performed to understand soot formation and car engine efficiency. Relevant reactions are summarized in [128] and a review of some of the kinetic issues can be found in [43]. The important role of stabilomers in the PAH formation route was elucidated by [119]. This was applied to PAH formation in AGB outflows by [35, 44] and further developed with realistic AGB wind models by [36]. Reference [79] reports the extraction of small PAHs with the same isotopic composition as their host stardust grains isolated from meteorites, linking (some) PAH formation firmly to AGB outflows. There is an extensive literature on the properties of interstellar dust. Particularly relevant here are the studies by [56, 57, 58] that elucidated the internal structure of soot and its dependence on physical conditions as well as extracted small PAHs from soot grains and studied their characteristics. This group also studied the observable characteristics of soot particles, see for example, [113]. The characteristics of high temperature carbon condensates have been studied by [58]. The role of radicals in the growth of PAHs has been advocated by [63]. Low temperature formation routes toward benzene and naphthalene have been studied experimentally by [91, 133]. The chemistry of pyridyl radicals (C_5NH_5) with small hydrocarbon compounds leading to quinoline-derivatives has been shown to occur rapidly even at low temperatures of the ISM [92, 93]. The formation of benzene in post-AGB objects has been investigated by [129]. Ion–molecule reactions leading to benzene ions and their derivatives are discussed by [24, 118]. Models for the formation of benzene in molecular clouds have been discussed by [15]. Shattering of soot particles as the origin of interstellar PAHs has been modeled by [49, 50, 62]. Experimental studies on the photochemical behavior of methylated PAHs have been reported by [137] and of hydrogenated amorphous carbon films by [1].

The conversion of large PAHs into cages and fullerenes has been introduced by [13] inspired by observations of the abundance behavior of PAHs and fullerenes in the reflection nebula, NGC 7023. Experimental evidence for the top-down conversion of PAHs to fullerenes is provided by [136].

The knock-out process has been studied by [102, 120]. The chemistry initiated by energetic ion–cluster interaction has been reviewed by [47]. Models for the destruction of PAHs by shocks, thermal plasmas, and cosmic rays are presented in [50, 80, 81, 82]. A general review of dust destruction by supernovae is presented by [83], see also [16, 17].

Clustering of PAHs in an astrophysical setting has been studied by [105]. The photochemical evolution of PAH clusters has been studied experimentally by [28, 134, 138]. Processing by energetic ions has been studied by [51, 61] and reviewed by [132].

12.10 Exercises

12.1 Abundances:

- The AIBs are perched on broad, 6−9 and 11−14 μm plateaus. Because of their spatial morphology, these plateaus are thought to be carried by PAH clusters containing a few hundred C-atoms. Assume that these clusters need to reach 500 K in order to emit in the 6–9 μm range after absorption of a single photon, calculate the typical size of the carrier.
- Typically, these plateaus carry about 8% of the observed infrared flux. Calculate the abundance of the carriers.
- As discussed in Section 12.5.3, the mid-IR flux of the cirrus is measured to be 1.6×10^{-24} erg/s/H-atom. Calculate the fraction of the carbon locked up in the carrier in the diffuse ISM.
- The C_{60} bands carry about 0.05% of the IR flux in the surface of the PDR associated with NGC 7023. Calculate the abundance of C_{60}.
- Take the abundances as listed in Table 12.3 and plot these as function of size. Take a fractional bin-size for each component of 0.3. Compare these abundances with the size distribution of interstellar carbon grains, $n\,(a)\,da = A_c a^{-3.5} da$ with $A_c = 6.9 \times 10^{-26}$ cm$^{2.5}$/H-atom.
- Calculate the total surface area of PAHs (per H-nuclei) and compare with that of interstellar dust grains.
- There seems to be a relationship between size and number density. Discuss possible astrophysical origins for such a relationship.

12.2 PAH molecules versus graphite grains:
Contrast the absorption and emission characteristics of a 50 C-atom PAH molecule with a spherical graphite dust particle with a radius $a = 10^3$ Å (and a specific density of 2.2 g cm^{-3}).

- The radiative equilibrium temperature of a graphite grain is given by $T_{gr} \simeq 14.3$ $(0.1\ \mu\text{m}/a)^{0.06}\ G_0^{1/5.8}$ with G_0 the intensity of the radiation field in units of the interstellar radiation field. Calculate the dust temperature in the Orion Bar ($G_0 = 5 \times 10^4$).
- If we assume that both the PAH and the dust grain are at 20 K, calculate the energy content (in eV) of each ($C_V = 3.84 \times 10^2\ VT^2$ erg K^{-1}).
- Calculate the UV absorption timescale for the graphite grain in the interstellar UV radiation field and compare to that of a 50 C-atom PAH (Chapter 8). Adopt an interstellar radiation field with 1.3×10^{-4} erg cm^{-2} s^{-1} sr^{-1}, a mean photon energy of 10 eV, and a UV absorption cross section for the dust given by $\sigma_{uv} = \pi a^2 \bar{Q}_{uv}$, with $\bar{Q}_{uv}$ the UV absorption efficiency, which we will set equal to unity.
- Assume that each absorbs a 10 eV photon. Calculate the temperature increase of each immediately after absorption.
- How small does the radiation field have to be for the grain to "behave" like a molecule?
- The (energy) cooling rate ($k_E = dE/dt$) is given by: $k_E = 4\pi \int \sigma(\nu)\, B(\nu, T)\, d\nu$ with B the Planck function at temperature T and frequency ν. The IR emission (absorption) cross section is given by, $\sigma(\nu) = \pi a^2 Q(\nu)$, with Q the absorption efficiency. Calculate the cooling timescale for the dust grain, adopting the expression for the Planck mean efficiency for graphite grains ($\bar{Q}_P = 3.2 \times 10^{-5}\ T^{1.8}\ (a/1\ \mu\text{m})^{0.06}$).
- Calculate the cooling timescale for the PAHs after absorption of a 10 eV photon, assuming that emission is dominated by one mode at 1600 cm^{-1} with an integrated strength of $\sigma = 4 \times 10^{-7}$ cm^{-2} Hz^{-1} (C-atom)$^{-1}$.
- Derive the expression relating the temperature cooling timescale (dT/dt) to the energy cooling rate (use the expression for the heat capacity above). Evaluate both expressions immediately after UV photon absorption.

12.3 Anomalous Microwave Emission:

- Describe the observational characteristics of the AME.
- Show that the intensity of the AME (Eq. (12.3)) is only a small fraction of the total energy emitted by PAHs.
- Derive Eq. (12.8) and simplify it to Eq. (12.9).
- Show that for high UV fields, the most probable J is J_{ir}.
- In a PDR, at what G_0/n_H ratio do collisions become important?
- Derive Eq. (12.17).

12.4 Calculate the charge distribution for circumcoronene in a diffuse interstellar cloud.

12.5 Describe the general spectral characteristics of PAHs and compare these with the observations. Include in this discussion the difference between CH and CC modes, the dependence on charge, and the influence of H-edge structure.

12.6 Describe the physical and chemical processes that affect the evolution of interstellar PAHs as material sojourns from its "birth site," AGB stars, to its "final" resting place, protostars and planetary systems.

12.7 PAH clusters:
Consider a "sandwich" of two 50 C-atom PAHs bonded by 50 meV per C-atom.

- Calculate the dissociation probability assuming a typical UV photon energy of 10 eV. Take $k_0 = 10^{16}$ s^{-1}.
- Calculate the photo-fragmentation rate as a function of distance from a star with a photon flux given by $G = 2600\,(40''/r)^2\,G_0$ ($G_o = 10^8$ photons cm^{-2} s^{-1}).
- Calculate the dimer formation rate if each collision leads to sticking. Assume a size of $0.9N_c^{1/2}$ Å, a temperature of 100 K, a H-nuclei density of $n_H = 3 \times 10^4$ cm^{-3}, and that 5% of the elemental C is in the form of these PAHs.
- At what distance would the dimer formation rate and the fragmentation rate balance?
- Consider only monomers and dimers. Plot the fraction of PAHs in dimers as a function of distance from the star.
- Consider a dense ($n_H = 3 \times 10^4$ cm^{-3}) PDR at a distance of 40″ from the star. In the PDR, the radiation field will be controlled by dust attenuation with a cross section of 10^{-21} cm^2/H-nuclei. Reevaluate the dimer/monomer ratio.

Bibliography

[1] Alata, I., Cruz-Diaz, G. A., Muñoz Caro, G. M., Dartois, E., 2014, *A & A*, 569, A119
[2] Ali-Haïmoud, Y., 2014, *MNRAS*, 437, 2728
[3] Allamandola, L. J., Tielens, A. G. G. M., Barker, J. R., 1985, *ApJ*, 290, L25
[4] Allamandola, L. J., Tielens, A. G. G. M., Barker, J. R., 1989, *ApJS*, 71, 733
[5] Allamandola, L. J., Tielens, A. G. G. M., 1989, *Interstellar Dust* Dordrecht: Reidel, IAU symp. 135
[6] Andrews, H., Peeters, E., Tielens, A. G. G. M., et al., 2018, *A & A*, 619, A170
[7] Bakes, E. L. O., Tielens, A. G. G. M., 1994, *ApJ*, 427, 822
[8] Bauschlicher, C. W., Jr., Peeters, E., Allamandola, L. J., 2008, *ApJ*, 678, 316
[9] Bauschlicher, C. W., Jr., Boersma, C., Ricca, A., et al., 2010, *ApJS*, 189, 341
[10] Bauschlicher, C. W., Jr., Peeter, E.,, Allamandola, L. J., 2009, *ApJ*, 697, 311
[11] Bauschlicher, C. W., Jr., Ricca, A., Boersma, C., Allamandola, L. J., 2018, *ApJS*, 234, 32
[12] Berné, O., Joblin, C., Rapacioli, M., et al., 2008, *A & A*, 479, L41
[13] Berné, O., Tielens, A. G. G. M., 2012, *PNAS*, 109, 401
[14] Bernstein, M. P., Sandford, S. A., Allamandola, L. J., 1996, *ApJ*, 472, L127
[15] Bettens, R. P. A., Herbst, E., 1997, *ApJ*, 478, 585
[16] Biscaro, C., Cherchneff, C., 2014, *A & A*, 564, A25
[17] Biscaro, C., Cherchneff, C., 2016, *A & A*, 589, A132
[18] Boersma, C., Mattioda, A. L., Bauschlicher, C. W., Jr., et al., 2009, *ApJ*, 690, 1208
[19] Boersma, C., Peeters, E., Martín-Hernández, N. L., et al., 2009, *A & A*, 502, 175

[20] Boersma, C., Bauschlicher, C. W., Allamandola, L. J., et al., 2010, *A & A*, 511, A32
[21] Boersma, C., Bauschlicher, C. W., Jr., Ricca, A., et al., 2011, *ApJ*, 729, 64
[22] Boersma, C., Bregman, J., Allamandola, L. J., 2013, *ApJ*, 769, 117
[23] Boersma, C., Bregman, J., Allamandola, L. J., 2016, *ApJ*, 832, 51
[24] Bohme, D., 1992, *Chem Rev*, 92, 1487
[25] Boulanger, F., Abergel, A., Bernard, J. P., et al., 1998, *Star Formation with the Infrared Space Observatory*, 15
[26] Boulanger, F., Abergel, A., Cesarsky, D., et al., 2000, ISO Beyond Point Sources: *Studies of Extended Infrared Emission*, 91
[27] Bréchignac,P., Garcia, G. A., Falvo, C., et al., 2014, *J Chem Phys*, 141, 164325
[28] Bréchignac, P., M. Schmidt, M., Masson, A., et al., 2005 *A & A*, 442, 239
[29] Cami, J., Bernard-Salas, J., Peeters, E., Malek, S. E., 2010, *Science*, 329, 1180.
[30] Campbell, E. K., Holz, M., Gerlich, D., Maier, J. P., 2015, *Nature*, 523, 322
[31] Campbell, E. K., Maier, J. P., 2017, *J Chem Phys*, 146, 160901
[32] Candian, A., Kerr, T. H., Song, I.-O., et al., 2012, *MNRAS*, 426, 389
[33] Candian, A., Sarre, P., Tielens, A. G. G. M., 2014, *ApJ*, 719, L10
[34] Cernicharo, J., Heras, A. M., Tielens, A. G. G. M., et al., 2001, *ApJL*, 546, L123
[35] Cherchneff, I., Barker, J., Tielens, A. G. G. M., 1992, *ApJ*, 401, 269
[36] Cherchneff, I., 2012, *A & A*, 545, A12
[37] Croiset, B. A., Candian, A., Berné, O., Tielens, 2016, *A & A*, 590, A26
[38] Dickinson, C., Ali-Haïmoud, Y., Barr, A., et al., 2018, *New Astron Rev*, 80, 1
[39] Doney, K. D., Candian, A., Mori, T., et al., 2016, *A & A*, 586, A65
[40] Foing B. H., Ehrenfreund, P., 1994, *Nature*, 369, 296
[41] Foing B. H., Ehrenfreund, P., *A & A*, 1997, 317, L59
[42] Fonfría, J. P., Cernicharo, J., Richter, M. J., Lacy, J. H., 2008, *ApJ*, 673, 445
[43] Frenklach, M., 2002, *Phys Chem Chem Phys*, 4, 2028
[44] Frenklach, M., Feigelson, E., 1989, *ApJ*, 341, 372
[45] Galliano, F., Madden, S. C., Tielens, A. G. G. M., Peeters, E., Jones, A. P., 2008, *ApJ*, 679, 310
[46] Galué, H. A., 2014, *Chem Sci*, 5, 2667
[47] Gatchell, M., Zettergren, H., 2016, *J Phys B: At Mol Opt Phys*, 49 162001
[48] Goto, M., Gaessler, W., Hayano, Y., et al., 2003, *ApJ*, 589, 419
[49] Hirashita, H., Kobayashi, H., 2013, *Earth, Planets, and Space*, 65, 1083
[50] Hirashita, H., and Murga, M. S., 2020, *MNRAS*, 492, 3779
[51] Holm, A. I. S., Zettergren, H., Johansson, H. A. B., Seitz, F., 2010, *Phys Rev Lett*, 105, 213401
[52] Hony, S., Van Kerckhoven, C., Peeters, E., Tielens, A. G. G. M., Hudgins, D. M., Allamandola, L. J., 2001, *A & A*, 370, 1030
[53] Hudgins, D. M., Allamandola, L. J., 1999a, *ApJ*, 513, L69
[54] Hudgins, D. M., Allamandola, L. J., 1999b, *ApJ*, 516, L41
[55] Hudgins, D. M., Bauschlicher, C. W., Jr, Allamandola, L. J., 2005, *ApJ*, 632, 316
[56] Jaeger, C.,Krasnokutski, A., Staicu, A., et al., 2006, *ApJS*, 166, 557
[57] Jaeger, C., Huiskens, F., Mutschke, H., et al., 2007, *Carbon*, 45, 2981
[58] Jaeger, C., Huiskens, F., Mutschke, H., et al., 2009, *ApJ*, 696, 706
[59] Joblin, C., Tielens, A. G. G. M., 2011, *PAHs and the Universe: A Symposium to Celebrate the 25th Anniversary of the PAH Hypothesis* (EAS publication series), v46
[60] Jochims, H. W., Baumgärtel, H., Leach, S., 1996, *A & A*, 314, 1003

[61] Johansson, H. A. B., Zettergren, H., Holm, A. I. S., et al., 2011, *Phys Rev A*, 84, 043201
[62] Jones, A. P., Tielens, A. G. G. M., Hollenbach, D. J., 1996, *ApJ*, 469, 740
[63] Jones, B. M., Zhang, F., Kaiser, R. I., 2011, *PNAS*, 11, 108
[64] Kuhn, M. Renzler, M. Postler, J., et al., *Nature Communication*, 7, 13550
[65] Langhoff, S. R., 1996, *J Phys Chem A*, 100, 2819
[66] Léger, A., d'Hendecourt, L., Boccara, N., eds., 1997, *Polycyclic Aromatic Hydrocarbons and Astrophysics* (Dordrecht: Reidel)
[67] Lemmens, A. K., Rap, D. B., Thunnisen, J. M. M., et al., 2019, *A & A*, 628, A130
[68] Linsky, J. L., Draine, B. T., Moos, H. W., et al., 2006, *ApJ*, 647, 1106
[69] Lu, J., Ren, X., Cao, L., 2016, *J. Energy Eng*, 142, 04015041
[70] Mackie, C. J., et al., 2015, *J Chem Phys*, 143, 224314
[71] Mackie, C. J., Candian, A., Huang X., et al., 2016, *J Chem Phys*, 145, 084313
[72] Mackie, C. J., Candian, A., Huang X., et al., 2018, *Phys Chem Chem Phys*, 20, 1189
[73] Mackie, C. J., Chen, T., Candian, A., et al., 2018, *J Chem Phys*, 149, 134302
[74] Malloci, G., Joblin, C., Mulas, G., 2007, *Chemical Physics*, 332, 353
[75] Maltseva, E., Petrignani, A. Candian, A., et al., 2015, *ApJ*, 814, 23
[76] Maltseva, E., Petrignani, A., Candian, A., et al., 2016, *ApJ*, 831, 58
[77] Maltseva, E., Petrignani, A. Candian, A., et al., 2018, *A & A*, 60, A65
[78] Matsuura, M., et al., 2014, *MNRAS*, 439, 1472
[79] Messenger, S., Amari, S., Gao, X., et al., 1998, *ApJ*, 502, 284
[80] Micelotta, E. R., Jones, A. P., Tielens, A. G. G. M., 2010, *A & A*, 510, A36
[81] Micelotta, E. R., Jones, A. P., Tielens, A. G. G. M., 2010, *A & A*, 510, A37
[82] Micelotta, E. R., Jones, A. P., Tielens, A. G. G. M., 2011, *A & A*, 526, A52
[83] Micelotta, E. R., Matsuura, M., Sarangi, A., 2018, *Space Sci Rev*, 214, 53
[84] Moutou, C., Leger, A., D'Hendecourt, L., 1996, *A & A*, 310, 297
[85] Moutou, C., et al., 1999, *A & A*, 347, 949
[86] Mulas, G., Malloci, Joblin, C., Toublanc, D., 2006, *A & A*, 460, 93
[87] Mulas, G., Falvo, C., Cassam-Chenaï, P., et al., 2018, *J Chem Phys*, 149, 144102
[88] Nagata, T., Tokunaga, A. T., Sellgren, K., et al., 1988, *ApJ*, 326,157
[89] Onaka, T., Mori, T. I., Sakon, I., et al., 2014, *ApJ*, 780, 114
[90] Pabst, C. H. M., Goicoechea, J. R., Teyssier, D., et al., 2017, *A & A*, 606, A29
[91] Parker, D. S. N., Zhang, F., Kim, Y. S., et al., 2012, *Proceedings of the National Academy of Science*, 109, 53
[92] Parker, D. S. N., Kaiser, R. I., Kostko, O., et al., 2015, *ApJ*, 803, 53
[93] Parker, D. S. N., Yang, T., Dangi, B. B., et al., 2015, *ApJ*, 815, 115
[94] Peeters, E., et al., 2002, *A & A*, 390, 1089
[95] Peeters, E., Tielens, A. G. G. M., Allamandola, L. J., Wolfire, M. G., 2012, *ApJ*, 747, 44
[96] Peeters, E., et al., 2017, *ApJ*, 836
[97] Piest, H., von Helden, G., Meijer, G., 1999, *ApJ*, 520, L75
[98] Pilleri, P., Montillaud, J., Berné, O., Joblin, C., 2012, *A & A*, 542, 69
[99] Pino, T., Dartois, E., Cao, A. T., et al., 2008, *A & A*, 490, 665
[100] Pirali, O., Vervloet, M., Mulas, G., et al., 2009, *Phys Chem Chem Phys*, 11, 3443
[101] Planck Collaboration, Ade, P. A. R., Aghanim, N., Arnaud, M., et al., 2011, *A & A*, 536, A20
[102] Postma, J., Bari, S., Hoekstra, R., et al., *ApJ*, 2010, 708, 435
[103] Puget, J. L., Leger, A., 1989, *Annu Rev Astron Astrophys*, 27, 161
[104] Rapacioli, M., Joblin, C., Boissel, P., 2005, *A & A*, 429, 193

[105] Rapacioli, M., Joblin, C., Boissel, P., 2006, *A & A*, 460, 519
[106] Ricca, A., Bauschlicher, C. W., Jr., Boersma, C., Tielens, A. G. G. M., Allamandola, L. J., 2012, *ApJ*, 754, 75
[107] Rosenberg, M. J. F., Berné, O., Boersma, C., 2014, *A & A*, 566, L4
[108] Roser, J. E., Ricca, A., Allamandola, L. J., 2014, *ApJ*, 783, 97
[109] Rouan, D., Léger, A., Omont, A., Giard, M., 1992, *A & A*, 253, 498
[110] Salama, F., Bakes, E. L. O., Allamandola, L. J., Tielens, A. G. G. M., 1996, *ApJ*, 458, 621
[111] Salas, P., Oonk, J. B. R., Emig, K. L., et al., 2019, *A & A*, 626, A70
[112] Schlemmer, S., Cook, D. J., Harrison, J. A., Wurfel, B., Chapman, W., Saykally, R. J., 1994, *Science*, 265, 1686
[113] Schnaiter, M., Henning, T., Mutschke, H., et al., 1999, *ApJ*, 687, 696
[114] Sellgren, K., 1984, *ApJ*, 277, 623
[115] Sellgren, K., Uchida K. I., Werner M. W., 2007, *ApJ*, 659, 1338
[116] Sloan, G., et al., 1997, *ApJ*, 474, 735
[117] Smith, J. D., et al., 2007, *ApJ*, 656, 770
[118] Soliman, A.-R., Hamid, A. M., Momoh, P. O., et al., 2012, *J Phys Chem*, A, 116, 8925
[119] Stein, S. E., Fahr, A., 1985, *J Phys Chem*, 89, 3715
[120] Stockett, M. H., Gatchell, M., Chen, T., et al., *J Phys Chem Lett*, 2015, 6, 22, 4504
[121] Szczepanski, J., Vala, M., 1993, *ApJ*, 414, 646
[122] Szczepanski, J., Wang, H., Vala, M., et al., 2006, *ApJ*, 646, 666
[123] Tielens, A. G. G. M., 2008, *Ann Rev Astron Astrophys*, 46, 289
[124] Tobita, S., Leach, S., Jochims, H. W., et al., 1994, *Can J Phys*, 72, 1060
[125] van Diedenhoven, B., Peeters, E., van Kerckhoven, C., et al., 2004, *ApJ*, 611, 928
[126] Verstraete, L., Leger, A., D'Hendecourt, L., Defourneau, D., Dutuit, O., 1990, *A & A*, 237, 436
[127] Wagner, D. R., Kim, H., Saykally, R. J., 2000, *ApJ*, 545, 854
[128] Wang, H., Frenklach, M., 1997, *Combustion and Flame*, 110, 173
[129] Woods, P. M., Millar, T. J., Zijlstra, A. A., Herbst, E., 2002, *ApJ*, 574, L167
[130] Yang, X. J., Li, A., Glaser, R., Zhong, J. X., 2017, *ApJ*, 837, 171
[131] Ysard, N., Verstraete, L., 2010, *A & A*, 509, A12
[132] Zettergren, H., 2017, *Nucl Instr Meth Phys Res B*, 408, 9
[133] Zhang, F., Parker, D., Kim, Y. S., et al., 2011, *ApJ*, 728, 141
[134] Zhang, W., Si, Y., Zhen, J., et al., 2019, *ApJ*, 872, 38
[135] Zhen, J., Paardekooper, D. M., Candian, A., Linnartz, H., Tielens, A. G. G. M., 2014, *Chem Phys Lett*, 592, 211
[136] Zhen, J., Castellanos, P., Paardekooper, D. M., Linnartz, H., Tielens, A. G. G. M., 2014, *ApJL*, 797, L30
[137] Zhen, J., Castellanos, P., Linnartz, H., Tielens, A. G. G. M., 2016, *Molecular Astrophysics*, 5, 1
[138] Zhen, J., Chen, T., Tielens, A. G. G. M., 2018, *ApJ*, 863, 128
[139] Zhen, J., Rodriguez Castillo, S., Joblin, C., et al., 2016, *ApJ*, 822, 113

Subject Index

Source Index

Index of Chemical Compounds

Compound	Chemical Formula	Page
acenaphthene	$C_{12}H_{10}$	302
acetaldehyde	CH_3CHO	31, 34, 227, 399, 400, 464, 523, 540
acetic acid	CH_3COOH	31, 35, 464, 471–478, 540
acetone	CH_3COCH_3	31, 34, 464
acetonitrile, see methyl cyanide		
acetylene (ethyne)	C_2H_2	18, 31, 32, 71, 76, 165, 174, 177, 235, 305, 369, 403, 464, 489, 490, 491, 502, 504, 597, 606–609
acetylene cation	$C_2H_2^+$	190, 330, 403, 526
alamine	$NH_2CH_3CHCOOH$	36
amino acetonitrile	NH_2CH_2CN	7
amino radical	NH_2	227, 235, 322, 333–335, 399, 414
amino radical cation	NH_2^+	334
ammonia	NH_3	7–9, 27, 30, 36, 71–73, 165, 174, 177, 179, 201, 223, 227, 235, 238, 239, 322, 334, 335, 369, 371–375, 378, 399, 401, 414, 424, 428, 430, 432, 434, 461, 471, 478, 490, 491, 508
ammonia cation	NH_3^+	333, 334, 401
aniline	$C_6H_5NH_2$	36
anthanthracene	$C_{22}H_{12}$	260
anthracene	$C_{14}H_{10}$	260, 262, 297, 302, 304, 613
argonium	ArH^+	322, 343
azulene	$C_{10}H_8$	33, 302
benzene	C_6H_6	26, 27, 33, 60, 243, 261, 263, 271–272, 304, 308, 605–607, 609
benzoic acid	C_6H_5COOH	35
benzonitrile	C_6H_5CN	399
Buckmunsterfullerene	C_{60}	33, 177, 258, 293, 301, 308–312, 593, 594, 598, 604, 610, 611
Buckmunsterfullerene, cation	C_{60}^+	258, 293, 322, 350–354, 594, 598
butadiynyl cation	C_4H^+	609
butadiynyl radical	C_4H	399, 523, 609
1-buten-3-ynyl	n-C_4H_3	607

Compound	Chemical Formula	Page
methylene	CH_2	165, 184, 197, 227, 235, 322, 330, 336, 526, 607
methylene cation	CH_2^+	189, 190, 197, 330, 331, 336, 526
methylene glycol	$CH_2(OH)_2$	239
methylidyne	CH	165, 184, 190, 197, 227, 235, 318, 319, 322, 324, 330, 332, 336, 338–340, 399, 523, 610
methylidyne cation	CH^+	165, 184–186,,190, 197, 201, 318, 319, 322, 330, 332, 336, 338–340, 403, 523–525
molecular hydrogen	H_2	25, 27, 28, 70, 72, 73, 88–91, 124–128, 143–145, 149, 150, 168–172, 177, 193, 197, 216, 217, 223, 227, 235, 236, 238, 322, 399, 532, 535, 538, 539
molecular hydrogen cation	H_2^+	178
molecular nitrogen	N_2	177, 179, 180, 193, 223, 227, 238, 243, 322, 334, 399, 428, 490, 491
molecular oxygen	O_2	28, 71, 165, 174, 177, 178, 179, 184, 193, 223, 227, 234, 236, 239, 395, 399, 425, 426, 438, 477
naphthyl	$C_{10}H_7$	609
napthalene	$C_{10}H_8$	33, 260, 262, 299, 302, 306, 308, 596, 607
nitric oxide	NO	227, 327, 399, 540
nitrogen dioxide	NO_2	237
nitrosyl hydride	HNO	227, 235
octatetraynyl radical	C_8H	399
ovalene	$C_{32}H_{14}$	260, 273, 292, 293, 299
ozone	O_3	223, 426, 477
pentadiyne	CH_3C_4H	399
pentadiynylidyne radical	C_4H	399, 526
perylene	$C_{20}H_{12}$	260
phenanthrene	$C_{14}H_{10}$	260, 302
phenol (carbolic acid)	C_6H_5OH	34
phenyl	C_6H_5	306, 607
phenyl cation	$C_6H_5^+$	609
2-phenyl ethenylium	$C_8H_7^+$	609
phosphene	PH_3	540
phosphorus mononitride	PN	540
phosphorus monoxide	PO	540
propane	C_3H_8	32
propane nitrile, see ethyl cyanide		
propargyl radical	C_3H_3	606
propenylidene	C_3H_2	322, 399, 432, 523, 526
propenylidene (linear)	C_3H_2	399, 526, 531
propyne	CH_3C_2H	399
propynylidene cation	C_3H^+	523, 526, 609
propynylidine	C_3H	399
protonated acetylene	$C_2H_3^+$	330, 403, 609

Compound	Chemical Formula	Page
water	H_2O	8, 27, 30, 60, 65, 66, 71, 73, 165, 174, 175, 177–179, 184, 190, 193, 197, 223, 227, 234, 235, 238, 239, 243, 322, 330, 331, 338, 341, 369, 395, 399, 400, 424–426, 430, 435, 438, 461, 464, 469–481, 502–505, 523, 535–537, 539, 540
water cation	H_2O^+	178, 197, 322, 330, 331, 344

Chemical Formula	Compound	Page
$(CH_2OH)_2$	ethylene glycol	461, 462, 464
ArH^+	argonium	322, 343
c-C_3D_2	doubly deuterated propenylidene	404
c-C_3H	cyclic propynylidyne	322, 399, 403, 523, 526
c-C_3H_2	cyclopropenylidene	399, 526, 531
c-C_3HD	deuterated propenylidene	404
c-C_5H_5	cyclopentadienyl	607
$C_{10}H_7$	naphthyl	609
$C_{10}H_8$	azulene	33, 302
$C_{10}H_8$	napthalene	33, 260, 262, 299, 302, 306, 308, 596, 607
$C_{12}H_{10}$	acenaphthene	302
$C_{14}H_{10}$	anthracene	260, 262, 297, 302, 304, 613
$C_{14}H_{10}$	phenanthrene	260, 302
$C_{150}H_{30}$	circumcircumcircumcoronene	582
$C_{16}H_{10}$	fluoranthene	302
$C_{16}H_{10}$	pyrene	177, 260, 273, 293, 302, 579, 580, 582, 583, 607, 613, 617
$C_{18}H_{12}$	chrysene	260, 263, 579, 580
$C_{18}H_{12}$	tetracene	263, 287
$C_{18}H_{12}$	triphenylene	263
C_2	dicarbon	165, 322, 323, 324, 330, 342, 343, 399
$C_{20}H_{10}$	corannulene	604
$C_{20}H_{12}$	perylene	260
$C_{22}H_{12}$	anthanthracene	260
C_{24}	C_{24}-fullerene	308
$C_{24}H_{12}$	coronene	177, 259, 260, 264, 268, 271, 273, 278, 282, 283, 293, 296, 299, 300, 302, 304, 306, 308, 594, 607, 613, 617
C_{28}	C_{28}-fullerene	308
C_2D	deuterated ethynyl	404
C_2H	ethynyl radical	83, 165, 322, 330, 341, 399, 403, 414, 523, 526, 607
C_2H^+	ethynyl cation	330, 526
C_2H_2	acetylene (ethyne)	18, 31, 32, 71, 76, 165, 174, 177, 235, 305, 369, 403, 464, 489, 490, 491, 502, 504, 597, 606–609
$C_2H_2{}^+$	acetylene cation	190, 330, 403, 526

Chemical Formula	Compound	Page
CH_3CH_2OH	ethanol	26, 31, 34, 174, 177, 239, 413, 461, 464, 540
CH_3CHO	acetaldehyde	31, 34, 227, 399, 400, 464, 523, 540
CH_3CN	methyl cyanide (acetonitrile)	31, 36, 177, 305, 374, 399, 461, 464, 466, 540
CH_3CNH^+	protonated methylcyanide	475
CH_3COCH_3	acetone	31, 34, 464
CH_3COOCH_3	ethyl formate	464
CH_3COOCH_3	methyl acetate	35
CH_3COOH	acetic acid	31, 35, 464, 471–478, 540
CH_3NCO	methyl isocyanate	464
CH_3NH_2	methyl amine	31, 36, 508
CH_3O	methoxy radical	227, 233, 235, 236, 477
CH_3OCH_3	dimethyl ether	18, 26, 31, 174, 177, 461, 462, 464, 471–478, 540
CH_3OH	methanol	17, 18, 34, 73, 174, 177, 223, 227, 232–236, 238, 239, 305, 322, 369, 374, 413, 424, 427, 430, 433, 435, 460, 461, 464, 466, 468–481, 492, 506, 523, 530
$CH_3OH_2{}^+$	protonated methanol	190, 474, 475
CH_3OOH	methyl peroxide	239
CH_3SH	methyl mercaptan	235
CH_4	methane	7–9, 18, 26, 27, 32, 65, 76, 174, 177, 193, 223, 227, 235, 239, 331, 369, 400, 403, 413, 424, 428, 430, 489, 490, 491
$CH_5{}^+$	protonated methane	190, 331
CHD_2COOH	doubly deuterated acetic acid	470
CHD_2O	deuterated methoxy radical	429
CHD_2OH	d2-methanol	233
CHD_2OH	doubly deuterated methanol (d2-methanol)	233, 429, 430, 469, 470
CHDOH	d1-hydroxymethyl radical	233
CN	cyano radical	165, 183, 227, 235, 318, 319, 322, 325, 334, 336, 399, 403, 413, 414, 522, 523
CO	carbon monoxide	7, 18, 60, 72, 73, 91–94, 165, 174, 177, 178, 193, 223, 227, 232–234, 237, 238, 243, 304, 322, 325, 330, 331, 333, 367, 369, 373, 375, 388, 395, 424, 426–428, 430, 432, 433, 435, 460, 464, 489, 492, 500, 502, 521, 523, 532, 535–537, 539, 540
$CO(NH_2)_2$	urea	7
CO^+	carbon monoxide cation	178, 197, 330, 523, 527
CO_2	carbon dioxide	7–9, 18, 27, 30, 60, 71, 73, 174, 180, 223, 237, 413, 424, 428, 430, 435, 464, 489, 502, 504
CS	carbon sulfide	72, 235, 322, 336, 369, 379, 399, 413, 414, 432, 433, 459–461, 464, 522, 523, 540

Chemical Formula	Compound	Page
H_3^+	protonated molecular hydrogen	178, 197, 322, 342, 343, 367, 400, 403, 407–409, 434, 459
H_3O^+	protonated water	178, 197, 322, 330, 331, 344
H_3S^+	protonated hydrogen sulfide	336
HC_3N	cyano acetylene	31, 116, 118, 119, 165, 174, 399, 403, 433, 464, 523, 540, 609
HC_3NH^+	protonated cyano acetylene	399, 403
HC_4N	cyano ethynyl methylene	399
HC_5N	cyano diacetylene	322, 351, 399, 540
HC_7N	cyano triacetylene	399
HC_9N	cyano tetraacetylene	399
HCl	hydrogen chloride	30, 35, 322, 338
HCl^+	hydrogen chloride cation	322, 338
HCN	hydrogen cyanide	7, 8, 31, 36, 72, 165, 174, 177, 227, 235, 238, 304, 322, 334, 369, 379, 390, 401, 413, 414, 433, 461, 464, 490, 491, 502–505, 508, 523, 540, 584, 606
$HCNH^+$	protonated hydrogen cyanide	399, 403, 475
HCO	formyl radical	165, 232–234, 237, 322, 399, 427, 472, 473, 523
HCO^+	formyl cation	72, 178, 197, 322, 330, 338, 340, 341, 399, 400, 409, 413, 428, 432, 433, 464, 490, 521–523, 527, 540
$HCONH_2$	formamide	227, 239, 464, 472, 507
HCOO	formyloxyl	227
$HCOOCH_3$	methyl formate	18, 31, 35, 461, 462, 471–478, 507, 540
HCOOH	formic acid	30, 35, 223, 227, 239, 399, 400, 424, 461, 464, 471–478, 506, 523, 540
HCS^+	protonated carbon sulfide	336
HCS^+	thioformyl cation	399, 464, 530
HD	deuterium hydride	177, 223, 322, 344–346, 405, 501
HD_2^+	protonated deuterium	404, 405, 434
HDCO	deuterated formaldehyde	404, 429, 469, 538, 540
HDCS	deuterated thioformaldehyde	404
HDO	deuterated water	469, 470, 530
HDS	deuterated hydrogen sulfide	404, 470, 540
HF	hydrogen fluoride	27, 322, 337, 338, 341
HNC	hydrogen isocyanide	227, 322, 334, 399, 403
HNC_3	iso cyano acetylene (iminipropadienylidene)	399
HNCO	isocyanic acid	227, 235, 464, 472, 523, 540, 541
HNO	nitrosyl hydride	227, 235
HO_2	hydroperoxy radical	227, 228, 234, 425, 437, 477
HOC^+	hydroxymethylidyne cation	178, 322, 523, 527
$HOCH_2CHO$	glycol aldehyde	464, 471–478, 540
HOCO	hydrocarboxyl radical	427
$HOCO^+$	hydrocarboxyl radical cation	540

Chemical Formula	Compound	Page
n-C_4H_3	1-buten-3-ynyl	607
N_2	molecular nitrogen	177, 179, 180, 193, 223, 227, 238, 243, 322, 334, 399, 428, 490, 491
N_2D^+	deuteronated dinitrogen	404
N_2H	dinitrogen monohydride	227
N_2H^+	protonated molecular nitrogen	334, 399, 400, 401, 414, 432, 434, 450, 460, 490, 502, 540
N_2H_2	diimide	227, 235, 427
N_2H_4	hydrazine	235
NCO	isocyanato radical	227, 235
ND	deuterated imidogen	404
ND_2H	doubly deuterated ammonia	404, 428
ND_3	triply deuterated ammonia	428
NH	imidogen	227, 322, 333–335, 399
NH^+	imidogen cation	202, 322
NH_2	amino radical	227, 235, 322, 333–335, 399, 414
NH_2^+	amino radical cation	334
NH_2CH_2CN	amino acetonitrile	7
NH_2CH_2COOH	glycine	7, 8, 31, 36, 508, 509
$NH_2CH_3CHCOOH$	alamine	36
NH_2D	deuterated ammonia	404, 428
NH_2OH	hydroxylamine	540
NH_3	ammonia	7–9, 27, 30, 36, 71–73, 165, 174, 177, 179, 201, 223, 227, 235, 238, 239, 322, 334, 335, 369, 371–375, 378, 399, 401, 414, 424, 428, 430, 432, 434, 461, 471, 478, 490, 491, 508
NH_3^+	ammonia cation	333, 334, 401
NH_4^+	protonated ammonia	333, 334
NO	nitric oxide	227, 327, 399, 540
NO_2	nitrogen dioxide	237
O_2	molecular oxygen	28, 71, 165, 174, 177, 178, 179, 184, 193, 223, 227, 234, 236, 239, 395, 399, 425, 426, 438, 477
O_3	ozone	223, 426, 477
OCN^-	cyanate anion	424
OCS	carbonyl sulfide	237, 399, 424, 464, 540
OH	hydroxyl radical	67–69, 165, 174, 178, 183, 184, 197, 201, 227, 234, 235, 238, 322, 331, 339–341, 344, 345, 399, 400, 502, 523, 526, 527, 535–537, 539
OH^+	hydroxyl cation	165, 178, 197, 322, 330, 331, 344, 345, 347, 523
PH_3	phosphene	540
PN	phosphorus mononitride	540
PO	phosphorus monoxide	540
SH	mercapto radical	322, 336, 337
SH^+	mercapto cation	322, 336, 339–341, 523–525